International Conference on Data and Information Science 2017

Journal of Physics: Conference Series Volume 971

Bandung, Indonesia
5 – 6 December 2017

ISBN: 978-1-5108-6121-3
ISSN: 1742-6588

Printed from e-media with permission by:

Curran Associates, Inc.
57 Morehouse Lane
Red Hook, NY 12571

Some format issues inherent in the e-media version may also appear in this print version.

Printed by Curran Associates, Inc. (2018)

Institute of Physics
Dirac House, Temple Back
Bristol BS1 6BE UK

Phone: 44 1 17 929 7481
Fax: 44 1 17 920 0979

techtracking@iop.org

Additional copies of this publication are available from:

Curran Associates, Inc.
57 Morehouse Lane
Red Hook, NY 12571 USA
Phone: 845-758-0400
Fax: 845-758-2633
Email: curran@proceedings.com
Web: www.proceedings.com

Table of contents

Volume 971

International Conference on Data and Information Science

5–6 December 2017, Telkom University, Indonesia

Accepted papers received: 15 February 2018
Published online: 5 April 2018

Preface

Papers

Data Science

Computational Linguistic

IOP Conf. Series: Journal of Physics: Conf. Series **971** (2018) 011001 doi:10.1088/1742-6596/971/1/011001

ICoDIS

In the name of Allah, The Most Gracious, The Most Merciful.

This inaugural conference is organized by School of Computing, Telkom University, and supported by Indonesia Data Science Society (IDSS) and Indonesia Association of Computational Linguistics (INACL). The advancement of today's computing technology has driven people to generate a vast amount of data with the size and variety that have never been experienced in the history of computing. The need to process and analyze big data attracts researchers interest to propose solutions. ICoDIS is organized to gather researchers to disseminate their relevant work on data science, computational linguistics, and information science.

We are very grateful to have 89 good research papers submitted from 7 countries. After having a rigorous review process, 56 papers were accepted. Therefore, the acceptance rate of this inaugural conference is 62.9%.

We would like to express our gratitude to invited speakers for contributing their time and effort to share their ideas in this conference. Last but not least, we thank all parties who have supported this conference.

Bandung, 5 December 2017

IOP Conf. Series: Journal of Physics: Conf. Series **971** (2018) 011001 doi:10.1088/1742-6596/971/1/011001

Fakultas Informatika
School of Computing
Telkom University

Telkom University

INACL
Indonesia Association for Computational Linguistics

INDONESIA DATA SCIENTIST SOCIETY

Sponsors:

Yayasan Pendidikan Telkom

trusted · IT · company
telkomsigma
by Telkom Indonesia

Peer review statement

All papers published in this volume of *Journal of Physics: Conference Series* have been peer reviewed through processes administered by the proceedings Editors. Reviews were conducted by expert referees to the professional and scientific standards expected of a proceedings journal published by IOP Publishing.

Class dependency of fuzzy relational database using relational calculus and conditional probability

Mohammad Deni Akbar[1 2], Yoshihiro Mizoguchi[3], Adiwijaya[1]

[1] School of Computing, Telkom University, Jl Telekomunikasi 01, Dayeuhkolot, Bandung-West Java 40257

[2] Graduate School of Mathematics, Kyushu University, 744 Motooka, Nishi-ku, Fukuoka 819-0395, Japan

[3] Institute of Mathematics for Industry, Kyushu University, 744 Motooka, Nishi-ku, Fukuoka 819-0395, Japan

Abstract. In this paper, we propose a design of fuzzy relational database to deal with a conditional probability relation using fuzzy relational calculus. In the previous, there are several researches about equivalence class in fuzzy database using similarity or approximate relation. It is an interesting topic to investigate the fuzzy dependency using equivalence classes. Our goal is to introduce a formulation of a fuzzy relational database model using the relational calculus on the category of fuzzy relations. We also introduce general formulas of the relational calculus for the notion of database operations such as 'projection', 'selection', 'injection' and 'natural join'. Using the fuzzy relational calculus and conditional probabilities, we introduce notions of equivalence class, redundant, and dependency in the theory fuzzy relational database.

1. Introduction

The relational database model was first introduced by Codd[3, 6]. One of the advantage of relational model is soundness of consistency of data. A procedure of data processing is sometimes described by dynamic sequences of operations which may have ambiguities in its implementation. Since a procedure in the relational database model is defined by a static formula, we can avoid inconsistencies in their implementations.

The relational algebra for the possibility-distribution-fuzzy-relational model of fuzzy databases was introduced by Umano and Fukami[12]. Further investigation of division operators are made by Nakata[8]. Our formula of fuzzy relational database model is slightly different since we are using a notion of relational calculus. Okuma[9] formulated the relational database model using a theory of relational calculus on the Dedekind Category. Since the Dedekind Category includes the category of fuzzy relations, we can formulate a fuzzy relational database model using the relational calculus on the category of fuzzy relations. Further, we introduce the notion of database operations using formulas of relational calculus. The advantage of our framework to approximate data reduction using fuzzy class clustering.

In traditional relational database, functional dependency (FD) is a set of constraints between two attributes in a relation. Functional dependency says that if two tuples have same values for attributes $A_1, A_2, \ldots, A_n$, then those two tuples must have to have same values for attributes $B_1, B_2, \ldots, B_n$. In fuzzy relational database, we can classification attribute value to class then we finding the dependency. Let see in Table 1 and 2. We classified the age 'young' from 19 to 22

Table 1. Traditional Relation Age and Salary

Age	Salary
19	125
19	100
22	100
22	125
34	150
38	150

Table 2. Equivalence Class Relation of Table 1

Age	Salary
{Young}	{Middle}
{Old}	{High}

and 'old' above 22. Also, the salary 'middle' from 100 to 125 and 'High' above 125. Using fuzzy dependency, we can reduce the data and also finding the rules that the salary depend with the age.

Fuzzy functional dependency is an important topic in the theory of fuzzy relational database. There are a lot of papers about data dependency analysis, based on such as similarity-based definition [11], proximity-based definition [2], possibility distribution based model[10], conformance-based definition [13] and semantic distance based definition [7]. But there are few investigations about an algorithm that would enable the identification of class of attribute relationships in fuzzy relations. In this paper, we introduce a definition of class dependency to define attribute dependency using a conditional probability method. We use the notions to reduce the data and finding the dependency then we can operate it using database operations.

2. Boolean relation theory

2.1. Basic notations

In this section, we introduce a notion of relational calculus. We follows the symbols and notations of relational calculus in [9, 5, 1].

Definition 2.1. *A* **relation** α *of a set* A *into another set* B *is a subset of the Cartesian product* $A \times B$ *and denoted by* $\alpha : A \rightharpoondown B$.

.

Definition 2.2. *The* **inverse relation** $\alpha^{\#} : B \rightharpoondown A$ *of* α *is a relation such that* $(b, a) \in \alpha^{\#}$ *if and only if* $(a, b) \in \alpha$.
The **composite** $\alpha\beta : A \rightharpoondown C$ *of* $\alpha : A \rightharpoondown B$ *followed by* $\beta : B \rightharpoondown C$ *is a relation such that* $(a, c) \in \alpha\beta$ *if and only if there exists* $b \in B$ *with* $(a, b) \in \alpha$ *and* $(b, c) \in \beta$.

As a relation of a set A into a set B is a subset of $A \times B$, the inclusion relation, union, intersection and difference of them are available as usual and denoted by $\sqsubseteq$, $\sqcup$, $\sqcap$ and $-$, respectively.The **identity relation** $\mathrm{id}_A : A \rightharpoondown A$ is a relation with $\mathrm{id}_A = \{(a, a) \in A \times A | a \in A\}$. The **empty relation** $\phi \sqsubseteq A \times B$ is denoted by $\mathbf{0}_{AB}$. The entire set $A \times B$ is called the **universal relation** and denoted by ∇_{AB}.The one point set $\{*\}$ is denoted by $\mathbf{I}$. We note that $\nabla_{II} = \mathrm{id}_I$.

2.2. Relational model

In the relational database theory as originally defined by E.F. Codd[3], a relation R is a set of tuples $(t_1, t_2, ..., t_n)$, where each element t_i is a member of, a data domain D_i. That is $R \sqsubseteq D_1 \times D_2 \times ... \times D_n$.

Lemma 2.3. *Let* $A = \{a_1, a_2, ..., a_n\}$ *be a finite set of attributes. For any attribute* $a_i \in A$,*there exists an attribute domain* D_{a_i}. *Let* X *be a subset of* A. D_X *is defined by* $D_X := \prod_{a \in X} D_a$. *We denote the projection map from* D_X *to* D_a *by* $\rho_{X_a} : \prod_{a \in X} D_a \to D_a \, (a \in X)$.

For $Y \sqsubseteq X$ we define a relation $\rho_{XY} : D_X \to D_Y$ as follows: $\rho_{XY} = \bigcap_{a\in Y} \rho_{X_a} \cdot \rho^{\#}_{Y_a}$. If $Y = \{a\}$ then it is trivial that $\rho_{X_{\{a\}}} = \rho_{X_a} (\forall a \in X)$, *and* $\rho_{XX} = \mathrm{id}_{D_X}$.

Lemma 2.4. *A database scheme* $R = (A, \{D_a \mid a \in A\})$ *is a pair of an attribute set and a class of attribute domains. An element t of* D_A *is called a tuple.*

Lemma 2.5. *a database relation r is defined as* $r : D_A \rightharpoondown D_A$ *with* $r \sqsubseteq \mathrm{id}_{D_A}$. *That is* $r = \{(t,t) \in D_A \times D_A\}$

Example 2.6. *We have relation between name, education and salary that is shown by Intan [4].*
Let $A = \{Name, Education, Salary\}$, *Domain of attribute 'Name'* $D_{Name} = \{N1, ..., N11\}$, $D_{Education} = \{Non, ES, JHS, SHS, BA, MS, PHD\}$, $D_{Salary} = \{0, ..., 500\}$.
where $D_{Name} = D_{Education} = N$ *and* $D_{Salary} = \Sigma$. $D_A \sqsubseteq D_{Name} \times D_{Education} \times D_{Salary}$. *Table 3 is a database relation* r_{ES} *is defined as* $r : D_A \rightharpoondown D_A$, *in first tuple in table* $(N1, Master, 400)$ *means* $((N1, Master, 400), (N1, Master, 400))$ *in* r_{ES}.

Table 3. Education and Salary

r_{ES}

Name	Education	Salary
N1	Master(MS)	400
N2	Senior High School(SHS)	150
N3	PHD	470
N4	Junior High School(JHS)	200
N5	Elementary School(ES)	125
N6	Senior High School(SHS)	250
N7	Master (MS)	420
N8	Senior High School(SHS)	175
N9	Master (MS)	415
N10	Senior High School(SHS)	275
N11	Non (N)	100

3. Fuzzy Relational Database

3.1. Basic notation

We follow the basic notations in Okuma[9]. A **fuzzy relation** α from a set A to another set B is a fuzzy subset of the Cartesian product $A \times B$. i.e. $\alpha : A \times B \to [0, 1]$. It is denoted by $\alpha : A \rightharpoondown B$. The **inverse fuzzy relation** $\alpha^{\#} : B \rightharpoondown A$ of α is a relation defined by $\alpha^{\#}(b, a) := \alpha(a, b)$. The **composite** $\alpha\beta : A \rightharpoondown C$ of $\alpha : A \rightharpoondown B$ followed by $\beta : B \rightharpoondown C$ is a fuzzy relation defined by $\alpha\beta(a, c) := \bigvee_{b\in B} (\alpha(a, b) \wedge \beta(b, c))$. As a relation $\alpha : A \rightharpoondown B$ is a fuzzy subsets of $A \times B$, the inclusion relation, union, intersection and difference of them are defined as fuzzy subset.

Let $\alpha : A \rightharpoondown B$ and $\beta : A \rightharpoondown B$ be fuzzy relations. We define the union as $(\alpha \sqcup \beta)(a, b) := \alpha(a, b) \vee \beta(a, b)$, intersection as $(\alpha \sqcap \beta)(a, b) := \alpha(a, b) \vee \beta(a, b)$, negation as $\bar{\alpha}(a, b) := 1 - \alpha(a, b)$ and difference as $(\alpha - \beta)(a, b) := \alpha(a, b) \wedge (1 - \beta(a, b))$.

Let $A = \{a_1, a_2, ..., a_n\}$ be a finite set of attributes. For any attribute $a_i \in A$, we define a set D_{a_i} called an attribute domain. Let X be a finite subset of A. D_X is the product set defined by $D_X := \prod_{a\in X} D_a$. A **database scheme** $R = (A, \{D_a\}_{a\in A})$ is a pair of an attribute set and a class of attribute domains. We denote the projection function D_X to D_a by ρ_a and D_X to D_Y by $\rho_{X,Y}$ where $a \in A$ and $Y \sqsubseteq X$.

A **database relation** r is a fuzzy relation $r : D_A \rightharpoondown D_A$ which satisfies $r \sqsubseteq \mathrm{id}_A$ where id_A is an identity function on A. We can consider a fuzzy relation r as a function $r : D_A \times D_A \to [0,1]$ and for a tuple $t \in D_A$, we call $r(t,t)$ the fuzzy value of t.

4. Database Operations

The fuzzy database operations model was introduced by Umano and Fukami[12]. They have defined 'projection', 'join', and 'selection'. In their paper, the formulas was used to compute the value of fuzzy relation, not to compute how to get table of 'projection', 'join', and 'selection' from the complete table. In our work, we use relational calculus to compute database operations from the complete relation($r : D_A \rightharpoondown D_A$). We use simpler and clearer formula to compute database operations.

(i) Let f and r be database relations $f, r : D_A \rightharpoondown D_A$. The **selection** $\sigma_f(r)$ of r with f is the database relation defined by $\sigma_f(r) = r \cdot f : D_A \rightharpoondown D_A$.

(ii) Let $X \sqsubseteq A$ and r a database relation $r : D_A \rightharpoondown D_A$. The **projection** $\pi_{A,X}(r)$ of r to X is the database relation defined by $\pi_{A,X}(r) = \rho^{\#}_{A,X} \cdot r \cdot \rho_{A,X} : D_X \rightharpoondown D_X$.

(iii) Let $X \sqsubseteq A$ and r_X a database relation $r_X : D_X \rightharpoondown D_X$. The **injection** $\eta_{X,A}(r_X)$ of r_X to A is the database relation defined by $\eta_{X,A}(r) = (\rho_{A,X} \cdot r_X \cdot \rho^{\#}_{A,X}) \sqcap \mathrm{id}_A : D_A \rightharpoondown D_A$.

(iv) Let $X, Y \sqsubseteq A$ and $Z = X \sqcup Y$. For database relations $r_X : D_X \rightharpoondown D_X$ and $r_Y : D_Y \rightharpoondown D_Y$, the **natural join** $r_X \bowtie r_Y$ is the database relation defined by $r_X \bowtie r_Y = \eta_{X,Z}(r_X) \cdot \eta_{Y,Z}(r_Y) : D_Z \rightharpoondown D_Z$.

5. Fuzzy Equivalence Class and Dependency

5.1. Fuzzy class partition

We begin this section by defining a fuzzy set using fuzzy relational database model. Let D_U be a classical set of objects D_X, called universe of discourse. An element of D_U is denoted by u.

Definition 5.1. *Let a fuzzy relation set be f, Domain of attribute X be D_X will be divided to n-class in domain of attribute class U, $D_U = \{u_1, u_2, \dots, u_n\}$. Let a collection attribute sets be $A = \{X, U\}$, with $D_A \sqsubseteq D_U \times D_X$. Hence, the fuzzy relation set f is denoted by the following $f : D_A \times D_A \to [0,1]$*

Example 5.2. *From Example 3, we would like to divide domain of attribute education and salary, $D_{Education}, D_{Salary}$. For $D_{Education}$ be divided in 3 class, $D_{LE} = \{High\ Education(HE),\ Middle\ Education(ME),\ High\ Education(HE)\}$ and D_{Salary} be divided in 3 class, $D_{LS} = \{High\ Salary(HS),\ Middle\ Salary(MS),\ High\ Salary(HS)\}$.*

Table 4. $f_{ClassEducation}$

f_1

Class Education	Education	Fuzzy Value
High Education	Non(N)	1
High Education	ES	0.8
High Education	JHS	0.5
Middle Education	ES	0.2
Middle Education	JHS	0.5
Middle Education	SHS	0.9
Middle Education	BA	0.2
Low Education	SHS	0.1
Low Education	BA	0.8
Low Education	MS	1
Low Education	PHS	1

Table 5. $f_{ClassSalary}$

f_2 Class Salary	Salary	Fuzzy Value
Low Salary	100	1
Low Salary	125	0.5
Middle Salary	125	0.5
Middle Salary	150	1
Middle Salary	175	1
Middle Salary	200	1
Middle Salary	255	0.9
Middle Salary	275	0.2
High Salary	255	0.1
High Salary	275	0.5
High Salary	400	1
High Salary	415	1
High Salary	420	1
High Salary	470	1

Table 4 showed relation $f_1 : D_{LE} \times D_{Education} \to [0,1]$, in first tuple we have seen $(HE, Non(N), 1)$ means that $f_1((HE, Non(N)), (HE, Non(N))) = 1$. Table 5 showed relation $f_1 : D_{LS} \times D_{Salary} \to [0,1]$, in first tuple we have seen $(LS, 100, 1)$ means that $f_1((LS, 100)), (LS, 100)) = 1$.

5.2. Degree of proximity relation

In this subsection, we extend the relational data model to handle fuzzy data, it is necessary to support integrity constraints expressing degree of proximity relation between member of class such that degree of 'approximately equal' or 'more or less equal'.

Definition 5.3. *Let domain of attribute X be D_X will be divided to n-class in domain of attribute class U, $D_U = \{u_1, u_2, \ldots, u_n\}$. Let a collection attribute sets be $A = \{X, U\}$, with $D_A \sqsubseteq D_U \times D_X$. Fuzzy relation be $f : D_A \times D_A \to [0,1]$. Degree of proximity relation between member of domain class u_i, u_j, $\mu_{PR}(u_i, u_j)$ follows:*
Consider that $u_i = t_i[U]$,
$\mu_{PR}(u_i, u_j) = 1 - \wedge\{f(t_i, t_i) \vee f(t_j, t_j) | \; t_i[U] \neq t_j[U] \wedge t_i[X], t_j[X]\}$ *and* $\mu_{PR}(u_i, u_i) = 1$

Example 5.4. *From table 4, we can compute* $\mu_{PR}(HE, ME) = 1 - \wedge\{(0.8 \vee 0.2), (0.5 \vee 0.5)\} = 0.5$.
From table 5, we can compute $\mu_{PR}(LS, MS) = 1 - \wedge\{(0.5 \vee 0.5)\} = 0.5$.

Table 6. μ_{PR} Class of Education

μ_{PR}	HE	ME	LE
HE	1	0.5	0
ME	0.5	1	0.2
LE	0	0.2	1

Table 7. μ_{PR} Class of Salary

μ_{PR}	HS	MS	LS
HS	1	0.5	0
MS	0.5	1	0.5
LS	0	0.5	1

6. Dependency Using Conditional Probability Method

6.1. Class dependency

There are some theory to find dependency relation in every tuple(Table8). Sozat and Yacizi [13] have defined conformance between tuple to compute dependency between tuple and attribute.

Table 8. Fuzzy implication operators.

Fuzzy implication	Corresponding formula for $X \leadsto^{\theta} Y$
Mamdani	$I_{MM} = min(x, y)$
KleeneDiense	$I_{KL} = max(1 - x, y)$
Early-Zadeh	$I_{EA} = max(1 - x, min(x, y))$
Willmot	$I_w = min[max(1 - x, y), max(x, 1 - x), max(y, 1 - y)]$

$$\mu_{x_i|y_j} = \frac{S(P[x_i \sqcap y_j](h))}{S(P[y_j](h)} \qquad \mu_{y_j|x_i} = \frac{S(P[x_i \sqcap y_j](h))}{S(P[x_i](h)}$$

In this research, we use conditional probability method to compute dependency of class every attribute.

Let X, Y be collection attribute of a relation. Consider $X, Y \sqsubseteq A$, $Z = X \sqcup Y$. Let $U_X \in X, U_Y \in Y$ be attribute classification. Domain of attribute U_X, $D_{U_X} = \{x_1, x_2, ..., x_i\}$ and $D_{U_Y} = \{y_1, y_2, ..., y_j\}$. For a database relation $f : D_X \rightharpoondown D_X$, $g : D_Y \rightharpoondown D_Y$ and $h = f \bowtie g$, $h : D_Z \rightharpoondown D_Z$. We define fuzzy value of x_i, y_j, $x_i \wedge yj$:
We abbreviate $P[U_X = x_i](h), P[U_Y = y_i](h), P[U_X = x_i \sqcap U_Y = y_j](h)$ by $P[x_i](h), P[y_i](h), P[x_i \sqcap y_j](h)$

$$\begin{aligned} P[x_i](h) &= \sigma_{U_X = x_i}(h) : D_Z \rightharpoondown D_Z \\ P[y_j](h) &= \sigma_{U_Y = y_i}(h) : D_Z \rightharpoondown D_Z, and \\ P[x_i \sqcap y_j](h) &= \sigma_{U_X = x_i \sqcap U_Y = y_i}(h) : D_Z \rightharpoondown D_Z \end{aligned}$$

Let $D_a \sqsubseteq R$, where R is a set of all real numbers. For database relation $h : D_Z \times D_Z \to [0, 1]$ we define the fuzzy sum $S(h)$ as follows:

$$S(h) = \sum\{h(t, t) \in [0, 1] \,|\, t \in D_a\}$$

So, we can define dependency between class $x_i \leadsto^{\mu_{x_i|y_j}} y_j$,

Example 6.1. *First, we transform table3 to fuzzy relational model,*$r_{ES} : D_A \times D_A \to [1]$. *After that we make natural between* r_ES, f_1, *and* f_2, *as we can see in table9*

From table 10,11,12, we can get class dependency $HE \leadsto HS$, $\mu_{HE|HS} = \frac{4.1}{4.6} = 0.891$, *and* $HS \leadsto HE$, $\mu_{HS|HE} = \frac{4.1}{4.5} = 0.911$.

Table 13 and table 14 is complete result of dependency between class of attribute. For example, in first tuple we got in table 13 $\mu_{LS|LE} = 0.6$ *and table 14* $\mu_{LE|LS} = 0.882353$, *it means 'Low Education' more determine than 'Low Salary'*

6.2. Attribute dependency

In this subsection, we would like to redefine dependency between attributes, we change conformance in fuzzy relational database with "degree" of class dependency that we have defined in the previous section.

Definition 6.2. *Let* U_X, U_Y *be class of attribute* X, Y, $U_X = \bigcup_{i=1}^{n} x_i$, $U_Y = \bigcup_{i=1}^{m} y_j$. *Fuzzy functional dependency* $X \leadsto^{\theta} Y$ *(reads X determines) if for* x_i *and* y_j *in r:*

$$\mu_{y_j|x_i} \geq min(\theta, \mu_{x_i|y_j})$$

where θ is real number within the range $[0, 1]$ describing the linguistic strength. When $\theta = 1$, fuzzy functional dependecy becomes functional dependency(Boolean), denoted as $r : D_X \times D_Y \to 0 \; or \; 1$.

Table 9. $r_{ES} \bowtie f_1 \bowtie f_2$

r_{LES}

Name	Class Education	Class Salary	Fuzzy Value
N1	High Education	High Salary	1
N2	High Education	Middle Salary	0.1
N2	Middle Education	Middle Salary	0.9
N3	High Education	High Salary	1
N4	Low Education	Middle Salary	0.5
N4	Middle Education	Middle Salary	0.5
N5	Low Education	Low Salary	0.5
N5	Low Education	Middle Salary	0.5
N5	Middle Education	Low Salary	0.2
N5	Middle Education	Middle Salary	0.2
N6	High Education	Middle Salary	0.1
N6	Middle Education	Middle Salary	0.9
N7	High Education	High Salary	1
N8	High Education	Middle Salary	0.1
N8	Middle Education	Middle Salary	0.9
N9	High Education	High Salary	1
N10	High Education	High Salary	0.1
N10	High Education	Middle Salary	0.1
N10	Middle Education	High Salary	0.5
N10	Middle Education	Middle Salary	0.5
N11	Low Education	Low Salary	1

Table 10. $P[HE]$

$P[HE]$

Name	CE	CS	$P[HE]$
N1	HE	HS	1
N2	HE	MS	0.1
N3	HE	HS	1
N6	HE	MS	0.1
N7	HE	HS	1
N8	HE	MS	0.1
N9	HE	HS	1
N10	HE	HS	0.1
N10	HE	MS	0.1
	S(P[HE])		4.5

Table 11. $P[HS]$

$P[HS]$

Name	CE	CS	$P[HE]$
N1	HE	HS	1
N3	HE	HS	1
N7	HE	HS	1
N9	HE	HS	1
N10	HE	HS	0.1
N10	ME	HS	0.5
	S(P[HE])		4.6

Table 12. $P[HE \sqcap HS]$

$P[HE \sqcap HS]$

Name	CE	CS	$P[HE]$
N1	HE	HS	1
N3	HE	HS	1
N7	HE	HS	1
N9	HE	HS	1
N10	HE	HS	0.1
	S(P[HE])		4.1

Example 6.3. *From table 13, and table 14, we can compute Education* $\rightsquigarrow^{\theta}$ *Salary:*

(i) (LE and LS) $\mu_{LS|LE} \geq min(\theta, \mu_{LE|LS}) = 0.6 \geq min(\theta, 0.82353)$

(ii) (LE and MS) $\mu_{MS|LE} \geq min(\theta, \mu_{LE|MS}) = 0.4 \geq min(\theta, 0.188679)$

(iii) (LE and HS) $\mu_{HS|LE} \geq min(\theta, \mu_{LE|HS}) = 0. \geq min(\theta, 0.)$

Table 13. $\mu_{ClassSalary|ClassEducation}$

CS	CEl	$\mu_{CS\|CE}$
Low Salary	Low Education	0.6
Low Salary	High Education	0.
Low Salary	Middle Education	0.0434783
Middle Salary	Low Education	0.4
Middle Salary	High Education	0.0888889
Middle Salary	Middle Education	0.847826
High Salary	Low Education	0.
High Salary	High Education	0.911111
High Salary	Middle Education	0.108696

Table 14. $\mu_{ClassEducation|ClassSalary}$

CE	CS	$\mu_{CE\|CS}$
Low Education	Low Salary	0.882353
Low Education	Middle Salary	0.188679
Low Education	High Salary	0.
High Education	Low Salary	0.
High Education	Middle Salary	0.0754717
High Education	High Salary	0.891304
Middle Education	Low Salary	0.117647
Middle Education	Middle Salary	0.735849
Middle Education	High Salary	0.108696

(iv) *(HE and LS)* $\mu_{LS|HE} \geq min(\theta, \mu_{HE|LS}) = 0. \geq min(\theta, 0.)$
(v) *(HE and MS)* $\mu_{MS|HE} \geq min(\theta, \mu_{HE|MS}) = 0.089 \geq min(\theta, 0.075)$
(vi) *(HE and HS)* $\mu_{HS|HE} \geq min(\theta, \mu_{HE|HS}) = 0.911 \geq min(\theta, 0.89)$
(vii) *(ME and LS)* $\mu_{LS|ME} \geq min(\theta, \mu_{ME|LS}) = 0.043 \geq min(\theta, 0.117)$
(viii) *(ME and MS)* $\mu_{MS|ME} \geq min(\theta, \mu_{ME|MS}) = 0.848 \geq min(\theta, 0.7358)$
(ix) *(ME and HS)* $\mu_{HS|ME} \geq min(\theta, \mu_{ME|HS}) = 0.11 \geq min(\theta, 0.11)$

from that condition we can conclude that minimum value of θ is less than 0.043. It means that X and Y are weak relation. But, using class dependency we can get conclusion that in the some level X and Y has strong relation.

Compare with Kiss Model:
Let $A = \{A_1, A_2, ..., A_n\}$ be collection of attributes sets, R be a fuzzy relation, $R : A_1 \times A_2 \times \ldots \times A_n \rightarrow [0, 1]$, t_1, t_2 be tuples such that $t_1, t_2 \in R$.
"Degree" dependency of fuzzy database by Kiss is:
$T_R(A_1, A_2) = 1 - \wedge\{R(t_1) \vee R(t_2))|\ t_1(A_1) = t_2(A_1) \wedge t_1(A_2) \neq t_2(A_2)$.
Using Kiss's formula we can compute degree dependency between attributes.
For example, we can compute degree dependency between "Class of Education$\rightsquigarrow$ Class of Salary" $T_R(LE,\ LS)$ from table 9. We got $ClassEducation \rightsquigarrow_{0.1} Class\ Salary$,
The Problem: Using Kiss's formula, we can compute dependency between attribute but can not compute "degree" dependency between class of attribute. As we can see in table 13 and table 14, degree dependency of "High Education" to "High Salary" different with degree dependency of "High Education" to "Low Salary". it means using conditional probability, we can compute and analyze trend data more precisely. Using conditional probability, we introduce formula degree of class of functional dependency. In table 13 and table 14, we got

Low Education $\rightsquigarrow_{0.88233}$ *Low Salary* and *Low Salary* $\rightsquigarrow_{0.6}$ *Low Education*, from the result we can conclude that "Low Education" more give effect to "Low Salary".

7. Conclusion and Future Works

Generally, our research has shown how to extract intelligent information from fuzzy relational databases. In order to to find sub-relations where criteria for the existence of fuzzy dependencies. We have tried to identify possible dependency between class of attributes of fuzzy database, and also we try to find dependency between attributes using the previous theorem of fuzzy implications. The most important contribution of this research is characterization of the theory of partial dependency or class dependency in the context of developing tools and algorithms that facilitate identification of relations between class of attributes in the observed relational schemes. Using fuzzy relational calculus, we make new formula of fuzzy database operations, class of dependency, find proximity relation between class of attributes and dependency between attributes. Hopefully, the notion is very useful to make database analysis, knowledge discovery application in fuzzy relational database. Using fuzzy relational calculus, we can compute easily the functional dependency using conditional probability method. Using class dependency, we can making analysis deeper and accurate, because every class has different dependency. In the next research, we would like to continue in soundness and completeness theorem and making simplification for implication problem in fuzzy database using deduction category and others. Fuzzy relational database is a combination of relational database and fuzzy theory.

References

[1] Mohammad Deni Akbar and Yoshihiro Mizoguchi. A formalization of a fuzzy relational database model using relational calculus. *Bulletin of Informatics and Cybernetics*, 48(6):83–104, 2016.

[2] F. Berzal, I. Blanco, D. Sánchez, J. M.Serrano, and M. A.Vila. A definition for fuzzy approximate dependencies. *Fuzzy Sets Syst.*, 149(1):105–129, January 2005.

[3] E. F. Codd. A relational model of data for large shared data banks. *Commun. ACM*, 13(6):377–387, jun 1970.

[4] Rolly Intan and Masao Mukaidono. Fuzzy conditional probability relations and their applications in fuzzy information systems. *Knowledge and Information Systems*, 6(3):345–365, May 2004.

[5] Yasuo Kawahara. Applications of relational calculus to computer mathematics. *Bulletin of Informatics and Cybernetics*, 23(1-2):67–78, 1988.

[6] A. Kiss. An application of fuzzy graphs in database theory. *Pure Mathematics and Applications*, 1(3-4):337–342, 1991.

[7] Wei-Yi Liu. The fuzzy functional dependency on the basis of the semantic distance. *Fuzzy Sets and Systems*, 59(2):173 – 179, 1993.

[8] Michinori Nakata. *Formulation of Division Operators in Fuzzy Relational Databases*, pages 144–156. Physica-Verlag HD, Heidelberg, 2000.

[9] H. Okuma and Y. Kawahara. Relational aspects of relational database dependencies. *Bulletin of Informatics and Cybernetics*, 16:91–104, 2000.

[10] K. V. S. V. N. Raju and Arun K. Majumdar. Fuzzy functional dependencies and lossless join decomposition of fuzzy relational database systems. *ACM Trans. Database Syst.*, 13(2):129–166, June 1988.

[11] Sujeet Shenoi, Austin Melton, and L. T. Fan. Functional dependencies and normal forms in the fuzzy relational database model. *Inf. Sci.*, 60(1-2):1–28, March 1992.

[12] Motohide Umano and Satoru Fukami. Fuzzy relational algebra for possibility-distribution-fuzzy-relational model of fuzzy data. *J. Intell. Inf. Syst.*, 3(1):7–27, February 1994.

[13] Mustafa lker Szat and Adnan Yazici. A complete axiomatization for fuzzy functional and multivalued dependencies in fuzzy database relations. *Fuzzy Sets and Systems*, 117(2):161 – 181, 2001.

Predictive modelling for startup and investor relationship based on crowdfunding platform data

Andry Alamsyah, Tri Buono Asto Nugroho
School of Economics and Business, Telkom University, Bandung, Indonesia

andrya@telkomuniversity.ac.id, tribuonoasto@student.telkomuniversity.ac.id

Abstract. Crowdfunding platform is a place where startup shows off publicly their idea for the purpose to get their project funded. Crowdfunding platform such as Kickstarter are becoming popular today, it provides the efficient way for startup to get funded without liabilities, it also provides variety project category that can be participated. There is an available safety procedure to ensure achievable low-risk environment. The startup promoted project must accomplish their funded goal target. If they fail to reach the target, then there is no investment activity take place. It motivates startup to be more active to promote or disseminate their project idea and it also protect investor from losing money. The study objective is to predict the successfulness of proposed project and mapping investor trend using data mining framework. To achieve the objective, we proposed 3 models. First model is to predict whether a project is going to be successful or failed using K-Nearest Neighbour (KNN). Second model is to predict the number of successful project using Artificial Neural Network (ANN). Third model is to map the trend of investor in investing the project using K-Means clustering algorithm. KNN gives 99.04% model accuracy, while ANN best configuration gives 16-14-1 neuron layers and 0.2 learning rate, and K-Means gives 6 best separation clusters. The results of those models can help startup or investor to make decision regarding startup investment.

1. Introduction

Crowdfunding platform is a place where startup exhibit and promote their project idea to public to get funded. The most funded crowdfunding platform for creative project is Kickstarter. Since its launch on April 28,2009, over 14 million people have commit to invest on many categories project, $3.3 billion has been invested, and 132,039 projects have been successfully funded [1]. In Kickstarter, startup promote their project by explain their idea, set project goal, deadline, and rewards. As consequence, people show their interest on the project, and finally fund the project. Crowdfunding platform is an efficient way for startup to get funded without traditional liabilities, such as in the conventional bank. There is various category project that startup can participate. A safety procedure is provided to ensure achievable low-risk environment. The startup promoted project must accomplish their funded goal target. If they fail to reach the target, then there is no investment activity take place. Regarding the crowdfunding platform rules, in this paper, we proposed models to predict and map the project successfulness to get funded.

There have been several references about crowdfunding platform and Kickstarter. Aslam [2] investigate the pattern of the relationship between startup and its investors based on network topology context using the small world characteristics. This study explores deeper about startup – investor

relationship on the model construction context using data mining framework. Greendberg et al [3] construct a classification models to predict project successfulness using static project data such as goal, category, pledged, video available or not, and some others. Etter et al [4] also built prediction model on Kickstarter with the model performance is 85% of correct predication, and propose Support Vector Machine model adopted on social attributes such as tweets with accuracy of 72%. Stam [5] also doing study based on Etter's work using static data project and social attributes on Twitter. He extends the study by adding the duration for a project to achieve their funding goal. This study adopts some of static attributes project data that have been studied before, with addition of comments and updates data attributes.
Based on the references above, we construct 3 models: First, the classification model using K-Nearest Neighbor (K-NN) to predict whether the project will be successful or failed. Second, the prediction model using Artificial Neural Network (ANN) to predict the number of successful project on each project category. Third, the model to map investor trend on investing the project using clustering methods based on K-means algorithm. The choice of each data mining model depends on data characteristics and the questions that investors and startups most asked. Through interviews, we investigate several scenarios, and narrowing down the results into three models above.

2. Theoretical Background

2.1. Crowdfunding
Crowdfunding is an initiative undertaken to raise money for new project proposed by someone by collecting small to medium-size investment from several other people. In the process of asking the public for donation that provide startup capital for new ventures, with this entrepreneurs and small business owners can bypass venture capitalist and angel investor entirely and instead pitch ideas straight to everyday internet users, who provided financial backing [6]. For those who invest their money will get rewards as appreciation for supporting their project.

2.2. Data Mining
Data mining is a scientific method in Knowledge Discovery in Database (KDD) process consisting of the application of data analysis and discovery of algorithm that, under acceptable computational efficiency limitation, to produce unknown data pattern [7].

2.3. Classification
Classification is the processing of finding set of models or function which describe and distinguish data classes or concepts [8]. For example, in the intrusion detection application, the classification algorithm gathers sufficient "normal" and "abnormal data, then produce classifier that can learn from those data classifier to label new unseen audit data whether belong to the "normal" or "abnormal" class [9]. One of classification methods based on non-parametric is KNN.

2.4. Prediction
Prediction is a model to predict the future output from data input based on certain criteria or formula [10]. One of the best blackbox algorithm that can be used is ANN, which imitate human brain system to simulates learning proses [11]. ANN prediction is useful when facing highly dimensional non-linear system.

2.5. Clustering
Clustering is model to group the objects of a dataset into meaningful cluster or subclass [12]. There are several methods to measure separation between data, some of them based on their distance, hierarchical, and data density. K-Means algorithm is clustering methods based on distance in vector space.

3. Methodology

3.1 Theoretical Framework

The research idea is based on the data exploratory analysis, thus construct models based on the data patterns. However, we also validate our findings with the most asked questions about business aspect by interviews the startups and investors. We come to conclusion that the previous 3 models we proposed answer the real-world problem. We explore crowdfunding data project, construct models and validate it to show models accuracy. Our research framework can be seen in Fig 1.

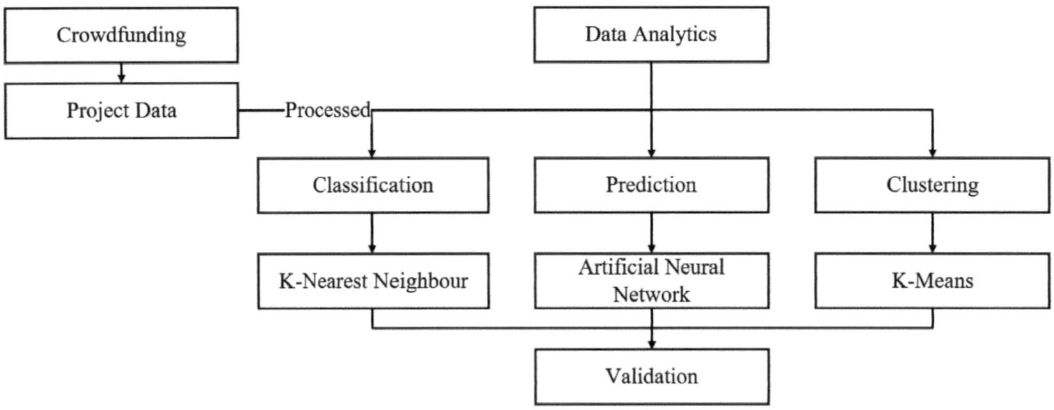

Figure 1 Research Framework

3.2 Dataset

We collect Kickstarter data through their open Application Programming Interface (API). The duration of data collection is from January 2016 to February 2017 which contains 45498 data projects. Table 1 shows dataset profile of Kickstarter.

Table 1 Kickstarter profile dataset

Attribute	Description	Type
Backers	Count backers that have backing on project	Features
Category	Category that project belongs to	Features
Comments	Total comment on project page	Features
Duration	Length of project campaign in day	Features
Goal	Goal of project in dollar	Features
Pledged	Total money of gathered in dollar	Features
State	Project state in the end of deadline	Target
Updates	Total updates on project page by creator	Features

3.3 Classification Process

The objective is to classify whether the project belong to success or fail class to get investment. We use state attribute as prediction target and the rest of the attributes as classifier set. We apply K-nearest neighbor (KNN) algorithm to classify which class a project belongs to, base on their distance closeness. We assign k=1 to define the project class of its nearest neighbor either “successful” or “failed” class. We validate the model based on 10 folds cross validation statistical performance.

3.4 Prediction Process

The objective is to construct the prediction model of number successful project per category. By using ANN algorithm, we collect sample from a multivariate value series dataset by windowing the input data per week. The windowing process is to transforms a given sample set containing series data into new example set containing single valued example. We run the methods on each project category to predict the number of successful project in each category. Dolezel et al [13] use same kind of dataset where X(k-1) and Y(k-1) as input and the output is Y(k). We train and test the data to estimate the performance of prediction model through validation process. ANN algorithm construct model by a feed-forward neural network trained by back propagation algorithm. Trial and error procedure applied to find the least error configuration by 1000 iterations. Model error rate measured using Root Mean Squared Error (RMSE) and Mean Absolute Error (MAE). Table 2, show the attributes that we use for prediction process.

Table 2 Dataset profile in time-series

Attribute	Description	Type
Historical successful project	Number of successful project per week per category	Features
Successful project	Number of successful project per category	Prediction

3.5 Clustering Process

The Objective is to map investor trend in investing on project. We construct clustering methods of pledged data that show total of money of investor to fund the project, and the backers data to show number of investor who fund the project. K-means algorithm is used to determine the similarity by means of the distance of objects data to each other, and assign the which cluster an object belongs to.

4. Result and Analysis

4.1. Classification Model

Figure. 2. shows the result of KNN classification model to classify whether a project belongs to "successful" or "failed" class. To classify those result, use the following attributes: pledged, backers, duration, goal, category, comments, and updates for social aspect. It also shows the result prediction of the actual data, in bottom left with blue dot shows prediction and actual project is failed. In bottom right, it shows prediction project is failed and the actual project is successful. In upper left, it shows prediction project is successful and the actual project data is failed. In upper right, it shows prediction and actual project is successful. The model accuracy can be seen in Table 3. It shows that prediction model has a high accuracy, class recall and class precision. It indicates that model has successfully predict most of the project entity.

By using this model for real world application, we have the information what category a project belongs to. To avoid a failed project, model characteristics inform us what aspect or data attributes needs to be improved. In real-world terms, we can call this model to predict a project likely to be successful or failed, but actually this model is classification problem.

Table 3 Classification Model Accuracy

Accuracy 99.04%	True Failed	True Successful	Class Precision
Pred. Failed	28389	214	99.25%
Pred. Successful	221	16673	98.69%
Class Recall	99.23%	98.73%	

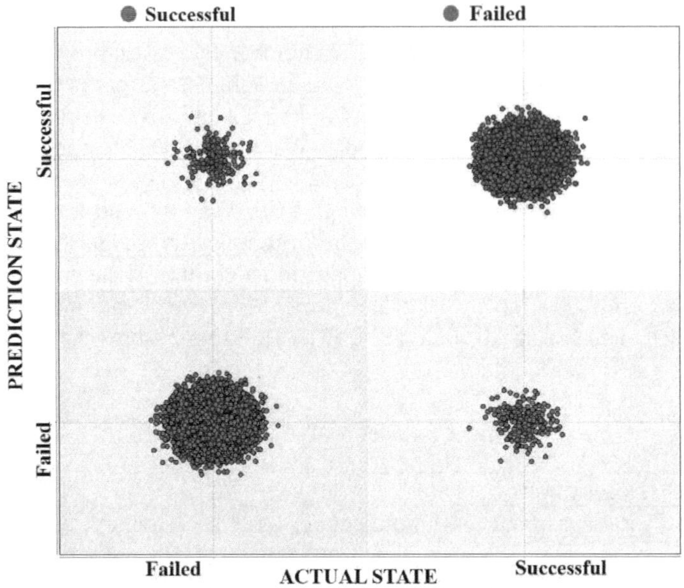

Figure 2 Classification Model Visualisation

4.2. Prediction Model

ANN algorithm predicts the number successful project per category. ANN contains Input Layer, Hidden Layer, and Output Layer of connected artificial neurons. The successful project data summarized based on 16 project categories. The output layer is one of the 16 project categories. We do trial and error to find the lowest error in hidden layer configuration. Table 4 shows the 4 lowest error. The result show that the lowest error and highest accuracy is in is Art category.

Table 4 Layer Configuration Error Test Result

Layer	RMSE	MAE
16-4-1	4.001	1.941
16-8-1	4.559	2.090
16-10-1	5.749	3.481
16-14-1	3.243	2.574

In Table 4 show that 16-14-1 configuration have the lowest error rate, we focus on building neural network based on this result. To measure the accuracy, we use three different value of Learning Rate 0.1, 0.2 and 0.3 while more than those values shows higher error. Table 5 shows the accuracy result of each learning rate. The 0.2% learning rate has the highest accuracy with 83.3%. It means predicting the number of successful project per category using ANN has high prediction rate. The ANN configuration of 16-14-1, can be seen in Figure 3.

Table 5 Learning Rate Test Result

Learning Rate	Accuracy
0.1	66.7%
0.2	83.3%
0.3	75%

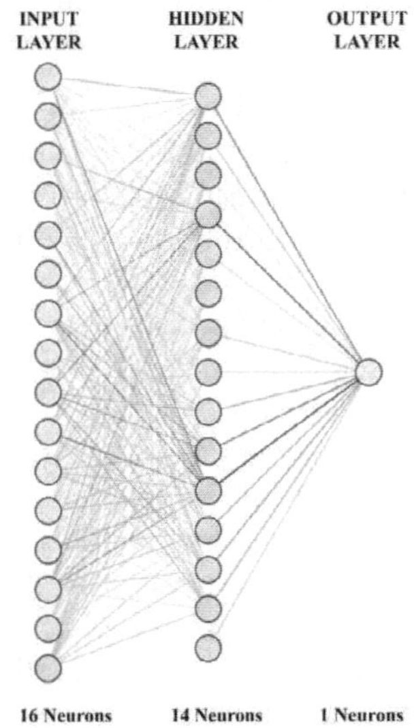

Figure 3 Neural Network Model

4.3. Clustering Model

To support investment activities exploratory analysis, we map number of investors and total investment values per project in Euclidean space. K-means algorithm is applied to the space. Based on elbow methods, number of clusters k=6 is the optimal configuration, thus we obtain 6 group / clusters. In Figure 4, it shows that cluster 1, cluster 3, cluster 4 and cluster 6 are normal scenario, where number of money invested and number of investor in project are both low. Cluster 5 has unique result, it shows total of invested money on projects is huge, while the number of investors are fairly low, which mean there are some investors that have high expectation on those projects. Cluster 2 shows different story, there are relatively small amount of total investment even though it attracts many investors.

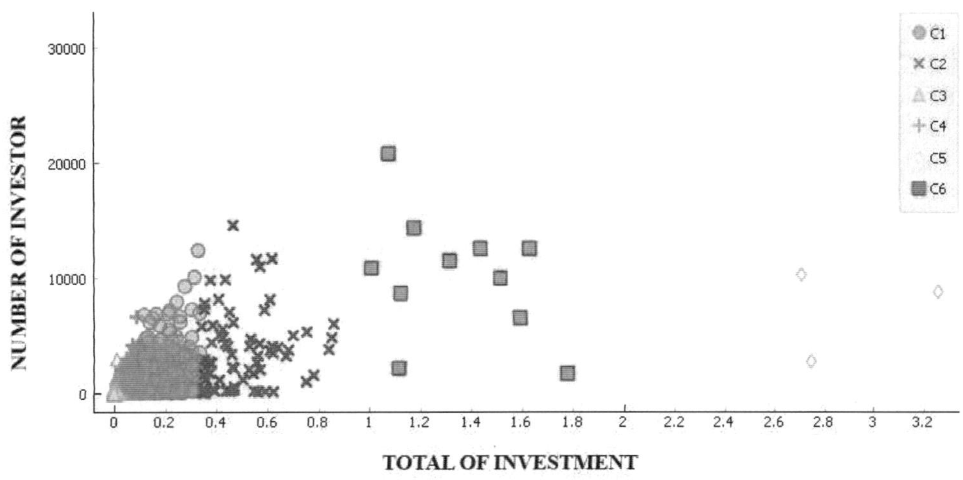

Figure 4 Clustering Model Visualization

IOP Conf. Series: Journal of Physics: Conf. Series **971** (2018) 012002 doi:10.1088/1742-6596/971/1/012002

Conclusion

In general, there are many possibilities of model construction to solve case scenario based on data pattern. It is not always guarantee that we can construct the model, since the dataset itself is not always give meaningful pattern. If we can find several patterns, thus we can construct many model, where each model answer different question. In this paper, our models answer most asked question real-world problem face by startups and investors.

This study presents how a dataset answer different questions or scenarios. First is about predicting project successfulness, where from classification model, we predict whether a project belong successful or failed class. The classification model has high accuracy 99.04%. From the prediction model, we predict the number of successful project on each project category. The prediction model has quite high accuracy learning rate of 83.3%. At last, the clustering model help us to map investor trend in investing their money. There are some projects that has fairly the same number of investor but have very high total of money invested. It shows some investor has high expectation in those projects.

Overall, the models can help startup and investor to understand comprehensively about their business nature. We found that Kickstarter represents crowdfunding platform business model, thus the model can closely represent real-world startup-investor relationship.

References

[1] Kickstarter, "Kickstarter About," [Online]. Available: https://www.kickstarter.com/about?ref=nav. [Accessed 30 September 2017].

[2] F. N. Aslam and A. Alamsyah, "The Small World Phenomenon and Network Analysis of ICT Startup Investment in Indonesia and Singapore," *The 7th Smart Collaboration for Business in Technology and Information Industry,* 2016.

[3] G. Michael D, P. Bryan, K. Hariharan and E. Gerber, "Crowdfunding support tools: predicting success & failure," *CHI'13 Extended Abstracs on Human Factors in Computing Systems,* pp. 1815-1820, 2013.

[4] V. Etter, M. Grossglauser and P. Thiran, "Launch Hard or Go Home! Predicting the Success of Kickstarter Campaigns," *Proceedings of the first ACM Conference on Online Social Networks (COSN'13),* pp. 177-182, 2013.

[5] M. Stam, *Crowdfunding Success Prediction: From Classification to Survival Regression and Back,* Amsterdam: University of Amsterdam, 2016.

[6] S. Steinberg, The Crowdfunding Bible: How to Raise Money for Any Startup, Video, Game, or Project, San Francisco: READ.ME, 2012.

[7] U. Fayyad, P.-S. G. and S. P, "Knowledge Discovery and Data Mining: Towards a Unifying Framework," *Proc. 2nd Int. Conference on Knowledge Discovery and Data Mining,* pp. 82-88, 1996.

[8] H. J. Parashar, S. Vijendra and N. Vasudeva, "An Efficient Classification Approach for Data Mining," *International Journal of Machine Learning and Computing,* vol. 2, no. 4, pp. 446-448, 2012.

[9] W. Lee, S. J. Stolfo and K. W. Mok, "A Data Mining Framework for Building Intrusion Detection Model," *Proceedings of the 1999 IEEE Symposium on Security and Privacy ,* pp. 120-132, 1999.

[10] D. Olshon and S. Yong, Intoduction to Business Data Mining, United States: McGraw-Hill Irwin, 2007.

[11] K. M, Data Mining: concept, models, methods, and algorthms, New Jersey: John Wiley, 2011.

[12] T. Sajana, C. M. Rani and K. V. Narayana, "A Survey on Clustering Techniques for Big Data Mining," *Indian Journal of Science and Technology,* vol. 9, no. 3, 2016.

[13] P. Dolezel, P. Skrabanek and L. Gago, "Weight Initialization Possibilities for Feedforward Neural Network with Linear Saturated Activation Functions," *AFAC-PaperOnline,* vol. 49, no. 25, pp. 049-054, 2016.

Support vector machine and principal component analysis for microarray data classification

Widi Astuti[1], Adiwijaya[2]

School of Computing, Telkom University, Bandung

[1]astutiwidi@telkomuniversity.ac.id; [2]adiwijaya@telkomuniversity.ac.id;

Abstract. Cancer is a leading cause of death worldwide although a significant proportion of it can be cured if it is detected early. In recent decades, technology called microarray takes an important role in the diagnosis of cancer. By using data mining technique, microarray data classification can be performed to improve the accuracy of cancer diagnosis compared to traditional techniques. The characteristic of microarray data is small sample but it has huge dimension. Since that, there is a challenge for researcher to provide solutions for microarray data classification with high performance in both accuracy and running time. This research proposed the usage of Principal Component Analysis (PCA) as a dimension reduction method along with Support Vector Method (SVM) optimized by kernel functions as a classifier for microarray data classification. The proposed scheme was applied on seven data sets using 5-fold cross validation and then evaluation and analysis conducted on term of both accuracy and running time. The result showed that the scheme can obtained 100% accuracy for Ovarian and Lung Cancer data when Linear and Cubic kernel functions are used. In term of running time, PCA greatly reduced the running time for every data sets.

1. Introduction

Cancer is a term used for diseases in which abnormal cells divide without control and are able to invade other tissues. Based on WHO cancer fact sheets [1], cancer is a leading cause of death worldwide, accounting for 8.2 million deaths in 2012. Conventional methods for monitoring and diagnosing cancers there are now very dependent on human observation to detect certain features. A cancer diagnosis is usually performed using imaging systems and analysis of morphological and clinical data. In recent decades Microarray take an important role in the diagnosis of cancer and improve the accuracy of cancer diagnosis compared to traditional techniques. With microarrays, it is possible to examine a gene expression within a single sample or to compare gene expressions within two tissue samples, such as in tumor and non-tumor tissues [2].

The characteristic of microarray data is small sample but huge dimension. Since that, there was a challenge for researcher to provide solutions for microarray data classification with high performance in both accuracy and running time. Some researches on cancer detection using microarray data are listed as follow. In 2007, Gavin and Talbot [3] proposed a modification for feature selection using sparse logistic regression [4] with Bayesian regularization which considered fast compared to previous method but the error rate is 17%. In 2012, Sahu et al [5] proposed a new approach for microarray data classification using feature selection which utilized Particle Swarm Optimization (PSO). The research showed that the usage of PSO on feature selection process could increase the accuracy. In 2014, Tran et all [6] did a review on PSO-based feature selection for data classification. Tran remarked that feature selection is a NP hard problem which is very challenging especially for high dimensional data. The use of PSO or another Evolutionary Computing method requires precision in represent the data into the searching space. Furthermore, Tran stated that on feature selection there is trade off conflict between increasing accuracy and reducing computational cost.

In 2015, Vanitha [7] proposed SVM along with mutual information gain for its feature selection. Vanitha used four different SVM namely linear SVM, SVM with RBF kernel, SVM with quadratic kernel, and SVM with polynomial kernel. Furthermore, Vanitha compared the proposed scheme result with KNN and ANN. Based on the result, the usage of MI-SVM gave better result than KNN and ANN and even hit 100% accuracy for some cases. The performance is low for colon cancer data set with 74.19% as its highest accuracy which is achieved from linear SVM and 38.70% accuracy achieved from SVM with quadratic kernel. Nurfalah [8] used PCA as feature selection for ANN based microarray data classification. Mollae [8] combined three feature selection methods and a feature extraction method called PSO-dICA before entering the SVM process. Mollae suggested that in future works, kernel functions are used. This research attempted to improve Vanitha [7] result and adopted [9] suggestion for better performance.

2. Microarray Data Classification

In our work, there are generally three steps which is illustrated on figure 1. First step is feature extraction using PCA. Next is classification using SVM with different kernels. Last step is performance evaluation using accuracy measurement.

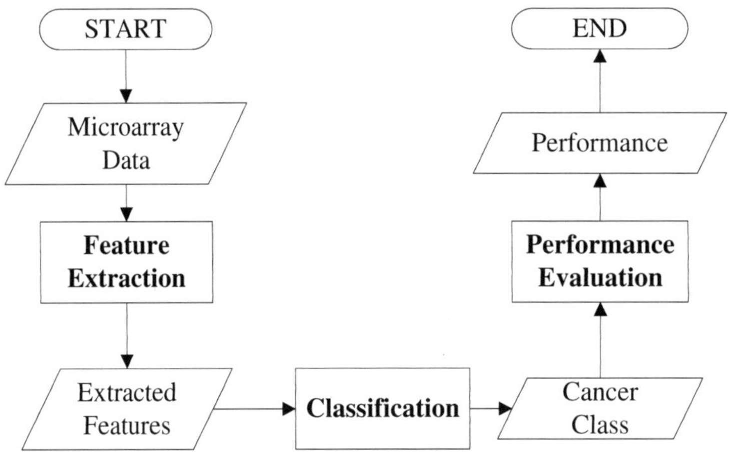

Figure 1. General Scheme.

2.1. Feature Extraction

As the dimension (number of features) increase, it will be harder to do the data mining task such as classification. To solve that problem, Feature Extraction namely PCA which extract original features into a new features using mapping function is used. PCA processes include: data centering, calculate covarian matrix, calculate Eigenvector and Eigenvalue, select top Eigenvector, and transformation data.

2.2. Classification

Support Vector Machine (SVM) is well known as a classifier which can model complex data, has good accuracy and less prone to overfitting. SVM works by searching the linear optimal separating hyperplane (decision boundary). The rationale is that decision boundary with large margin is better when handling unseen data compared to decision boundary with small margin. When the data are not linearly separable, SVM transform original data into a higher dimension using a nonlinear mapping to obtain the separating hyperplane.

IOP Conf. Series: Journal of Physics: Conf. Series **971** (2018) 012003 doi:10.1088/1742-6596/971/1/012003

2.3. Performance Evaluation

In this study, there are two type of experiments conducted. First experiment type is conducted to see the effect of the number of component chosen in PCA. The first experiment will be conducted on colon cancer data set. Colon cancer data set is chosen due to its position as performance comparison with previous research. Second experiment is conducted to see the proposed method's accuracies and running time. The experiment is conducted on seven data set, namely: Breast Cancer, Nervous System Cancer, Colon Cancer, Lung Cancer, Leukemia, Ovarian Cancer, and Prostate Cancer data. For each data set, 12 scenarios of experiment which explained on Table 1 will be performed.

Table 1. Scenarios for second experiment

Scenario #	Reduction Dimension	Kernel in SVM
1	None	Linear
2	None	Quadratic
3	None	Cubic
4	None	Fine Gaussian
5	None	Medium Gaussian
6	None	Coarse Gaussian
7	PCA	Linear
8	PCA	Quadratic
9	PCA	Cubic
10	PCA	Fine Gaussian
11	PCA	Medium Gaussian
12	PCA	Coarse Gaussian

IOP Conf. Series: Journal of Physics: Conf. Series **971** (2018) 012003 doi:10.1088/1742-6596/971/1/012003

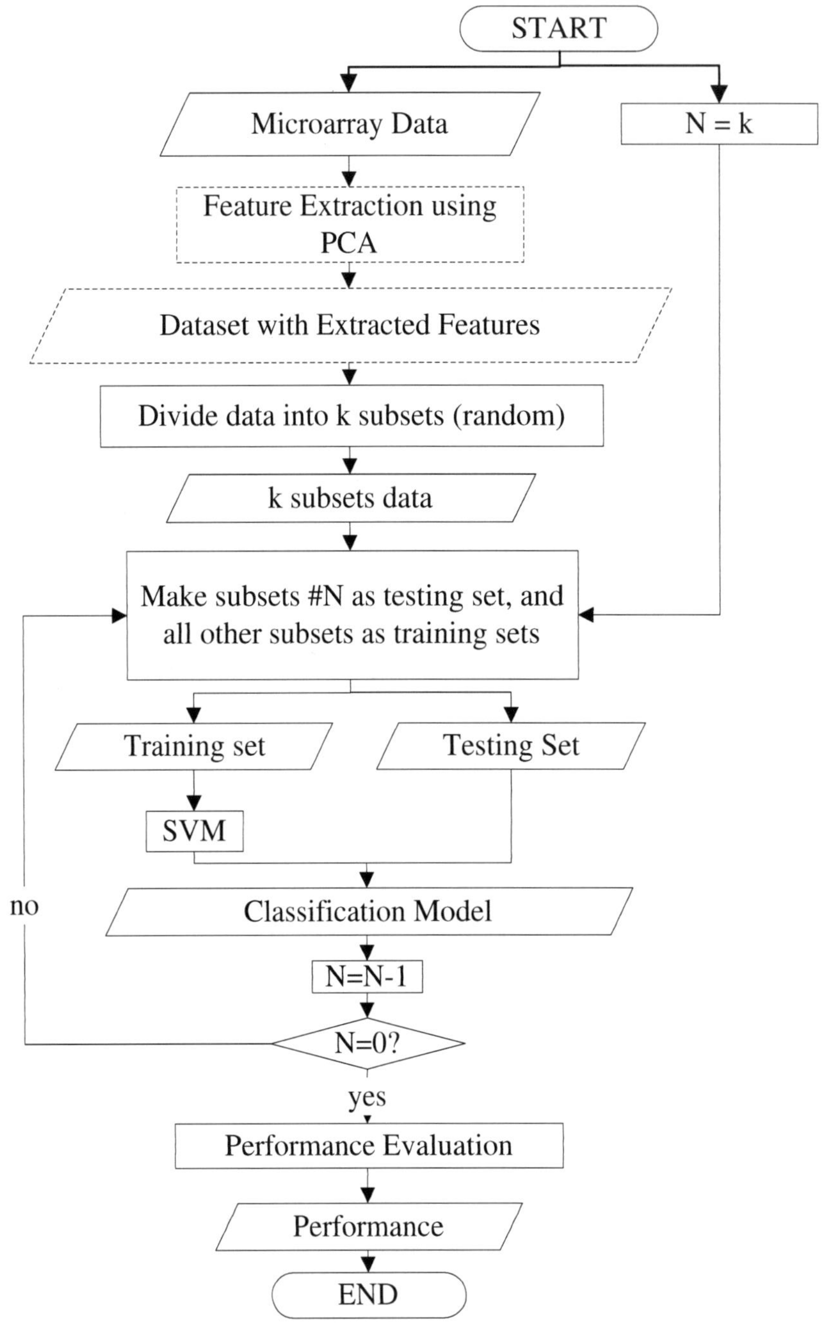

Figure 2. Detailed General Scheme.

Microarray data that have very small sample can complicate the evaluation to generated model so that evaluation results can over fit to only one subset of data. We apply the k-fold cross validation technique to resolve this problem, k-fold cross validation technique ensure that evaluation of the resulting model is not over fit to only one data so that model selection is more valid. By using k-fold cross validation, model selected and evaluated through k times iteration. The diagram of experiment process in this study which using k-fold cross validation can be seen in Figure 2. The number of k which are used in this study is 5.

3. Testing and Analysis

As explained on 2.3, an experiment on Colon Cancer Data is conducted to see the effect of the number of component chosen in PCA. Colon Cancer Data is chosen due to its position as performance comparison with previous research. Colon Cancer Data contains 62 samples with 2,000 features. After the explained variances are discovered, the colon cancer data is transformed into new data set which has lower dimension than original data. Nine different experiment scenarios are conducted and the accuracies are measured for six different SVM. The nine scenarios are chosen using explained variance threshold. The nine thresholds are 25%, 45%, 65%, 80%, 90%, 95%, 97.5%, 98.75% and 100%. It means first scenario would pick a minimum number of component which satisfies variance kept ≥ 25%, and so on. After the PCA conducted, the results are put into six different SVMs. The accuracies are measured for each classification results. The accuracies are presented on Table 2.

From Table 2, it can be seen that for Linear SVM, the best accuracy obtained when the number of component chosen is 8 or 16. From Quadratic and Cubic SVM, best accuracy is obtained when the number of component is 25 which keep 95% variance during PCA process. Best accuracy for Coarse Gaussian is obtained when the number of component after PCA is 43. Furthermore, best accuracy for Fine Gaussian and Medium Gaussian is 67.7% which is obtained when the number of component kept is 4 and 43 respectively.

From accuracy average for each different scenarios, it can be seen that the higher explained variance, the accuracy also increases. This trend stopped on 95% threshold and decreases after that. Best result of all scenario of choosing number of component and 6 different SVM obtained by Quadratic SVM with 95% variance kept after PCA.

Table 2. Accuracies for PCA parameter.

Number of Components after PCA		1	2	4	8	16	25	34	43	61
Variance Kept after PCA (%)		36.1	48.44	66	80.14	90.34	95.08	97.52	98.85	100
		Accuracy (%)								
Kernel Used	Linear	64.5	66.1	75.8	83.9	83.9	79.0	72.6	69.4	66.1
	Quadratic	62.9	62.9	74.2	79.0	82.3	85.5	80.6	79.0	75.8
	Cubic	45.2	72.6	79.0	72.6	79.0	82.3	79.0	80.6	75.8
	Fine Gaussian	59.7	67.7	67.7	64.5	64.5	62.9	64.5	64.5	64.5
	Medium Gaussian	62.9	66.1	66.1	64.5	62.9	64.5	64.5	67.7	64.5
	Coarse Gaussian	64.5	64.5	62.9	62.9	64.5	64.5	64.5	64.5	64.5
Average		59.95	66.65	70.95	71.23	72.85	**73.12**	70.95	70.95	68.53

The experiment result for second experiment is presented on Table 3 and Table 4. Table 3 shows the accuracies of each scenarios for each data set while Table 4 shows the running time of each scenarios for each data set.

Table 3. Accuracies for proposed method.

Scenario #	Accuracy (%)						
	Breast	**Nervous**	**Colon**	**Lung**	**Ovarian**	**Leukemia**	**Prostate**
1	56.41	61.70	83.90	**100.00**	**100.00**	87.70	**93.10**
2	**65.41**	61.70	**85.50**	**100.00**	99.60	87.70	92.20

3	60.25	60.00	83.90	96.90	99.60	87.70	90.20
4	56.41	65.00	64.50	50.00	64.00	40.40	65.70
5	58.97	65.00	79.00	90.60	99.20	87.70	87.30
6	56.41	65.00	64.50	56.30	92.10	66.70	63.70
7	56.41	**68.30**	79.00	93.80	100.00	**96.50**	88.20
8	57.69	**68.30**	80.60	90.60	98.40	93.00	88.20
9	51.28	61.70	82.30	90.60	51.30	93.00	90.20
10	56.41	65.00	62.90	75.00	56.40	68.40	51.00
11	56.41	65.00	56.40	75.00	56.40	68.40	51.00
12	56.41	65.00	56.40	75.00	56.40	68.40	51.00
Best Result	65.41	68.30	85.50	100.00	100.00	96.50	93.10

For breast cancer data, it can be seen that accuracy from scenario without using PCA is better than scenario with PCA process, especially for quadratic and cubic kernel, but relatively similar for others kernel. For Nervous cancer, surprisingly accuracy from scenario which used PCA is better than scenario without PCA process in 3 kernels, and similar in 3 other kernels. Accuracy for colon data from scenario without dimension reduction is better for SVM with Medium Gaussian Kernel, but relatively similar for other kernels. In lung cancer case, the accuracy of the scenario without dimensional reduction process is slightly larger for SVM with linear, quadratic, cubic, and Gaussian kernels. However, the accuracy of the scenario using PCA is better for SVM with the Fine Gaussian and Coarse Gaussian kernels. Furthermore, the mean of scenario accuracy using PCA is greater than the mean of scenario accuracy without using PCA. The accuracy of the scenario without dimensional reduction process is better for SVM with the Gaussian Medium and Coarse Gaussian kernels, but relatively similar for other kernel types in ovarian cancer data. For leukemia data, in terms of accuracy, scenarios using dimensional reduction processes result in higher performance than scenarios without dimensional reduction for 5 SVM kernels other than the Medium Guassian kernel. Lastly, for prostate cancer data, accuracy of scenarios without dimensional reduction processes is generally higher than scenarios using PCA. In term of running time, scenarios with PCA constantly give better running time.

Table 4. Running Times for proposed method.

Scenario #	Breast	Nervous	Colon	Lung	Ovarian	Leukemia	Prostate
1	520.32	50.40	8.99	103.07	157.41	102.43	101.45
2	493.48	52.13	5.64	97.74	157.03	98.98	99.73
3	493.62	46.80	5.55	96.75	156.61	95.65	99.34
4	357.08	48.37	5.63	96.83	159.91	95.96	96.84
5	357.08	48.73	5.40	96.48	157.52	99.06	101.52
6	356.44	53.20	5.423	97.66	158.71	122.96	100.02
7	164.10	29.89	8.10	54.36	80.327	62.32	58.10
8	162.80	27.88	7.23	49.24	81.794	60.32	51.41
9	162.39	28.15	7.04	48.82	162.39	51.63	69.85
10	147.13	28.00	7.20	49.98	147.13	58.63	50.58
11	155.82	27.85	7.17	48.87	155.82	51.52	57.68
12	146.54	27.68	7.09	49.04	146.54	51.14	50.69

4. Conclusion

From experiment result, it can be concluded that proposed scheme for microarray data classification which utilize Principal Component Analysis as dimension reduction method and SVM with kernel functions as classifier can surpass performance of previous research [7]. The usage of Principal Component Analysis proven to reduce the running time. Compared to [7] which also used colon cancer data set, the proposed scheme which utilizes PCA produces better accuracy by 4.81% for Linear SVM and 46.80% for Quadratic SVM. Various kernel function had been explored in this research along with suggestion from [9]. The result showed that from 6 from 7 highest accuracy for each data are obtained when kernel function is not used except for Breast Cancer Data when Quadratic Kernel Function increase the accuracy by 1.6%. We can concluded that the microarray data which had been used are linearly separable.

References

[1] World Health Organization, Cancer fact sheet, Retrieved from : http://www.who.int/mediacentre/factsheets/fs297/en/, Accessed : February 2016.
[2] Sarhan Ahmad M., Cancer Classification Based on Microarray Gene Expression Data Using DCT and ANN, Journal of Theoretical and Applied Information Technology, 2009.
[3] Cawley GC, Nicola L, Talbot C. Gene selection in cancer classification using sparse logistic regression with Bayesian regularization. Bioinformatics 2006;22(19).
[4] Shevade S.K., Keerthi S.S.. A simple and efficient algorithm for gene selection using sparse logistic regression, Bioinformatics,2003, vol. 19.
[5] Sahu B., Mishra D., A Novel Feature Selection Algorithm using Particle Swarm Optimization for Cancer Microarray Data, Procedia Engineering, 2012 (38).
[6] Tran B., Xue B., Zhang M. (2014) Overview of Particle Swarm Optimisation for Feature Selection in Classification. In: Dick G. et al. (eds) Simulated Evolution and Learning. SEAL 2014. Lecture Notes in Computer Science, vol 8886. Springer, Cham
[7] Vanitha, C.D.A., Devaraj, D. and Venkatesulu, M.. 2015. Gene Expression Data Classification Using Support Vector Machine and Mutual Information-based Gene Selection. Procedia Computer Science, 47, pp.13-21.
[8] Nurfalah A., Adiwijaya, Suryani A.S.. 2016. Far East Journal of Electronics and Communications. 16 (2). pp. 269-281.
[9] Mollaee, M., & Moattar, M. H. (2016). A novel feature extraction approach based on ensemble feature selection and modified discriminant independent component analysis for microarray data classification. Biocybernetics and Biomedical Engineering, 36(3), 521-529

On the classification techniques in data mining for microarray data classification

Husna Aydadenta and Adiwijaya

School of Computing, Telkom University, Jl. Telekomunikasi No.1, Bandung, 40257, Indonesia

E-mail: aydadenta@student.telkomuniversity.ac.id and adiwijaya@telkomuniversity.ac.id

Abstract. Cancer is one of the deadly diseases, according to data from WHO by 2015 there are 8.8 million more deaths caused by cancer, and this will increase every year if not resolved earlier. Microarray data has become one of the most popular cancer-identification studies in the field of health, since microarray data can be used to look at levels of gene expression in certain cell samples that serve to analyze thousands of genes simultaneously. By using data mining technique, we can classify the sample of microarray data thus it can be identified with cancer or not. In this paper we will discuss some research using some data mining techniques using microarray data, such as Support Vector Machine (SVM), Artificial Neural Network (ANN), Naive Bayes, k-Nearest Neighbor (kNN), and C4 .5, and simulation of Random Forest algorithm with technique of reduction dimension using Relief. The result of this paper show performance measure (accuracy) from classification algorithm (SVM, ANN, Naive Bayes, kNN, C4.5, and Random Forets).The results in this paper show the accuracy of Random Forest algorithm higher than other classification algorithms (Support Vector Machine (SVM), Artificial Neural Network (ANN), Naive Bayes, k-Nearest Neighbor (kNN), and C4.5). It is hoped that this paper can provide some information about the speed, accuracy, performance and computational cost generated from each Data Mining Classification Technique based on microarray data.

1. Introduction

Microarray technology is one tool used by many biologists to monitor the genome level of gene expression within a particular organism [1]. Microarray is usually a glass slide where DNA molecules are installed regularly in certain locations called spots (features). Microarray contains thousands of spots and each spots contain millions of copies of identical DNA molecules that uniquely fit the genes (Figure 1).

Because the microarray contains many spots (features) to perform cancer detection using microarray data requires the help of data mining. Data mining is a series of processes to extract the added value of information that has been unknown manually from a database to manipulate data into more valuable information obtained by extracting and recognizing important or interesting patterns of data contained in the database [2].

Detecting cancer with data mining can be accomplished by using classification technique, this technique is used to predict the group membership of data instances [2]. To do the classification using microarray data is the biggest challenge in data mining. This is due to the small number of samples on microarray data while the number of features on microarray is a lot (high dimensional). The higher the dimension in the data and the number of fixed

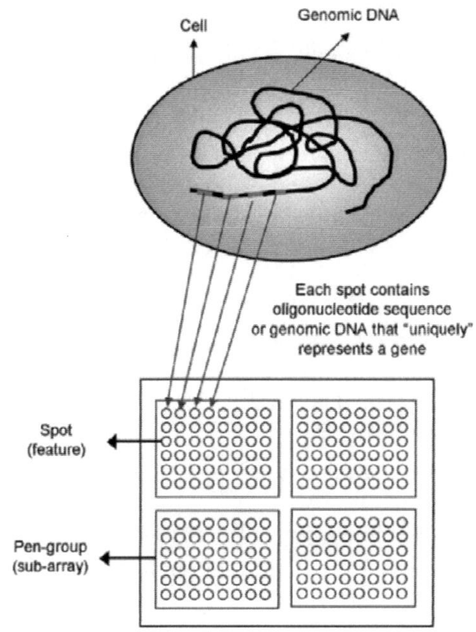

Figure 1: A microarray may contain thousands of spots. Each spot contains many copies of the same DNA sequence that uniquely represents a gene from an organism. Spots are arranged in an orderly fashion into Pengroups.

observations then the value of accuracy on the classification at a certain point will decrease. To overcome this problem, we can use the process of reduction.

Dimensional reduction can eliminate redundancy in the data thus the features used in the classification are features that have a high correlation to the class. There are two types of dimensional reduction: feature selection and feature extraction. The feature selection works by removing irrelevant features and redundancy [3]. The purpose of feature selection is to get rid of irrelevant and noisy genes from the input data set, to speed up the processing of data by reducing the dimensionality, and to avoid over fitting of the classifier [4] While extraction features work by transforming the original data into a new representation. Feature extraction has the same goal as feature selection that eliminates irrelevant or noise features in the data and removes redundancy in the data in order to increase the value of classification accuracy.

2. Literature Survey

Kohbalan Moorthy and Mohd Saberi Mohamad [5] in their paper classified microarray data using Random Forest algorithm on microarray data. The data used were Adenocarcinoma, Brain, Breast2, Breast3, Colon, Leukemia, Lymphoma, NCI60, Prostate, SRBCT. For the classification using Random Forest they tested the OOB error rate with conditions, ie the lowest OOB error rate and the fewest subsets, the lowest OOB error rate and the largest number of subsets and standard processes or default random forest algorithm. From the research, the accuracy of the standard classification or the default random forest algorithm for gene selection was always lower than the lowest OOB error rate and the lowest number of subsets and the lowest OOB error rate and the largest number of subsets.

Devi Arockia Vanitha, Devaraj, and Venkatesulu M [6] in their research used Support Vector Mechine (SVM) algorithm to classify two microarray data namely Colon cancer and Lymphoma. Then, they also used Mutual Information (MI) between genes and class labels to identify informative genes. The selected genes were used for learning using SVM and the evaluation in this study was using the Leave-one-Out Cross Validation (LOOCV) method. The proposed method of research conducted was able to reduce the feature dimension by identifying the most informative subset of genes as well as improving classification accuracy.

Nurfalah, A., Adiwijaya, and Suryani, A.A. [7], did research which aimed to explore the use of Neural Networks (ANN) and feature extraction using PCA. Futhermore, Tsun-Chen Lin, Ru-Sheng Liu, Ya-Ting Chao, and Shu-Yuan Chen [8] did a research which aimed to explore the use of gene expression data in differentiating heterogeneous types of cancer. They used Genetic Algorithm hybrid learning methodology (GA) and Neural Networks (ANN) to find an optimal subgroup of genes for tissue cancer classification. The proposed method was tested on two microarray datasets namely NCI60 and GCM. The results of the research showed that GA and ANN methods not only excel in the prediction approach but also able to reduce the number of predictive genes required in the analysis of classification.

C.Lavanya, M.Nandihini, R.Niranjana, and C.Gunavathi [9] built a comprehensive framework that combined the features of selection and classification technique, in their research. In which classification technique used was Naive Bayes algorithm. As for the feature selection using T-Test measurement, Chi-Square Test was used to select the possibility of cancer-related genes from large microarray gene expression data. From the classification framework built, it successfully classified new samples as infected or normal.

Sampath Deegalla and Henrik Bostrom [10] conducted classification research on microarray data using KNN algorithm. The dataset used among others were Colon Tumor, Leukemia, Central Nervous, SRBCT, Lymphoma, Brain, NCI60, and Prostate. Prior classification they did feature reduction which included several method of features reduction used, namely Principal Component Analysis (PCA), Random Projection (RP), Partial Least Squares (PLS) and Information Gain (IG). The results showed that PCA and PLS achieved the best accuracy with fewer components than the other two methods. And the main conclusion of this research was that the option of the feature reduction method can be very important when classifying the microarray using the KNN algorithm.

Hong Hu, Jiuyong Li, Ashley Plank, Hua Wang, and Grant Daggard [11] in their research, did a classification using several methods such as LibSVMs, C4.5, BaggingC4.5, AdaBoostingC4.5, and Random Forest in seven data collection of Microarray cancer (Breast Cancer, Lung Cancer, Lymphoma, Leukemia, Colon, Ovarian, and Prostate) . The results of the study showed that all ensemble methods outperform C4.5. The experimental results also showed that these five methods benefit from preprocessing data, including gene selection and discrimination, in classification accuracy.

3. Classification Techniques for Microarray Data

In this section includes the following subsections, namely: (1) Comparison of Classification Techniques; (2) Random Forest on the Microarray Data Classification

3.1. Comparison of classification techniques

In this subsection we provide a strength and weakness of some classification techniques for microarray data classification, namely Support Vector Machine, Artificial Neural Network, Naive Bayes, k-Nearest Neighbor, and C4.5.

3.1.1. Support Vector Machine (SVM) is a learning machine algorithm first introduced by Vapnik in 1992 [12]. The algorithm works with the principle of Structural Risk Minimization (SRM) with the aim of finding the best hyperplane that separates the two classes in the input space. The basic principle of SVM is linear classifier, and further developed to work on nonlinear problems. by incorporating the kernel trick concept in high-dimensional workspaces.

3.1.2. Artificial Neural Network (ANN) is a classification algorithm that represents the behavior of human neural network. The basic principle of ANN is that a number of parameters as inputs are processed in such a way as in the hidden layer (multiplication, addition, division, etc.), then processed again in the output layer to produce an output.

3.1.3. Naive Bayes Naive Bayes is a classification algorithm using a simple probability that applies Bayes Theorem with high independent assumptions. Bayes theorem is a theorem used in statistics to calculate the probability of a class of each attribute group present, to determine which class is optimal.

3.1.4. k-Nearest Neighbor (k-NN) is a supervised learning algorithm in which the results of a new instance query are classified by the majority of the categories in KNN. Working system of k-NN algorithm based on the shortest distance from query instance to training sample to determine its KNN. Training samples are projected into many-dimensional spaces, where each dimension represents a feature of the data. This space is divided into sections based on training sample classification. A point in this space is marked by class c if class c is the most common classification of the k nearest neighbor of the point. Nearby or the distance of neighbors are usually calculated based on Euclidean Distance.

3.1.5. C4.5 is one of the most popular classification algorithms in the decision tree algorithm group. At the learning stage of the train data, the C4.5 algorithm constructs a decision tree. Then at the classification stage, the decision tree is used to predict the class of a sample whose class is not yet known.

Table 1: Strength and weakness of classification techniques for microarray data

No	Algorithm	Strength	Weakness
1	Support Vector Machine (SVM)	- Generally generates accurate classification. - Less overfitting, robust to noise	- Requires long training time. - Because it takes long learning thus it will produce expensive computing.
2	Artificial Neural Network (ANN)	- Relatively easy to use. - Good for complex issues like image recognition	-It takes a long process time for large neural networks. - ANN architecture is different from the microprocessors architecture so it takes time to imitate it.
3	Naive Bayes	- It only requires a small amount of training data to determine the estimated parameters required in the classification process. - Able to handle quantitative and discrete data. - Handle missing values by ignoring agencies during an opportunity . -Fast and space efficiency. - Sturdy on irrelevant attributes	- Requires expensive computing. - Not applicable if the conditional probability is zero, if zero then the predicted probability will be zero as well. - Assume independent variables.
4	k-Nearest Neighbor (k-NN)	- Able to generate strong or clear data (especially if using a quadratic multiplication derivative at a distance magnitude). - Effective if used for large data.	- Requires a value of k as a parameter. - Computational price calculation is very high because this experiment requires calculating the distance from multiple queries for all experimental data.

5	C4.5	- The area of decision making that was previously complex and very global, can be transformed into simple and specific. - Elimination of unnecessary calculations, because when using the decision tree method the example is tested only on the basis of certain criteria or classes.	- Overlap occurs especially when the classes and criteria used are numerous. - Difficult in optimal decision tree design. - Decision-quality results obtained from the decision tree method depend largely on how the tree is designed.

3.2. Random forest algorithm on the microarray data classification

In the research conducted by Ramn Daz-Uriarte and Sara Alvarez de Andrs about "*Gene selection and classification of microarray data using random forest*" said that if Random Forest Algorithm is suitable for use in Microarray Datasets this is because Random Forest algorithm can be used when the number of variables are more than the number of samples, as it is known that it is characteristic of microarray. In addition the Random Forest algorithm can also be used on data that has 2 classes or multiple classes, has good predictive performance even when most are predictive noise variables, and therefore RF does not require feature selection, and able to handle categorical and continuous mixtures Predictors [13].

The Random Forest algorithm is a classification algorithm developed by Leo Breiman [14] which uses an ensemble classification tree. The ensemble method is a classification method that produces more than one model, thus the random forest consists of more than 1 tree. Then, each of the classification trees is built using a boostrap sample of data at random. Classes resulting from this classification process are derived from the most classes generated by the decision trees present in the Random Forest. By voting on available decision trees making accuracy of Random Forest increases [13]. For the working process of random forest can be seen in the following Figure 2

One of the hallmarks of the random forest algorithm is that it can be applied to a dataset that has a larger number of variables than the sample, but by applying random forest as gene selection, it is still very difficult, it is necessary to pre-process the removal dimension by sorting strong gene as a feature subset of the classification process. And for gene selection, this research used method of relief. Relief is one feature selection methods that utilizes weighting technique to measure the significance of a feature in the context of classification and features that have a weight value above the threshold used will be selected. Relief Weights are continuous values and allow features to be classified by relevance. Relief is also an interesting algorithm in Feature Selection because it has efficient computing.

As in Figure 2, the proposed method in this research consists of two processes, namely preprocessing and processing. In preprocessing, do normalization, it aims to speed up the computation process, then feature selection using relief method, using relief method will generate the rank for each feature, then selected top ranking as feature subset to be used for processing, it aims to remove redundancy and features that are no relevance. Then, the preprocessing is classified using a random forest algorithm, where the data subset used for learning is a feature subset of top ranking results in pre-processing. The random forest algorithm will produce more than one classifiers according to the number of trees in use. To make predictions using this algorithm using voting, where the most class of voting is what will be the predicted result on the model. Furthermore, after the model is obtained will be evaluated by using confusion matrix, and this process will be obtained accurately from the proposed method.

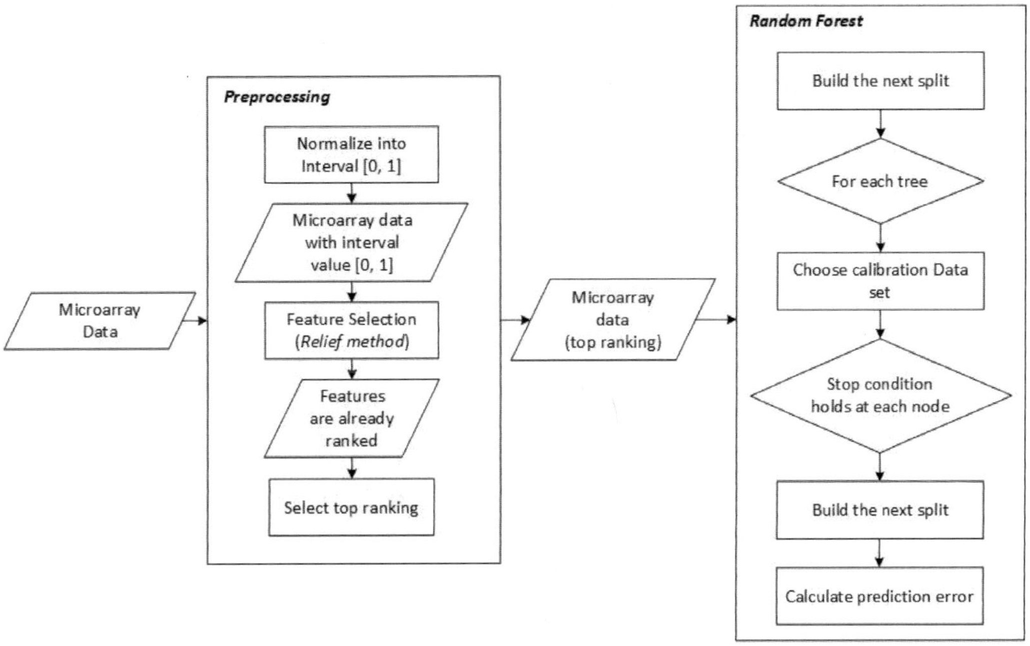

Figure 2: The proposed method

4. Results and Analysis

In this research, it was conducted Random Forest algorithm simulation using microarray data, which used data taken from `http://www.gems-system.org/`, in this simulation, the data used were amounting to 6, among others: Table 3 shows the accuracy produced using Random Forest

Table 2: Microarray Data

Data	Number of class	Sample	Feature
Colon	2	62 (22 Positif, 40 Negatif)	2001
Ovarian	2	253 (91 Normal, 162 Cancer)	15155
Central Nervous	2	60 (21 Class1, 39 Class0)	7129
Lung Cancer	2	181 (31 Mesothelioma, 150 ADCA)	12533
Tumor Prostate	2	136 (77Tumor, 59 Normal)	12600

algorithm for gene selection and classification and using Relief for gene selection & Random Forest algorithm for classification.

The simulation was done using two scenarios, using Random Forest for classification and gene selection, then before using Random Forest for classification, the data was done preprocessing i.e. gene selection using Relief method. When the random forest algorithm is used as a gene selection as in research from Kohbalan Moorthy and Mohd Saberi Mohamad [5], it is performed by bootstrap by selecting as much as N at random with the replacement of all N available in training data. Bootstrap is a training data used to form a tree. Bootstrap consists of randomly selected data with returns. The data selection was done as much as the original data set, and the bootstrap was formed according to the number of trees to be built on the Random Forest method. Unselected data on bootstrap for each tree is used as Out-Of-Bag (OOB) data, this is

Table 3: Accuracy of Random Forest algorithm simulation

Data	Tree	Random Forest	k	Random Forest & Relief	Data	Tree	Random Forest	k	Random Forest & Relief
Colon	20	75%	10	91.67%	Central Nervous	20	58.33%	10	75%
			30	83.33%				30	66.67%
			50	83.33%				50	66.67%
	50	66.67%	10	83.33%		50	58.33%	10	66.67%
			30	100%				30	66.67%
			50	91.67%				50	75%
	100	75%	10	83.33%		100	41.67%	10	75%
			30	83.33%				30	66.67%
			50	91.67%				50	66.67%
Leukemia	20	85.71%	10	100%	Lung Cancer	20	77.22%	10	94.40%
			30	100%				30	97.22%
			50	92.86%				50	100%
	50	85.71%	10	100%		50	80%	10	100%
			30	100%				30	97.22%
			50	100%				50	100%
	100	80.00%	10	100%		100	80%	10	97.22%
			30	92.86%				30	97.22%
			50	100%				50	100%
Ovarian	20	80%	10	96%	Tumor Prostate	20	57.78%	10	85.19%
			30	98%				30	88.89%
			50	100%				50	85.19%
	50	78%	10	96%		50	57.78%	10	85.19%
			30	100%				30	81.48%
			50	100%				50	100%
	100	78%	10	100%		100	68.89%	10	96.30%
			30	100%				30	88.89%
			50	100%				50	92.59%

because the data selection is done randomly with a return. OOB data will be used in internal model testing using OOB error rate estimation. In the research Kohbalan Moorthy and Mohd Saberi Mohamad [5], the development of the OOB process, by comparing three scenarios, is a selection of the smallest subset of genes with lowest OOB error rates, selection of the biggest subset of genes with lowest OOB error rates and setting the minimum number of genes to be selected.

However, in this research tried to do the gene selection by using the method of relief. The relief method works by ranking for each gene, then selected top 10, 30, and 50 rankings and is used as a feature subset of the random forest process for the Bootstrap process, and tree building.

Then, on the simulation of Random Forest algorithm it was also tested with three parameters of the number of trees ie 20 trees, 30 trees, and 50 trees. In addition to the parameter of the number of trees it was also used three parameters k, k value namely the number of variables used in the process of classification, the number of variables in total of three, ie 10, 30, and 50.

From the results in Table 3, it can be seen the accuracy of classification using Random Forest algorithm which before the classification process has been in dimension reduction using Relief method always higher than just using Random Forest classification algorithm. From the simulation results we can conclude that Random Forest algorithm is good for microarray data, but Random Forest also has some weaknesses, such as yielding unstable accuracy, this is influenced by input parameter in random forest, and random process of taking variable of each development trees, and large number of trees can make this algorithm slow for real-time predictions. Furthermore, this study also compared the performance of several classification algorithms on Tabel 4. Central Nervous has a sample size of 60 and a total of 7129 features, and has 2 values in the target class ie class1 and class0. The results of comparing some of the classification algorithms can be seen in Table 4.

Table 4: The results of comparing some of the classification algorithms

Method	Accuracy
Random Forest	75%
Support Vector Mechine (SVM)	66.67%
Artificial Neural Network (ANN)	66.67%
Naive Bayes	41.67%
k-Nearest Neighbor (k-NN)	58.33%
C4.5	58.33%

From Table 4 we compared some [15]classification algorithms and used Relief method for gene selection process with parameter k for Relief method of 50. Parameter k is number of features taken in gene selection to be subset feature. Then, other required parameters depending on the classification algorithm used. In this research, the parameter of tree number of random forest algorithm used was 30 trees, the number of neutrons in ANN 5 algorithm, and the number k on k-NN was 5. From the 6 algorithms the Random Forest algorithm produced higher accuracy than the other five classification algorithms.

5. Conclusion

In this research it was done survey on some data mining classification techniques using microarray data. This study analyzed several classification algorithms, such as Random Forest, Support Vector Machine (SVM), Artificial Neural Network (ANN), Naive Bayes, k-Nearest Neighbor (kNN), and C4.5. Some classification algorithms provided results based on the speed, accuracy, performance and resulting computational costs. From the research that has been done by using some of the classification algorithms produced good accuracy, and RF algorithm gave higher accuracy than other algorithms, but the high or low accuracy on the classification on microarray data can also be influenced by dimensional reduction used, such as in the simulation that has been done in this research using Relief method. And in this paper, it provides some surveys about data mining classification techniques using microarray data as well as the performance of the algorithm with the result of accuracy obtained. And for further research, it may be possible to add the compare time complexity required for some classification algorithms using microarray data.

References

[1] A. Snchez and M. C. R. de Villa, "A tutorial review of microarray data analysis," Departament d'Estadstica. Universitat de Barcelona., 2008.

[2] P. ning Tan, M. Steinbach, and V. Kumar, *Introduction to Data Mining*, Pearson, 2005.

[3] Z. M. Hira and D. F. Gillies, "A review of feature selection and feature extraction methods applied on microarray data," in *Advances in Bioinformatics*, 2015.

[4] A. P. K and P. V, "Review on feature selection techniques of dna microarray data," *Computer Applications* , 2013.

[5] K. Moorthy and M. S. Mohamad, "Random forest for gene selection and microarray data classification," in *Springer-Verlag Berlin Heidelberg*, 2012.

[6] D. Arockia, Vanitha.C, Devaraj.D, and Venkatesulu.M, "Gene expression data classification using support vector machine and mutual information-based gene selection," in *Procedia Computer Science (ScienceDirect)*, **47**, p. 13 21, 2015.

[7] A. Nurfalah, Adiwijaya, and A. Suryani, "Cancer detection based on microarray data classification using pca and modified back propagation.," *Far East Journal of Electronics and Communications* **16(2)**, p. p.269, 2016.

[8] T.-C. Lin, R.-S. Liu, Y.-T. Chao, and S.-Y. Chen, "Multiclass microarray data classification using ga/ann method," in *Springer-Verlag Berlin Heidelberg*, p. 1037 1041, 2006.

[9] C.Lavanya, M.Nandihini, R.Niranjana, and C.Gunavathi, "Classification of microarray data based on feature selection method," in *International Journal of Innovative Research in Science, Engineering and Technology*, **3**, February 2014.

[10] S. Deegalla and H. Bostrom, "Classification of microarrays with knn: Comparison of dimensionality reduction methods," in *Springer-Verlag Berlin Heidelberg*, pp. 800–809, 2007.

[11] H. Hu, J. Li, A. Plank, H. Wang, and G. Daggard, "A comparative study of classification methods for microarray data analysis," in *Conferences in Research and Practice in Information Technology (CRPIT)*, **61**, 2006.

[12] L. Auria and R. A. Moro, *Support Vector Machines (SVM) as a Technique for Solvency Analysis*, German Institute for Economic Research, 2008.

[13] R. Daz-Uriarte and S. A. de Andrs, "Gene selection and classification of microarray data using random forest," *BMC Bioinformatics* , 2006.

[14] L. Breiman, "Random forest–random feature," in *Statistics Department University of California Berkeley*, 1999.

[15] C. Sekhar, Minal, and E. Madhu, "Multimodal choice expression modelling using random forest decision trees," in *ijtte*, 2016.

Classification of polycystic ovary based on ultrasound images using competitive neural network

R M Dewi[1], Adiwijaya[2], U N Wisesty[3], Jondri[4]

[1,2,3,4]School of Computing, Telkom University, Telekomunikasi 1st, West java, ID

E-mail: riamaydewi@students.telkomuniversity.ac.id[1], adiwijaya@telkomuniversity.ac.id[2], untarinw@telkomuniversity.ac.id[3], jondri@telkomuniversity.ac.id[4]

Abstract. Infertility in the women reproduction system due to inhibition of follicles maturation process causing the number of follicles which is called polycystic ovaries (PCO). PCO detection is still operated manually by a gynecologist by counting the number and size of follicles in the ovaries, so it takes a long time and needs high accuracy. In general, PCO can be detected by calculating stereology or feature extraction and classification. In this paper, we designed a system to classify PCO by using the feature extraction (Gabor Wavelet method) and Competitive Neural Network (CNN). CNN was selected because this method is the combination between Hemming Net and The Max Net so that the data classification can be performed based on the specific characteristics of ultrasound data. Based on the result of system testing, Competitive Neural Network obtained the highest accuracy is 80.84% and the time process is 60.64 seconds (when using 32 feature vectors as well as weight and bias values respectively of 0.03 and 0.002).

1. Introduction

Fertility is one of the factors that may affect the integrity of a household. While infertility is a problem that occurs in the reproduction system of women or men. Inhibition of follicular maturation process can interfere with women ovulation, causing the number of follicles called polycystic ovaries (PCO). PCO symptoms can be detected early through hormone examination. Because such examinations are very expensive, many people turn to ultrasonography (USG) examination that produces an ultrasound image as shown in Figure 1.

According to the Rotterdam conference [1], a patient can be said has PCOS if she meets two of the three symptoms: (1) failure of ovulation, (2) high androgen hormones, or (3) the presence of polycystic ovaries circumstances. Morphologically, it can be said if the presence of polycystic ovaries are twelve or more follicles 2-9 mm in diameter either and or the occurrence of ovarian volume increase by more than $10cm^3$ [2]. Ultrasound image is checked manually by a doctor by counting the number and size of follicles in the ovaries. However, the investigation takes a long time and needs high accuracy to detect whether the patient has polycystic ovary syndrome or not.

The problem can be overcome by using an automated system that can detect PCO by ultrasound images. In general, the follicles can be detected using a system with machine learning approach (feature extraction and classification) and stereology approach. This research used machine learning approach with Gabor Wavelet method for feature extraction and Competitive

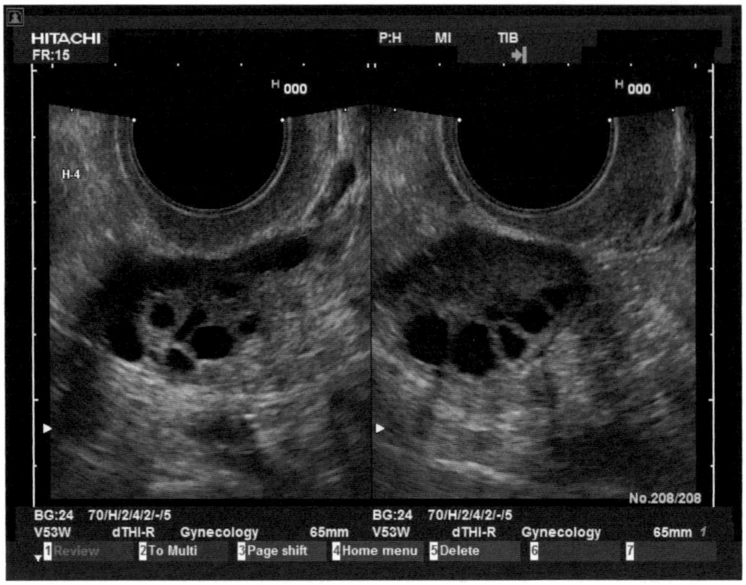

Figure 1. Ultrasound image

Neural Network methos for classification. So, this research expected to help doctors to detect PCO easier by ultrasound images.

2. Related Works

Follicles detection can be done in two ways, namely stereology and a sequencial process of feature extraction and classification. In the paper entitled Follicle Ultrasound Detection on the Images to Support Determination Polycystic Ovary Syndrome [3], polycystic ovary can be identified using stereology to calculate the number and size of each follicle and euclidean distance to measure the diameter of the follicle. In the other paper, entitled Modified Backpropagation Algorithm for Polycystic Ovary Syndrome Detection based on Ultrasound Images [4], the research conducted by using Gabor Wavelet [5] for feature extraction and Lavenberg-Marquardt Optimization and Conjugate Gradient-Fletcher Reeves as a modification of backpropagation to classify PCO. In [6], the segmentation of the follicles using Particle Swarm Optimization (PSO) to make modifications to the fitness function. In [7], the research using three classifier namely (1) Neural Network - Learning Vector Quantization (LVQ), (2) KNN - Euclidean Distance, and (3) SVM - RBF Kernel.

In [8], follicles can be segmented using region growing method. This method tests whether the neighbor of initial seeds should be added to segmentation region. In the previous research, the segmentation process of ultrasound image has been done in [9, 10] that uses edge detection as segmentation process. In [9], a median filter can be used to remove noise form image. It is based on storing and updating the gray level histogram of the picture elements in the window. The main idea of the median filter is to find a median in a specific window. The center window will be updated with a median of the window. Meanwhile, otsu global threshold [10] of the image to is the way to find the pixel similarity to its neighbors. Otsu's thresholding method involves iterating through all the possible threshold values and calculating a measure of spread for the pixel levels each side of the threshold, i.e. the pixels that either fall in foreground or background. As well-known, Canny has introduced an edge detection method in [11]. In the Canny edge detection, a computational approach to edge detection was presented which is

IOP Conf. Series: Journal of Physics: Conf. Series **971** (2018) 012005 doi:10.1088/1742-6596/971/1/012005

using a smoothing process to the image with a Gaussian, optimizing the trade-off between noise filtering and edge localization. It can be used to detect follicles in the ultrasound image which is diagnosed PCO.

3. Proposed Schema

The data used in this research is ovarian ultrasound image derived from Klinik Bersalin Permata Bunda Syariah, Cirebon and the data has been validated by a gynecologist. Based on Figure 2, the ultrasound image has four step process. First, preprocessing on the image has been done, then segmentation has been performed to obtain follicle and feature extraction by using Gabor Wavelet. The final process is classifying the follicle using Competitive Neural Network to get the final result is "1" or "2", where "1" is a non-PCO and "2" is PCO. Figure 2 describes general scheme of PCO detection system.

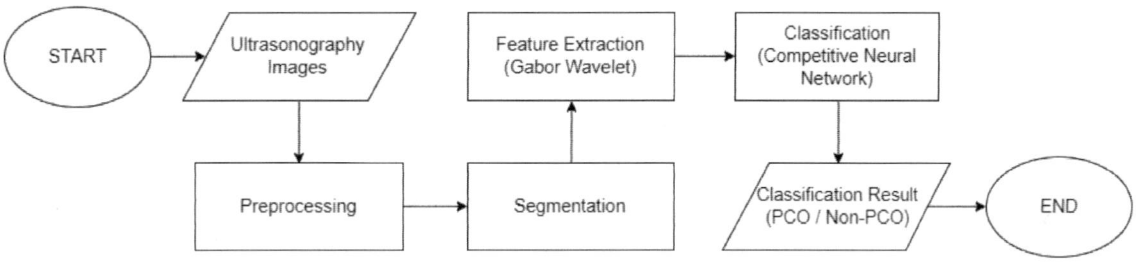

Figure 2. System Flowchart

Preprocessing is an important step before the data is processed. This stage will produce data that has better quality and important information contained in the image to be more easily obtained. Preprocessing has several processes, namely grayscalling, histogram equalization, image binarization, morphology, invert image, and data cleaning. Grayscalling is used to convert color images into grayscale images. Histogram equalization is used to capture the difference pixels that have a low frequency. Output obtained in this process is an image that has the same histogram values so that the important information contained in the images are not lost. Binary image is a process to change the image matrix values to 1 or 0, which means the image turns to black and white. Morphology used are erosion and dilation. Erosion is used to attenuate objects that are considered noise in binary image so that the object will disappear. Then the dilation process is used to thicken binary image objects that have been lost because of the erosion process. Invert image is a process to change black into white and vice versa, because the object is more easily detected when the object's color is black and the background is white. The final is data cleaning process to remove the objects that are not important.

Segmentation is the step to separate an object with the background. This step has two processes, namely follicle edge detection and cropping. Edge detection aims to recognize objects contained in the image. In this research, the object is the follicle and edge detection conducted using modern algorithms (canny method). Furthermore, each follicle has been detected will be labeled and every labeled follicle will be cropped to produce a new image that will be used in the next process.

Feature extraction is the step to extract feature/information of the object in an image to be distinguished from other objects. This research uses a Gabor Wavelet feature vector which results from each follicle. Figure 3 describes the flowchart of Gabor Wavelet process.

Classification is the last step that will be used to classify the data into several classes. In this research, the data will be classified into class "1" which means non-PCO or class "2" which means PCO. The method used is Competitive Neural Network (CNN) which has a single layer neural

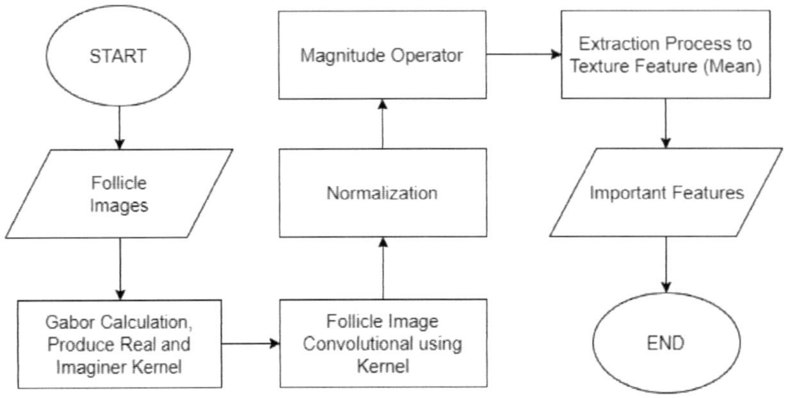

Figure 3. Gabor Wavelet Process

network architecture. This neural network only has input and output layer. CNN learning systems using unsupervised learning method which is no target output. CNN was selected because this method combined Hemming Net and The Max Net, so data classification can be performed based on the specific characteristics of ultrasound data. The process of learning on CNN using the WTA (winners take all) systems. The whole neurons that exist on CNN will be competing and they will produce only one winning neuron. The training and testing process of competitive neural network can be seen in Figure 4.

CNN classification consists of two processes, namely:

3.1. Training Process

The training process is done by using a training dataset and started by initializing the entire weights randomly. Then, performed two CNN training's phase that are done repeatedly weight value. Both phases are:

3.1.1. Competing Phase In this phase, all neurons will compete to be a winning neuron in rewarding phase. The process done in this phase are:

(i) Choosing the input vector randomly from the entire data.

(ii) Calculating output on all output nodes based on the following formula:

$$o_j = p_l \cdot w_j \tag{1}$$

with an output nodes o, p is input vectors, and w is the weight vector of all the input vector.

(iii) Determining the winning neurons of all output nodes.

3.1.2. Rewarding Phase The purpose of this phase to give a reward to one neuron that has been won in competing phase. Weight vector in the neurons will be updated based on the Equation 2 so that the winner can be closer to the input vector:

$$w_j(n+1) = w_j(n) + \xi(n)(p_i - w_j) \tag{2}$$

by w is the weight vector of all the input vector, ξ is a competitive learning rate (bias), and p is input vectors.

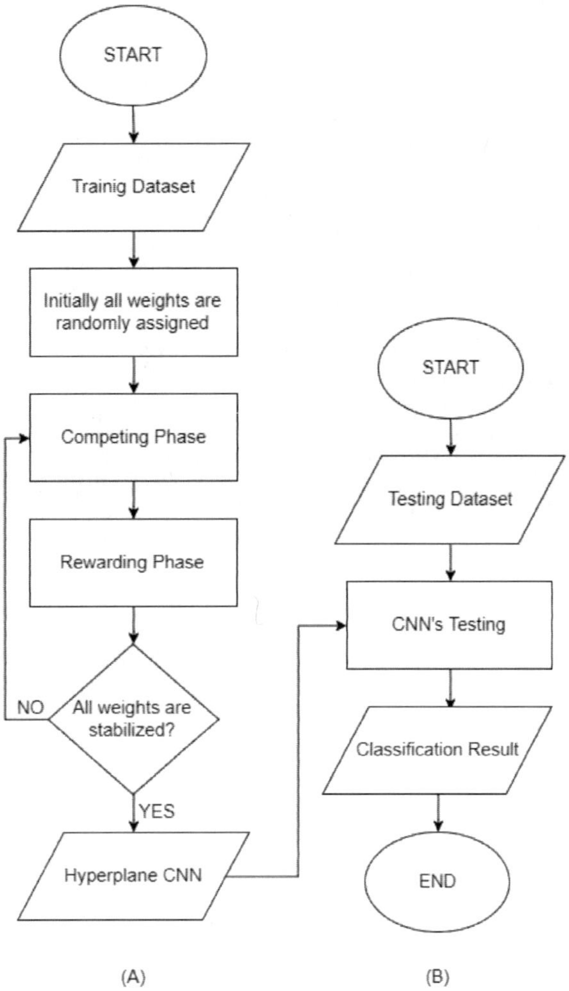

Figure 4. (A) Training of Competitive Neural Network, (B) Testing of Competitive Neural Network

3.2. Testing Process
This process used a testing dataset and it was conducted testing using hyperplane of the training process to get the classification of follicles detected "1" (non-PCO) or "2" (PCO).

4. Result and Discussion

4.1. Testing of Weight Parameter
This test aims to analyze the influence of weight parameter on accuracy. Weight parameter is used in the competing phase to obtain neutal input. There are 10 variations of weight parameters used are 0.01, 0.02, 0.03, 0.04, 0.05, 0.1, 0.15, 0.2, 0.25, and 0.3.

The best accuracy is 80.84% with a weight of 0.03, 0.15, 0.2, and 0.25 at 32 feature vectors. Figure 5 shows that the accuracy obtained from each weight parameter produce value almost stable at 16, 24 and 32 feature vectors.

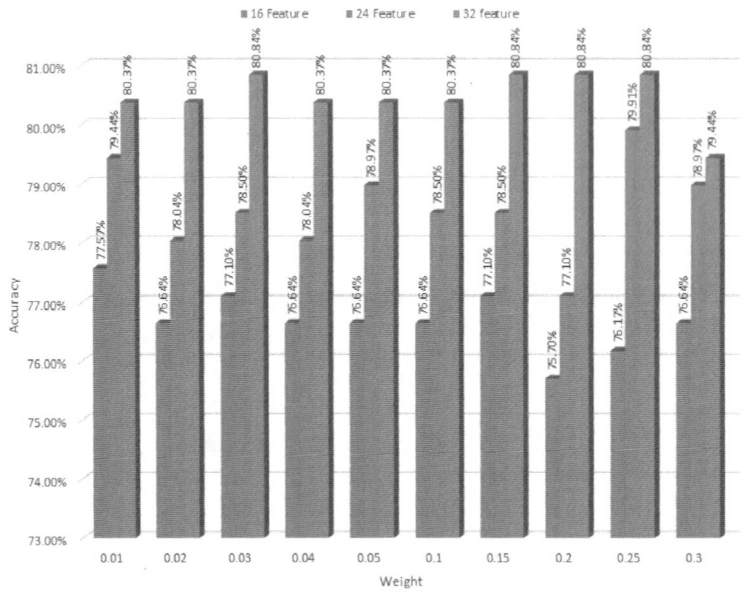

Figure 5. Comparison of Accuracy based on Weight Parameter

4.2. Testing of Bias Parameter

This test aims to analyze the influence of bias parameter on accuracy. Bias parameter in the competitive neural network is a learning rate that affects the speed of neural network to learn the pattern of training data. There are 10 variations of bias parameter used namely 0.001, 0.002, 0.003, 0.004, 0.005, 0.01, 0.02, 0.03, 0.04 and 0.05.

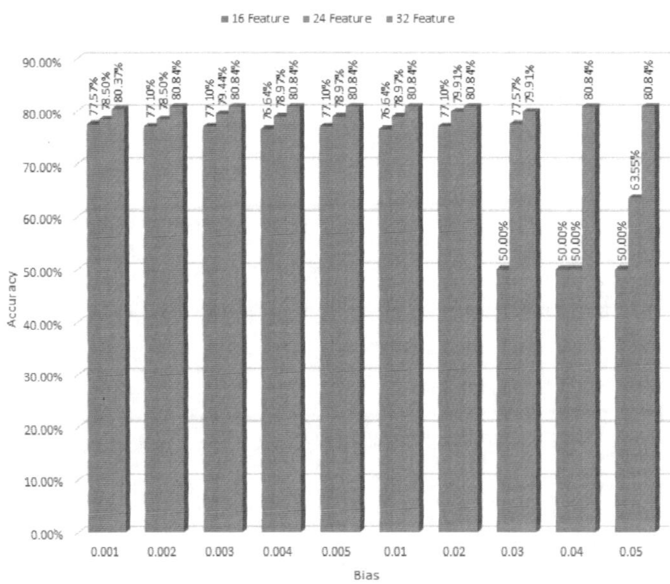

Figure 6. Comparison of Accuracy based on Bias Parameters

Based on Figure 6, the fastest time is 57 seconds with bias parameter of 0.01, 0.02, 0.03, 0.04, and 0.05 at 32 feature vectors. It means the greater value of bias, the faster training process done. This is because the value of bias affecting the neural network got pattern from training data.

4.3. Testing of The Number of Feature Vectors

This test aims to analyze the influence of the number of feature vectors of Gabor Wavelet on the accuracy and computational time on the use of Competitive Neural Network (CNN) method. The number of feature vectors tested are 16, 24, and 32 features.

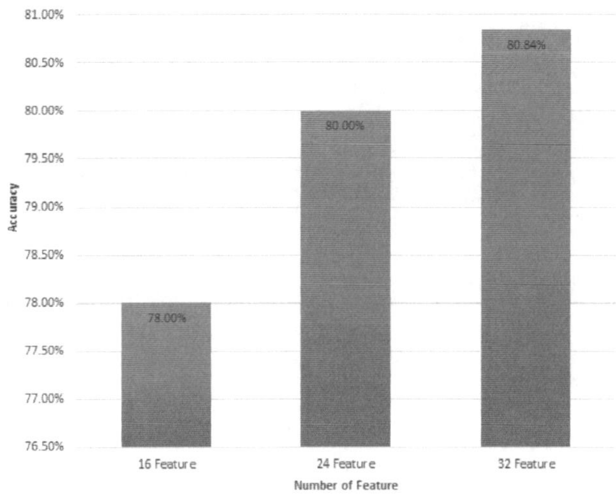

Figure 7. Comparison of Accuracy based on The Number of Feature Parameters

Based on Figure 7, the best accuration is 80.84% at 32 feature vector. The number of feature vectors in Gabor Wavelet process is directly proportional to the accuracy of PCO classification. The greater number of feature vectors, the greater accuracy obtained to perform PCO classification based on ultrasound image.

5. Conclusion

Classification of polycystic ovaries can be done by using machine learning approach (feature extraction and classification). One of the classification method is Competitive Neural Network. The best accuracy is 80.84% when using 32 feature vector (time process is 60.64 seconds). It can be happened because the number of feature vectors is directly proportional to the accuracy of classification using Competitive Neural Network.

References

[1] National Institutes of Health. 2012. Evidence-based Methodology Workshop on Polycystic Ovary Syndrome. U.S. Department of Health and Human Services

[2] Lawrence M J, Eramian M G, Pierson R A and Neufeld E 2007 Computer assisted detection of polycystic ovary morphology in ultrasound images *Fourth Canadian Conference on Computer and Robot Vision (CRV 2007), 28-30 May 2007, Montreal, Quebec, Canada* pp 105–112 URL `https://doi.org/10.1109/CRV.2007.18`

[3] Adiwijaya, Purnama B, Hasyim A, Septiani M, Wisesty U and Astuti W 2015 Follicle detection on the usg images to support determination of polycystic ovary syndrome *Journal of Physics: Conference Series* vol 622 (IOP Publishing) p 012027

[4] Wisesty U N, Nasri J and Adiwijaya 2016 Modified backpropagation algorithm for polycystic ovary syndrome detection based on ultrasound images *Recent Advances on Soft Computing and Data Mining - The Second International Conference on Soft Computing and Data Mining (SCDM-2016), Bandung, Indonesia, August 18-20, 2016, Proceedings* pp 141–151 URL `https://doi.org/10.1007/978-3-319-51281-5`$_1$5
[5] Adiwijaya, Maharani M, Dewi B, Yulianto F and Purnama B 2013 digital image compression using graph coloring quantization based on wavelet-svd *Journal of Physics: Conference Series* vol 423 (IOP Publishing) p 012019
[6] Setiawati E, Adiwijaya and Tjokorda A 2015 Particle swarm optimization on follicles segmentation to support pcos detection *Information and Communication Technology (ICoICT), 2015 3rd International Conference on* (IEEE) pp 369–374
[7] Purnama B, Wisesti U N, Adiwijaya, Nhita F, Gayatri A and Mutiah T 2015 A classification of polycystic ovary syndrome based on follicle detection of ultrasound images *Information and Communication Technology (ICoICT), 2015 3rd International Conference on* (IEEE) pp 396–401
[8] Setiawati E, Adiwijaya, Wirayuda T A and Astuti W 2016 *Advanced Science Letters* **22** 1997–2001
[9] Mehrotra P, Chakraborty C, Ghoshdastidar B, Ghoshdastidar S and Ghoshdastidar K 2011 Automated ovarian follicle recognition for polycystic ovary syndrome *Image Information Processing (ICIIP), 2011 International Conference on* (IEEE) pp 1–4
[10] Rihana S, Moussallem H, Skaf C and Yaacoub C 2013 Automated algorithm for ovarian cysts detection in ultrasonogram *Advances in Biomedical Engineering (ICABME), 2013 2nd International Conference on* (IEEE) pp 219–222
[11] Canny J F 1986 *IEEE Trans. Pattern Anal. Mach. Intell.* **8** 679–698 URL `https://doi.org/10.1109/TPAMI.1986.4767851`

The comparison between SVD-DCT and SVD-DWT digital image watermarking

Kurniawan Wira Handito, Zulfikar Fauzi, Firda Aminy Ma'ruf, Tanti Widyaningrum, Kemas Muslim Lhaksmana
School of Computing, Telkom University

handitowira@student.telkomuniversity.ac.id, zulfikarfauzi@student.telkomuniversity.ac.id, firdamaruf@student.telkomuniversity.ac.id, aprilia_zooili@yahoo.com, kemasmuslim@telkomuniversity.ac.id

Abstract. With internet, anyone can publish their creation into digital data simply, inexpensively, and absolutely easy to be accessed by everyone. However, the problem appears when anyone else claims that the creation is their property or modifies some part of that creation. It causes necessary protection of copyrights; one of the examples is with watermarking method in digital image. The application of watermarking technique on digital data, especially on image, enables total invisibility if inserted in carrier image. Carrier image will not undergo any decrease of quality and also the inserted image will not be affected by attack. In this paper, watermarking will be implemented on digital image using Singular Value Decomposition based on Discrete Wavelet Transform (DWT) and Discrete Cosine Transform (DCT) by expectation in good performance of watermarking result. In this case, trade-off happen between invisibility and robustness of image watermarking. In embedding process, image watermarking has a good quality for scaling factor < 0.1. The quality of image watermarking in decomposition level 3 is better than level 2 and level 1. Embedding watermark in low-frequency is robust to Gaussian blur attack, rescale, and JPEG compression, but in high-frequency is robust to Gaussian noise.

1. Introduction

Development of computer technology will always be followed by the development of internet. Because of this, the long distance would not be a big problem since information in form of digital data can easily be accessed even though they are originally from far places. This ease of data exchange makes the content of data cannot be completely monitored. Therefore, various ideas are made to create an owner mark of the digital data possessed. It can be implemented by using Steganography technique called watermarking.

Watermarking is one of method to hide or give certain information into a digital data[5]. One application of watermarking is copyright-labelling[2,6]. There are various types of watermarking methods which each of them has its points of strength and weakness [2,6]. The watermarking technique that is used must be robust to the manipulation means, and this technique is called Robust Watermarking [15]. Based on the concept of insertion, watermarking is divided into two, which are spatial domain and transform domain. On this research [16], it is mentioned that for watermarking with the objective of data concealing (including copyright-labelling) will be more suited using transform domain.

There are many methods that can be applied on transform domain, such as Singular Value Decomposition (SVD), Discrete Cosine Transform (DCT), Discrete Fourier Transform (DFT), Discrete

Wavelet Transform (DWT), Contourlet Tranform (CT). There have been many research using transform domain method for watermarking either by one method or combining more than one methods. Singular Value Decomposition is a technique to decompose a matrix which is often used in various application like image compression[1], watermarking[2,4,6], and other applications. In this paper, watermarking performance using SVD-DCT and SVD-DWT will be compared. The watermarking term here is reconciled with the non-blind watermarking.

2. Related Works

As mentioned earlier, the development in computer technology makes it easier to retrieve data from any places, without exception, the digital medical images and medical records of patients. But confidentiality is the most important thing for medical record, since only the patient and authorized person is allowed to see it, and with this rapid technology development, it is kind of hard to find a secure network. Therefore, Widi Astuti's research[7], implemented steganography to solve this problem. The technique itself embed secret data into the image to conceal the secret data. The research itself provide an implementation of data hiding scheme using Vector Quantization and graph coloring based on evolutionary algorithms which are Genetic Algorithm (GA) and Particle Swarm Optimization (PSO). The result of this research showed that GA provide better performance than PSO in terms of computational time, but both of this algorithm approximately gave the same performance in the hiding capacity.

Prathiwi et al.[8] proposed a watermark scheme where signature watermark is embedded into frequency domain (Integer Wavelet Transform). To improve robustness, Reed-Muller Codes is applied to detect and fix if there is a manipulation attack on the digital medical images. Authenticity Control in this system is sing Hash Block Chaining from hash MD5. This proposed scheme has a good performance for authenticity and integrity of digital medical images, where based on testing results, use of Reed-Muller on embedding process can improve the robustness of signature watermark from attack like Gaussian noise, sharpening, blurring and JPEG Compression.

Adiwijaya et al.[9] proposed a scheme of reversible watermark using a modified LSB and Huffman compression to detect and recover the manipulated medical image. This Watermarking scheme shows a good detection performance, with the testing results as follows. The watermark scheme is able to detect attacks with up to 100% accuracy and can perform recovery with up to 98% accuracy for some attacks.

Rosiyadi et al.[10] proposed copyright protection scheme using watermarking for e-government document which combine Discrete Cosine Transform (DCT) and Singular Value Decomposition (SVD) using a control parameter to avoid the false-positive problem. From the experimental result, this proposed scheme can improve the image quality GA-based evolution and that this approach is robust under several types of attacks.

Horng et al.[11] proposed a blind watermarking scheme using DCT-SVD and GA, it is found that this scheme has robustness and offers a high imperceptibility to various attacks. While Kurniawan et al.[12] proposed a watermarking technique which applied Reed-Solomon code for robust watermark in wavelet domain and SHA-256 for fragile watermark in Hash Block Chaining. The proposed technique can be implemented simultaneously on an image so that the integrity control and authenticity of the image detection can be applied at once.

Adiwijaya et al[13] stated that the primary applications of watermarking are to protect copyright and integrity verification (authentication). The principle watermarking of this research is inserting digital data (text or image) into host medical image to maintain authority of ownership and to detect authenticity of medical image. Adiwijaya divide medical image into two parts, namely Region of Interest (ROI) and Region of Non-Interest (RONI). The reference watermark which is used to detect the authenticity of medical images (integrity control) that is embedded in the ROI image. Meanwhile, the signature watermark which is used for authority of ownership (proprietary rights) is embedded in the RONI image on the wavelet domain. This research used Reed-Muller Code method for signature watermark embedding using error correcting code, while Hash Block Chainning for reference watermark embedding. With a host of type JPEG image, the quality of watermarked images was generated on each color component that has the same PSNR values. It is showed that PSNR values for each color component of this value is not much different from the value of PSNR as bitmap image of the host type.

Adiwijaya et al.[14] proposed a graph coloring technique in the quantization process on image compression scheme based on Wavelet-SVD. The system shows a good performance based on PSNR and compression ratio, with the testing results as follows. The average compression ratio generated by the system stands between 50-60%, while the average PSNR stands between 40-80dB.

This research proposed the use of Singular Value Decomposition (SVD) algorithm, because it has proven to give good performance on processing image. SVD gave good performance on image compression compared to JPEG2000[14]. Therefore, that Singular Value Decomposition will be suitable to be used on watermarking process.

3. Watermarking Using SVD Based on DCT and DWT

This research provided a scheme that will embed watermark image into host image that produce watermarked image. In addition, this scheme also extracts the embedded image from the watermarked image, so the image can be recovered. The scheme receives input consisting of host image (.bmp format), watermark image (.bmp format), input parameters decomposition level, embedding subband, scale factor and block for DCT. The output consisting of watermaked image, image disturbance result (.bmp format) and etraction image (.bmp format), along with value of PSNR and correlation coefficient for each image.

In a watermarking scheme using SVD based on DCT/DWT, there are two big actions; embedding process and extraction process. The figure below is the flow of SVD-DCT Insertion scheme in watermarking process to an image.

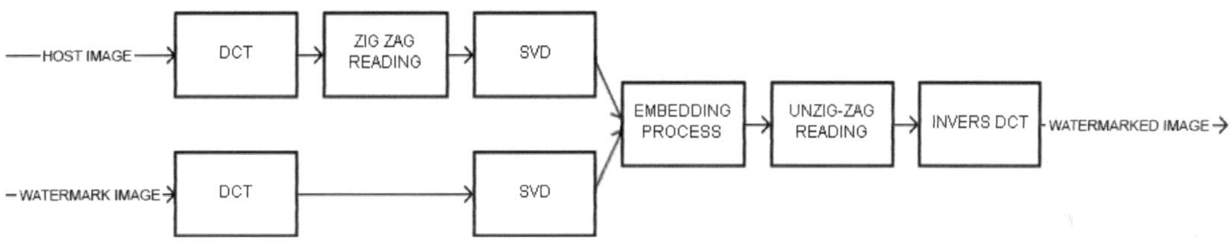

Figure 1. The Flow of SVD-DCT Insertion

Steps of watermark insertion process above are:

1. Applying DCT upon host image (A).
2. Zig-zag reading is aimed to make a vector asserting series of DCT coefficients from the low to the high frequency. Then, they are divided into 4 subbands of B1, B2, B3, and B4.
3. Determining the value of SVD subband chose for insertion location.

$$A^k = U_A^k S_A^k V_A^{kT} \tag{1}$$

   where

   k : chosen subband

   A : host image
4. Applying DCT upon watermark image (W) which will be inserted.
5. Determining SVD value of watermark image (W).

$$W = U_W S_W V_W^T \tag{2}$$

6. Singular value manipulation process is:

$$\lambda_i^{*k} = \lambda_i^k + \alpha_k \lambda_{Wi} \tag{3}$$

   where

   λ_i^{*k} : element of the diagonal of manipulated subband singular matrix (S_A^{*k}).

λ_i^k : element of the diagonal of host image singular matrix on subband where watermark is inserted (S_A^k).

λ_{Wi} : element of the diagonal of watermark image singular matrix (S_W).

α_k : scale factor of watermark insertion on a subband k.

Scale factor determines how big the singular value used in insertion.

7. Determining the new matrix of modified image, by using:

$$A^{*k} = U_A^k S_A^{*k} V_A^{kT} \quad (4)$$

where

A^{*k}: matrix of watermarking image in frequency domain

S_A^{*k}: matrix of modified singular subband

8. Doing position replacement using un-zig-zag reading of the new modified image matrix.
9. Applying inverse DCT after doing replacement to the original position.

The steps of extraction process are provided in Figure 2.

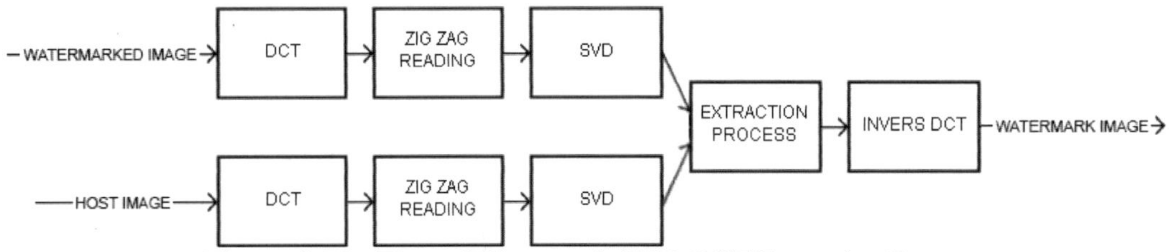

Figure 2. The Flow of Watermark SVD-DCT Extraction Process

To show the comparison of watermarking process, the insertion and extraction scheme in watermarking process using SVD based on DWT is provided.

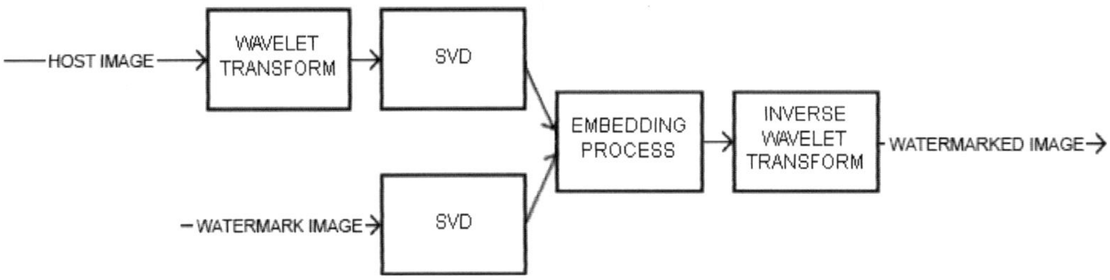

Figure 3. Embedded Process of Image Watermarking

The steps of watermark embedding process are:

1. Wavelet transform is charged on host image and produce a media image in frequency domain (there are 4 subband: LL, LH, HL, and HH)
2. One of the subband is selected as the embbeding point, then determining the SVD value

$$A^k = U_a^k S_v^k V_a^{kT} \quad (5)$$

where, k = 1,2,3,4 which indicate LL, LH, HL, and HH

A = host image

3. Calculate SVD value of watermark image

$$W = U_W S_W V_W^T \quad (6)$$

where, W = watermarked image

4. In the embedding process, there is a manipulation process of the subband singular matrix from orignial image (S_a^k) with singular matrix from watermark image (S_w). The manipulation process is as follows:

$$\lambda_i^{*k} = \lambda_i^k + \alpha.\lambda_w^k \tag{7}$$

where, λ_i^{*k}= diagonal entry singular matrix of the manipulated subband (S_a^{*k}).
λ_i^k = dagonal entry singular matrix of original image subband (S_a^k).
λ_w^k = diagonal entry singular matrix of watermarked image subband (S_w).

5. Produce a modified image matrix

$$A^{*k} = U_a^k S_a^{*k} V_a^{kT} \tag{8}$$

where, A^{*k} = singular matrix of watermarked image in frequency domain.
S_a^{*k} = Singular matrix of watermaked imge

6. Inverse wavelet transform is charged on matrix A^{*k} and produce a watermaked image.

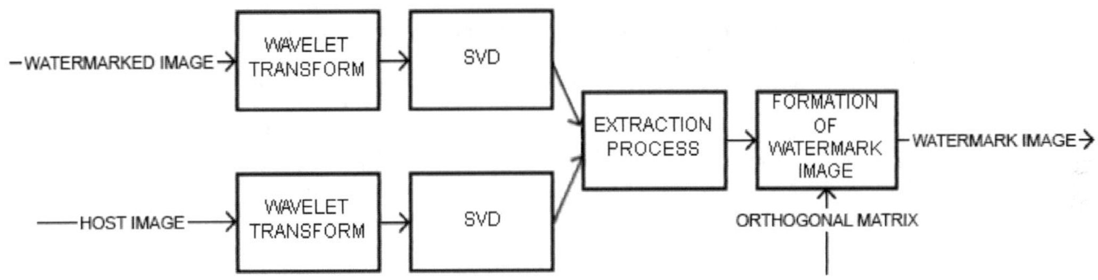

Figure 4. Extraction Process of Watermarked Image

The steps of watermark extraction process are:

1. Wavelet transform is charged on watermarked image (A^*), producing 4 subband : LL*, LH*, HL*, and HH*
2. Wavelet transform is charged on original image (A), producing 4 subband : LL, LH, HL, and HH
3. Compare the coefficient value of watermarked image subband with coefficient value of original image, so it detects in which subband watermark is embedded.
4. Calculate singular matrix of watermarked image subband (S_a^{*k})

$$A^{*k} = U_a^k S_a^{*k} V_a^{kT} \tag{9}$$

5. Calculate singular matrix of original image suband (S_a^k)

$$A^k = U_a^k S_a^k V_a^{kT} \tag{10}$$

6. In the extraction process, singular matrix of watermarked image subband (S_a^{*k}) is reduced by singular matrix of original image (S_a^k), and producing a singular matrix of watermark image (S_w). The reduction operation is as follows :

$$\lambda_i^w = \frac{(\lambda_i^{*k} - \lambda_i^k)}{\alpha} \tag{11}$$

7. Producing a matrix watermark image by multiplying the singular matrix watermark with its 2 orthogonal matrices.

$$W = U_w S_w V_w \tag{12}$$

4. Outcome and Test Analysis

The test is done by using a host image that showed on figure 5(a), the size is 512x512 which in .bmp format and the watermark image is showed on figure 5(b), the size is 128x128 which in *.bmp fomat and doing insertion on low frequency. It is for insertion on low frequency resulting extracted image which will be more robust towards several noises. This research provides correlation coefficient graphs showing the influence of scala factor on invisibility and robustness value if the watermark image is inserted in different subband and difference decomposition level of DWT and also difference block of DCT.

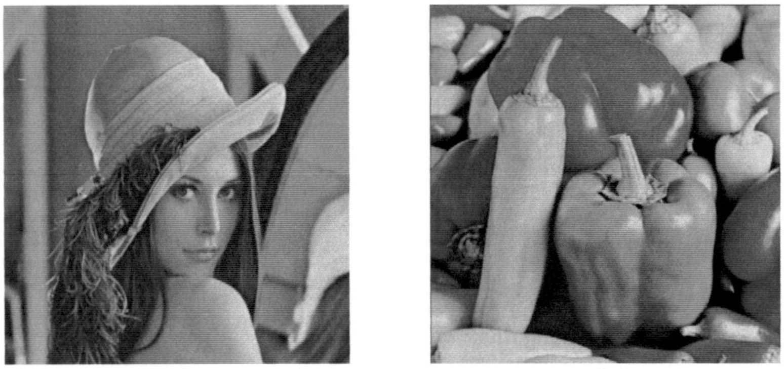

A b

Figure 5. (a) Original host image, (b) watermark image

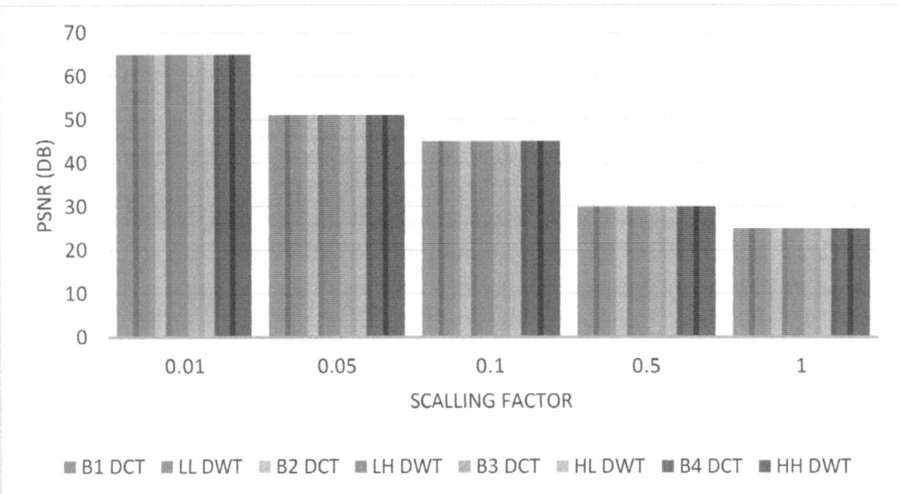

Figure 6. Influence of Scala factor on Invisibility

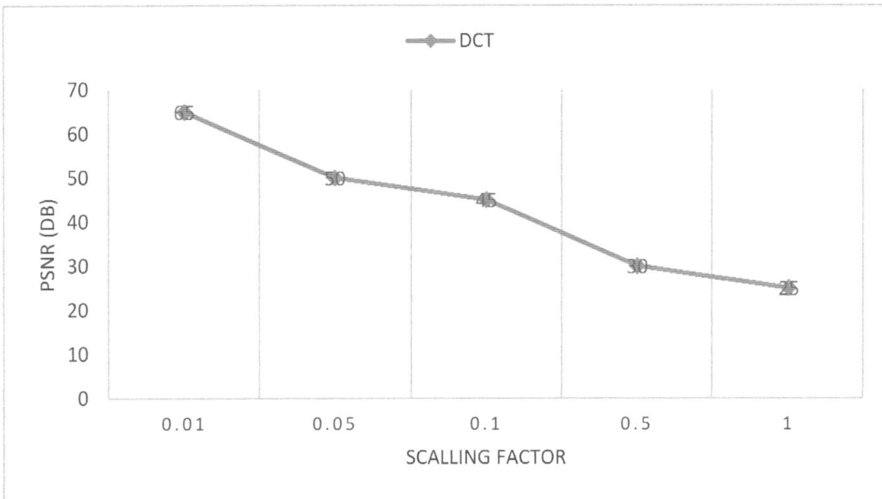

(a)

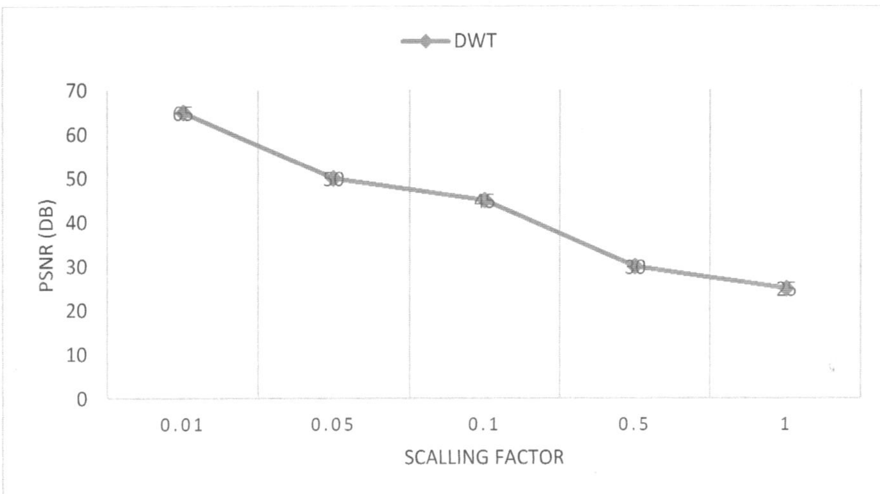

(b)

Figure 7. PSNR Invisibility of Watermarked Image (a) DCT (b) DWT

Figure 6 and 7 show that the increasing of the scale factor will increase the robustness of watermark against attacks, but it will increase the distortion of the watermarked image. That is why the PSNR value is decreasing. The decreasing of the scale factor will decrease the robustness of watermark against attacks, but it will also decrease the distortion of the watermarked image. Therefore, the selection of the scale factor must preserve the balance between robustness and the perceptual invisibility (distortion of watermarked image). The bigger scale factor used, the bigger image singular value inserted; hence the alteration watermarking image is bigger. The bigger alteration between watermarking image and original image will increase the error rate. It is shows that in scaling factor up to 0.1, watermarked image still has a good PSNR value (>30db) for all subband and all decomposition levels and all blocks.

Figure 7 (a) and (b) shows comparison between watermarking using DCT-SVD and DWT-SVD where the quality of watermarking image using DCT-SVD is equal with using DWT-SVD.

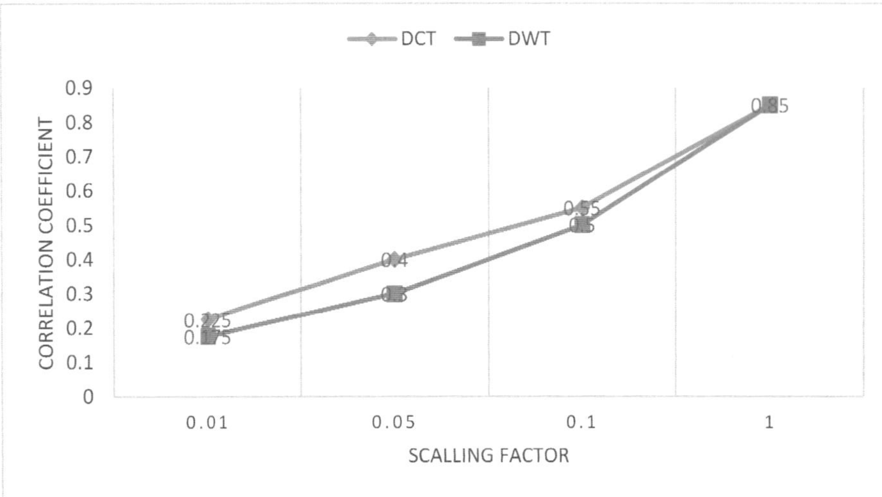

Figure 8. Robustness of Watermarking

Figure 8 shows that the bigger scale factor used, the bigger robustness of inserted image to a noise. Meanwhile, Figure 6 and 7 show that PSNR invisibility value is decreasing along with increasing on scale factor; but this increasing causes better robustness on watermark image. Figure above also proves trade-off between invisibility and robustness.

Correlation coefficient is a value that indicates the degree of connectivity between two variables. In general, it can be said that if the value of correlation coefficient is huge, then it means that the connectivity of the variables is strong and vice versa. The formula of correlation coefficient represents data as $\rho(x,y) = \frac{K\sum_{i=1}^{K} X_i Y_i - (\sum_{i=1}^{K} X_i)(\sum_{i=1}^{K} Y_i)}{\sqrt{[\sum_{i=1}^{K} X_i^2 - (\sum_{i=1}^{K} X_i^2)][\sum_{i=1}^{K} Y_i^2 - (\sum_{i=1}^{K} Y_i)]}}$, where X is host image and Y is watermarked image. Therefore, if the correlation coefficient is huge between those 2 variables, then there's a strong connectivity between them. If there's a strong connectivity between the host image and the watermarked image, then the less damage that will occur to the watermarked image (not many watermarked image information lost), therefore the more data contained within can be restored to its original form

Scale factor is a measure of how many data that will2 be inserted into an image, in this case is a singular matrix of watermark image. For example, if the scale factor is 0.05 then the amount of data that is inserted into the image is 5%. Therefore, scale factor shows the capacity of the inserted data.

Noise in watermarked image will decrease the quality of that image and influenced the durability of the image. But this decrease is not only influenced by the selection of subband insertion place, it is also influenced by the intensity of the interference.

This research provides scale factor effect to inserted image robustness. Disturbances given are noise, rescale, and JPEG compression.

IOP Conf. Series: Journal of Physics: Conf. Series **971** (2018) 012006 doi:10.1088/1742-6596/971/1/012006

- **Gaussian Noise**

Below is a correlation coefficient of watermark image robustness on the host image against JPEG compression

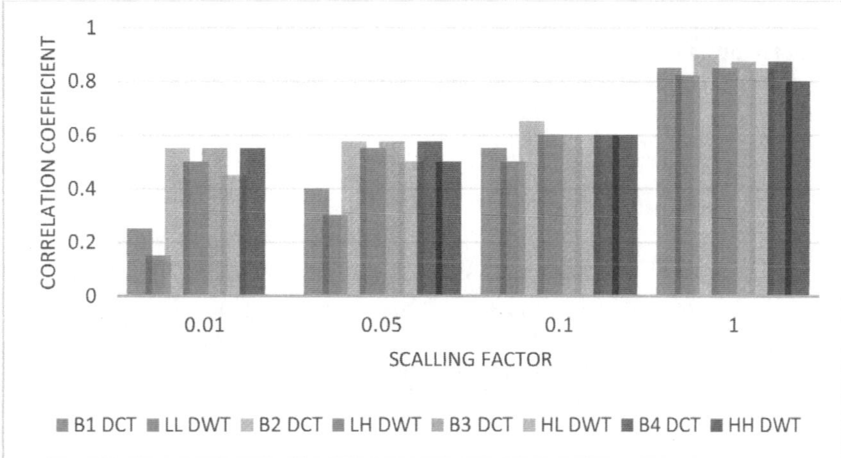

Figure 9. Robustness of watermaked image from Noise

Showed on figure 9, the bigger scale factor used, inserted image will be more robust to the noise. Using DCT results better robustness on inserted image than using DWT. It is also known that the inserted watermark image on subband LL couldn't withstand Gaussian noise for all scalling factor. This is due to the high coefficient that subband LL has compare to the others. The sum of noise matrix directly on subband LL coefficient will cause a huge change on subband coefficient, so at the time of extraction there are some components of the original low frequency image that get into high frequency range, so that only a little watermark image on the LL subband that can be extracted.

- **Rescale**

Below is a correlation coefficient of watermark image robustness on the host image against JPEG compression

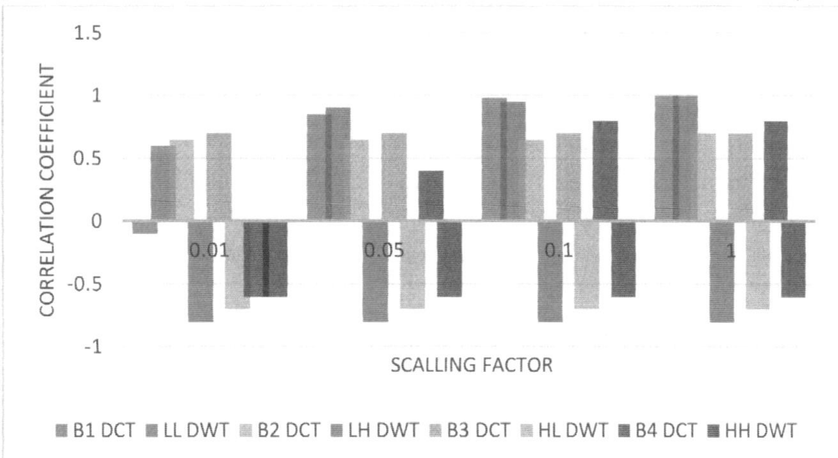

Figure 10. Robustness of Watermarked Image from Rescale

Figure 10 shows that the bigger scale factor used, inserted image will be more robust to rescale. Using DCT relatively results better robustness on inserted image than using DWT. It is known that watermark image durability towards rescale disturbance is determined by decomposition level, subband insertion place and wavelet filter that is used and also the image characteristic. This happened because rescale is a type of disturbance on spatial domain, that is by changing the size of the image resulted from watermarking to be a smaller size and then the size of the image will be returned to normal, so that there would not be any duplicated image component. If the watermark image is inserted on frequency image, it will cause inconsistency on an image durability towards rescale disturbance on specific level and wavelet filters.

- **JPEG Compression**

Below is a correlation coefficient of watermark image robustness on the host image against JPEG compression

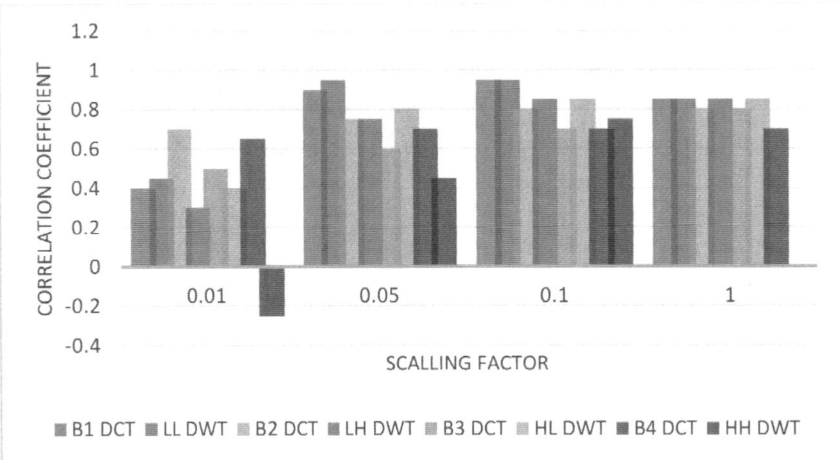

Figure 11. Robustness of Watermarked Imaged from JPEG Compression

On figure 11, the bigger scale factor used, then the inserted image will be more robust to compression, the correlation between the watermark image from extraction process and original watermark image is better, this is because of the higher scalling factor of JPEG Compression then the less information of watermarked image lost and the data contained can be restored. Using DWT relatively results better robustness on inserted image than using DCT. Besides, insertion on low frequency is relatively more robust than insertion on high frequency subband. It is also known that watermark image has a good durability if it is inserted on LL subband for all scalling factor. This is because LL subband has subband approximation, therefore LL subband coefficient has an important meaning on representing an image. While JPEG compression using DCT transformation that transforms an image from spatial domain to frequency domain. In DCT, the coefficient is sorted from the most important to the least important, the most important value will be on the top-left, and getting to the bottom-right it is getting less important. It also causes durability to scalling factor for DCT lower than DWT.

Compression is done by quantizing and removing the value from the bottom-right to the top-left to the desired quality limit. If the data is inserted on LL subband, then the image component that keep that data would not be wasted too much, so there will be many data that can be restored. The greater the JPEG compression quality used, the better the watermark image on the LL subband.

From Figure 9, 10, 11 above, it can be concluded that watermark image durability towards certain types of disturbance are influenced by the magnitude of the scale factor used and the use of a combination of input parameters (subband, decomposition level on DWT and block on DCT).

5. Conclusion

The bigger scale factor used, the more visible of alteration on the image; but the watermark image inside is more robust to noise. Furthermore, scale factor selection should consider the subband of insertion. High frequency insertion is better using smaller scale factor than in low frequency insertion. Watermark image robustness to noise is influenced by the type and intensity of noise, and insertion location on subband. Insertion on low frequency subband will be robust to From the experiment results, combination of SVD-DCT method result a better watermark robutsness from combination of SVD-DWT method. The correlation coefficient is related with robustness of watermark method, from the robust watermarking method shows that if watermark image after extraction and attackking process is compared with original watermark image, it has a good similarity (correlation coefficient). The proposed future works can be listed as follows:

a. The proposed watermarking scheme can be implemented on other digital medias such as audio and video;
b. The watermarking technique can be used and implemented on different kind of fields such as fingerprint, tamper detection, owner identification, data authentication and etc. Therefore, the proposed watermarking scheme can be implemented on other fields beside copyright-labelling;
c. The proposed watermarking scheme is a non-blind watermarking, therefore in the near future, a research can be done for the blind watermarking scheme using the proposed method.
d. A combination of watermarking methods can also be explored more in the near future and some optimization algorithms can also be used to maximize the result of the research.

IOP Conf. Series: Journal of Physics: Conf. Series **971** (2018) 012006 doi:10.1088/1742-6596/971/1/012006

References

[1] Abdi, Herve.(2007). *Singular Value Decomposition (SVD) and Generalized Singular Value Decomposition (GSVD)*. Encyclopedia of Measurement and Statistics.Another reference

[2] Ganic Emir, M. Eskicioglu Ahmet.(2005), "Robust Embedding of Visual watermarks Using DWT-SVD", Department of Computer and Information Science CUNY Brooklyn College.

[3] Gonzales,Rafael C.(1993). Digital Image Processing. Addison-Wesley Publishing Company Inc.

[4] Khayam,Syed Ali.(2003). *The Discrete Cosine Transform(DCT): Theory and Application*. Department of Electrical and Computer Engineering Michigan State University.

[5] Supangkat,Suhono H, Kuspriyanto, Juanda.(2000). *Watermarking sebagai Teknik Penyembunyian Label Hak Cipta pada Citra Digital*. Departemen Teknik Elektro Institut Teknologi Bandung.

[6] Sverdlov,A., Dexter,S., dan Eskicioglu, A..(2005). *Robust DCT-SVD Domain Image Watermarking For Copyright Protecyion: Embedding Data In All Frequencies.* http://www.sci.brooklyn.cuny.edu/~sdexter/Pubs/EUSIPCO 2005-WM.pdf

[7] Astuti, W, Adiwijaya.(2015). *Graph Coloring Based on Evolutionary Algorithms to Support Data Hiding Scheme on Medical Images.* Procedia Computer Science.

[8] Prathiwi, D. S., Astuti, W., Adiwijaya, and Wirayuda, T. A. B. (May, 2015). *Watermarking Scheme for Authenticity and Integrity Control of Digital Medical Image Using Reed-Muller Codes and Hash Block Chaining*. In 2015 3rd International Conference on Information and Communication Technology (ICoICT) pp. 23-29.

[9] Adiwijaya, Faoziyah, P. N., Permana, F. P., Wirayuda, T. A. B., and Wisesty, U. N. (Sept, 2013). *Tamper Detection and Recovery of Medical Image Watermarking Using Modified LSB and Huffman Compression*. In 2013 Second International Conference on Informatics Applications (ICIA) pp. 129-132.

[10] Rosiyadi, D., Horng, S. J., Fan, P., Wang, X., Khan, M. K., and Pan, Y. (July, 2012). *Copyright Protection for E-Government Document Images*. IEEE MultiMedia, 19(3), 62-73.

[11] Horng, S.-J., Rosiyadi, D., Li, T., Takao, T., Guo, M., and Khan, M. K. (2013). *A Blind Image Copyright Protection Scheme for E-government*. Journal of Visual Communication and Image Representation. 24(7), 1099-1105.

[12] M. T. Kurniawan, Adiwijaya and W. Agung.(2012). *Multiple Watermarking on Digital Medical Images for Tamper Detection and Integrity Control*. 2nd International Conference on Uncertainty Reasoning and Knowledge Engineering, Jakarta. pp. 145-148.

[13] Adiwijaya, T. A. B. Wirayuda, S. D. Winanjuar, and U. Muslimah. (2014). *The Multiple Watermarking on Digital Medical Image for Mobility and Authenticity*. pp. 457–462, Springer International Publishing, Cham, 2014.

[14] Adiwijaya, M.Maharani, B.K.Dewi, F.A.Yulianto, and B.Purnama. (2013). *Digital Image Compression Using Graph Coloring Quantization Based on Wavelet-SVD.* Journal of Physics: Conference Series 423.

[15] Tao, H., Chongmin, L., Zain, J. M., and Abdalla, A. N. (2014) Robust Image Watermarking Theories and Techniques: A Review. Journal of Applied Research and Technology, 12(1), 122-138.

[16] Prathiwi, D. S., Astuti, W., Adiwijaya, and Wirayuda, T. A. B. (May,2015) Watermarking scheme for authenticity and integrity control of digital medical image using Reed-Muller Codes and Hash Block Chaining. In 2015 3rd International Conference on Information and Communication Technology (ICoICT) pp. 23-29.

Handling imbalance data in churn prediction using combined SMOTE and RUS with bagging method

Eka Pura Hartati[1], Adiwijaya[2], Moch Arif Bijaksana[3]

School Of Computing, Telkom University, Indonesia

ekapura@telkom.co.id[1], adiwijaya@telkomuniversity.ac.id[2], arifbijaksana@telkomuniversity.ac.id[3]

Abstract. Customer churn has become a significant problem and also a challenge for Telecommunication company such as PT. Telkom Indonesia. It is necessary to evaluate whether the big problems of churn customer and the company's managements will make appropriate strategies to minimize the churn and retaining the customer. Churn Customer data which categorized churn Atas Permintaan Sendiri (APS) in this Company is an imbalance data, and this issue is one of the challenging tasks in machine learning.
This study will investigate how is handling class imbalance in churn prediction using combined Synthetic Minority Over-Sampling (SMOTE) and Random Under-Sampling (RUS) with Bagging method for a better churn prediction performance's result. The dataset that used is Broadband Internet data which is collected from Telkom Regional 6 Kalimantan.
The research firstly using data preprocessing to balance the imbalanced dataset and also to select features by sampling technique SMOTE and RUS, and then building churn prediction model using Bagging methods and C4.5.

1. Introduction

Customer churn has become a significant problem and also a challenge for Telecommunication company. It is necessary to evaluate whether the big problems of churn customer and the company's managements will make appropriate strategies to minimize the churn from its own data and retaining the customer to save the revenue.
Churn prediction is considered as one of data mining application that reflecting imbalanced problems. Based on data from PT. Telkom Indonesia, the average churn rate Atas Permintaan Sendiri (APS) for Internet broadband customer of PT.Telkom Indonesia Regional 6 Kalimantan was less than 1% per month. The imbalanced data-set problem occurs when one class, usually the one that refers to the concept of interest (positive or minority class), is underrepresented in the data-set and the number of negatives (majority) instances have outnumbers of positive class instances [1].
The objective of this research is to handle imbalanced data problems using combine sampling SMOTE and RUS, and measure the performance of churn prediction using a classifier C4.5 with bagging approach. The dataset used in this research is dataset from PT. Telkom Indonesia Regional 6 Kalimantan which has 1,08% churn data (imbalanced dataset). A higher performance has been obtained by applying combine sampling technique SMOTE and RUS. The implementation of bagging approach in classification is able to improve the F-score of a single classifier (C4.5).

2. Literature Review

2.1. Related Work

Many researchers have attempted to address the imbalanced data. Because this learning task is quite challenging over the past 15 years. A lot of techniques have been developed for dealing with imbalanced data according to their categories approach. There were researchers who focus on algorithm level, called Internal approach, some others focus on the data level, called external approach and the last cost-sensitive learning falls between data and algorithm approach.

Based on three approaches above, algorithm level tries to adapt the existing classifier to bias the learning toward the minority class [2]. On data level approach aims to rebalance the class distribution by resampling the data [1], [3]. These techniques were independent of the classifier used and usually more versatile. The last cost-sensitive level incorporates both data level transformation by adding cost to an instance and modify the algorithm so accepts the additional cost [4].

Ensembles based methods that used in this research was bagging. This technique consists in a combination of bagging and data level approach to preprocess the data before training each classifier.

2.2. Customer Churn Management

Churning Customers can be divided into two groups, Voluntary and Non-Voluntary churner (Mattison, 2005). Non-voluntary churners are easier to identify, They have had their service withdrawn by Company. This group includes Customers that churned for fraud and non-payment Customer. Voluntary churners are more difficult to identify because these churners occur when the customer makes a decision to terminate their service to a provider. Voluntary churners can be sub-divided into two groups, an incidental churner, and a deliberate churner. an incidental churner usually happens when there is a change in circumstances prevents the Customer from further requiring the provided service. Deliberate churner happens for reason of technology (there is a new technology), economics (customer's packet price), service quality, psychological factor and other reasons.

2.3. Methods and Classifier

2.3.1. Random Under Sampling, SMOTE and Bagging. Sampling is a popular methodology to handle the problem of class imbalance. There are three types of sampling methods, namely Random Under-Sampling (RUS), Random Over Sampling and Hybrid methods that combine both sampling methods. RUS method, is a nonheuristic method that aims to balance class distribution through the random elimination of majority class examples. The majority class instances are discarded at random until a more balanced distribution is reached. Its major drawback is that it can discard potentially useful data, which could be important for the induction process. Random Over Sampling methods, which create a superset of the original data-set by replicating minority class instances but it can also increase the likelihood of occurring the overfitting, since it makes exact copies of existing instances. SMOTE is an oversampling method which the main idea is to create new minority class examples by interpolating several minority class instances that lie together. Its create instances by randomly selecting one or more of the nearest neighbors (kNN) of a minority class instance and the generation of the new instance value from a random interpolation of both instances. Thus, the overfitting problem is avoided and causes the decision boundaries for the minority class to be spread further into the majority class space. Bagging is one of ensemble methods which manipulate the training data set by sampling with replacement. Each new training data set has the same size with the original data training. Bagging methods use diverse training samples (bags) to train independent base learner capable of handling imbalanced data in a parallel way. Bagging method is very suitable for unstable classifier [1], through their ability technique to do sampling with a replacement which can make lower variance of dataset.

2.3.2. Classifier This research used C4.5 Decision Tree as the weak learner, because C4.5 gives a good performance for binary classification [1]. C4.5 is an improvement of ID3 which it is a possibility to use continuous data, using unknown (missing) values, having the ability to use attributes with different weights, pruning the tree after being created, pessimistic prediction error, and also sub-tree raising [2]. Moreover, its classification model is easy to understand and have high precision but C4.5 is also an unstable classifier like regression tree, artificial neural network, and rule-base classifier. This classifier is very sensitive to the change of data training, if the data is changed so does the classifier while ensemble learning can effectively improve its stability and generalization performance [3].

3. Material and Proposed Method

3.1. Proposed Scheme

The combined sampling SMOTE with RUS was proposed in this research to balance the dataset and using bagging approach to classify the churn class. The design process of this research was as described below.

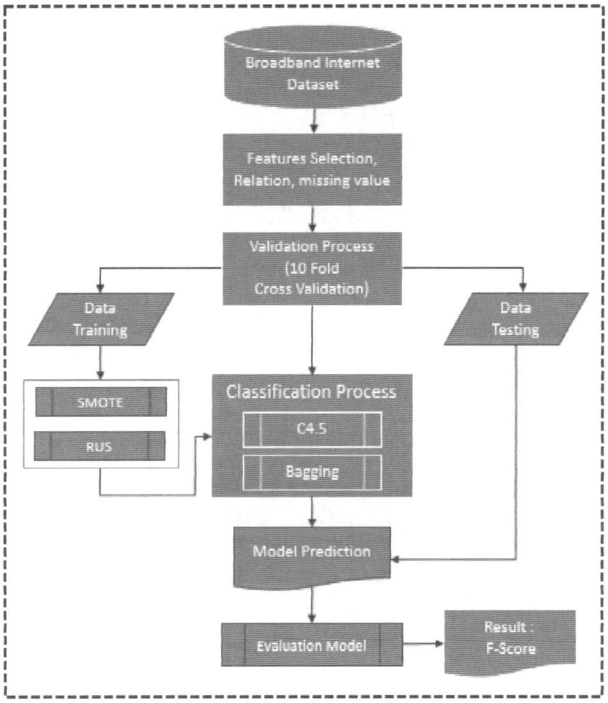

Figure 1. Block Diagram of Research Design

3.1.1. Data Preparation. After collecting data from database resource then those data were combined and selected the fitting attributes. These were seven attributes that were used in this process, namely CUST_OLD, CUST_CAT, CUST_BILL, PACKET_NAME, KWAD, TROUBLE_TIC, STAT_CHURN. The Attribute with a missing value and had redundant information was not used in this research and then all dataset transformed into numerical type for the next step.

3.1.2. Data Preprocessing. Data preprocessing was a process before data were processed. Sampling became one of standard approach for improving classification. The training data was altered to create more balanced class distribution before they were used in the learning stage. This research used combined RUS and SMOTE as sampling methods. Therefore the dataset could be balanced by neither loosing information too much (i.e., under sampling too many majority class instances), nor suffering from over-fitting (i.e., oversampling too heavily).

3.1.3. Learning Stage. There were two processes in this stage, they were training and testing. Bagging was one of ensemble methods which manipulate the training dataset by sampling with replacement. Each new training dataset has the same size with the original data training. Bagging methods use diverse training samples (bags) to train independent base learner capable of handling imbalanced data in a parallel way. Bagging method is very suitable for an unstable classifier, through their ability technique to do sampling with a replacement which can make lower variance of dataset. It also used voting to choose the best prediction that was resulted from basic classifiers. In this stage, C4.5 was used for a basic classifier to predict the churn. The decision trees were built, using a set of a training dataset. At each node of the tree, C4.5 taken one attribute of the data that most effectively splits its set of samples into subsets enriched in one class or the other. After the churn model has built then it was tested using prepared data test by 10-fold cross-validations technique.

4. Experiments

4.1. Parameter Measurement

Precision is the number of True Positives (TP) divided by the number of True Positives (TP) and False Positive (FT). It is the number of positive prediction divided by the total number of positives class value predicted (1). Recall is the number of True Positives (TP) divided by the number of True Positives (TP) and the number of False Negatives (FN). It is the number of positive prediction divided by the number of positive class values in the test data (2).

Table 1. Confusion Matrix for Two-Class

ACTUAL CLASS	PREDICTIVE CLASS	
	POSITIVE (CHURN)	NEGATIVE (NOT CHURN)
POSITIVE (CHURN)	True Positive (TP)	False Negative (FN)
NEGATIVE (NOT CHURN)	False Positive (FP)	True Negative (TN)

$$\text{Precision} = TP/(TP+FP) \quad (1)$$

$$\text{Recall} = TP/(TP+FN) \quad (2)$$

$$\text{F-Score} = (2\ (\text{precision} \times \text{recall}))\ /\ ((\text{precision}+\text{recall})\ ) = (2 \times TP)\ /\ (2 \times TP + FP + FN) \quad (3)$$

Precision and recall can be summarized into another metric known as the F-measure. It is also called the F Score. The F Score conveys the balance between the precision and the recall (3).

4.2. Experiment's Scenario

The data set that use in this reseach is data broadband Internet from PT. Telekomunikasi Indonesia Area Regional 6 Kalimantan (Kalsel, Kalbar, Kalteng, Balikpapan, Samarinda, and Tarakan) in October 2016.

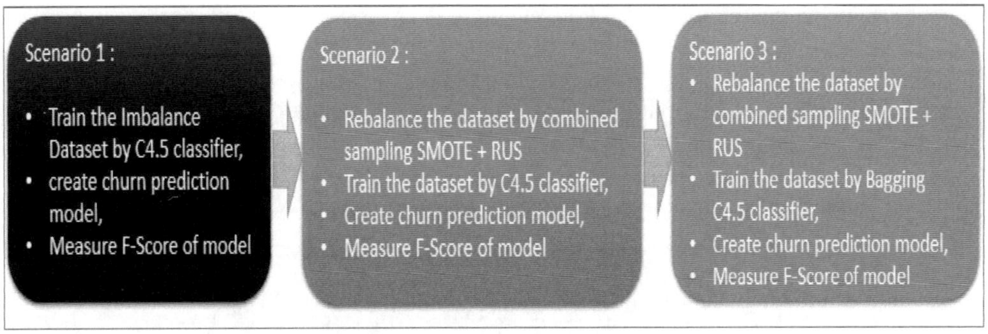

Figure 2. Experiment's Scenario

There were three steps of Experiment's Scenario:
1) Scenario 1: the first scenario aims to show the performance of classification without using data sampling and only use single classifier C4.5. F-Score was measured to see the performance of churn prediction model.
2) Scenario 2: the second scenario shows the performance of classification using data preprocessing by sampling with RUS and SMOTE in the learning process. It was also using single classifier C4.5. The F-Score was measured with combined SMOTE (find the best N% SMOTE) and RUS (find the best N% RUS). Then the F-score performance was evaluated.
3) Scenario 3: the last scenario aims to show the performance of classification with data preprocessing combined SMOTE and RUS, and using bagging methods in classification. This scenario used C4.5 for its classifier and investigate how many bags (Nmodel) that showed the best performance.

4.3. Experiment's Result and Discussion

The first scenario aims to show the performance of classification without using data sampling and only use single classifier C4.5. F-Score was measured to see the performance of churn prediction model with F-score.

Table 2. F-Score Value of Scenario 1

Measurements	Number of Experiments									
	1	2	3	4	5	6	7	8	9	10
TESTING (F-Score)	0,011	0,016	0,019	0,014	0,015	0,020	0,019	0,014	0,017	0,017
TIME (sc)	14,710	13,539	14,257	13,922	13,809	13,611	13,437	14,201	13,454	13,434

There were not any input parameters at this fist scenario because it just wants to show the real result of C4.5 as a single classifier to build a churn model (base line) and to measure the F-Score value. The average F-Score of Training in this scenario was 0.053 and Testing was 0.013. The result of this scenario still had a small value of F-score. The small value of F-Score because the dataset that used was still imbalanced and in the next scenario, the dataset will be rebalanced by using sampling process.

The second scenario showed the performance of the classification using data preprocessing by sampling with SMOTE and RUS in the learning process. It also used single classifier C4.5. The F-Score was measured with combined SMOTE (find the best N% SMOTE) and RUS (find the best N% RUS). The amount of minor data were 1.502 and the variable N SMOTE used in this experiments were N = 5x, N = 10x and N = 20x, thus, the minority class was generated as many as 5x (minor data will become 7.510),

10x (minor data will become 15.020) and 20x (minor data will become 30.040) in the experiment. By SMOTE processing, the nearest neighbors (kNN) with k = 3 was used to avoid overfitting problem and caused the decision boundaries for the minority class to spread further into the majority class space. The N RUS variables that follow this experiments were N = ½ and N = ¾, the amount of major data were 136.404. It then reduced the major data by N= ½ means the major data became 68.202. The F-score performance was evaluated. The results were shown in the tables below:

Table 3. F- Score Value of Scenario 2A, N RUS=1/2, N SMOTE=20x

Measurements	Number of Experiments					AVERAGE
	1	2	3	4	5	
F-SCORE TESTING	0,0444	0,0452	0,0450	0,0443	0,0448	0,045
CH RATE TESTING (%)	0,5921%	0,5792%	0,5860%	0,5921%	0,5869%	
TIME (sc)	343,499	344,739	342,649	343,485	341,581	343,629

Table 4-3 shows there is an increase of F-Score value, with average value is 0,045 with N SMOTE = 20x and N RUS = ½.

Table 4. F- Score Value of Scenario 2A, N RUS=1/2, N SMOTE=10x

Measurements	Number of Experiments					AVERAGE
	1	2	3	4	5	
F-SCORE TESTING	0,081	0,081	0,082	0,080	0,085	0,081
CH RATE TESTING (%)	0,8140%	0,8173%	0,8121%	0,8143%	0,8129%	
TIME (sc)	340,342	342,646	327,057	343,395	341,700	336,682

In the table 4-4, the N SMOTE = 10x and N RUS = ½, the F-Score value gets a higher value with average 0,081.

Table 5. F- Score Value of Scenario 2A, N RUS=1/2, N SMOTE=5x

Measurements	Number of Experiments					AVERAGE
	1	2	3	4	5	
F-SCORE TESTING	0,087	0,090	0,091	0,086	0,093	0,089
CH RATE TESTING (%)	0,8339%	0,8300%	0,8268%	0,8316%	0,8267%	
TIME (sc)	338,688	349,440	337,294	338,936	339,061	341,807

Table 4-5 shows the F-Score value still increase with average 0,089 and comparing to the tables above, there is a significant improvement of F-Score especially with N SMOTE = 5x and N RUS = ½. The average value of baseline F-Score was 0,013 and had increased to 0,089. Since the implementation of SMOTE and RUS, the churn rate value also increased the churn rate. It indicated that there was a decreasing number of majority data in order to balance the dataset. As the resume, the values of F-Score Testing of N RUS = ½ were presented by the graphic below.

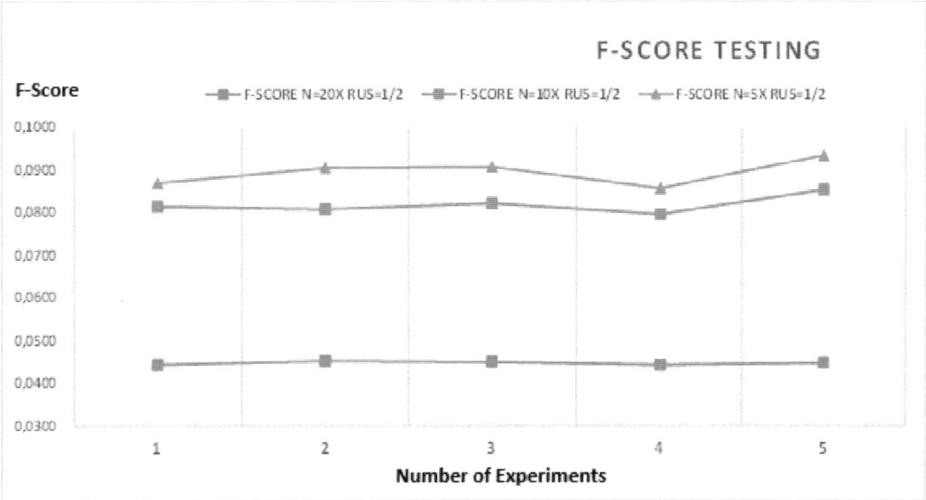

Figure 3 F-Score Testing of N RUS = 1/2

From variable N = ½ , the research continued to check another value of N RUS = ¾. The major data reduced to 102.303 and N SMOTE = 20x, 10x and 5x.

Table 6. F-Score Value of Scenario 2B, N RUS=3/4, N SMOTE=20x

Measurements	Number of Experiments					AVERAGE
	1	2	3	4	5	
F-SCORE TESTING	0,066	0,068	0,069	0,072	0,071	0,068
CHURN RATE TESTING	0,8132%	0,8123%	0,8118%	0,8098%	0,8126%	
TIME (sc)	375,355	374,932	375,345	375,965	375,888	375,211

Table 6 shows there is an increase of F-Score value, with average value is 0,068 with N SMOTE = 20x and N RUS = ¾.

Table 7. F-Score Value of Scenario 2B, N RUS=3/4, N SMOTE=10x

Measurements	Number of Experiments					AVERAGE
	1	2	3	4	5	
F-SCORE TESTING	0,089	0,086	0,087	0,092	0,088	0,087
CHURN RATE TESTING	0,8264%	0,8279%	0,8284%	0,8237%	0,8272%	
TIME (sc)	387,647	375,395	376,106	379,773	372,092	379,716

In the table 7, the N SMOTE = 10x and N RUS = 3/4, the F-Score value gets a higher value with average 0,087.

Table 8. F-Score Value of Scenario 2B, N RUS=3/4, N SMOTE=5x

Measurements	Number of Experiments					AVERAGE
	1	2	3	4	5	
F-SCORE TESTING	0,093	0,089	0,089	0,094	0,090	0,090
CHURN RATE TESTING	0,8302%	0,8332%	0,8324%	0,8302%	0,8382%	
TIME (sc)	372,364	371,603	371,809	372,068	372,518	371,925

Table 8 shows the F-Score value still increase with average 0,090 and comparing to the three tables above, there is a significant improvement of F-Score specially with N SMOTE = 5x and N RUS = 3/4. SMOTE created instances by interpolating minor data in 5x ratio and RUS decreased the major data by N= ¾, these best parameters would be used in the next scenario 3 to obtain the better F-Score value with bagging approach. The F-Score's testings of scenario 2B were figured in the graphic below as the resume from the tables above.

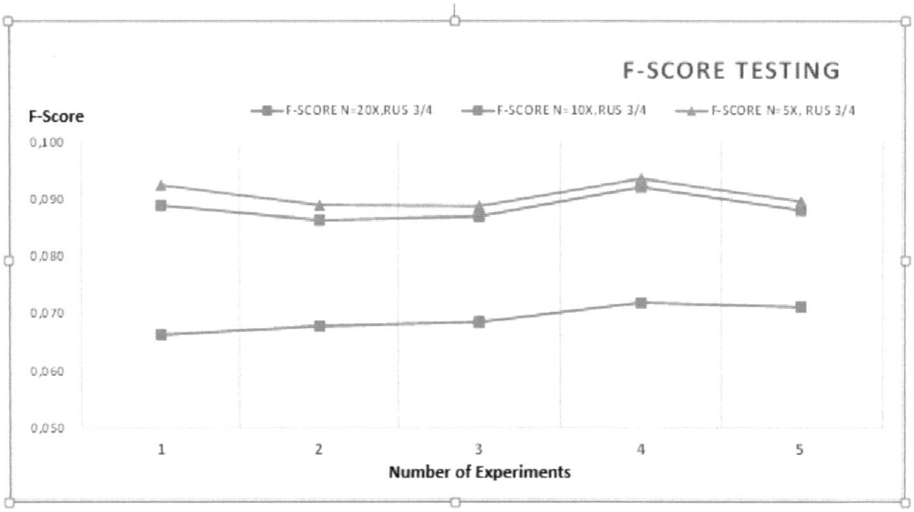

Figure 4. F-Score Testing of N RUS = 3/4

The combination of SMOTE and RUS had been able to reduce the probability of overfitting problem and rebalance the data distribution of majority class without loosing its substantial data. The best result from those parameters will be used in the next scenario 3 to aims the better F-Score value with bagging approach.

The testing's result with bagging compared to without bagging and with a number of bags = 5 or 7. It was shown the better F-Score resulted from bagging process with numbers of bags 7 than 5. The best average F-Score value was 0,140 as explained in the table below.

Table 9. F-Score Value of Scenario 3, N RUS=3/4, N SMOTE=5x, Bags=5

Measurements	Number of Experiments					AVERAGE
	1	2	3	4	5	
F-SCORE TESTING	0,159	0,138	0,118	0,126	0,126	0,138
CH RATE TESTING (%)	0,8308%	0,8308%	0,8315%	0,8351%	0,8336%	
TIME (sc)	1.875,62	1.878,40	1.872,86	1.727,26	1.882,81	1.875,63

Table 9 shows there is an increase of F-Score value, with average value is 0,138 with N SMOTE = 5x, N RUS = ¾ and bags = 5.

Table 10. F-Score Value of Scenario 3, N RUS=3/4, N SMOTE=5x, Bags=7

Measurements	Number of Experiments					AVERAGE
	1	2	3	4	5	
F-SCORE TESTING	0,178	0,146	0,132	0,141	0,136	0,152
CH RATE TESTING (%)	0,8328%	0,8329%	0,8351%	0,8301%	0,8322%	
TIME (sc)	2.640,42	2.627,27	2.645,73	2.628,35	2.629,72	2.637,80

In the table 4-10, the N SMOTE = 5x, N RUS = ¾ and bags = 7, the F-Score value gets a higher value with average 0,152.

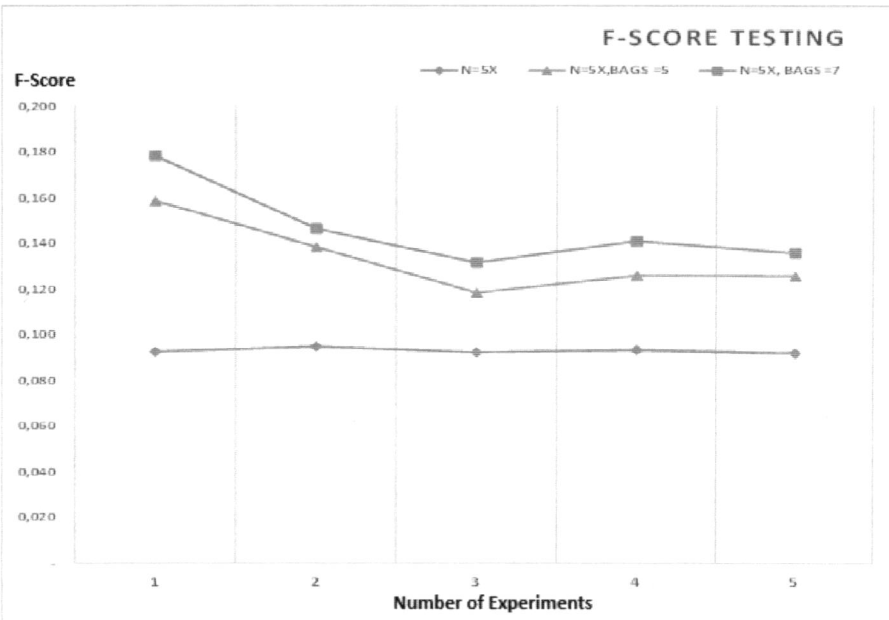

Figure 5. F-Score Testing of Scenario 3

The graphic as described in figure 4 conveys that the more bags used in the research the better F-Score we had since the new dataset was formed to train each single classifier of C4.5 by randomly drawing (with replacement) instances from the original dataset and the best prediction from each bag then tobe chosen to infer the class.

5. Conclusions & Future Work

5.1. Conclusions

Based on the result this research, the performance of F-score is quite good, but it still needs more research to obtain the increase of F-score. The combined SMOTE and RUS were able to handle the imbalanced data and improved the churn prediction performance. SMOTE was used to generate the synthetic data from the churn class in order to increase the probability of drawing the churn data (churn is the minority class). RUS was used to reduce the probability of overfitting problem caused by the oversampling in SMOTE. This combined method improved the F-Score to 571 % from dataset without sampling. The Average F-Score Value was 0,090. The implementation of sampling and bagging methods in prediction process by C4.5 improved the performance of F-Score. The F-Score increased about 56 % from without bagging. The average F-Score was 0,152, this combined method was very suitable for unstable classifier like C4.5, through their ability technique to do sampling with replacement to reduce the variance of dataset. The number of bags also influenced the improvement of F-Score since there were more option values for the voting process. The best F-Score resulted from a number of bags = 7.

5.2. Future Work

What most attributes that can influence customer is also needed to be investigated in attempt to identify why they churn and also an additional comparison for other ensemble methods such as boosting is needed to be evaluated

References

[1] N. V. Chawla, N. Japkowicz and A. Kotcz, "Editorial : special issue on learning from imbalanced data sets," *ACM Sigkdd Explorations Newsletter,* pp. 1-6, 2004.

[2] G. M. Weiss, "Mining with rarity: a unifying framework," *ACM SIGKDD Explorations Newsletter,* p. 7–19, 2004.

[3] H. Haibo, M. Yuqian and others, "Imbalanced Datasets : from Sampling to Classifiers," *Imbalanced Learning, Foundation, Algorithms, and Application,* pp. 52-69, 2013.

[4] E. Dwiyanti, Adiwijaya and A. Ardiyanti, "Handling Imbalanced Data in Churn Prediction Using RUSBoost and Feature Selection," *International Conference Soft Computeing and Data Mining,* pp. 376-385, 2016.

[5] Tan, P. Ning, M. Steinbach and others, "Introduction on Datamining," *Boston, Pearson Addison Wesley,* 2006.

[6] S. Kotsiantis, D. Kanellopoulos, P. Pintelas and others, "Handling imbalanced datasets: A review," *GESTS International Transactions on Computer Science and Engineering,* pp. 25-36, 2006.

[7] J. Tang, S. Alelyani and H. Liu, "Feature selection for classification: A review," *Data Classification: Algorithms and Applications,* p. 37, 2014.

[8] Q. R.J, "Bagging, Boosting, and C4.5," in *in AAAI/IAAI : Proceeding of the 13th National Conference on Artificial Intelegence and 8th Innovative Applications of Artificial Intelligence Conference*, Portland, Oregon, 1996.

[9] H. Haibo, M. Yuqian and others, Imbalanced Learning, Foundations, Algorithms, and Applications, Wiley, 2013, pp. 22-51.

[10] L. S. Salzberg, "C4. 5: Programs for machine learning by j. ross quinlan. morgan kaufmann publishers, inc., 1993," *Machine Learning,* vol. 16, pp. 235-240, 1994.

Process mining in oncology using the MIMIC-III dataset

Angelina Prima Kurniati[1,2], Geoff Hall[3,4], David Hogg[1], Owen Johnson[1]

[1]School of Computing, University of Leeds, UK
[2]School of Computing, Telkom University, Bandung, Indonesia
[3]School of Medicine, University of Leeds, UK
[4]Leeds Institute of Cancer and Pathology, St James's University Hospital, Leeds, UK
Corresponding author's e-mail address: scapk@leeds.ac.uk

Abstract. — Process mining is a data analytics approach to discover and analyse process models based on the real activities captured in information systems. There is a growing body of literature on process mining in healthcare, including oncology, the study of cancer. In earlier work we found 37 peer-reviewed papers describing process mining research in oncology with a regular complaint being the limited availability and accessibility of datasets with suitable information for process mining. Publicly available datasets are one option and this paper describes the potential to use MIMIC-III, for process mining in oncology. MIMIC-III is a large open access dataset of de-identified patient records. There are 134 publications listed as using the MIMIC dataset, but none of them have used process mining. The MIMIC-III dataset has 16 event tables which are potentially useful for process mining and this paper demonstrates the opportunities to use MIMIC-III for process mining in oncology. Our research applied the L* lifecycle method to provide a worked example showing how process mining can be used to analyse cancer pathways. The results and data quality limitations are discussed along with opportunities for further work and reflection on the value of MIMIC-III for reproducible process mining research.

1. Introduction

Cancer can affect any part of the body [1] and is recognized as a large group of diseases with at least 65 recognised types of cancer [2]. The complex nature of this disease makes choices in cancer care pathways particularly challenging. Process mining offers the opportunity to develop a deeper understanding of this complexity and, if applied to cancer patient records, may help improve cancer care pathways and outcomes for cancer patients.

Process mining [3] is an emerging approach for discovering and analysing business process models which uses data extracted from the event logs in information systems. An event log is a record of timestamped activities automatically generated by the system as it is used to support the business activities of an organisation. These records can provide insights into the real-world effectiveness of business processes. Process mining has been applied to the analysis of healthcare processes with the aims of improving quality of care, patient safety, and optimization of resources [4]. Process mining is especially useful to analyse highly complex and flexible patient care processes (care pathways), as happen in cancer patient treatments.

Process mining has been used in healthcare to describe what happened, why it happened, what will happen, and what is the best that can happen [5]. Rojas et al. [6] reviewed previous studies using process mining in healthcare, and found that the most process mining case studies in healthcare were in oncology. A previous study by the authors [7] found 37 peer reviewed papers describing process mining based research in oncology with a regular complaint being the limited availability and accessibility of suitable fine grained datasets with suitable information for process mining. Most of the papers (24 of 37) focused on gynaecological cancer, mainly because there was a Business Process Improvement (BPI) challenge held using this dataset [8]. We conclude that the availability of data for researchers is a key enabler for process mining research. Specifically, the availability and accessibility of sufficiently large volumes of detailed patient data is due to the sensitivity and confidentiality of such personal data.

Publicly available datasets are one option and this paper describes the potential to use MIMIC-III for process mining in oncology. The MIMIC-III dataset is an open access dataset from a hospital in the USA with a large number (n=46,520) of de-identified patient records. This paper presents a worked example to explain how MIMIC-III can be used to analyse patient data using process mining tools. We describe how selection criteria were used to construct event logs from cancer patient treatment records and then analysed to gain insight into treatment pathways. Our experiments are reproducible using the available information from the MIMIC-III database and all SQL queries, figures, resulting process models and supporting resources have been made available in a GitHub repository.

2. Background

2.1. Healthcare processes

The processes during a patient's time in a hospital consists of many different activities, including administrative (admission, discharge, transfer to a ward, etc.) and clinical activities (triage, test and scans, diagnosis, therapy, etc.) [9]. These activities are performed by different clinical roles (doctors, nurses, technical specialist, etc.) and vary from one healthcare organization to another [5]. Healthcare processes are recognized as nontrivial because the steps involved are nonlinear, complex and unpredictable, such processes do not always follow standard sequences [10].

Many healthcare organizations are now using electronic health record (EHR) systems to record administrative and clinical information about their patients and track the treatments provided. These EHR systems evolved from paper-based physician notes and the requirement to structure records more formally has increased as health organisations have grown in size and complexity [11]. Patient-level information, including demographic data and some clinical information (e.g. allergies, long-term conditions) is supplemented by time-stamped records recording observations, diagnosis, prescriptions, treatment and administrative processes such as admission and discharge [12]. Event data will generally be a mixture of coded variables and natural language text logged against the date, time, user id and type of event. These EHR systems therefore contain longitudinal data that can be explored using process mining techniques [4] and, as EHR systems mature, the opportunities to find and analyse process information about treatment pathways are growing.

Our paper focuses on cancer patient treatment and the sequence of administrative steps in the care pathways. Because we work only in the administrative steps, we hope that the approach and the data are appropriate for more general use.

2.2. Process mining

Process mining is an emerging research discipline which combines computational intelligence, data mining, process modelling and analysis approaches. The idea of process mining is to discover, monitor, and improve real processes by extracting knowledge from the event logs that can be extracted from business information systems [3]. Process mining is complementary with more traditional process modelling performed by business analysts.

There are three types of process mining: discovery, conformance checking, and enhancement [13]. Discovery takes an event log and creates a graphical business process model. Conformance checking can be used to check if reality, as recorded in the log, conforms to the model and vice versa. In process modelling a process can be represented as a directed graph structure with nodes representing transitions (i.e. events that may occur, e.g. chemotherapy, follow up) and places (i.e. conditions). The directed arcs determine which places are pre- and/or post conditions for which transitions. This structure for modelling a business process can be shown graphically following notations such as Petri nets [14] or

IOP Conf. Series: Journal of Physics: Conf. Series **971** (2018) 012008 doi:10.1088/1742-6596/971/1/012008

Business Process Modelling Notation (BPMN) [15]. The model can be used for further analysis, such as detecting deviations, deadlocks, and repairing or enhancing the model. Enhancement extends or improves the process model using additional information, such as resources, decision rules, and performance data [16]. In our literature review, 36 out of 37 studies analysed the control-flow (the ordering of activities) and performance perspectives. Those studies implemented process discovery and conformance checking to check model conformance against the event log. Most of the studies (24 of 37) used the ProM software tool framework (www.promtools.org). This paper presented the implementation of a methodology called the L* lifecycle model [16] covering process discovery and conformance checking in both control-flow and time perspectives.

2.3. The MIMIC-III Dataset

Medical Information Mart for Intensive Care III (MIMIC-III) is a database comprising EHR information related to patients admitted to critical care units at the Beth Israel Deaconess Medical Centre (BIDMC), in Boston, USA. The dataset version used in this research is MIMIC-III v1.4, released on September 20th, 2016 [17]. It contains data such as: vital signs, medications, laboratory measurements from within the hospital (i.e. in-patient) and from clinics (i.e. out-patient), charted observations during a patient's stay in the intensive care unit, and de-identified notes regarding the patient's stay, including nursing notes, physician notes and discharge summaries [18].

MIMIC-III consists of 26 relational tables, where 16 of them contain timestamped event information. Tables are linked by identifiers: SUBJECT_ID refers to a unique patient and HADM_ID refers to a unique admission. This study will focus on using the 16 event tables, but other tables provide supporting information. For example, when we used `chartevents` table, we would need to refer to `d_item` table to get the label of `item_id` specified in the `chartevents` table. Diseases and procedures in the MIMIC-III are encoded using the International Classification of Diseases version 9 (ICD-9) codes, and the mapping can be found in `diagnoses_icd` and `procedures_icd` tables.

Time in the MIMIC-III database is stored with one of two suffixes: TIME (down to the minute) and DATE (down to the day). Most data are recorded with a time indicating when the event took place (CHARTTIME) and when it was validated (STORETIME). In this study, the event logs were created using CHARTTIME attributes, as this is the best match to the time of actual measurement. All patient data in the MIMIC-III database has been de-identified and all dates have been randomly shifted to the future so that dates are internally consistent for the same patient but inconsistent across patients. The handling of dates and times presents issue for process mining which we will discuss.

3. Methodology

This study was exploratory in nature, with the goal of understanding cancer treatment processes in the MIMIC-III dataset as a publicly available data. The methodology in this research is the L* lifecycle model, suggested by van der Aalst et al [16] for typical process mining projects. Stage 1 (Extraction) was enriched with adoption of *data processing* stage in Process Mining Project Methodology (PM^2)[19]. Note that Stage 1 of the L* life-cycle model does not detail the extraction activities. The PM^2 does provide helpful guidance and these were added in Stage 1. Note also that the Stage 4 of the L* life-cycle model covers explicit operational support and was therefore beyond the scope of this particular study.

In Stage 0 (Plan and justify), planning involved identifying research questions as a starting point for investigation. We drew our research questions from our literature review of cancer pathways.

Stage 1 (Extract) involved extraction to build the log by applying the selection criteria to collect records of treatment cases for patients diagnosed with cancer. Our approach included four preprocessing activities described in the PM^2: *creating views, aggregating events, enriching logs, and filtering logs*. Initial investigations revealed a large number of cases and our approach was to create simplified extracts of the dataset at three different level of abstractions and also select specific cancer types to ensure more homogeneous subset of cases.

In Stage 2 (Create control-flow model and connect event log), the extracted data subsets were analysed using several process mining approaches. This was related to discovery and conformance checking. In this paper, discovery was performed using the Inductive Visual Miner [20] algorithm. Inductive Visual Miner was used since it is robust, user friendly, and feature rich, can deal with noise and exceptions, and enabled the focus to be on the main process flow instead of on every detail of the behaviour appearing in the process log.

Stage 3 (Create integrated process model) extended the process models discovered in the Stage 2 with an additional perspective. The analysis was performed using Petri Net as a process modelling standard, to investigate the activities with the longest waiting times.

The tools used in this study were PostgreSQL (through PgAdminIII graphical interface), Python, and ProM 6.5.1. PostgreSQL was used as the database management system which allowed SQL-based queries though PgAdminIII to extract and select data from MIMIC-III database. Python was used to create more structured way of data processing. The ProM provided process mining toolset for discovery, conformance checking, and enhancement of process models. The SQL queries and resulting models are available for reuse in Github repository (https://github.com/angelinast3/mimic3cancerpromin).

4. Results

4.1. Stage 0: Plan and Justify

Planning for this study was started with a list of frequently posed questions suggested by Mans et all for process mining in healthcare [21]. Those questions were adapted in this study:

Q1. What are the most followed paths and what exceptional paths are followed?
Q2. Are there differences in care paths followed by different patient groups?
Q3. Where are the long waiting time activities in the process?

4.2. Stage 1: Extract

The extract stage in this study started with selecting records of patients diagnosed with cancer. The cancer codes in ICD-9 are 140x-239x [22], and can be grouped based on the cancer types. These codes can be found in `diagnoses_icd` table.

In total, 7,361 patients were found to have at least one diagnoses related to cancer. All of these were selected for this study. Note that a patient might have more than one type of cancer, which make this patient fall into more than one group. Total patients in this study were recorded against 13 cancer types, with the highest number in cancer type 7 (2,846 patients), cancer type 2 (1,400 patients), and cancer type 8 (1,110 patients). The results are summarized in Table 1.

Table 1. Summary of Cancer Types

#	Description	Patients	Admissions
1	Malignant neoplasm of lip, oral cavity, and pharynx (140-149)	87	135
2	Malignant neoplasm of digestive organs and peritoneum (150-159)	1400	**2012**
3	Malignant neoplasm of respiratory and intrathoracic organs (160-165)	1056	1540
4	Malignant neoplasm of bone, connective tissue, skin, and breast (170-175)	279	337
5	Kaposi's sarcoma (176-176)	14	19
6	Malignant neoplasm of genitourinary organs (179-189)	722	1025
7	Malignant neoplasm of other and unspecified sites (190-199)	2846	**3950**
8	Malignant neoplasm of lymphatic and hematopoietic tissue (200-209)	1110	1692
9	Neuroendocrine tumors (209-209)	26	38
10	Benign neoplasm (210-229)	1215	**2036**
11	Carcinoma in situ (230-234)	47	65
12	Neoplasms of uncertain behavior (235-238)	675	1065
13	Neoplasms of uncertain nature (239)	60	105
	Any type of cancer (140 - 239)	7361	10857

Extraction of all cancer patient data was done by selecting records from each of the 16 event tables in MIMIC-III. A summary of the extracted records is presented in Table 2.

Table 2. Summary of Extracted Tables

#	Table Name	Patients	Activities	Rows
1	`admissions`	7361	8	35843
2	`callout`	4197	6	27402
3	`chartevents`	7359	2580	38766594
4	`cptevents`	3327	1	19310
5	`datetimeevents`	5648	148	925542
6	`icustays`	7345	2	22976
7	`inputevents_cv`	3924	756	1833886
8	`inputevents_mv`	3850	251	664209
9	`labevents`	7351	556	6912233
10	`microbiologyevents`	3457	47	11670
11	`noteevents`	5351	584	129712
12	`outputevents`	7278	415	824665
13	`prescriptions`	6900	2697	685648
14	`procedureevents_mv`	3853	114	53440
15	`services`	7357	18	15657
16	`transfers`	6902	8	29708
	Total	7361	7929	50958495

The resulting `allevents` table contained 50,958,495 rows. Using this table directly for process mining resulted in a "spaghetti" process model which is impossible to analysed and it is not presented here. Further transformation and creation of a simplified version of the dataset was essential.

Transformation was performed in several steps, as shown in Fig. 1. A transaction table was created consisting of `subject_id`, `activity`, `category`, `tablename`, `charttime` and records were inserted from all tables extracted before.

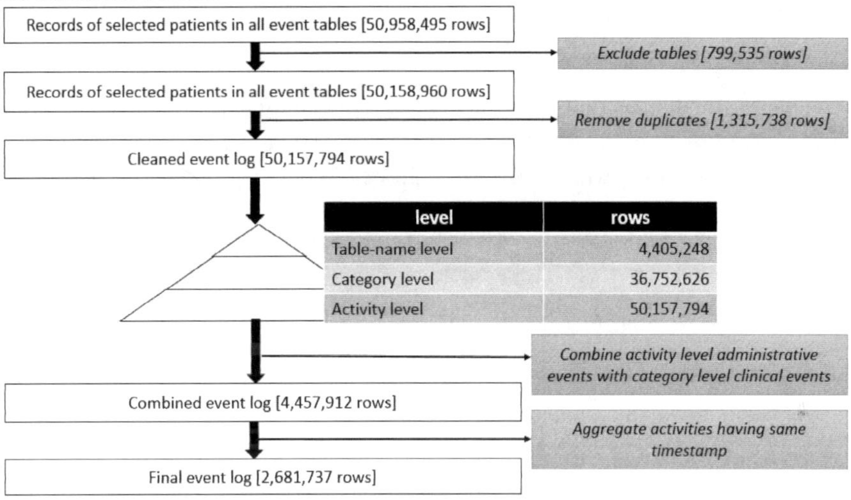

Fig. 1. The sequence of data extraction and transformation in Stage 1

Each row in the table was transformed into an event log record with the following preprocessing based on PM^2 methodology:

1) *Filtering log*
   a. Three tables were excluded because they provide timestamps down to the day only. These were `callout`, `cptevents`, and `prescription` tables.
   b. Records which did not have a timestamp or an activity name were excluded.
   c. Duplicate records were reconciled by keeping one record and deleting other duplicate records. For example: `admission` in the `admissions` table, `Hospital Admit Date` in `datetimeevents` table, and `admit` in `transfers` table are refering to the same event.
2) *Enriching log* by creating three levels of details with three different values for activity labels, which are the original activity names (activity level), category names from `d_items` and `d_labitems` tables (category level), and table name (table-name level). Those three different levels are needed to make it possible to analyse the data in three different level of details, based on information available in the dataset.
3) *Creating views* was done by deciding which level of detail needed in the next stage. This was done based on two general types of events recorded in the tables, which are administrative and clinical events. In the MIMIC-III, administrative events are recorded in `admissions`, `icustays`,

`services`, and `transfers` tables. A general suggestion to get a high level process model from all events in a group of patients would be to combine activity level of administrative events and middle or high level of clinical events.

4) *Aggregating events*. There were a small number of activities with the same timestamp. This need to be aggregated to get the correct sequence of events in the model.

The final event log was saved in a standard comma separated values (.csv) format file, which is readable in ProM. All codes and results of the transformation adjustments are available in the github repository for reusability purpose.

Loading, the final step in this stage was done by importing the file to ProM, converting it to the XES standard, and processing it to discover process models using the available algorithms/ plug-in modules. Unless otherwise stated, all plug-ins in ProM were applied with default parameter settings.

4.3. Stage 2: Create control flow model and connect event log

The next stage was to create control flow models and link these to the event log, using selected plugins in ProM. This study uses BPMN Miner as the main plugin for process model discovery. The discovered process models provided answers to some of the research questions specified before. The event log was filtered and adapted based on insight gained from the preliminary model (e.g. removing rare activities or outlier cases, focusing on specific activities, etc.).

The first research question was Q1 (*What are the most followed paths and what exceptional paths are followed?*). This could be answered by using event logs from three different levels. For example, to find the most followed admission and Intensive Care Unit (ICU) stay paths, records in the `admissions` and `icustays` tables can be used, which resulted in process model as shown in Fig. 2.

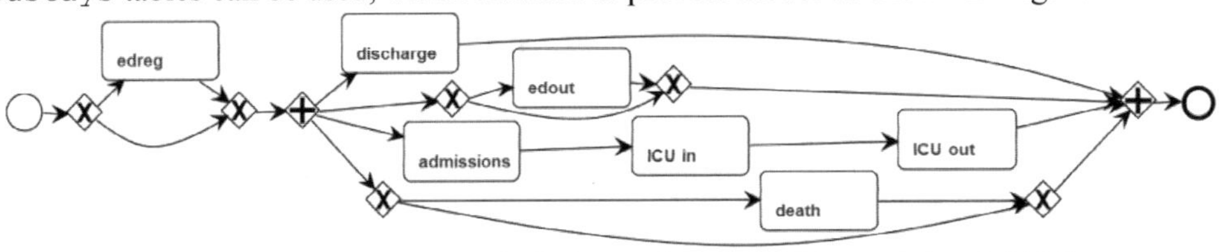

Fig. 2. Process model of admissions (fitness 0.971, precision 0.8808)

The process model in Fig. 2 was created with the "`BPMN Miner`" plug-in [23] in ProM, representing the 10,843 admission of 7,350 cancer patients having complete pathways. The evaluation was performed to the generated model with "`Replay a log for conformance checking`" plugin [24] in ProM. The generated model was evaluated using fitness and precision measures. Fitness quantifies the extent to which the discovered model can accurately reproduce the cases recorded in the log. Precision quantifies the fraction of the behaviour allowed by the model that is not seen in the event log. High fitness (0.971) and high precision (0.8808) means that the discovered model allows for the behaviour seen in the event log and does not allow for behaviour unrelated to what was seen in the event log.

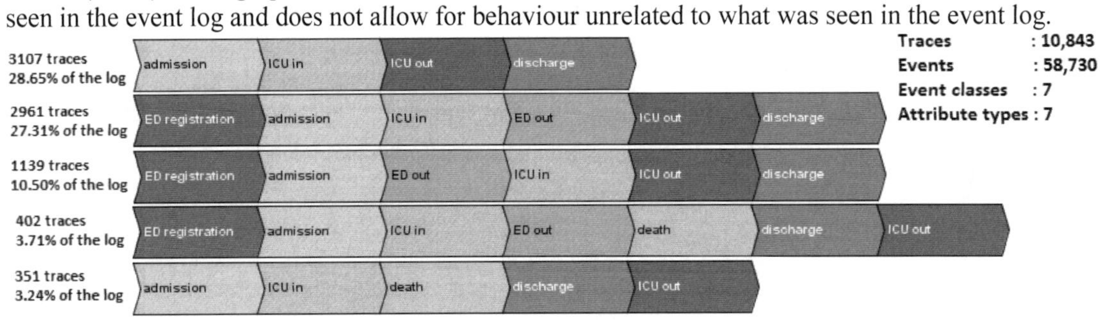

Fig. 3. The five most common trace variants representing 73.41% of traces

IOP Conf. Series: Journal of Physics: Conf. Series **971** (2018) 012008 doi:10.1088/1742-6596/971/1/012008

The most common trace variants are presented in Fig. 3. The pathways start with `admission` or `ED registration` and end with `discharge`. When a patient is admitted to the hospital, they could be admitted to a standard admission or registered through emergency department (ED). It can be seen that the most frequent activities are: `admission` and `discharge` activities. Other possible events are `ED registration`, `ED out` and `death`. We could also see that some variants are different because of different administrative steps, such as in the second and third variant (see Fig. 3). ICU in and ED out happened in the different order, while other events are the same, which indicate the different work order in cases where patients are moving from ED to the ICU.

Additional evaluation was done by 5-fold cross-validation. In this step, the original event log was randomly partitioned into 5 equal sized sub-eventlog. Of the 5 sub-eventlog, a single sub-eventlog was used as the validation data for testing the model, and the remaining 4 sub-eventlogs were used as training data. The cross-validation process was then repeated 5 times (the folds), with each of the 5 sub-eventlogs used exactly once as the validation data. The 5 results from the folds were then be averaged to produce a single estimation. From this step, the resulted fitness was 0.968644 and precision was 0.79726. The validation shows that both measures were expectedly lower than the conformance to itself, but both are still representing high values of fitness and precision.

Following the same method, we created process models from groups of patients with different cancer types to answer Q2 (*Are there differences in care paths followed by different patient groups?*). For example, the pathways of the three most common groups are presented in Fig. 4.

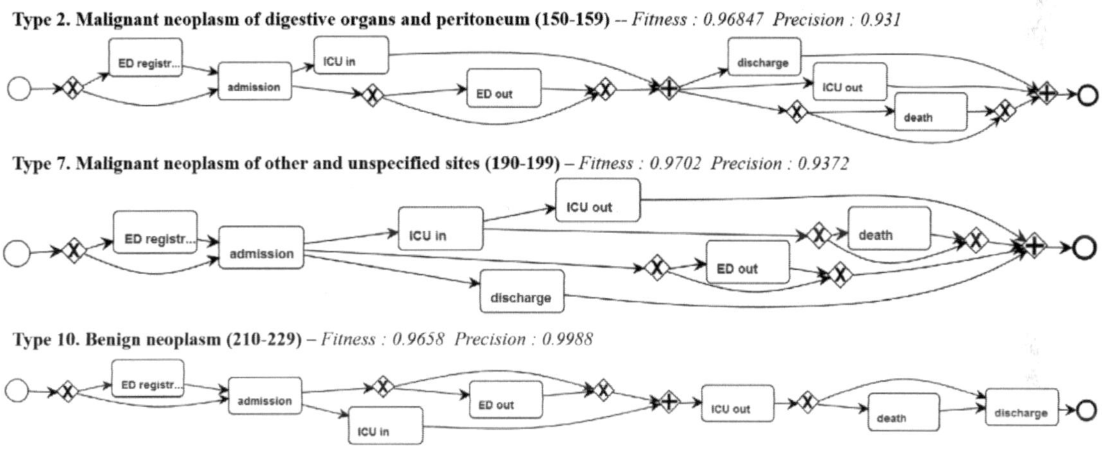

Fig. 4. Process models of three different cancer types

Our clinical domain expert reviewed the models based on their visual utility by comparing process models of cancer type 2, type 7, and type 10. The findings are: (1) `ICU in`, `ICU out`, `admission` and `discharge` are always happened in all three cancer types regardless of the sequence; (2) `ED registration`, `ED out` and `death` are possible events in all three types; (3) `admission` happened as the first event or after `ED registration` in all three types; (4) Some patients with cancer type 7 have been discharged after admission without other major events but type 2 and type 10 cancer patients have had more extensive care.

4.4. Stage 3: Create integrated process models

In stage 3, the models could be extended with other perspectives (e.g. date, time, and resources). In this study, the time information was used to analyse the waiting times for the admission pathways of all cancer patients. This was done to answer Q3 (*Where are the long waiting time activities in the process?*).

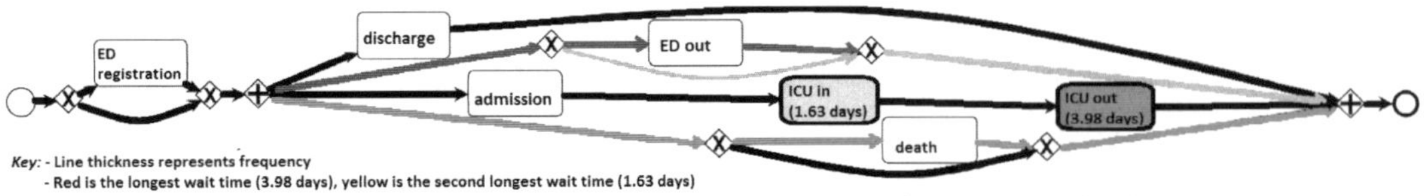

Fig. 5. Waiting time analysis (combined output from ProM plugins)

The BPMN in Fig. 2 was analysed with the "Replay a Log on Petri net for Performance/ Conformance Analysis" in ProM and the result was combined with the original BPMN, as shown in ***Fig.*** *5*. The analysis revealed that the longest waiting time is in ICU out (3.98 days in average), while the second longest waiting time is in ICU in (1.63 days in average). The long waiting times in ICU in and ICU out give an insight to dig deeper in the lower level of activities between ICU in and ICU out to understand which activities contribute to the long waiting time in ICU.

5. Discussion

The L* lifecycle was suitable for process mining in oncology using the MIMIC-III dataset, with the limitation that Stage 4 (Explicit operational support) was not applicable in this study due to the de-identified nature of the dataset. Stage 0 (Plan and justify) were performed by listing research questions based on frequently posed questions for process mining project in healthcare, which could be adapted in any clinical domains including oncology. All three questions listed have been answered through the experiments.

One important step during the Stage 1 (Extraction) was the transformation, which included the preprocessing of the dataset. Preprocessing should be based on a good understanding of the data and this study demonstrates how several preprocessing approaches were used to simplify the MIMICIII dataset. Preprocessing requires care, excluding a table might result in incomplete analysis and be insufficient to answer the research questions in the study.

Stage 2 (Creating control flow models) and Stage 3 (Create integrated process models) have been done in this study (Section 4) to answer some standard research questions. Process model comparisons were performed in a simple way in this study. Data quality issues were detected during these stages, and merit further analysis. This could be done by selecting events in the `allevents` table at three different levels.

6. Conclusion

This paper focuses on the applicability of process mining within healthcare domain, specifically cancer pathway, using the MIMIC-III dataset. MIMIC-III is representative of ICU and hospital datasets and its use can support reproducible research due to it being publicly available. This paper focused on gaining insights from the patient flow by implementing L* lifecycle using the control-flow and time perspective. It was shown that it is possible to mine complex hospital processes with existing techniques to discover and analyse process models.

Future work will explore several aspects. The first will be improving data quality with using data cleaning approaches. This will include the approaches to define a method to select events to include in the process discovery. The second area of focus will apply different algorithms for process mining to find process models with high conformance to the event log. While using only the Inductive Miner in this study, we recognise that there are many other algorithms available in ProM, where each of them are suitable for different characteristics of event log. This might lead to choosing the most suitable algorithm for this specific case study, or developing a new algorithm. Lastly, we will focus on advanced analysis with more clinical based research questions and comparison studies with other real-life datasets.

Acknowledgment
This research is supported by ClearPath Connected Cities Project. This is also a part of a PhD study funded by the Indonesia Endowment Fund for Education (LPDP).

References

[1] American Cancer Society, "The History of Cancer," 2011. [Online]. Available: www.cancer.net/patient/Advocacy and Policy/Treatment_Advances_Timeline.pdf. [Accessed: 09-Aug-2016].
[2] Cancer Research UK, "Your cancer type," 2014. [Online]. Available: http://www.cancerresearchuk.org/about-cancer/type/. [Accessed: 09-Aug-2016].
[3] W. M. P. van der Aalst, "Process Mining: Data Science in Action," in *Process Mining*, 2nd ed., Springer-Verlag Berlin Heidelberg, 2016, pp. 3–23.
[4] R. S. Mans, M. H. Schonenberg, M. Song, W. M. P. van der Aalst, and P. J. M. Bakker, "Application of Process Mining in Healthcare – A Case Study in a Dutch Hospital," *Proc. BIOSTEC 2008*, vol. 25, pp. 425–438, 2008.
[5] R. Mans, W. M. P. van der Aalst, and R. Vanwersch, *Process Mining in Healthcare: Evaluating and Exploiting Operational Healthcare Processes*, 1st ed. Springer International Pub, 2015.
[6] E. Rojas and J. Munoz-Gama, "Process mining in healthcare: A literature review," *J. Biomed. Inform.*, vol. 61, pp. 224–236, 2016.
[7] A. P. Kurniati, O. Johnson, D. Hogg, and G. Hall, "Process Mining in Oncology: a Literature Review," in *The 6th ICICM, IEEE*, 2016.
[8] B. W. Boudewijn van Dongen, Diogo R. Ferreira, "Business Process Intelligence Challenge (BPIC)," *IEEE Task Force on Process Mining*, 2011. [Online]. Available: http://www.win.tue.nl/bpi/doku.php?id=2011:challenge. [Accessed: 22-Apr-2016].
[9] R. Lenz and M. Reichert, "IT support for healthcare processes – premises, challenges, perspectives," *Data Knowl. Eng.*, vol. 61, no. 1, pp. 39–58, 2007.
[10] M. Poulymenopoulou, F. Malamateniou, and G. Vassilacopoulos, "Specifying Workflow Process Requirements for an Emergency Medical Service," *J. Med. Syst.*, vol. 27, no. 4, 2003.
[11] O. Johnson, P. S. Hall, and C. Hulme, "NETIMIS : Intelligent Simulation of Health Economics Outcomes using Big Data," *Pharmacoeconomics*, vol. 34, no. 2, pp. 107–114, 2015.
[12] K. HAYRINEN, K. SARANTO, and P. NYKANEN, "Definition, structure, content, use and impacts of electronic health records: A review of the research literature," *Int. J. Med. Inform.*, vol. 77, no. 5, pp. 291–304, May 2008.
[13] W. M. P. van der Aalst, *Process mining: discovery, conformance and enhancement of business processes*, 1st ed. Springer-Verlag Berlin Heidelberg, 2011.
[14] W. M. P. van der Aalst, "The application of Petri nets to workflow management," *J. circuits, Syst. Comput.*, vol. 8, no. 1, pp. 21–66, 1998.
[15] P. Wohed, W. M. P. van der Aalst, M. Dumas, A. H. M. ter Hofstede, and N. Russell, "Pattern-based Analysis of BPMN," *Lect. Notes Comput. Sci.*, vol. 4102/2006, pp. 161–176, 2006.
[16] W. M. P. van der Aalst, A. Adriansyah, A. K. A. de Medeiros, F. Arcieri, T. Baier, T. Blickle, J. C. Bose, P. van der Brand, R. Brandtjen, J. Buijs, A. Burattin, J. Carmona, M. Castellanos, J. Claes, and J. Cook, "Process Mining Manifesto," *BPM Work.*, vol. 99, pp. 169–194, 2011.
[17] MIT Laboratory for Computational Physiology, "MIMIC-III v1.3," 2015. [Online]. Available: https://mimic.physionet.org/about/releasenotes/. [Accessed: 03-Aug-2016].
[18] A. E. W. Johnson, T. J. Pollard, L. Shen, L. H. Lehman, M. Feng, M. Ghassemi, B. Moody, P. Szolovits, L. A. Celi, and R. G. Mark, "Data Descriptor : MIMIC-III , a freely accessible critical care database," *Sci. Data*, pp. 1–9, 2016.
[19] M. L. Van Eck, X. Lu, S. J. J. Leemans, and W. M. P. Van Der Aalst, "PM2: A process mining project methodology," *Lect. Notes Comput. Sci. (including Subser. Lect. Notes Artif. Intell. Lect. Notes Bioinformatics)*, vol. 9097, pp. 297–313, 2015.
[20] S. J. J. Leemans, "Inductive visual Miner manual," pp. 1–16, 2017.
[21] R. S. Mans, W. M. P. van der Aalst, R. J. B. Vanwersch, and A. J. Moleman, "Process mining in healthcare: Data challenges when answering frequently posed questions," *LNCS (including Subser. Lect. Notes Artif. Intell. Lect. Notes Bioinformatics)*, vol. 7738 LNAI, pp. 140–153, 2013.
[22] "Online ICD9/ICD9CM codes - Neoplasms." [Online]. Available: http://icd9cm.chrisendres.com/index.php?action=child&recordid=1059. [Accessed: 16-Mar-2017].

[23] R. Conforti, M. Dumas, L. García-Bañuelos, and M. La Rosa, "BPMN Miner: Automated discovery of BPMN process models with hierarchical structure," *Inf. Syst.*, vol. 56, 2016.
[24] A. Adriansyah, "Replay a Log on Petri Net for Conformance Analysis plug-in.pdf | Algorithms | Logarithm."

IOP Conf. Series: Journal of Physics: Conf. Series **971** (2018) 012009 doi:10.1088/1742-6596/971/1/012009

Telkom UData sentiment analysis using crowdsourcing and trust

Edvya Noer[1], Dana Sulistyo Kusumo[2] and Yanti Rusmawati[3]

Telkom University, Bandung, Indonesia

Abstract. Microblogging sites have millions of people sharing their thoughts daily because of its characteristic short and simple manner of expression. Sentiments analysis are often being used to analyse the user customer opinions regarding brand images or products. For some reasons, not all sentiment generated using this existing machine-based algorithms yields satisfying results. This is mostly due to the uniformity of the informal language used in the social media sentences. This condition also occurs in Telkom UData on our preliminary study, where the machine-based provided less then optimal results in analysing the sentiment. This research offers concepts with human interaction using crowdsourcing where people are involved to analyse sentiments, while forming the new training dataset at the same time. From the research results found that sarcastic and contradictory sentences can be recognized by humans, to be utilized as new training datasets for further machine learning. From this experiments, that approach are likely increase the accuracy of the sentiments in UData from neutral to become positive or negative polarized up to 39%. We do as well simulated trust concept through sociometric to ensure the crowdsource workers are trusted and capable enough in analysing the sentiments on social media.

1. Introduction

Since 1990s, the internet has been growing so far and becoming a powerful media to gain information from. Some of which is used to obtain the voice of customers by using microblogging channels known as social media. Customers now have enough trusted information from their global peers such that they can select the best product and the best service provider available to them every single time. Every single action on a social network is a proactive expression of what a customer sees, thinks and feels. Smart companies will embrace social support and social selling [1]. Referring to the statistic presents the social network penetration in Indonesia. As of the fourth quarter 2016, 40% of the total Indonesian population were active with social media users. The most popular social platforms are YouTube, Facebook, Tweeter and Instagram with more than 50% of penetration rate [2].

In Telecom industry, provider such as Telkom Indonesia is facing the period where competition leads to creativity, no longer to the price war. The most creative products will win the battle field. The creative products are the products that can fit customer's needs and attract their interest. The creativity should be achieved by looking for new product ideas and perceptions from Telkom product user: the customers.

With the rise of social networking era, there has been a surge of user generated content. Microblogging sites have millions of people sharing their thoughts daily because of its characteristic short and simple manner of expression. Sentiment analysis usually used as an integral part of social

[1] edvyanoer@student.telkomuniversity.ac.id
[2] danakusumo@telkomuniversity.ac.id
[3] rusmaway@telkomuniversity.ac.id

listening. Sentiments analysis are often being used to analyze the user customer opinions regarding brand images or products.

The main problem in identifying sentiment in sentences, is that sometimes the meaning of opinions are not clearly implied. Despite the high usage of the sentiment analysis application, there are still rooms for improvements. One of the problems that still become a challenge in sentiment analysis is sarcasm, contradictory. Research has been conducted to identify sarcasm in sentences especially in Bahasa Indonesia, where their observation found that in Indonesian social media people tend to criticize something using sarcasm for certain topics [3]. The results of the study found that the additional features for recognizing sarcasm can increase the accuracy of up to 6%, by using negativity and interjection words. In spite of the accuracy did not increase rapidly, this research did provide encouraging results to reduce the inability of computers to recognize sentences that have double meaning.

Another way to solve the inability of computers to recognize those kind of sentences is by proposing human interaction into sentiments identification. Involving human works for online analyzing over large scale data is called crowdsourcing. Crowdsourcing is an online and distributed problem-solving and production model that has emerged in recent years. Crowdsourcing involves human interaction as volunteers or hired workers to find specific solution of a problem [4]. Notable examples of the model include Wikipedia iStockphoto, Inno-Centive, and Amazon's Mechanical Turk (Mturk) - an online market place for work that require human intelligence. While the latest in Indonesia is applied on www.kawalpemilu.org, which showed the power of netizen voluntarily providing their effort to validating the final results of Indonesian Presidential Election in 2014. Since the sentiments retrieved from the internet are large-scale data, human involved in this progress should be plenty [5].

Some researches has been done involving crowdsourcing approach. Barbier [6] stated that crowds of people can solve some problems faster than individuals or small groups. A crowd can also rapidly generate data about circumstances affecting the crowd itself. This crowdsourced data can be leveraged to benefit the crowd by providing information or solutions faster than traditional means. Other research such as Djelassi [7] focused on customer's participation in a product development process through crowdsourcing practices. Crowdsourcing generates a win-win relationship, creating value for both firms and customers. The results suggest the need to establish an open business model based on crowdsourcing. On the same way Mukherjee [8] used Amazon Mechanical Turk to crowdsource fake hotel reviews. Truthful reviews were obtained from the TripAdvisor Web site. They tried several classification approaches which have been used in related tasks such as genre identification, psycholinguistic deception detection, and text classification.

Telkom itself has started to utilize social media as a great resource for its customers to obtain useful information for the sustainability of their business. One of them is by providing free service on Telkom UData website at http://socmed.udata.id. This portal provides free facilities for public to obtain sentiments based on certain keywords that will be very useful to analyze a particular issue that is related especially to customer.

In this research, the sentiments gathered from UData will be used as pre-processing data for crowdsourcing purpose. There are need for some improvements in machine learning while improving the accuracy of NLP intelligence especially in Indonesian languange are still about to plan. With current conditions, some tweets containing sarcasm, and contradictions are not completely recognized. For instance when Telkom observes the sentiments as a product owner.

We are proposing crowdsourcing in this experiment since UData is a third party built-in-app and creating new training dataset is the only way to improve its perform. Another reason for proposing crowdsourcing is in order to find another alternative methods of sentiment analytics other than machine-base algorithm that has been done by many researches.

As an interesting example, when we enter the keyword "indihome" then we get the result of sentiment with the composition as shown by Figure 1, where the neutral percentage (contains no polarized sentiment) has the greatest composition 46%.

IOP Conf. Series: Journal of Physics: Conf. Series **971** (2018) 012009 doi:10.1088/1742-6596/971/1/012009

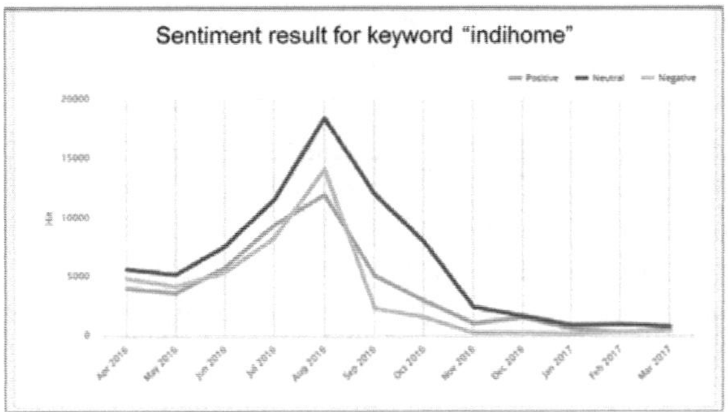

Sentiment	Count	%
negative	42,273	26%
neutral	75,314	46%
positive	46,985	29%
Total	164,572	100%

Most tweets are labeled as neutral because the computer cannot recognize any sarcasm or contradictory in the sentences

Figure 1. Tweets with keyword ‘indihome’ retrieved through Apr 2016 – Apr 2017

Then we look further particularly for those sentences polarized as neutral, for instance as shown in Figure 2 below. There comes curiosity where the sentence “Bisa mohon dibantu untuk proses berhenti berlangganan indihome via twitter?” or in English “*Would you please help me to stop subscribing indihome via twitter?*” should be negatively polarized, and UData did not recognize it appropriately. With Telkom's point of view as a product owner, this is a sample of losing customer loyalty which is not explicitly stated.

Figure 2. Sentence polarized as neutral

Regarding to the background stated above, we identify the research question as follows:

1. How to increase the accuracy information of UData, in this case sentiment analysis obtained from social network (Twitter), by involving human interaction via crowdsourcing approach.
2. How to identify trust of the crowdsource workers, in order to improve the trustworthiness of sentiments obtained.

The hypothesis of this research is by adapting crowdsourcing method in constructing the training dataset can improve the sentiment obtained from the machine-based algorithm. Words found in Twitter mostly used are unstructured and informal. The crowdsourcing result should produce more accurate sentiment of opinion from customers. The result can be useful for monitoring customer needs, analysis of future business growth, predicting the future business trends, as well as more accurate decision making, to get more fresh ideas or to find a new business opportunity. Moreover, we can also compute trust weight of each crowdsource workers to further strengthen the credence of the sentiment they propose.

As a delimitation this research limit the scope only for Telkom product (i.e. indihome), and the crowdworkers involved are only for Telkom employee. In this case all tweets obtained from twitter are filtered for sentences that has only keyword ‘indihome’ in them, to focus on sentiment of that specific Telkom product. Since the crowdworkers involved are restricted to only in Telkom Group, this crowdsourcing model is known as Enterprise Crowdsourcing.

2. Literature Review

With more and more social media emergence in this era, people are now faced with curiosity to know what is being discussed in social media in real time. Sentiment analysis plays an important role in mapping the growing issue among netizens, so there is a need to ensure that the sentiments produced by the machine are quite accurate. This section provide an overview of terminology, methods, and limitations related to the existing systems.

2.1. Sentiment Analysis on UData

Sentiment analysis, or sometimes called opinion mining, is used to extract and analyze the emotions conveyed in texts [9]. Numerous research from various domains takes advantage of the sentiment analysis; for instance, sentiment analysis is applied to detecting political stances, characterizing personality, measuring happiness, and even predicting stock market. With the growing use of the Worldwide Web, the demands for sentiment analysis are increasing.

We use already collected data of tweets from Telkom UData that can be freely accessed at https://socmed.udata.id. UData is a social media analytics portal developed by Telkom and already has sentiment analytics in Bahasa Indonesia. The result of sentiments could be retrieved as positive, negative, or neutral labeled tweets.

2.2. Crowdsourcing

Crowdsourcing is a term, sometimes associated with Web 2.0 technologies, that describes outsourcing of tasks to a large often anonymous community [10]. It is an online, distributed problem-solving and production model that has emerged in recent years. Crowdsourcing involves human interaction as volunteers or hired workers to find specific solution of a problem. Notable examples of the model include Wikipedia iStockphoto, Inno-Centive, and Amazon's Mechanical Turk (Mturk). While the latest in Indonesia existed on www.kawalpemilu.org, which presented the power of netizen voluntarily providing their effort to validate the final result of the Indonesian Presidential Election in 2014.

Crowdsourcing has a great potential to verify the sentiment result by implicating human involvement. This would also be useful in validating the sentiment results whether actually accurate or not. In this research, we built Crowdsourcing mechanism to improve the sentiment of tweets to have better analyzing and more accurate sentiment. Every crowdsource workers (or crowdworkers) will be given some tweets they can verify. For each collected verification, the workers will be awarded points.

Crowdworkers consist of volunteers choosen from Telkom employees, based on consideration that they are familiar enough about Telkom product and will have better understanding on customer perception related to the product.

2.3. Trust

According to Kim et.al [11] we cannot judge human work only by the concept of quality, trust needs to complement it. Predictive evaluation of data based on trust is important, as if we cannot verify data we have to trust the provider. Understanding the interaction between quality of data and trust of source is most important. Incorporating more trust mechanisms, especially in crowdsourced practices will helpful to filter out low quality data. In recent trust and reputation concepts, there has been significant number of elements that have been researched. Such as dimensions like reliability, honesty assessment. However to be implemented in open community system like crowdsourcing there should be other parameter added to improve the quality of judgements and recommendations, such as anonymous detection, socio metric between agents, etc. Situm has already categorized existing algorithms used for computing trust among friends, Table 1 below represents a review of some of the currently available literature about existing approaches in trust computation on social networks or peer-to-peer networks [12].

Table 1. Research of Trust

	Application Domain	*Trust Evaluation*	*Information Collection*	*Trust / Distrust*	*Trust Representation*
Eigen-Trust	P2P	Graph Based	Experience	Trust	Probabilistic
Peer-Trust	P2P	Graph Based	Experience	Trust	Gradual
Yu	P2P	Graph Based	Experience	Trust	Probabilistic
Golbeck	Social Networks	Graph Based	Experience	Trust	Gradual
Josang	P2P	Graph Based	Experience	Trust and Distrust	Probabilistic

Power-Trust	P2P	Graph Based	Experience and Behavioral	Trust	Gradual
Chaverlee	Social Networks	Interactions	Behavioral	Trust	Probabilistic
Liu	Social Networks	Interactions	Behavioral	Trust	Gradual
Nepal	Social Networks	Interactions	Behavioral	Trust	Gradual
Proposed Method	P2P	Graph Based	Experience	Trust and Distrust	Gradual

Those algorithms are categorized to algorithms which are implemented in P2P domain problem or social network domain problem. First five algorithms are graph based and use propagative and aggregative properties of the trust to compute trust values. Last three algorithms use only direct interactions among members to compute trust. Interaction-based algorithms also use behaviors to compute trust, while mostly all peer-to-peer algorithms use peer's experiences of communication with other peers.

This experiment is proposing the trust approach similar to Josang with domain in peer-to-peer interaction, graph based evaluation and have distrust/ less-trusted judgement, however we are using gradual representation in differ to accommodate the human interaction in crowdsourcing. According to Haydar [13], gradual representation of trust is more similar to the human way in expressing trust, whereas probabilistic representation is more meaningful mathematically.

3. Research Methodology

3.1. Problem Identification

This reseach has identified potential improvement in identifying sentiment in sentences especially those in social-medias, by combining with crowdsourcing mechanism that can improve the sentiment analysis result. Commonly used methods of determining sentiments such as Naïve Bayes and Support Vector Machine (as well as those used in UData). The main problem in identifying sentiment in sentences with those methods is that some opinions are not only positive or negative. Despite the high usage of the sentiment analysis application, there are still rooms for improvements. One of the problems that still become a challenge in sentiment analysis is sarcasm and contradictory. So in this research another mechanism proposed to validate whether the sentiment are accurate, by involving human interaction and using crowdsource platform. The final result is expected to affirm more confident sentiments from the trusted crowdsource workers.

3.2. Model Design

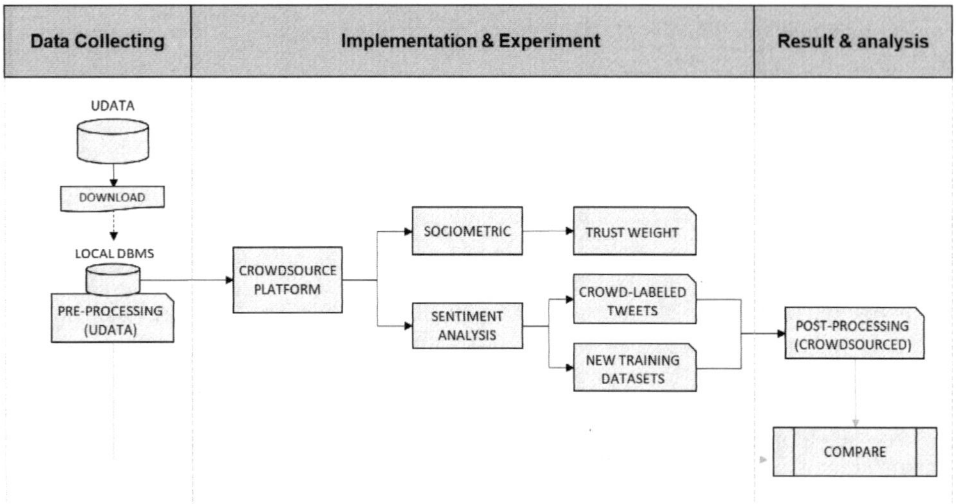

Figure 3. Model Design

In Figure 3, the basic step for design implementation to validate sentiment over human interaction are as follow.

1. The first objective of this research is to retrieve tweets with sentiments from online social media (Twitter). For this research the sentiments data is retrieve from Telkom UData. We use keyword "indihome" to limit the tweets data only for this specific Telkom product. The tweets already labeled with positive, negative and neutral tags by machine algorithm. This sentiments data then saved to local database, we call the sentiments gained from UData as pre-processing data.
2. For crowdsourcing mechanism, we focus on the neutral-labeled tweets because these are the tweets that cannot recognized by the machine so further analysis is needed by crowdsourcing. At this part, crowdsource work means labeling the neutral sentences that cannot recognized by machine (which may include slank, sarcasm, contradictory etc.). Each sentences will be labeled by at least 20 people [5].
3. Parallel in sociometric session, each crowdsource worker will be judged for their trust by using sociometric experiment among them, by giving trust weight to other workers. This concept will determine which workers have the highest level of trust.
4. On the result & analysis stage, we compare the crowdsourced result (as post-processing data) and UData (as pre-processing data) to find out the increasing performance of sentiment analysis gained from crowdsourcing.

3.3. Data Collecting and Processing

The next step is data collection and processing which collect trust data from the crowdworkers by using sociometric experiment. The experiment is collecting trustworthiness each other among the crowdworkers. The collected data then run in to expected format dataset process to calculate the socio trust matrix.

3.4. Experiment Design

The initial step of the experiment is determined by choosing the appropriate workers to get the job done. As stated before this experiment uses convenience sampling. Convenience sampling is a statistical method by selecting people who are conveniently available to participate in study because of their availability or easy access. The crowdworkers involved in this experiment are chosen based on consideration as follows:

1. The crowdworkers are the Telkom employees with mostly educated as Bachelor and Master Degree, and should have job allocation related to customer based oriented. For example: Account Manager, Call Center Agent, Solution, Technician and Networking.
2. The crowdworkers can access the experiment platform only during free time on their working hours.

Crowdworkers will be shown other participant names who has previously been defined initially from the database as member of crowdsource workers in this platform. They should give their level of confidence to each of these peers so that will construct the weight of socio metric for all crowdworker peers.

3.5. Weighting Model

Sociometric is a quantitative method for measuring social relationships. In this research sociometric is used to give ability to the crowdsource workers to rate for each other's trustworthiness. Every worker will be able to rate for other worker trustworthiness based on their social interaction. The more trustiness a worker obtains, will affect to the result of his validation of sentiment. The sample of the sociomatrix is as Table 2 which is constructed to clarify weights and introduce terminology to illustrate the strength of sociometry [15].

Table 2. The Sociomatrix

Sociometric Choice Matrix							
Chooser	Choosen						
	A	**B**	**C**	**D**	**E**	**F**	**G**
A		+		+		-	-
B	+				-	-	+
C				+	+		
D	-	-	+		+		
E		-	+	+			-
F		-	-	+	+		
G	-		+	+		-	
+ Choices Received	**1**	**1**	**3**	**5**	**3**	**0**	**1**
- Choises Received	**2**	**3**	**1**	**0**	**1**	**3**	**2**

4. Experiment and Analysis

4.1. Data Source

As mention in section 2, the initial datasets are taken from portal Telkom UData. We also select tweets that posted on range of period between Jan 2017 – Apr 2017. The retrieved tweets are based on search with keyword "indihome" to limit and focus only analyzing to that kind of Telkom product. The data are downloadable from UData and already have prior sentiment labeled automatically using machine-based algorithm. From the collected data of total 3.951 tweets, we found about 41% (1.601 tweets) are labeled positive, 25% (1.004 tweets) are labeled negative while other 34% (1.346 tweets) are labeled neutral.

Each crowdworkers are pushed to analyze over 300 neutral tweets that presented one by one to the workers, as we choose with randomly unique tweets. This amount of tweets are based on calculation of sample size formula, by assuming there are 1.346 tweets, with confidence level of 95% and Margin of Error 5%.

4.2. Data Preparation

After having datasets collected, then the datasets are filtered manually to reduce non-related customer-based opinion, for example in this case tweets from telkom user itself (@telkomcare, @telkomsolution, @telkompromo, @indihomewitel, etc). Mostly tweets from those users are related to promotion or advertisement that will obscure the purpose of the experiment to perceive real customer opinion. Datasets that used for experiment are only the ones with neutral-labeled, by assumption those ones are the tweets cannot recognized well by the machine.

4.3. Profile of the Crowdworkers

Prior in the previous planning, the crowdworkers that will be involved in this experiment were about 24 workers, but only 21 workers are ready to join this experiment. This was due to the experiment was held on office hours, and the crowdworkers mostly were still doing their main duties. About 81% of the crowdworkers are male (17 workers), while the other 19% are female (4 workers). 52% of them are 20-30 years old, and 48% are 30-40 years old. None of them is above 40 years old. All of them are full time job employee of PT Telkom, with vary expertise such as Account Manager, Call Center, Networking, Solution and Technician. The assignment is done by considering that they are frequently get involved with customer, familiar enough in using social media, understand the terms commonly used by netizens, and also understand Telkom products especially Indihome.

4.4. Experiment Results

The experiment are started by giving the crowdworkers to login to the platform first. The system will log the time they start to record how long they use the crowdsource platform. After login the crowdworkers will be shown a sociometric form, each of them has to pass it by giving answer their trust on the other workers. Then the form continues with crowdsource platform. In this form, the crowdworkers are intended to analyze the sentence whether it is polarized positive or negative, while at the same time they have to tag keyword that polarize the sentence. The keyword found is also meant for the new training data set against data tweets are still neutral. From the proposed crowdsource results we have collected 110 keyword as new training dataset for UData.

We also found some tweets (in Indonesian language) that contains anomaly which did not detected normally by machine based algorithm as in Table 3 and 4. Keyword is word tagged by crowdworkers in which according to their assumptions the word that make the sentence polarized positive or negative, and contra is the word found in the sentence that known generally has a normal meaning. For instance word "gangguan" or "trouble" in English, usually polarized as negative. However since the sentence also contain word "tahan" or "stand" in English, the crowdworkers decide the overall meaning of the sentence is positive. This is a contradictory example found, and cannot be recognized by UData.

Table 3. Tweets contain contradictory to keywords from Crowdsourcing result

NO	TWEETS	CONTRA	KEYWORD	CROWDSOURCED SENTIMENT
1	TelkomCare: andara_santoso Hai Kak Andara. IndiHome Fiber lebih tahan dalam kondisi cuaca apapun seperti serangan petir dan gangguan elektr… *(TelkomCare: andara_santoso Hai Sis Andara. IndiHome Fiber is more resistant in any weather conditions such as thunder and electrical trouble)*	gangguan *(trouble)*	tahan *(resistant)*	Positive: 100% Negative: 0% Neutral: 0%
2	@TelkomCare Tanggal 16/3 lusa genap 3 bulan saya menikmati free indihome. Bgmn apabila saya ingin berhenti agar tdk dikenai tagihan? *(@TelkomCare on date 16/3 its completely 3 months for me*	nikmati *(enjoying)*	berhenti *(quit)*	Positive : 0% Negative : 100% Neutral: 0%

	enjoying free indihome. How if I want to quit so I won't be charged later?)			
3	@TelkomCare Bisa mohon dibantu untuk proses berhenti berlangganan indihome via twitter? *(Please help me to stop subscribing indihome via twitter?)*	bantu, langganan *(help, subscribing)*	berhenti *(stop)*	Positive : 0% Negative : 100% Neutral: 0%
4	@TelkomCare Okee saya mau registrasi min. Tadi via app indihome tidak bisa min force close app nya *(@TelkomCare Okee I'd like to register.I could not do that via indihome the app was forced close)*	tidak bisa *(could not)*	registrasi *(register)*	Positive: 100% Negative: 0% Neutral: 0%

Table 4. Tweets contain sarcasm

NO	TWEETS	KEYWORD	CROWDSOURCED SENTIMENT
1	Indihome itu bisa murah karena mereka gak spend budget on customer-service things *(Indihome is cheap because they don't spend budget on customer-service things)*	Murah *(cheap)*	Positive: 0% Negative:100% Neutral: 0%
2	@TelkomCare Mudah2an segera baik2 saja ya min….. Saya butuh indihome segera on nih *(@TelkomCare Hope will be good fixed soon, I need indihome is on soon)*	baik *(good)*	Positive: 0% Negative: 100% Neutral: 0%
3	Ktnya unggul kq malah rto ini gan... RT @TelkomCare: Sobat, simak keunggulan internet fiber dari IndiHome Fiber ini yuk :) *(It is said superior then why rto now... RT @TelkomCare: Friends, check out the fiber internet superiority of Indihome Fiber :))*	unggul *(superior)*	Positive: 0% Negative: 100% Neutral: 0%

After some adjustments and filtering works we finally have nice fact. The result generates more polarized sentiments than previous one. The table below shows us the results before and after the crowdsourcing process. There are a significant about 39% of neutral sentiment slightly changed to positive or negative polarized.

Table 5. Comparison before and after the crowdsource process

SENTIMENT	*UDATA*	*PROPOSED CROWDSOURCING*	*% OF INCREASE*
negative	1.004	1.242	24%
positive	1.601	2.372	48%
total pos + neg	**2.605**	**3.614**	**39%**
neutral	1.346	337	-75%

As we found that most significant changes in sentiment found by crowdsource platform is because the lack of dataset for unknown keywords in UData. The crowdworkers have provided considerable input for the new training dataset and caused an enormous differentiation between UData result and proposed crowdsource result.

Paired Samples Test

	Paired Differences					t	df	Sig. (2-tailed)
				95% Confidence Interval of the Difference				
	Mean	Std. Deviation	Std. Error Mean	Lower	Upper			
Pair 1 VAR00001 - VAR00002	-.25538	.43613	.00694	-.26898	-.24178	-36.806	3950	.000

Figure 4. Paired Sample Test

In Paired Sample T-Test each subject or entity is measured twice, resulting in pairs of observations. As in this experiment where we have two observation sources, from UData (Pre-Processing) and Proposed Crowdsourcing (Post-Processing). The Paired Samples Correlation table adds the information that pre-process and post-process result are significantly positively correlated (.425).

From the Paired Sample T-Test result (Figure 18) we found that the probability value Sig.2 (tailed) = .000 < α = 0.05, it means there are significant changes between pre and post process result. Our Hypothesis stated there are significant changes in sentiment of UData after the post-process result are acceptable.

5. Conclusion

Experiment shows the significant changes of UData performance improvement after the training dataset which derived from crowdsource collaboration. By using crowdsourcing approach, we can be sure that sarcastic and contradictory sentences can be recognized by humans, to be utilized as new training datasets for further machine learning. And from this experiments, that approach are likely increase the accuracy of the sentiments in UData from neutral to become positive or negative polarized up to 39%. Sociometric is used in this experiment to identify trustiness among the crowdsource workers, in order to make sure the workers are trusted for validating the sentiments. The index result obtained from historical sociometric can be useful later to determine which of crowdworkers are capable enough and trusted by workers to be assign for future crowdsource works.

For a recommendation, UData is a sentiment analytics for public and global platform, not devoted only for Enterprise Platform. There are some considerations whether it is applicable or not when we train its dataset using corporate point of view or product based perception. There should be further work to separate the function or privileged, so it can be utilized both for global and enterprise purpose.

References

[1] Sprinklr, "Journal of Customer Experience Vol 1 Spring 14," Sprinklr New York, 10001 USA, 2014.

[2] Statistica Portal, "Penetration of leading social networks in Indonesia as of 4th quarter 2016," [Online]. Available: https://www.statista.com/statistics/284437/indonesia-social-network-penetration. [Accessed March 2017].

[3] E. Lunando and A. Purwarianti, "Indonesian Social Media Sentiment Analysis with Sarcasm Detection," International Conference on Advanced Computer Science and Information Systems (ICACSIS), pp. 195-198, 2013.

[4] D. C. Brabham, "Crowdsourcing as A Model for Problem Solving," Convergence: The International Journal of Research into New Media Technologies, 2008.

[5] A. Carvalho, S. Dimitrov and K. Larson, "How many crowdsourced workers should a requester hire?," Annals of Mathematics and Artificial Intelligence, vol. 78, no. 1, pp. 45-72 , 2016.

[6] G. Barbier, R. Zafarani, H. Gao, G. Fung and H. Liu, "Maximizing benefits from crowdsourced data," Computational & Mathematical Organization Theory, vol. 18, no. 3, pp. 257-279, 2012.

[7] D. Souad and I. Decoopman, "Customers' participation in product development through crowdsourcing: Issues and implications," Industrial Marketing Management., vol. 42, pp. 683-692, 2013.

[8] A. Mukherjee and V. Venkataraman, "What Yelp Fake Review Filter Might Be Doing?," Proceedings of the 7th International Conference on Weblogs and Social Media, ICWSM, pp. 409-418, 2013.
[9] A. Kumar, "Sentiment Analysis on Twitter," IJCSI International Journal of Computer Science Issues, vol. Vol. 9, no. Issue 4, No 3, 2012.
[10] A. Brew, D. Greene and P. Cunningham, "Using Crowdsourcing and Active Learning to Track Sentiment in Online Media," Proceedings of the 2010 conference on ECAI 2010: 19th European Conference on Artificial Intelligence, pp. 145-150, 2010.
[11] H. Kim, J. Kim and A. T. Chen, "Building Trust Mechanisms in Crowdsourcing Application," F1000 Research Open for Science, 2015.
[12] M. Situm, Analysis of Algorithms for Determining Trust Among Friends on Social Networks, Faculty of Electrical Engineering and Computing, University of Zagreb, 2014.
[13] C. Haydar, A. Roussanaly and A. Boye, "Individual Opinions vs Collective Opinions in Trust Modelling," OTICS 2013, The Third International Conference on Social Eco-, pp. 92-99, 2013.
[14] M. Adriani, J. Asian and N. Nazief, "Stemming Indonesian: A Confix-Stripping Approach," ACM Transactions on Asian Language Information Processing (TALIP), vol. 6, no. 4, 2007.
[15] W. A. D. Jr, International Encyclopedia Of The Social Sciences, 2nd Edition, USA: MacMillan Reference, 2008.
[16] Social Media Marketing University (SMMU) , "Brands Are Not Prepared to Manage Complaints in Social Media," 2014. [Online]. Available: http://www.smmu.com.
[17] Y. Sravya and N. D. Prasad, "Twitter Sentiment Classifier," Dept. of C.S.E., GVP College of Engineering (Autonomous),Andhra Pradesh, IndiaInternational Research Journal of Engineering and Technology (IRJET), 2016.
[18] B. L. Bayus, "Crowdsourcing New Product Ideas Over Time: An Analysis of Dell's IdeaStorm Community," Management Science, vol. 59 , no. 1, 01, pp. 226-244, 2013.
[19] Y. Wang and J. Vassileva, "Bayesian Network-Based Trust Model," WI '03 Proceedings of the 2003 IEEE/WIC International Conference on Web Intelligence, p. 372, 2003.
[20] E. Seltzer and D. Mahmoudi, "Citizen Participation, Open Innovation, and Crowdsourcing: Challenges and Opportunities for Planning," Journal of Planning Literature, vol. 28, no. 1, pp. 3-18, 2012.
[21] S. Croon, "Developing new products and services in cooperation with customers in social media platforms," 2013. [Online]. Available: http://essay.utwente.nl/63700.
[22] J. M. Wing and V. Gligor, "Towards a theory of trust in networks of humans and computers," SP'11 Proceedings of the 19th international conference on Security Protocols, pp. 223-242, 2011.
[23] D. Artz and Y. Gil, "A survey of trust in computer science and the Semantic Web," Web Semantics: Science, Services and Agents on the World Wide Web, vol. 5, no. 2, pp. 58-71, 2007.
[24] A. Rachmat and Y. Lukito, "Klasifikasi Sentimen Komentar Politik dari Facebook Page Menggunakan Naive Bayes," Jurnal Informatika dan Sistem Informasi Universitas Ciputra, vol. 2, p. 26, 2016.
[25] S. H. Bao, Y. Deng, H. L. Guo, Q. Hu, J. A. Laredo, R. Vaculin and M. Vukovic, "Using Crowdsourcing to Improve Sentiment Analytics". Patent US 9330420 B2, 3 May 2016.
[26] Z. Noorian and M. Ulieru, "The State of the Art in Trust and Reputation Systems: A Framework for Comparison," Journal of Theoretical and Applied Electronic Commerce Research, vol. 5, no. 2, pp. 97-117, 2010.
[27] Statistica Portal, "Most popular social media of mobile internet users in Indonesia," [Online]. Available: https://www.statista.com/statistics/279776/preferred-netizen-social-media-in-indonesia-by-age. [Accessed May 2017].
[28] J. Nolte, The Philosophy, Theory and Methods of J. L. Moreno, New York: Routledge Taylor &

IOP Conf. Series: Journal of Physics: Conf. Series **971** (2018) 012009 doi:10.1088/1742-6596/971/1/012009

Francis Group, 2014.
[29] C. E. Hollander, An Introduction to Sociogram Construction, Snow Lion Press, 1978.
[30] J. McGready, "The Paired t-test and Hypothesis Testing," Johns Hopkins University, 2006.
[31] J. E. Bartlett, "Organizational Research: Determining Appropriate Sample Size in Survey Research," Information Technology, Learning, and Performance Journal, vol. 9, no. 1, 2001.
[32] P. Del Siegle, "Confidence Intervals and Levels," 2017. [Online]. Available: http://researchbasics.education.uconn.edu/confidence-intervals-and-levels. [Accessed 2017 May].

Prediction of DHF disease spreading patterns using inverse distances weighted (IDW), ordinary and universal kriging

S S Prasetiyowati, Y Sibaroni

School of Computing, Telkom University, Bandung

srisuryani@telkomuniversity.ac.id, yuliant@telkomuniversity.ac.id

Abstract. Dengue hemorrhagic disease, is a disease caused by the Dengue virus of the Flavivirus genus Flaviviridae family. Indonesia is the country with the highest case of dengue in Southeast Asia. In addition to mosquitoes as vectors and humans as hosts, other environmental and social factors are also the cause of widespread dengue fever. To prevent the occurrence of the epidemic of the disease, fast and accurate action is required. Rapid and accurate action can be taken, if there is appropriate information support on the occurrence of the epidemic. Therefore, a complete and accurate information on the spread pattern of endemic areas is necessary, so that precautions can be done as early as possible. The information on dispersal patterns can be obtained by various methods, which are based on empirical and theoretical considerations. One of the methods used is based on the estimated number of infected patients in a region based on spatial and time. The first step of this research is conducted by predicting the number of DHF patients in 2016 until 2018 based on 2010 to 2015 data using GSTAR (1, 1). In the second phase, the distribution pattern prediction of dengue disease area is conducted. Furthermore, based on the characteristics of DHF epidemic trends, i.e. down, stable or rising, the analysis of distribution patterns of dengue fever distribution areas with IDW and Kriging (ordinary and universal Kriging) were conducted in this study. The difference between IDW and Kriging, is the initial process that underlies the prediction process. Based on the experimental results, it is known that the dispersion pattern of epidemic areas of dengue disease with IDW and Ordinary Kriging is similar in the period of time.

1. Introduction

Dengue disease or Dengue Hemorrhagic Fever (DHF), is a disease caused by the Dengue virus of the Flavivirus genus Flaviviridae family. DHF is transmitted to humans through the bite of Aedes mosquitoes that have been infected with dengue virus. DHF is commonly found in tropical and sub-tropical regions. WHO data indicated that ASIA ranks first in terms of the number of DHF patients each year. While Indonesia is the country with the highest case of dengue in Southeast Asia [11].

In Indonesia, for the last 41 years, from 1968 to 2009, there has been an increasing number of dengue endemic provinces and cities, from 2 provinces increasing to 32 provinces with an increase rate about 97%, and from 2 cities increasing to 382 cities with an increase rate about 77%. In addition, there was also an increase in the number of dengue cases, in 1968 only 58 cases to 158,912 cases in 2009 [11].

Some cases of dengue fever cause high mortality. The high number of deaths is caused by several factors including delayed medical treatment and low public awareness of prevention. Many people are not aware of the existence of this disease, one of them caused by the lack of information about the spread of disease they get. The availability of information about the spread of dengue disease area, will make it easier for public health office to know areas that have great potential for dengue fever. So, that preventive actions can be done earlier to the society and the death rate due to dengue fever in a region can decreased.

Several studies related to DHF have been conducted by some researchers but still not focused on the spread of DHF in a region [5], [6], [8], [15]. Although studies in dengue fever which consider spatial factors have been carried out by a number of researchers but the approach used has not used a spatial analysis approach [13], [16], whereas Puspitasari and Irwan [12] indicate that there is spatial influence in the spread of this DHF. Based on this result, the focus of this research is to determine the pattern of dissemination Disease dengue in a region appropriately using spatial analysis approach.

The reason we use three methods of spatial analysis, namely IDW, Ordinary Kriging (OK) and Universal Kriging (UK) to describe the pattern of dengue fever spread, is based on the characteristic pattern of increasing the number of dengue cases, that is decreasing, increasing the number of cases or tend to remain from year to year. The pattern is in accordance with the characteristics of the method used, namely the possibility of distribution pattern is stationary or there is a trend up or down. The formulation of the problem presented in this paper are

- How is the pattern of the spread of dengue fever area in the city of Bandung in 2016 until 2018, considering West Java is the sixth order based on the number of cases
- How the implementation of science data in solving this problem, given the availability of information about the pattern of the spread of dengue fever areas for the coming years
- There are many methods used to create a regional dispersion pattern map, in this paper will examine the implementation of three interpolation methods in spatial analysis used for mapmaking ie IDW, OK, and UK, since data on the number of cases per region is spatial data and number patterns the case varies

It is expected that the results of this study can obtained a clear picture of the pattern of the spread of DHF in each village in the city of Bandung until 2018. So, that the parties concerned can immediately take a policy to address the spread of a very significant in some villages in the city of Bandung.

2. Related works

Dengue disease in Indonesia is a contagious disease that is still a national problem in the field of health. This is because the patient number of this disease tends to increase and more widely spread. Research to obtain the relation between dengue disease with climate factor, which includes temperature, wind speed, humidity, and rainfall in Padang city done by Mangguang [8] . From the result of the research, it can be concluded that there is no significant correlation between climate factor and DHF incidence.

Puspitasari and Irwan [12], conducted a study that aims to analyze the pattern of dengue fever dispersion spatially. Spatial linkages in dengue fever dispersion are measured through spatial autocorrelation using the Moran index. The pattern of dengue fever dispersion are examined using ANN (Average Nearest Neighbor), while mapping of areas with high risk dengue fever dispersion is done by using the Kernel density estimation. The study showed that there is spatial autocorrelation in the dengue dispersion and the study also show pattern type of dengue fever occurrence is clustered pattern. Furthermore, the Kernel density estimation can show mapping of areas with high risk dengue fever dispersion.

In other research, Latifah N et. al [10] presents infectious disease data in the form of maps to know the dispersion pattern and its relation to environmental conditions, and also determine the level of vulnerability. The distribution pattern of infectious diseases have in common characteristic, which has a diffuse pattern with random grouping level. Most of the parameters relating to environmental conditions didn't have a significant linkage to the spread of DHF. The risk of transmission in some areas of Semarang city based on qualitative analysis show that 25% areas are classified as very high risk level and 25% other areas are classified as high risk level.

Similar research is also conducted by Dom C. N., et.al [5] using temporal-spatial to determine the severity and magnitude of outbreak transmission. Spatial models are used to determine high risk of dengue fever spread, which is done by measuring three characteristics of temporal risk, namely frequency, duration and intensity. Meanwhile Rosa et.al. [13] examined the temporal and spatial distribution of dengue fever by considering factors such as dengue coefficient, viral location index, population density and income level. The results showed that the main risk area of transmission of this disease is the city center, where the population has the highest income. Spatial analysis used for DHF notification is ploting census data such as age, sex, population, income, and number of cases with gradient maps. The results obtained are graphs and gradient maps, which are used as DBD disease notification, so they can be integrated with vector control programs that transmit disease.

Dom C. N., et.al [6] analyzed the pattern of dengue dispersion monthly that included spatial deployment and hotspot identification. Estimates were performed using Nearest Neighboring Averages (ANN) and Kernel Density (KD), to assess spatial spread of dengue cases and detect dengue hotspots. The results suggest that spatial patterns for dengue cases over a 5-year period Is a clustered spatial (with a value R, 1) based on monthly frequency data.

Arrowiyah [2], examined the description and map of the spread of DHF incidence in the period 2006-2009, and compared the two analyzes produced by the moran's index method I with Geary's C. From this study it was found that the results released by the moran's index were more sensitive than Geary's. In addition, in this research can produce classification of disease incidence rate in Surabaya City. As for not yet done in this study is the use of distance weights in identifying patterns of spread of dengue disease in the city of Surabaya.

While the research that specifically discuss about the comparison of methods between IDW and kriging, has been done by Setianto A, et al, [14]. In the research said that the technique used for better selection of methods, not only on performance type and data concentration, but also on implementation for various spatial issues. Based on the methodology used, the IDW pattern on its surface display is smoother and sharper than the kriging. Similarly from surface analysis factors, the IDW method is more accurate than kriging, since the presence of minor surfaces and less than maximal displays is useful as a surface-surface interpretation force. While the pattern of kriging shows the breadth of the observed area so that interpolation is less accurate.

Based on the research that has been done by researcher above, hence we do research by making map of pattern of spreading area of DHF disease in Bandung City, taking into account the weight of distance, and then compare between IDW method with kriging method by using statistical test.

3. Methodology

The data used in this study is the number of DHF disease patients in 152 villages in the city of Bandung from 2010 to 2015 obtained from the health department of Bandung. Prediction of disease spread pattern is done through 2 stages, first is to predict the number of patients per year using GSTAR and continued by predicting of dengue fever pattern dispersion area using IDW and Kriging Interpolation.

3.1. Prediction number of DHF patients per year

The prediction of the DHF patients number is done to determine the number of DHF patients for each urban village in the region of Bandung from 2016 - 2018. The method used is GSTAR, by taking into account the spatial and time. GSTAR is one of the models used in the prediction of time series data and location. In the GSTAR models, parameter changes at each location forming a diagonal matrix with parameter $\emptyset_{10}^{(i)}$ and $\emptyset_{11}^{(i)}$. ,Borovkova et al, [3], the GSTAR model (1: 1) for the number of DHF patients at each location i = 1,2, ..., N and time (t) can be seen in equation (1),

$$\hat{Z}_i(t) = \emptyset_{10}^{(i)} Z_i(t-1) + \emptyset_{11}^{(i)} \sum_j^N W_{ij} Z_j\,(t-1) + \varepsilon_i(t) \quad (1)$$

3.2. Prediction of DHF disease spread pattern

After the prediction model in 2016 until 2018 was obtained, the next step is formation a spatial model based on prediction data. The purpose of this spatial modeling is to obtain an area dispersion model of dengue fever. Spatial models used in this research are IDW and Kriging interpolation. Based on the characteristic pattern of the number of sufferers, then the appropriate Kriging interpolation method used is Ordinary Kriging and Universal Kriging. In IDW method it is assumed that each point has local influence decreases with distance. The points have closer distance to the location estimation, would be given greater weight. Case number estimation at position $\boldsymbol{S_0}$ using IDW method by Watson et al, [7] can be seen in equation (2)

$$\widehat{\boldsymbol{Z}}(\boldsymbol{S_0}) = \frac{\sum_{i=1}^{n} \boldsymbol{Z(S_i)} \cdot \boldsymbol{d_{i0}^{-p}}}{\sum_{i=1}^{n} \boldsymbol{d_{i0}^{-p}}} \tag{2}$$

Where $\widehat{\boldsymbol{Z}}(\boldsymbol{S_0})$ the estimated value at the point $\boldsymbol{S_0}$, $\boldsymbol{S_0}$ is location estimation, n is the number of nearest neighbors, $\boldsymbol{Z(S_i)}$ is case number of the sample location, with i = 1,2, ..., n, d is the distance from the sample location $\boldsymbol{S_i}$ to location predictions $\boldsymbol{S_0}$ and p is the exponent that determines the value of the weight for each prediction.

While the case number estimate at point S_0 for Ordinary Kriging and Universal Kriging is obtained from Amstrong, [1] and expressed by equation (3)

$$\hat{Z}(S_0) = \sum_{i=1}^{n} \lambda_i Z(S_i) \tag{3}$$

Where λ_i is weight value of i-th data, $Z(S_i)$ is value of the number of dengue fever patients in each village, and n is number of samples in the estimation process.

In Universal Kriging [17], function f_1 of the coordinate s, which is considered as deterministic, because they were known to each domain location. Random function Z(s) is composed of a deterministic component (m (s)) and Y (s) is stationary random function of second order. Formula of Z(s) can be seen in equation (4)

$$Z(s) = m(s) + Y(s) \tag{4}$$

It is assumed that Y (s) is a second order stationary with mean is equal zero and covariance function C (h), so

$$E[Z(s)] = m(s) \tag{5}$$

[17] assumed that drift m (s) can be represented as a linear combination of deterministic function f_1 with coefficients α_l(not zero)

$$m(s) = \sum_{l=0}^{L} \alpha_l f_l(s) \tag{6}$$

Where function $f_0(s)$ is defined as constant and have value is 1, then equation of Kriging is presented by [17] as linear combination $Z^* = \sum_{\alpha=1}^{n} w_\alpha Z(s_\alpha)$, So $\sum_{l=0}^{L} \alpha_1 (f_1(s_0) - \sum_{\alpha=1}^{n} w_\alpha f_1(s_\alpha)) = 0$ and if α_1 not zero, then value of w_α is

$$\sum_{\alpha=1}^{n} w_\alpha f_l(s_\alpha) = f_l(s_\alpha) \quad for \;\; l = 0, \dots, L, \tag{7}$$

Equation (7) is referred to as universal **condition**. While for constant function $f_0(s)$, t he condition is expressed as $\sum_{\alpha=1}^{n} w_{\alpha=1}$. By using Lagrange parameter μ_1 and minimizing, [16] will obtain Universal Kriging (UK) system equation :

$$\begin{cases} \sum_{\beta=1}^{n} w_\beta C(s_\alpha - s_\beta) - \sum_{l=0}^{L} \mu_l f_l(s_\alpha) = C(s_\alpha - s_0) \\ \qquad\qquad for \; \alpha = 1, \dots, n \\ \sum_{\beta=1}^{n} w_\beta f_l(s_\beta) = f_l(s_0) \\ \qquad\qquad for \; l = 0, \dots, L \end{cases} \tag{8}$$

Equation (8) can be expressed in matrix in equation (9)

$$\begin{pmatrix} C & F \\ F^T & 0 \end{pmatrix} \begin{pmatrix} W \\ -\mu \end{pmatrix} = \begin{pmatrix} C \\ f \end{pmatrix} \tag{9}$$

Where $F = \{f_0, \ldots., f_L\}$ and F is a collection of linear function stating coordinate location, $Y(s)$ is stationery random function of 2-th order, $(x\,, y)$ is a location coordinate, w is weight factor for point i, μ is lagrange multiplier used to minimize the possibility of estimation error, and α is the trend local coefficient [17].

3.3. *Determining the appropriate model*

The predicted results accuracy of dengue fever dispersion with either IDW or Kriging will depend on the appropriate of the selected model. Determining the right model is based on the results of the validation and verification using a statistical measure of RMSE (Root Mean Square Error). In the selection of optimum weight on IDW and best variogram for Kriging, residual analysis was performed between observation data and estimation data using the RMSE criteria. RMSE is the root of MSE (Mean Square Error). While The MSE formula can be seen in equation 10, [9]

$$\text{MSE} = \frac{1}{n}\sum_{i=1}^{n}(\hat{\boldsymbol{Z}}(\boldsymbol{S}_0) - \boldsymbol{Z}(\boldsymbol{S}_i))^2 \quad (10)$$

Where $\hat{\boldsymbol{Z}}(\boldsymbol{S}_0)$ = estimation value and $\boldsymbol{Z}(\boldsymbol{S}_i)$ = actual / observation value of i-th data. The smaller the value of RMSE, then the selection of models and predicted results will be better. It shows how close the observation data to the predicted value.

4. Design of system

The interpolation process to predict the dispersion pattern of dengue disease areas, with IDW and kriging methods (Ordinary and Universal Kriging) is described in Figures 1 and 2. And the interpolation process steps are described as follows

4.1. *Input of sample data*

User input sample data, ie spatial data contains data on the number of cases of DBD and the parameters required during the calculation process.

4.2. *Prepocessing sample data*

If there is missing sample data from the regular pattern, then the missing sample value does not need to be interpolated by taking its mean value or replacing it with a zero value

4.3. *Interpolation process*

4.3.1. *IDW*

Local/ global interpolation

Deterministic interpolation techniques can be divided into two groups: global interpolation and local interpolation. The global technique calculates the predictions using all the data sets whereas the local technique calculates the prediction of the size points that exist within the neighborhoods area, ie smaller spatial areas within larger study areas.

After determining the interpolation technique, then calculate the predicted value using Inverse Distance Weighted (IDW) method of formula 2.

4.3.2. *Kriging methoda (Ordinary and Universal Kriging)*

4.3.2.1. Experimental semivariogram calculations. An experimental semivariogram was used to determine the possibility of anisotropy. The semivariogram is calculated in various directions. When each direction gives the same parameter value is isotropic, meaning semivariogram depends only on distance, h. If the semivariogram depends on the distance h and direction θ then it is called anisotopic. To investigate anisotropic is usually selected at least four directions.

4.3.2.2. Fitting of theoretical semivariogram model. After the experimental semivariogram model is plotted, the fitting results of the experimental semivariogram plot with the theoretical semivariogram.

IOP Conf. Series: Journal of Physics: Conf. Series **971** (2018) 012010 doi:10.1088/1742-6596/971/1/012010

Then select the theoretical semivariogram model, input range, sill, and nugget values based on the experimental semivariogram obtained.

4.3.2.3. Validation of theoretical semivariogram model. After the theoretical semivariogram model is obtained, then the validation process of theoretical semivariogram model. Model validation is the testing process of the selected model with respect to the smallest RMSE value

4.3.2.4. Estimated number of dengue cases by kriging method. After the theoretical semivariogram model is valid, then the model is used to estimate the number of dengue cases with ordinary kriging and universal kriging.

4.3.2.5. Kriging interpolation. After the estimation process, kriging interpolation is done by mapping on the contour map

4.4. Implementation

At this stage of implementation, in addition to data on the number of cases of dengue fever, it takes some other inputs ie spatial data or location information in the form of coordinates of both geographic coordinates (latitude and longitude), XYZ coordinates, including datum and projection information, conversion to Universal Transverse Mercator format (UTM) into meters. In its implementation to the software, it takes a shapefile file of urban village of Bandung, which consists of vector data of an attribute. Vector data is the shape of the earth that is represented into a set of lines, dots, and areas.

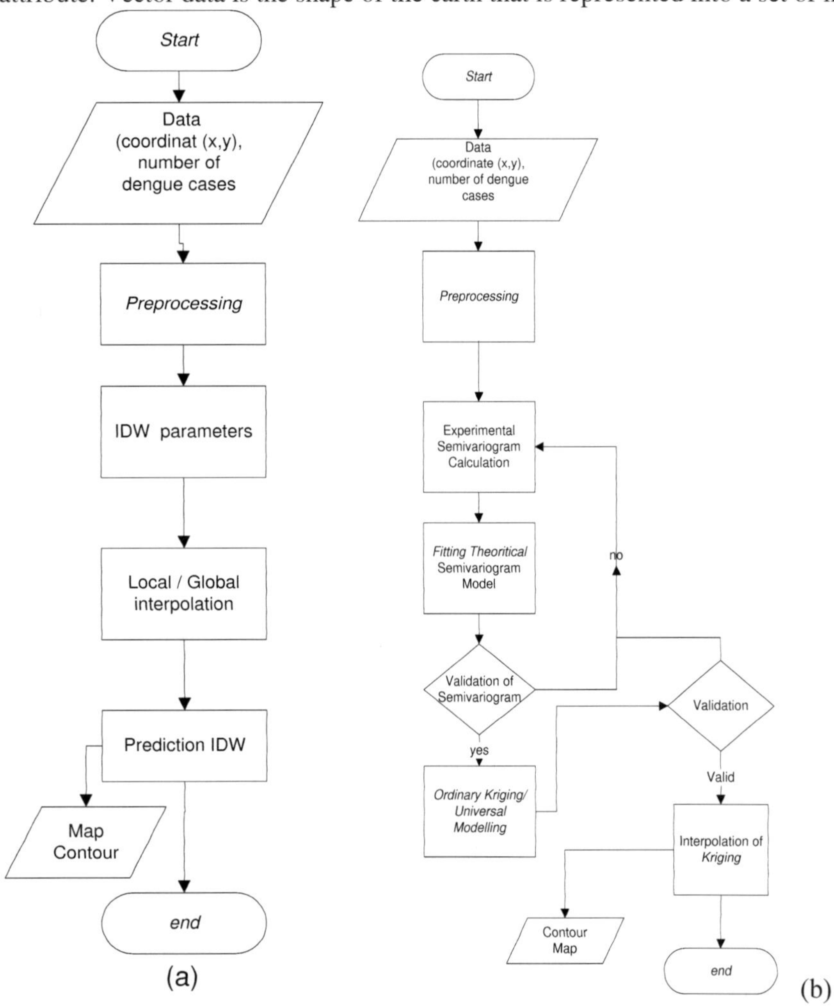

Figure 1. (a) Diagram of distribution pattern Prediction using IDW, (b) Diagram of distribution pattern Prediction using Kriging

5. Result and discusion

5.1. Prediction of Distribution Pattern Using IDW

Inverse distance weighted (IDW) is a deterministic spatial interpolation method, using exact calculations. In this case, the deterministic interpolation technique used is the local interpolation. Prediction calculations are taken from the points size is within the neighborhoods, in a smaller spatial area, and in the larger study area. After the interpolation is done, then the predicted value is calculated by Inverse Distance Weighted (IDW). Table 1 contains the RMSE values for the three predicted values ranging (p=1, p=1.5, and p=2) with a minimum of neighborhoods is 10. The sample data is taken from 2010 to 2015.

Table 1. RMSE value of IDW from 2010 to 2015

Year	RMSE		
	P = 1	P = 1.5	P = 2
2010	17.17	22.38	17.89
2011	19.54	19.81	20.41
2012	22.81	23.33	24.19
2013	28.54	29.00	29.94
2014	14.39	14.41	14.65
2015	16.71	16.75	15.98

Table 2. Estimation value of RMSE from 2016 to 2018

Year	RMSE		
	P = 1	P = 1.5	P = 2
2016	18.56	18.86	18.99
2017	9.53	9.68	9.97
2018	20.91	21.25	21.87

Based on Table 1 and Table 2, it can be concluded that the smallest RMSE values obtained when the value of p = 1. Hence for p values used in IDW interpolation is 1. The results of the regional dispersion patterns resulting from the IDW interpolation based on sample data from 2010 to 2015 can be seen in Figure 2

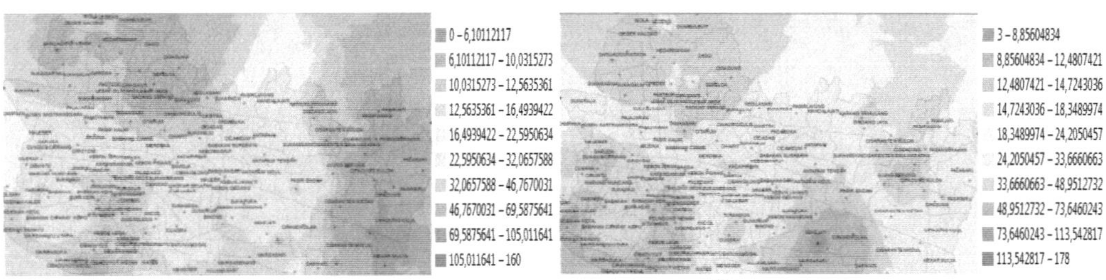

(a) (b)

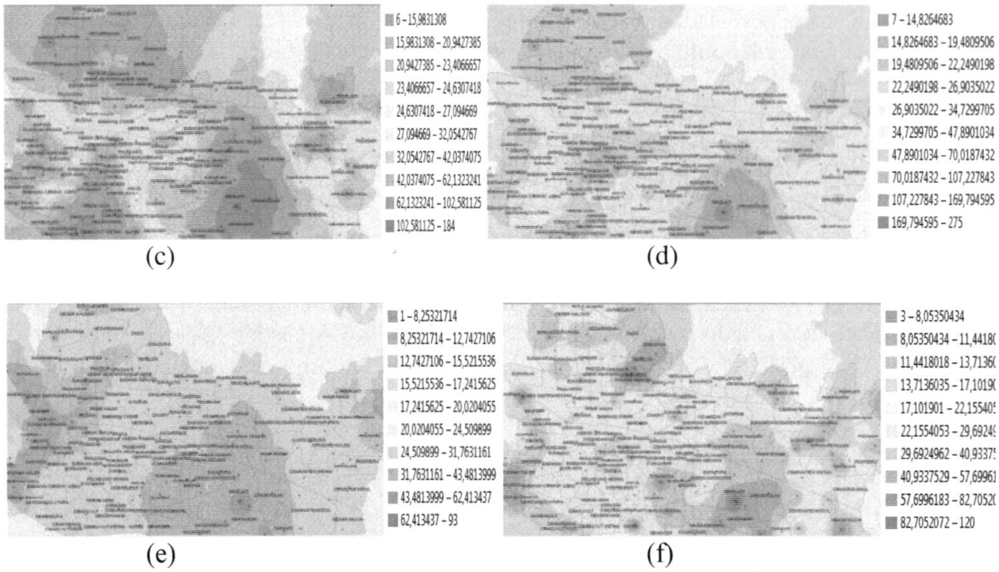

Figure 2. The region dispersion patterns of DHF diseases at (a) 2010, (b) 2011, (c) 2012, (d) 2013, (e) 2014, and (f) 2015 using IDW

By considering the pattern of DHF spread in the Figure 2, it can be informed that the number of DHF patients in the northern region of Bandung from 2010 until 2013 tended to constant around 70 to 105 and decreased from 2014 and 2015 around 40 to 50. At the western region of Bandung, the pattern of distribution tends to decrease from 2010 to 2014 around the number 30. Then in 2015, the number of patients of dengue disease tends to fall again but not significant to below 23. While the number of patients in the southern, central part and east region of Bandung, starting in 2011 tends to rise then constant at around 105 in 2013. The number of patients is decreased in 2014 and 2015 at a rate of about 40 to 60 per year. In contrast to other region of Bandung, in middle region of Bandung since 2010 until 2011, the number of patients around 25, in 2012 to 2015 tends to constant not exceeding 30.

While the pattern of spread of dengue disease territories of the data predicted results, using GSTAR for 2016 and 2018 can be seen in Figure 3.

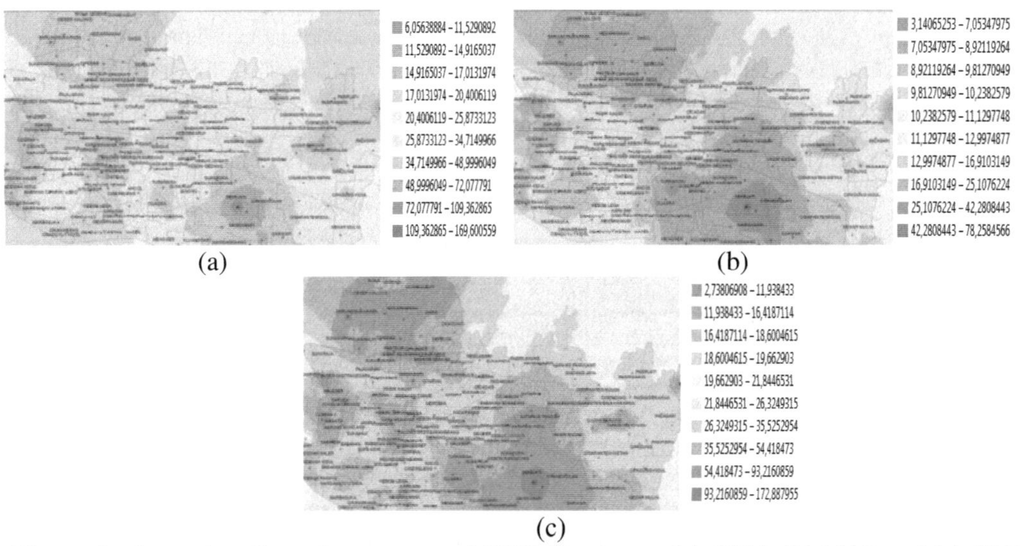

Figure 3. The region dispersion patterns of DHF diseases at (a) 2016, (b) 2017, and (c) 2018 using IDW

Based on the area spread pattern of dengue disease from GSTAR prediction data from 2016 to 2017, it shows that for the north and south region of Bandung tends to fall, but in 2018 rose to 94. Similarly, the pattern in the western, central and eastern region of Bandung, distribution patterns tend to fall in 2017, but rose in 2018.

5.2. Prediction of distribution pattern using kriging

Unlike the IDW method, before the Kriging interpolation method is implemented, a semivariogram measurement is computed first, to determine the most suitable theoretical semivariogram model. The theoretical semivariogram model is then used as an input model on Kriging interpolation. In this study, it is only used three options semivariogram theoretical models, namely *Spherical, Exponential* and *Gaussian.* Determination of the best theoretical semivariogram model is based on the smallest RMSE value. Table 3 presented RMSE values of each model semivariogram theoretical for data preprocessing and data predicted results GSTAR from 2010 to 2015 and from 2016 to 2018, where the shading cells in the table show the model used in the Kriging process.

Table 3. RMSE for each semivariogram models

Year	Ordinary Kriging			Universal Kriging		
	Eksp	Gauss	Spher	Eksp	Gauss	Spher
2010	18.03	17.57	17.34	17.65	17.73	17.47
2011	20.24	20.01	19.75	20.31	21.21	20.45
2012	23.96	23.25	23.4	23.76	24.53	24.21
2013	29.03	28.29	28.49	28.57	28.74	28.43
2014	14.38	14.44	14.35	15.25	15.11	15.21
2015	15.81	16.15	15.74	15.82	16.42	15.84
2016	19.25	18.97	18.96	19.08	19.19	19.11
2017	9.85	9.70	9.67	9.81	9.89	9.90
2018	21.71	21.26	21.34	22.39	22.16	23.06

The spread pattern of dengue disease area generated using Ordinary Kriging interpolation is presented in Figure 4

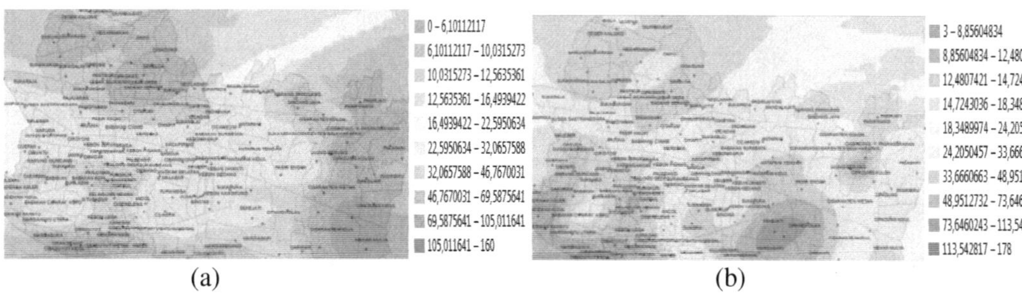

(a) (b)

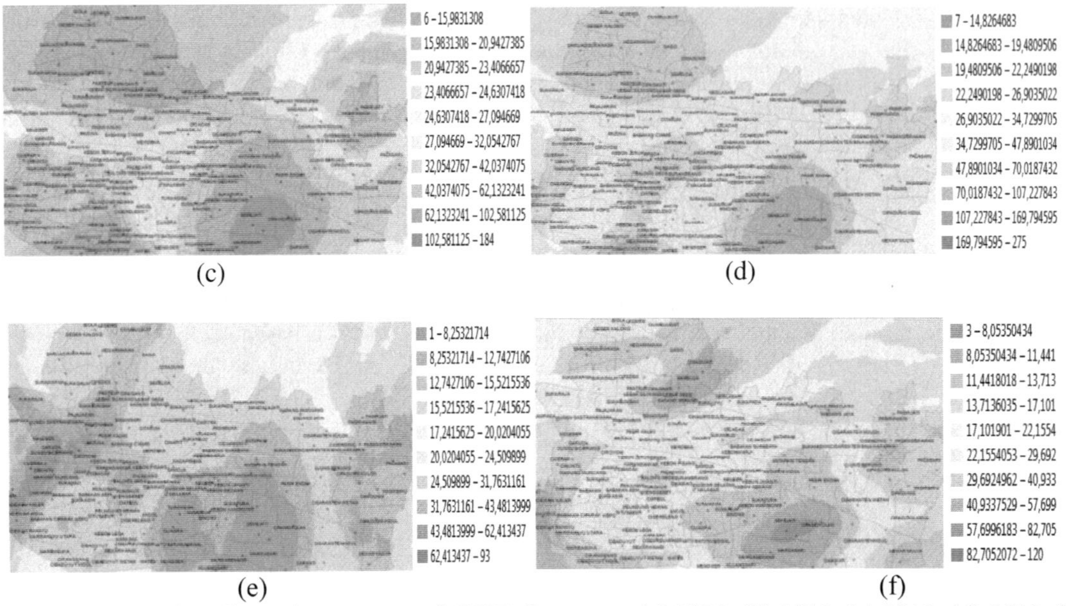

Figure 4. The region dispersion patterns of DHF diseases at (a) 2010, (b) 2011, (c) 2012, (d) 2013, (e) 2014, and (f) 2015 using Ordinary Kriging

Figure 4 shows that the number of DHF patients in the northern region of Bandung in 2010 about 32 to 70 people, then decreased in 2011, 2014 and 2015 under 50 Although in 2012 and 2013 had risen to reach 70 people. In the southern region of Bandung there is an increase from 2011 to 2015 with the number of patients reached 114 people. This condition extends to the eastern region of Bandung in 2012, but then becomes stable in subsequent years under 23. While in the western and central region of Bandung tend to have a constant pattern that is under 23 patients per year.

While the pattern of spread of dengue disease area predicted results using Ordinary Kriging interpolation GSTAR from 2016 to 2018 are presented in Figure 5.

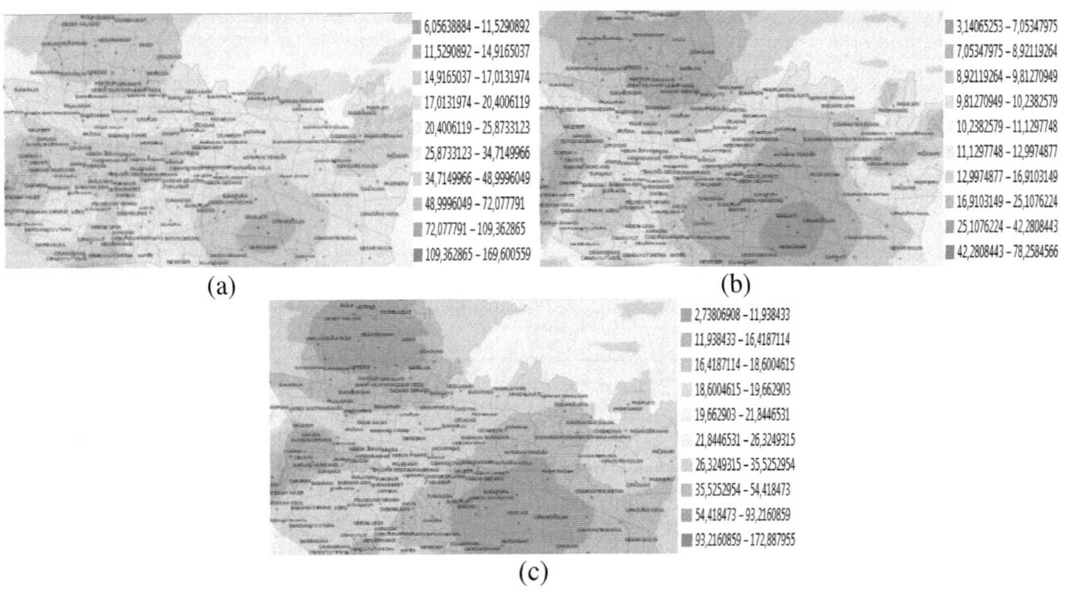

Figure 5. The region dispersion patterns of DHF diseases at (a) 2016, (b) 2017 and (c) 2018 using Ordinary Kriging

Based on Figure 5, it appears that the number of patients in the northern region of Bandung in 2017 decreased significantly to below 26, compared to number of patients in 2016 which reached 50. But in 2018, it is predicted this number will increase drastically reaches 94. In the south region of Bandung from 2016 to 2018, the pattern tends to be constant with the number of DHF patients per year ranged from 50 to 73. The condition is also not much different than the eastern region of Bandung in 2017 and 2018. Like the results obtained with IDW, on western and central region of Bandung from 2016 to 2018 are predicted to tend to be below 26.

While the spread pattern of dengue disease territory acquired from Universal Kriging interpolation can be seen in the Figure 6.

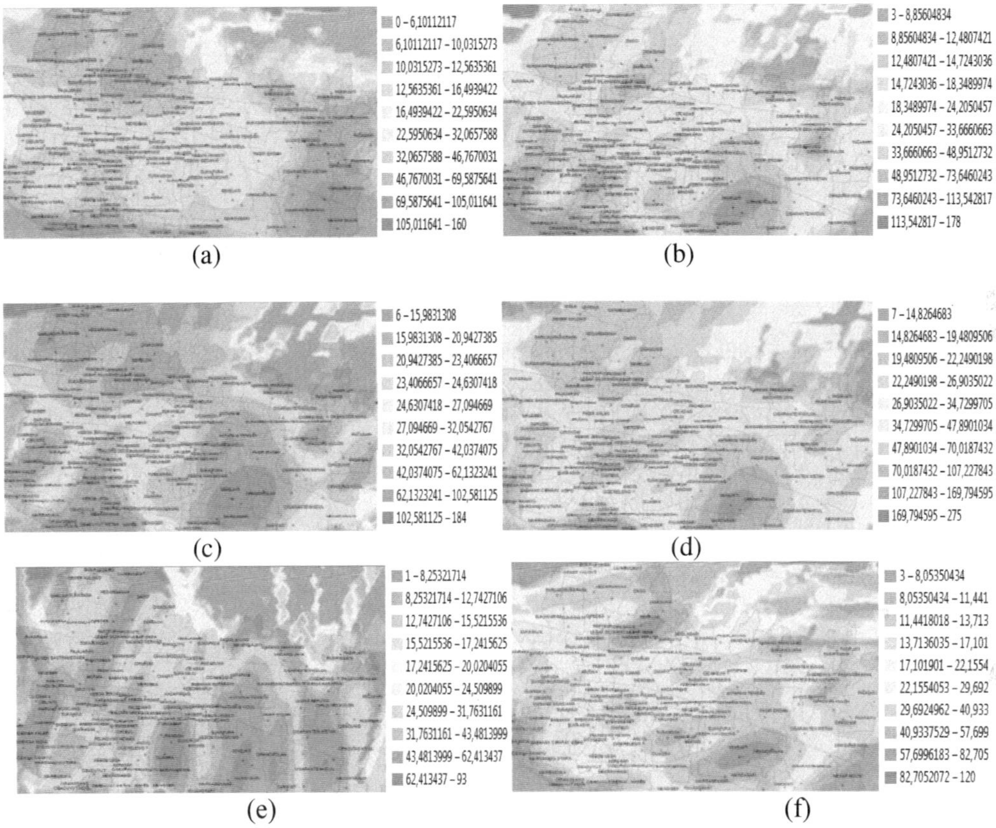

Figure 6. The region dispersion patterns of DHF diseases at (a) 2010, (b) 2011, (c) 2012, (d) 2013, (e) 2014, and (f) 2015 using Universal Kriging

Figure 6 shows that DHF patients in the northern part of Bandung in 2010 around 70, then tends to rise to 103 in 2011 to 2013. Then from 2013 to 2015 has decreased below 30. In the southern part of Bandung, the number of patients from 2011 to 2015 tends to increase to 80 per year, where previously in 2010 under 23 patients with DHF. While the number of patient in western and central part of Bandung tend to be constant under 30, from 2010 until 2015. Similarly, the pattern on the eastern city of Bandung, there is an increase in the number of patients about 30% in 2012.

The spread pattern prediction of dengue disease by using GSTAR, and Universal Kriging interpolation from 2016 to 2018 can be seen in Figure 7.

IOP Conf. Series: Journal of Physics: Conf. Series **971** (2018) 012010 doi:10.1088/1742-6596/971/1/012010

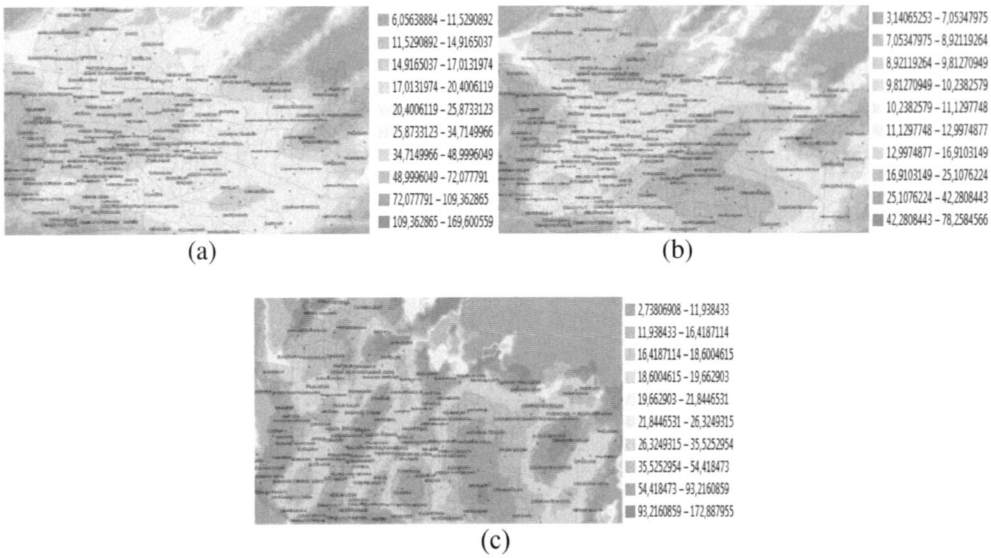

Figure 7. The region dispersion patterns of DHF diseases at (a) 2016, (b) 2017 and (c) 2018 using Universal Kriging

Based on the spread pattern of dengue disease area as seen in Figure 7, it is estimated until the end of 2016, the number of patients in most areas of Bandung city is low, that is below 35, even patient number in the western and central part of Bandung, the number is lower around 18. This condition tends to decline steadily in 2017 in all areas of Bandung. But in 2018, it is estimated that the average increase again reaches 94, even some villages in the northern part of Bandung reaches 173 patients per year.

5.3. Test of compare of IDW and kriging methods

To find out which method gives the best predicted results from the three methods used by comparison of RMSE values. Tables 4 and 5 contain the RMSE values of each method for preprocessing data from 2010 to 2015 and predicted results with GSTAR for 2016 through 2018.

Table 4. Value of RMSE for 3 methods from 2010 to 2015

Year	RMSE		
	IDW	Ordinary Kriging	Universal Kriging
2010	17.17	17.34	17.47
2011	19.54	19.75	20.31
2012	22.81	23.25	23.76
2013	28.54	28.29	28.43
2014	14.39	14.35	15.11
2015	15.98	15.74	15.82

Table 5. Value of RMSE for 3 methods from 2016 to 2018

Year	RMSE		
	IDW	Ordinary Kriging	Universal Kriging
2016	18.56	18.96	19.08
2017	9.53	9.67	9.81

2018	20.91	21.26	22.16

To compare the three methods used above, based on the RMSE values obtained from each method from 2010 to 2018, we will use the t-Student test. While the data used as input on the t-Student test is the value of the RMSE. The test results obtained are presented in Table 6.

Table 6. t-Student Testing

test	Value of t calculated	T value on t-student table	Conclusion
IDW vs OK	0.34	2.14	IDW & OK is not different significantly
IDW vs UK	2.09	2.14	IDW & UK is not different significantly
OK vs UK	3.92	2.14	OK & UK is different significantly

In table 6. ie on the line IDW vs OK and IDW vs UK, it appears that the value of t calculated is smaller than the t value on t-student table. So it can be concluded that the IDW method does not differ significantly with kriging method (OK and UK). While on the line OK vs UK, it appears that the value of t calculated is greater than t value on t-student table, which means that Ordinary Kriging method is significantly different with universal kriging method. Based on the test results, it can be concluded that in this study, the IDW method is not influenced by the characteristic data, while the method of Ordinary Kriging and Universal Kriging is very sensitive to the characteristics of data that is stationary or there is a trend.

Based on the test with t-student statistics on the three methods, presented in table 6, the results showed that the pattern of the spread of dengue disease area, produced by IDW method did not differ significantly with kriging method. But precisely the method of ordinary kriging with universal kriging method gives significantly different results on the pattern of the spread of dengue disease area in the city of Bandung. And based on the results of the above research, the prediction results obtained from the IDW and kriging methods are not much different. The pattern of the spread of dengue fever in Bandung until 2018, still has the same pattern that is the increase in the number of cases is quite large occurred in the city of Bandung in the north and south.

For the researcher, the future research plan that will be conducted, based on the above results, will examine the factors that influence the spread of dengue fever in the region mentioned above by using Genetic Algorithm.
While the proposals submitted to the government of Bandung, associated with the results of this study, the government of Bandung, especially the Department of Health should be as early as possible to disseminate about the spread of dengue fever and how the management of this disease to urban villages in the northern and southern Bandung.

6. Conclusion

Based on the above results and discussion can take several conclusions as follows

- In general, the pattern of the spread of dengue disease area in northern and southern Bandung changed significantly i.e. from 2010 to 2015 tends to fall. The pattern will still occur until 2017, but the number of people with DHF will rise in 2018. In the western, central and east region of Bandung is relatively constant until the year 2018.

- Implementation of the three-spatial interpolation method combined with GSTAR to predict the dispersion pattern of dengue disease in Bandung area, give information that IDW and Kriging methods did not differ significantly, but Ordinary Kriging and Universal Kriging method give different result.
- IDW method is not influenced by the characteristic data, while the method of Ordinary Kriging and Universal Kriging is very sensitive to the characteristics of data that is stationary or there is a trend.

Acknowledgment
The authors wish to thank Telkom University for the support of funds and infrastructure, so that this research can be completed.

References

[1] Armstrong M 1998 *Basic Linear Geostatistic* Berlin Springer-Verlag
[2] Arrowiyah Sutikno *Spatial Pattern Analysis Kejadian Penyakit Demam Berdarah Dengue Untuk Informasi Early Warning Bencana di Kota Surabaya* unpublish
[3] Borovkova S Hendrik P Budi Nuraini R 2008 *Consistency and Asymptotic Normality of Least Squares Estimators in Generalized STAR Model* Statistica Neerlandica Vol 62 No. 4 pp 482 – 508
[4] C.D Lloyd 2007 *Local Models for Spatial Analysis* London Taylor & Francis Group
[5] Dom C N Ahmad A H Latif Z A Ismail R Pradhan B 2012 *Manifestation of GIS Tool for Spatial Pattern Distribution Analysis of Dengue Fever Epidemic in the City of Subang Jaya, Malaysia,* Journal Environment Asia, 82 – 92, The Society of Higher Education Institutes on Environment
[6] Dom C N Ahmad A H Latif Z A Ismail R 2013 *Measurement of Dengue Epidemic Spreading Pattern Using Density Analysis Method Retrospective Spatial Statistical Study of Dengue in Subang Jaya Malaysia 2006 – 2010* The Royal Society of Tropical Medicine & Energy
[7] D. Watson and P G M 1985 *A Refirement of Inverse Distance Weighted Interpolation Geo-Processing 2* pp 315 – 327
[8] Mangguang M Dt 2010 *Analisis Epidemologi Penakit Demam Berdarah Dengue Melalui Pendekatan Spasial Temporal dan Hubungannya dengan Faktor Iklim di Kota Padang tahun 2008 – 2010* Fakultas Kesehatan Masyarakat, Universitas Andalas Padang
[9] M Suprajitno 2005 *Pengantar Geostatistik* Jakarta Universitas Indonesia
[10] Latifah N Saraswati E Widayani P *Pemetaan Data Penyakit Menular di Kota Semarang (Studi Kasus : Penyakit DBD, Diare, Pneumonia, dan TB Paru +)* unpublish
[11] Pusat Data dan Surveilans Epidemiologi Kementrian Kesehatan RI 2010 *Demam Berdarah Dengue di Indonesia tahun 1968 – 2009* Buletin Jendela Epidemiologi Vol 2 ISSN 2087 – 1546
[12] Puspitasari R Irwan S 2011 *Analisis Spasial Kasus Demam Berdarah di Sukoharjo Jawa Tengah dengan Menggunakan Indeks Moran* Prosiding Seminar Nasional Matematika dan Pendidikan Matematika Wackernagel, H., 1998, Multivariate Geostatistica An Introduction with Application, Springer, Verlag Berlin Heidelberg New York ISBN 978-979-16353-6-3
[13] Rosa Freitas M G Tsouris P Sibajev A Tetiani E Ubirajara A Lopes R Francisco J 2003 *Exploratory Temporal and Spatial Distribution Analysis of Dengue Notification in Boa Vista Roraima Brazilian Amazon 1999 – 2001* Dengue Bulletin Roraima Federal University Vol 27
[14] Setianto A Triandini T 2013 *Comparison of Kriging and Inverse Distance Weighted (IDW) Interpolation Methods in Lineament Extraction and Analysis* J. SE Asian Appl Geol Vol 5(1) pp 21 – 29
[15] Suryani S Yuliant S 2007 *Pemodelan Penyebaran Penyakit Demam Berdarah Dengan Model SIR*. Jurnal Saintifika Sekolah Tinggi Teknologi Telkom

IOP Conf. Series: Journal of Physics: Conf. Series **971** (2018) 012010 doi:10.1088/1742-6596/971/1/012010

[16] Tran A Depuris X Dussart P Morran J Raburison P Remy F Polidori L Gardon J 2004 *Dengue Spatial and Temporal, French Guiana 2001* Emerging Infectious Diseases, Volume 10 No. 4

[17] Wackernagel H 1998 *Multivariate Geostatistica An Introduction with Application* Springer Verlag Berlin Heidelberg New York

IOP Conf. Series: Journal of Physics: Conf. Series **971** (2018) 012011 doi:10.1088/1742-6596/971/1/012011

Implementation of mutual information and bayes theorem for classification microarray data

Mahendra Dwifebri Purbolaksono, Kurnia C Widiastuti, Mohamad Syahrul Mubarok, Adiwijaya, Firda Aminy Ma'ruf

School of Computing, Telkom University

febhius@gmail.com, kurniabahariawan@gmail.com, msyahrulmubarok@telkomuniversity.ac.id, adiwijaya@telkomuniversity.ac.id, marufirda@gmail.com

Abstract. Microarray Technology is one of technology which able to read the structure of gen. The analysis is important for this technology. It is for deciding which attribute is more important than the others. Microarray technology is able to get cancer information to diagnose a person's gen. Preparation of microarray data is a huge problem and takes a long time. That is because microarray data contains high number of insignificant and irrelevant attributes. So, it needs a method to reduce the dimension of microarray data without eliminating important information in every attribute. This research uses Mutual Information to reduce dimension. System is built with Machine Learning approach specifically Bayes Theorem. This theorem uses a statistical and probability approach. By combining both methods, it will be powerful for Microarray Data Classification. The experiment results show that system is good to classify Microarray data with highest F1-score using Bayesian Network by 91.06%, and Naïve Bayes by 88.85%.

1. Introduction

Cancer is the leading cause of death for everyone around the world. According to WHO's data, there are about 14 million new cases just in 2012 which can increase next year. Early observation of disease is needed to prevent cancer growth. Furthermore, cancer detection is also important. There are several ways to detect cancer, such as physical detection, blood test and X-*ray*. But these could take a long time and containing human errors during detection. In 2013, *British Medical Journal* (BMJ) released that medical error caused 251.454 death every year in United States of America. Therefore, it is required to have technologies that can solve the problem. One of them is Gene Expression Technology that involves Microarray technology for cancer detection [1]..

Microarray technology is a developing technology used to study the expression of many genes at once [2]. It is the groundwork for many cancer detection researches. However, the main problem of Microarray data is processing the data itself since it contains many number of insignificant and irrelevant attributes [2] which affect the classification process and result. Therefore, it needs a method to reduce the dimension of Microarray data without eliminating important information in every attribute. After reducing the dimension, the data will be classified based on the existing classes.

The problem of data classification will be solved using Bayesian methods. For classifying Microarray data, the methods have attracted a lot of attentions from researchers to study this topic, such as Liying Yang [3], Devi Arockia [4] and Luis M. de Campos [5]. In reference [5] it explained the detail about Microarray, such as data distribution and approach to solve Microarray classification problems. Many different approaches have been used, one of them is feature selection approach using Mutual Information (MI) and Support Vector Machine (SVM) for the classifier [4]. In the research, it showed that MI has good performance for dimensional reduction. However, for classification process, this research has a deficiency for classifying multi-class data. While in research [5], it used a variety of Bayes Theorem such as Naive Bayes and Bayesian Network that have good performance in this topic. But in research [5], it has deficiency for data dimensional reduction process. That happened because Certainty Factors method can only process maximum 2 data. If it is more than two data, the system needs more time to execute it.

Therefore, this research uses Mutual Information for feature selection and it is expected to produce a better result after dimension gets reduction. While for classification, it uses Naïve Bayes and Bayesian Network since they have ability to get pattern and dependency among variables using probability approach.

2. Study of Literature

Two main focuses in this research are the usage of Mutual Information and Bayes Theorem. The Bayesian Theorem that we used are Naïve Bayes and Bayesian Network. For the system, there are three main stages namely Discretization, Feature Selection and Learning of Bayesian.

2.1. Discretization

We used K-Means algorithm to transform continue data into discrete data. K-Means works by dividing data into several groups, also called clustering. Firstly, K-Means randomly distribute center points (centroids) for clusters representation. Because of randomly distribution, it is possible that some centroids are not properly located. Therefore, they need to be updated as follows,

$$\mu_k = \frac{1}{N_k}\sum_{q=1}^{N_k} x_q \quad (1)$$

where μk is centroid of kth cluster, Nk is number of data belongs to k^{th} cluster, and axis q^{th} data on k^{th}-cluster. Updating centroids is done until centroids do not change.

2.2. Feature Selection

The problem in Microarray data is about large dimension of the gene. Therefor, we need to reduce the number of dimension. One of the solution is using Mutual Information. In this work, Mutual Information is used to select informative gene from the original gene expression profile [6]. To compute Mutual Information, the probability distribution of genes is needed which in practice are not known, and the best we can do is using histogram of data. The steps involved in computing the Mutual Information from the histogram of the training data are given as follows,

a. The data set is arranged in an ascending order based on the output.
b. The output class label (Y) is divided into two groups and the initial entropy H(Y) is calculated using
$$H(Y) = -\sum_{j=1}^{Ny} P(Y\text{–}i).\log(P(Y\text{–}i)) \quad (2)$$
c. The input genes (X) are divided into ten levels and their conditional entropies H(Y|X) are evaluated using
$$H(Y|X) = -\sum_{i=1}^{Nx} P(X\text{–}i).\sum_{j=1}^{Ny} P(Y\text{–}j|X\text{–}i).\log(P(Y\text{–}j|X\text{–}i)) \quad (3)$$
d. Next, the mutual information of each gene with respect to the output is computed using
$$I(Y;X) = H(Y) - H(Y|X) \quad (4)$$
Where:
X-i = Input number-i { $X_1 : 2.5, X_2 : 1.5 \ldots.. X_n : 2.25$}
Y-i = Class number-i {$Y_1 : ALL, Y_2 : AML$}
e. The result of Mutual Information will be sorted ascending. First attribute is the highest value and will be selected as the most informative gene.

2.3. Bayes Theorem

Bayes Theorem is a theorem which is used in statistic for calculating probability of hypothesis. Bayes Optimal Classifier [7] calculates probability from each group of an existing class and for deciding the most optimal class. Naïve Bayes is the first considered as Bayesian Classifier.
Naïve Bayes or can be called as Bayes Theorem with the assumption that the attribute independency is a classification that depends on two assumptions. Those assumptions are as follow: each attribute will be independent given class attribute and all attributes have influence to a class. Assumption of attribute independency will eliminate the needs of amount training data from multiplication of Cartesian by all of attribute which used to classify a data [8]. In a Directed Acyclic Graph (DAG) which represents the Naive Bayes (NB) structure, all of the bows will be aimed from attribute class to all attributes like Figure 1.

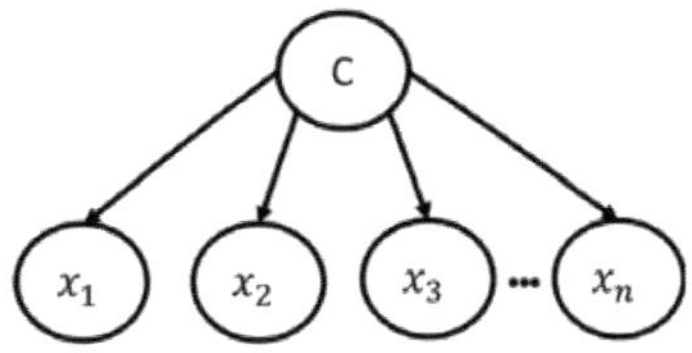

Figure 1. Example of Naïve Bayes Structure

Where:
C = class $\{C_{Leukimia\ ALL-A} \ : \ ALL, AML\ \}$
$_1$ = attribute number-1 $\{x_1 : \ 250\}$

The development of Naïve Bayes method is accomplished by using many methods [13]. One of it begins from the basic topology, for example by connecting each attributes to one another. This method is one of the Bayes theorem methods which is also knows as Bayesian Network. For an example, Bayesian Network with a connection between each attribute like Figure 2.

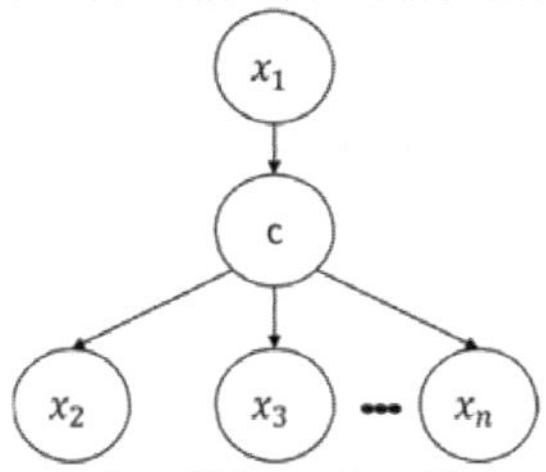

Figure 2. Example of Bayesian Network Structure

Where:
C = class
x_n = attribute

From all of the methods, Bayes Theorem is using this equation [9]:

$$P(A \mid B) = \frac{P(B \mid A)\, P(A)}{P(B)} \tag{5}$$

Where:
A is data hypothesis A (specific class)
B is data with unknown class
P(A | B) is probability hypothesis A given B condition (posterior | probability)
P(B | A) is probability B given hypothesis A condition
P(A) is probability of hypothesis A (prior probability)
P(B) is probability of B

IOP Conf. Series: Journal of Physics: Conf. Series **971** (2018) 012011 doi:10.1088/1742-6596/971/1/012011

The first step of this schema is to establish the Naïve Bayes or Bayesian Network structure using MAP estimation (Maximum A Posterior).

To calculate the MAP, given r_i which signifies cardinality from X_i, and q_i that represent cardinality from parent set X_i. After that, the conditionally probability $P(X_i|pa(X_i))$ can be represented into $\theta_{ijk} = P(X_i = k|pa(X_i) = j)$, where $\theta_{ijk} \in \theta$, $1 \leq i \leq n$, $1 \leq j \leq q_i$ and $1 \leq k \leq r_i$. Then assumed that $D = \{D_1, D_2, \ldots, D_n\}$ as the observed dataset [10].

$$\theta_{ijk} = \frac{N_{ijk} + \alpha_{ijk}}{N_{ij} + \alpha_{ij}} \tag{6}$$

Where N_{ijk} is the amount of data in D for X_i obtained from k value and parent pa(X_i) which is obtained from j value. N_{ij} is the amount of data from D for each $pa(X_i)$ obtained from j value. With α_{ijk} and α_{ij} value as follows.

$$\alpha_{ijk} = \frac{\alpha}{r_{ijk}\, q_{ijk}} \tag{7}$$

$$\alpha_{ij} = \frac{\alpha}{r_{ij}\, q_{ij}} \tag{8}$$

Where α value is a hyper parameter as a number which works for smoothing or avoiding being divided by zero. In this system α value will be 0.001.

3. System Design

The main system flow is divided into two, which are training and testing. On the training flow, the input data is all of training data. The purpose of this process is to obtain the parameter from NB or BN based on the distribution of training data. The flow will be explained in Figure 3.

On the testing flow, it is more or less the same with training flow. The difference is the learning process of Bayes. That process is not required, but what is needed is the retrieval of previously trained Bayes model data to be interfered using joint probability for existing data testing. The form of testing input data is the same with training input data.

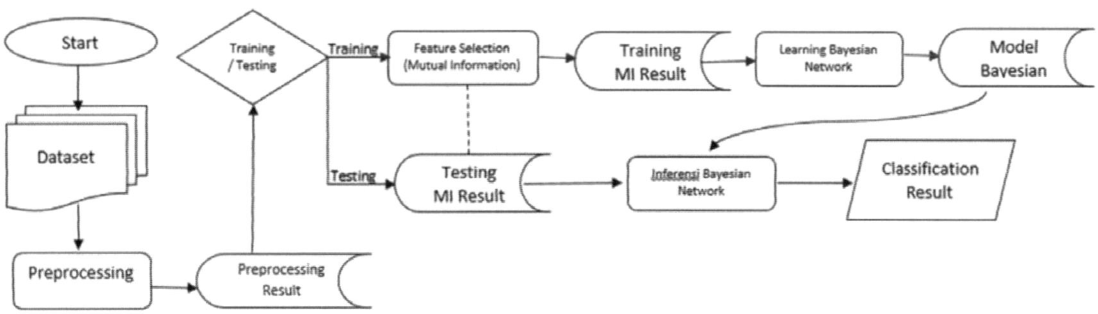

Figure 3. Flow of System Design

The system performance can be measured when the testing schema is done. Which means the model performance can be used for the new data, in this case the testing data. The model performance will be measured using F1-score.

$$Precision = \frac{True\ Positive}{Total\ Predicted} \tag{9}$$

$$Recall = \frac{True\ Posotive}{Total\ Target} \tag{10}$$

$$F1 = 2 \times \frac{Precision \times Recall}{Precision + Recall} \tag{11}$$

4. Testing and Analysis

The dataset used is Kent Ridge Bio-medical [11], which is consisting of training data and testing data.

Table 1. Distribution of dataset [11]

No	Data sets	Number of Instances	Number of Genes	Number of Classes
1	Colon Cancer	62	2000	2
2	Ovarian Cancer	253	15154	2
3	Leukemia ALL-AML	72	7129	2
4	Breast Cancer	97	24481	2
5	Lung Cancer	181	12533	2

The objective is to classify the microarray data accurately using a system that is built by combining different algorithms. To accomplish that objective, there are two scenarios used, which are the observation on the effect of dimensional reduction and the second scenario is the observation on the effect of K value in K-Means.

4.1. Scenario 1: Dimensional Reduction

In this scenario, the feature selection will be done using Mutual Information. Mutual Information method is one of the methods that is useful to reduce the dimension of microarray data. Therefore, all of the attributes can be observed by looking at the importance of each attribute. This scenario is important because it will solve the problem of microarray data which is its high dimension.

It has been observed that Colon Cancer with K=10, Ovarian Cancer with K = 8, Leukemia ALL-AML with K=10, Breast Cancer with K=10, and Lung Cancer with K=12. There are six K-Means combinations for each data with conditions mentioned in Table 2.

Table 2. Distribution of MI value in every data sets

Combination	Range of MI Score				
	Colon Cancer	Ovarian Cancer	Leukemia ALL-AML	Breast Cancer	Lung Cancer
1	0.6544-0.6353	0.6541-0.6394	0.6016-0.6017	0.6849-0.6742	0.6931-0.6931
2	0.6544-0.6274	0.6541-0.6364	0.6016-0.5973	0.6849-0.6703	0.6931-0.6873
3	0.6544-0.6203	0.6541-0.6311	0.6016-0.4587	0.6849-0.6679	0.6931-0.6817
4	0.6544-0.6130	0.6541-0.6278	0.6016-0.3870	0.6849-0.6655	0.6931-0.6814
5	0.6544-0.6055	0.6541-0.6256	0.6016-0.3614	0.6849-0.6633	0.6931-0.6758
6	0.6544-0.5961	0.6541-0.6215	0.6016-0.3407	0.6849-0.6617	0.6931-0.6752

Those values are referring to the range of MI score which is every attribute score. Therefore, based on the value distribution, those results will be used in first scenario with five datasets and six combinations (1-6) such as Table 3.

Table 3. Comparison Performance of Various Attributes Number

Combination	Range of MI Value									
	Colon Cancer		Ovarian Cancer		Leukemia ALL-AML		Breast Cancer		Lung Cancer	
	NB	BN	NB	BN	NB	BN	NB	BN	NB	BN
1	0.6667	0.7333	0.9677	0.9677	0.588s 2	0.7647	0.5789	0.5789	0.9731	0.9731
2	0.7333	0.8	0.9677	0.9677	0.7647	0.7647	0.7368	0.7368	0.9799	0.9801
3	0.7333	0.8	0.9839	0.9839	0.8823	0.8823	0.7895	0.7895	0.9799	0.9801
4	0.7333	0.8	0.9677	0.9677	0.8823	0.8823	0.7895	0.8420	0.9866	0.9876
5	0.7333	0.8	0.9677	0.9677	0.8823	0.8823	0.7895	0.8420	0.9866	0.9876

6	0.8	0.8667	0.9516	0.9516	0.8823	0.8823	0.7895	0.8420	0.9866	0.9876

The overall analysis obtained from analyzing each of the datasets:

a. Increasing the range value of MI does not necessarily improve the performance result, this is shown by the performance results from using Ovarian Cancer and Leukemia dataset, where the performance results were not affected by the increment of the value range.
b. There is a max point in some of the dataset, like Breast Cancer, Colon Cancer and Lung Cancer. At first the results increased until it reached max point but then it decreased.

The analysis mentioned above are caused by different amount of attributes which become a unique characteristic of each datasets and also the diverse data distribution.

4.2. Scenario 2: Distribution Value of K in K-Means

Scenario 2 is done by clustering the data using K-Means. K-means method is one of methods that can be used for discretization. Thus all values on attributes that originally have a large distribution of data can be minimized. This scenario will be able to solve the problem of calculating Mutual Information, which is the different distribution between variables to class. The performance results comparison of the various K values is written at Table 4.

Table 4. Performance Comparison of Various K Value

Evaluation		**K=2**	**K=8**	**K=10**	**K=12**	**K=16**
Colon Cancer	NB	0.6667	0.8	0.7333	0.7333	0.7333
	BN	0.684	0.631	0.84	0.631	0.631
Ovarian Cancer	NB	0.9508	0.9839	0.9839	0.9839	0.5484
	BN	0.950	0.967	0.984	0.984	0.598
Leukemia ALL-AML	NB	0.8823	0.7941	0.8823	0.5882	0.5882
	BN	0.882	0.765	0.765	0.882	0.588
Breast Cancer	NB	0.6842	0.6316	0.7895	0.6316	0.6316
	BN	0.6842	0.6316	0.8420	0.6316	0.6316
Lung Cancer	NB	0.9799	0.9799	0.9799	0.9866	0.9731
	BN	0.973	0.973	0.980	0.980	0.980
Average	NB	0.8328	0.8379	0.8738	0.7847	0.6949
	BN	0.846	0.842	0.887	0.857	0.705

From all of the analysis of each dataset, it can be concluded that increasing the K value will not necessarily improve the performance result. However, each of dataset will have its own pattern of value. In Ovarian, Lung Cancer, Breast Cancer and Colon Cancer data sets have maximal score when K value is 10, but it did not apply in Leukemia by Bayesian Networks. These results occurred because every dataset has different characteristics or data distribution to each other.

4.3. Comparison of Maximal Average Result

After all of the experiments are finished, we got the maximum result of every data sets in Table 5.

Table 5. Average and Comparison of Maximum F1-Score

Evaluation		**Colon Cancer**	**Ovarian Cancer**	**Leukemia ALL-AML**	**Breast Cancer**	**Lung Cancer**	**AVG**
Max Accuracy	NB	0.8	0.9839	0.8823	0.7895	0.9866	0.8885
	BN	0.867	0.984	0.882	0.84	0.980	0.9106

IOP Conf. Series: Journal of Physics: Conf. Series **971** (2018) 012011 doi:10.1088/1742-6596/971/1/012011

Based on the results shown in Table 5, it proves that Mutual Information for feature selection method and Bayesian for classifier method can classify microarray data with the average score of Naïve Bayes is 0.8885 and the Bayesian Network is 0.9106. The results were obtained from the analysis of dimensional reduction and the distribution value of K in K-Means in every data set. In this research, Bayesian Network has the biggest average because the structure of Bayesian Network allows connection between each and every attributes.

5. Discussion and Conclusion

In microarray case, the first thing that should be noticed is the type of the data, whether it is continuous data or discrete data, both of them have wide data distributions. Therefore, clustering is required to solve that problem. This research used K-Means for discretization or clustering method. Amount of K for K-Means has different trend in every data, depends on the amount of random attribute or data sets type.

For feature selection, this research used Mutual Information method. From the feature selection process, it can be concluded that increasing the value range of MI will increase the number of random attribute. However, it does not guarantee that the performance will increase as well. Each data set has its own boundary to get its maximum *F1-score*. It would be better if the next observation focuses more on using the other discretization methods and feature selection methods. It will gain a new trend of data which can improve the performance.

The machine learning approach using Bayes Theorem has proven to successfully classify microarray data with good performance results. Based on one of the Bayes Theorem methods used which is the Bayesian Network, it is proven to acquire better performance in classifying the microarray data than Naïve Bayes method. Bayesian Network obtained 0.9106 average score. That score is good for classification of microarray data and used to detect cancer. Bayesian Network can be improved again with optimized DAG by Learning Structure. The future research itself is not limited in exploring the use of Learning Structure method only, but it can also be further expanded by looking into the other methods of feature selection, because it will help to gain new attributes.

References

[1] Nurfalah, Adiyasa; Adiwijaya; Suryani, Arie Ardiyanti, "Cancer Detection Based On Microarray Data Classification Using PCA and Modified Backpropagation," *Far East Journal of Electronics and Communications,* vol. 16, no. 2, pp. 269-281, 2016.

[2] Singh, Rabindra Kumar; Sivabalakrishman, Dr. M.,, "Feature Selection of Gene Expression Data For Cancer Classifiaction: A Review," in *in 2nd International Symposium on Big Data dan Cloud Computing (ISBCC'15)*, Vellore, India, 2015.

[3] Liying Yang; Zhimin Liu; Xiguo Yuan; Jianhua Wei; Junying Zhang, "Random Subspace Aggregation for Cancer Prediction with Gene Expression Profile," *BioMed Research International,* vol. 2016, p. 10, 2016.

[4] C, Devi Arockia Vanitha; D, Devaraj; M, Venkatesulu, "Gene Expression Data Classification using Support Vector Machine and Mutual Information-based Gene Selection," in *n Graph Algorithms, High Performance Implementations and Application (ICGHIA2014)*, Tamil Nadu, India, 2015.

[5] Luis M. de Campos; Andres Cano; Javier G. Castellano; Seraffin Moral, "Bayesian Networks Classifiers for Gene-Expression Data," in *Department of Computer Science and Artificial Intelligence. Granada University.*, Spain, 2011.

[6] C, Devi Arockia Vanitha; D, Devaraj; M, Venkatesulu, "Gene Expression Data Classification using Support Vector Machine and Mutual Information-based Gene Selection," in *Graph Algorithms, High Performance Implementations and Application (ICGHIA2014)*, Tamil Nadu, India, 2015.

[7] T. M. Mitchell, Machine Learning, New York: McGraw-Hill, 1997.

[8] Berson, A; J, Smith S., Data Warehousing, Data Mining, & OLAP, New York: McGraw-Hill, 2001.

[9] Almira Syawli; M. Ali Fahmi; Silvia Ari Santhy; Zulkarnaen, "Diagnosa Penyakit Diabetes Mellitus Berbasis Desktop Application," *Jurnal Diabetes Mellitus,* vol. 3, no. 2, pp. 50-62, 2010.

[10] Yun Zhou; Norman Fenton; Martin Neil, "Bayesian Network Approach to Multinomial Parameter Learning Using Data and Expert Judgments," *International Journal od Approximate Reasoning,* vol. 55, pp. 1252-1268, 2014.

[11] Bio-medical, Kent Ridge, "Kent Ridge Bio-medical Dataset," 2003. [Online]. Available: http://datam.i2r.a-star.edu.sg/dataset/krbd. [Accessed 12 2016].

[12] W. H. Organization, "Cancer fact sheet," Februari 2015. [Online]. Available: http://www.who.int. [Accessed 1 November 2016].

[13] Mubarok, MS; Adiwijaya; Aldhi, MD; “Aspect-based sentiment analysis to review products using Naive Bayes”. *In AIP Conference Proceedings* 1867, 020060, 2017.

IOP Conf. Series: Journal of Physics: Conf. Series **971** (2018) 012012 doi:10.1088/1742-6596/971/1/012012

Comparison of ANN and SVM for classification of eye movements in EOG signals

Lim Jia Qi, Norma Alias

Ibnu Sina Institute, Faculty of Science, Universiti Teknologi Malaysia

norma@ibnusina.utm.my

Abstract. Nowadays, electrooculogram is regarded as one of the most important biomedical signal in measuring and analyzing eye movement patterns. Thus, it is helpful in designing EOG-based Human Computer Interface (HCI). In this research, electrooculography (EOG) data was obtained from five volunteers. The (EOG) data was then preprocessed before feature extraction methods were employed to further reduce the dimensionality of data. Three feature extraction approaches were put forward, namely statistical parameters, autoregressive (AR) coefficients using Burg method, and power spectral density (PSD) using Yule-Walker method. These features would then become input to both artificial neural network (ANN) and support vector machine (SVM). The performance of the combination of different feature extraction methods and classifiers was presented and analyzed. It was found that statistical parameters + SVM achieved the highest classification accuracy of 69.75%.

1. Introduction

In recent years, there is growth in interest among researchers in pattern recognition, which is a study of how machines, instead of human can learn and identify patterns (classes) of interest, and make precise decision on the classes identified [14]. Pattern recognition has been applied in numerous scientific and engineering disciplines such as biology (medical diagnosis), artificial intelligence (AI), remote sensing and etc. HCI, which is based on pattern recognition is helpful in assisting users with various tasks. The research on HCI can greatly improve the quality of life, especially for those severely disabled person that suffer from spinal cord injuries, Locked-in Syndrome, muscular dystrophy, Amyotrophic Lateral Sclerosis (ALS) and etc. These patients have difficulty in conveying their intentions because they lack control of their voluntary muscle. However, they are still conscious and are able to control their eye movements.

There are several devices that are used to record activities of human eyes, however most of them have their own limitations. Video-oculography is costly, and has to be calibrated repeatedly to obtain accurate measurements [10]. Magnetic search coil, being another eye tracking device can cause irritation and discomfort to subjects because magnetic search coil has to be inserted into a subject's eye. In view of this, EOG signals can be another feasible alternative. EOG is measure of changes in corneo-retinal potential (CRP) according to eye movement. This potential difference arises from positive poles of cornea and negative poles of retina. Due to its relatively larger amplitude reading compared to other biomedical signals like electroencephalogram (EEG) and electromyogram (ECG), EOG signals is easier to be recorded. Therefore, EOG is a good candidate to be chosen as the input of eye movement classification system despite the fact that EOG is seldom deterministic. In fact, many HCIs using EOG as input have been proposed such as electrical wheelchair control [5], mobile robot control [17], eye writing recognition [1] [24], activity recognition [7] and so on.

As for the classification scheme, two of the most popular classification models, ANN and SVM were employed in this study. ANN can be defined as computing model that originated from the understanding of the structure and function of biological neural networks. ANN can be configured for specific mathematical purposes like data classification, curve fitting, mathematical modeling and etc. On top of that, ANN is able to derive relations between inputs and outputs which are too complex for human. Nowadays, ANN is a very important and flexible classification algorithm in designing novel pattern recognition systems.

SVM is another classifier that has been studied extensively after its introduction by Bernhard E. Boser et al. at COLT-92 conference [6]. Originally, it was designed for binary classification problem. SVM is famous for its excellent classification performance in various applications. It is not affected by the curse of dimensionality, which means it can handle large dimension input variables classification problems by projecting the input data into higher dimensional space. In addition, tradeoff between classifier complexity and error can be adjusted explicitly by manipulating certain parameters.

The primary motivation of this study is to assess and compare the performance of different clasifiers (ANN and SVM) in terms of classification accuracy and training time. This study also explores different types of feature extraction method on the preprocessed EOG signals. It is hope that this paper can help in development of robust and user friendly algorithm for classification of EOG signals.

2. Materials and Methods

2.1. Data Acquisition

EOG signal was collected from 5 volunteers with the age in between 22 and 26. The details of the 5 subjects are listed in Table 1. EOG measurements were taken using Biopac MP36R data acquisition unit. The sampling rate is set to be 100 Hz for both channels.

Table 1. Participants' details.

Subjects	Ages	Gender	Vision
1	24	Male	Glasses
2	25	Male	Normal
3	25	Female	Glasses
4	26	Male	Glasses
5	22	Male	Normal

(1) Before the start of the experiment, subjects were instructed to perform eye movements until they are familiar with the tasks. The subjects was asked to stay relaxed sitting on a chair. Disposable lead electrodes were attached to positions around the eyes as shown in Figure 1.

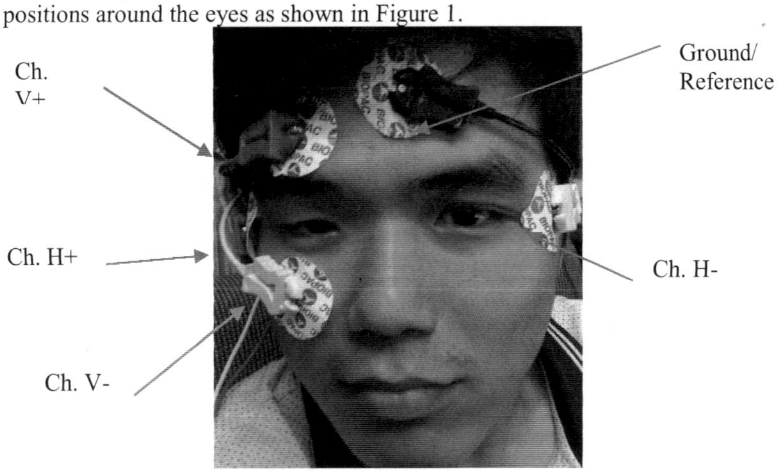

Figure 1. Placement of surface EOG electrodes. Ch. V+ and Ch. V- measure the vertical eye movement, while Ch. H+ and Ch. H- measure the horizontal eye movement.

(2) A simple eye movement interface using animation features of Microsoft Power Point 2013 was designed prior to the experiment. Before the start of the recording of EOG signals, subjects were instructed to keep their eyes on the ball that appear on the laptop screen. The ball followed the movement as illustrated in Fig. 2. Each eye movements (target actions) was repeated for 5 times per session. A total of 10 session was conducted for each subject. The participants were asked to blink only when the ball was at its central position. During the experiment, the eyes position was aligned with the central position of the ball and the distance was fixed at 30 cm.

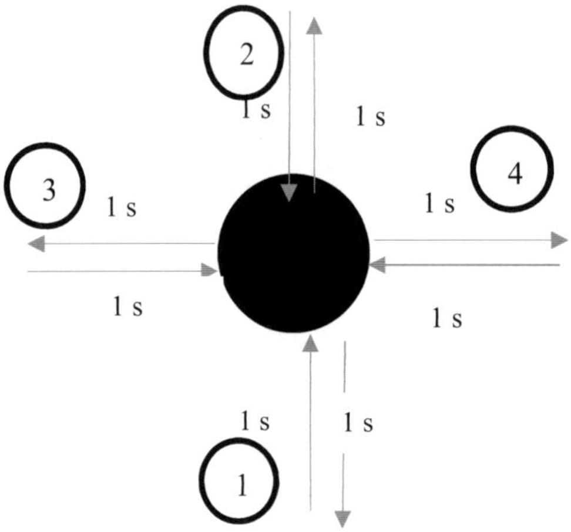

Figure 2. The ball will move downwards in 1 second and back to its original central position in 1 second. The ball will be stationary at the central position for 5 seconds before moving upwards according to the order shown by the circled numbers. This cycle will repeat 5 times each session. 10 session were conducted for each subjects.

(3) Data preprocessing, feature extraction and supervised classification algorithm thereafter was performed with the aid of Matlab (R2013a).

2.2. Data Preprocessing

EOG signals are frequently contaminated with noises and other interferences, such as eye blinks, powerline interference, head movements and so on. Knowing that the effective EOG signals is in between 0.1Hz (DC)-50Hz, Chebyshev 4th order bandpass filter of the particular frequency range was used. This is to attenuate and simultaneously suppress power supply interference. The baseline of raw EOG signals is removed using:

$$X_i = S_i - \mu(x) \tag{1}$$

where $\mu(x)$ is the mean of raw EOG signal, S_i is EOG signal data points.

The EOG signals is normalized after baseline removal. The normalization is carried out with the following formula:

$$x_i = \frac{X_i}{Max(abs(X_i))} \quad (2)$$

Figure 3 shows the EOG signals before and after preprocessing.

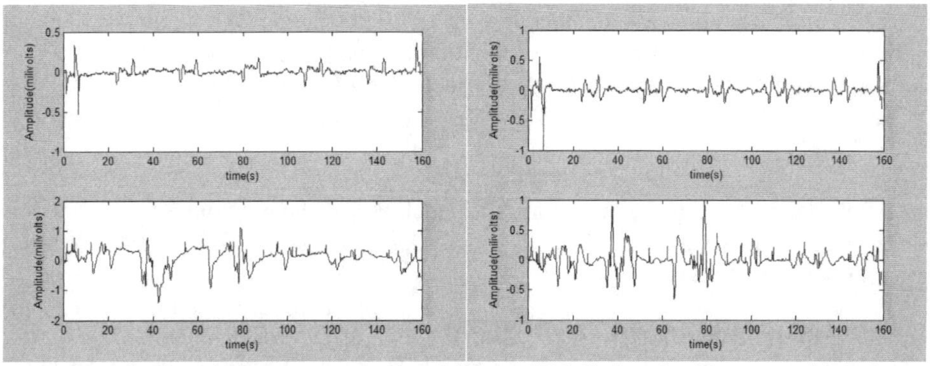

Figure 3. Before (left) and after preprocessing (right) of EOG signals. (Source: subject 1 session 1).

2.3. Feature Extraction

Classification using raw EOG signals would be time consuming and ineffective because of the signal noise originated from eye blinks, EMG from facial response, incorrect electrodes placement, influence of luminance and etc. Another major drawback would be high computational cost due to enormous input data size. Nonetheless, EOG signals can be represented by means of some attributes. Such attributes can be used as features to characterize the signals. The method to obtain these features is called feature extraction. Feature extraction is pivotal in deriving unique pattern from the original dataset without loss of key information. In other words, feature extraction reduces dimensionality of data but maximizes inter-class separability and intra-class similarity at the same time. In the present work, statistical features [23], AR coefficients derived from Burg method and PSD estimation using Yule-Walker method [3].

2.3.1. Statistical Parameters. The following ten statistical features of each EOG channels (horizontal & vertical eye movement) are used as parameters to characterize the EOG signals:

i) Minimum, x_{min}
ii) Maximum, x_{max}
iii) Mean
iv) Median
v) Mode
vi) First quartile
vii) Third quartile
viii) Interquartile range
ix) Standard deviation

$$S = \sqrt{\frac{\sum_{i=1}^{N}(x_i - \mu)^2}{N-1}} \quad (3)$$

x) Kurtosis coefficients

$$k = \frac{E(x-\mu)^4}{\sigma^4} \quad (4)$$

$$k = \frac{\frac{1}{N}\sum_{i=1}^{N}(x_i - \mu)^4}{\left(\frac{1}{N}\sum_{i=1}^{N}(x_i - \mu)\right)^2} \tag{5}$$

2.3.2. Autoregressive (AR) model coefficients derived using Burg algorithm. In an AR model, the variable of interest is predicted using linear combination of past values of the variable. It is commonly used to model time series data. The AR model of order k, AR(k) is as follow:

$$x_t = \sum_{i=1}^{k} a_i x_{t-i} + \varepsilon_t \tag{6}$$

where a_i is AR coefficients, x_t is the output of AR model, x_{t-i} is the previous known data points, ε_t is the residual term, or so called Gaussian white noise. In order to estimate a_i, Burg method [8]was utilized. In this paper, AR(4) model is chosen

2.3.3. Parametric Power Spectral Density (PSD) Estimation using Yule-Walker method. Parametric or model based method of spectral estimation assume that the signal can be estimated by model with known functional form. Once the coefficients in the assumed model can be estimated with certain algorithm, the signal's spectral characteristics are then derived from the estimated model. Common models available are AR model, moving average (MA) model, and autoregressive moving average (ARMA) model. Since AR model is appropriate in representing spectra with narrow peaks [25], so AR(4) model is chosen and Yule-Walker method is used to estimate the coefficients by solving equation (7).

$$\begin{pmatrix} x(0) & \cdots & x(-n+1) \\ \vdots & \ddots & \vdots \\ x(n-1) & \cdots & x(0) \end{pmatrix} \begin{pmatrix} a_1 \\ \vdots \\ a_n \end{pmatrix} = -\begin{pmatrix} x(1) \\ \vdots \\ x(n) \end{pmatrix} \tag{7}$$

The power spectral density, $S_x(f)$ of the time series, x_t is given in equation (8).

$$S_x(f) = |H(f)|^2 S_\varepsilon(f) = \frac{1}{f_s} \frac{\sigma^2}{\left|1 - \sum_{i=1}^{k} a_i e^{-ji(2\pi f)T}\right|^2} \tag{8}$$

where H(f) is the transfer function where a_i are estimated by Yule-Walker equation, $S_\varepsilon(f)$ is power spectrum of white noise with variance, σ^2 and sampling frequency, f_s. In this paper, I used 65-point discrete approximation of the power spectrum.

2.4. Eye Movement Classification using ANN

ANN has already achieved a huge number of practical applications [18][22][4]. ANN is more superior to other artificial intelligence (AI) system because it is flexible and adaptive which means the network can adjust itself to new environment. Another justification for this is ANN user does not need to have access to domain-specific knowledge. However, ANN algorithm is iterative, which can sometimes suffer from convergence problem. In this research, multilayer feedforward ANN [15] is chosen because it is suitable for classification problem where supervised learning is implemented with Levenberg-Marquardt (LM) algorithm [12]. ANN was implemented in Matlab version R2013a with neural network toolbox).

There are two main steps in the algorithm of feed forward ANN. First, the neurons (processing units) calculate the weighted sum of its inputs and then apply activation function to this sum to derive its outputs [21]. Second, ANN will undergo backpropagation. The purpose behind this step is to adjust the connection weights of multilayer perceptron (MLP) based on the errors of outputs produced. An MLP consists of three layers: input layer, hidden layer and output layer. Each layer is composed of predefined number of neurons. Figure 4 shows basic architecture of multilayer feed forward ANN. Figure 5 shows Matlab generated schematic diagram of neural network when statistical parameters is the input.

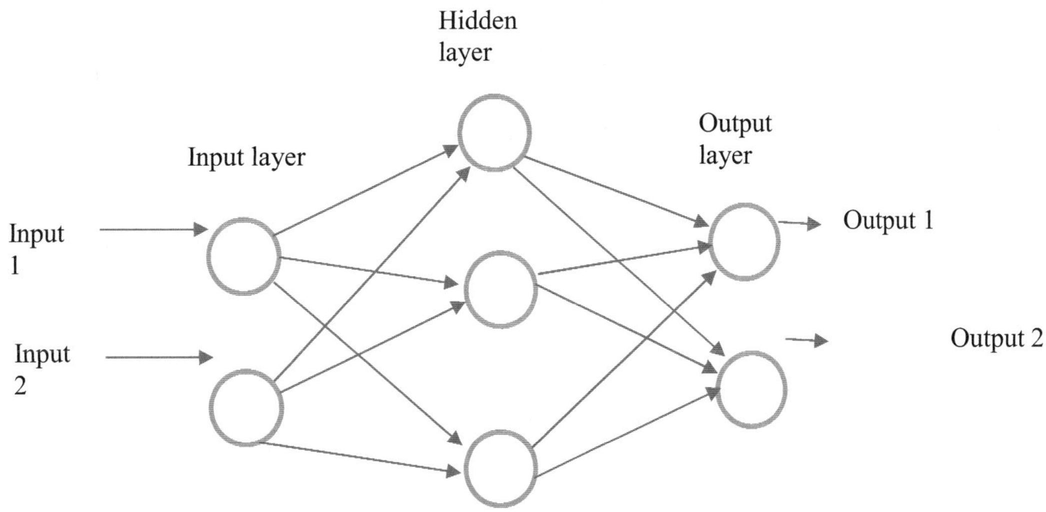

Figure 4. Architecture of feed forward neural network. The number of nodes of input layer is equal to the number of input feature for each data point. In hidden layer, the number of nodes is user-defined. We choose 10 hidden nodes for statistical features and PSD estimation, while 5 hidden nodes for AR coefficients. The number of nodes in output layer corresponds to number of classes (targets).

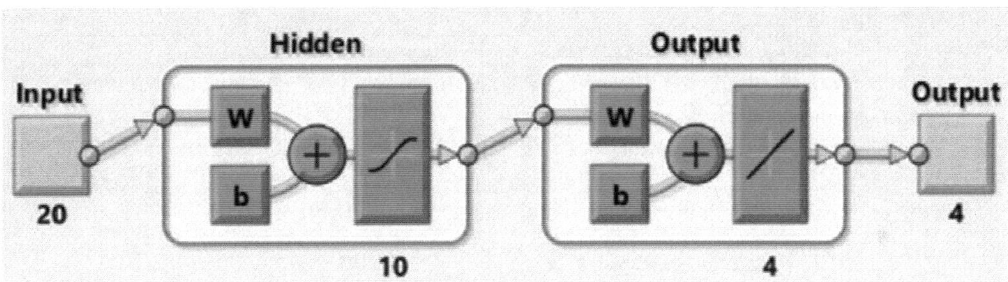

Figure 5. Schematic diagram of neural network.

2.5. Eye Movement Classification using SVM

Undoubtedly, SVM is an effective classification algorithm in various research fields. Its application encompasses text categorization [16], face detection [11], diagnosis of breast cancer [20] and etc.

So, how SVM works? SVM is a binary classifier that tries to find the optimal hyperplane. Optimal hyperplane means a linear decision surface that splits n-dimensional space ($\mathbb{R}^n$) into two parts and simultaneously the distance of decision boundary from the data of both classes should be as far away as possible. This is shown schematically in Fig. 6.

With the introduction of slack variables and kernel functions, SVM can be applied in non-linearly separable problem with high dimensional input data. In this study, the following SVM formulation will be used:

Training phase:

minimize $\frac{1}{2}\sum_{i=1}^{n}\sum_{j=1}^{n}\alpha_i\alpha_j\, y_i y_j \mathrm{K}(\boldsymbol{x}_i, \boldsymbol{x}_j) - \sum_{i=1}^{n}\alpha_i$

subject to $0 \leq \alpha_i \leq C$, $C > 0$ (user-defined parameter)

$$\sum_{i=1}^{n}\alpha_i\, y_i = 0$$

Testing phase:

Classify the unseen objects, **x** using the function below:

$$f(\boldsymbol{x}) = sign(\sum_{i=1}^{n} \alpha_i y_i \, K(\boldsymbol{x}_i, \boldsymbol{x}) + b) \quad (9)$$

The SVM algorithm is carried out in MATLAB (R2013a) using LIBSVM (version 3.20) package available at http://www.csie.ntu.edu.tw/~cjlin/libsvm/. Radial Basis Function (RBF) was chosen for classification because it is widely used and usually the first choice compared to other kernel functions in the literature [9].

3. Results and Discussions

3.1. Construction of different feature sets

Out of 10 sessions of EOG reading for each subject, last 3 sessions of the EOG signals are selected as the testing data. The remaining seven are used as the training data. Different features are extracted from these data. The dimension of training dataset and testing dataset is shown in Table 2. The training vectors are used to train ANN and SVM classifiers, whereas the accuracy and efficacy of the trained models were evaluated using testing vectors.

Table 2. Dimensions of training and testing data for different feature extraction methods.

Feature extraction methods	Dimensions	
	Training data	Testing data
Statistical parameters	120x20	80x20
AR coefficients	120x10	80x10
PSD features	120x130	80x130

3.2. Comparison between ANN and SVM

In this paper, the performance of classifiers is evaluated subject by subject. Performance evaluation and comparison is presented with the use of testing accuracy and training time. The classification accuracy is calculated as proportion of actual positives which are predicted positive [2]:

$$\text{Classification accuracy}(\%) = [\frac{NT}{(NT + NF)}] \times 100 \quad (10)$$

where NT is the number of trials with true prediction and NF is the number of trials with false prediction.

Based on the results from Table 3 to Table 8, under the same features data as input, both classifiers produces roughly similar results. ANN is slightly better than SVM in terms of classification rate when AR and PSD features were used, but takes longer training time. On the other hand, SVM outperforms ANN in both aspects when statistical parameters was used. When classification accuracy of the three methods are compared, statistical method achieves the best accuracy: 66.5% (ANN), 69.75% (SVM), in contrast with AR coefficients accuracy: 54% (ANN), 53.75% (SVM) and PSD accuracy: 50.25% (ANN), 49% (SVM). It can be seen from the table that SVM with statistical parameters as the input produces the best results, with accuracy of 69.75% and fairly low training time of 0.158s.

In terms training time of both classifiers, SVM outperforms ANN in multiclass classification. This can be explained by how classification model is trained under SVM. As a robust supervised learning algorithm, SVM attempts to maximize the distance (margin) between the closest training points (support vectors) from either class in order to achieve better generalization performance on test data [13]. The solution is based on support vectors. Thus, it reduces the number of operation needed to compute a model (not all features vector is included in the calculation). Unlike SVM, the central idea of ANN is to sum up linear combinations of all input data and bias value in each neuron, and then model the output as a nonlinear function of these summation values, which increases the number of operation as well as computational complexity and cost. Fig. 6 and Fig. 7 shows the summary of results in terms of classification accuracy and training time.

Table 3. Classification results for statistical parameters as features using ANN as classifier.

Subjects	Classification accuracy (%)				Overall classification accuracy (%)	Duration for training (s)
	Down[a]	Up[b]	Left[c]	Right[d]		
1	55	65	90	90	75	1.161
2	30	60	45	80	53.75	0.412
3	70	50	45	85	62.5	0.611
4	85	85	65	75	77.5	0.445
5	35	70	75	75	63.75	0.407
Mean	55	66	64	81	66.5	0.607

Note:
Down[a] represents event which eyes move centre-down-up.
Up[b] represents event which eyes move centre-up-down.
Left[c] represents event which eyes move centre-left-right.
Right[d] represents event which eyes move centre-right-left.

Table 4. Classification results for statistical parameters as features using SVM as classifier.

Subjects	Classification accuracy (%)				Overall classification accuracy (%)	Duration for training (s)
	Down	Up	Left	Right		
1	55	85	95	95	82.5	0.129
2	45	90	30	80	61.25	0.129
3	40	85	65	80	67.5	0.142
4	90	85	50	85	77.5	0.247
5	60	50	55	75	60	0.144
Mean	58	79	59	83	**69.75**	**0.158**

Table 5. Classification results for AR coefficients as features using ANN as classifier.

Subjects	Classification accuracy (%)				Overall classification accuracy (%)	Duration for training (s)
	Down	Up	Left	Right		
1	75	80	75	30	65	0.355
2	40	60	65	70	58.75	0.296
3	20	55	70	25	42.5	0.345
4	60	55	60	50	56.25	0.309
5	65	15	60	50	47.5	0.323
Mean	52	53	66	45	54	0.326

Table 6. Classification results for AR coefficients as features using SVM as classifier.

Subjects	Classification accuracy (%)				Overall classification accuracy (%)	Duration for training (s)
	Down	Up	Left	Right		
1	80	80	60	50	67.5	0.115
2	30	70	75	50	56.25	0.091
3	20	60	45	45	42.5	0.058
4	60	25	35	70	47.5	0.248
5	65	35	70	45	53.75	0.246

Mean	51	54	57	52	53.5	0.152

Table 7. Classification results for PSD as features using ANN as classifier.

Subjects	Classification accuracy (%)				Overall classification accuracy (%)	Duration for training (s)
	Down	Up	Left	Right		
1	50	75	15	90	57.5	6.519
2	75	35	20	90	55	5.124
3	50	90	50	40	57.5	4.121
4	75	30	50	25	45	5.05
5	45	30	25	50	36.25	3.871
Mean	59	52	32	59	50.25	4.937

Table 8. Classification results for PSD as features using SVM as classifier.

Subjects	Classification accuracy (%)				Overall classification accuracy (%)	Duration for training (s)
	Down	Up	Left	Right		
1	80	45	60	55	60	0.066
2	30	80	60	45	53.75	0.056
3	20	65	50	45	45	0.065
4	45	45	30	55	43.75	0.082
5	80	50	10	30	42.5	0.129
Mean	51	57	42	46	49	0.080

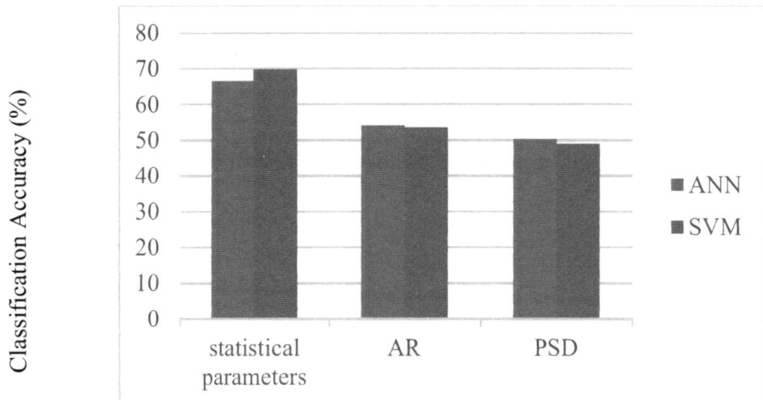

Figure 6. Classification accuracy of different feature inputs and classifiers.

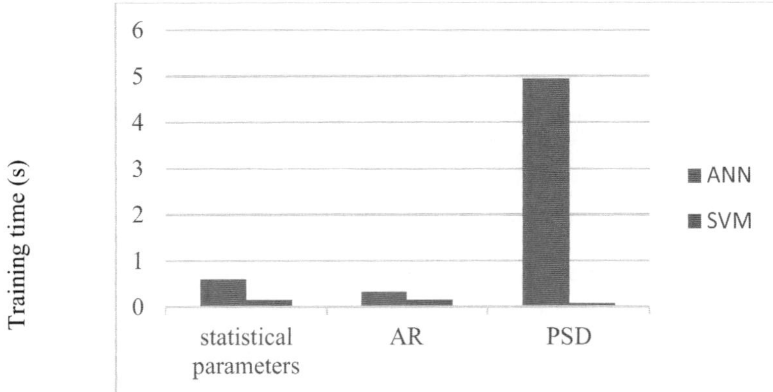

Figure 7. Training time of different feature inputs and classifiers.

Various approaches and findings and their respective classification accuracies for different types of eye movements have already been presented in the literature. Anwashe et al.10 achieved 80% classification accuracy for six types of eye movements by using AR+PSD as input for feed forward neural network. In Deng et al.2 work, 90% of accuracy was achieved for four directional eye movements using fuzzy distinction rule. Merino et al. proposed an algorithm that works with derivative and amplitude level and obtained 94% accuracy [19].

4. Conclusion

EOG signals have huge potentials to be applied in communication support system, such as eye-controlled wheelchair, virtual keyboard and etc. Features extracted from dataset plays vital role in the subsequent classification stage of EOG signals. This paper has taken into account some strategies to improve the analysis and classification EOG signals.

The introduced method of extracting statistical parameters from the time series of EOG data was found to be superior to other alternatives (AR coefficients and PSD features) regarding classification accuracy. PSD estimation is not a good feature extraction method, compared to other feature extraction method. This finding actually contradicts with the work of Anwashe et al.10 The combination of statistical feature extraction method and SVM was proved to be better than using ANN as classifier due to high generalization performance of SVM with classification accuracy of 69.75% and lower training time.

Future research will be focused on more combination of feature extraction methods and classifiers. State-of-the-art feature extraction algorithm like discrete wavelet transform (DWT) and newly emerged classifiers such as Extreme Learning Machine (ELM) can be good options for the improvement of the classification accuracy and training time.

References

[1] A.B. Usakli, S. G. F. A., et. al. (2009). On the Use of Electrooculogram for Efficient Human Computer Interfaces. *Computational Intelligence and Neuroscience.*

[2] Abdelkader, N. B., Yoshimura, N. & Shin, D. (2013). Classification of Four Eye Directions from EEG Signals for Eye-Movement-Based Communication Systems. *Journal of Medical and Biological Engineering, 34*, 581-588.

[3] Anwashe, B., Datta, S., Pal, M. & et.al. (2013). Classifying Electrooculogram to Detect Directional Eye Movements. *Procedia Technology, 10*, 67-75.

[4] Aysegul, G. K., S. (2006). Classification of electro-oculogram signals using artificial neural network. *Expert Systems with Applications, 31*, 199-205.

[5] Barea, R. B., L. Lopez. E. (2002). System for Assisted Mobility Using Eye Movements based on electrooculography. *IEEE Trans. Neural. Syst. Rehabil., 4*, 209-218.

[6] Bernhard E. Boser, I. M. G., V. N. Vapnik. (1992). *A Training Algorithm for Optimal Margin*

Classifiers. Paper presented at the Proceedings of the Fifth Annual Workshop on Computational Learning Theory 5, Pittsburgh.

[7] Bulling, A., Ward. J.A., Gellersen. H, et al. (2011). Eye Movement Analysis for Activity Recognition Using Electrooculography. *IEEE Trans. Pattern Analysis Machine Intelligence, 33*, 741-753.

[8] Burg, J. (1975). *Maximal Entropy Spectral Analysis*. Stanford University.

[9] Chih-Wei, H., Chang, C.-C. & Lin, C.-J. (2010). A Practical Guide to Support Vector Classification. http://www.csie.ntu.edu.tw/~cjlin

[10] Deng, L. Y., Hsu, C.-L., Lin, T.-C. & et. al. (2010). EOG-based Human–Computer Interface System Development. *Expert Systems with Applications*(37), 3337-3343.

[11] E. Osuna, R. F., F. Girosit. (1997). *Training support vector machines: an application to face detection*. Paper presented at the Computer Vision and Pattern Recognition, 1997. Proceedings., San Juan, Puerto Rico, USA, USA. http://ieeexplore.ieee.org/document/609310/

[12] Hao Yu, B. M. W. (2010). Levenberg-Marquardt Training (pp. 1-16).

[13] Hastie, T. T. R. F. J. (2001). *The Elements of Statistical Learning*: Springer New York Inc.

[14] J. K. Basu, D. B., T. Ki,. (2010). Use of Artificial Neural Network in Pattern Recognition. *International Journal of Software Engineering and Its Applications, 4*(2).

[15] Jain, A., Mao, J. & Mohiuddin, K. (1996). Artificial Neural Networks: A Tutorial. *Computer (Long Beach Calif)*, 31-44.

[16] Joachims., T. (2005). Text categorization with Support Vector Machines: Learning with many relevant features. *Machine Learning: ECML-98, 1398*, 137-142.

[17] Kim, Y., D. N.L., Y. Y., et. al. (2007). Robust Discrimination Method of the Electrooculogram Signals for Human-Computer Interaction Controlling Mobile Robot. *Intelligence Automation Software Computing, 13*, 319-336.

[18] Lin, H., Hou, W., Zhen, X. & Peng, C. (2006). *Recognition of ECG Patterns Using Artificial Neural Network*. Paper presented at the Sixth International Conference on Intelligent System Design and Applications.

[19] Merino, M., Rivera, O. & Gómez, I. (2010). *A Method of EOG Signal Processing to Detect the Direction of Eye Movements*. Paper presented at the 2010 First International Conference on Sensor Device Technologies and Applications (SENSORDEVICES), Venice, Italy. http://ieeexplore.ieee.org/document/5632116/

[20] Nasser H. Sweilam, A. T. N. A. M. (2010). Support vector machine for diagnosis of cancer disease: A comparative study. *Egyptian Informatics Journal, 11*, 81-92.

[21] Russell, S. J., Norvig, P., & Davis, E. (2010). *Artificial intelligence: A modern approach*: Prentice Hall.

[22] Shahrin, A. N., Omar, N., Jumari, K. F. & Khalid, M. (2007). *Face Detection using Artificial Neural Networks Approach*. Paper presented at the First Asia International Conference on Modelling & Simulation.

[23] Siuly, L., Y. & Wen, P. (2011). Clustering technique-based least square support vector machine for EEG signal classification. *Computer Methods and Programs in Biomedicine, 104*, 358-372.

[24] Tobias Wissel, R. P. (2013). Considerations on Strategies to Improve EOG Signal Analysis.. In: Investigations Into Living Systems. *Artificial Life, and Real-world Solutions*, 204-218.

[25] U¨beyli, E. D. G. l. (2005). Feature extraction from Doppler ultrasound signals for automated diagnostic systems. *Computers in Biology and Medicine, 35*(9), 735-764.

IOP Conf. Series: Journal of Physics: Conf. Series **971** (2018) 012013 doi:10.1088/1742-6596/971/1/012013

Visualization of time series statistical data by shape analysis (GDP ratio changes among Asia countries)

Yukari Shirota[1], Takako Hashimoto[2] and Riri Fitri Sari[3]

[1]Gakushuin University, Tokyo, Japan
[2]Chiba University of Commerce, Chiba, Japan
[3]Universitas Indonesia, Depok, Indonesia

yukari.shirota@gakushuin.ac.jp

Abstract. It has been very significant to visualize time series big data. In the paper we shall discuss a new analysis method called "statistical shape analysis" or "geometry driven statistics" on time series statistical data in economics. In the paper, we analyse the agriculture, value added and industry, value added (percentage of GDP) changes from 2000 to 2010 in Asia. We handle the data as a set of landmarks on a two-dimensional image to see the deformation using the principal components. The point of the analysis method is the principal components of the given formation which are eigenvectors of its bending energy matrix. The local deformation can be expressed as the set of non-Affine transformations. The transformations give us information about the local differences between in 2000 and in 2010. Because the non-Affine transformation can be decomposed into a set of partial warps, we present the partial warps visually. The statistical shape analysis is widely used in biology but, in economics, no application can be found. In the paper, we investigate its potential to analyse the economic data.

1. Introduction

We would like to analyse a deformation. For a long time, it has been of great importance to measure shapes of objects such as shapes in morphology. In shape analysis, it is important how to compare different shapes; they have different sizes and orientations as well as different shapes. The process of transforming different sets of data into one coordinate system is called image registration by register marks. That is a tough problem and the image registration problem had been discussed for a long time. However, the research conducted by University of Leeds members and others made remarkable progresses in shape analysis to solve the problem[1]. The research field is called "statistical shape analysis" or "geometry driven statistics."[2, 3]. The field of statistical shape analysis involves methods for the study of the shapes of objects where location, rotation and scale information can be removed to compare the shapes[4]. To compare the two shapes, we use landmarks which illustrate the corresponding pairs. The landmarks are translated, rotated, and scaled so that they lined up and matched as closely as possible. The procedure is known as a generalized Procrustes analysis[3]. The shape analysis applications include medical imaging and morphometrics for biology [5-11]. Although there are many papers on biology and medical imaging fields, as far as we know, there are no economics analysis paper by this statistical shape analysis except [12]. We would like to apply the statistical shape analysis methods for the economics analysis[13, 14]. In the paper, we analyse the GDP ratio changes from 2000 to 2010 in Asia between the agriculture and industry, value added (percentage of GDP). We handle the data as a set of landmarks on a two dimensional image to see the deformation using the principal components. In the next section, we shall explain the GDP data we use. In Section 3, we present the principal warps of the given formation. In Section 4, concerning the deformation, the resulting partial warp data and others are visually presented. In Section 5, we shall discuss the potential and possibilities of the statistical shape analysis to analyse the economic data. Finally we conclude the paper in Section 6.

2. GDP Data

In the section, we shall explain the GDP data we use. The data are the GDP ratio changes from 2000 to 2010 in Asia between the agriculture and industry, value added (percentage of GDP) cited from World Bank statistical data (http://data.worldbank.org/indicator/NV.IND.TOTL.ZS). The value added of an industry is the contribution of a private industry or government sector to overall GDP. Value added equals the difference between an industry's gross output (consisting of sales or receipts and other operating income, commodity taxes, and inventory change) and the cost of its intermediate inputs (including energy, raw materials, semi-finished goods, and services that are purchased from all sources) (https://www.bea.gov/faq/index.cfm?faq_id=184). Agriculture, value added (percentage of GDP) includes forestry, hunting, and fishing, as well as cultivation of crops and livestock production. Value added is the net output of a sector after adding up all outputs and subtracting intermediate inputs (http://data.worldbank.org/indicator/NV.IND.TOTL.ZS). The target countries are ten countries: they are China, India, Indonesia, Thailand, Philippines, Malaysia, Singapore, Japan, Korea, and Pakistan. The data is presented in Figure 1. The individual country is denoted as one landmark on the formation. We handle the data as a set of landmarks on a two dimensional image. In Singapore, Japan and Malaysia, only the industry percentage decrease are seen. On the other hand, China and Indonesia, and India show only an agriculture decline tendency. To compare the two formations, we have to make landmarks translated, and re-scaled. The formation after the translation and scaling is called a "pre-shape." The pre-shapes of Figure 1 is shown in Figure 2. The scaling makes the axis have no dimension. Therefore, in all figures from Figure 2 to 6, the axis has no dimension. The scaling factor for a pre-shape is the centroid size which is the square root of the sum of squared Euclidean distances from each landmark to the centroid [1].

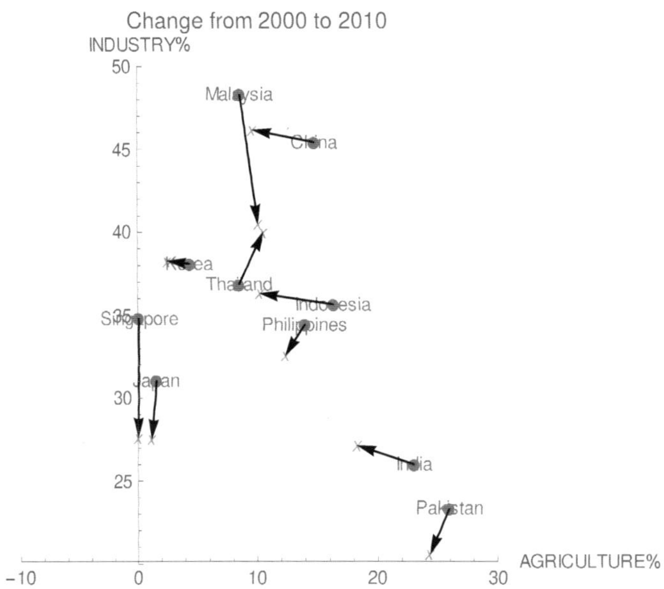

Figure 1. Asian countries' Agriculture and Industry, value added (percentage of GDP) in 2000 and 2010. The arrow illustrates each country's difference from the 2000 position to the 2010 position.

IOP Conf. Series: Journal of Physics: Conf. Series **971** (2018) 012013 doi:10.1088/1742-6596/971/1/012013

We would like to analyse the differences separately by the Affine and non-Affine transformations. By global differences we mean large scale trends, such as overall Affine transformation. Local differences are on a smaller scale, for example highlighting changes in a few nearby landmarks. Global differences are smooth changes between the figures, whereas local changes are the remainder of the components of a deformation and are less smooth [1]. When we consider the meaning of "a few nearby landmarks", for example, in skull data of animals, the meaning is the physically connection on the skull.

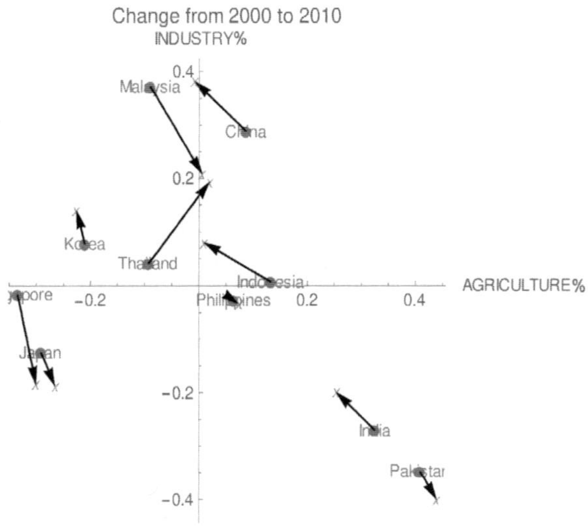

Figure 2. Pre-shapes of Asian countries' agriculture and industry, value added (percentage of GDP) in 2000 and 2010. The arrow illustrates each country's difference from the 2000 position to the 2010 position. The two axes have no dimension because we obtained pre-shapes by scaling. In the scaling, the original Euclidian distance is divided by the Centroid size.

On the other hand, in the GDP data, there is no geographical relationships among nearby countries; the deformation in ten years depends only on the country's characteristics. In the next section, we introduce a thin-plate interpolation among landmarks. In a skull data, there is a real physical surface. In biology, we can conduct reasoning of the cause and effect relationship of the deformation. However, in the GDP data, there is no real thin-plate. There, we can see only the resulting changes; the causes of the change are considered from each country's characteristics. The paramount issue of the economical shape analysis is the interpretation of the thin-plate. Concerning this, we have to study continuously.

3. Principal Warps

In the section, we will present the principal warps of the given pre-shape of the 2000 data which we represent by T. The principal warp is defined just on one shape. For example, given the shape T, the set of principal warps is defined. For the shape T, a bending energy matrix can be calculated. Then when we can do an eigen-decomposition of the matrix, the obtained eigenvectors are called principal warps. After an eigen-decomposition of the bending energy matrix, we can get (k-3) non-zero eigenvalues and eigenvectors. In this example, as there are 10 landmarks (countries), the number of the principal warps is seven. Figure 3 illustrates the first and second principal warps of T among the 10 principal warps. The spline surface is determined so that the bending energy is minimized. We know theoretically that the i-th partial warp is defined as an inner product between the i-th principal warp eigenvector of T and the basis function vector $s(t) = \left(\sigma(t - t_1), \dots, \sigma(t - t_k)\right)^T$ where $\sigma(t - t_k)$ is the interpolation base function of which centre is (t_k), the k-th landmark position.

The first principal warp shows the Philippine's difference relative to Indonesia and Thailand (See Figure 4). The second principal warp reflects the difference of Indonesia, Japan and Korea relative to Thailand, Philippines, and Singapore (See Figure 5). The eigenvalues are as follows: 109, 76, 38, 28, 18, 11 and 3. That means that the first and second principal warps are dominant ones. Then let us consider the meaning of the principal warp. In general, when a matrix is given and we may get its eigenvectors, each eigenvector shows the principal characteristic of the matrix, which is the PCA (Principal Component Analysis). Like that, given the formation T, the principal warp means the principal characteristic of the shape T which is interpolated with the thin-plate.

4. Partial Warps of the Deformation

In the section, we analyse the deformation from the 2000 pre-shape (T) to the 2010 pre-shape (Y). As shown in Figure 5, The deformation is decomposed into the Affine transformation and the non-Affine transformation. The non-Affine transformation is decomposed into the seven partial warps. The number of partial warps is equal to the number of principal warp eigenvectors, which is the theory. In other words, to express the deformation T to Y, we use the principal components of the given T where the partial warp is a projection mapping of the pre-shape Y to the principal warp eigenvectors of T. In the example, because the number of principal warps was seven, the number of partial warps is also seven. The original pre-shape T and the shape after the Affine transformation of T are shown in Figure 5. The Cartesian transformation grids are called transformation grids. The Affine transformation is expressed the skewed transformation grids. The Affine transformation shows a global change tendency. In the example, we can see from the Affine transformation a globally positive correlation between agriculture and industry and growth of both. An industrious growth is a bit bigger than an agricultural one. We suppose that a thin-plate exists according to the transformation grids. The total non-Affine transformation is shown in Figure 5. The transformation grids are changed as if the thin-plate is folded.

Next, we will divide the non-Affine transformation to seven partial warp eigenvectors. That is a principal component of the deformation T to Y. The first, second and third partial warps are presented in Figure 6. In each partial warp figure pair, the right figure shows the change of the transformation grids. The grid change shows the local changes and is not smooth. The difference/change on each country can be expresses by a vector. A set of the vectors are plotted from the origin (0, 0) (See the oval area in the right graph). The left figure is a magnification one of the oval area in the right figure. In the left figure, instead of the circle marks without country names, country names are plotted around the origin (0, 0). The country name in the left figure is used there as a point marker. Theoretically the vector points are listed on a straight line as shown in the right and left figures. Although we see the decline angles of the two corresponding ovals are not the same, this is because the axis scaling is not the same. If we use the same axis scaling, the decline angles become the same.

In the first partial warp, Indonesia's change direction is opposite to the direction of Philippines' one (See the left figure). As shown in the right figure of the first partial warp, the transformation grids show the local changes in the neighbourhood around Philippines and Indonesia; there the transformation grid lines are skewed. Then in the second partial warp, there are a local change on Korea and Japan and another local change on India and Pakistan. The two changes are the opposite directional changes as shown in Figure 5. This change is mainly caused by the same tendency in the second principal warp of T as shown in Figure 4. Because the partial warp is the mapping of Y to the principal warp of T, the same tendency can be seen as the base principal warp.

In the third partial warp, India shows the opposite directional change to ones of Singapore and Malaysia. In Figure 6, the transformation grids by the third partial warp shows the big wave-like change in the vicinity of India; on the other hand, the vicinity of Singapore shows the opposite directional grid change.

5. Discussion

In the section, we shall consider the effects of the shape analysis method in economic data analysis. In this GDP example, finally we found that in the first partial warp Indonesia and Philippines offer the opposite directional changes, concerning the growth of industry percentage, although Indonesia and Philippines are located in the neighbourhood in the 2000 pre-shape. Secondly, the second principal local change is that Korea and Japan and India and Pakistan show the opposite directional changes concerning the growth of agricultural percentage. Like this, the statistical shape analysis method can describe the local changes of each country. We discussed, in Section 2, that economics data has no real thin-plate among landmarks. Although the real thin-plate does not exist among countries, the partial warps are helpful to see the local change of each country. As the aim of this paper is to demonstrate the shape analysis method on economics data, we cannot say the reason of each local change in detail. We can, however, say that the statistical shape analysis method has the power of decomposition of the complicated local changes. In other words, we can extract principal components of the complicated local movements of each country on the deformation.

The statistical shape analysis extracts principal components of the non-Affine transformation part using the bending energy matrix. Concerning the principal component related analysis, in economics data analysis, we mainly use PCA (Principal Component Analysis) and SVD (Singular Value Decomposition) (See [15]). For example, time series data such as stock prices are analysed by using Random Matrix Theory which uses SVD [16-18]. The purpose of the existing studies by PCA and SVD is to extract principal components for groping of objects such as companies and countries. However, we use the shape analysis to extract principal components of their local movements. Then we focus on each country's peculiar movement. The advantage of the shape analysis is that we can examine the non-Affine transformation and that we can divide the non-Affine transformation to several principal components. Almost all data deformation analysis in economics have mainly examined the Affine transformation. However the shape analysis method enabled us to conduct another approach of analysing time series data analysis. The shape analysis method should be used more widely for the economics data analysis.

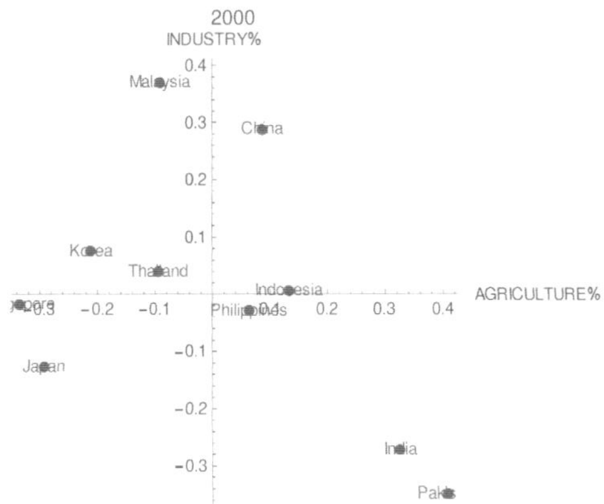

Figure 3. The pre-shape of the 2000 data and its first and second principal warps.

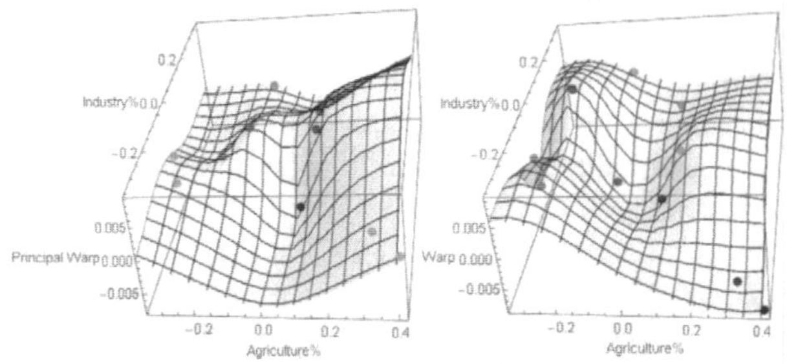

Figure 4. In each principal warp eigenvector of the 2000 pre-shape, each country has its value in the bar chart. The value indicates the contribution of the country to the thin-plate interpolation surface. The y axis shows a value of each country in the principal warp eigenvector and the y axis has no dimension.

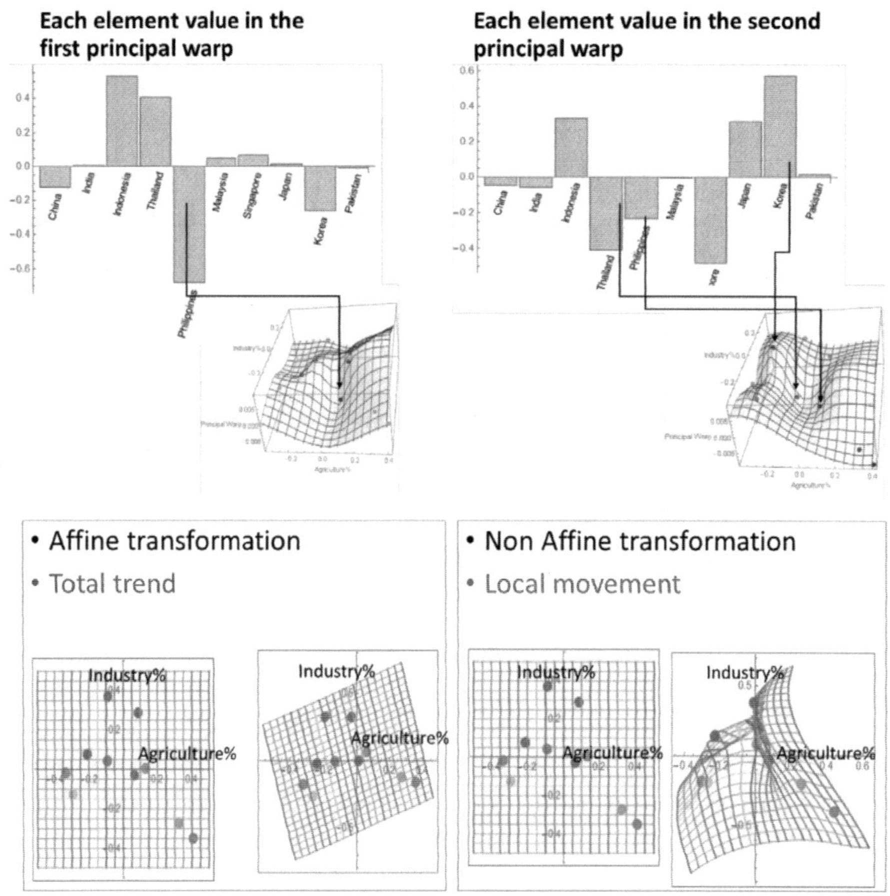

Figure 5. The deformation is divided to Affine and non Affine transformations.

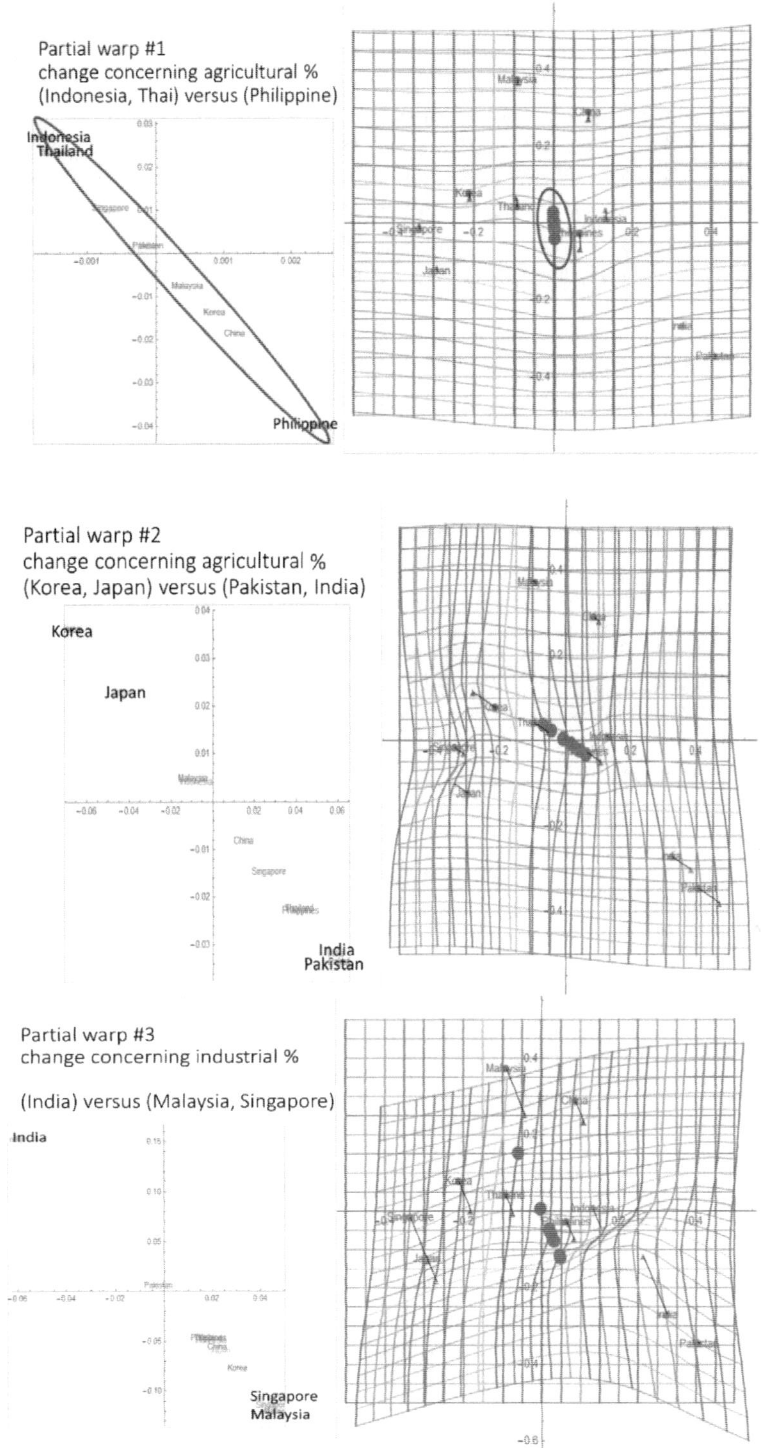

Figure 6. The first, second and third partial warps of the deformation from 2000 to 2010. Only the difference vectors of each countries are plotted in the oval areas around the origin (0, 0).

6. Conclusion

In the paper we explained the way of applying the statistical shape analysis method on economics data. We used as an example the agriculture and industry value added percentage of GDP data in Asia between in 2000 and in 2010. We analysed the deformation of the growth ratio among the Asian 10 countries. There a country is expressed as a landmark. A set of 10 landmarks forms a shape. To analyse the deformation between the two shapes, the deformation is decomposed into the Affine transformation and the non-Affine transformation. When we are not interested in the total change tendency and we are interested in each country's peculiar movement, we would like to extract only the non-Affine transformation. The non-Affine transformation can be decomposed into a set of partial warps. Each partial warp features the character of the local changes. For example, the first partial warp shows the first principal component of the local change. In the data case, the first partial warp shows that Indonesia and Philippines show the opposite directional changes concerning the growth of industry percentage, although they are located in the neighbourhood in 2000.

Almost all data deformation analysis in economics have mainly examined the Affine transformation part. However the shape analysis method enabled us to conduct another approach of analysing time series data analysis. The shape analysis method should be used more widely for the economics data analysis. We would like to apply the shape analysis method to many kinds of economic data.

References

[1] I. L. Dryden and K. V. Mardia, *Statistical shape analysis* vol. 4: J. Wiley Chichester, 1998.
[2] I. L. Dryden and K. V. Mardia, *Statistical shape analysis with Applications in R (Second Edition)*: J. Wiley Chichester, 2016.
[3] I. L. Dryden and J. T. Kent, *Geometry Driven Statistics*: Wiley Online Library, 2015.
[4] K. Mardia, F. Bookstein, and J. Kent, "Alcohol, babies and the death penalty: Saving lives by analysing the shape of the brain," *Significance,* vol. 10, pp. 12-16, 2013.
[5] H. Bayanati, R. E. Thornhill, C. A. Souza, V. Sethi-Virmani, A. Gupta, D. Maziak, *et al.*, "Quantitative CT texture and shape analysis: Can it differentiate benign and malignant mediastinal lymph nodes in patients with primary lung cancer?," *European radiology,* vol. 25, pp. 480-487, 2015.
[6] F. Nemmi, U. Sabatini, O. Rascol, and P. Péran, "Parkinson's disease and local atrophy in subcortical nuclei: insight from shape analysis," *Neurobiology of aging,* vol. 36, pp. 424-433, 2015.
[7] M. Burgaleta, A. Sanjuán, N. Ventura-Campos, N. Sebastian-Galles, and C. Ávila, "Bilingualism at the core of the brain. Structural differences between bilinguals and monolinguals revealed by subcortical shape analysis," *NeuroImage,* vol. 125, pp. 437-445, 2016.
[8] S. Srivastava, S. B. Lal, D. Mishra, U. Angadi, K. Chaturvedi, S. N. Rai, *et al.*, "An efficient algorithm for protein structure comparison using elastic shape analysis," *Algorithms for Molecular Biology,* vol. 11, p. 27, 2016.
[9] F. L. Bookstein, *The measurement of biological shape and shape change* vol. 24: Springer Science & Business Media, 2013.
[10] A. McIntosh, F. Bookstein, J. V. Haxby, and C. Grady, "Spatial pattern analysis of functional brain images using partial least squares," *Neuroimage,* vol. 3, pp. 143-157, 1996.
[11] F. L. Bookstein, "Integration, disintegration, and self-similarity: characterizing the scales of shape variation in landmark data," *Evolutionary biology,* vol. 42, pp. 395-426, 2015.
[12] A. K. Laha, "Big Data Analytics in Management: Some Statistical Challenges and Opportunities," in *The 33rd Leeds Annual Statistical Research Workshop*, pp. 107-108.
[13] Y. Shirota and T. Hashimoto, "Visual Explanation of Deformation Theories in Shape Analysis," *Gakushuin Economics Papers,* vol. 54, p. in printing, 2017.

[14] C. Apriono, R. F. Sari, Y. Yano, and Y. Shirota, "Economic Indicator Evaluation Based on Shape Deformation Analysis of Indonesian Provinces Statistics," *Gakushuin Economics Papers,* vol. 54, pp. 1-22, 2017.

[15] J. Friedman, R. Tibshirani, and T. Hastie, "The elements of statistical learning: data mining, inference, and prediction," Springer, 2013.

[16] V. Plerou, P. Gopikrishnan, B. Rosenow, L. A. N. Amaral, and H. E. Stanley, "A random matrix theory approach to financial cross-correlations," *Physica A: Statistical Mechanics and its Applications,* vol. 287, pp. 374-382, 12/1/ 2000.

[17] V. Plerou, P. Gopikrishnan, B. Rosenow, L. A. N. Amaral, T. Guhr, and H. E. Stanley, "Random matrix approach to cross correlations in financial data," *Physical Review E,* vol. 65, p. 066126, 06/27/ 2002.

[18] Y. Yano and Y. Shirota, "SVD and Text Mining Integrated Approach to Measure Effects of Disasters on Japanese Economics: Effects of the Thai Flooding in 2011" in *Neural Information Processing (LNCS 9949)*, A. Hirose, S. Ozawa, K. Doya, K. Ikeda, M. Lee, and D. Liu, Eds., ed: Springer, 2016, pp. 20--29.

Spoofing detection on facial images recognition using LBP and GLCM combination

F Sthevanie, K N Ramadhani
Multimedia Laboratory, School of Computing, Universitas Telkom, Jl. Telekomunikasi No. 1, Terusan Buah Batu Bandung 40257, Indonesia

sthevanie@telkomuniversity.ac.id, kurniawannr@telkomuniversity.ac.id

Abstract. The challenge for the facial based security system is how to detect facial image falsification such as facial image spoofing. Spoofing occurs when someone try to pretend as a registered user to obtain illegal access and gain advantage from the protected system. This research implements facial image spoofing detection method by analyzing image texture. The proposed method for texture analysis combines the Local Binary Pattern (LBP) and Gray Level Co-occurrence Matrix (GLCM) method. The experimental results show that spoofing detection using LBP and GLCM combination achieves high detection rate compared to that of using only LBP feature or GLCM feature.

1. Introduction

Nowadays, the face detection system has been widely implemented in daily life. The face detection system usually used in security system because of the unique features on every human faces. However, the face detection system has some drawbacks such as image spoofing. Facial image spoofing is an attack that deceive the face detection system using imprinted images. Spoofing attack occurs when someone try to impersonate a registered user by forging the face image and taking advantage of such illegal access[1]. The biometric system becomes vulnerable without the spoofing detection mechanism.

Original facial images and imprinted facial images reflect light in different ways. Original face is a complex 3D-object where the imprinted face image is rigid planar object. Other differences is that human face has pigment while imprinted face image does not have any pigment. Using those differences, we can build a system that can detect spoofing attack on facial biometric system.

Texture analysis is one of the fundamental research in computer vision and is an important component in the pattern recognition and classification. Texture analysis is commonly used for image classification, content-based image retrieval, 3D image reconstruction and image segmentation. There were a lot of research used texture analysis to detect spoofing attack in facial biometric system. The method used were Local Binary Pattern (LBP), Gabor Wavelet, Haboor Wavelet, Histogram of Oriented Gradient (HOG), Spatiotemporal Local Binary Pattern[2,3,4,5,6]. The ability of LBP to extract image features was used to distinguish the original human face images and the imprinted face images. Another feature extraction used to detect spoofing was GLCM which able to extract the contrast level of an image. The contrast level was used to distinguish spoof or non-spoof images. This study compares the performance of Local Binary Pattern and Gray-Level Co-Occurrence Matrix methods as a feature extraction method to classify the image into two classes i.e. spoof and non-spoof classes. We combine the LBP and GLCM method to extract the features used to detect the spoofing images. The LBP and GLCM are based on texture feature, but GLCM uses statistical function so the extraction result represents the overall image texture features. By combining the

LBP histogram and GLCM statistical function, we expect to obtain a better result compared to by using only one feature function.

2. Related works

The attacks on biometric can be divided into several scenarios. The attack can be a forcing to a registered users, registering to the system using deceased person, using truncated body parts with a genetic clone, or by spoofing the data of registered user. The biometric characteristics of each individual registered user must be different even for identical twins. Using only algorithm is not strong enough to counter those differences[7]. This research is focused on spoofing attack.

Spoofing occurs when someone wants to fool the system by faking the existing data which in this case is the face image of registered user. The attacker uses an imprinted facial image of registered user to enter the biometric security access and then gain access of the system. Without the anti-spoofing mechanism, the recognition system is very vulnerable to be attacked. Anti-spoofing mechanism can be built using liveness detection. The spoofing detection can be built using some liveness sign to differentiate between the original face and the imprinted image. Human face has a complex 3D features that are not rigid, while the imprinted face image is a rigid planar object. We can use texture features to extract those differences. Thus, we can use texture analysis to detect spoofing attacks on human face images[8].

3. Local Binary Pattern (LBP)

Local Binary Pattern (LBP) method was proposed to describe the texture of an image. LBP was used in many researchs in computer vision such as face recognition, facial experession recognition, motion and action modeling and in medical image analysis. The LBP is derived from the neighboring radius of a given pixel. The LBP value is calculated by comparing the intensity of neighboring pixel to the intensity of the central pixel starting from a certain angle and repeat it on the next neighboring pixel[9]. The comparison result is a binary number of 1 and 0. A value 1 is given if the center pixel intensity is less than the intensity of the neighboring pixel, whereas a value of 0 is given if the center pixel intensity is more than the intensity of the neighboring pixel. The most commonly used sample is $P = 8$ with the circle radius $R = 1$. $P = 8$ means that we conduct the LBP on 8 neighboring pixel around the center pixel. $R = 1$ means that the distance between the center pixel and neighboring pixel is 1. However, other number of neighboring pixel and radius values such as $P = 16$ and $R = 2$ can be used. After that, we concate the binary value of the neighboring pixel clockwise or counterclockwise. The binary value is then converted to decimal value. This process is conducted for every pixel in the image. The illustration of LBP process can be seen on Figures 1 and 2.

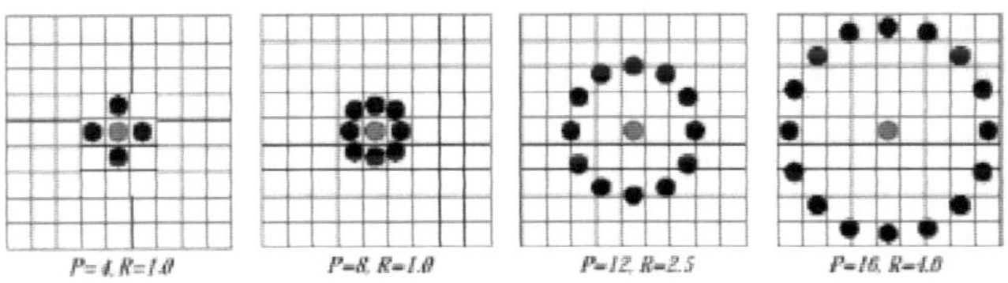

Figure 1 LBP Illustration for different P and R values

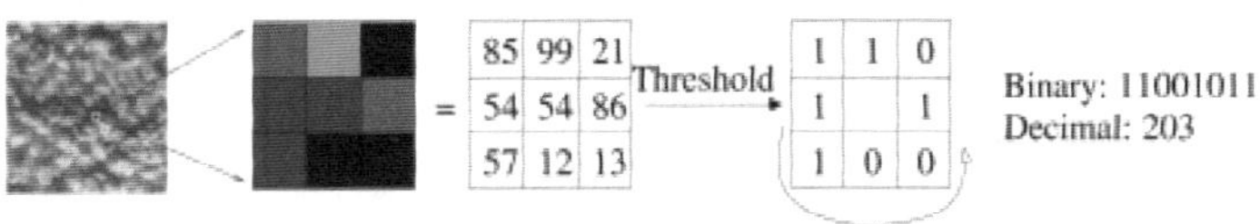

Figure 2 LBP Encoding Example

The LBP code of a pixel (x_c, y_c) [10] can be calculated by:

$$LBP_{P,R} = \sum_{P=0}^{P-1} s(g_p - g_c)2^P, \tag{1}$$

where g_c is the intensity value of the central pixel (x_c, y_c), g_p is the intensity value of the neighboring pixel P in radius R, and s is defined as the thresholding function as:

$$s(x) = \begin{cases} 1, if\ x \geq 0; \\ 0, otherwise. \end{cases} \tag{2}$$

The feature extraction using LBP is divided into two stages.

1. Build LBP matrix by generating every LBP value for every pixel. The example of LBP matrix can be seen on Figure 3.

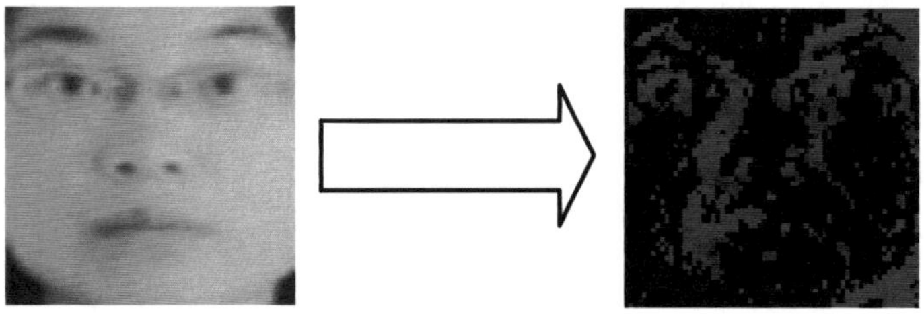

Figure 3 The Example of LBP Matrix (right) from original image (left)

2. Calculate histogram from LBP matrix. On this stage, we build histogram that contain the number of pixel for every LBP intensity level. The number of LBP intensity level is depend on the number of neighboring pixel (P value), that is 2^P. On this research, we use P=8, so the number of LBP intensity level is 256. The example for LBP histogram can be seen on Figure 4.

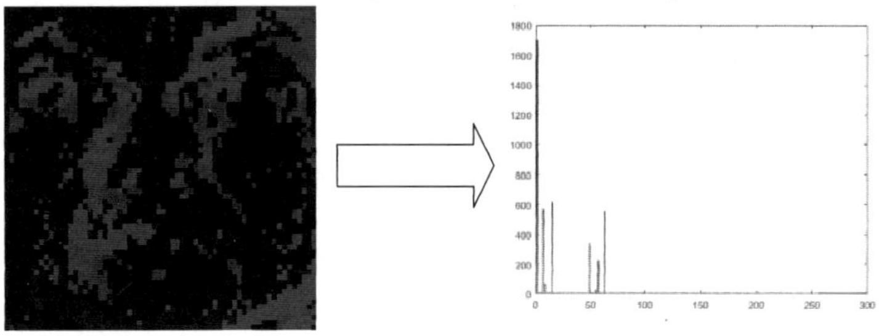

Figure 4 Example of LBP Histogram (right) from LBP Matrix (left)

4. Gray-Level Co-Occurrence Matrix (GLCM)

The texture analysis approach has been widely developed in recent decades, and can be classified into four categories such as statistical, geometrical, model-based and signal processing. Gray Level Co-occurrence Matrix (GLCM) has been shown to be one of the most efficient approaches to texture analysis among other statistical methods[11]. GLCM describes the spatial distribution of the gray value of an image and the frequency of gray occurrences pattern with a specified angles and distance. GLCM can be used for recognition and texture classification. Texture analysis using Gray Level Co-occurence Matrix (GLCM) was introduced by Haralick[12]. GLCM is a matrix sized NxN where N is the grayscale intensity level. Each of matrix element value located on (i,j) is calculated from the number pattern from original grayscale image that contain a specified pattern. The specified pattern contains the gray pixel i neighboring with gray paxel j for distance d and angle θ. The parameter d represents the distance of two neigboring pixel in the grayscale image and θ is the size of the discrete angle (0^0, 45^0, 90^{0}, 135^0). The commonly angle value used in GLCM calculations is θ = 45⁰ or the mean of the four corners (θ = 0⁰, 35⁰, 90⁰, 135⁰). The illustration of GLCM can be seen on Figure 5.

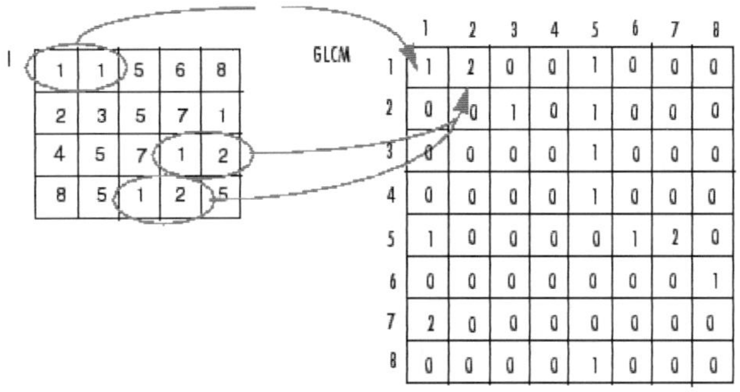

Figure 5 The Illustration of GLCM method

Based on the GLCM definition, Haralick defines several functions to describe the textual features from GLCM, such as[11]:

1. *Angular Second Moment* (ASM),

$$f_1 = \sum_i \sum_j \{p(i,j)\}^2 \quad (3)$$

2. *Contrast,*

$$f_2 = \sum_{n=0}^{N_g-1} n^2 \left\{\sum_{i=1}^{N_g} \sum_{j=1}^{N_g} p(i,j)\right\}, |i-j| = n \quad (4)$$

3. *Correlation,*

$$f_3 = \frac{\sum_i \sum_j (ij) p(i,j) - \mu_x \mu_y}{\sigma_x \sigma_y} \quad (5)$$

4. *Variance,*

$$f_4 = \sum_i \sum_j (i - \mu)^2 p(i,j) \quad (6)$$

5. *Inverse Difference Moment,*

$$f_5 = \sum_i \sum_j \frac{1}{1+(i-j)^2} p(i,j) \quad (7)$$

6. *Sum Average,*

$$f_6 = \sum_{i=2}^{2N_g} i p_{x+y}(i) \quad (8)$$

7. *Sum Variance,*

$$f_7 = \sum_{i=2}^{2N_g} (i - f_8)^2 p_{x+y}(i) \quad (9)$$

8. *Sum Entropy,*

$$f_8 = -\sum_{i=2}^{2N_g} p_{x+y}(i) \log\{p_{x+y}(i)\} \quad (10)$$

9. *Entropy,*

$$f_9 = -\sum_i \sum_j p(i,j) \log(p(i,j)) \quad (11)$$

10. *Difference Variance,*

$$f_{10} = variance\ of\ p_{x-y} \quad (12)$$

11. *Difference Entropy*

$$f_{11} = -\sum_{i=0}^{N_g-1} p_{x-y}(i) \log\{p_{x-y}(i)\}. \quad (13)$$

$p(i,j)$ is the normalized value of GLCM element located on (i,j), $p(i,j) = P(i,j)/R$. $\mu_x, \mu_y, \sigma_x, \sigma_y$ are the mean values and variance values for P_x dan P_y. μ is the mean value for image P. $p_x(i)$ is the value of marginal-probability matrix calculated by summing the row value from $p(i,j), = \sum_{i=1}^{N_g} P(i,j)$. N_g is the *gray levels* used.

$$P_y(j) = \sum_{i=1}^{N_g} p(i,j) \tag{14}$$

$$P_{x+y}(k) = \sum_{i=1}^{N_g}\sum_{j=1}^{N_g} p(i,j),\ i+j=k;\ k=2,3,\dots,2N_g \tag{15}$$

$$P_{x-y}(k) = \sum_{i=1}^{N_g}\sum_{j=1}^{N_g} p(i,j),\ |i-j]=k;\ k=0,1,\dots,N_g-1 \tag{16}$$

5. K-Nearest Neighbour

K-Nearest Neighbors (K-NN) is one of the simplest algorithms and has long been used in classification. In some cases, K-NN method provides better results compared to other classifier. The K-NN method classifies the unlabeled vector (testing data) by assigning the most-appearing labels between the nearest training distance k around the testing data. Therefore, the performance of this method is highly dependent on the distance matrix used to identify its nearest neighbor[3]. The distance function usually used in K-NN are euclidean distance, minkowski distance, mahalanobis distance, city block distance, hamming distance, and mahalanobis distance. The distance function commonly used for continuous variables is the euclidean distance. For discrete variables, such as text-based data, other distance calculation methods can be used, such as Hamming Distance. The distance calculation using the euclidean distance is obtained by comparing each value in the training data vector with each value in the testing data vector as can be seen in:

$$d(a,b) = \sqrt{\sum_{i=1}^{n}(a_i - b_i)^2} \tag{17}$$

6. Research methodology

This research consist of two steps. The first step is modelling. The modelling step is conducted to build the model for texture feature of the training data (spoof and non-spoof) and insert the features to the database. The output of this step is the database of texture feature of training data. The flowchart of modelling step can be seen on Figure 6.

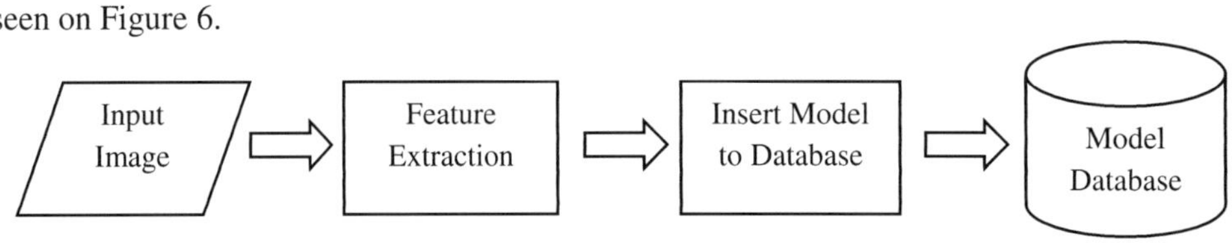

Figure 6 Modelling Step

The second step is testing step. The testing step is an identification process to classify the testing data to spoof or non-spoof label. The step start with input image of testing data. The feature extraction process is conducted on training data to obtain the feature vector. Then, the classification process using K-NN algorithm is conducted. The classification process uses the model database built from modelling step. The output of the classification process is the conclusion whether the data is labelled spoof or non-spoof. The process can be seen on Figure 7.

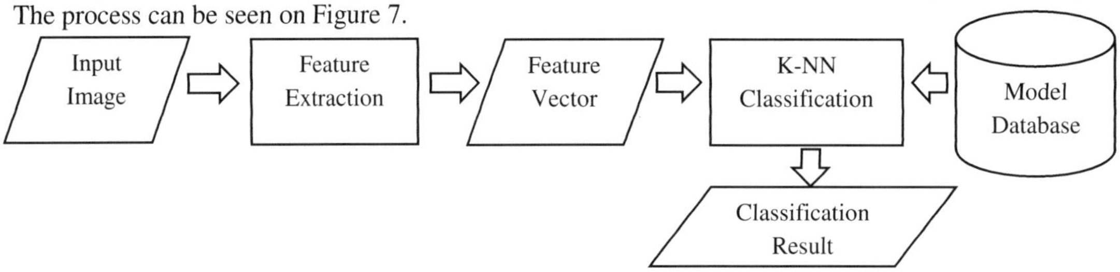

Figure 7 Testing Step

The dataset used in this research are from NUAA dataset. The NUAA database images consist of 15 user images and the number of overall images are 12.614 images of real face and impostor face images. The images were resized to 64 x 64 pixels. The color representation of the images are grayscale with 256 levels.

7. Testing result

There are three scenario for testing process. The first scenario is spoofing detection using LBP algorithm. The second scenario is spoofing detection using GLCM algorithm. The final scenario is spoofing detection by combining LBP and GLCM algorithm.

7.1 Spoofing Detection using LBP Algorithm.

On this step, we conduct testing on spoofing detection using LBP algorithm to determine the best R value from 1 to 5 using value $P = 8$. The result can be seen on Figure 8.

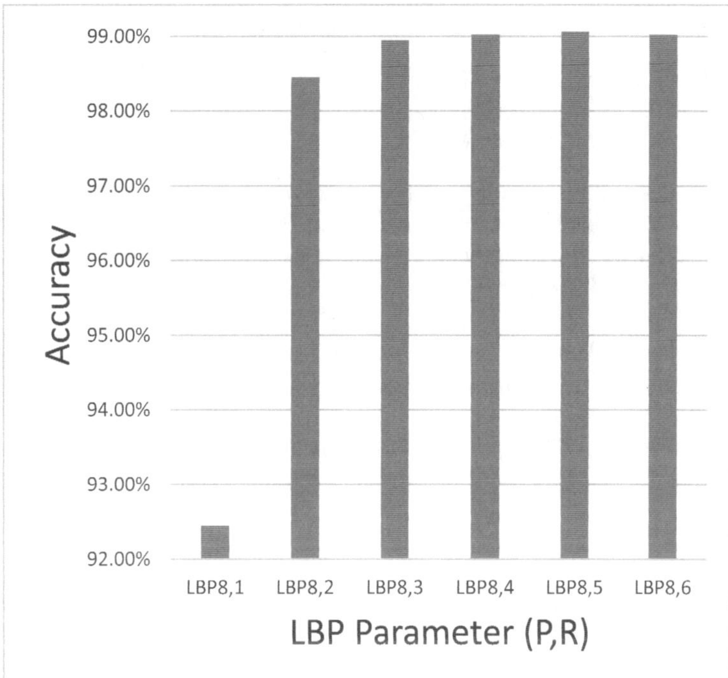

Figure 8 LBP Testing Result

The x-axis shows the different value for LBP radius (R). The y-axis shows the accuracy of spoofing detection using given LBP radius. From Figure 8, can be seen that the overall performance of LBP parameter configuration were more than 90%. The best accuracy was obtained by using R=5 where the accuracy was 99.07%.

7.2 Spoofing Detection using GLCM Algorithm.

On this step, we conduct testing on spoofing detection using GLCM algorithm. We use two mechanism on GLCM algorithm.

1) GLCM1 is implemented by using GLCM on every part of human face image.
2) GLCM2 is implemented by using GLCM only on eyes and nose area. We use only the area that has a contour difference on face image.

The result can be seen on fig 9.

IOP Conf. Series: Journal of Physics: Conf. Series **971** (2018) 012014 doi:10.1088/1742-6596/971/1/012014

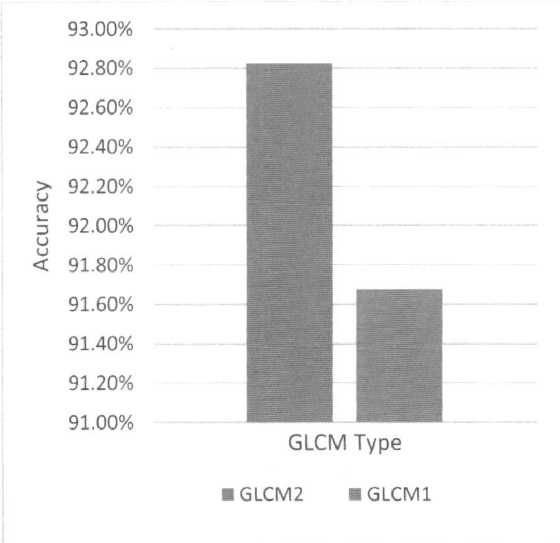

Figure 9 GLCM Testing Result

From Figure 9, we can see that the accuracy of the overall performance spoofing detection using GLCM for the two configuration obtained the accuracy more than 90%, but the performance of GLCM still less than the performance of LBP algorithm. The best accuracy was 92.82% using GLCM2 configuration (using GLCM only in eyes and nose area).

7.3 Spoofing Detection using LBP and GLCM Combination

We combine the LBP and GLCM feature using several scenario.

1) Combination 1. We combine the result of K-NN algorithm on LBP features and GLCM features.
2) Combination 2. We conduct GLCM algorithm on LBP matrix.
3) Combination 3. Similar to combination 2, but the LBP matrix is constructed only from eyes and nose area.
4) Combination 4. Similar to combination 1, but the LBP matrix and GLCM matrix constructed only from eyes and nose area.

The best accuracy from four combination was obtained using combination 4. The best accuracy was 99.26% obtained by combining the K-NN classification result on LBP and GLCM features. The detail of testing result can be seen on Figure 10.

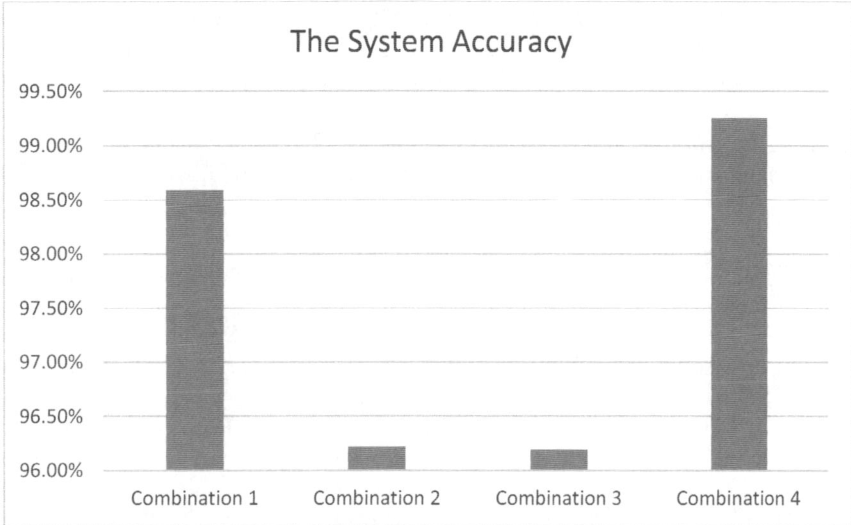

Figure 10 Testing Result on LBP and GLCM Combination

Furthermore, we compared the result of the best combination scenario and the result of LBP algorithm. We compared the error rate for the two scenario. The result can be seen on Figure 11.

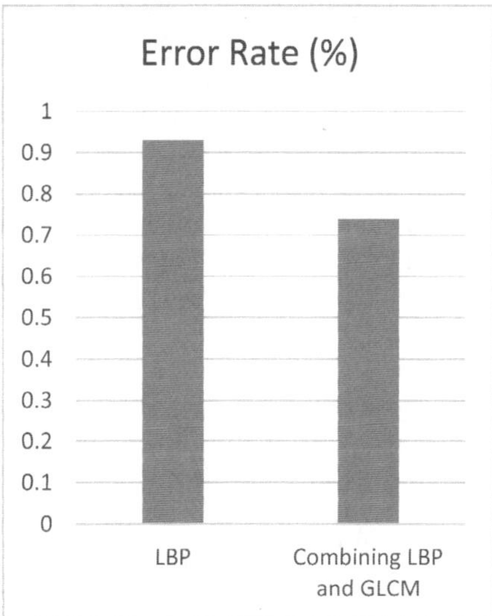

Figure 11 Comparing Error Rate Between LBP and Combining LBP+GLCM

The error rate of spoofing detection using LBP and GLCM combination feature extraction is less than that of feature extraction using only LBP. By using GLCM, we reduced the error rate on LBP feature from 0.93% to 0.74% or the error rate reduced to 20.4%. Thereby, we conclude that the spoofing detection using combination of LBP and GLCM feature extraction is better than using only one feature extraction method, such as LBP and GLCM.

8. Conclusion

In this research, we implemented the spoofing detection on facial biometric using texture features. We implemented two texture feature extraction algorithm, those were LBP and GLCM. The research showed that the spoofing detection using LBP algorithm has a better accuracy than that of using GLCM algorithm.

The LBP algorithm gave the accuracy 99.07% while the GLCM algorithm only gave accuracy 92.82%. The research also showed that the eyes and nose area has an important feature to detect the spoofing on facial biometric. The accuracy of using only eyes and nose area was 92.82% while the accuracy of using every part on human face was 91.67%. By using the combination of LBP and GLCM algorithm for feature extraction algorithm, the spoofing detection system accuracy was 99.26%. That performance was better than that of using only a single feature extraction algorithm, such as LBP and GLCM.

References

[1] Maatta, J., Hadid, A., & Pietikainen, M. (2012). Face spoofing detection from single images using texture and local shape analysis. *Biometrics, IET, 1*(1), 3-10.

[2] X.Tan, Y.Li, J.Liu and L.Jiang. Face Liveness Detection from A Single Image with Sparse Low Rank Bilinear Discriminative Model, In: Proceedings of 11th European Conference on Computer Vision (ECCV'10), Crete, Greece. September 2010

[3] Komulainen, J., Hadid, A., & Pietikäinen, M. (2012, November). Face spoofing detection using dynamic texture. In *Computer Vision-ACCV 2012 Workshops*(pp. 146-157). Springer Berlin Heidelberg.

[4] Määttä, J., Hadid, A., & Pietikainen, M. (2011, October). Face spoofing detection from single images using micro-texture analysis. In *Biometrics (IJCB), 2011 international joint conference on* (pp. 1-7). IEEE.

[5] Wen, D., Han, H., & Jain, A. K. (2015). Face spoof detection with image distortion analysis. *Information Forensics and Security, IEEE Transactions on,10*(4), 746-761.

[6] de Freitas Pereira, T., Anjos, A., De Martino, J. M., & Marcel, S. (2012, November). LBP– TOP based countermeasure against face spoofing attacks. In *Computer Vision-ACCV 2012 Workshops* (pp. 121-132). Springer Berlin Heidelberg.

[7] Schuckers, S. A. (2002). Spoofing and anti-spoofing measures. *Information Security technical report, 7*(4), 56-62.

[8] Mansfield, A. J., & Wayman, J. L. (2002). *Best practices in testing and reporting performance of biometric devices* (pp. 1-36). Teddington, Middlesex, UK: Centre for Mathematics and Scientific Computing, National Physical Laboratory.

[9] Lahdenoja, O., Poikonen, J., & Laiho, M. (2013). Towards understanding the formation of uniform local binary patterns. *ISRN Machine Vision, 2013.*

[10] Hadid, A. (2008, November). The local binary pattern approach and its applications to face analysis. In *Image Processing Theory, Tools and Applications, 2008. IPTA 2008. First Workshops on* (pp. 1-9). IEEE.

[11] Haralick, R. M., Shanmugam, K., & Dinstein, I. H. (1973). Textural features for image classification. Systems, Man and Cybernetics, IEEE Transactions on, (6), 610-621.

[12] Hu, Y., Zhao, C. X., & Wang, H. N. (2008, December). Directional analysis of texture images using gray level co-occurrence matrix. In Computational Intelligence and Industrial Application, 2008. PACIIA'08. Pacific-Asia Workshop on (Vol. 2, pp. 277-281). IEEE.

[13] Jakkula, V. (2006). Tutorial on support vector machine (svm). School of EECS, Washington State University.

[14] Cristianini, N., & Shawe-Taylor, J. (2000). An introduction to support vector machines and other kernel-based learning methods. Cambridge university press.

Social network analysis using k-Path centrality method

Natya Taniarza[1], Adiwijaya[2] and Warih Maharani[3]

School of Computing, Telkom University, Bandung, 40257, Indonesia

[1]natyataniarza@gmail.com, [2]adiwijaya@telkomuniversity.ac.id, [3]wmaharani@gmail.com

Abstract. k-Path centrality is deemed as one of the effective methods to be applied in centrality measurement in which the influential node is estimated as the node that is being passed by information path frequently. Regarding this, k-Path centrality has been employed in the analysis of this paper specifically by adapting random-algorithm approach in order to: (1) determine the influential user's ranking in a social media Twitter; and (2) ascertain the influence of parameter α in the numeration of k-Path centrality. According to the analysis, the findings showed that the method of k-Path centrality with random-algorithm approach can be used to determine user's ranking which influences in the dissemination of information in Twitter. Furthermore, the findings also showed that parameter α influenced the duration and the ranking results: the less the α value, the longer the duration, yet the ranking results were more stable.

1. Introduction

Social networks can be defined as finite sets of users and relationships between other users [1]. Users and relationships can be measured and mapped in graphs by interaction pattern analysis or defined as Social Network Analysis (SNA) where the user is represented as a node and the relation is represented as edge [2]. One of the most commonly used SNA applications is centrality measurement [3].

Centrality measurement is used to measure the ability of a node in flowing information. In general, the centrality measurement consists of 3 points of view: Degree centrality, Betweenness centrality and Closeness centrality [4]. By using the Betweenness centrality method, node most often located on the shortest path is the most influential node in the dissemination of information. While shortest paths are not always an optional path of information [5], enabling influential nodes in the flow of information is a node that is not on the shortes path.

The problem can be solved by using k-Path centrality method with random algorithm approach, where in determining path or path of a node does not use shortest path but by using random algorithm, so it will look for possible combinations of path of information or path. Therefore, in this paper use k-Path centrality method to know the ranking of users who have influence on Twitter network.

The remainder of this paper is structured as follows: section 2 briefly presents the methodology, section 3 is system design and implementation, section 4 reports the experiment and results from the case study conducted, and section 5 provides conclusion and suggestion for further research.

2. Methodology

2.1. Social Network Analysis
Social Network Analysis (SNA) is a mathematical method that allows to map and measure interaction paths formed between nodes on a network. SNA aims to understand how groups of individuals relate and how the consequences, and understand how they behave [3]. Scott [6] defines SNA is a set of methods used to investigate relationships in the data structure. So with SNA we can analyze the interactions within a group by looking at the social actors (nodes) with the relationships between them. Otte and Rousseau [7] shows that SNA can be used for implicit information retrieval purposes, including interaction and friendship relationships between users.

2.2. Centrality Measurement
The measurement of centrality is one example of applying SNA. Used to measure the ability of each node in streaming information. To date, many development methods can be used to measure centrality. Each method has its own advantages and disadvantages. Selection of methods can be tailored to the needs. Of the various methods, the most widely used according to Otte and Rousseau [7] are Degree centrality, Betweenness centrality and Closeness centrality.

2.3. k-Path Centrality
k-Path centrality ($\boldsymbol{C_k}$) is the development of Betweenness centrality, where the same general concept is to emphasize the relationship of a node to another node. But on the calculation of this centrality without specifying the shortest paths as Betweenness does, according to Tharaka [8] shortest paths are not always the path through which information passes, in other words it allows nodes that have a high centrality value instead of nodes that are in the path of shortest paths. Tharaka also revealed that nodes that have a high $\boldsymbol{C_k}$ then have a high $\boldsymbol{C_B}$ as well [5,8]. Centrality calculation formula by using k-Path centrality method on a node v for weightless graph that is [5,8]:

$$C_k(v) = \sum_{s \neq v} \sum_{1 \le \ell \le k} \frac{p_v(s;\ell)}{p(s;\ell)} \quad (1)$$

Where $p_v(s;\ell)$ is the number of paths from node s to another node along k that passes through the node v as much as ℓ and $p(s;\ell)$ is the number of paths from node s to another node along k as much as ℓ.

2.4. Random Algorithm Approach
The random algorithm approach is used to estimate the k-Path centrality value of each node with the concept of retrieving the nodes randomly using a uniform distribution. Because using the uniform distribution then the chances of each node to be drawn together. Calculation of k-Path centrality value in random algorithm approach [5,8] becomes:

$$\boldsymbol{C_k(v) = B(v)\frac{k.n}{\ell}} \quad (2)$$

$$\boldsymbol{\ell = 2\, k^2 n^{1-2\alpha} \ln n} \quad (3)$$

$$\boldsymbol{k = \ln\,(n + m)} \quad (4)$$

Where B(v) is the weight value of node v, the weight here means the number of nodes v taken. ℓ is number of paths generated, $\boldsymbol{k}$ is positive integer number which is the number of nodes in a single path. n is number of nodes while m is number of edges and α is real number $(0, \frac{1}{2})$.

3. System Design and Implementation

IOP Conf. Series: Journal of Physics: Conf. Series **971** (2018) 012015 doi:10.1088/1742-6596/971/1/012015

In this section, the system design of the current research will be elaborated. The steps of the process are as shown in Figure 1.

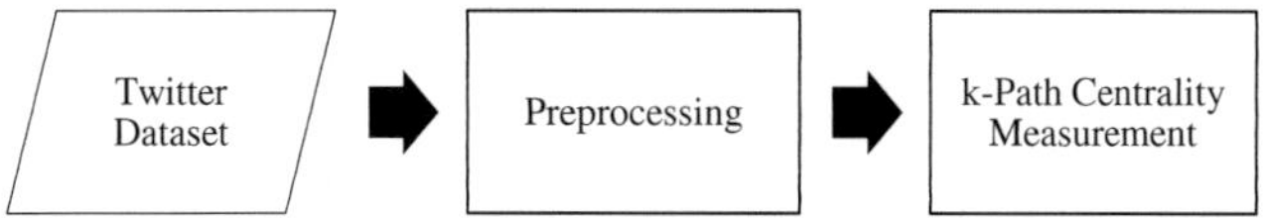

Figure 1. Overall Process.

3.1. Twitter Dataset

Twitter data is a data set derived from social media Twitter taken using NodeXL software. The data obtained is a table of relations consisting of user names and relationships between users. Data set used in this experiment comes from social media Twitter taken from Twitter @natyataniarzaa users. Data in the form of friendship relation between user and other user. The relations used are follower and followed / following relations. From the retrieval of data obtained by the user as much as 377 and 17006 relation.

3.2. Preprocessing

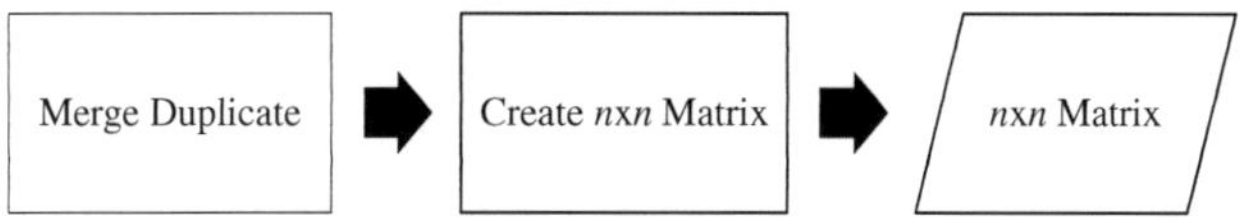

Figure 2. Preprocessing Process.

Preprocessing is done to get the matrix ready to be input to the system, this process is done in NodeXL software. The detail steps in preprocessing are as shown in Figure 2. The first is the process of merge duplicate on the relation table, this is done because on Twitter data allows the existence of the same data or duplicate. For example when user A performs the following to user B with user A being a user follower B, this data is what will be merged. Results table of relationships formed on Twitter user @natyataniarzaa can be seen in Figure 3.

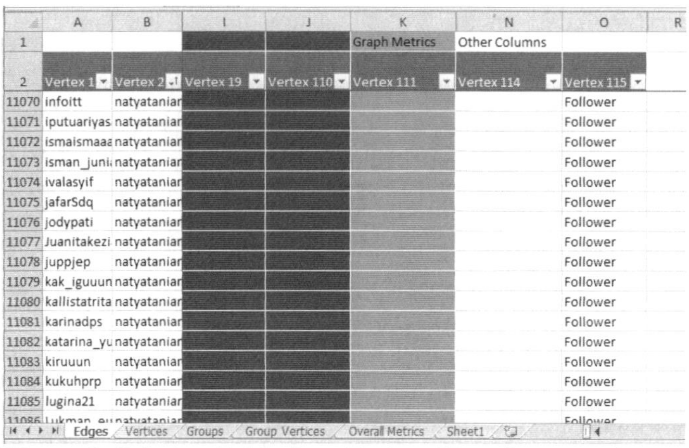

	A	B	I	J	K	N	O	R
1					Graph Metrics	Other Columns		
2	Vertex 1	Vertex 2	Vertex 19	Vertex 110	Vertex 111	Vertex 114	Vertex 115	
11070	infoitt	natyataniar					Follower	
11071	iputuariyas	natyataniar					Follower	
11072	ismaismaaa	natyataniar					Follower	
11073	isman_juni	natyataniar					Follower	
11074	ivalasyif	natyataniar					Follower	
11075	jafarSdq	natyataniar					Follower	
11076	jodypati	natyataniar					Follower	
11077	Juanitakezi	natyataniar					Follower	
11078	juppjep	natyataniar					Follower	
11079	kak_iguuun	natyataniar					Follower	
11080	kallistatrita	natyataniar					Follower	
11081	karinadps	natyataniar					Follower	
11082	katarina_yu	natyataniar					Follower	
11083	kiruuun	natyataniar					Follower	
11084	kukuhprp	natyataniar					Follower	
11085	lugina21	natyataniar					Follower	
11086	Lukman_eu	natyataniar					Follower	

Edges / Vertices / Groups / Group Vertices / Overall Metrics / Sheet1

Figure 3. Relation Table from Twitter User @natyataniarzaa.

The non-duplicate relation table is then represented in the form of *nxn* matrix, where n is the number of users retrieved. The example of *nxn* matrix results that formed on Twitter @natyataniarzaa user can be seen in Figure 4.

	A	B	C	D	E	F	G	H	I	J	K	L
1		yogifahmi	natyatani‹	ucilfauzi	Egiiiilang	imam_mu	esty_p	mutiaanu	kiruuun	dinnurilla	nunikania	iendrabay a
2	yogifahmi	0	1	1	1	0	0	0	0	0	0	0
3	natyatani‹	1	0	1	1	1	1	1	1	1	1	1
4	ucilfauzi	1	1	0	1	0	1	0	0	0	0	0
5	Egiiiilang	1	1	1	0	0	0	1	0	0	0	0
6	imam_mu	0	1	1	0	0	0	1	1	1	1	1
7	esty_p	0	1	1	0	0	0	0	0	0	0	0
8	mutiaanu	0	1	0	1	1	0	0	0	0	1	1
9	kiruuun	0	1	0	0	1	0	0	0	1	1	1
10	dinnurilla	0	1	0	0	1	0	0	1	0	1	1
11	nunikania	0	1	0	0	1	0	1	1	1	0	1
12	iendrabay	0	1	0	0	1	0	1	1	1	1	0
13	alfathxp	0	1	0	0	1	0	1	1	1	1	1
14	anggiama	0	1	0	0	1	0	1	1	1	1	1
15	adbudsan	0	1	0	0	1	0	1	1	1	1	1
16	dhanfatha	0	1	0	0	0	0	0	0	0	0	0
17	luthfydelf	0	1	0	0	1	0	0	0	1	0	1
18	FabianRes	0	1	0	0	0	0	0	0	0	0	0
19	yudhisriar	0	1	0	0	0	0	0	0	0	0	0
20	satwikapc	0	1	1	1	0	0	0	0	0	0	0

Sheet1 Sheet2 Sheet3

Figure 4. nxn Matrix from Twitter User @natyataniarzaa.

3.3. k-Path Centrality Measurement

To facilitate the calculation of k-Path centrality is done in 3 stages, namely initialization, path determination and accumulation (see Figure 5).

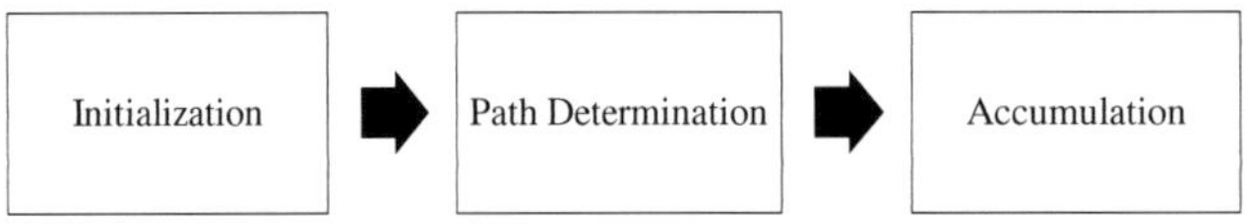

Figure 5. k-Path Centrality Measurement Process.

In the first process is the initialization of the required parameters values when the path determination and accumulation. The second process is the process of determining the path, in this process produced a combination of simple paths taken at random. The path determination process aims to give the weight of each node that is picked up or missed. Each retrieval of values and nodes is taken randomly with a uniform distribution approach, where each value or node has the same chance of being retrieved. The last process is the process of accumulation, this process aims to get the value of the centrality of each node based on the weight obtained from the previous process included in equation 1.

4. Experimental Result

The result of the experiment is intended to implement and analyze the effect of parameter α on processing time and ranking result. This experiment aims to know the rank of the influential user in spreading information on Twitter. The detail of how the procedure is conducted will be detailed in the following subsections.

4.1. Experimental Setup

Based on the objectives to be achieved, tested the value of parameter α. In the process of testing is done changes the value of α as much as 49 times that starts from 0.01 and increased 0.01 each test to 0.49.

The dataset used in this experiment comes from Twitter social media taken from Twitter @natyataniarzaa users with the help of NodeXL software. Data in the form of friendship relation between user and other user. The relations used are follower and followed/following relations. From the retrieval of data obtained by the user as much as 377 and 17006 relations.

4.2. Analysis Parameter α on The Processing Time

Testing conducted 49 times on Twitter data, where each test is done as much as 30 times then generated the average time in seconds. In order to more easily understand the effect of α then given the visualization in graphical form as follows:

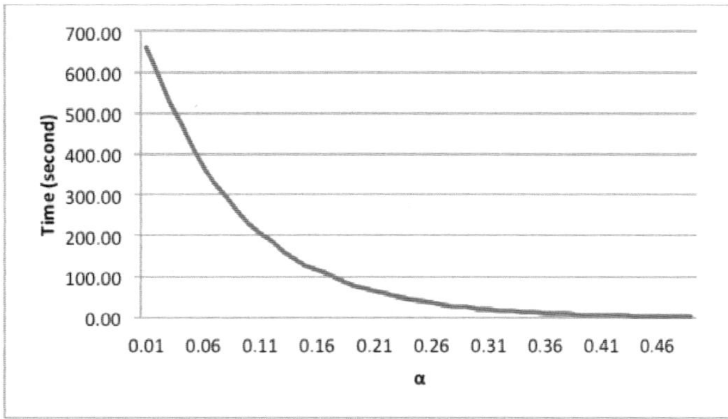

Figure 6. Change The Value α to Processing Time

Based on Figure 6 can be seen that the graph is a non-linear graphic where the smaller the value of α used then the time required to obtain the value of centrality longer. This can be due to the smaller value of α used, the more combinations of paths are generated. To find the combination of these paths depends on the duration of the calculation of centrality.

4.3. Analysis Parameter α on Ranking Result

In this experiment is done 49 times where each test is done as much as 10 times. In this paper is shown 3 tables of test results that can represent the overall test.

Based on Table 1 it can be seen that the 'hmifitt' and 'johan_jojo' nodes are exchanged positions when the 8th and 10th tests can be caused by two things, first because the difference in the centrality value of the two nodes is small and the second the number of each node picked up may change as a result of random path determination to allow the number of 'johan_jojo' nodes to be retrieved in the 8th and 10th tests more than the number of captured 'hmifitt' nodes.

Table 1. Rank Result Where α = 0.01.

rank	1st test	2nd test	3rd test	4th test	5th test	6th test	7th test	8th test	9th test	10th test
1	1	1	1	1	1	1	1	1	1	1
2	2	2	2	2	2	2	2	2	2	2
3	3	3	3	3	3	3	3	3	3	3
4	4	4	4	4	4	4	4	4	4	4
5	5	5	5	5	5	5	5	5	5	5
6	6	6	6	6	6	6	6	6	6	6
7	7	7	7	7	7	7	7	7	7	7
8	8	8	8	8	8	8	8	9	8	9
9	9	9	9	9	9	9	9	8	9	8
10	10	10	10	10	10	10	10	10	10	10

1. natyataniarzaa 2. infobdg 3. infoitt 4. detikcom 5. sidiik 6. bem_itt 7. Kuliner_Bandung 8. hmifitt 9. johan_jojo 10. Adevaoktoveri

Based on Table 2 it can be seen that the 'edwinugroho' and 'permib' nodes can fit into the top 10, this can be due to the centrality of a node varying each test. The value of this centrality can be variable because in determining the combination path is done randomly, thus allowing the path in first test will be different from the second test.

Table 2. Rank Result Where α = 0.2.

rank	1st test	2nd test	3rd test	4th test	5th test	6th test	7th test	8th test	9th test	10th test
1	1	1	1	1	1	1	1	1	1	1
2	2	2	2	2	2	2	2	2	2	2
3	3	3	3	3	3	3	3	3	3	3
4	4	4	4	4	4	4	4	4	4	4
5	5	5	5	5	5	5	5	5	5	5
6	6	6	6	6	6	6	6	6	6	6
7	7	7	7	7	7	7	7	7	7	7
8	9	9	8	12	8	8	8	9	9	8
9	8	11	9	9	9	9	12	8	8	9
10	10	8	10	8	10	11	10	10	10	10

1. natyataniarzaa 2. infobdg 3. infoitt 4. detikcom 5. sidiik 6. bem_itt 7. Kuliner_Bandung 8. hmifitt 9. johan_jojo 10. Adevaoktoveri 11. edwinugroho 12. permib

Based on Table 3 it can be seen that only 'natyataniarzaa' and 'infobdg' nodes that do not change rank, this can happen for two reasons. First because the centrality value of the two nodes is far adrift from the other nodes, so that no node can defeat the centrality of the two nodes, and the second because all nodes have relation with the natyataniarza node so nat 'natyataniarzaa' is a strong link in the deployment information, this is what causes node 'natyataniarzaa' always in the top position.

Table 3. Rank Result Where $\alpha = 0.4$.

rank	1st test	2nd test	3rd test	4th test	5th test	6th test	7th test	8th test	9th test	10th test
1	1	1	1	1	1	1	1	1	1	1
2	2	2	2	2	2	2	2	2	2	2
3	3	3	3	3	3	3	5	3	3	3
4	4	4	4	4	5	4	3	4	4	5
5	5	6	6	5	4	5	4	5	5	4
6	6	5	5	7	7	7	6	6	6	7
7	8	7	7	6	6	6	7	8	7	6
8	7	9	11	8	11	9	9	7	9	9
9	9	8	9	12	9	13	12	9	12	8
10	10	10	10	9	10	8	11	10	14	10

1. natyataniarzaa 2. infobdg 3. infoitt 4. detikcom 5. sidiik 6. bem_itt 7. Kuliner_Bandung 8. hmifitt 9. johan_jojo 10. Adevaoktoveri 11. edwinugroho 12. permib 13. diskayunita 14. anggiamanda

Based on Table 1, 2, and 3 it can be concluded that the value of α affects the ranking results, where the greater α used the more changes in the ranking results. Changes in ranking results because of the greater the value of α then the combination of paths taken less, so the combination of each test path will be much different than the combination of paths taken more.

4.4. Result

From the previous analysis based on two angle of view that is time and result of ranking in taking value of α used, hence at this test value of α used is 0,2, where time used not too long and rank yielded stable enough. On this Twitter data is tested 30 times then averaged to get top 10 users who have the highest centrality value. Here is displayed the value of centrality, the number of follower relations and relations following each node:

Table 4. Top 10 Rank Result Twitter User

Rank	Node	Number of follower relation	Number of following relation	Centrality value
1	natyataniarzaa	232	320	1788.55
2	infobdg	252	41	293.49
3	infoitt	232	1	231.65
4	detikcom	151	0	191.17
5	sidiik	176	192	173.41
6	bem_itt	153	1	158.18

IOP Conf. Series: Journal of Physics: Conf. Series **971** (2018) 012015 doi:10.1088/1742-6596/971/1/012015

7	Kuliner_Bandung	122	0	145.57
8	hmifitt	128	12	122.65
9	johan_jojo	123	126	121.70
10	adevaoktoveri	132	134	115.89

Based on Table 4 it can be seen that the number of relations following a node has no effect on the ranking result of the node. While the number of follower relationships is quite influential on the node ranking results, but that does not mean that the node has the lowest number of follower relation ranked the lowest as well. For example, the 'Kuliner_Bandung' node has the lowest follower relation among the other 9 nodes, but the 'Kuliner_Bandung' node is not in the lowest rank. This is because the position of node 'Kuliner_Bandung' is a strong link in the dissemination of information in the network. The strong connector here means it lies between a set of nodes. To compare the location of the 'Kuliner_Bandung' node can be seen in Figure 7.

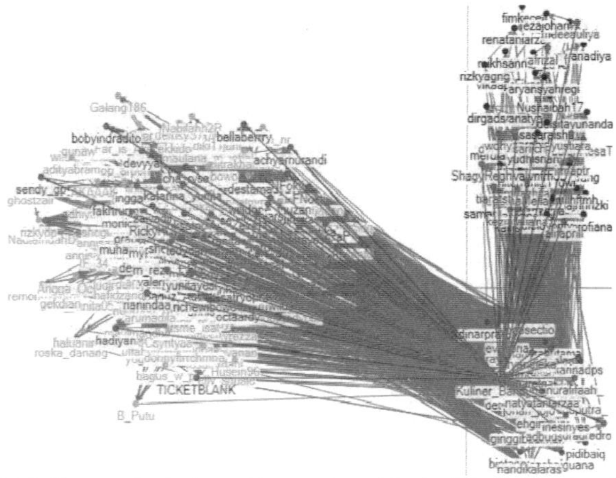

Figure 7. Follower Relations Formed on Node 'Kuliner_Bandung'.

5. Conclusion

From the results of experiments conducted found that in the calculation of centrality is influenced by the value of α, the smaller the value of α used then takes longer time calculation, but the ranking results tend to be more stable. The best α value parameter in this paper is 0.2, where the time required is not too long and the ranking result is quite stable. In addition to the number of followers, the ranking results are influenced by the user's position. The stronger the user as a liaison of a group of users the more influential the user is in the dissemination of information.

For further research can perform an accuracy analysis of k-Path centrality method, for example by kuisoner or survey. It can also implement the weighting effect on the k-Path centrality method for the weighted graph, by adding the mention and reply relations. Analyzing the influence of data characteristics, for example with the diversity of data amount or depth of graph used.

References

[1] R. Rabade, N. Mishra, S. Sharma. (2014). "Survey of Influential User Identification Techniques in Online Social Networks". Rajiv Gandhi Proudyogiki Vishwavidyalaya: School of Information Technology.

[2] Hopkins, Adelaide. (2010). *Graph Theory, Social Networks and Counter Terrorism.* University of Massachusetts: Department of Mathematics.

[3] K. Hildrun and K. Theo. (2010). "A New Centrality Coefficient for Social network Analysis Applicable to Bibliometric and Webometric Data". Humboldt-University Berlin: Department of Library and Information Science.

[4] E. Santos, E. D. Sotelino, Y. Cao, E. Brown, E. Santos. (2007). "Effective and Efficient Methodologies for Social Network Analysis".

[5] A. Tharaka, T. Rahul, K. Nicolas, S. Ramanuju, I. Andrian. (2010). "K-Path Centrality: A New Centrality Measure in Social Network". University of South Florida.
[6] Scoot, John. (2011). "Social network Analysis Theory and Application".
[7] E. Otte and R. Rosseau. (2002). "Social Network Analysis: a Powerful Strategy, also for the Information Science".
[8] Alohakoo, Tharaka. (2010). "Path centrality: A new centrality measure in networks". University of South Florida.
[9] Maharani, W., Adiwijaya and Gozali, A.A., 2014, October. Degree centrality and eigenvector centrality in twitter. In Telecommunication Systems Services and Applications (TSSA), 2014 8th International Conference on (pp. 1-5). IEEE.
[10] Rachman, Z.A. Maharani, W. and Adiwijaya, 2013, March. The analysis and implementation of degree centrality in weighted graph in Social Network Analysis. In Information and Communication Technology (ICoICT), 2013 International Conference of (pp. 72-76). IEEE.

An implementation of Elman neural network for polycystic ovary classification based on ultrasound images

I F Thufailah[1], Adiwijaya[2], U N Wisesty[3] and Jondri[4]
[1,2,3,4] School of Computing, Telkom University, Telekomunikasi 1 St, West Java, ID.

E-mail: vlouchi@student.telkomuniversity.ac.id[1], adiwijaya@telkomuniversity.ac.id[2], untarinw@telkomuniversity.ac.id[3], jondri@telkomuniversity.ac.id[4]

Abstract. Polycystic Ovary Syndrome (PCOS) is a reproduction problem that causes irregular menstruation period. Insulin and androgen hormone have big roles for this problem. This syndrome should be detected shortly, since it is able to cause a more serious disease, such as cardiovascular, diabetes, and obesity. The detection of this syndrome is done by analyzing ovary morphology and hormone test. However, the more economical way of test is by identifying the ovary morphology using ultrasonography. To classify whether one ovary is normal or it has polycystic ovary (PCO) follicle, the analysis will be done manually by a gynecologist. This paper will design a system to detect PCO using Gabor Wavelet method for feature extraction and Elman Neural Network is used to classify PCO and non-PCO. Elman Neural Network is chosen because it contains context layer to recall the previous condition. This paper compared the accuracy and process time of each dataset, then also did testing on elman's parameters, such as layer delay, hidden layer, and training function. Based on tests done in this paper, the most accurate number is 78.1% with 32 features.

1. Introduction

Polycystic Ovary Syndrome (PCOS) is a reproduction problem that causes irregular menstruation cycle of woman in her reproductive age [1]. The main cause of PCOS is still uncertain. However, the resistance of insulin and androgen has roles for this problem [2]. A patient can be declared of having PCOS if she acquires two of three diagnostic criteria established in Rotterdam Conference: (1) Oligo/an ovulation, (2) clinical or biochemical symptoms from having too much androgen, (3) ovary with polycystic cyst [3]. From morphological point of view, an ovary is assumed as polycystic ovary if there are, at least, 12 follicles with 2-9 mm of diameter, or there is an increase of ovary volume that exceeded 10 cm^3.

PCOS is not only causing irregular menstruation cycle, but, it is also causing a sterile (unable to have child) in woman, cardiovascular disease, diabetes, and obesity [2]. To be able to detect whether a woman is having PCOS or not, there are two kinds of tests; hormonal test and ovary morphological test. However, from economical point of view, the cost of ovary morphology is cheaper because it uses ultrasonography equipment. Later, the ultrasound image will be analyzed by gynecologist. The analysis of ultrasound image is done manually by gynecologist. It takes time, for sure. Figure 1 is an ultrasound image that has been classified into polycystic

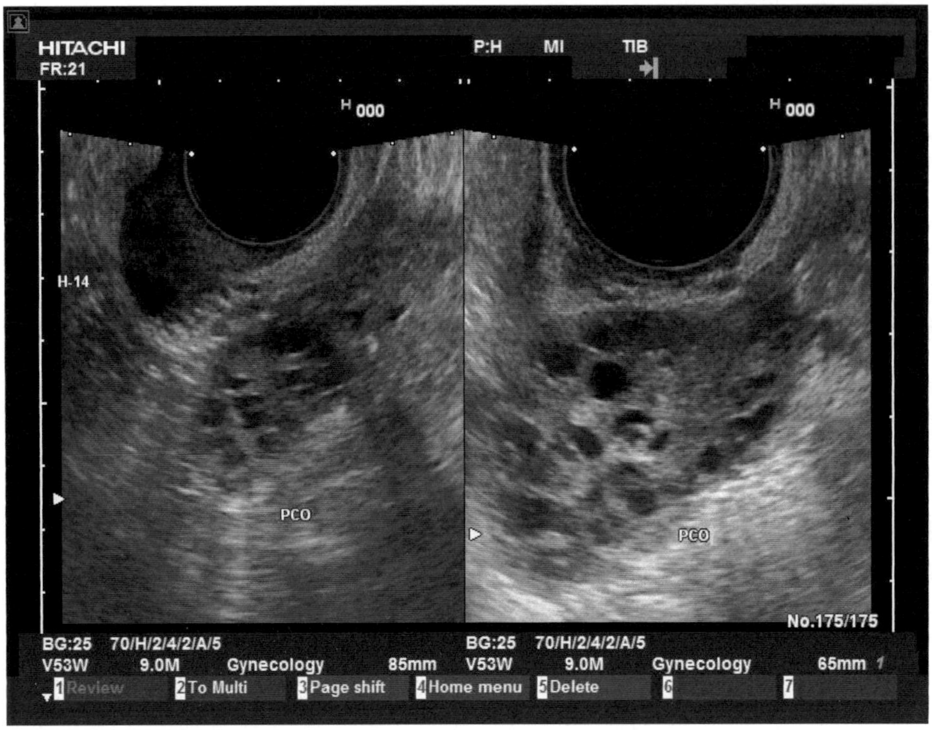

Figure 1. Ultrasound Image

ovary by gynecologist . If there are numbers of images, and the result must be handed to the patient soon, it will influence the thoroughness and accuracy of the gynecologist in analyzing the images. For that reason, a system has been built to help the detection of PCO by using Gabor Wavelet method for feature extraction and Elman Neural Network for classification.

2. Related Works

In this paper [4] Setiawati et al. used Particle Swarm Optimization to cluster ultrasound image. This paper proposed a non-parametric fitness function to improve the convergence solution of PSO clustering. Using contrast enchancement was expected to increase the extracted follicles size as close as prossible as the real follicles size. This paper also suggested to use machine learning approach to classify the follicle.

Meanwhile Purnama et al. [5] proposed preprocessing, segmentation, feature extraction, and classification as his detection polycystic method. He used gabor wavelet as the segmentation method and three machine learning approaches: Neural Network-Learning Vector Quantization (LVQ), KNN - Euclidean distance, and Support Vector machine (SVM) - RBF Kernel. The best accuracy gained from SVM-RBF Kernel on C was 40.

Wisesty et al. stated [6] there were two approaches to detect PCO follicles: 1) streteolgy, and 2) feature extraction and classification. Wisesty himself used the second approach. He then used gabor wavelet as the extraction method and modified backpropagation as the classification method. His modified backpropagation were called Lavenberg-Marquadt optimization and COnjugate Gradient-Fletcher Reeves, which is able to improve the convergance rate.The best accuration reached using Levenberg-Marquardt was 93.925%, which was gained from 33 neurons and 16 vector features.

3. Proposed Scheme

Data used in this paper is an ultrasound image (Example: Figure 1) that come from Permata Bunda Syariah Maternity Clinic Cirebon. This data is already validated by a gynecologist. Based on the Figure 2 ultrasound image data will go through several processes:

(1) Preprocessing
(2) Segmentation
(3) Feature Extraction
(4) Classification.

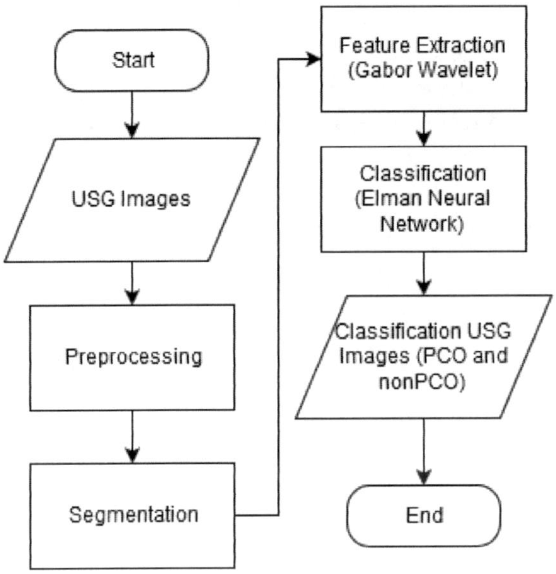

Figure 2. Flowchart of System Design.

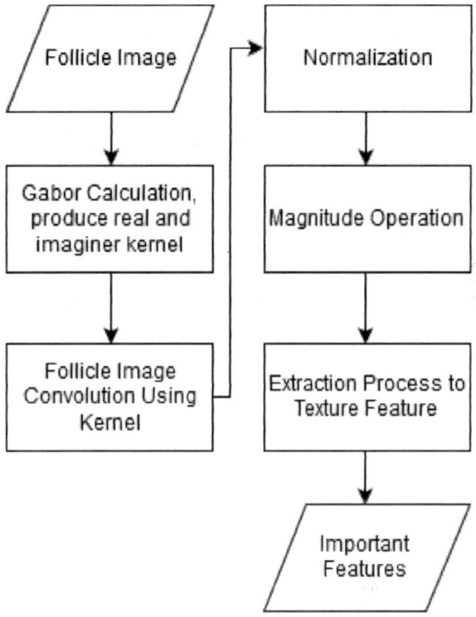

Figure 3. Flowchart of Gabor Wavelet

3.1. Preprocessing

In this preprocessing stage there are several stages such as grayscalling, histogram equalization, image binarization, morphology, invert image, and data cleaning. Grayscalling is aimed to change the images color from RGB to gray. Histogram equalization is to evenly distribute the number of pixels that is owned by every degree of grayish color in ultrasound image. Image binarization is a process to change the color of image to be only two colors, 0 for black, and 1 for white. In preprocessing phase, there is also process of morphology; erosion and dilation. Erosion is a process to reduce objects color in the image, so noise in that image is decreased. Dilation is process to improve objects color in the image. Invert image is process to flip the color of image; from black to white, and vice versa. It is important because white object is easier to detect. And the final process of preprocessing is data cleaning. Data cleaning is aimed to omit unimportant object.

3.2. Segmentation

Segmentation is a phase to separate object from background. This phase has two processes, they are: edge detection and, follicle cutting. Edge detection is used to recognize object in the image. This paper uses canny operator for edge detection. After follicle in the ultrasound image is detected, next, it is labeled and cut into some different images and shows each follicle.

3.3. Feature extraction
Feature extraction is a process to take information from an object in the image to be recognized and differentiated from other object. This paper uses Gabor Wavelet method. Result of Gabor wavelet method is feature vector of each follicle. Figure 5 shows the block diagram of the feature extraction process.

3.4. Classification
Elman Neural Network was invented by Jeff Elman. Elman Neural Networks is built using a feedforward network architecture and adds connections to the previous layer (extra nodes). This extra node contains a representation of the contents of one layer that existed when the pattern was previously trained [7]. This extra set is called a network context unit. Figure 4 is the architecture of Elman Neural Network, while Figure 5 is a process flow. The learning process is following [8]:

(1) the context unit is set to 0; $t = 1$;

(2) pattern x^t clamped, the forward calculations are performed once;

(3) the backpropagation learning rule is applied;

(4) $t \leftarrow t + 1$; go to 2;

The context unit in step t thus always has the hidden unit activation value in step t-1.

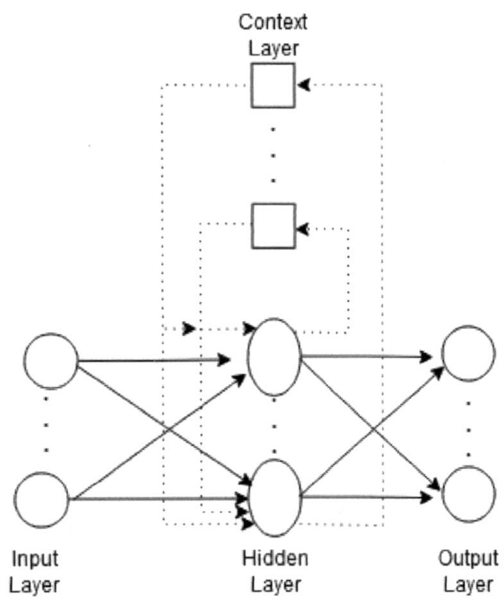

Figure 4. Structure of Elman Neural Network [9].

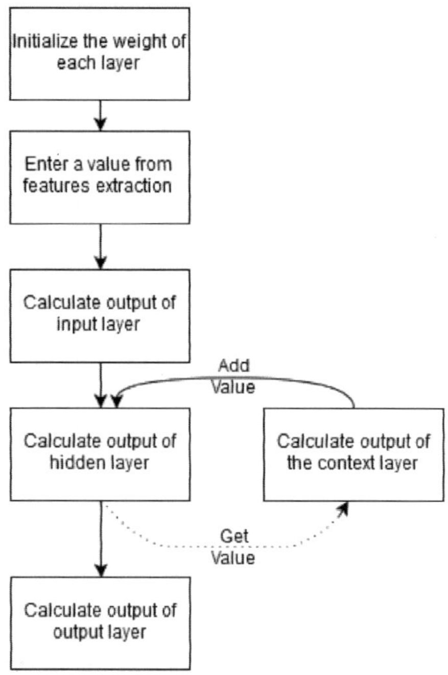

Figure 5. Flowchart Elman Neural Network

4. Result and Discussion

4.1. Accuracy and elman neural network process duration

In this paper, extraction results of Gabor Wavelet used are 16, 24, and 32 feature. This feature are classified using Elman Neural Network. There are 3 different parameters that are tested in that feature set. The parameters are layer delays, hidden sizes, and training function. Figure 6 shows that accuracy of Set data 16 feature is 68.8%, accuracy of Set data 24 feature is 66.7% and accuracy of 78.1% for set data with 32 features.

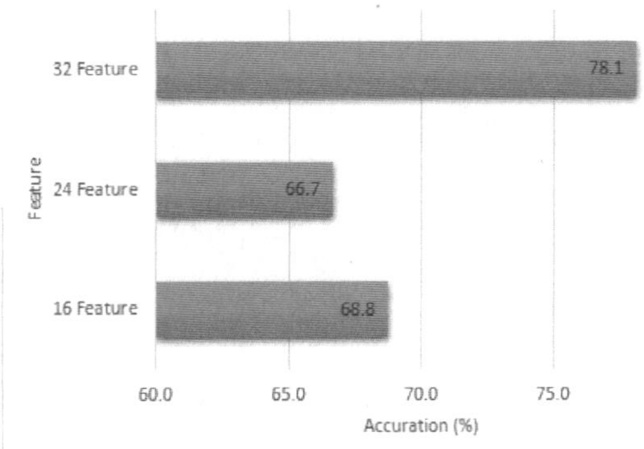

Figure 6. Accuracy of Elman Neural Network

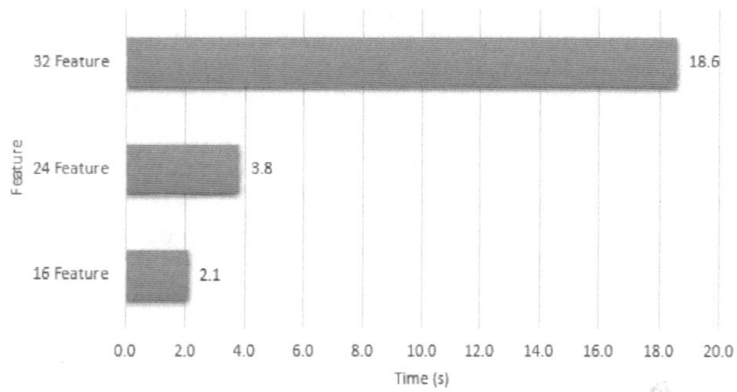

Figure 7. Time Process of Elman Neural Network

Figure 7 shows a comparison of process duration of Elman Neural Network for feature in data 16, 24 and 32. The longest duration is data 32. It is because numbers of data processed are more than the others data set.

4.2. Parameter layer delay test

This test is aimed to analyze the influence of parameter layer delay parameter to data accuracy that is trained and tested by Elman Neural 1:2, 1:3, 1:4, 1:5. The following graph shows that accuracy from every layer delay variation in the data with 16, 24 and 32 features. There are four layer delay variations, they are 1:2, 1:3, 1:4, 1:5. The following graph shows that accuracy in every layer delay variation in the data with 16, 24, and 32 features. Figure 8 shows that the

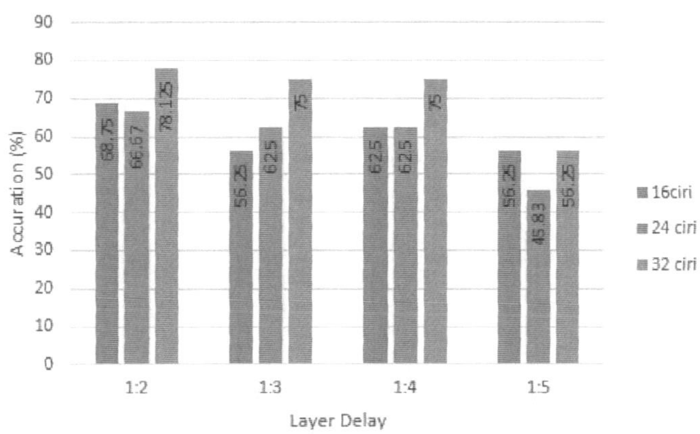

Figure 8. Comparison of Accuracy and Layer Delay Variation

biggest accuracy of layer delay 1:2 is 78.125% for 32 data features. In delay of 1:3, the biggest accuracy is 75% for 32 data features. But, for layer delay 1:5, the highest accuracy is 56.25% for 16 and 32 data features. From the result, layer delay variation has influence level of data accuracy using Elman Neural Network.

4.3. Parameter hidden layer test

This test aims to analyze the effect of hidden layer parameters on the accuracy of the data being trained and tested with Elman Neural Network. There are 5 variations of hidden layer parameters that are 10, 15, 20, 25, 30. The graph below shows the accuracy of each variation of the hidden layer on the data of 16 characteristics, 24 characteristics, and 32 features. The number of hidden layers gives little change to the accuracy of each data. For the highest hidden layer 10 accuracy of 78.125%, the greatest accuracy with 15 hidden layers is 78.125%, for 20 and 25 the largest hidden layer accuracy is 75% and for 30 hidden layers is 78.125%. All of the greatest accuracy is for data with 32 features. From the results of this test is known that the number of hidden layers affect the accuracy of the Elman Neural Network.

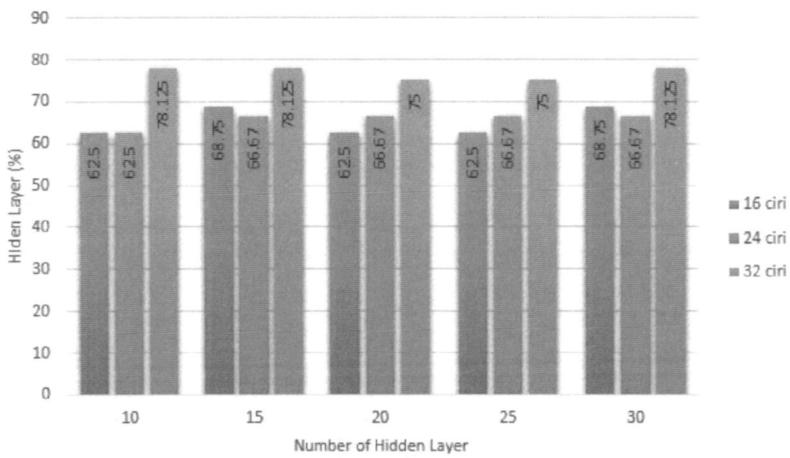

Figure 9. Comparison of Accuracy and Hidden Layer

4.4. Test of training function parameter

This test is aimed to analyze the influence of training function parameter to data accuracy that are trained and tested using Elman Neural Network. There are 2 training function, Lavenberg Marquardt and BFGS quasi-Newton. Figure 10 shows that accuracy from every hidden layer variation, and layer delay in data with 16, 24, and 32 features to training function type, have the same accuracy.

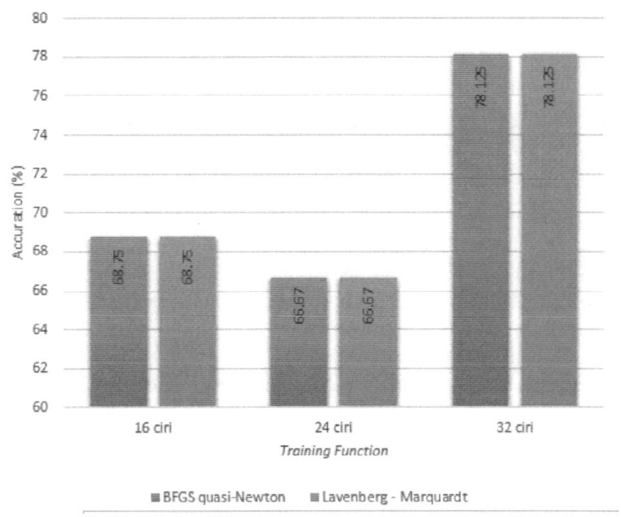

Figure 10. Comparison between Accuracy and Training Function Type

4.5. Result classification elman neural network and other method

Table 1. Result Classification

Features	Method	Accuration (%)	Time (s)
16	SVM Linear	83.2	0.4
	SVM Polynomial	100.0	0.2
	SVM RBF	99.5	0.1
	Elman Neural Network	68.8	2.1
24	SVM Linear	85.5	0.2
	SVM Polynomial	100.0	0.1
	SVM RBF	100.0	0.1
	Elman Neural Network	66.7	3.8
32	SVM Linear	83.6	0.5
	SVM Polynomial	100.0	0.4
	SVM RBF	99.1	0.1
	Elman Neural Network	78.1	18.6

Table 1 shows the result classification from ovaries USG using Elman Neural Network and compared with Support Vector Machines. Based on table 1 shows that best classification is using SVM Polynomial with accuration is 100% and process time is 0.2 second for data with 16 features. For Support Vector Machine processing time required is much shorter than Elman Network. The SVM Process takes time in range 0 to 1 second, while Elman Neural Network process time reaches 18 second. This happens because on Elman Neural Network had hidden layers and layers delay.

5. Conclusion

Elman Neural Network can be used to classify ovary problem. The best accuration (78.1%) using Elman Neural Network when using data 32 features. Using more data for training, the accuration increase proportionaly. But more data for training will affect the time process. For data set with 32 feature the process duration is 18.6 second. The number of hidden layers gives little change to the accuracy of each data set. The training function do not give any effect on accuracy. Nevertheless, SVM has a better performance than Elman for PCO classification based on feature which is obtained by Gabor Wavelet.

References

[1] Sinclair C 2010 *Jakarta: EGC*

[2] Lawrence M J, Eramian M G, Pierson R A and Neufeld E 2007 Computer assisted detection of polycystic ovary morphology in ultrasound images *Fourth Canadian Conference on Computer and Robot Vision (CRV 2007), 28-30 May 2007, Montreal, Quebec, Canada* pp 105–112 URL `https://doi.org/10.1109/CRV.2007.18`

[3] National Institutes of Health. 2012. Evidence-based Methodology Workshop on Polycystic Ovary Syndrome. U.S. Department of Health and Human Services

[4] Setiawati E, Adiwijaya and Tjokorda A 2015 Particle swarm optimization on follicles segmentation to support pcos detection *Information and Communication Technology (ICoICT), 2015 3rd International Conference on* (IEEE) pp 369–374

[5] Adiwijaya, Purnama B, Hasyim A, Septiani M, Wisesty U and Astuti W 2015 Follicle detection on the usg images to support determination of polycystic ovary syndrome *Journal of Physics: Conference Series* vol 622 (IOP Publishing) p 012027

[6] Wisesty U N, Nasri J and Adiwijaya 2016 Modified backpropagation algorithm for polycystic ovary syndrome detection based on ultrasound images *Recent Advances on Soft Computing and Data Mining - The Second International Conference on Soft Computing and Data Mining (SCDM-2016), Bandung, Indonesia, August 18-20, 2016, Proceedings* pp 141–151 URL `https://doi.org/10.1007/978-3-319-51281-5`$_1$5

[7] Balkin S D 1997 Using recurrent neural networks for time series forecasting *International Symposium on Forecasting,Barbados*

[8] Kröse B and van der Smagt P 1993 *An introduction to neural networks* (The University of Amsterdam)

[9] Sundaram N M, Sivanandam S and Subha R 2016 *International Journal of Applied Engineering Research* **11** 1835–1840

Predicting Jakarta composite index using hybrid of fuzzy time series and support vector regression models

Rian Febrian Umbara, Dede Tarwidi, Erwin Budi Setiawan

Telkom University, Bandung, Indonesia

rianum@telkomuniversity.ac.id, erwinbudisetiawan@telkomuniversity.ac.id, dedetarwidi@telkomuniversity.ac.id

Abstract. The paper discusses the prediction of Jakarta Composite Index (JCI) in Indonesia Stock Exchange. The study is based on JCI historical data for 1286 days to predict the value of JCI one day ahead. This paper proposes predictions done in two stages., The first stage using Fuzzy Time Series (FTS) to predict values of ten technical indicators, and the second stage using Support Vector Regression (SVR) to predict the value of JCI one day ahead, resulting in a hybrid prediction model FTS-SVR. The performance of this combined prediction model is compared with the performance of the single stage prediction model using SVR only. Ten technical indicators are used as input for each model.

1. Introduction

Stock price index is an indicator or a reflection of stock price movements. Index is one of the guidelines for investors to invest capital in the capital market, especially stocks. Stock indices derived from stock prices with high market capitalization can provide an overall picture of the economy and depend on several factors [1].

Predicting stock price index and movement has been considered one of the most challenging applications of time series prediction. While there are many empirical studies dealing with issues in predicting stock price indices, most of the research is done in countries with advanced financial markets. An accurate prediction of stock price index movements is essential for developing an effective market strategy [2]. Thus, investors can minimize the market risk and the opportunity to make a profit with stock index trading [3]. Predicting stock and derivatives prices have been done by many scholars around the world. Patel, Shah, Thakkar, & Kotecha predict the stock price index CNX Nifty and S & P Bombay Stock Exchange (BSE) using SVR-ANN, SVR-RF, SVR-SVR algorithms [1]. While C.P. Lee, W.C. Lin, and C.C Yang [4] used fuzzy time series combined with least square support vector regression and bootstrap method to forecast option prices.

In this study, prediction of Jakarta Composite Index (JCI) is done in two stages. Predictions done in two stages are expected to result in smaller errors than the one-stage prediction. The first stage of the prediction is to calculate ten technical indicators, which are commonly used by analysts to analyze stock price movements, based on JCI data at time t. Next we will use the Fuzzy Time Series method to predict the value of the ten indicators at time t + 1. The predicted value of the ten indicators will then be used to predict the JCI closing value at time t + 1 using the Support Vector Regression method. Support Vector Regression is the most widely used algorithm to predict stock prices and stock price indices proposed by Vladimir N Vapnick in 1996.

2. Construction of Prediction Model

The prediction model to be constructed is a hybrid of Fuzzy Time Series model and Support Vector Regression (FTS-SVR). The prediction process involved 10 technical indicators. Here's an explanation of Fuzzy Time Series, Support Vector Regression, and 10 Technical Indicators..

2.1. Fuzzy Time Series

Fuzzy time series is one of soft computing methods that has been used and applied in time series data analysis. This method was proposed by Song and Chissom. The main purpose of fuzzy time series is to predict time series data that can be widely used in any real time data, including capital market data.

In the last few decades, research on time series has progressed in handling predictions. But in real life, people tend to face many random fuzzy sequences that contain noise or disturbance. Predictions based on traditional time series data do not seem to be able to deal with this problem. But fuzzy mathematics has an advantage in solving this problem. Therefore, Song and Chissom introduced the concept of fuzzy mathematics into time series data and proposed a fuzzy time series concept.

Let ***U*** universe of discourse, where $U = \{u_1, u_2, \dots, u_n\}$. A fuzzy set defined in ***U*** can be represented as follow: $A = f_A(u_1)/ u_1 + f_A(u_2) / u_2 + \dots + f_A(u_n)/u_n$, where f_A is a membership function of the fuzzy set *A*, $f_a: U \rightarrow [0,1]$, and $f_A(u_i)$ represents the membership degree of u_i belongs to fuzzy set A, $f_A(u_i) \in [0,1], and\ 1 \le i \le n$.

Here are some steps to construct Fuzzy Time Series model [5]:

1. Define the scope of the universe on the basis of data over a predetermined period of time. Let D_max and D_min be the maximum and minimum values of the data value, then the interval of the universal set is U = $[D_{min}, D_{max}]$.
2. Divide U into several intervals $\{u_1, u_2, \dots, u_n\}$ with the same length.
3. Define fuzzy sets A_t based on the interval formed and the time series data is fuzzified. The fuzzy set A_t defines the linguistic variables of the time series data..
4. Identify fuzzy relationships using fuzzyfication data. If the time series variable *F*(*t*-1) is fuzzyfied to A_k and *F*(*t*) becomes A_m, then A_k has a relationship with A_m. Denote the relation with $A_k \rightarrow A_m$, where A_k is the *current state* of the original value dan A_m is the next *current state*. The entire fuzzy relation will be saved and if there is a repetition of the same fuzzy relation more than once, the same fuzzy relation is stored only once.
5. Insert fuzzy relation into fuzzy relation group. If the fuzzy set relates to more than one set, then another relation is placed on the right of the other set. For example, A_1 relates to A_1, dan A_2, then the resulting fuzzy relation group is $A_1 \rightarrow A_1, A_2$. The results are entered and stored into the ith fuzzy relation group.
6. Defuzzyfy the prediction results with rules:
   a. If there is a one-to-one relationship within the group from A_j, such as $A_j \rightarrow A_k$, and the highest membership degree of A_k is in the interval u_k, then the prediction result from F(t) is the middle value of the interval u_k.
   b. If there is an empty relation, such as A_j than has no relation, or $A_j \rightarrow \emptyset$, and the highest membership degree of lies between the interval ranges u_j, then the prediction result obtained is the middle value of u_j.
   c. If there is a one-to-many relation within the relation group of A_j, such as $A_j \rightarrow A_1, A_2, \dots, A_n$, and the highest membership degrees lies in the intervals $u_1, u_2, \dots, u_n$, then the prediction result obtained is the average of the middle values $m_1, m_2, \dots, m_n$ of the intervals $u_1, u_2, \dots, u_n$, that is $\frac{m_1 + m_2 + \dots + m_n}{n}$.

2.2. *Support Vector Regression*

Support Vector Regression (SVR) is a method first introduced by Vladimir N. Vapnik, Harris Drucker, Christopher JC Burges, Linda Kaufman and Alexander J. Smola in 1996. The model introduces Support Vector Classification which relies on a subset of training data. SVR is an application of Upport Vector Machine (SVM) used for regression cases, whose output is a real or continuous number. SVR is a method that can overcome the overfitting, so it will produce a good performance with small error [6]. Overfitting is a condition in which a model does not describe the main relationship between input and output variables but rather describes random error or noise. This condition will result in poor predictions. SVR builds the linear regression function equation as follows

$$\boldsymbol{f(x,w) = w^T x + b} \quad (1)$$

Vapnik creates a formula to minimize the function of the linear Ɛ-insensitivity loss function, that is

$$|y - f(x,w)|_\varepsilon = \begin{cases} 0, & jika\ |y - f(x,w)| \le \varepsilon \\ |y - f(x_i,w)| - \varepsilon, & lainnya \end{cases} \quad (2)$$

Based on the above equation, the linear regression of $f(x,w)$ is estimated by minimizing $||w||^2$ and the sum of linear ε-insensitivity losses. Where the factor $||w||^2$ is called regularization. Minimizing $||w||^2$ will make the function as small as possible, so it can control the function of capacity.

$$R = \frac{1}{2}||w||^2 + c(\textstyle\sum_{i=1}^{m} |y - f(x,w)|_\varepsilon \quad (3)$$

$$R = \frac{1}{2}||w||^2 + c\textstyle\sum_{i=1}^{m}(\xi + \xi^*) \quad (4)$$

$$(w^T x_i + b) - y_i \le \varepsilon + \xi_i \quad (5)$$

$$y_i - (w^T x_i + b) - y_i \le \varepsilon + \xi_i \quad (6)$$

$$\xi_i, \xi_i^* \ge 0, i = 1,2, \ldots, m \quad (7)$$

where ξ_i dan ξ_i^* are slack variables, one to exceed the target value by more than ε and other tfor being more than ε below the target. The value of c serves to minimize risk. It controls a trade-off between an approximation error and the norm of weight vector ||w||.

For optimization problems it uses langrange theory and Karush-Kuhn-Tucker condition to get the desired regression function weights. SVR can predict the stock price index value at the time of closing by using the kernel assistance and 10 technical indicators as input. The kernel function used is

$$RadialBasisFunction \quad : K(x_i, x_j) = exp(-\gamma||x_i \cdot x_j||^2) \quad (8)$$

where γ is constant of radial basis function

2.3. Technical Indicators

Ten technical indicators are some indicators or basis for calculating stock index movement [7]. Here are the indicators

Table 1. Ten technical indicators

No	Indicators	Formula
1.	Simple n-day moving average (SMA)	$\frac{C_t + C_{t-1} + \cdots + C_{t-(n-1)}}{n}$
2.	Weighted moving average (WMA)	$\frac{(n)C_t + (n-1)C_{t-1} + \cdots + (1)C_{t-(n-1)}}{n + (n-1) + \cdots + 1}$
3.	Stochastic K% (STK)	$\frac{C_t - LL_{t-(n-1)}}{HH_{t-(n-1)} - LL_{t-(n-1)}} x100$
4.	Stochastik D% (STD)	$\frac{\sum_{i=0}^{n-1} K_{t-i}}{10}\%$

5.	Relative Strength Index (RSI)	$100 - \frac{100}{1 + (\sum_{i=0}^{n-1} \frac{UP_{t-i}}{n}) / (\sum_{i=0}^{n-1} \frac{DW_{t-i}}{n})}$
6.	Moving average convergence divergence (MACD)	$MACD(n)_{t-1} + \frac{2}{n+1} \times (DIFF_t - MACD(n)$
7.	Commodity channel index (CCI)	$\frac{M_t - SM_t}{0.015D_t}$
8.	Larry William's R% (LWR)	$\frac{H_t - C_t}{H_t - L_t} \times 100$
9.	A/D oscillator (AD)	$\frac{H_t - C_{t-1}}{H_t - L_t}$
10.	Momentum (MOM)	$C_t - C_{t(n-1)}$

where :

1. C_t = close price of the stock index at time t
2. L_t = lowest price of the stock index at time t
3. H_t = highest price of the stock index at time t
4. LL_t = lowest price of the stock index in time inteval t
5. HH_t = highest price of the stock index in time interval t
6. $DIFF_t$ = $EMA(12)_t - EMA(26)_t$
7. $EMA(k)_t$ = $EMA(k)_{t-1} + \alpha \times (C_t - EMA(k)_{t-1})$
8. α = $\frac{2}{k+1}$
9. k = k-days period for EMA
10. M_t = $\frac{H_t+L_t+C_t}{3}$
11. SM_t = $\frac{(\sum_{i=1}^{n} M_{t-i+1})}{3}$
12. DM_t = $\frac{(\sum_{i=1}^{n} |M_{t-i+1} - SM_t|)}{n}$
13. UP_t = the increase of the stock price index at time t
14. DW_t = the decrease of the stock price index at time t
15. n = number of stocks

2.4. The Proposed Prediction Model

The proposed prediction model is a hybrid model of FTS-SVR, which performs predictions in two stages. The first stage is to predict the value of 10 technical indicators at time t + 1 based on data of 10 technical indicators at time t using Fuzzy Time series. The second stage is to predict the closing value of JCI at time t + 1 based on the prediction of 10 technical indicators generated in the first stage by using Support Vector Regression. The performance of the proposed model will then be compared to its performance with a one-stage prediction model, which predicts the value of JCI at time t + 1 based on 10 technical indicators data at time t by using Support Vector Regression.

3. Prediction Results

Before the prediction model of FTS-SVR was established, JCI closing value data was collected from January 12, 2012 until May 19, 2017. The closing value data was then used to calculate the value of 10 technical indicators per day using the formulas listed in Table 1. Thus we obtained daily value of 10 technical indicators starting from March 1, 2012 to May 19, 2017. In the first stage, each technical indicator is then divided into 105 intervals to form fuzzy time series models. The formation of Fuzzy Time Series models for each technical indicator is performed using the historical data of each indicator from March 1, 2012 until April 21, 2016. So we get 10 models of Fuzzy Time Series for each indicator. The models obtained are then used to predict the value of 10 technical indicators one day ahead (t + 1) starting from April 22, 2016 to May 19, 2017. The performance of the models is measured by calculating prediction errors, using Mean Absolute Percentage Error (MAPE) and Root Mean Square Error (RMSE), with the following formulas

$$MAPE = \frac{1}{n}\sum_{t=1}^{n}\frac{|A_t - F_t|}{|A_t|} \times 100 \tag{9}$$

$$RMSE = \left(\sum_{i=1}^{n}(A_t - F_t)^2/n\right)^{1/2} \tag{10}$$

where:

A_t= actual value of the stock price index at time t
F_t= predicted value of the stock price index at time t
n = number of data
t = time index

Here are the RMSE and MAPE values from the predictions of each technical indicator using Fuzzy Time Series.

Table 2. Error prediction of each technical indicator at time t + 1 with Fuzzy Time Series

No	Prediction of Indicators	MAPE	RMSE
1.	Prediction of SMA	0.8804	88.456
2.	Prediction of WMA	0.7701	74.86
3.	Prediction of STK	86.34	19.18
4.	Prediction of STD	7.5	4.77
5.	Prediction of RSI	311.191	4173.403
6	Prediction of MACD	67.95	5.54
7.	Prediction of CCI	223.76	66.25
8.	Prediction of LWR	157.763	29.49
9	Prediction of AD	384.736	0.53
10	Prediction of MOM	298.1	61.07

In the second stage, the prediction result of 10 technical indicators at t + 1 then becomes the input for SVR process to predict the JCI closing value at t + 1. The predicted indicators values are then divided into two groups, training data from April 22, 2016 until February 13, 2017 (200 business days), and testing data from February 14, 2017 to February 13, 2017 (62 business days). The kernel used in SVR is a radial basis function with parameter values C = 91556.41, ε = 6.0026, and $\gamma = 740.1296$. Prediction errors for training and testing are as follows.

Table 3. Prediction errors of the JCI closing value

No	Process	MAPE	RMSE
1.	Training	0.459	37.047
2.	Testing	2.282	165.393

The following figures show the comparison between the actual JCI closing values and the predicted JCI closing values, for both training and testing data.

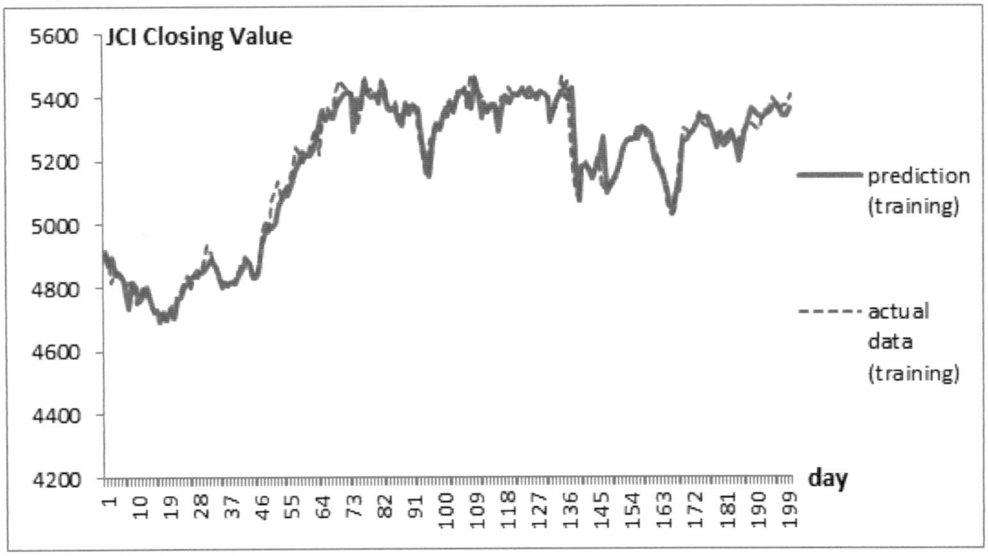

Figure 1. Comparison of FTS-SVR model prediction results with actual data for training data prediction

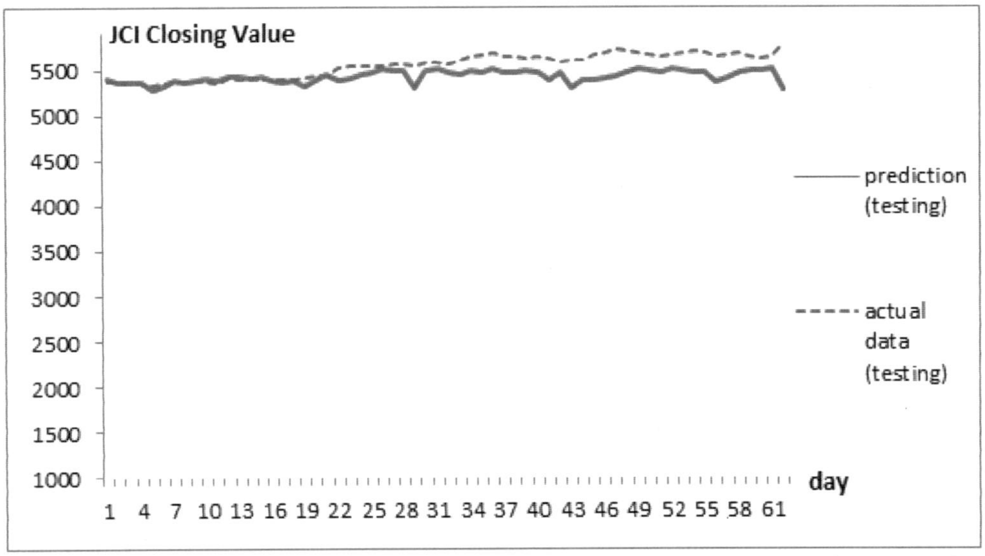

Figure 2. Comparison of FTS-SVR model prediction results with actual data for testing data prediction

JCI closing price prediction results from hybrid model of FTS-SVR are then compared with prediction results from one-stage prediction model using SVR. As the inputs are the actual 10 technical indicators at time t, and the output is the prediction of JCI closing value at t + 1. Training data are the actual values of the ten indicators from April 22, 2016 to February 13, 2017 (200 working days) and testing data are the actual values of the ten indicators from February 14, 2017 to February 13, 2017 (62 working days). Figures 3 and 4 show a comparison of the results of the close price predictions with their actual values for training and testing.

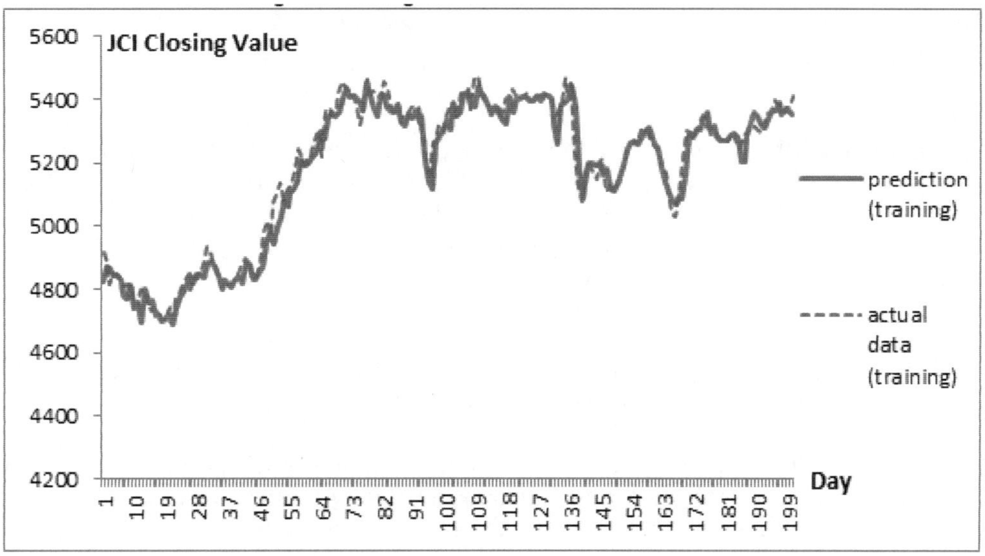

Figure 3. Comparison of SVR model prediction results with actual data for training data prediction

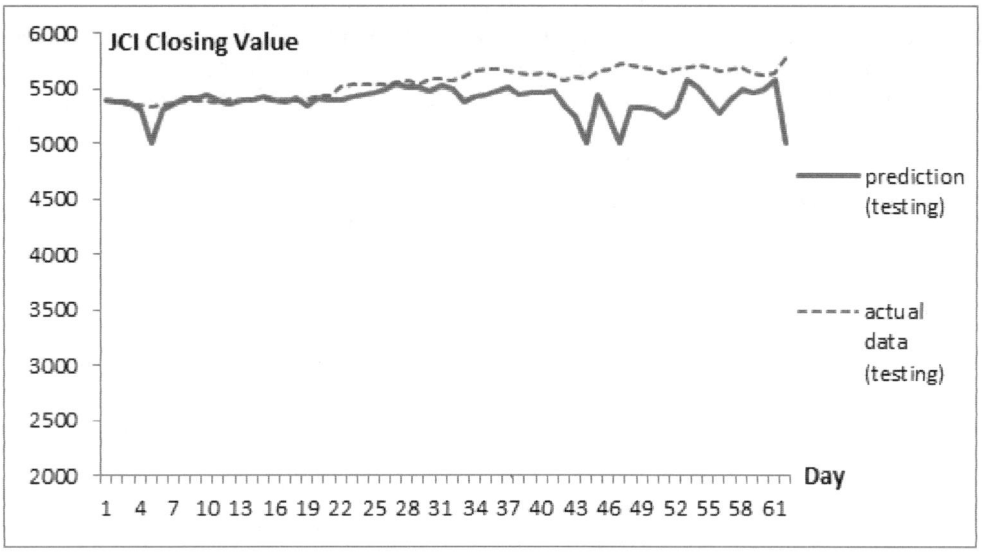

Figure 4. Comparison of SVR model prediction results with actual data for testing data prediction

Table 4 shows the comparison of prediction errors from the FTS-SVR model and the SVR model.

Table 4. Comparison of MAPE and RMSE between predicted values resulted from FTS-SVR model and SVR model

Prediction Model	MAPE for training data prediction	MAPE for testing data prediction	RMSE for training data prediction	RMSE for testing data prediction
FTS-SVR	0.459561173	2.281763222	37.0468724	165.3924739
SVR	0.524795171	3.003133976	41.6945237	242.4911956

It can be seen that the error value (MAPE and RMSE) of predicted JCI closing values resulted from FTS-SVR model is smaller than those of SVR model.

4. Conclusions

The results have shown that the hybrid FTS-SVR model gives a better prediction of JCI closing value than the SVR model. This can be seen from the MAPE and RMSE values of the FTS-SVR models which are smaller than those of the SVR model, both for training data and testing data. For future work, fuzzy time series method can be combined with optimization algorithm, such as genetic algorithm, to get optimal intervals for prediction.

References

[1] Patel, Jigar., Shah, Sahil., Thakkar, Priyank., & Kotecha, K. (2015). Predicting stock market index using fusion of machine learning techniques. Expert Systems with Applications 42 (2015) 2162-2172.

[2] Leung, M. T., Daouk, H., & Chen, A. S. (2000). Forecasting stock indices: A comparison of classification and level estimation models. International Journal of Forecasting, 16, 173–190.

[3] Manish, K., & Thenmozhi, M. (2005). Forecasting stock index movement: A comparison of support vector machines and random forest. In Proceedings of ninth Indian institute of capital markets conference, Mumbai, India.

[4] Lee, C. P., Lin, W. C., & Yang, C. C. (2014). A strategy for forecasting option prices using fuzzy time series and least square support vector regression with a bootstrap model. Scientia Iranica. Transaction D, Computer Science & Engineering, Electrical, 21(3), 815.

[5] Jens Runi Poulsen. (2009). Fuzzy Time Series Forecasting. Aalborg University Esbjerg.

[6] Parrella. F. (2007). Online support vector regression (Master's thesis). Italy: Department of Information Science, University of Genoa.

[7] Pring, Martin.(2001), Introduction to Technical Analysis. McGraw-Hill. Singapore.

IOP Conf. Series: Journal of Physics: Conf. Series **971** (2018) 012018 doi:10.1088/1742-6596/971/1/012018

Leukemia and colon tumor detection based on microarray data classification using momentum backpropagation and genetic algorithm as a feature selection method

Untari N Wisesty[1], Riris S Warastri[2] and Shinta Y Puspitasari[3]

[1,2,3] School of Computing, Telkom University, Bandung, Indonesia
[1] untarinw@telkomuniversity.ac.id

Abstract. Cancer is one of the major causes of mordibility and mortality problems in the worldwide. Therefore, the need of a system that can analyze and identify a person suffering from a cancer by using microarray data derived from the patient's Deoxyribonucleic Acid (DNA). But on microarray data has thousands of attributes, thus making the challenges in data processing. This is often referred to as the curse of dimensionality. Therefore, in this study built a system capable of detecting a patient whether contracted cancer or not. The algorithm used is Genetic Algorithm as feature selection and Momentum Backpropagation Neural Network as a classification method, with data used from the Kent Ridge Bio-medical Dataset. Based on system testing that has been done, the system can detect Leukemia and Colon Tumor with best accuracy equal to 98.33% for colon tumor data and 100% for leukimia data. Genetic Algorithm as feature selection algorithm can improve system accuracy, which is from 64.52% to 98.33% for colon tumor data and 65.28% to 100% for leukemia data, and the use of momentum parameters can accelerate the convergence of the system in the training process of Neural Network.

1. Introduction

Cancer is one of the major causes of morbidity and mortality worldwide, with 14 million new cases and 8.8 million cancer-related deaths by 2015. It is estimated that the number of new cases will increase by about 70% over the next 2 decades [1]. Detection of cancer can be done by obtaining information from the patient's DNA data in the form of microarray data that has thousands of attributes. Data that has many attributes (dimensionality), will take a lot of time and take advantage of excessive computing efforts that are difficult to handle. Therefore, a dimension reduction method is needed to select the optimal attributes so that computational effort can be performed efficiently [2].

Vanitha D.A. [3] use Mutual Information (MI) as feature selection and classification methods, namely ANN with 50.94% accuracy and SVM with 67.74% accuracy on Colon Tumor data. Nurfalah.A [4] construct a scheme using PCA as dimensional reduction and Backpropagation with Conjugate Gradient as microarray data classification, resulting an accuracy of 76.92% for Colon Tumor and 97.14% for Leukemia data. In both studies, it is difficult to determine the number of optimal attributes produced by MI and PCA.

In previous research, microarray data classification process using Evolutionary Artificial Neural Network (EANN) which is a combination of Genetic Algorithm (GA) as feature selection and Artificial Neural Network (ANN) as cancer classification. From the merger is produced very good performance. This is because EANN to get a solution does not require expert knowledge and there is a selection of attributes in the evolutionary process that can avoid the occurrence of ANN network structures that are too large which can lead to an increase in the computation process, resulting in lower performance [5]. In addition, other studies related to GA use for feature selection in the classification process i.e. GA and Backpropagation Neural Network (BPNN) combined resulted in an average accuracy of 93.8% [2].

BPNN is one of the learning algorithm of Multi-Layer Neural Network. BPNN provides an efficient computational method for converting weights in a forward network feed with differentiated activation in the function unit. In previous research, BPNN was used to detect breast cancer by producing higher accuracy value than Support Vector Machine (SVM) and Decision Tree, where 99% accuracy was obtained for BPNN, 96.19% for SVM and 92.38% for Decision Tree [6]. Other studies related to BPNN have an accuracy of 93% for colon tumor dataset and 95% in leukemia dataset, but the result of accuracy obtained is still not maximal because it still processes from all attributes, so that research can be continued by applying GA for feature selection process [7]. However, in another study said that BPNN still has a lack of convergence is slow and requires a long training process. This can be overcome by adding momentum to the BPNN used on each iteration, so that it will speed up the training process. Momentum is a parameter used to prevent a system from converging to a local minimum and can increase the convergence speed of the system [8].

In this research built a system that can classify and detect a person whether the patient suffered from cancer or not, using GA and Momentum BPNN algorithm in the case of high-dimensional data. Leukemia and colon tumor patients' data were from Kent Ridge Bio-medical Dataset, where data of colon tumor patients with the number of attributes 2000 [9] and in leukemia patients had the number of attributes of 7129 [10]. With the detection model is expected to help overcome the problems in the medical world.

2. Research Scheme

In this section, we will describe the general system design that will be used to detect Leukemia and Colon Tumor in high-dimensional data. In detecting Leukemia and Colon Tumor and getting good performance, dimensional reduction process is needed to obtain optimal attribute. The method used is a combination of Genetic Algorithm and Momentum Backpropagation (Momentum BPNN). Genetic Algorithm serves to perform dimensional reduction process and Momentum BPNN to perform the process of classification. Data used in this study are leukemia and colon tumor data taken from Kent Ridge Bio-medical Dataset. The design of Leukemia and Colon Tumor detection system that is built can be seen in Figure 1.

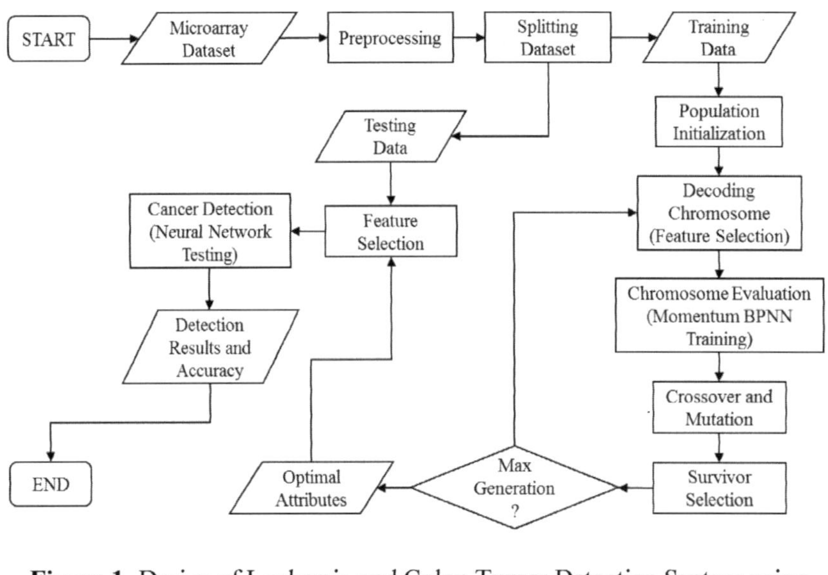

Figure 1. Design of Leukemia and Colon Tumor Detection System using Genetic Algorithm and Momentum BPNN.

2.1. Dataset

Data used in this research is data of colon tumor and Leukemia disease. The data were obtained from the Kent Ridge Bio-medical Dataset [9, 10]. The data available on the repository are gene expression data, protein profiling data and genomic sequence data. The data used included into high-dimensional data that having 2000 attributes for colon tumor data and 7129 attributes for Leukemia data, contains 62 samples for colon tumor data and 72 samples for Leukemia data. Colon tumor and Leukemia sample data can be seen in Table 1 and 2.

Table 1. Colon Tumor Data Samples

No	Attribute 1	Attribute 2	Attribute 3	Attribute 4	Attribute 5	...	Attribute 2000	Class
1	8589.42	5468.24	4263.41	4064.94	1997.89	...	28.70	Negative
2	9164.25	6719.53	4883.45	3718.16	2015.22	...	16.77	Positive
3	3825.71	6970.36	5369.97	4705.65	1166.55	...	15.16	Negative
4	6246.45	7823.53	5955.84	3975.56	2002.61	...	16.09	Positive
5	3230.33	3694.45	3400.74	3463.59	2181.42	...	31.81	Negative
....	....	....	....	....	....	...	....	....
62	7472.01	3653.93	2728.22	3494.48	2404.67	...	39.63	Positive

Table 2. Leukimia ALL-AML Data Samples

No	Attribute 1	Attribute 2	Attribute 3	Attribute 4	Attribute 5	...	Attribute 7129	Class
1	-214	-153	-58	88	-295	...	-37	ALL
2	-139	-73	-1	283	-264	...	-14	ALL
3	-76	-49	-307	309	-376	...	-41	ALL
4	-135	-114	265	12	-419	...	-91	ALL
5	-106	-125	-76	168	-230	...	-25	ALL
....	...	...	...	...	...	...	...	...
72	-135	-186	-70	337	-407	...	-10	AML

2.2. Data Preprocessing

At the pre-processing stage will be done some processing steps. The first process is the dataset will be processed using the normalization method. The normalization method is used to convert data containing real numbers into range [0,1]. The second process is to divide the data that has been normalized into two parts, namely testing and training data using K-Fold Cross Validation, where K=10.

2.3. Feature Selection using Genetic Algorithm

The feature selection stage is done after going through the pre-processing and data division. In the feature selection process uses training data as input. Training data will enter at population initialization process by generating a number of chromosomes with a random value of 0 and 1 with a gene length equivalent to the number of attributes entered. Next is the decoding stage that is the selection of attributes from the result of randomly generated population initialization, where the attribute of value 0 is not used and the attribute of value 1 will be used in the next step. The attribute reduction is done by using the feature selection method to generate the optimal attribute. Examples of chromosome representation results can be seen in Figure 2.

1	0	1	0	0	1

Figure 2. Example of Chromosome Representation Results

The next stage is the process of individual evaluation by applying the process of learning and validation using the Momentum BPNN algorithm which will result training accuracy where the training accuracy results are used as the fitness value to enter at the next stages of GA. The highest fitness value obtained will be used as the best solution.

The elitism process is done to store the chromosome with the best fitness value obtained from the previous process, so that the best fitness value or solution ever obtained is not lost [11]. The next stage is every individual through the process of parent selection using Tournament Selection. Tournament Selection method is used because according to the result of literature study which has been done, where the method of Fitness Proportionate Selection (FPS) with Roulette Wheel and Rank Based Selection method still has weakness that is the bottleneck where more computation process is needed to complete the selection process due to the size of the population used [12]. The parent selection process will produce an individual parent who will be recombined on each of the chromosomes. Recombination is done to produce new and more varied individuals, where each child's chromosome has genes from their parents. The recombination process uses 1 point crossover.

New individuals generated from the recombination process will then undergo the mutation process. In the process of mutation begins with the generation of random numbers. If the random number generated is less than the mutation probability then the gene will be replaced by the value of the opposite. Selection of survivors is the process of replacing old chromosomes with new chromosomes. The survivor selection used is the generational replacement model. Then check whether the system has reached the maximum generation given or not. If not, then the process will return again in the decoding stage, but if it has reached the maximum value it will produce the most optimal attributes. These attributes will then continue in the process of Leukemia and Colon Tumor detection.

2.4. Training Process of Momentum BPNN

The data used in the training stage of Momentum BPNN is the training data that has been through the feature selection stage. Then enter at the training stage using Momentum BPNN. Where in the first stage is the initialization of the parameters that will be used in the construction of network architecture in Momentum BPNN. Then going into the second stage is a random process of weight and bias that will be used in the training stage. Next will be checked for epoch value, if epoch value is more than maximized epoch value that has been initialized at the beginning then the process is done, but if epoch value less than maximum epoch value specified it will go to next stage [13]. In the Backpropagation algorithm there are two stages, namely feed forward phase and backward phase. In the feed forward phase there are three processes, namely the calculation of the output of the hidden layer and output layer and put it into the activation function in the form of sigmoid function [14], as well as calculating the error value (MSE). With the calculation of the error value then the feed forward stage has been completed.

The next stage is the backward phase that there is a process of calculating the factors of weight and bias changes to generate new weight and bias values based on errors that have been obtained before. In the calculation of weight change factors added momentum parameters to avoid premature convergence, as in equations 1 and 2 [15]. The last stage of calculating the new weights and bias based on the old weights and bias as well as weight change factor. Next will be done looping on the feed forward and backward process in accordance with the epoch value used. The optimal weight and bias with the smallest MSE will be stored for use in the testing phase.

$$\Delta w_{jk}(t+1) = \alpha.\delta_k.z_j + mc.\Delta w_{jk}(t) \quad (1)$$
$$\Delta v_{ij}(t+1) = \alpha.\delta_j.z_j + mc.\Delta v_{ij}(t) \quad (2)$$

Where:

δ_k : Error information for k^{th} output unit
δ_j : Error information for hidden unit Z_j.
z_j : Hidden unit j^{th}.

α : *Learning rate*
mc : Momentum
Δw_{jk} : The weight change factor of the j-hidden units and k-output units.
Δv_{ij} : The factor changes the weight of the i-input unit to the j-hidden unit.

2.5. Testing Process of Momentum BPNN

The testing process is used to obtain Leukemia and Colon Tumor detection results in data testing. In the testing stage used only the feed forward process without using the backward stage, so there is no weight update process. At the testing stage we use the best weights and biases that have been obtained in the training process. The classification results are obtained from the output of the neuron at the output layer, where the outpur value has been converted into 0 or 1. The number 1 states detected cancer, and vice versa. Furthermore, the results obtained will be used in the calculation process of accuracy by using a comparison between the actual data with classification results. In the testing process produces an output of accuracy along with the label of each data.

3. Analysis and Experimental Results

The purpose of system testing to be conducted in this study is to obtain optimal attributes that can detect whether a person is positively infected with cancer or not on microarray data of a high dimension. In the test, things that can affect system performance are structures in network construction on Momentum Backpropagation (Momentum BPNN), as well as parameters of the number of individuals in a population (population size), mutation probability (Pm), and crossover probability (Pc) in GA.

3.1. Influence Analysis of Momentum Parameters on BPNN Algorithm

In the BPNN algorithm has several variations that can be used, one of them is with the addition of parameters momentum (mc) on the process of weight and bias improvement. In this research, the system test using standard BPNN and BPNN using momentum (mc) or commonly known as momentum BPNN, where this test is done to know the influence of the momentum parameter in BPNN. The momentum parameter value (mc) used in momentum BPNN is 0.9, while for other parameter values both in BPNN and in BPNN momentum is the same, as in Table 3. This is done in order to know the difference of standard BPNN with BPNN using momentum.

Table 3. BPNN Testing Parameters With and Without Momentum Parameters

BPNN Parameters			GA Parameter			
Epoch	*Learning Rate*	*Hidden Layer*	Max Generation	Population Size	*Pc*	*Pm*
100	0.01	5	20	50	0.65	0.01

Figure 3 show the value of MSE generated takes a long time to reach convergent point, where at epoch reach 100-point value of MSE that have not reached convergence point with accuracy value obtained equal to 83.33% for data colon tumor and 91.42% for leukemia data. While in Figure 4, it can be seen that adding momentum to BPNN can accelerate convergence time and increase accuracy value 94.99% for colon tumor and 98.57% data for leukemia data, this is because the momentum parameter can make a significant leap toward MSE value which is produced closer to the minimum position, this is done by taking the weight value on the previous iteration and adding the current weighting value to improve the learning on the BPNN algorithm. In the presence of the momentum parameter, the convergence point can be reached when it is at the epoch value 10.

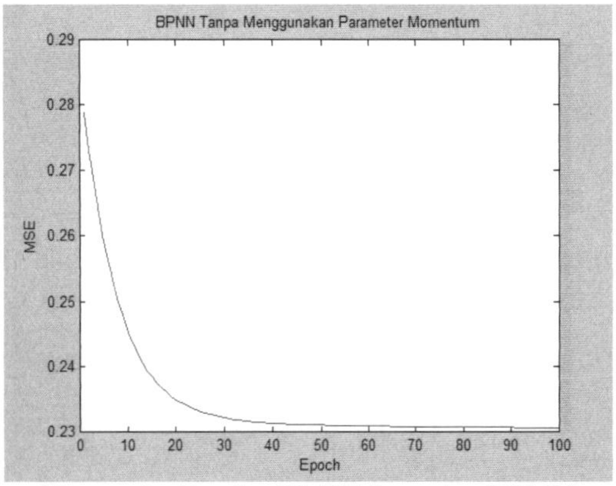

Figure 3. Graph of MSE Learning BPNN Without Momentum

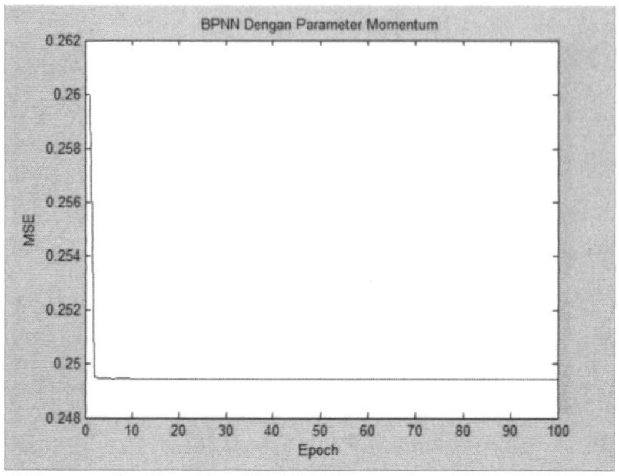

Figure 4. Graph of MSE Learning BPNN With Momentum

Thus, it can be concluded that if using momentum on BPNN can accelerate the time to reach the converging point in the training process. So with a faster convergence process will accelerate also in the training process. This proves the results of the analysis ever done that the addition of momentum BPNN can overcome the weakness of the BPNN algorithm is a slow convergence and require a long training process.

3.2. Testing of Number of Hidden Neurons on the Momentum BPNN Algorithm

This test is used to determine the effect of number of hidden neurons to the accuracy produced. The number of hidden neurons used are 5, 10 and 15. The average accuracy obtained from these three values can be seen in Figure 4.

Table 4. Result of Number of Hidden Neurons Experiment

Number of Hidden Neurons	Accuracy (%)	
	Colon Tumor	Leukimia
5	94.62	95.71
10	94.62	95.71
15	94.62	95.71

In Table 4, it appears that the average accuracy value obtained yields the same result for the different number of hidden layer neurons. In the data colon tumor generated an average accuracy of 94.62% for the number of hidden neurons 5, 10 and 15. While the leukimia data yields an average accuracy of 95.71% for the number of hidden neurons 5, 10 and 15. It can be concluded that in this test the number of hidden neurons does not affect the accuracy of results obtained. This is because the hidden layer is used to build the network architecture that will be used in Momentum BPNN so that the number of neurons in the hidden layer depends on the uniqueness of each dataset pattern used.

3.3. Testing Momentum and Learning Rate Parameters on Momentum BPNN Algorithm

This test is used to determine the effect of learning rate and momentum parameter values on the resulting accuracy. The value of momentum used is 0.9, 0.5 and 0.1, while the value of learning rate used is 0.01, 0.1 and 0.2. Accuracy value obtained from the combination of momentum and learning rate in colon tumor data can be seen in Table 5 and leukemia data in Table 6.

From the Table 5 and Table 6, can be seen that the colon tumor data get the highest accuracy value is by using a combination of momentum parameter value (mc) 0.9 and the value of learning rate parameter (lr) 0.2 with an accuracy of 98.33%. While the leukemia data obtained results have similarities with the data Colon Tumor is the highest accuracy value obtained on the combination of momentum parameter value (mc) 0.9 and the value of learning rate (lr) 0.2 with 100% accuracy value obtained.

Table 5. Testing Result of Momentum and Learning Rate Parameters In Colon Tumor Data to System Accuracy

Momentum	Learning Rate		
	0.01	0.1	0.2
0.9	94.99%	96.66%	98.33%
0.5	96.66%	94.99%	96.66%
0.1	83.33%	93.32%	96.66%

Table 6. Testing Result of Momentum and Learning Rate Parameters In Leukimia Data to System Accuracy

Momentum	Learning Rate		
	0.01	0.1	0.2
0.9	98.57%	97.14%	100%
0.5	95.71%	92.85%	98.57%
0.1	91.42%	94.28%	92.85%

From the results of both data accuracy can be concluded that the greater the value of momentum (mc) and learning rate (lr) used can achieve optimum accuracy. This is because if the greater the value of momentum used it will accelerate the network weight changes to achieve the minimum error or reach the convergent point, so that the faster in achieving the minimum error and convergence will affect the value of accuracy obtained. The larger momentum parameter will avoid the convergence at local minimum, so we can get the optimum weight and bias value. With the acceleration of the training process, the accuracy of the value obtained will also be higher. However, if the value of learning rate is too large, it will cause the learning process of Momentum BP not get optimum convergence. This makes the system unable to produce the expected accuracy value.

3.4. Testing of Genetic Algorithm Parameters

Based on the results of the test, the combination of GA parameters used are maximum generation, population size, crossover probability (Pc), and mutation probability (Pm) by using pre-processing process first then will be entered into the system with the data division of training and testing using 10-Fold Cross Validation. The resulting output is a classification class and accuracy of the class that is detected correctly.

Table 7. Result of Population Size Experiment

Population Size	Accuracy (%)	
	Colon Tumor	Leukimia
50	91.78	97.36
100	94.53	97.83
200	94.87	98.12

In Table 7, it can be seen that in colon tumor data using population size at GA that is 50 produce accuracy 91.78% and with population size 100 produce accuracy 94.53%, while using population size 200 obtained highest accuracy is 94.53%. In leukemia data obtained the test results with 97.26% accuracy for population size 50 and 97.83% for population size 100. While the highest accuracy obtained from the test results using population size 200. So it can be concluded that the greater the size of the population the greater the accuracy obtained. This is because when the size of the population gets bigger, then the number of samples taken as a candidate solution will also be more and more. So that will increase the probability of getting the best solution among the large solution space.

The second parameter used in GA is the crossover probability (Pc) used to determine what percentage of the probability of doing crossover in order to obtain a good individual. The effect of crossover probability can be seen in Table 8.

Table 8. Result of Crossover Probability Experiment

Crossover Probability	Accuracy (%)	
	Colon Tumor	Leukimia
0.65	91.80	97.39
0.9	95.65	98.15

In Table 8, it can be seen that in colon tumor data using crossover probability value (Pc) 0.65 resulted 91.80% accuracy and 95.65% accuracy using crossover probability value (Pc) 0.9. Leukemia data obtained 97.39% accuracy by using probability value of crossover (Pc) 0.65 and get 98.15% accuracy using crossover probability value (Pc) 0.9. So the test results from both data, it can be concluded that the higher the probability value of crossover (Pc) used then the higher the accuracy produced. This is because the greater probability of crossover (Pc) will cause greater chromosomal chances of crossover, thus increasing the chances of getting the best individual that will have an effect on the accuracy obtained.

Table 9. Result of Mutation Probability Experiment

Mutation Probability	Accuracy (%)	
	Colon Tumor	Leukimia
0.65	91.80	97.39
0.9	95.65	98.15

In Table 9, it can be seen that in colon tumor data using mutation probability value (Pm) 0.01 yield accuracy of 93.92% and accuracy of 97.53% if using mutation probability (Pm) of 0.3. While on leukemia data, on graph of test result can be seen that by using mutation probability value (Pm) 0.01 yield accuracy 93.54% and 98.01% if using mutation probability value (Pm) 0.3. Can be concluded from the test results on both data is generated the highest accuracy if the value of mutation probability (Pm) is used greater. Where mutation probability (Pm) is used to control how many chromosomes will be mutated. So the greater the probability value of mutation (Pm) the more chromosomes will undergo the mutation process. This causes more new chromosomes to be generated from the mutation process and used in subsequent generations, so that the individual generated will also be more varied and can prevent the system from being trapped in the local optimum.

3.5. Testing of Feature Selection Effect on System Performance

Based on the results of tests that have been done on colon tumor data with the number of attributes 2000 and leukimia data with the number of attributes 7129 done classification with Momentum BPNN using GA and Momentum BPNN without using GA, where GA is used as feature selection process for optimal attribute selection. The result of comparison on detection of cancer patient by using Momentum BPNN with GA and without GA can be seen in Table 10.

Table 10. Comparison Results of GA Use in Momentum BPNN

Algorithm	*Colon Tumor*		*Leukimia*	
	Number of Attributes	Accuracy (%)	Number of Attributes	Accuracy (%)
Momentum BPNN	2000	64.516%	7.129	65.277 %
Momentum BPNN and GA	968	98.33%	3526	100%

Based on Table 10, accuracy obtained from testing of detection process on cancer patient using Momentum BPNN without using GA that is in colon tumor data have accurate value 64,516% with attribute count used is 2.000 attributes, while leukemia data has accurate value 65.277% with attributes used as many as 7,129 attributes. Both accuracy values are still low, this is because all attributes that exist in both data are still processed as a whole, so the attribute is less informative also used in the process of classification.

While in Momentum BPNN by using GA get accuracy value equal to 98.33% with optimal attribute that used is 968 for colon tumor data, while for leukimia data yield 100% accuracy value with optimal attribute count is 3,526 attributes. In the test results it can be seen that Leukemia and Colon Tumor detection using Momentum BPNN and GA has a higher accuracy value compared to Momentum BPNN without using GA, this is because the attributes processed in the classification process has been reduced so that only the optimal attributes used in the classification process.

It can be concluded that the use of Momentum BPNN algorithm using GA has a higher accuracy value than without using GA. Increased accuracy is due to GA can be used for dimensional reduction process through the feature selection stage to select the optimal attribute without reducing the information from the data. So with dimension reduction with GA as feature selection which only chooses the optimal attribute which will be used in the classification process, then the attribute is processed only the selected attribute is not the whole of the attribute. Whereas, if without using GA will cause the overall attributes contained in the data used in the classification process, so the presence of attributes that do not provide information to be processed in the process of classification. This may affect the accuracy and performance results of the system being built because not all attributes have the information used for the classification process.

4. Conclusion

Based on the system testing that has been done, it can be concluded that the system built has been very good in detecting colon tumor and leukemia disease with the best accuracy of 98.33% for colon tumor data and 100% for leukemia data. The feature selection process using Genetic Algorithm has an effect on improving system accuracy, from 64.52% to 98.33% for colon tumor data and 65.28% to 100% for leukemia data. This is due to the feature selection process so as to produce the optimal attributes that will be used in the classification process. The existence of momentum parameters at BPNN can accelerate the time to reach the convergent minimum point in the training process. In BPNN test results without using momentum parameters convergent point were achieved at epoch 90 with an accuracy of 83.33% for colon tumor data and 91.42% for leukemia data, whereas when the momentum parameters were added to BPNN, convergent points could be achieved at epoch 10, with accuracy obtained for 98.33% in colon tumor data and 100% in leukemia data.

Acknowledgments

The authors would like to thank Telkom University for financially supporting in this research.

References

[1] *World Health Organization* 2016 Available Online: http://www.who.int/mediacentre/factsheets/fs297/en/. [Accessed 22 October 2016]

[2] Khobragade V P and Vinayababu D A 2012 A Classification of Microarray Gene Expression Data Using Hybrid Soft Computing Approach *International Journal of Computer Science Issues* vol 9 no 6 pp 246-255

[3] Vanitha C D A Devaraj D and Venkatesulu M 2015 Gene Expression Data Classification Using Support Vector Machine and Mutual Information-based Gene Selection *Procedia Computer Science 47* pp 13- 21

[4] Nurfalah A Adiwijaya and Suryani A A 2016 Cancer Detection Based on Microarray Data Classification Using PCA And Modified Back Propagation *Far East Journal of*

Electronics and Communications 16(2) pp 269-281
[5] Kim K J and Cho S B 2004 Prediction of Colon Cancer Using An Evolutionary Neural Network *Neurocomputing* pp 361-379
[6] Pawar P S and Patil D 2012 Breast Cancer Detection Using Backpropagation Neural Network with Comparison Between Different Neuron *IEEE International Conference on Parallel, Distributed and Grid Computing* pp 170-173
[7] Satpathy S and Mahapatra P 2014 Microarray Classification Using Intelligent Techniques *International Journal of Scientific & Engineering Research* vol 5 no 7 pp 1663-1667
[8] Yu C C and Liu B D 2002 A Backpropagation Algorithm With Adaptive Learning Rate and Momentum Coefficient *IEEE* pp 1218-1223
[9] *Kent Ridge Bio-medical Dataset* Available Online: http://datam.i2r.a-star.edu.sg/datasets/krbd/ColonTumor/ColonTumor.html [Accessed 22 October 2016]
[10] *Kent Ridge Bio-medical Dataset* Available Online: http://datam.i2r.a-star.edu.sg/datasets/krbd/Leukemia/ALLAML.html [Accessed 22 October 2016]
[11] Suyanto 2008 Evolutionary Computation Komputasi Berbasis "Evolusi" dan "Genetika" *Bandung: Informatika*
[12] Suyanto 2008 Soft Computing "Membangun Mesin Ber-IQ Tinggi" *Bandung: Informatika*
[13] Noertjahyana A and Yulia Studi Analisa Pelatihan Jaringan Syaraf Tiruan Dengan dan Tanpa Algoritma Genetika
[14] Andrian Y and Putra P H 2014 Analisis Penambahan Momentum Pada Proses Prediksi Curah Hujan Kota Medan Menggunakan Metode Backpropagation Neural Network *Seminar Nasional Informatika* pp 165-172
[15] Hermawan A 2006 Jaringan Saraf Tiruan, Teori dan Aplikasi *Yogyakarta: Andi Offset*

IOP Conf. Series: Journal of Physics: Conf. Series **971** (2018) 012019 doi:10.1088/1742-6596/971/1/012019

Steady state numerical solutions for determining the location of MEMS on projectile

K Abiprayu, M F F Abdigusna and P H Gunawan

School of Computing, Telkom University, Jalan Telekomunikasi No. 1 Terusan Buah Batu, Bandung 40257, Indonesia.

E-mail: krishajarr@gmail.com; firly.feisal@gmail.com; phgunawan@telkomuniversity.ac.id

Abstract. This paper is devoted to compare the numerical solutions for the steady and unsteady state heat distribution model on projectile. Here, the best location for installing of the MEMS on the projectile based on the surface temperature is investigated. Numerical iteration methods, Jacobi and Gauss-Seidel have been elaborated to solve the steady state heat distribution model on projectile. The results using Jacobi and Gauss-Seidel are shown identical but the discrepancy iteration cost for each methods is gained. Using Jacobi's method, the iteration cost is 350 iterations. Meanwhile, using Gauss-Seidel 188 iterations are obtained, faster than the Jacobi's method. The comparison of the simulation by steady state model and the unsteady state model by a reference is shown satisfying. Moreover, the best candidate for installing MEMS on projectile is observed at point $\mathbf{T(10,0)}$ which has the lowest temperature for the other points. The temperature using Jacobi and Gauss-Seidel for scenario 1 and 2 at $\mathbf{T(10,0)}$ are 307 and 309 Kelvin respectively.

1. Introduction

Microelectromechanical systems or MEMS is a combination of electrical and mechanical components in a miniature sized device. Generally, this technology has been developed since the last few decades and its limited applications have been used. For instance, several applications of MEMS can be found in MIT microengine, electrostatic micromotor, thermal inkjet printhead, digital micromirror device, and etc [1, 2]. Moreover, this technology is used currently for military purposes. One of examples is MEMS can be integrated on projectile for obtaining the accuracy of a shot to the target. Indeed, the technology MEMS brings high level technology in military forces [3, 4, 5].

Several research and studies about the installation MEMS on projectile have been done [6, 7]. The fact that, the technology MEMS can work properly in a maximum temperature 71° C or 344 K. Meanwhile, the temperature of projectile can reach a maximum temperature 267° C or 540 K [8, 9]. Therefore, the installation of MEMS on the projectile is not an easy task. Since the MEMS might malfunction if the installation place on the surface of projectile has a temperature more than 344 K.

In the papers of [8, 9], the numerical simulation of unsteady heat distribution is elaborated. The numerical method for solving the unsteady model is quite complicated since the discrete time and space should be considered. For instance, in the paper of [8], the unsteady model is approximated by finite element method which has complicated discrete property. Nevertheless, this method is known as a robust method for complex geometry domain. In this paper, the steady state heat distribution model will be considered on fix domain of projectile. This idea appears to be emphasized to reduce the complexity of numerical scheme such as a stability from the relation of discrete time and space.

Here, steady model will be discretized by a finite difference method which has uniform grid spacing. This method is simple and straightforward method. Thus, no complicated discrete property will be considered. By using the steady state model, the system algebraic equations will be obtained. Indeed, the approximation of solution for linear system equations should be calculated. Here, two numerical iterative methods for solving linear system equations will be used. The methods are known as Jacobi and Gauss-Seidel method. The comparison of results by unsteady model in a reference and steady model also will be considered in this paper.

2. Mathematical Model

In this paper, the steady state of temperature distribution model will be considered. The 2D mathematical model of steady state distribution of heat over entire domain of projectile $\Omega + \partial\Omega$ is given as

$$\frac{\partial^2 T(x,y)}{\partial x^2} + \frac{\partial^2 T(x,y)}{\partial y^2} = 0, \quad (x,y) \in \Omega, \tag{1}$$

$$T(x,y) = \beta(x,y), \quad (x,y) \in \partial\Omega \tag{2}$$

where $T(x,y)$ is the temperature distribution function, $\beta(x,y)$ is the known boundary function, x and y are the spatial coordinate of domain. Here, the $\partial\Omega$ denotes the boundary of domain. The illustration of projectile domain can be seen in Fig 1 .

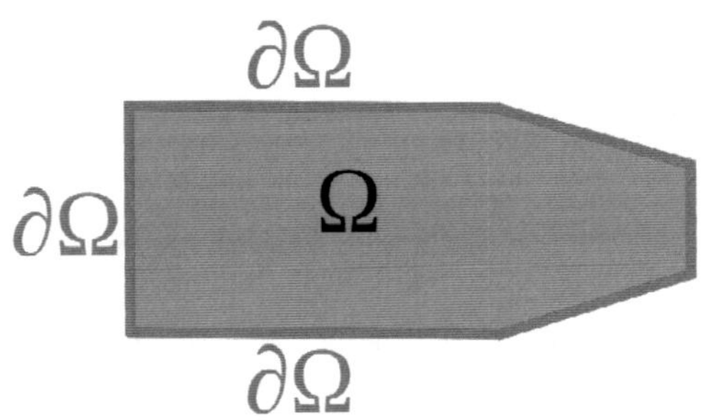

Figure 1. The configuration of projectile domain.

In order to discretize (1-2), the domain (including boundary) can be discretized by several points in a mesh (see Figure 2 for more detail). Assume a point in a mesh is in a set $\mathcal{M} = \{1, 2, \cdots, N\}$, $N \in \mathbb{Z}^+$, thus the discrete form of equation (1) can be given as

$$\frac{(T_a - 2T_i + T_b)}{\Delta x^2} + \frac{(T_c - 2T_i + T_d)}{\Delta y^2} = 0, \quad i, a, b, c, d \in \mathcal{M} \tag{3}$$

The grids spacing Δx and Δy are set to be uniform and similar. Thus the equation can be rewritten as

$$T_i = \frac{T_a + T_b + T_c + T_d}{4} \tag{4}$$

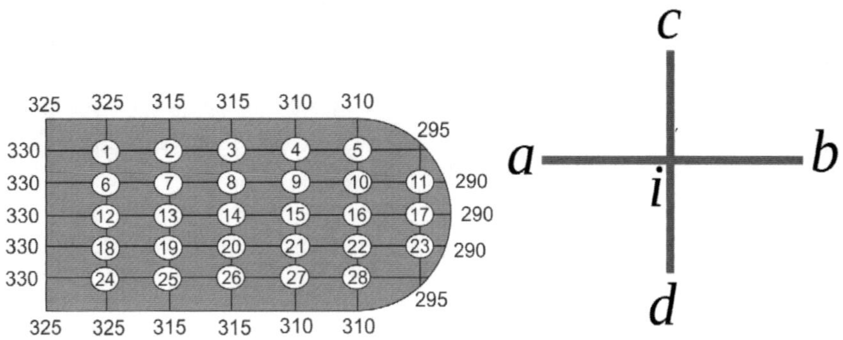

Figure 2. A mesh generated over a domain (left). The stencil for point i (right).

Thus from Figure 2, the system of equations can be formed as follow,

$$
\begin{aligned}
T_1 &= \frac{325 + 330 + T_2 + T_6}{4}, & T_2 &= \frac{315 + T_1 + T_3 + T_7}{4}, \\
T_3 &= \frac{315 + T_2 + T_4 + T_8}{4}, & T_4 &= \frac{310 + T_3 + T_5 + T_9}{4}, \\
T_5 &= \frac{310 + T_4 + 295 + T_{10}}{4}, & T_6 &= \frac{330 + T_1 + T_7 + T_{12}}{4}, \\
T_7 &= \frac{T_2 + T_6 + T_8 + T_{13}}{4}, & T_N &= \frac{\cdots + \cdots + \cdots + \cdots}{4}.
\end{aligned} \tag{5}
$$

Indeed now the problem is to solve this system of equations (5). In this paper, two numerical methods Jacobi and Gauss-Seidel iterative method for approximating the solutions of (5) will be used.

3. Numerical Methods and Algorithms

3.1. Jacobi iterative method

Jacobi iterative method is a method for solving the linear system of equations $A\mathbf{x} = \mathbf{b}$. Indeed the system (5) can be transformed into $A\mathbf{T} = \mathbf{b}$. The formula of Jacobi iterative method is given as

$$T_i = \frac{1}{a_{ii}}\left[b_i - \sum_{j=1, j\neq i}^{N} (a_{ij}T_j)\right], i \in \mathcal{M}. \tag{6}$$

The equation (6) will be solved iteratively using the initial guess $T_i^{(0)}$ of $\mathbf{T}^{(0)}$. The algorithm of Jacobi is given in Algorithm 1. In the Algorithm 1, the iterative method can be interrupted using the relation of $||\mathbf{T}^{(k)} - \mathbf{T}^{(k-1)}|| < TOL$. Here, TOL is the tolerance number which indicates the steady state is tolerated at a fix distance. The second stopping criteria in this iterative is used the maximum number of iteration $Kmax$. Since in some cases, the distance $||\mathbf{T}^{(k)} - \mathbf{T}^{(k-1)}||$ is never reach the TOL value due to the initial guess $\mathbf{T}^{(0)}$.

3.2. Gauss-Seidel iterative method

$$T_i^{(k)} = \frac{1}{a_{ii}}\left[b_i - \sum_{j=1}^{i-1}\left(a_{ij}T_j^{(k)}\right) - \sum_{j=i+1}^{N}\left(a_{ij}T_j^{(k-1)}\right)\right]. \tag{7}$$

Algorithm 1 The algorithm of Jacobi's method.

Step 0. Start.

Step 1. Set $k = 1$, TOL: tolerance, $Kmax$: maximum number of iterations, $\mathbf{T}^0$: initial approximation.

Step 2. While ($k \leq Kmax$) do Steps **3-6**

  Step 3. For $i \in \mathcal{M}$

$$T_i^{(k)} = \frac{1}{a_{ii}} \left[b_i - \sum_{j=1, j \neq i}^{N} \left(a_{ij} T_j^{(k-1)} \right) \right].$$

  Step 4. If $||\mathbf{T}^{(k)} - \mathbf{T}^{(k-1)}|| < TOL$

    the OUTPUT ($\mathbf{T}^{(k)}$).

    (The procedure was successful.) STOP

  Step 5. Set $\mathbf{T}^{(k-1)} = \mathbf{T}^{(k)}$

  Step 6. Set $k = k + 1$

Step 7. OUTPUT ('Maximum iterations exceeded.')

Step 8. End.

Algorithm 2 The algorithm of Gauss-Seidel method.

Step 0. Start.

Step 1. Set $k = 1$, TOL: tolerance, $Kmax$: maximum number of iterations, $\mathbf{T}^0$: initial approximation.

Step 2. While ($k \leq Kmax$) do Steps **3-6**

  Step 3. For $i \in \mathcal{M}$

$$T_i^{(k)} = \frac{1}{a_{ii}} \left[b_i - \sum_{j=1}^{i-1} \left(a_{ij} T_j^{(k)} \right) - \sum_{j=i+1}^{N} \left(a_{ij} T_j^{(k-1)} \right) \right].$$

  Step 4. If $||\mathbf{T}^{(k)} - \mathbf{T}^{(k-1)}|| < TOL$

    the OUTPUT ($\mathbf{T}^{(k)}$).

    (The procedure was successful.) STOP

  Step 5. Set $\mathbf{T}^{(k-1)} = \mathbf{T}^{(k)}$

  Step 6. Set $k = k + 1$

Step 7. OUTPUT ('Maximum iterations exceeded.')

Step 8. End.

The second method is called the Gauss-Seidel iterative method. This method is a modification of Jacobi's method [10]. The modification on acting on the iterative process. The modified iteration method is given in Equation (7), for all $i \in \mathcal{M}$.

Further, Algorithm 2 shows the algorithm of Gauss-Seidel method. Here, the modification to calculate $a_{ij} T_j^{(k-1)}$ in Jacobi's method is considered. Indeed, for $i > 1$, the solutions $T_1^{(k)}, \cdots, T_{i-1}^{(k)}$ have been calculated and thus the remain $T_{i+1}^{(k-1)}, \cdots, T_N^{(k-1)}$ should be used.

4. Numerical simulations

In this paper, the investigation of finding the best position for installing MEMS on projectile will be elaborated. The investigation of the observation points on projectile can be seen in Figure 3. This observation points can be used as a candidate for installing MEMS on projectile.

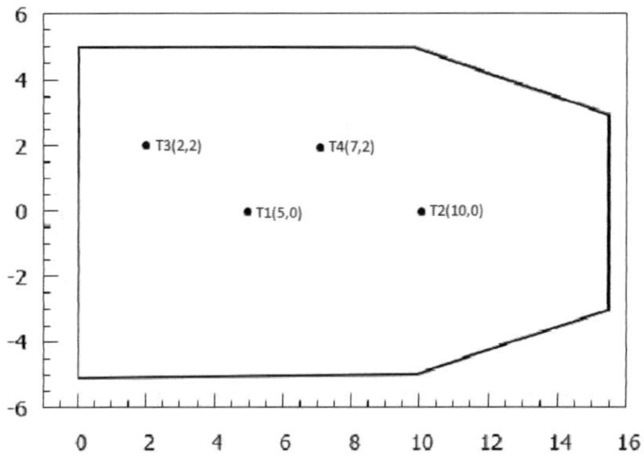

Figure 3. The candidate places of points on MEMS. Four points are chosen in order to validate the proposed methods.

Here, two scenarios are elaborated based on the input of initial data at boundary. First scenario, the boundary condition is set as the results of unsteady heat distribution model in [8]. The example of the application interface for Jacobi and Gauss-Seidel can be seen in Figure 4.

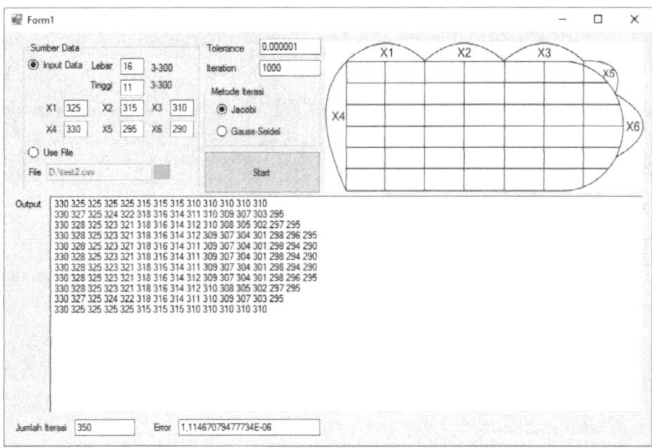

Figure 4. The example of application interface with Jacobi results.

In this simulation, the other conditions are given as follow

$$\mathbf{T}^{(0)} = 280 \text{ Kelvin, (for inner domain)},$$
$$TOL = 10^{-6},$$
$$Kmax = 10^{4}.$$

The boundary conditions of this simulation in the area $X1$, $X2$, $X3$, $X4$, $X5$ and $X6$ are described in Figure 4. These values of boundary conditions are obtained from the paper of Akbar A, et al., [8].

The results of first scenario can be seen in Figure 5. From Figure 5, the heat map shows the highest temperature is located on the back side of projectile (330 K). Meanwhile the lowest temperature is shown in the front of projectile (290). Using previous initial conditions, the iteration by Jacobi to solve this scenario is 350 iterations. Meanwhile, by Gauss-Seidel the final iteration is obtained 188 iterations.

For the second scenario, the modification of initial boundary condition from first scenario is made. Here, the boundary conditions is obtained from [8], but in more detail than the approximation as shown in Figure 4. The result of second scenario is shown in Figure 6. In this results, the highest temperature is not on entire of back side projectile. However, the highest temperature is located at the back corners of projectile only.

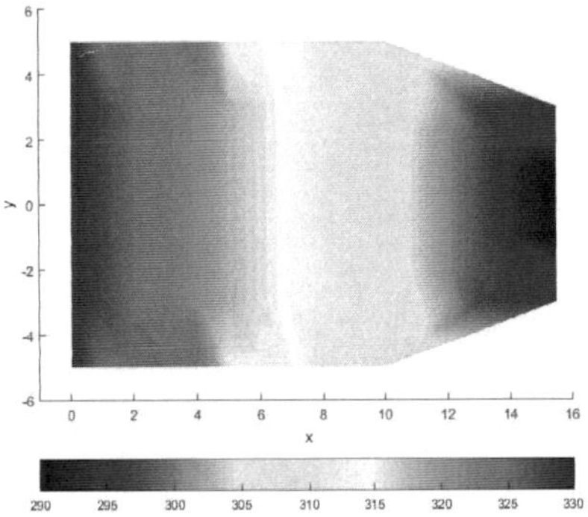

Figure 5. The results of steady state heat distribution on projectile for first scenario.

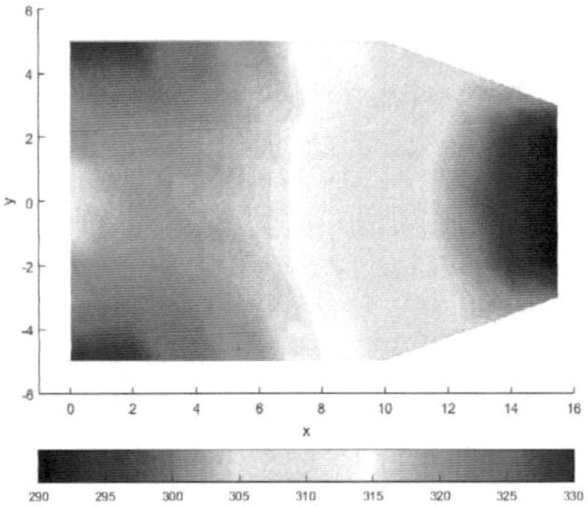

Figure 6. The results of steady state heat distribution on projectile for second scenario.

In addition, the comparison of numerical result of steady state heat distribution model using Jacobi and Gauss-Seidel methods and unsteady model using finite element method is shown in Figure 7.

From Figure 7, the result using scenario 2 is close to the numerical result by [8]. Indeed, the close solution to the reference is depend on the initial boundary conditions. As in Figure 3, the candidate for installing MEMS on projectile should be investigated. The investigation results for each observation points are shown in Table. 1.

Table 1. The comparison of temperature for each observation points.

Points	Scenario 1	Scenario 2	Results in [8]
$T1(5,0)$	318	317	315
$T2(10,0)$	307	309	305
$T3(2,2)$	325	320	318
$T4(7,2)$	314	315	315

From Table 1, it appears that temperature for four points are still under heat tolerance of MEMS which is 344 Kelvin. And it is concluded that the location placement of MEMS can be done in every observation points. Moreover, from Table 1, the best location to install MEMS is obtained at point $T2$ which has a lowest temperature than the other points.

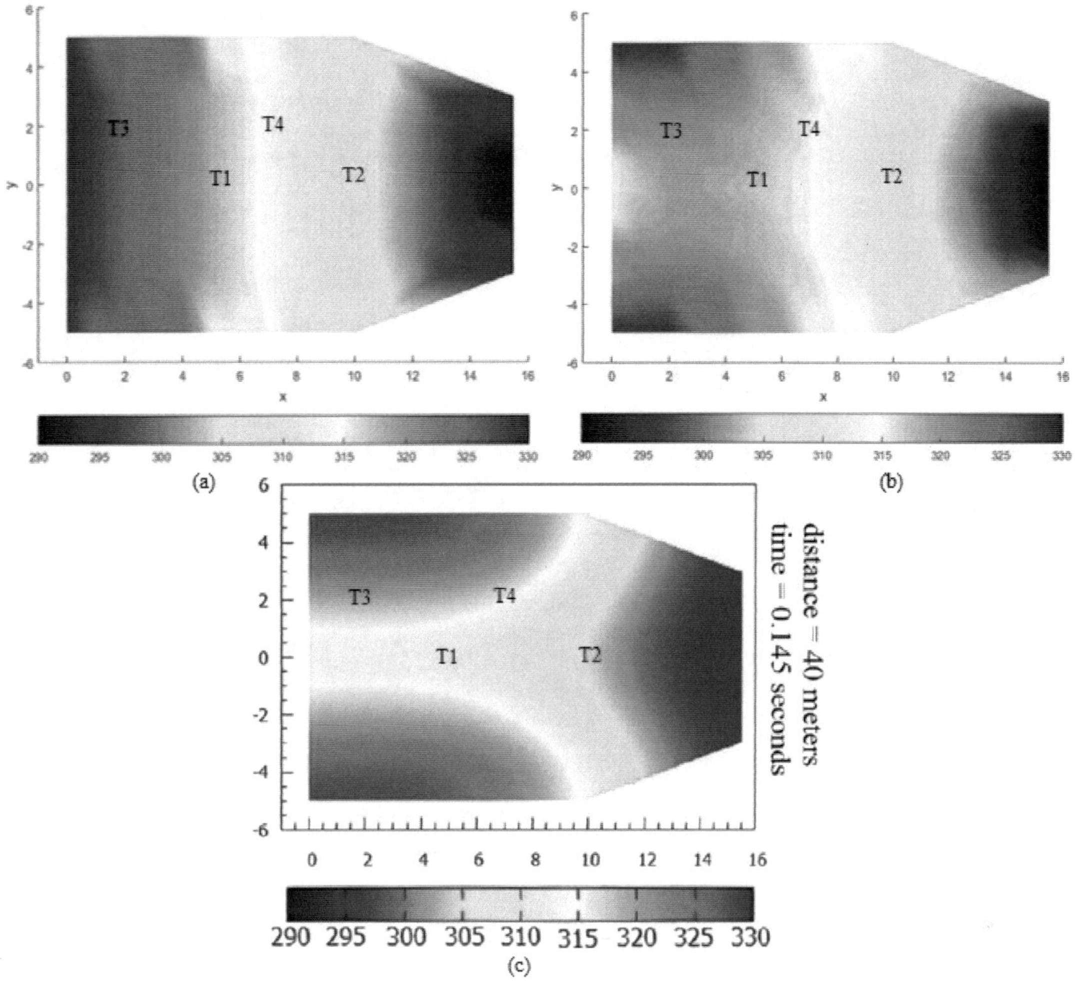

Figure 7. The results of steady state heat distribution using Jacobi's (a) and Gauss-Seidel (b) method and simulation by [8] (c) on projectile.

5. Conclusion

The Jacobi and Gauss-Seidel iteration methods have been elaborated to solve the steady state heat distribution model on projectile. The results using Jacobi and Gauss-Seidel is similar but they have different iteration cost. Using Jacobi's method, the steady state is obtained by 350 iterations. Meanwhile, using Gauss-Seidel the steady state is obtained by 188 iterations. The satisfied comparison by the simulation result in [8] is achieved. Moreover, from two scenarios, the best candidate for installing MEMS on projectile is observed at point $T2(10, 0)$ which has the lowest temperature.

References

[1] Gardner J W, Varadan V K and Awadelkarim O O 2001 *Microsensors, MEMS, and smart devices* vol 1 (Wiley Online Library)
[2] Ho C M and Tai Y C 1998 *Annual Review of Fluid Mechanics* **30** 579–612
[3] Brown T G 2003 *Sensors, 2003. Proceedings of IEEE* vol 2 (IEEE) pp 753–760
[4] Schadow K 2004 *Proceedings of CANEUS 2004 Conference on Micro-Nano-Technologies* vol 10 pp 6–2004

[5] Gripton A 2002 *Position Location and Navigation Symposium, 2002 IEEE* (IEEE) pp 28–35
[6] Davis B S 1998 *Position Location and Navigation Symposium, IEEE 1998* (IEEE) pp 594–601
[7] Davis B S, Denison T and Kuang J 2006 *Shock and Vibration* **13** 127–135
[8] Akbar A, Tarwidi D and Setiawan E 2016 *International Journal on Information and Communication Technology*
[9] M B T and L D 2010 *Finite element modeling of transient temperatures in a small-caliber projectile* (J. Eng. Applied Sci)
[10] Burden R L and Faires J D 2001 *Numerical analysis (7th)* (Prindle Weber and Schmidt, Boston)

IOP Conf. Series: Journal of Physics: Conf. Series **971** (2018) 012020 doi:10.1088/1742-6596/971/1/012020

Staggered grid implementation of 1D Boussinesq model for simulating dispersive wave

D Adytia[1, a)], D Tarwidi[1], S A Kifli[1] and S R Pudjaprasetya[2],

[1]School of Computing, Telkom University, Jl. Telekomunikasi 1 Bandung 40257, Indonesia
[2]Industrial & Financial Mathematics Research Group, Faculty of Mathematics & Natural Sciences, Institut Teknologi Bandung, Jl. Ganesha 10, Bandung 40132, Indonesia

adytia@telkomuniversity.ac.id

Abstract. In this paper, a numerical implementation of 1D Variational Boussinesq (VB) wave model in a staggered grid scheme is discussed. The staggered grid scheme that is used is based on the idea proposed by Stelling & Duinmeijer (2003) who implemented the scheme in a non-dispersive Shallow Water Equations in a conservative form. Here, we extend the idea of the staggered scheme to be applied for VB wave model. To test the accuracy of the implementation, we test the numerical implementation of VB wave model for simulating propagation of solitary wave against analytical solution. Moreover, to test dispersiveness of the model, we simulate a standing wave against analytical solution. Results of simulations show a good agreement with analytical solutions.

1. Introduction

Water wave modelling has been a favorite topic among coastal engineers since the last 3 decades; especially when it is used as a tool to understand wave propagation and behavior for design engineering of an offshore as well as coastal structure. Over several physical aspects of water wave, the most import aspects are dispersion property and nonlinearity. Dispersion in water wave means that longer wave (or low frequency waves) travels faster than shorter waves (high frequency waves). Boussinesq-type models (BTMs) are the most favorite wave model for researchers to simulate wave propagation, for long wave such as tsunami, as well as short wave such as wind wave. The original Boussinesq model was introduced by J.V. Boussinesq in 1872 [7] is valid only for long waves above flat bottom. The original model is then extended by Peregrine in 1967 [14] to be able to simulate wave above an uneven bottom. Two most popular BTMs are the Boussinesq of Madsen & Sorensen in 1992 [12] and the Boussinesq of Nwogu in 1993 [13]. Since then, these BTMs are extended for simulating highly dispersive and highly nonlinear wave propagation; see [8] for review of the development of BTMs.

Numerical implementation to be chosen for the BTMs is also important to solve the model numerically. Most popular numerical implementations among these BTMs are Finite Difference Method (FDM), Finite Volume Method (FVM) and Finite Element Method (FEM). These numerical implementations are mostly using collocated grid, i.e. all variables are defined in the same grid points. For continuous problems the collocated grid gives good performance, while in boundary condition such wet-dry condition, the collocated grid may lead to a stability problem where an artificial damping is required (see [11]). A different approach is a staggered grid approach, where grids of surface elevation and velocity are in different locations. Stelling & Duinmeijer in 2003 [15] propose a staggered grid scheme for solving Shallow Water Equations (SWE) that performs relative good for rapidly varied shallow water flows. The staggered scheme is then extended for simulating dispersive Non-Hydrostatic wave model in [16].

In this paper, we use a Boussinesq type model that is derived by using a variational approach, i.e. that is so-called the Variational Boussinesq (VB) model. The model was implemented by using spectral method in [9] and by using Finite Element Method (FEM), see [1, 2, 4, 5, 6] and later is extended to a fully nonlinear model by also using FEM [3]. Both implementations are using collocated grid, which give challenges when dealing with wet-dry boundary conditions. Here, we implemented numerically the VB model in a staggered grid scheme as proposed by Stelling & Duinmeijer in [15]. Accuracy of the implementation is tested for simulating two cases, i.e. a dispersive standing wave in a close basin and a propagation of a solitary wave.

The structure of this paper is as follows. In the next section, we describe the VB model as it is introduced in [1]. This section is followed by a staggered grid implementation of the VB model using the basic idea in [15]. In Section 4, we test the performance of the implementation for simulation two test cases against analytical solution. Finally, some conclusions and discussions are described in the last section.

2. Variational Boussinesq Model

In this paper, we use the 1D Variational Boussinesq (VB) model that is introduced by Klopman ea. in 2010 [9], for simulating dispersive wave. The VB model is derived based on variational formulation proposed by Luke in 1967 [10]. By assuming the water as an ideal fluid, i.e. inviscid, incompressible and the flow is assumed to be irrotational, the dynamic of wave can be exactly discribed by a Hamiltonian system, see [17]. The Hamiltonian is the total energy, i.e. sum of the potential and the kinetic energy. As derived in [1], the 1D VB model is described by two dynamic equations as follows

$$\partial_t \eta = -\partial_x\big((h+\eta)\partial_x \phi\big) - \partial_x(\beta \partial_x \psi) \tag{2.1}$$

$$\partial_t \phi = -g\eta - (\partial_x \phi)^2/2 \tag{2.2}$$

and an additional elliptic equation

$$-\partial_x(\alpha \partial_x \psi) + \gamma\psi = \partial_x(\beta \partial_x \phi) \tag{2.3}$$

where $\eta(x,t)$ and $\varphi(x,t)$ are the canonical variables of the Hamiltonian system, i.e. surface elevation and surface potential, respectively. Here, *h(x)* is the water depth and $g = 9.81 m/s^2$ is the gravitational acceleration. A spatially dependent variable $\psi(x)$ is an auxiliary variable that is used in the approximation of the VB model, see [1, 5], which has to calculated in every time step by solving the elliptic equation in (2.3). The coefficients α, β & γ are the VB coefficients. For parabolic vertical profile the coefficients are given by

$$\alpha = \frac{2}{15}H^3; \quad \beta = -\frac{H^2}{3}; \quad \gamma = \frac{H}{3} \tag{2.4}$$

where $H = h + \eta$ is the total depth.

Note that the VB model in (2.1)-(2.3) are written in canonical variables $\eta(x,t)$ and $\varphi(x,t)$. Most Boussinesq type of model as well as non-hydrostatic model are written in $\eta(x,t)$ and $u(x,t)$, where *u(x,t)* is horizontal velocity. As shown in [9], the VB model can be written in surface horizontal velocity *u(x,t)* by using the fact that *u* can be written as $u = \partial_x \phi$. Therefore, the VB model can be rewritten as

$$\partial_t \eta = -\partial_x(Hu) - \partial_x(\beta \partial_x \psi) \tag{2.5}$$

$$\partial_t u = -g\partial_x \eta - u\partial_x u \tag{2.6}$$

$$-\partial_x(\alpha \partial_x \psi) + \gamma\psi = \partial_x(\beta u) \tag{2.7}$$

Notice that equation (2.5) and (2.7) are actually the Shallow Water Equations (SWE) with an additional term in the continuity equation (2.5), i.e. $-\partial_x(\beta\partial_x\psi)$, and the equation (2.6) is an additional equation for searching ψ in every time step. As described in [3] the VB model (2.4)-(2.6) can be extended for simulating highly dispersive and strongly nonlinear waves. In the next section we describe the staggered grid scheme that is proposed by [15] for implementing numerically the equation (2.4)-(2.6).

3. Staggered Grid Scheme

To design an efficient and accurate wave code, not only an accurate wave model is needed but it is also important to choose a stable, simple, and efficient numerical implementation. In this paper we implemented the VB model in (2.4)-(2.6) numerically in a staggered grid scheme proposed in [15]. Staggered grid means that eq. (2.4) and (2.5) are approximated on different cells. Here, η and ψ are calculated at full grid points notated by x_i, $i = 1, 2, \cdots, N$ and u is calculated at half grid points $x_{i+1/2}$, $i = 1, 2, \cdots, N+1$ as illustrated in Figure 1. Just as η, the depth h and the total depth H is also in the full grid. As consequences, the coefficient α, β and γ are also in the full grid.

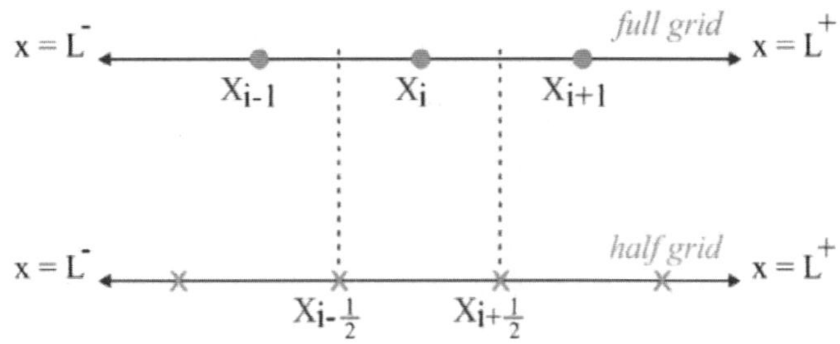

Figure. 1. Illustration of staggered grid scheme.

In the staggered scheme discretization proposed by Stelling & Duinmeijer (2003) [15], the spatial derivatives are approximated by second order center difference and first order upwind scheme. The approximation of the continuity equation of VB model (2.5) is given by the following relation

$$\frac{\eta_i^{n+1}-\eta_i^n}{\Delta t} = -\left(\frac{{}^*H^n_{i+\frac{1}{2}}u^n_{i+\frac{1}{2}} - {}^*H^n_{i-\frac{1}{2}}u^n_{i-\frac{1}{2}}}{\Delta x}\right) - \left(\frac{{}^*\beta^n_{i+\frac{1}{2}}(\psi^n_{i+1}-\psi^n_i) - {}^*\beta^n_{i-\frac{1}{2}}(\psi^n_i-\psi^n_{i-1})}{(\Delta x)^2}\right) \tag{2.8}$$

Here, $\eta(x_i, t_n)$ is denoted by η_i^n, $u(x_{i+1/2}, t_n)$ is denoted by $u^n_{i+1/2}$, just as well as other variables, i.e. H, u, β, ψ. The length spatial grid is denoted by Δx and the length of time discretization is denoted by Δt. In the eq. (2.8), the values of $^*H$ at $\left(i+\frac{1}{2}\right)$ and $\left(i-\frac{1}{2}\right)$at are undefined, these terms are indicated by the superscript *. Following [15], the values of $^*H$ is approximated using first-order upwind method as follows.

$$ {}^*H^n_{i+\frac{1}{2}} \begin{cases} H^n_i, & if\ u^n_{i+\frac{1}{2}} \geq 0 \\ H^n_{i+1}, & otherwise \end{cases} \tag{2.9}$$

The condition (2.9) above states that, when the flow is going to the right or $u_{i+\frac{1}{2}} \geq 0$, the information for $^*H_{i+\frac{1}{2}}$ values can be obtained from H_i. The other way around, when the flow is to the left or $u_{i+\frac{1}{2}} < 0$, the information for $^*H_{i+\frac{1}{2}}$ values is obtained from H_{i+1}. Since the coefficients α, β, dan γ are function of H, therefore the condition (2.9) are also applied for these coefficient for obtain values for $*\alpha$, $*\beta$, dan $*\gamma$.

Similar to the idea in [14], the approximation of the momentum equation of VB model (2.6) is given by the following relation

$$\frac{u^{n+1}_{i+\frac{1}{2}}-u^{n}_{i+\frac{1}{2}}}{\Delta t} = -g\left(\frac{\eta^{n+1}_{i+1}-\eta^{n+1}_{i}}{\Delta x}\right) - (u\partial_x u)^n_{i+\frac{1}{2}} \qquad (2.10)$$

Rather than directly subtitute the values of u in the advection term $\partial_x u$, instead, the term is calculated by using the following relation

$$u\partial_x u = \frac{1}{h}\left(\frac{\partial(qu)}{\partial x} - u\frac{\partial q}{\partial x}\right)$$

where $q = hu$ is the horizontal momentum. A consistent approximation for the advection term as suggested in [15] is as follows

$$(u\,\partial_x u)_{i+1} = \frac{1}{\bar{H}_{i+\frac{1}{2}}}\left(\frac{\bar{q}_{i+1}\cdot{}^{*}u_{i+1}-\bar{q}_i\,\cdot{}^{*}u_i}{\Delta x} - u_{i+\frac{1}{2}}\frac{\bar{q}_{i+1}-\bar{q}_i}{\Delta x}\right) \qquad (2.11)$$

where the bar sign denotes a simple interpolation on a point based on two points besides it, i.e.

$$\bar{H}_{i+\frac{1}{2}} = \frac{1}{2}\left(H_i + H_{i+1}\right), \qquad \bar{q}_i = \frac{1}{2}\left(q_{i+\frac{1}{2}} + q_{i-\frac{1}{2}}\right), \qquad q_{i+\frac{1}{2}} =.{}^{*}H_{i+\frac{1}{2}} + u_{i-\frac{1}{2}}$$

Just as $^*H$ in eq. (2.8), the $^*u$ is calculated using upwind approximation as

$$^{*}u^n_i \begin{cases} u^n_{i-\frac{1}{2}}, & if\ \bar{q}^n_i \geq 0 \\ u^n_{i+\frac{1}{2}}, & otherwise \end{cases} \qquad (2.12)$$

The last equation (2.7), i.e. the elliptic equation for solving ψ in every time step, is approximated by the following relation

$$\frac{^{*}\alpha_{i+\frac{1}{2}}(\psi_{i+1}-\psi_i) - {}^{*}\alpha_{i-\frac{1}{2}}(\psi_i-\psi_{i-1})}{(\Delta x)^2} +.{}^{*}\gamma_i\psi_i = \frac{^{*}\beta_{i+\frac{1}{2}}u^n_{i+\frac{1}{2}} - {}^{*}\beta_{i-\frac{1}{2}}u^n_{i-\frac{1}{2}}}{\Delta x} \qquad (2.13)$$

Note that in eq. (2.13) leads into a linear matrix system that has to be solved every time step in order to obtain value of ψ for a given value of u. The resulting matrix system is a tridiagonal matrix system that can be solved efficiently using Thomas' algorithm.

4. Test Cases

To test the accuracy and consistency of numerical implementation of VB model as proposed in the previous section, we use two test cases, i.e. a dispersive standing wave in a close basin and a propagation of solitary wave. Results of simulation of both cases will be compared with analytical solutions. To show the importance of dispersive effects, we compare the simulations performed with VB model with non-dispersive Shallow Water Equations (hydrostatic model) in Stelling & Duinmeijer (2003) [15].

4.1. Standing wave in a close basin

The first case to test consistency of dispersive effect of VB model, we simulate a standing wave in a close basin. Computational domain that is used is $x \in [0,20]\ m$ with a hardwall boundary condition on both sides. In the domain, we use an initial wave condition

$$\eta(x, t = 0) = 0.1\cos(k_0 x) \qquad (2.14)$$

With wave number $k_0 = \pi/20$ above depth of $h_0 = 10m$. Notice that the amplitude is small compared to the depth, therefore this case is actually a weakly nonlinear problem. On the other hand, this is a dispersive case, since $k_0 h_0 = \pi/2 = 1.57$. The exact solution of the standing wave is given in the following expression

$$\eta(x, t = 0) = 0.05[\cos(k_0(x - ct)) + \cos(k_0(x + ct))] \tag{2.15}$$

where c is the exact phase velocity, defined as

$$c = \frac{\omega}{k_0} = \sqrt{\frac{g}{k_0}\tanh(k_0 h_0)}$$

For computation of the standing wave, we use $\Delta x = 0.05m$ and $\Delta t = 0.01s$. The grid size is chosen to represent the wave to be simulated. Snapshot of the simulation at t=*0s, 1s, 1.5s* and *2.5s* are shown in Figure 2.

To compare results of simulation with analytical solution as well as non-dispersive SWE, we extract a signal at $x=1m$. The comparison is shown in Figure 3. It can be seen that the simulation of VB model give a good agreement with the analytical solution in (2.15), whereas the simulation with non-dispersive SWE gives wrong result. The SWE has faster phase velocity, i.e. $c_0 = \sqrt{gh_0}$, which causing the propagation of standing wave of SWE is too fast.

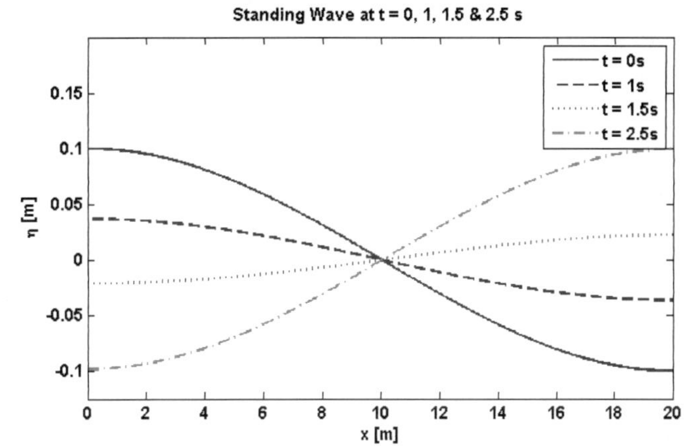

Figure 2. Snapshot of surface elevation at various times for standing wave simulation.

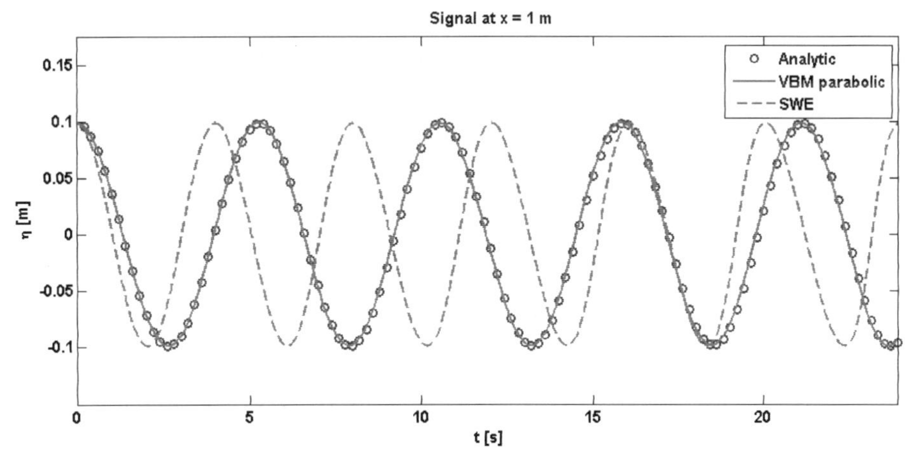

Figure 3. Comparison of signal at $x=1m$ of analytical solution (dotted blue circles), VB model (solid red line) and SWE (dashed red line) for standing wave simulation.

4.2. Solitary wave propagation

The second test case is a propagation of a solitary wave. Solitary wave is a wave that propagates with a non-disturbed wave form, which occurs when the effect of dispersion is balanced with the nonlinear effect. Therefore, to be able to simulate solitary wave accurately, the wave model (as well as its numerical implementation) should has a correct dispersion as well as nonlinearity. The exact solitary wave is given by the following expression

$$\eta(x,t) = A_0 \operatorname{sech}^2(\zeta(x - \lambda t)) \ , \ \lambda = \sqrt{g(h_0 + A)}, \ \zeta = \sqrt{\frac{3A_0}{4h_0(h_0+A_0)}}$$

Here, A_0 is the amplitude of the solitary wave, λ is the speed of the solitary wave. The solitary wave above is obtained from analytical solution of KdV equation.

We consider a domain of $x \in [0,450]\ m$ that is discretized with $\Delta x = 1m$. The length of time discretization is $\Delta t = 0.01s$. In Figure 4, snapshot of solitary wave propagation is at *t*=14*s*, 21*s* and 28*s*, for SWE model (left plot) and for VB model (right plot). Both numerical simulations are compared with the analytical solution. As the SWE has no dispersion effect, the nonlinearity of SWE become too dominant, as a result the solitary wave becomes steeper, nearly breaking. On the other hand, results of simulation with VB model show a good agreement with analytical solution. We also compare signals that are extracted at *x*=250*m* as shown in Figure 5. Just as in the snapshot plot, similar patterns are also seen in these signals, i.e. the SWE result shows steeper wave as the dispersion of the model is absent.

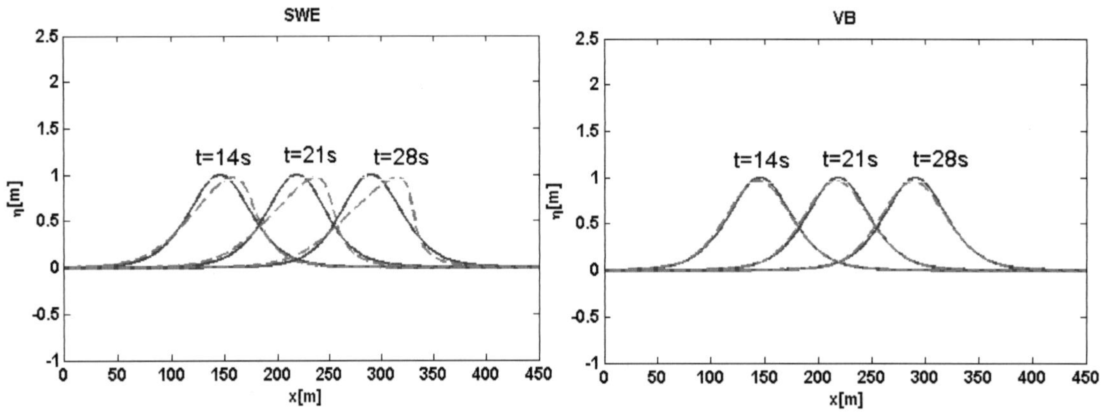

Figure. 4. Snapshot of simulation with SWE model (left plot, denoted by red dashed line) and with VB model (right plot, denoted by red dashed line) with analytical solution (solid blue line) at *t*=14s, 21s, and 28s.

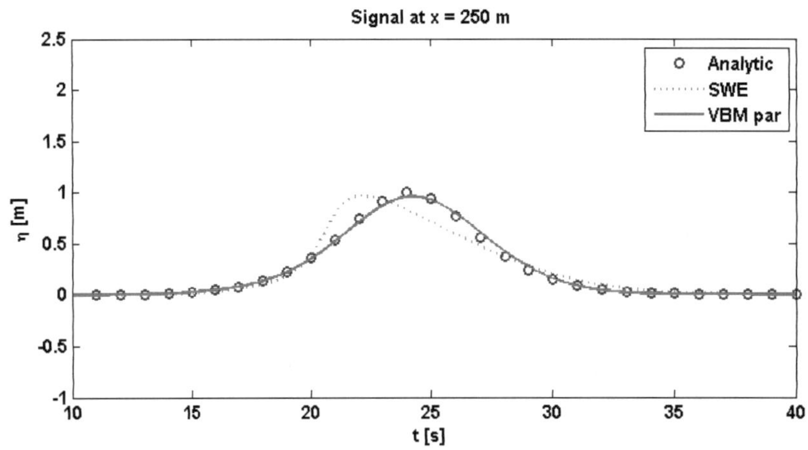

Figure. 5. Comparison of signal at *x*=250*m* of analytical solution of solitary wave (blue circles), simulation with SWE (dotted red line) and simulation with VB model (solid red line).

5. Conclusion and discussion

The original idea of the staggered grid introduced by Stelling & Duinmeijer (2003) [15] is extended for solving numerically the dispersive wave model, i.e. the Variational Boussinesq (VB) model. An additional elliptic equation in the VB model leads to a tridiagonal matrix system that is solved efficiently by using Thomas' algorithm. Two test cases are presented to test consistency of numerical implementation of the VB model, i.e. standing wave simulation and solitary wave propagation. Both cases show the importance of dispersion property of wave model. The comparison of the results of simulation with staggered grid VB model show a good agreement with analytical solution, whereas the non-dispersive staggered grid Shallow Water Equations (SWE) cannot simulate the proposed phenomena accurately. The staggered grid implementation for the 1D VB model here can be extended into 2D problem.

References

[1] Adytia D and van Groesen E 2012 *Coast. Eng.* 64 139–150
[2] Adytia D 2014 *Proc.24th Int. Ocean and Polar Engineering Conference* ISOPE 2014 912–918
[3] Adytia D and Lawrence 2016 *Proc. of the ASME 2016 35th International Conference on Ocean, Offshore and Arctic Engineering* vol 7
[4] Adytia D and van Groesen E 2011 *Proc. International Conference on Developments in Marine CFD*, in Chennai, India. RINA 25–29
[5] Adytia D 2012 *Coastal zone simulations with Variational Boussinesq Modelling*, Ph.D thesis, University of Twente, The Netherlands
[6] Adytia D and van Groesen E 2010 *Proc. Of Asian and Pacific Coasts 2009* 122-128
[7] Boussinesq J V 1872 *J. Math. Pures Appl.* 17 55–108
[8] Brocchini M 2013 *Proc R Soc A* 469 20130496
[9] Klopman G van Groesen E and Dingemans M 2010 *J. Fluid Mech.* 65 36–63
[10] Luke J 1967 *J. Fluid. Mech.* 27 395–397
[11] Lynett P J Wu T and Liu P L -F 2002 *Coast. Eng.* 46 89–107
[12] Madsen P A and Sorensen O R 1992 *Coast. Eng.* 18 183–204
[13] Nwogu O 1993 *J. Waterw. Port Coast. Ocean Eng.* 119 (6) 618– 638
[14] Peregrine D H 1967 *J. Fluid Mech.* 27 815–827
[15] Stelling G S and Duinmeijer S P A 2003 *Int. J. Numer. Methods Fluids A* 43 1329–1354
[16] Stelling G S and Zijlema M 2003 *Int. J. Numer. Methods Fluids* 43 1–23
[17] Zakharov V 1968 *J. Eppl. Mech.Tech. Phys.*, 9 (2) 190–194

Mapping online transportation service quality and multiclass classification problem solving priorities

Andry Alamsyah[1], Imam Rachmadiansyah[2]
School of Economics and Business, Telkom University, Bandung, Indonesia
[1] andrya@telkomuniversity.ac.id, [2]imamrachmadian@gmail.com

Abstract. Online transportation service is known for its accessibility, transparency, and tariff affordability. These points make online transportation have advantages over the existing conventional transportation service. Online transportation service is an example of disruptive technology that change the relationship between customers and companies. In Indonesia, there are high competition among online transportation provider, hence the companies must maintain and monitor their service level. To understand their position, we apply both sentiment analysis and multiclass classification to understand customer opinions. From negative sentiments, we can identify problems and establish problem-solving priorities. As a case study, we use the most popular online transportation provider in Indonesia: *Gojek* and *Grab*. Since many customers are actively give compliment and complain about company's service level on *Twitter*, therefore we collect 61,721 tweets in *Bahasa* during one month observations. We apply *Naive Bayes* and *Support Vector Machine* methods to see which model perform best for our data. The result reveal *Gojek* has better service quality with 19.76% positive and 80.23% negative sentiments than *Grab* with 9.2% positive and 90.8% negative. The *Gojek* highest problem-solving priority is regarding application problems, while *Grab* is about unusable promos. The overall result shows general problems of both case study are related to accessibility dimension which indicate lack of capability to provide good digital access to the end users.

1. Introduction

Online transportation service is known for its accessibility, transparency, and tariff affordability. These points make online transportation have advantages over the existing conventional transportation service. Online transportation is an example of disruptive technology that makes customers have better bargaining position than the companies. It changes the way how customers and companies interact towards conclude transactions [1]. Co-occurrence with online transportation increasing popularity, we can easily find those topics discussed on social media, such as *Twitter, Facebook,* and *Instagram.*
Gojek and *Grab* are the most popular online transportation company in Indonesia. They have over 500.000 users in 2017, based on playstore unique users. Both provide variety of service, such as delivery, financial, and public transportation. Customers are actively give compliment and complain about both company's service level on social media. Those data are extracted, analyze, and finally turn into knowledge that can help to improve their service level[1]. For this paper, we investigate which social media are the most popular to hold the discussion about both company. From table 1, we see that on average number of discussion on *Twitter* is higher than others, thus we pick *Twitter* as representative media to get the data. The previous approach of identifying problem priorities mostly based on interviews or questionnaires, which basically more time-consuming effort compared to collecting online opinions.

Table 1. Average Number of Discussion in Each Social Media during Observation Time

Company	*Social Media*		
	Twitter	*Facebook*	*Instagram*
Gojek	761/week	140/week	73/week
Grab	402/week	105/week	239/week

We apply transportation service quality [2] measurement to map each dimension proportion to the global domain problem and sentiment analysis [3] method to classify people's feeling to phenomenon. The purpose of those methods is to identify service satisfaction, to improve service quality, and to increase accuracy of decision support making. To classify service problems, we use multiclass classification method[4]. The most used multiclass classification methods are *Naive Bayes* (NB) and *Support Vector Machine* (SVM). Both classification methods are applied in this research to evaluate multiclass classification performance.
Our global objectives are first to map problems of online transportation companies and second to set problem-solving priorities. We apply both sentiment analysis and multiclass classification to understand customer thought of each dimension or in other word dimension performance. As the result, we can identify problems from set of negative opinions, then build problem solving priorities based on that information. The problem-solving priorities help companies to decide what problems should be dealt first. Problems-solving priorities arranged based on amount of negative opinions in each dimension.

2. Theoretical Background

Service quality is a conceptual model to assess services based on its performance's perception. There are 5 general service quality dimensions [4]: *tangibles, reliability, responsiveness, assurance*, and *empathy*. However, in this research, since our object is about online transportation, thus, we apply transportation service quality [2] to identify problems classification. Each opinion has contextual similarities to others classified into the same class. Multiclass classification method help to determines customer satisfaction and opinion in each class. Thus, it makes us easier to achieve research objectives. The transportation service quality dimension shown in table 2:

Table 2. Transportation Service Quality Dimensions

Service Quality Dimension	*Description*
Availability	The availability of transportation services everytime, everywhere, and in any condition.
Accessibility	The application service ease of use in certain time, condition, and area.
Information	Information availability, well informed customer such as travel fee before making the journey
Time	The detail information about departure time, arrival time, and travel duration.
Customer Service	The capabilities of company to handle complaints, suggestions; The capabilities of company to response customer inquiry in reasonable time; The information about promotional activities.
Comfort	Company effort to provide comfort to customer, such as all-weather protection, vehicles hygiene, and driving style
Safety	Company effort to provide safety and security, such as driver preparation, driving attributes, route knowledge, traffic condition awareness
Environment	Vehicle noise, and vehicles contribution in gas emission

Multiclass classification is a classification method which has more than two target classes. The target result shows classes of tweets contain specific issue related to dimensions in Table 2. Sentiment analysis is a method to analyze the user opinion or sentiment by classification. Sentiment analysis performed automatically to classify opinion based on words or phrase that formed during learning procedure. The automated sentiment analysis is important to reduce human error and processing time to identify, extract and summarize text [3]. The previous research [5] address many advantage of applying multiclass classification problem in customer relationship management area, such as faster identification, considerably good accuracy, and support real-time customer complaints. In the face of unstructured nature of online data, text mining based multiclass classification methods is preferable. The learning process become advantage in identifying future complains automatically.
To perform sentiment analysis methods, we need preprocessing procedure which consist of the several following processes [5]. First is data cleaning to remove useless items for the study, such as emoji and noise words. Second is tokenization process to separate the sentences into words, phrases, symbols called tokens. Third is filtering process to remove punctuation, hyphens, characters and high frequency intermediate words, such as *'dari', 'untuk', 'belakang', 'atas'* and *'depan'*. In addition, this step also

transform size letters into lowercase. Fourth is stemming process to remove incremental word or turn word into a basic word [6], like *'bepergian'* to *'pergi'*. Table 3 shows examples of each process in preprocessing procedure

Tabel 3. Preprocessing Example

Raw Data	Kenapa ya susah banget dapet gocar padahal armadanya banyak di sekitarku, @Go-jekindonesia? So saaad :(
Tokenization	Kenapa ya susah banget dapet gocar padahal armadanya banyak di sekitarku, @Go-jekindonesia? So saaad :(
Filtering	susah banget dapet gocar armada so saad
Stemming	susah sekali dapat gocar armada sangat sedih

To measure how important a word to a document we use TF-IDF [7] procedure. The main function of weighting process is to enhance term retrieval effectiveness. There are 2 effective factors for effective retrieval. First, terms likely to be relevant to the user's need must be retrieved. Second, terms likely to extraneous must rejected [8]. This process is done before we can classify user sentiments. First step is to give weight to each word in document using with TF-IDF formula.

$$W_{d,t} = TF_{d,t} \quad (1)$$

$tf_{d,t}$ is the term occurrences of the word *t* in documents *d.* The calculation of $tf_{d,t}$ is known to increase the recall value but does not increase the precision value. To deal with this problems terms with high frequency are usually removed.

$$idf_t = log\frac{N}{df_t} \quad (2)$$

N is the number of words contained in the text and *df* is the number of texts containing the word *t*. df_t is the number of text that contain the term t. It means Idf_t is the inverse of term t in number that existed in the documents. Idf_t represent term specificity which expected to improve precision. So, by using $W_{d,t}$ with $tf_{d,t}$ and idf_t together, the method performs better, because it addresses both aspects of recall and precision. The better recall and precision give us better overall performance. Then TF-IDF will be described as an equation.

$$W_{t,d} = tf_{t,d}\ X\ idf_{t.} \quad (3)$$

3. Research Methodology

The overall research workflow can be seen in Figure 1. We set the research objective to map service quality and problem classification to determine each opinion class category. The following step is finding the best method to perform the research. In data collection process, we decide to use *Twitter* for the reason explained in Introduction chapter. Our focus to choose the best sentiment analysis methods, between NB and SVM methods in term of the prediction accuracy multiclass classification problem regarding service quality. Both methods have similar performance in computation time, thus we can ignore this aspect.

We collect all *Twitter* data concerning *@gojekindonesia* which represent *Gojek* and *@GrabID* which represent *Grab* during the data collection time. Following the data collection process, we do data preprocessing procedure, which consist removal irrelevant tweet; noise removal by tokenizing, stemming, filtering; weighting the words with TF-IDF. After preprocessing procedure, we build training and testing data before applying NB and SVM methods.

Model evaluation are done by measuring recall, precision, f-measure, and kappa value of each methods. The last step is to analyze the result and determine the best way to achieve research goals to establish problem-solving priorities.

Data collecting process is done for the period of one month in March 2017. We collect in total 61,721 tweets, consist of 49,947 tweets associated with *@Gojekindonesia* and 11,774 tweets associated with *@GrabID*. After preprocessing step, we obtain 1,463 tweets for *@Gojekindonesia* and 1,098 tweets for *@GrabID*. We classify those tweets into 8 different dimensions of transportation service quality. The examples of tweet classification based on transportation service quality dimension is in Table 4. The example of sentiment classification tweets is in Table 5. The example of preprocessing to of raw data into clean data is in Table 6.

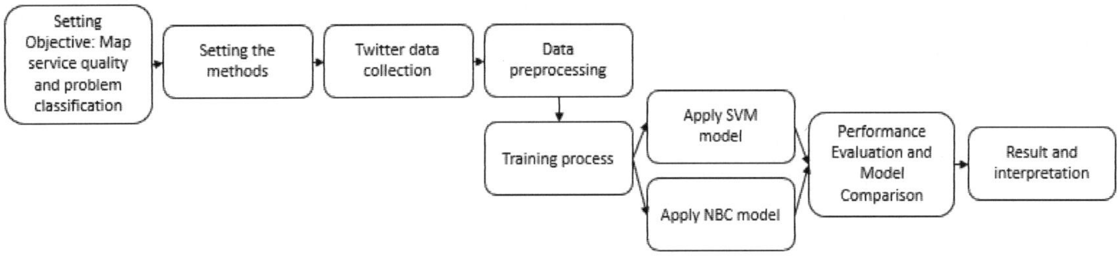

Figure 1. Research Workflow

Table 4. Example of Tweet Classification based on Transportation Service Quality Dimension

Example Tweets	*Dimension*
Susahnya Go-jek pada jam jam mau berangkat :"	*Availability*
@Go-jekindonesia min saya gagal verifikasi kode akun Go-jek	*Accessibility*
Karena go-mart cukup mahal	*Information*
Pesen Go-jek abangnya lama banget huhu padahal ga terlalu jauh dari lokasi kasian ibu gue nungguin :(	*Time*
Thanks to gopay gratisan dari Go-jek	*Customer Service*
Pertama kali ini Go-jek ada yg cakep plus wangiiiii...sukaaa	*Comfort*
Termengerikan adalah naik Go-jek yg super aneh kayak baru belajar naik motor takut	*Safety*
Go-jek parkir semprawut gak jelas sampai di trotoar	*Environment*

Table 5. Example of Tweet Classification by Sentiment

Tweet	Sentiment
Terima kasih bapak Go-jek baik hati yang mau antri lama demi saya ngidam ini	Positive
Go-jek ini Tarifnya gak masuk akal	Negative

Table 6. Preprocessing Example

Raw Data	*Clean Data*
App Gojek giliran lagi dibutuhin error hhhhh sebhul.	Aplikasi error sebal

We apply TF-IDF procedure to weight the words using *Rapidminer* application. The result determines each word into classes and omit words to increase the multiclass classification performance, for example, *'aplikasi'* has 0,134 weight in *accessibility* dimension, while *'bayar'* has 0 weight in *accessibility* and *information* dimension, so *'bayar'* should be deleted to increase the performance of multiclass classification.

Problem-solving priorities applied to see what problem must be dealt first in order. It intended to allocate companies' resource efficiently. We use multiclass classification and sentiment analysis to detect problems by each class negative opinion quantities. The largest negative quantities classes assigned to be the highest prioritized dimension, while the largest negative quantities of words in classes assigned to be the highest prioritized problems. For example, *availability* dimension attains 142 negative sentiment, *accessibility* dimension attains 210 negative sentiment, *information* dimension attains 134 negative sentiment and *time* dimension attains 78 negative sentiment. So, the top 3 prioritized dimension are *accessibility, availability* and *information*. Moreover, as example, *time* dimension has *'late pickup'* with 130 times mentioned while *'slow travel duration'* with 145 times mentioned. The highest prioritized problem should be *'slow travel duration'*.

4. Result and Analysis

NB classification methods has good accuracy for both object. The value of both kappa is around 0.80 which predicated a nearly perfect agreement of its classification model, while precision and recall value reach between 80-95%, with F-measure 91.46% and 89.36% as harmonization of precision and recall value indicating the classification process is going well. NB determine words naively.

SVM classification model has higher accuracy than NB as shown in Table 7 and Table 8. It shows that SVM kappa and F-measure for is better than NB. As the result, for this study SVM is more suitable method to do the classification task. SVM model consider each word class before determining the word included in specific class.

Table 7. Naive Bayes Text Classification Performance

Company	Accuracy	Kappa	Precision	Recall	F-measure
Gojek	93,92%	0.823	88.64%	94.47%	91,46%
Grab	96,58%	0.786	90.77%	88.00%	89,36%

Table 8. Support Vector Machine Text Classification Performance

Company	*Accuracy*	*Kappa*	*Precision*	*Recall*	*F-measure*
Gojek	97,26%	0.910	97,66%	93,66%	95,61%
Grab	97,26%	0.809	98.54%	85.00%	91,27%

Both *GoJek* and *Grab* are actively engaged in interaction with customers through social media. These interactions crawled and mapped to determine customer perceived quality. On Table 9, we can see the sentiment proportion of both companies. *Gojek* obtain better opinion for their service than Grab, with 19.76% against 9.2% positive.

Table 9. The Tweet Sentiment Proportion

Company	*Total Tweet*	*Positive*	*Negative*
Grab	1462	134 (9.2%)	1328 (90.8%)
Gojek	1098	217 (19.76%)	881 (80.23%)

Table 10. *Gojek* Dimension Proportion

Sentiment	*Availability*	*Accessibility*	*Information*	*Time*	*Customer Services*	*Comfort*	*Safety*	*Environment*
Positive	25.3%	1.3%	17.9%	9.6%	18.8%	21.1%	5.5%	0.5%
Negative	24.1%	34.7%	13.6%	8.4%	6.7%	7%	4.9%	0.6%

The top-3 negative dimension in *Gojek* are *accessibility, availability* and *information. Gojek* problems in *accessibility* dimension are application problems, where 273 times mentioned about '*aplikasi*', '*error*', '*sistem*', and '*server*'; difficult to logging in or verification problem, where 182 times mentioned about '*akun*', '*log*', '*masuk*', '*keluar*', '*password*', '*kode*' and '*verifikasi*'; ordering transportation service, where 72 times mentioned about '*pesan*' and '*invalid*'. Problems in *availability* dimension are driver ignorance and difficult to get transportation service, where 244 times mentioned about '*ambil*', '*dapat*', '*susah*', '*pesan*', '*tiada*', '*lama*', '*jemput*', and '*invalid*'; unilateral cancellation or driver's cancellation request, where 30 times mentioned about '*batal*' and '*tolak*'. Problem in *information* dimension are the expensive tariff mentioned 77 times about '*mahal*', '*bayar*' and '*tarif*'; gopay truncated due unilateral cancellation from driver mentioned 33 times about '*batal*', '*potong*', and '*ambil*'.

We use the result above to set *Gojek* problem-solving priorities. There are 3 problems occurred to be the most mentioned from 2 different dimension. So, the table 11 below are the problems that should be dealt first.

Table 11. Gojek Problem-Solving Priorities

No	*Problems*	*Mentioned*	*Solution*
1	Application problems	273 Times	• Improve application performance and system optimization • Conduct research about most used phone in the market and user habit
2	Driver ignorance and difficult to get transportation service	244 Times	• Create reward and punishment system for the driver

3	Difficult to logging in or verification problem	182 Times	• Create auto login from same device • Integrate the application with social media service to get alternatives code access verification • Send login confirmation to email

Table 12. *Grab* Dimension Proportion

Sentiment	*Availability*	*Accessibility*	*Information*	*Time*	*Customer Services*	*Comfort*	*Safety*	*Environment*
Positive	1.5%	3.8%	21.4%	6.6%	13.3%	45.3%	7.4%	0.7%
Negative	10.3%	23%	18.4%	3.4%	33.7%	7.1%	3.6%	0.5%

The top-3 negative dimension in *Grab* are *customer service, accessibility* and *information*. The problems in *customer service* dimension are unusable promos, where 375 times mentioned about '*promo*', '*diskonseharian*' and '*kode*'; no response from admin, where 134 times mentioned about '*admin*' and '*komplain*'; no *grabpay* bonus received, where 364 times mentioned about '*bonus*', '*grabpay*', '*top*', '*isi*', '*masuk*', and '*gratis*'. The problems in *accessibility* dimension are unusable grabpay credit, where 341 times mentioned about '*masuk*', '*grabpay*', '*isi*', '*top*', '*bayar*', and '*lama*'; application error, where 107 times mentioned about '*aplikasi*', '*error*', and '*update*'; difficult to order and cancel, where 64 times mentioned about '*dapat*', '*batal*', '*invalid*' dan '*pesan*'. The problems in *information* dimension are *grabpay* value reduction due improper cancellation, where 136 times mentioned about '*grabpay*', '*potong*', '*batal*', '*pesan*', and '*jemput*'; high tariff, where 55 times mentioned about '*word*', '*mahal*', '*harga*', and '*tarif*'.

We use the result above to set *Grab* problem solving priorities. There are 3 problems occured to be the most mentioned from 2 different dimension. So, the table 14 below are the problem that should be dealt first.

Table 13. Grab Problem-Solving Priorities

No	*Problems*	*Mentioned*	*Solution*
1	Unusable promos	375 Times	• Arrange the promo frequency wisely • Make sure the promos work exactly as term made • Inform it to customer about the time and usage limit
2	No grabpay bonus received	364 Times	• Improve server capacity to build faster response and prevent overload request • Anticipate current and prospective customer before work the bonus plan.
3	Grabpay credit refill process taking long time	341 Times	• Improve server capacity, faster response, and prevent overload request. • Co-operate with reliable company, to prevent any obstacle company, such as reload grabpay balance delayed because bad access

Conclusion and Recommendation

Both multiclass classification, NB and SVM are able to classify the datasets well, but SVM performance is better than NB measured by kappa, precision, recall, f-measure and accuracy. So, it is better to use SVM as multiclass classification model for Indonesian *Twitter* dataset.

Gojek sentiment rating with 19.76% positive and 80.23% negative are better than Grab with 9.2% positive and 90.8% negative. The proportion of dimensions in *Gojek* are *accessibility, availability, information, comfort, customer service, time, safety* and *environment*. While *Grab* are *customer service, accessibility, information, comfort, availability, safety, time* and *environment*.

The main problems of *Gojek* are application problems, difficult to get transportation service, and difficult to logging in. The main problem of *Grab* are unusable promos, no response from admin, and no *grabpay* bonus received. Both companies suggested that the prioritized top-3 problems should be dealt as soon as possible, before negative opinion among customer get out of control. Overall, arising problems of both online transportation companies are related to *accessibility* which indicate lack of capability to provide good digital access to the society. Hence, we conclude sentiment analysis and multiclass classification are capable to establish problem-solving priorities.

References

[1] Wilson, T. D. 2013. The e-book phenomenon: a disruptive technology. *Proceeding of the International conference Publishing – trends and contents.* UDC: 655:[002:004](485)=111.

[2] Poliakova, A. 2010. Application of the Customer Satisfaction Index(CSI) to transport services. *eJournal of Slovak University of Technology in Bratislava*. **5** 208-215.

[3] Liu, B. 2012. *Sentiment analysis and opinion mining*. (California: Morgan & Claypool Publisher).

[4] Parasuraman, A., Berry, L. L., and Zeithaml, A. 1991. Perceived Service Quality as a Customer-Based Performance Measure: An Empirical Examination of Organizational Barriers Using an Extended Service Quality Model. *Human Resource Management.* **30** 335-364.

[5] Arusada, M. D. N., Putri, N. A. S., and Alamsyah, A. 2017. Training data optimization strategy for multiclass text classification. *The 5th International Conferences on Information and Communication Technology.* ISBN: 978-1-5090-4911-0.

[6] Han, J., Kamber, M., and Pei, J. 2012 . Data Mining Concepts and Technique. 3rd Edition (Waltham: Elsevier).

[7] Tokunaga, T. and Iwayama, M. 1994. Text categorization based on weighted inverse document frequency. *Technical Report of Departemen of Computer Science Tokyo Institute of Techonolgy*. ISSN 0918-2802

[8] Salton, G. and Buckley, C. 1988. Term-weighting approaches in automatic text retrieval. *Information processing & management* **24** 513-523.

IOP Conf. Series: Journal of Physics: Conf. Series **971** (2018) 012022 doi:10.1088/1742-6596/971/1/012022

Analysis OpenMP performance of AMD and Intel architecture for breaking waves simulation using MPS

M N A Alamsyah[1,*], A Utomo[2] and P H Gunawan[3]

School of Computing, Telkom University, Jalan Telekomunikasi No. 1 Terusan Buah Batu, Bandung 40257, Indonesia.

E-mail: [1]alamsyah@student.telkomuniversity.ac.id;
[2]ardhitoutm@student.telkomuniversity.ac.id;
[3]phgunawan@telkomuniversity.ac.id.
*Corresponding author

Abstract. Simulation of breaking waves by using Navier-Stokes equation via moving particle semi-implicit method (MPS) over close domain is given. The results show the parallel computing on multicore architecture using OpenMP platform can reduce the computational time almost half of the serial time. Here, the comparison using two computer architectures (AMD and Intel) are performed. The results using Intel architecture is shown better than AMD architecture in CPU time. However, in efficiency, the computer with AMD architecture gives slightly higher than the Intel. For the simulation by 1512 number of particles, the CPU time using Intel and AMD are 12662.47 and 28282.30 respectively. Moreover, the efficiency using similar number of particles, AMD obtains 50.09 % and Intel up to 49.42 %.

1. Introduction

Making a mathematical model for solving a real life problem is needed, since it will be unpractical and time consuming to be solved by direct experiment. For instance in wave problems, obtaining solutions with a simulation is more viable than doing an experiment. This is due to experiments are depended by many unnecessary factors, such that it is hard to focus on the wave's movement only.

In every atmospheric models, Navier-Stokes equation is used because it is a non-linear differential equation which can describe the flowing of fluid substances such as a gas or water. One of the example in a branch of atmospheric model is the wave problem. This problem can be approximated by using either grid or particle methods. Generally, both of those methods have their own characteristics. In the grid method, a model is constructed based on the value of macroscopic variables of its domain. An example of grid method is known as a robust method which is called a Finite Volume Method (FVM), see in [1] for more detail. Unlike grid method, the particle method provides the solutions in a form of particles, which gives more detail and comprehensive result. This is because of the deformation of the topology can be analyzed on each of particles. Several particle methods are available in some references, for instance, a Finite Pointset Method (FPM)[2], Smoothed Particle Hydrodynamics (SPH)[3], and Moving Particle Semi-Implicit (MPS) [4].

Based on the given solution, particle method is separated into two types, probabilistic and deterministic model. The probabilistic model works by seeing each particle statistics in a big picture, which means the accurate average value is gained from seeing the overall particles alteration. Whereas, deterministic particle model calculation depends on each particle in every loop, which decreases the computation load and memory usage.

Particle method is a numerical method of Lagrangian model, which is commonly used to approximate atmospheric models [5]. Here, Lagrangian model works by determining the movement of particles by calculating wind effect, buoyancy force, and turbulence effect [6]. This model indeed in contrast with the Eulerian model, which works with a grid system that monitors atmospheric properties such as temperature, pressure, and chemical concentration. Moreover, Eulerian model may ensued errors in spatial discretization that can be avoided by using the Lagrangian method [6].

The calculation time of particles movement takes a long time simulation, since the number of iteration is immense and also the use of a large amount of particles. Therefore in this paper, the parallel computing will be implemented to accelerate the computational time. Here, multicore architecture which is OpenMP platform will be used to parallelize code in the implementation of MPS method.

2. Moving Particle Semi-Implicit Method

In this section, the MPS method will be elaborated trough the governing equation, particle dynamics and kernel function.

2.1. Governing equation

Navier-Stokes equations are the main foundation in every atmospheric model. These equations, which are used in this paper, express the conservation of two forms: mass and momentum[7]. These equations are written as,

$$\partial_t \rho + \nabla \cdot (\rho u) = 0, \tag{1}$$

$$\partial_t u + u(\nabla \cdot u) = -\frac{1}{\rho}\nabla p + F + \frac{\mu}{\rho}\nabla^2 u. \tag{2}$$

where ρ is the density function, and u is the velocity function, p is the pressure function, F is the force, and μ is the viscosity of the fluid. Equations (1) and (2) are called a mass and momentum equation respectively. Here, the solution of Equations (1) and (2) will be approached using Lagrangian model. In the next section, the particle dynamic based on Equations (1) and (2) will be given.

2.2. Particle dynamics

The fluid used in this simulation is an incompressible fluid, which means the density of the fluid is unlikely to change over time, in other word $\partial_t \rho = 0$. Then it leaves Equation (1) becomes $\nabla \cdot (\rho u) = 0$. Hence, Equation (2) can be rewritten as

$$\frac{Du}{Dt} = -\frac{1}{\rho}\nabla p + F. \tag{3}$$

According to Equation (3), there are two sources of particles acceleration, vector gradient of pressure and external force. In MPS method, the solving process of Equation (3) is separated into two steps. The steps can be seen as the predictor-corrector steps.

Particles velocity caused by an external force is calculated first explicitly with Equation (4) which is given as

$$u_i^* = u_i^n + \Delta t F, \tag{4}$$

where u_i^* is the temporal velocity of particle-i, u_i^n is the velocity of particle-i at time step n, and Δt is the time step interval [8]. Then, the vector gradient of pressure is taken into account after the pressure is solved implicitly as described in Equation (5):

$$u_i^{n+1} = u_i^* - \frac{1}{\rho}\nabla p, \tag{5}$$

where u_i^{n+1} is the velocity of particle-i at time step $n+1$.

In particle term, the vector gradient of pressure in Equations (2),(3), and (5) is a vector gradient between particle-i and particle-j of scalar function p and is defined as

$$\nabla p_i = \frac{p_j - p_i}{|r_j - r_i|^2}(r_j - r_i). \tag{6}$$

However, the interaction between particles can occur with any combination of two particles, since a particle has several neighboring particles. Therefore, the accumulated vector gradient of pressure in particle-i is formulated as

$$\nabla p_i = \frac{d}{n_0}\sum_{j\neq i} \frac{p_j - p_i}{|r_j - r_i|^2}(r_j - r_i)\cdot w(|r_j - r_i|), \tag{7}$$

where d is the number of spatial dimensions, n_0 is the particle number density, and w is the kernel function which weights each neighboring particle influence and depends on the distance between particle-i and particle-j [4]. This kernel function can be found in Subsection 2.3 for more detail.

2.3. Kernel function

Kernel function is a function controlled by a condition either a particle is ranged by the effect of its surrounding or not. The condition is mainly affected by an area bounded by $r_e \in \mathbb{R}^+$ around a particle. If some particles are inside an area of r_e of a particle, the existence of those particles will affect that particle. Thus, the formulation of kernel function is given as

$$w(r) = \begin{cases} \dfrac{r_e}{r} - 1, & 0 \leqslant r < r_e \\ 0, & r \geqslant r_e \end{cases} \tag{8}$$

From Equation (8), it is explained that a particle can not be placed at the same spot as another particle, because the gap of particle-i and -j, $r = |r_j - r_i|$, of both particles is going to be 0 and the weighting function to be infinity.

3. Parallel Algorithm and Implementation

3.1. OpenMP and shared memory parallelism

Parallel computing is a technique in the computational field which some tasks in a program are processed concurrently by many processors such that the execution time does not explode despite the huge computation load. In general, there are three models of parallelism commonly used based on the architecture of CPUs and memories:

(i) Shared memory model,

(ii) Distributed memory model,

(iii) Hybrid model.

Each of models has some Application Program Interfaces (APIs) which can be implemented and are still developed. The shared memory model has POSIX Threads and OpenMP as its API. Meanwhile, there is Message Passing Interface (MPI) as a standard API for distributed memory model. As for hybrid model, it uses a combination of an API of each model. For more details see [9]. In this paper, the shared memory parallelism with OpenMP as the API will be used since its simplicity [10, 11, 12, 13, 14, 15].

Shared memory parallelism is implemented in this paper because of its memory addressing and accessing concept is easy and user-friendly from the programmer's perspective. Moreover, data sharing between processes is fast and uniform due to the proximity of memory to CPUs [9]. In the other side, OpenMP is used as the API because of the facts that it can be implemented on many different platforms (multi-platform). Moreover, it is compiler directive based, thus it easier to develop a parallel code from serial directly with a full control over parallelization to programmers [16].

3.2. Algorithm

The algorithm of MPS in this paper is classified into nine main processes. Six of them can be parallelized entirely, one of them partially, and the rest cannot be parallelized at all. The general overview of processes and its order is shown at Figure 1. The description and the explanation of each process are described as follows,

- **Initial Position**: generating particles by initializing their coordinates, velocity, and pressure value.
- **Particle Density**: calculating particle concentration of each particle with the formula in Equation (9):

$$n_i = \sum_{j \neq i} w(|r_j - r_i|), \tag{9}$$

  where n_i is the number density of particle-i and w is the kernel function which is described in Equation (8) [4, 8]. This process is done at the beginning of the simulation and every time iteration.
- **Determine Boundary**: determining particles which become the free surface of the fluid with condition $n_i < n_0\beta$, where n_0 is the initial number density and β is the boundary parameter in range $(0, 1)$ since particles at free surface have less neighboring particles[4][8]. This process also is done at the beginning of the simulation and every time iteration.
- **Set dt**: calculating the optimal time step interval of every iteration, so that the stability of the numerical simulation is conserved, yet the number of total iteration needed does not blow up. The interval is bounded by particles velocity and fluid viscosity.
- **Explicit Step**: calculating the velocity of each particle caused by source terms and updates its position.
- **Collision Step**: calculating the momentum of colliding particles then updates their velocity and position.
- **Poisson Solver**: generating the linear equations system of pressure of particles.
- **LU Decomposition**: is a sub-process of Poisson Solver to solve the equations.
- **Implicit Step**: calculating particles velocity caused by pressure gradient then updates their velocity and position.
- **Print Particle Data**: writing particles position and pressure into text files as the data output.

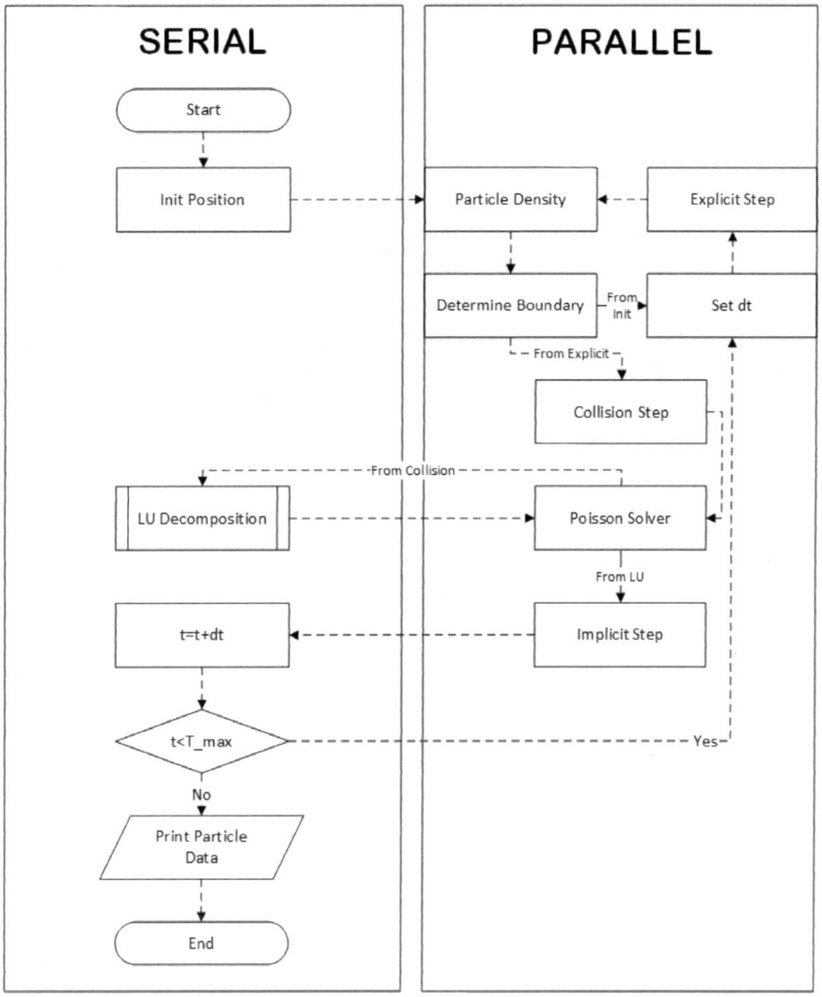

Figure 1. Flowchart of MPS implementation in serial and parallel part.

Note that each processor executes the parallel part in Figure 1 with the number of particles is defined in maximum $\lceil Total\ particles/p \rceil$ where p denotes the number of processors. Moreover, each processors in parallel part is executed in same time.

4. Numerical Results and Parallel Performances

4.1. Breaking waves problem

Wave Breaking is a phenomenon of fluid dynamics commonly seen at ocean and coastal area. This problem is simulated often due to its great impact to abrasion, man-made structures, and even the ocean current itself [17]. Here, the problem is about to see the profile of water surface at a breaking wave. The program simulates a wave tank in which breaking waves are happened. The configuration of the wave tank initial condition is shown in Figure 2. In order to generate waves, a wave paddle is given in the left boundary, which oscillates in x direction with its velocity depends on time $f(t) = 0.2\pi \sin(4\pi t)$. The wave tank also has a topography with slope $m = \frac{1}{2}$ in the right side. Here, the simulation is observed for 5 seconds and the wave paddle has oscillated 10 times since the frequency is 2 Hz based from $f(t)$.

Several simulation for different numbers of particles (489, 646, 833, 1037, 1263, and 1512

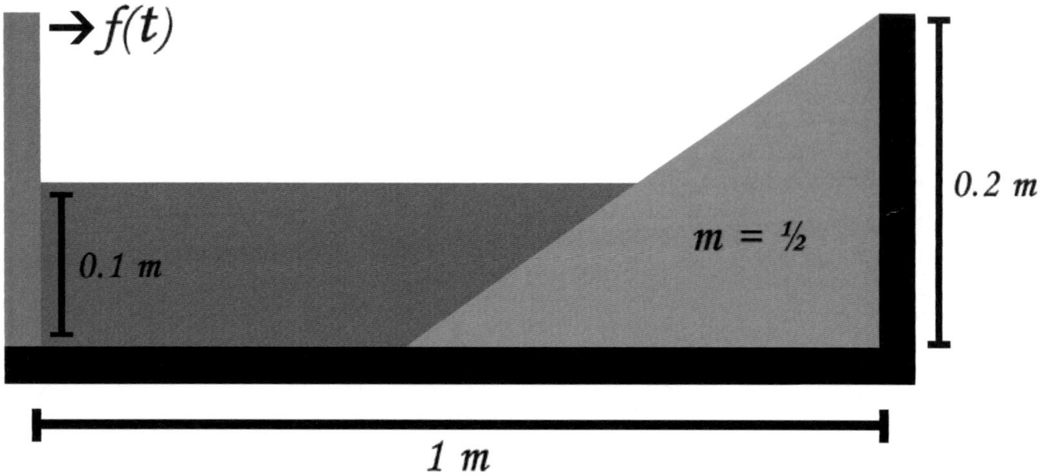

Figure 2. The wave tank setup used for breaking wave simulation.

particles) are given in this paper. Those numbers of particles are depend on the particle resolution. The particle resolution of 489 particles simulation is 50, which means there are 50 particles in a line along a unit length. Then, particle resolutions for particle numbers 646, 833, 1037, 1263, and 1512 are 60, 70, 80, 90, 100 respectively. Results of some simulations are shown in Figure 3.

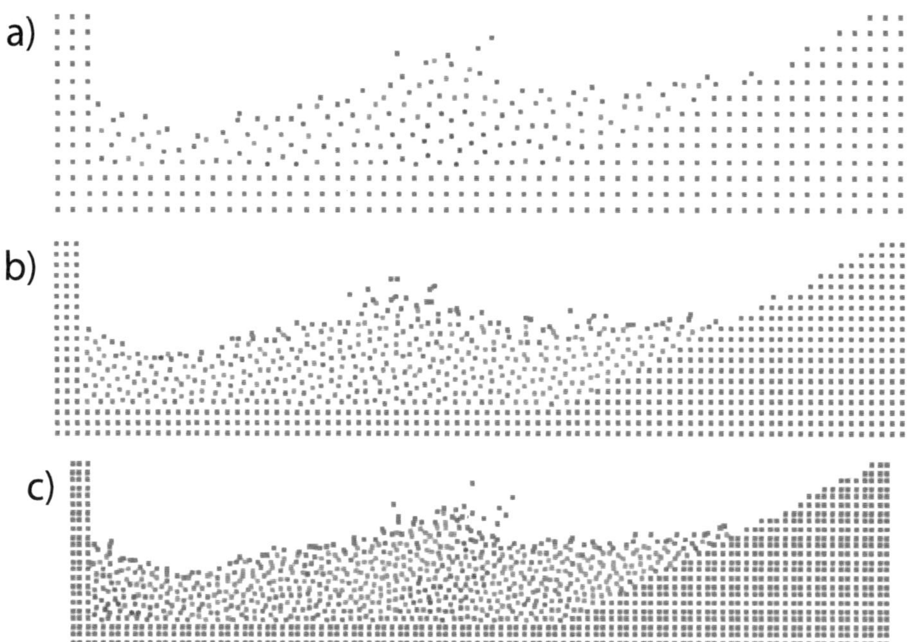

Figure 3. Breaking wave simulations using MPS method at $t = 5$ seconds with different numbers of particles: a) 489 particles, b) 1037 particles, and c) 1512 particles.

From Figure 3, the resolution of the results is depend on the number of particles. Using number of particles 1512 produces higher resolution than using 489 particles. However, at the same time, the water profile is shown similar for each simulation. In other side, using a large amount of particles, the computational time will increasing even to get the small final time of simulation. The analysis of parallel performance to handle the simulation time will be explained in the next subsection.

4.2. Computation details

In this experiment, the simulations are run on two computers with different CPU architecture, Intel and AMD. Details of the specification is shown at Table 1. Although both of them have different number of processors, these simulations only use 4 processors for the sake of relevance of this comparison analysis.

Table 1. The specifications of computer experiment.

Model Name	**Intel(R) Core(TM) i5-2500**	**AMD Opteron 6376**
Number of processor(s)	4	16
Clock frequency (MHz)	1600	2600
RAM size (GB)	8	8
Cache size (KB)	6144	512
Operating System	Ubuntu 12.04.3 LTS	Ubuntu 14.04.3 LTS

In parallel performance analysis, two benchmark tools are used which are speedup and efficiency. Speedup shows how much the parallel implementation affects the execution time. Speedup number is obtained by dividing the serial execution time by the parallel execution time. Meanwhile, efficiency shows how well the parallel implementation utilizes the resource. Efficiency number is obtained by dividing the speedup number by the total number of processors used, in this case is 4 processors. Details of speedup and efficiency of these simulations is shown in Table 2.

It can be noticed that according to the execution time in serial and parallel, the execution time of Intel processor is far better than the AMD one. The Intel processor only needs half time the AMD processor needed to run the simulation. However, from Table 2, the speedup and efficiency of AMD processor has a slightly higher than the Intel processor in general.

Note that the execution time using Intel is observed higher than AMD, since from Table 1, cache size of Intel is given larger than AMD. The size of cache in Intel is proportional to the size of the total particles in simulation, thus high speed of memory interaction is obtained.

5. Conclusion

The parallel implementation of MPS for breaking wave simulation has been elaborated. In the term of algorithm, the efficiency of parallel implementation of MPS stays 50.66% in average, since not all processes can be parallelized, even some processes that can be entirely parallelized still have a large chunk of serial code in each processor. Moreover, MPS method has a linear equations system which needs to be solved and the solver algorithm used in this paper is LU Decomposition which has an immense computation load. Then, in the term of CPU architecture, AMD gives slightly better result in efficiency of parallelization with 50.09% compared with Intel which has 49.42% for 1512 particles. However, Intel gives far better result in CPU time of parallel code with 12662.47 seconds of execution time, less than half of the execution time needed by AMD with 28282.30 seconds.

Table 2. Details of speedup and efficiency of simulations for each number of particles.

Processor	Number of Particles	Execution Time (s)		Speedup	Efficiency (%)
		Serial	Parallel		
Intel	489	209.62	102.04	2.05	51.36
	646	616.30	310.88	1.98	49.56
	833	1770.36	880.20	2.01	50.28
	1037	4545.67	2321.00	1.95	48.96
	1263	11037.56	5765.20	1.91	47.86
	1512	25031.81	12662.47	1.98	49.42
AMD	489	577.90	257.47	2.24	56.11
	646	1726.40	788.23	2.19	54.76
	833	4794.59	2314.30	2.07	51.79
	1037	11201.86	5839.13	1.92	47.96
	1263	25392.28	12763.27	1.99	49.74
	1512	56671.11	28282.30	2.00	50.09

References

[1] Eymard R, Gallouët T and Herbin R 2000 *Handbook of numerical analysis* **7** 713–1018
[2] Onate E, Idelsohn S, Zienkiewicz O and Taylor R 1996 *International journal for numerical methods in engineering* **39** 3839–3866
[3] Liu G R and Liu M B 2003 *Smoothed particle hydrodynamics: a meshfree particle method* (World Scientific)
[4] Koshizuka S and Oka Y 1996 *Nuclear science and engineering* **123** 421–434
[5] GOTOH H and SAKAI T 1999 *Coastal Engineering Journal* **41** 303–326
[6] Lagzi I, Meszaros R, Gelybo G and Leelossy A 2013 *Atmospheric Chemistry* (hgpu.org)
[7] Temam R 1984 *Navier-stokes equations* vol 2 (North-Holland Amsterdam)
[8] Koshizuka S, Nobe A and Oka Y 1998 *International Journal for Numerical Methods in Fluids* **26** 751–769
[9] Barney B 2012 *Available on https://computing. llnl. gov/tutorials/parallel comp*
[10] Gunawan P H 2016 *Information and Communication Technology (ICoICT), 2016 4th International Conference on* (IEEE) pp 1–5
[11] Mulyani, Putri N D and Gunawan P H 2017 *Information and Communication Technology (ICoIC7), 2017 5th International Conference on* (IEEE) pp 1–5
[12] Pahlevi M R and Gunawan P H 2017 *Information and Communication Technology (ICoIC7), 2017 5th International Conference on* (IEEE) pp 1–4
[13] Juliati S and Gunawan P H 2017 *Information and Communication Technology (ICoIC7), 2017 5th International Conference on* (IEEE) pp 1–5
[14] Alamsyah M N A, Simanjuntak C A, Bagustara B A R H, Pradana W A and Gunawan P H 2017 *Information and Communication Technology (ICoIC7), 2017 5th International Conference on* (IEEE) pp 1–6
[15] Iryanto and Gunawan P H 2017 *Information and Communication Technology (ICoIC7), 2017 5th International Conference on* (IEEE) pp 1–5
[16] Barney B 2008 *Website: https://computing. llnl. gov/tutorials/openMP*
[17] Cokelet E 1977 *Nature* **267** 769–774

Simulating dam-break over an erodible embankment using SWE-Exner model and semi-implicit staggered scheme

M D Ambara[1,*] and P H Gunawan[2]

School of Computing, Telkom University, Jalan Telekomunikasi No. 1 Terusan Buah Batu, Bandung 40257, Indonesia.

E-mail: [1]`makotoda@student.telkomuniversity.ac.id`;
[2]`phgunawan@telkomuniversity.ac.id`
`*Corresponding author`

Abstract. The impact of a dam-break wave on an erodible embankment with a steep slope has been studied recently using both experimental and numerical approaches. In this paper, the semi-implicit staggered scheme for approximating the shallow water-Exner model will be elaborated to describe the erodible sediment on a steep slope. This scheme is known as a robust scheme to approximate shallow water-Exner model. The results are shown in a good agreement with the experimental data. The comparisons of numerical results with data experiment using slopes $\Phi = 59.04$ and $\Phi = 41.42$ by coefficient of Grass formula $Ag = 2 \times 10^{-5}$ and $Ag = 10^{-5}$ respectively are found the closest results to the experiment. This paper can be seen as the additional validation of semi-implicit staggered scheme in the paper of Gunawan, et al (2015).

1. Introduction

Floods from heavy rains can create flood waves that can damage or even destroy river dikes in the form of sediments. The propagation of flood waves that pass through such erosion-like areas are greatly influenced by water flow and sediment. The interaction of flood with the sediments causes damage to the surrounding environment, even can threaten human life. Therefore, a simulation is needed to find out more about the influence of water waves on the sediment such that the possible impacts can be detected and prevented earlier.

Experimental activities to find out more about the impact of water flow on sediments have been discussed and conducted at the LIA Lab (Laboratory of Water Engineering), University of Cassino and Southern Lazio, 2016 [1, 2]. The configurations of this experiment is shown in Figure 1.

In the paper of Di Cristo et al [1], numerical solutions using mixed cell-centered finite-volume (CCFV) and node-centred finite-volume (NCFV) discretization methods have been proposed. The results show that the model allows for a logical simulation of free surface elevation experimental trends independently of the geofailure operator. In this paper, the another model will be used to simulate this experiment. This model is known as the SWE-Exner model which has been applied in many applications such as erodible damreak, subcritical steady and transcritical flow over a bump, etc.

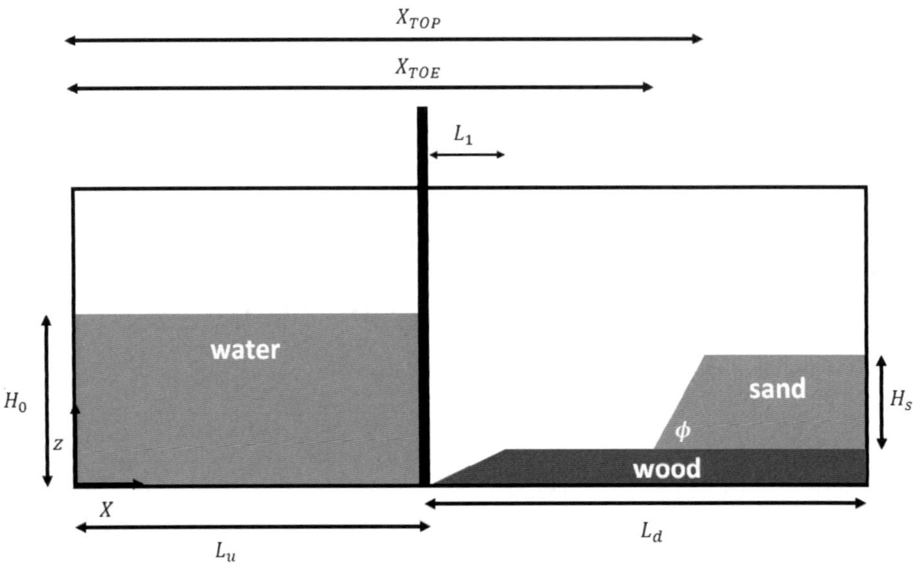

Figure 1. The lateral view of the experiment with the parameters $H_0 = 0.20$ m, $H_S = 0.10$ m $L_u = 3.00$ m, $L_d = 3.03$ m and $L_1 = 1.00$ m. The value of Φ are depend on X_{TOP} and X_{TOE}.

Moreover, in order to approximate the solution of SWE-Exner model, a robust semi-implicit staggered grid scheme will be used (this scheme is introduced in [3] in detail). This scheme is a very appropriate method for estimating the Exner-shallow water equation for bedload sediment modeling. Several numerical tests in [3] are shown in good convergence properties to analytical solutions and match well-designed data experiments in dam-break on an erodible embankment.

2. The 1D SWE-Exner and Staggered Scheme

2.1. Water-bedload sediment model

Sediment transport model by fluid flow can be described with a couple of hydrodynamic and morphological sediment change model. Hydrodynamic component can be used to simulate the fluid flow, in this case shallow water equations can be considered. Meanwhile, the morphological sediment change can be given by transport equation.

In hydrodynamic model, the shallow water equations are written as follows

$$\frac{\partial h}{\partial t} + \frac{\partial q}{\partial x} = 0, \tag{1}$$

$$\frac{\partial q^2}{\partial t} + \frac{\partial}{\partial x}\left(\frac{q^2}{h} + \frac{1}{2}gh^2\right) + gh\frac{\partial(z-H)}{\partial x} + S_f = 0, \tag{2}$$

where $h(x,t)$ is the water depth, $q(x,t) = h(x,t)u(x,t)$ the discharge, $u(x,t)$ the average velocity, H the fix bedrock bottom, $z(x,t)$ the sediment, $S_f = g\eta^2|u|uR_h^{-\frac{4}{3}}$ the friction term, and R_h is hydraulic ratio that can be approached by h [4]. The numerical methods for approximating the solution of Equations (1) and (2) are available in several references [5, 6, 7, 8].

Here, the morphodynamic model is given as follows:

$$\frac{\partial z_b}{\partial t} + \xi\frac{\partial q_b}{\partial x} = 0, \tag{3}$$

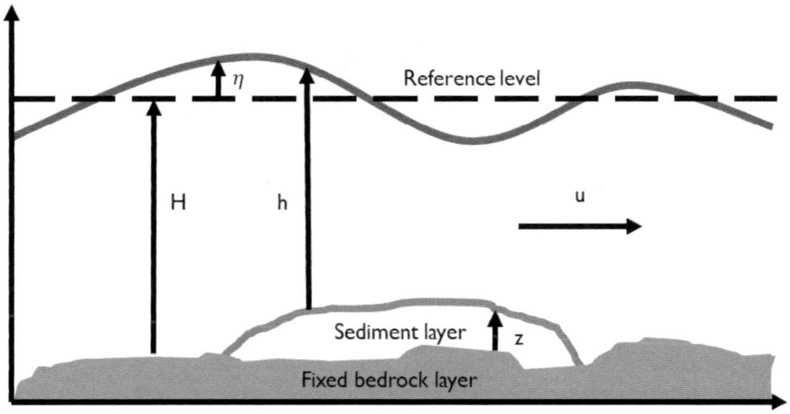

Figure 2. Sketch of 1D shallow water equations and their primitive variables.

where Q is solid transport discharge, where ξ obtained from $\xi = (1-\phi)^{-1}$, where ϕ is the size of the empty space between the basic sediments (bed sediment porosity)[3].

There are many empirical formulas from some researchers, one of them is called Grass formula,

$$q_b = A_g \frac{q}{h} \left| \frac{q}{h} \right|^{m_g - 1}, \qquad 1 \leq m_g \leq 4, \tag{4}$$

where m_g value commonly used is $m_g = 3$, so that

$$Q = A_g u^3. \tag{5}$$

where A_g range from 0 and 1. The interaction between the sediment and the fluid becomes smaller if the value of A_g decreases, and vice versa.

2.2. Semi-implicit staggered grid scheme

In this paper, the semi-implicit staggered grid scheme will be used to approximate the solution of water-sediment model (1-3). This numerical scheme is introduced in the paper of Gunawan P. H. et al [3] and Doyen et al [8]. Following are the discrete properties of staggered grid scheme for SWE-Exner model:

- Time domain: the time interval $(0, T)$ is divided by time step Δt such that the discrete time t^n is obtained for all $n \in 0, ..., N_t, t^n := n\Delta t$.
- Space domain: the domain $\Omega := (0, L)$ is divided into N_x cells by uniform discrete space step $\Delta x = L/N_x$.
- Full grid: variables h_i and z_i are defined in the full grids $i \in \mathcal{M} := \{1, ..., N_x\}$.
- Half grid: the variables u_i and q_i are defined in the half grids $i \in \varepsilon_{int} := \{1, ..., N_x - 1\}$.
- The rest of sets is defined as $\varepsilon_b := \{0, N_x\}$, and $\varepsilon := \varepsilon_{int} \cup \varepsilon_b$.

The space discretization can be shown in Figure 3.

Giving the average of the initial conditions by,

$$h_i^0 = \int_{x_{i-\frac{1}{2}}}^{x_{i+\frac{1}{2}}} \frac{h_0(x)dx}{\Delta x}, \qquad \forall i \in \mathcal{M}, \tag{6}$$

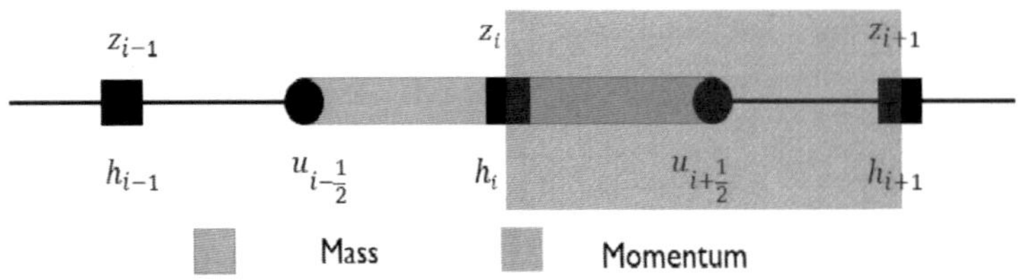

Figure 3. 1D *staggered grids*. The orange area represents volume control for mass conservation, association with unknown h_i and z_i. Volume control for momentum conservation ($u_{i+\frac{1}{2}}$) shown in red area.

$$z_i^0 = \int_{x_{i-\frac{1}{2}}}^{x_{i+\frac{1}{2}}} \frac{z_0(x)dx}{\Delta x}, \quad \forall i \in \mathcal{M}, \tag{7}$$

and

$$u_{i+\frac{1}{2}}^0 = \int_{x_i}^{x_{i+1}} \frac{u_0(x)dx}{\Delta x}, \quad \forall i \in \varepsilon_{int}. \tag{8}$$

The discretization of mass and momentum conservation is given as

$$h_i^{n+1} - h_i^n + \frac{\Delta t}{\Delta x}(q_{i+\frac{1}{2}}^n - q_{i-\frac{1}{2}}^n) = 0, \quad \forall i \in \mathcal{M}, \tag{9}$$

and

$$\begin{aligned} h_{i+\frac{1}{2}}^{n+1} u_{i+\frac{1}{2}}^{n+1} - h_{i+\frac{1}{2}}^n u_{i+\frac{1}{2}}^n + \frac{\Delta t}{\Delta x}[q_{i+1}^n \hat{u}_{i+1}^n - q_i^n \hat{u}_i^n + \frac{1}{2} g((h_{i+1}^{n+1})^2 - (h_i^{n+1})^2) \\ + g h_{i+\frac{1}{2}}^{n+1}(z_{i+1}^n - z_i^n) - g h_{i+\frac{1}{2}}^{n+1}(H_{i+1} - H_i)] + \Delta t S_{fi+1/2}^n = 0, \quad \forall i \in \varepsilon_{int}, \end{aligned} \tag{10}$$

respectively. The momentum balance equation is discretized with explicit upwind fluxes for the convection term and implicit centered fluxes for the pressure and topography term. The fluxes $q_{i+\frac{1}{2}}^n$ and q_i^n can be found in the paper [3] in detail.

The friction term $S_{fi+1/2}^n$ form is given in the case of Manning's friction (see [9]) by

$$S_{fi+1/2}^n = k^2 |u_{i+\frac{1}{2}}^n| u_{i+\frac{1}{2}}^{n+1} \left(h_{i+\frac{1}{2}}^{n+1} \right)^{-\frac{4}{3}} \tag{11}$$

and at the case of Darcy-Weisbach's friction by

$$S_{fi+1/2}^n = f |u_{i+\frac{1}{2}}^n| u_{i+\frac{1}{2}}^{n+1} \left(8 g h_{i+\frac{1}{2}}^{n+1} \right)^{-1} \tag{12}$$

At the sediment equation, the discretization of Exner equation is given as

$$z_i^{n+1} - z_i^n + \frac{\Delta t}{\Delta x} \xi (Q_{i+\frac{1}{2}}^{n+1} - Q_{i-\frac{1}{2}}^{n+1}) = 0, \quad \forall i \in \mathcal{M}. \tag{13}$$

where $Q_{i+\frac{1}{2}}^{n+1} = Q_s(h_{i+\frac{1}{2}}^{n+1}, u_{i+\frac{1}{2}}^{n+1})$.

Therefore, since sediment transport has no effect on fluid flow, sufficient stability conditions are obtained by determining conditions Courant-Friedrichs-Lewy (CFL) only on fluid part. For the Courant number $\nu \geq 1$, this condition is expressed as

$$\max_{i\in\mathcal{M}}\left(\sqrt{gh_i^n}+\frac{\left|q_{i-\frac{1}{2}}^n+q_{i+\frac{1}{2}}^n\right|}{2h_i^n}\right):=\frac{\nu\Delta x}{\Delta t}. \tag{14}$$

3. Flowchart algorithm

This is the flowchart algorithm of the simulation. There are some steps that must be done so the simulation works well. It can be seen in Figure 4.

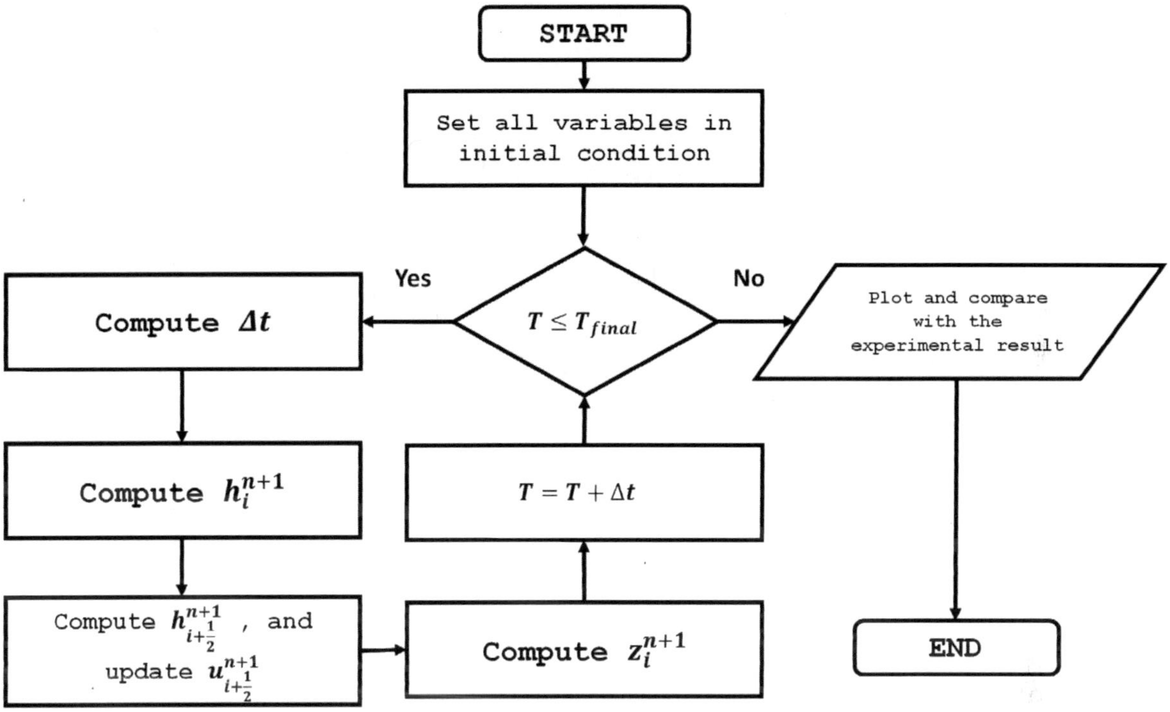

Figure 4. Flowchart algorithm of Dam-break over an erodible embankment Simulation.

Following is the explanations of algorithm which can be found in Figure 4.

- The first step is to set the value of variables for the initial conditions. There are 3 values of A_g and 2 slopes Φ which will be compared. Other variables are set according to Figure 1.
- After all variables initialized, compute the value of Δt at the beginning of all iterations by using (14). The value of this Δt may be different for each iterations depend on the stability criteria.
- The next step is to compute the value of h_i^{n+1} that can be computed by using the Δt and other variables as in water conservation equation (9).
- After that, compute the value of $h_{i+\frac{1}{2}}^{n+1}$ and update the value of $u_{i+\frac{1}{2}}^{n+1}$. This is the part of semi-implicit scheme because to compute the value of $u_{i+\frac{1}{2}}^{n+1}$, the value updated water height $h_{i+\frac{1}{2}}^{n+1}$ is used instead of old water height $h_{i+\frac{1}{2}}^{n}$. This momentum balance equation can be seen in (10).
- After all iteration done, the last step is to plot those variables and compare it to the experimental result.

IOP Conf. Series: Journal of Physics: Conf. Series **971** (2018) 012023 doi:10.1088/1742-6596/971/1/012023

4. Numerical result vs data experiment

Here, the numerical results of SWE-Exner model using staggered scheme will be compared with the data experiment obtained in [1].

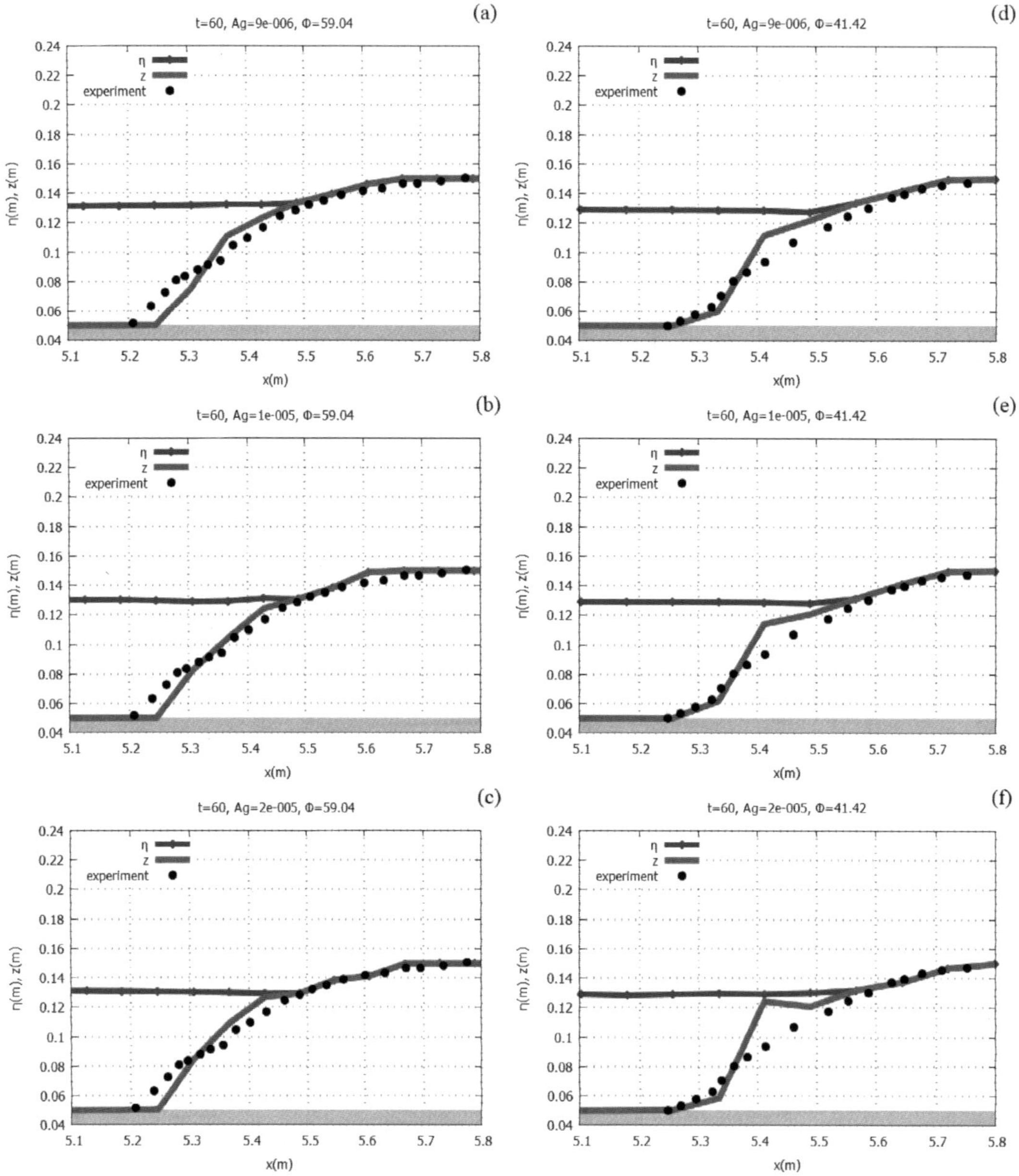

Figure 5. The comparison of numerical result and data experiment dambreak over an erodible embankment [1]. Figures (a), (b) and (c) are the comparison results in slope $\Phi = 59.04$ using $A_g = 9 \times 10^{-6}$, $A_g = 10^{-5}$ and $A_g = 2 \times 10^{-5}$ respectively. Figures (d), (e) and (f) are the comparison results in slope $\Phi = 41.42$ using $A_g = 9 \times 10^{-6}$, $A_g = 10^{-5}$ and $A_g = 2 \times 10^{-5}$ respectively.

In this simulation, the final time $t = 60$ s is used. Figures 5(a), (b), and (c) are obtained using various A_g for a slope of initial sediment $\Phi = 59.04^o$. From the results, the numerical results of sediment are shown in a good agreement with the data experiment. Moreover, the discrete norm error for each simulations is given in Table 1. The error is calculated by

$$||Error|| = \sum_{k=1}^{N_x} |Z_k - Z_{experiment}(k)|,$$

where $Z_{experiment}(k)$ is the sediment profile of experiment.

Figure 5(c) is shown the best result compared with the data experiment within slope $\Phi = 59.04^o$. The discrete norm error of Figure 5(c) is obtained 0.093389 (see Table 1).

In another comparison, Figures 5(d), (e), and (f) show the comparison of numerical simulations and data experiment for slope $\Phi = 41.42^o$. The best result is observed using $A_g = 1 \times 10^{-5}$ which is the closest profile to the experimental results. Here, the discrete norm error using $A_g = 1 \times 10^{-5}$ in Table 1 is given 0.073335.

Table 1. The discrete norm error for the results in Figure 5.

No	Figure 5	$\|\|Error\|\|$
1	(a)	0.128325
2	(b)	0.095005
3	(c)	0.093389
4	(d)	0.084301
5	(e)	0.073335
6	(f)	0.094042

5. Conclusion

The numerical simulation of erodible dambreak embankment in two different slopes are given. The results are obtained using SWE-Exner equation which is discretized by semi-implicit staggered scheme. The results show that the sediment profile of numerical result are match very well with the data experiment. Here, the value of coefficient Grass formula A_g becomes the main influence of the results. In the simulations with both slopes $\Phi = 59.04^o$ and $\Phi = 41.42^o$ the closest results are obtained using $A_g = 2 \times 10^{-5}$ and $A_g = 1 \times 10^{-5}$ respectively. The discrete norm error for using slope $\Phi = 59.04^o$ and $A_g = 2 \times 10^{-5}$ is obtained 0.093389. Meanwhile, using slope $\Phi = 41.42^o$ and $A_g = 1 \times 10^{-5}$ the discrete norm error is observed 0.073335.

References

[1] Di Cristo C, Evangelista S, Greco M, Iervolino M, Leopardi A and Vacca A 2017 *Journal of Hydraulic Research* 1–15
[2] Greco M, Di Cristo C, Evangelista S, Leopardi A, Iervolino M and Vacca A 2016 *River Flow 2016, Eighth International Conference on Fluvial Hydraulics* pp 9781315644479–77
[3] Gunawan P H, Eymard R and Pudjaprasetya S 2015 *Computational Geosciences* **19** 1197–1206
[4] Díaz M C, Fernández-Nieto E D and Ferreiro A 2008 *Computers & Fluids* **37** 299–316
[5] Stelling G and Zijlema M 2003 *International Journal for Numerical Methods in Fluids* **43** 1–23
[6] Stelling G S and Duinmeijer S A 2003 *International Journal for Numerical Methods in Fluids* **43** 1329–1354
[7] Pudjaprasetya S R and Magdalena I 2014 *East Asian J. Appl. Math.(EAJAM)* **4** 152–165
[8] Doyen D and Gunawan P H 2014 *Finite Volumes for Complex Applications VII-Methods and Theoretical Aspects* (Springer) pp 227–235
[9] Cao Z, Pender G, Wallis S and Carling P 2004 *Journal of hydraulic engineering* **130** 689–703

Context-aware recommender system based on ontology for recommending tourist destinations at Bandung

L Rizaldy Hafid Arigi, Z K Abdurahman Baizal, Anisa Herdiani

School of Computing, Telkom University, Bandung

[1]rizaldo.arigi@gmail.com, [2]baizal@telkomuniversity.ac.id,
[3]anisa@telkomuniversity.ac.id

Abstract. Recommender System is software that is able to provide personalized recommendation suits users' needs. Recommender System has been widely implemented in various domains, including tourism. One approach that can be done for more personalized recommendations is the use of contextual information. This paper proposes a context aware recommender based ontology system in the tourism domain. The system is capable of recommending tourist destinations by using user preferences of the categories of tourism and contextual information such as user locations, weather around tourist destinations and close time of destination. Based on the evaluation, the system has accuracy of of 0.94 (item recommendation precision evaluated by expert) and 0.58 (implicitly from system-end user interaction). Based on the evaluation of user satisfaction, the system provides a satisfaction level of more than 0.7 (scale 0 to 1) for speed factors for providing liked recommendations (PE), informative description of recommendations (INF) and user trust (TR).

1. Introduction

Improvements in the mobile technology and communications networks field made mobile services available for access anywhere. In order for this mobile service to match the state of the user, the services that are available should be responsive to the context around the user [1]. Context that can be used as an object's characteristic is defined as any information [2].

The majority of studies related to the recommender system still focus on recommending items to users based on user profiles or preferences without taking contextual information such as user locations, weather around tourist destinations and local time into account. Contextual information will become an especially important aspect of decision making for mobile app users [3].

There are several approaches to recommender system algorithms such as content-based, collaborative filtering and knowledge based Recommender systems with content-based approaches recommend items which are similar to what the users liked in the past. Similar to content-based, collaborative filtering recommends items that have been favored by other users with the same tastes. Because of relying on user history, both content-based and collaborative filtering have cold start problem. This problem does not appear on the recommender system that does not rely on user history such as knowledge-based.

Systems with a knowledge-based approach recommend items based on the knowledge of how an item characteristic can meet user needs and preferences [4].

One of the models that can be used to represent the domain of knowledge is ontology. Ontology offers good expressiveness and support reasoning task better than key-pair value, object oriented and logic based on contextual-based case [5].

Context-aware recommender system topic has been widely studied. One study proposed a recommender system model to recommend venues based on what users liked in the past and take advantage of user location and local time. The division of venue categories according to the category used by Foursquare [6]. While other studies propose a model of recommender system that utilizes contextual information to predict information categories that may be interesting to users. Items from category prediction results are evaluated with a model named distance punisher that utilizes the user's location, the weather in the destination, and the user's mode of transport [3]. However, the models mentioned only cover the general category of point of interest. Meanwhile, there are many tourist destination categories that have not been covered by the tourism destinations classification proposed by Inskeep [7].

In this paper we propose a context-aware recommender system method that is able to recommend tourist destinations by utilizing ontology developed in accordance with the principles of development and tourism planning, as well as taking user preferences and contextual information, such as user location, weather in tourist destinations and tourist destination's open time, into account. The proposed context-aware recommender system method is evaluated based on recommendation accuracy and user perceptions to application.

2. Ontology

The recommender system we developed using ontology as a representation of knowledge on the domain of tourism category. Ontology is built on the results of consultations with experts (academics and officials of cultural and tourism government services) and literature studies with reference to Inskeep's work [7].

On the ontology developed there are tourism destination categories and subcategories that become ontology class. One subcategory can be one or more subclasses of the tourism destination categories. The destination becomes direct individual from the tourist destination subcategories. One tourist destination can be an individual of one or more subcategories. Figure 1 is a visualization of some part of ontology we develop.

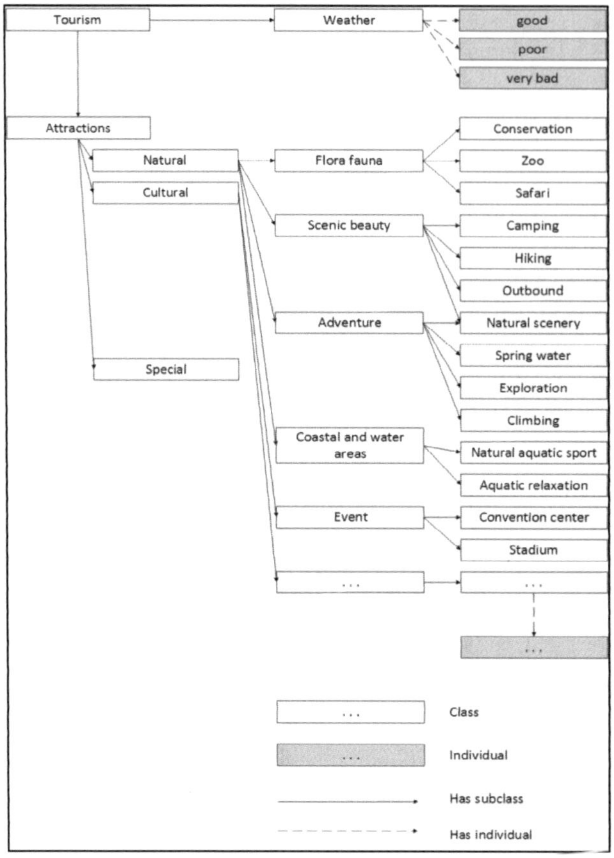

Figure 1. Part of ontology visualization.

All tourist destination individuals have attributes related to the processing of contextual information such as latitude, longitude and open-close time. In addition there are also supporting attributes such as description of tourist destinations. The latitude and longitude attribute serves to indicate the position of the destination on the map. This is required for calculating the distance from the user to the destination.

Tourist destination's open time becomes fourteen different attributes because every day there are two times when a tourist destination changes it's status, those changes are from close to open and from open to close. Open and close time values are not stored in the standard format (hh: mm: ss), but are stored in real number. Converting to a real number is required to make searching for a destination with queries easier. To convert the time of the standard format to the real value, the following equation 1 is used:

$$tn = \frac{60h + m}{1440} \tag{1}$$

while h is the hour and m is the minute. The value of tn will be in the range of zero to one. 1440 is the constant number of minutes per day. Table 1 is an example of a tourist destination specification with the mentioned attributes.

Table 1. Individual Specification Example

Specification	Description
Name	Alam wisata cimahi
Latitude	-6.841073f
Longitude	107.54805f
Monday open	0.3334f
Monday closed	0.875f
--	--
Sunday closed	0.875f
Class	Kuliner umum, outbound, taman bermain
Description	Tourist destination that unites several concepts of tourism such as, resorts, culinary, outbound and natural scenery

In addition to categories and subcategories of tourist destinations, the weather is also a class of its own. The weather class has an individual that represents the weather level. There are three levels of weather that is good, poor weather and very bad weather.

Each individual of the weather class has two attributes namely the weather value and the weather code. The weather value is the weather weight. The worse the weather, the greater the weight. While the weather code is a code of representation of weather conditions according to OpenWeatherMap standards. One individual can have multiple attributes of weather code. Table 2 is individual specification of good weather individual from weather class.

Table 2. Good Weather Individual Specification

Specification	Description
Name	Good weather
Weather code	801, 802, 803, 804, 952, 953
Weather value	1

3. Recommendation Model

We propose a recommendation model of destination based on user preferences on the category of tourist and user contextual information. Contextual information such as location, weather around the tourist destination and the open time of the tourist destination, is taken into account.

There are two steps in this model:

- Search for tourist destination that is not closed now. This is performed by utilizing the open and close time attributes on each individual destination.
- Evaluate each selected destination from the first step by assigning a utility value. Utility values are calculated based on user distance to destination, weight of weather value around destination and user preferences.

The first stage in this method relies on SPARQL queries. Results from SPARQL queries are URIs of currently open tourist destinations. The example of SPARQL queries to search for destinations that open on Monday at 12 noon, is as follows,

```
PREFIX rdf: <http://www.w3.org/1999/02/22-rdf-syntax-ns#>
PREFIX owl: <http://www.w3.org/2002/07/owl#>
PREFIX rdfs: <http://www.w3.org/2000/01/rdf-schema#>
PREFIX xsd: <http://www.w3.org/2001/XMLSchema#>
PREFIX ns:
<http://www.semanticweb.org/dell/ontologies/2017/4/tourism#>
SELECT ?destination
WHERE {?destination ns:senin_buka ?open; ns:senin_tutup
?closed

FILTER( (?open < 0.5 && 0.5 < ?closed)
|| ((?closed > 0.5 || ?open < 0.5) && (?closed < ?open)) )}
```

The second stage begins by calculating the distance from the user's location to each destination. To calculate the distance, euclidean equation can be used [8]:

$$d_i(u,i) = \sqrt{(x_u - x_i)^2 + (y_u - y_i)^2} \tag{2}$$

while x is latitude and y is longitude. Where u is the origin location and i is the destination location. After calculating the distance, next task is calculate similarity value between user preferences to the *i-th* tourism destination category by using equation (3) which is a modification of Multi Attribute Utility Theory [9]:

$$p_i = \sum_{j \in m} m_j f(i,j) \tag{3}$$

while m is the vector of interest level of the user to the category of tourist destinations and $f(i,j)$ is the relationship between the destinations to the i to category j. The value of $f(i,j)$ will always be more than equal to zero.

Each tourist destination does not become a direct individual of the category, but becomes an individual of the subcategories that are grouped by a category, so to calculate $f(i,j)$ equations (4) is used.

$$f(i,j) = \sum_{k \in S_j} r_{ik} \tag{4}$$

while r_{ik} is the value of the relationship between the *i-th* tourist destination with the *k-th* subcategory that becomes the subclass of the *j-th* category. The value of the relationship becomes one if the individual tourist destination is indeed an individual of the subclass of that category and zero if it is not.

Next step is to assess how reasonably visiting tourist destinations are based on the weather. To do so, there are two ways that can be done. First, the system must know the actual weather conditions in the destination by requesting direct weather data through the *OpenWeatherMap* API one by one for each destination. Second, by using a weather reference location.

The weather reference location is the location to which the *OpenWeatherMap* weather data is retrieved. Each tourist destination will be associated to a weather reference location using equation (2). Table 3 is a list of weather reference locations used for weather data retrieval in Bandung.

Table 3. Good Weather Individual Specification

No	Location Code	Latitude	Longitude	Geocoded
1	1650357	-6.9039	107.6186	Kota Bandung
2	8059762	-7.2026	108.586	Kota Cimahi

From the weather condition code obtained in the weather reference location data, the weather conditions code will be used to determine the level of weather conditions, whether good, poor or very bad. This is achieved by searching in all individuals from weather class, which individuals have the rated weather conditions coded. If there are individuals who have the weather condition code, then the weather weight score will be taken from the attribute of the individual weather value as w_i. The weight value for each weather level can be seen at table 4.

Table 4. Good Weather Individual Specification

Weather Level	Weight Value
Good	1
Poor	5
Very Bad	9

The last step is to calculate the utility value of a tourist destination. The calculation of utility values is performed using equation (5), modification of the proposed distance punisher model is as follows,

$$util_i = e^{-(d_i w_i)} p_i \quad (5)$$

Weather values can be determined freely as long as weather weight values follow constraint $w_{good} < w_{poor} < w_{very\_bad}$.

The difference between original distance punisher model with equation (5) is the multiplication with p_i as addition to the original equation. Calculation by equation (5) is performed on all open destinations. After that the system will take ten tourist destinations with the highest utility value.

It is important to remember that equation (5) will cause utility score drops with high w_i (worsening weather) and high d_i (distance between user current location and tourist destination take further). Equation (5) also cause utility score rises when similarity between interest level of the user to the category of tourist destination also higher.

To summarize the recommendation model step, Figure 2 shows aforementioned steps is applied

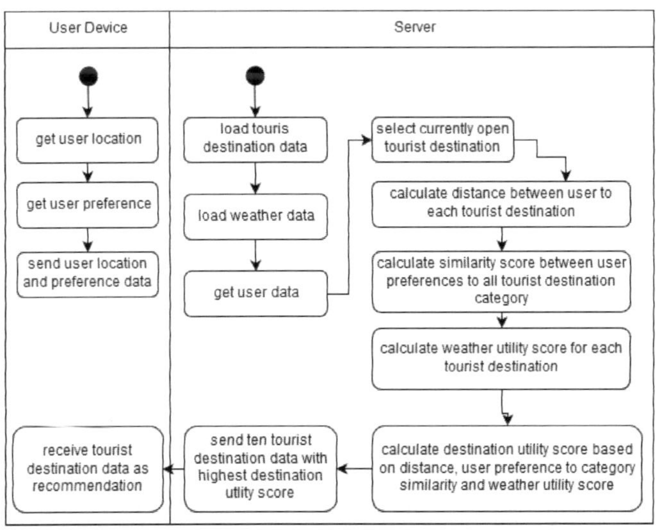

Figure 2. Recommendation model visualization

4. Eliciting User Preference

In order to obtain a user preference for a category of travel destinations, the user explicitly states an interest in a category of travel destination. There are four levels of user interest, those are not interested, less interested, interested enough and not interested. Each preference level has their own weighted value. Table 5 shows the weighted value of each level of interest in the tourism category. Meanwhile, Figure 3 shows the graphical user interface to explicitly insert the reference.

Table 5. Preference Level Weighted Value

Level	Value
Not interested	0
Less interested	0.33
Interested	0.67
Very interested	1

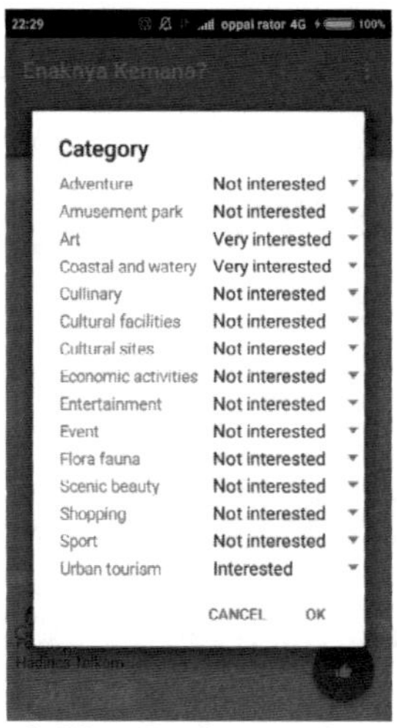

Figure 3. User Preference Input Menu

Users choose the appropriate level of interest through all tourist destination categories which are a direct subset of the natural, cultural and special classes. Users can view the description of categories of tourist destinations by tapping text tap categories. The front-end application will send location data and the level of user interest through the HTTP protocol.

5. Performance and User Satisfaction Evaluation

Evaluation were conducted on tourism experts and end users, so the parameters evaluated in both evaluation were different. The purpose of evaluating to experts and users is to evaluate the level of accuracy of the system from the expert and users' point of view and the level of user satisfaction.

5.1. Recommendation Accuration Evaluation

Evaluations to get the accuracy of recommendations provided involve tourism experts and regular users. Scenarios used in accuracy evaluating involving experts with regular users. System accuracy evaluation with experts was performed using precision parameters to measure the relevance of recommendations [10] [11]. Equation (6) is used to measure recommendation precision.

$$precision = \frac{TP}{TP + FP} \quad (6)$$

while TP (true positive) is the right tourist destination with full category travel preference and FP (false positive) is a less suitable tourist destination. Evaluation scenario is shown on Table 6.

Table 6. Evaluation Scenario With Expert

No	User location	Local time	Weather	Preference Evalution No.
1	Alun-alun cimahi	09:00	Good	1
2	Alun-alun cimahi	09:00	Poor	1
3	Alun-alun cimahi	09:00	Very Bad	1
...	...	...	...	...
13	Jl. Setiabudhi Bandung	13:00	Good	2
...	...	...	...	...
27	Jl. Buah Batu Bandung	18:00	Very Bad	3

There are 27 cases of system accuracy evaluating with experts. 27 cases that use three different locations as the location of the user when using the application. In each location there are three times of application usage evaluating, which is 9 am, 1 pm and 6 pm and at each evaluation time there are three times the weather changes.

The preference evaluation number is the number in the preference case table that is the input of the system. Preference evaluation number corresponds to application time usage, e.g evaluation number 1 for clock 9, evaluation number 2 for clock 13 and so on. Preference evaluation number refers to table 7. Table 7 is a table of user preference cases used as input.

Table 7. Preference Evaluation Cases

Preference Evaluation No.	Tourist destination category	Preference level
1	Adventure	Very interested
	Flora fauna	Interested
	Scenic beauty	Interested
	Coastal and water areas	Interested
	Economic activity	Less interested
2	Entertainment	Interested
	Event	Interested
	Sport	Interested
	Shopping center	Interested
	Culinary	Very interested
	Theme parks	Very interested
	Cultural facilities	Very interested
	Arts	Very interested
	Cultural sites	Very interested
3	Entertainment	Very interested
	Sport	Interested
	Shopping center	Very interested
	Culinary	Interested
	Art	Interested
	Urban tourism	Very interested

The steps taken for evaluating recommender system with experts are as follows,

- We ran the system according to scenario at Table 6 and get system output data.
- We submitted the system output data to the expert.
- Expert evaluated system output data.
- Evaluation result was processed with equation (6).

The precision mean from evaluating 27 system output results is 0.94. Precision based on weather system is shown in Figure 4.

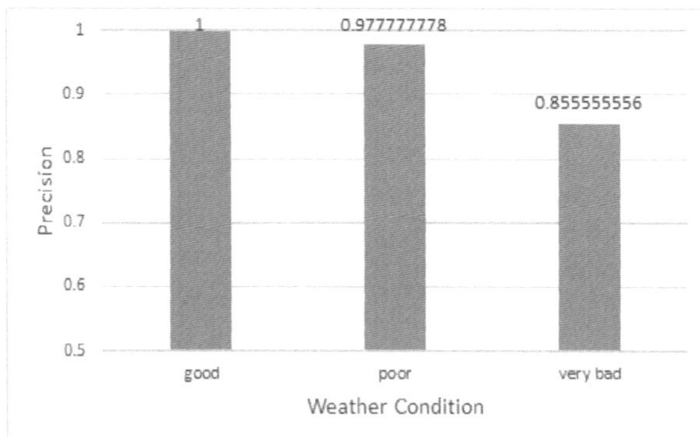

Figure 4 Precision Mean Group by Weather Condition

Precision of tourist destination recommendation is declining along with worsening weather. This is happened because the result of equation (5) is decreased if the wi value increases. However, the precision recommendations of tourist destinations during very bad weather can still be said to be quite high. Logically, the precision of the very bad weather should be lower, as searching for a destination that fits the user's preferences and the weather at the destination is still good is difficult when the actual weather is categorized as very bad. While the mean system precision based on application usage time is shown in Figure 5.

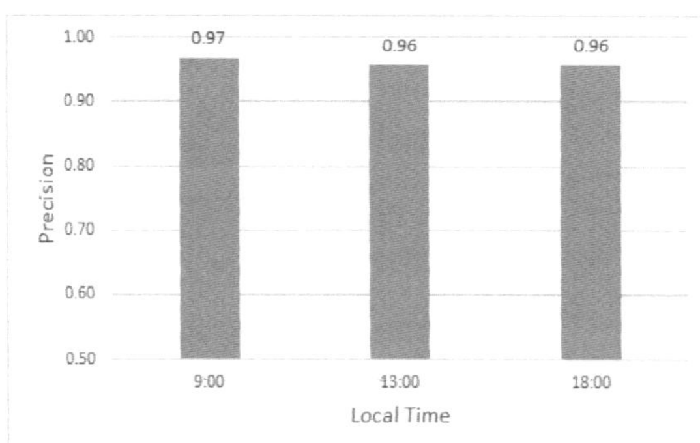

Figure 5. Precision Mean Group by Local Time

Recommendation precision declined with the passing of the day, although not significant. The cause of insignificant precision decline is due to the adjustment of the preference evaluation number to the local time. For example, preference evaluation number one and two are preference case for tourist destination category that have the majority of tourist destinations open in the morning until late afternoon. While the preference evaluation number three is preference case skewed towards tourist destination categories that have majority of tourist destination open until the evening.

System accuracy evaluation involving the ordinary user is done implicitly, it means the accuracy obtained through analysis of user activity [12]. The calculation of the recommendation accuracy to the user is calculated by the equation (7) [12].

$$accuracy = \frac{number\ of\ successful\ recommendations}{number\ of\ recommendations} \tag{7}$$

Each user requests travel attraction recommendations through the app at least once. The application will count the number of times the user requested a recommendation and how many times the user chose the tourist destination in the recommendation list. The number of recommendation requests will be incremented each time the user requests a recommendation and the number of successes will increase by one if the user selects one of the destinations from the list of recommended travel destination recommendations

5.2. User Satisfaction Evaluation

To measure the level of user satisfaction, there are several evaluation parameter was used. The evaluation parameters used are perceived recommendation quality (PRQ), usability (USA), informative (INF), perceived efficiency (PE) and trust (TR). The five user satisfaction parameters have been used in previous studies that examined the recommender system for recommending smartphones [12].

In user satisfaction evaluating, users will be asked to fill out a questionnaire after requesting a destination recommendation through an app at least once. Table 8 contains a list of statements in the questionnaire given to the user after using the application.

Table 8. Questionnaire Statement List [12]

No	Factor	Statement
1	PRQ	I love the tourist destination I chose from the recommendation list.
2	PRQ	I like the application interaction.
3	INF	I think the tourist destination description on recommendation list is quite informative.
4	USA	I have difficulty while using the application
5	PE	I can find a tourist destination i like quickly.
6	TR	I will use this app if I want to travel in Bandung.

There were 63 respondents using front-end application we developed. Respondents come from various backgrounds ranging from students, lecturers to ordinary citizens. Sampling of respondents was done randomly. User satisfaction evaluation results can be seen in Figure 6.

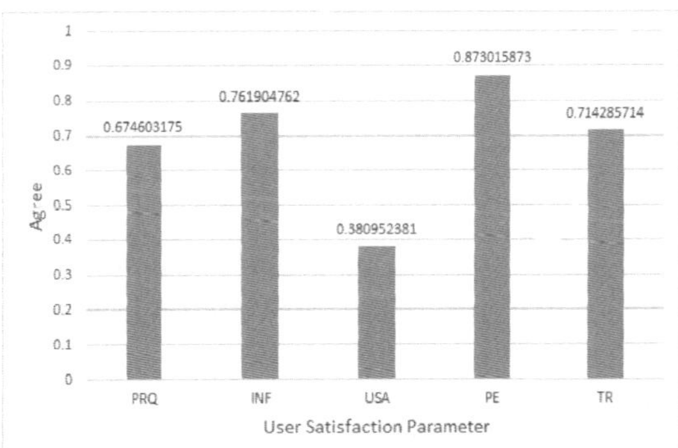

Figure 6. User Experience Questionnaire Result

On the perceived recommendation quality (PRQ) measurements, 67% of users agree that they like the tourist destinations on the recommendation list. PRQ score is still low because there are most users are not quite familiar with the delivery of tourist destination description. It is seen in the evaluation results of INF parameters. Not much different from the PRQ parameter evaluation results, 76% of users agreed that the description of the destination is informative enough (INF). This result is close to the result of PRQ parameter evaluation. Front-end application is still not easy to use. It is proved that 38% of users still encounter difficulties when using the application, the results of usability (USA) parameter evaluation.

Users can find the preferred tourist destination quickly. It is proven that 87% of users stated that they can find their preferred travel destination quickly (perceived efficiency / PE). This parameter gets the highest result because this application only takes three steps to get recommendation of destination. The three steps are clicking the user preference button, select the level of preference for a particular tourist destination category and choose a recommended travel destination. The result of the user trust (TR) parameter evaluation is 71% of users stated that they want to use this app if they want to travel in Bandung.

6. Conclusion

The general classification system in the tourism domain is discussed in the main reference [7] can be converted into a functional ontology model to represent knowledge in the tourism domain. The accuracy of the system is based on evaluating with high experts (overall reaching 0.94) so that the built system has been able to provide appropriate recommendations according to user preferences, but the accuracy of the system based on evaluating with the user is much lower (only 0.58) because it is influenced by high user dissatisfaction with the application (38% of users said the app is hard to use).

References

[1] Alhazbi S, Lotfi L, Ali R and Suwailih R 2013 Ontology-based model in tourism context-aware in *ICT Convergence (ICTC), 2013 International Conference on*, Jeju Island,.

[2] Abowd G, Dey A and Davies N 1999 Towards a better understanding of context and context-awareness in *Handheld and ubiquitous computing*, New York,.

[3] Passimier T D, Dooms S and Martens L 2014 Context-aware recommendations through context, *Multimedia Tools and Applications,* p 2925-2948,.

[4] Ricci F, Rokach L and Saphira B 2015 *Recommender systems handbook*, New York: Springer

[5] Strang T and Popien C L 2004 A context modeling survey in *Workshop on advanced context modelling, reasoning and management, UBICOMP 2004*, Nottingham,.

[6] Jakkhupan W, Rattanothai S and Kaewonjing J 2015 A context-aware personalized venue recommender system on smart device in *Wireless and Mobile (APWiMob), 2015 IEEE Asia Pacific Conference on*, Bandung.

[7] Inskeep E 1991 Tourism planning: an integrated and sustainable development approach, New York: Wiley

[8] Lee C S, Chang Y C and Wang M H 2009 Ontological recommendation multi-agent for Tainan City travel, *Expert Systems with Applications,* vol. 36, no. 3, p 6740-6753,.

[9] Beibei H and Aufaure M A 2013 A query refinement mechanism for mobile conversational search in smart environments in *Proceedings of the IUI 2013 on Second Workshop on Interacting with Smart Objects*, Santa Monica,.

[10] Arnett D K and Baizal Z K A 2015 Recommender system based on user functional requirements using Euclidean fuzzy in *Information and Communication Technology (ICoICT), 2015 3rd International Conference on*, Denpasar,.

[11] Chen R C, Huang Y H, Bau C T and Chen S M 2012 A recommendation system based on domain ontology and SWRL for anti-diabetic drugs selection, *Expert Systems with Applications,* vol. 39, no. 4, p 3995-4006

[12] Baizal Z A and Murti Y 2017 Evaluating functional requirements-based compound critiquing on conversational recommender system. *In Information and Communication Technology* (ICoIC7), 2017 5th International Conference p. 1-6. IEEE

[13] Noy N F and McGuinness L D 2017 Protege Stanford, Stanford University, [Online]. http://protege.stanford.edu/publications/ontology_development/ontology101-noy-mcguinness.html. [Accessed 17 August 2017].

[14] Fernandez M Perez A G and Juristo N 1997 Methontology: From ontological art towards ontological engineering, American Asociation for Artificial Intelligence

IOP Conf. Series: Journal of Physics: Conf. Series **971** (2018) 012025 doi:10.1088/1742-6596/971/1/012025

Breaking Megrelishvili protocol using matrix diagonalization

Muhammad Arzaki[1], Danang Triantoro Murdiansyah[2], Satrio Adi Prabowo[1]

[1]Computing Laboratory, School of Computing, Telkom University, Indonesia
[2]High Performance Computing Laboratory, School of Computing, Telkom University, Indonesia

E-mail: {arzaki, danangtri}@telkomuniversity.ac.id, prabowosatrioadi@students.telkomuniversity.ac.id

Abstract. In this article we conduct a theoretical security analysis of Megrelishvili protocol—a linear algebra-based key agreement between two participants. We study the computational complexity of Megrelishvili vector-matrix problem (MVMP) as a mathematical problem that strongly relates to the security of Megrelishvili protocol. In particular, we investigate the asymptotic upper bounds for the running time and memory requirement of the MVMP that involves diagonalizable public matrix. Specifically, we devise a diagonalization method for solving the MVMP that is asymptotically faster than all of the previously existing algorithms. We also found an important counterintuitive result: the utilization of primitive matrix in Megrelishvili protocol makes the protocol more vulnerable to attacks.

1. Introduction

A reliable communication between two or more participants is an essential feature in modern cryptography. The protection of the communication channel can be attained in many ways, one of them is by constructing a mutual secret key for the rightful parties. The Diffie-Hellman key exchange (DHKE) protocol in [1] is one of the most commonly used technique for creating such key. It allows the participants to concur on a specific key via an open channel. DHKE has also been extensively used as a building block for many new protocols. One of them is the Megrelishvili protocol whose theoretical notion was first presented in [2].

Megrelishvili protocol is a linear algebra-based modification of the conventional DHKE. The mutual secret key in this protocol is a vector over a particular finite field. To construct the key, the protocol employs matrix exponentiation and left-multiplication of a matrix by a vector. These two linear algebra operations cause the security of Megrelishvili protocol is not directly associated with a specific discrete logarithm problem in a particular group. Hence, Megrelishvili key exchange procedure presents an alternative solution for the creation of the common key between two participants.

The unique characteristics of Megrelishvili protocol has attracted numerous investigations to explore its mathematical and computational properties [3, 4, 5, 6, 7, 8, 9]. Despite these numerous studies, until recently the practical application of this protocol has not been widely implemented. Nevertheless—from theoretical viewpoint—the comprehensive theoretical security analysis of this protocol still needs to be carried out. In this paper we study the Megrelishvili vector-matrix problem (MVMP [7, 8, 9, 10]) as a mathematical problem that strongly correlates with the security of Megrelishvili protocol. Specifically, we focus our investigation regarding the computational complexity of the MVMP that involves diagonalizable public matrix.

We formulate efficient algorithms that solve the MVMP which involves diagonalizable public matrix in Section 4. We also prove that these algorithms are asymptotically faster than the other existing attack methods. Additionally, we also discuss some of the experimental results in Section 5 to support our theoretical analysis and to evaluate the practical efficiency of the proposed technique.

2. Finite Fields and Linear Algebra over Finite Fields

The working of Megrelishvili protocol relies heavily on mathematical theories of matrices and vector spaces over finite fields. We first briefly discuss some of our basic mathematical notations and terminologies throughout this paper. We use the standard mathematical notations as in [7, 8, 9, 10]. We write $\mathbb{F}_q$ to represent the finite fields of q elements, $\mathbb{F}_q^n$ to denote an n-dimensional vector space over $\mathbb{F}_q$, and $GL_n(q)$ to signify the collection of all nonsingular (invertible) matrices over $\mathbb{F}_q$. We write vectors with bold lowercase letters (e.g. $\mathbf{v}$) and matrices using bold capital letters (e.g. $\mathbf{M}$). The vectors are considered and handled as row or column matrices. Furthermore, we treat the product of a $1 \times n$ vector $\mathbf{v} \in \mathbb{F}_q^n$ and an $n \times 1$ vector $\mathbf{w} \in \mathbb{F}_q^n$ (i.e., $\mathbf{vw}$) as a scalar value in $\mathbb{F}_q$.

Given a finite field $\mathbb{F}_q$, the *order* of any nonzero element $\beta \in \mathbb{F}_q$, denoted by $|\beta|$, is defined as the least positive integer ℓ such that $\beta^\ell = 1$. It is well-known that $1 \leq |\beta| \leq q-1$ and $|\beta|$ divides $q-1$ [11, 12, 13]. A nonzero element $\alpha \in \mathbb{F}_q$ is called a *primitive element* if and only if $|\alpha| = q-1$. This condition also implies that for any nonzero $\beta \in \mathbb{F}_q$ there exists an integer $t \in [1, q-1]$ such that $\alpha^t = \beta$. The problem of determining the exponent t that satisfies $\alpha^t = \beta$ where α is the primitive element of $\mathbb{F}_q$ and β is a nonzero element in $\mathbb{F}_q$ is called the *discrete logarithm problem* (DLP) in $\mathbb{F}_q$. We refer the reader to [14] for more detailed discussion regarding this problem.

A polynomial $f(x)$ of *degree* $n \in \mathbb{N}$ over $\mathbb{F}_q$ is an expression $\sum_{i=0}^{n} \alpha_i x^i$ where $\alpha_i \in \mathbb{F}_q$ for all $0 \leq i \leq n$. We denote the set of all polynomials over $\mathbb{F}_q$ by $\mathbb{F}_q[x]$. The *degree* of a nonzero polynomial $f(x)$, denoted by $\deg(f(x))$, is defined as the greatest integer k such that the coefficient of x^k is nonzero. A polynomial $f(x) \in \mathbb{F}_q[x]$ is *reducible* if and only if $f(x) = g(x)h(x)$ for some $g(x), h(x) \in \mathbb{F}_q[x]$ with $0 < \deg(g(x)), \deg(h(x)) < \deg(f(x))$; otherwise, it is *irreducible*. An irreducible polynomial $f(x)$ is called a *primitive polynomial of degree d* over $\mathbb{F}_q$ if it is the polynomial with the least positive degree such that $f(\alpha) = 0$ where α is a primitive element of $\mathbb{F}_{q^d}$.

It is well-known that any finite field must contain a prime power number of elements, that is, for any field $\mathbb{F}_q$, we have $q = p^m$ for some positive prime p and positive integer m. The finite field $\mathbb{F}_{p^m}$ can be constructed by choosing an irreducible polynomial $f(x) \in \mathbb{F}_p[x]$ of degree m and constructing the quotient ring $\mathbb{F}_p[x]/(f(x))$, i.e., the ring of polynomials over $\mathbb{F}_p$ where the addition and multiplication are defined modulo $f(x)$ [11, 12, 13]. As a result, the finite field $\mathbb{F}_{p^m}$ is isomorphic to $\mathbb{F}_p/(f(x))$. In this paper, we consider both of these finite fields as identical fields.

For any nonsingular $n \times n$ matrix $\mathbf{M}$ over $\mathbb{F}_q$, the (*multiplicative*) *order* of $\mathbf{M}$, written by $\text{ord}(\mathbf{M})$, is defined as the smallest positive integer s such that $\mathbf{M}^s = \mathbf{I}$, where $\mathbf{I}$ is the $n \times n$ identity matrix. From [15], we have $1 \leq \text{ord}(\mathbf{M}) \leq q^n - 1$. Furthermore, we call $\mathbf{M}$ as a *primitive matrix* of size n over $\mathbb{F}_q$ whenever $\text{ord}(\mathbf{M}) = q^n - 1$. Given a square matrix $\mathbf{M}$ of size n over $\mathbb{F}_q$, an *eigenvector* of $\mathbf{M}$ is a nonzero vector $\mathbf{v} \in \mathbb{F}_q^n$ that satisfies $\mathbf{Mv} = \lambda\mathbf{v}$ for some scalar λ. The scalar λ is called an *eigenvalue* of $\mathbf{M}$ *corresponding to the eigenvector* $\mathbf{v}$. Moreover, a scalar λ is an eigenvalue of $\mathbf{M}$ if and only if $\det(\lambda\mathbf{I} - \mathbf{M}) = 0$. The polynomial $p_{\mathbf{M}}(x) = \det(x\mathbf{I} - \mathbf{M})$ is called the *characteristic polynomial* of $\mathbf{M}$. Therefore, we can find all eigenvalues of $\mathbf{M}$ by solving the *characteristic equation* $p_{\mathbf{M}}(x) = 0$ for x. However, this equation does not always have solutions in $\mathbb{F}_q$. When this condition happens, we need to extend our field of consideration to the *splitting field* of $p_{\mathbf{M}}(x)$ over $\mathbb{F}_q$, that is, the smallest field extension of $\mathbb{F}_q$ over which $p_{\mathbf{M}}(x)$ decomposes to linear factors [12, 13]. Suppose $p_{\mathbf{M}}(x)$ is a monic polynomial of degree n over $\mathbb{F}_q$ which can be factored as $p_{\mathbf{M}}(x) = \prod_{i=1}^{n}(x - \lambda_i)$ where each λ_i is a root of $p_{\mathbf{M}}(x)$, then the splitting field of $p_{\mathbf{M}}(x)$ over $\mathbb{F}_q$ is denoted by $\mathbb{K}(\lambda_1, \lambda_2, \ldots, \lambda_n)$. In this situation, elements of a splitting field can be treated as polynomials or vectors over $\mathbb{F}_q$.

An $n \times n$ matrix $\mathbf{M}$ over $\mathbb{F}_q$ is said to be *diagonalizable* if there exists a nonsingular matrix $\mathbf{P}$ such that $\mathbf{P}^{-1}\mathbf{MP}$ is a diagonal matrix. Suppose $\mathbf{\Lambda} = \mathbf{P}^{-1}\mathbf{MP}$ is a diagonal matrix, then it is easy to show that the diagonal entries of $\mathbf{\Lambda}$ are the eigenvalues of $\mathbf{M}$. Moreover, if we write $\mathbf{\Lambda} = \text{diag}(\lambda_1, \lambda_2, \ldots, \lambda_n)$ and $\mathbf{P} = [\mathbf{p}_1 \mid \mathbf{p}_2 \mid \cdots \mid \mathbf{p}_n]$ where each $\mathbf{p}_i$ is a column vector, then each $\mathbf{p}_i$ is the eigenvector of $\mathbf{M}$ associated with eigenvalue λ_i for all $1 \leq i \leq n$. Hence, an $n \times n$ matrix $\mathbf{M}$ is diagonalizable if and only if it has n linearly independent eigenvectors [16, Theorem 1.3.7]. This condition is equivalent to the statement that $\mathbf{M}$ is diagonalizable if and only if the *geometric multiplicity* of each eigenvalue of $\mathbf{M}$ is identical to its *algebraic multiplicity*. Furthermore, from [16, Theorem 1.3.9] we know that $\mathbf{M}$ is diagonalizable if all of its eigenvalues are distinct.

3. The Standard Megrelishvili Protocol and Megrelishvili Vector-Matrix Problem (MVMP)

The standard two-party Megrelishvili protocol is an example of linear algebra-based variant of the generic Diffie-Hellman key agreement. The theoretical notion of this protocol was first presented in [2]. It uses matrix exponentiation and left-multiplication of a matrix by a row vector that are performed over a specific finite field. Throughout this paper, we utilize the formal description of Megrelishvili protocol as explained in [6, 7, 8, 9]. The protocol differs from the typical DHKE because its security does not straightforwardly relate to a specific variation of the DLP in a finite group. For the discussion pertaining to the algorithm analysis of Megrelishvili protocol and its comparison to other variants of Diffie-Hellman protocol, the reader is referred to [7].

Unlike the conventional DHKE, the common confidential keys of the participants involved in Megrelishvili protocol is a (row) vector in $\mathbb{F}_q^n$. Suppose there are two participants, namely participant 1 and participant 2 who want to construct a common secret key using the standard Megrelishvili protocol. Initially, both participants need to agree on several public parameters which consists of a finite field $\mathbb{F}_q$, a (large) integer n, a nonzero (row) vector $\mathbf{v} \in \mathbb{F}_q^n$, and an invertible matrix $\mathbf{M} \in GL_n(q)$. Each participant i then constructs their respective secret matrix by raising the matrix $\mathbf{M}$ to a power of a secret integer t_i. The result of this exponentiation is a secret matrix $\mathbf{S}_i$ which is exclusively known by participant i itself. For the key exchange, each participant i left-multiplies $\mathbf{S}_i$ by the public nonzero vector $\mathbf{v}$ and sends the result (i.e., $\mathbf{v}\mathbf{S}_i$) to another participant via an open channel. Finally, each participant i retrieves the mutual secret vector by left-multiplying $\mathbf{S}_i$ by the vector obtained from its counterpart. Observe that we have $\mathbf{S}_1\mathbf{S}_2 = \mathbf{M}^{t_1}\mathbf{M}^{t_2} = \mathbf{M}^{t_2}\mathbf{M}^{t_1} = \mathbf{S}_2\mathbf{S}_1$, and thus $\mathbf{v}\mathbf{S}_1\mathbf{S}_2 = \mathbf{v}\mathbf{S}_2\mathbf{S}_1$, which signifies that both participants possess identical secret vectors at the end of the exchange. We summarize this standard two-party Megrelishvili key exchange procedure in Table 1.

Table 1: Generic Megrelishvili key exchange protocol for two participants according to [6, 7, 8, 9].

Public parameters of Megrelishvili protocol	
A reliable third party picks and announces: a finite field $\mathbb{F}_q$, a (large) integer n, a nonzero (row) vector $\mathbf{v} \in \mathbb{F}_q^n$, and a matrix $\mathbf{M} \in GL_n(q)$.	
Participant 1	**Participant 2**
Secret matrices generation	
Choose an integer t_1. Compute $\mathbf{S}_1 = \mathbf{M}^{t_1}$.	Choose an integer t_2. Compute $\mathbf{S}_2 = \mathbf{M}^{t_2}$.
Exchange of vectors via an open channel	
Compute $\mathbf{a}_1 = \mathbf{v}\mathbf{S}_1$. Send $\mathbf{a}_1$ to participant 2.	Compute $\mathbf{a}_2 = \mathbf{v}\mathbf{S}_2$. Send $\mathbf{a}_2$ to participant 1.
Mutual key retrieval for each participant	
Compute $\mathbf{a}_1' = \mathbf{a}_2\mathbf{S}_1$	Compute $\mathbf{a}_2' = \mathbf{a}_1\mathbf{S}_2$.
The common secret key is $\mathbf{a}_1' = \mathbf{a}_2'$.	

By observation, it is desirable if the values of $\mathbf{v}\mathbf{M}^t$ are all distinct for as many integer t as possible. However, since for any $\mathbf{M} \in GL_n(q)$ we have $\mathrm{ord}(\mathbf{M}) \leq q^n - 1$, then $\mathbf{v}\mathbf{M}^t$ has at most $q^n - 1$ different values in $\mathbb{F}_q^n$. There is an algorithm in [8] that provides a method for constructing a public matrix $\mathbf{M}$ whose order is precisely $q^n - 1$. The public matrix $\mathbf{M}$ is generated using the companion matrix of a primitive polynomial of degree n over $\mathbb{F}_q$. In such condition, $\mathbf{M}$ is also a *primitive matrix* of size n over $\mathbb{F}_q$. Moreover, this algorithm also entails that the values of $\mathbf{v}\mathbf{M}^t$ are all distinct for every $0 \leq t < q^n - 1$ for all nonzero vectors $\mathbf{v}$, which means that $\mathbf{v}\mathbf{M}^t$ can be any nonzero vector in $\mathbb{F}_q^n$ [8, Theorem 3].

The exchange of vectors in Megrelishvili protocol is performed over an open channel which is possibly being monitored by a wiretapper. In this step, participant i sends the vector $\mathbf{a}_i = \mathbf{v}\mathbf{P}_i = \mathbf{v}\mathbf{M}^{t_i}$ to its counterpart. Since the values of $\mathbf{a}_i$, $\mathbf{v}$, and $\mathbf{M}$ are public, a wiretapper may intercept the message $\mathbf{a}_i$ and unravel the value of the secret integer t_i by solving the equation $\mathbf{a}_i = \mathbf{v}\mathbf{M}^{t_i}$ for t_i. Consequently—from theoretical standpoint—the security of Megrelishvili protocol relates to the hardness of computing the value t from the equation $\mathbf{w} = \mathbf{v}\mathbf{M}^t$, where $\mathbf{v}, \mathbf{w} \in \mathbb{F}_q^n$ and $\mathbf{M} \in GL_n(q)$. This problem is called the *Megrelishvili vector-matrix problem* and it is formally described in Definition 1.

Definition 1 ([7, 8, 9, 10]) *The Megrelishvili vector-matrix problem (MVMP) is a linear algebra problem of determining an integer t (if such integer exists) from the equation $\mathbf{w} = \mathbf{v}\mathbf{M}^t$ where $\mathbf{v}$ and $\mathbf{w}$ are row vectors in $\mathbb{F}_q^n$ and $\mathbf{M} \in GL_n(q)$.*

The MVMP is similar to the well-known discrete logarithm problem that underlies the security of the conventional Diffie-Hellman protocol. However, it should be noted that the values on both sides of the equation in the MVMP are vectors, which are not elements of finite multiplicative group. From Table 1, a wiretapper can obtain the mutual secret vector of both participants by solving two instances of the MVMP, i.e., recovering the value t_1 from $\mathbf{a}_1 = \mathbf{v}\mathbf{M}^{t_1}$ and recovering the value t_2 from $\mathbf{a}_2 = \mathbf{v}\mathbf{M}^{t_2}$. Then, the mutual vector can be retrieved simply by computing $\mathbf{v}\mathbf{M}^{t_1+t_2}$ which can be done in polynomial number of scalar operations in $\mathbb{F}_q$. From [7, Theorem 5], we know that the MVMP in $\mathbb{F}_q^n$ can be solved using exhaustive search (brute-force) approach in $\mathcal{O}(T_{\text{mat}}(n) \cdot \text{ord}(\mathbf{M}))$ number of scalar operations, where $T_{\text{mat}}(n)$ is the time complexity of the preferred matrix multiplication algorithm. Furthermore, the fact that $T_{\text{mat}}(n) = \mathcal{O}(n^3)$ [17, Chapter 4] and $\text{ord}(\mathbf{M}) \leq q^n - 1$ gives us an $\mathcal{O}(n^3 q^n)$ upper bound for the number of scalar operations required to solve the MVMP. This exhaustive search approach also gives an $\mathcal{O}(n)$ space complexity bound for storing the vectors [9].

If the public matrix $\mathbf{M}$ is constructed using Algorithm 2 in [8], then the values of $\mathbf{v}\mathbf{M}^t$ are all distinct for all integers $0 \leq t \leq q^n - 2$. In this case, brute-force attack is not the only admissible technique for solving the MVMP. Using the adaptation of Shanks's baby-step giant-step algorithm in [18], there exists collision algorithms which solve the MVMP in $\mathbb{F}_q^n$ using $\mathcal{O}(n^4 \cdot q^{n/2} \cdot \log q)$ scalar operations[1] [9]. In other words, these collision algorithms solve the MVMP in $\mathbb{F}_q^n$ faster than the brute-force method by a factor of $\mathcal{O}((1/n \log q) \cdot q^{n/2})$. Additionally, the numerical experiment in [9] also indicates that the collision algorithms practically work faster than the exhaustive search attack, with the crossover points as low as $n = 3$ and $q = 2$. Despite their superior speed, both theoretical and experimental results show that the memory consumption for the collision algorithms in [9] are larger than those for the brute-force approach[2].

4. Diagonalization Methods for Breaking Megrelishvili Protocol

4.1. General diagonalization method for solving the MVMP

We first present a general approach for solving the MVMP using matrix diagonalization method. The idea of our technique is inspired by the research concerning the DLP in $GL_n(q)$ in [19, 20]. In both of these papers, the authors exploit Jordan canonical forms of matrices—which can be considered as generalized forms of diagonal

[1] This quantity is obtained under the assumption that $T_{\text{mat}}(n) = \mathcal{O}(n^3)$ and $\text{ord}(\mathbf{M}) = q^n - 1$.

[2] More precisely, the memory usage for the collision algorithms in [9] is larger than that for the brute-force attack by a factor of up to $\mathcal{O}(q^{n/2})$.

matrices. However, here we restrict our interest to a special case when the public matrix $\mathbf{M}$ is diagonalizable.

Given a public matrix $\mathbf{M} \in GL_n(q)$, a public vector $\mathbf{v} \in \mathbb{F}_q^n$, and a known vector $\mathbf{w} \in \mathbb{F}_q^n$ that is transmitted over an open channel, our objective is to find an integer t (if such value exists) such that

$$\mathbf{v}\mathbf{M}^t = \mathbf{w}. \tag{1}$$

If $\mathbf{M}$ is diagonalizable, then we can write $\mathbf{M} = \mathbf{P}\boldsymbol{\Lambda}\mathbf{P}^{-1}$, where $\boldsymbol{\Lambda}$ is a diagonal matrix whose diagonal entries are the eigenvalues of $\mathbf{M}$ and $\mathbf{P}$ is an invertible matrix whose columns correspond to the eigenvectors of $\mathbf{M}$. Consequently, (1) becomes

$$\mathbf{v}\left(\mathbf{P}\boldsymbol{\Lambda}\mathbf{P}^{-1}\right)^t = \mathbf{w} \tag{2}$$

$$\mathbf{v}\left(\mathbf{P}\boldsymbol{\Lambda}^t\mathbf{P}^{-1}\right) = \mathbf{w} \text{ (by mathematical induction on } t) \tag{3}$$

$$\mathbf{v}\mathbf{P}\boldsymbol{\Lambda}^t = \mathbf{w}\mathbf{P} \text{ (by right-multiplying both sides by } \mathbf{P}). \tag{4}$$

To ease our subsequent analysis, we write $\boldsymbol{\Lambda} = \operatorname{diag}(\lambda_1, \lambda_2, \ldots, \lambda_n)$,where each λ_i for $1 \le i \le n$ is the eigenvalue of $\mathbf{M}$. Since $\boldsymbol{\Lambda}$ is a diagonal matrix, from basic facts in matrix theory, we have $\boldsymbol{\Lambda}^t = \operatorname{diag}\left(\lambda_1^t, \lambda_2^t, \ldots, \lambda_n^t\right)$. Suppose we write $\mathbf{P} = [\mathbf{p}_1 \mid \mathbf{p}_2 \mid \cdots \mid \mathbf{p}_n]$ where each $\mathbf{p}_i$ is the eigenvector of $\mathbf{M}$ associated with eigenvalue λ_i for $1 \le i \le n$. By considering the vectors $\mathbf{v}$ and $\mathbf{w}$ as $1 \times n$ row vectors and the vectors $\mathbf{p}_i$ for all $1 \le i \le n$ as $n \times 1$ column vectors, we have

$$\mathbf{v}\mathbf{P} = \mathbf{v}[\mathbf{p}_1 \mid \mathbf{p}_2 \mid \cdots \mid \mathbf{p}_n] = [\mathbf{v}\mathbf{p}_1 \ \mathbf{v}\mathbf{p}_2 \ \cdots \ \mathbf{v}\mathbf{p}_n] \text{ and} \tag{5}$$

$$\mathbf{w}\mathbf{P} = \mathbf{w}[\mathbf{p}_1 \mid \mathbf{p}_2 \mid \cdots \mid \mathbf{p}_n] = [\mathbf{w}\mathbf{p}_1 \ \mathbf{w}\mathbf{p}_2 \ \cdots \ \mathbf{w}\mathbf{p}_n], \tag{6}$$

where each $\mathbf{v}\mathbf{p}_i$ and $\mathbf{w}\mathbf{p}_i$ are scalar values for all $1 \le i \le n$. For brevity, we write $\mathbf{v}\mathbf{p}_i = \hat{v}_i$ and $\mathbf{w}\mathbf{p}_i = \hat{w}_i$ for all $1 \le i \le n$ and

$$\mathbf{v}\mathbf{P} = [\mathbf{v}\mathbf{p}_1 \ \mathbf{v}\mathbf{p}_2 \ \cdots \ \mathbf{v}\mathbf{p}_n] = [\hat{v}_1 \ \hat{v}_2 \ \cdots \ \hat{v}_n] = \hat{\mathbf{v}} \text{ and} \tag{7}$$

$$\mathbf{w}\mathbf{P} = [\mathbf{w}\mathbf{p}_1 \ \mathbf{w}\mathbf{p} \ \cdots \ \mathbf{w}\mathbf{p}_n] = [\hat{w}_1 \ \hat{w}_2 \ \cdots \ \hat{w}_n] = \hat{\mathbf{w}}. \tag{8}$$

Using the previously mentioned notations, (4) becomes

$$\hat{\mathbf{v}}\boldsymbol{\Lambda}^t = \hat{\mathbf{w}} \tag{9}$$

$$[\hat{v}_1 \ \hat{v}_2 \ \cdots \ \hat{v}_n] \operatorname{diag}\left(\lambda_1^t, \lambda_2^t, \ldots, \lambda_n^t\right) = [\hat{w}_1 \ \hat{w}_2 \ \cdots \ \hat{w}_n] \tag{10}$$

$$\left[\hat{v}_1\lambda_1^t \ \hat{v}_2\lambda_2^t \ \cdots \ \hat{v}_n\lambda_n^t\right] = [\hat{w}_1 \ \hat{w}_2 \ \cdots \ \hat{w}_n]. \tag{11}$$

All values $\hat{v}_i\lambda_i^t$ and $\hat{w}_i$ for all $1 \le i \le n$ are scalars in a particular extension field of $\mathbb{F}_q$. This extension field definitely must contain all eigenvalues λ_i for all $1 \le i \le n$. Hence, it suffices to choose the extension field as the splitting field of $p_{\mathbf{M}}(x)$ over $\mathbb{F}_q$, where $p_{\mathbf{M}}(x)$ is the characteristic polynomial of $\mathbf{M}$. We denote this splitting field by $\mathbb{K}(\lambda_1, \lambda_2, \ldots, \lambda_n)$. Clearly $\hat{v}_i\lambda_i^t$ and $\hat{w}_i$ for all $1 \le i \le n$ are elements of $\mathbb{K}(\lambda_1, \lambda_2, \ldots, \lambda_n)$. From the previous analysis, the values of $\hat{v}_i$, $\hat{w}_i$, and λ_i can be identified by a wiretapper by determining the eigenvalues of the public matrix and their corresponding eigenvectors.

Since for all $1 \le i \le n$ the quantities $\hat{v}_i$ and $\hat{w}_i$ are elements of $\mathbb{K}(\lambda_1, \lambda_2, \ldots, \lambda_n)$, the quantities $\hat{w}_i \cdot (\hat{v}_i)^{-1}$ are well-defined and are scalars in $\mathbb{K}(\lambda_1, \lambda_2, \ldots, \lambda_n)$. Moreover, because the finite field multiplication is commutative, we can represent $\hat{w}_i \cdot (\hat{v}_i)^{-1}$ by $\hat{w}_i/\hat{v}_i$ and accordingly denote this value by $\hat{u}_i$. Therefore, from (11), we obtain

$$\hat{v}_i \lambda_i^t = \hat{w}_i \text{ for all } 1 \le i \le n \tag{12}$$

$$\lambda_i^t = \hat{w}_i \cdot (\hat{v}_i)^{-1} = \hat{w}_i/\hat{v}_i = \hat{u}_i \text{ for all } 1 \le i \le n, \tag{13}$$

where $\hat{u}_i$ for all $1 \le i \le n$ are elements in the splitting field $\mathbb{K}(\lambda_1, \lambda_2, \ldots, \lambda_n)$. Consequently, the integer t that satisfies (1) can be determined by solving n instances of the generic DLP in $\mathbb{K}(\lambda_1, \lambda_2, \ldots, \lambda_n)$ of the forms $\lambda_i^t = \hat{u}_i$ for all $1 \le i \le n$. In each of the DLP instances, every possible value of t is unique modulo $|\lambda_i|$. Suppose the solutions of the n instances of the DLP are

$$\begin{array}{c} t \equiv t_1 \,(\text{mod}\, |\lambda_1|) \\ t \equiv t_2 \,(\text{mod}\, |\lambda_2|) \\ \vdots \\ t \equiv t_n \,(\text{mod}\, |\lambda_n|) \end{array} \tag{14}$$

or in general $t \equiv t_i \,(\text{mod}\, |\lambda_i|)$ with $1 \le t_i \le |\lambda_i|$ for all $1 \le i \le n$. The value t that satisfies (1) can be obtained by applying the generalized Chinese Remainder Theorem for non-coprime moduli to (14). From [21], we know that the solution to (14) exists if and only if $t_i \equiv t_j \,(\text{mod}\, \gcd(|\lambda_i|, |\lambda_j|))$ for all $1 \le i < j \le n$. Furthermore, if the solution exists, then it is unique modulo $\text{lcm}(|\lambda_1|, |\lambda_2|, \ldots, |\lambda_n|)$.

Suppose the characteristic polynomial of a public matrix $\mathbf{M} \in GL_n(q)$ is a polynomial $p_{\mathbf{M}}(x)$ over $\mathbb{F}_q$ of degree n. The aforesaid analysis shows that whenever the public matrix $\mathbf{M}$ is diagonalizable, then the MVMP in $\mathbb{F}_q^n$ can be reduced to at most n instances of the DLP in the splitting field of $p_{\mathbf{M}}(x)$. From the facts in finite field theory, this splitting field is a subfield of the extension field $\mathbb{F}_{q^n}$. Hence, whenever the public matrix is diagonalizable, the MVMP in $\mathbb{F}_q^n$ is reduced to at most n instances of the DLP in $\mathbb{F}_{q^n}$.

4.2. Efficient piecewise algorithm for solving the MVMP

The method in Section 4.1 stipulates that it is mathematically feasible to solve the MVMP using matrix diagonalization approach. However, mathematical theory alone is inadequate for asserting that matrix diagonalization method outperforms the other existing algorithms. From computational perspective, the existence of a rigorous and comprehensive algorithm that represents the approach in Section 4.1 is necessary. The previous mathematical analysis in Section 4.1 states that whenever the public matrix $\mathbf{M} \in GL_n(q)$ is diagonalizable, the MVMP $\mathbf{vM}^t = \mathbf{w}$ for $\mathbf{v}, \mathbf{w} \in \mathbb{F}_q^n$ can be reduced to at most n instances of the DLP in the splitting field of the characteristic polynomial of $\mathbf{M}$. This condition ostensibly requires the wiretapper to work with an enormous extension field of $\mathbb{F}_q$ at once. In such circumstance, the algorithm would expect an immense time and storage requirement.

In this section we propose a rigorous and comprehensive piecewise method for solving the MVMP. Let $\mathbf{M} \in GL_n(q)$ be a public matrix and $p_{\mathbf{M}}(x) \in \mathbb{F}_q[x]$ be its characteristic polynomial. We assume that $p_{\mathbf{M}}(x) = \prod_{i=1}^{n} (x - \lambda_i)$ where $\lambda_1, \lambda_2, \ldots, \lambda_n$ are the roots (counting multiplicity) of $p_{\mathbf{M}}(x)$. Instead of working with all roots of $p_{\mathbf{M}}(x)$

simultaneously, we may expect the diagonalization method to work more efficient if we handle each of the eigenvector one at a time. In such condition, we only need to consider the extension field $\mathbb{K}(\lambda_i)$ whenever we deal with the eigenvalue λ_i.

We describe the steps of our piecewise approach as follows. Initially, we need to determine the characteristic polynomial $p_{\mathbf{M}}(x)$ of the public matrix $\mathbf{M} \in GL_n(q)$. For obtaining $p_{\mathbf{M}}(x)$, the authors in [19] suggest Hessenberg algorithm—which requires polynomial number of scalar operations in $\mathbb{F}_q$. Although other efficient algorithms do exist (see, e.g.: [22, 23]), it should be noted that the formation of $p_{\mathbf{M}}(x)$ for $\mathbf{M} \in GL_n(q)$ typically requires at most $\mathcal{O}(n^3)$ scalar operations in $\mathbb{F}_q$.

The next step is the factorization of $p_{\mathbf{M}}(x)$ to linear factors which is required for determining the eigenvalues of $\mathbf{M}$. There are several techniques to perform polynomial factorization in $\mathbb{F}_q[x]$. Three prominent methods are: Berlekamp algorithm, Cantor–Zassenhaus algorithm, and von zur Gathen–Shoup algorithm [24]. The complexities of these algorithms are measured in terms of the size of the finite field (denoted by q) and the degree of the polynomial (denoted by n). For sufficiently large values of q and n, von zur Gathen–Shoup algorithm is the asymptotically fastest technique for factoring polynomials over $\mathbb{F}_q$ of degree n. This algorithm is expected to use $\mathcal{O}(n^2 \log^2 n \log\log n)$ number of scalar operations in $\mathbb{F}_q$ [24]. In terms of memory consumption, Berlekamp algorithm requires $\mathcal{O}(n^2)$ space for storing the elements of $\mathbb{F}_q$, while Cantor–Zassenhaus and von zur Gathen–Shoup algorithms respectively need $\mathcal{O}(n)$ and $\mathcal{O}(n^{3/2})$ space for storing the elements of $\mathbb{F}_q$.

After the factorization is completed, we assume that $p_{\mathbf{M}}(x)$ can be factored as $\prod_{i=1}^{n}(x-\lambda_i)$ where $\lambda_1, \lambda_2, \ldots, \lambda_n$ are the roots (counted with multiplicity). The subsequent important step is to check the diagonalizability of $\mathbf{M}$. This can be done by examining whether the geometric multiplicity of each eigenvalues of $\mathbf{M}$ is the same as its algebraic multiplicity. To do so, we first consider the distinct roots of $p_{\mathbf{M}}(x)$ and determine each of their algebraic multiplicities. For simplicity, the algebraic multiplicity of an eigenvalue λ is written by $m_a(\lambda)$. If the roots of $p_{\mathbf{M}}(x)$ are all distinct, then $\mathbf{M}$ is diagonalizable. Otherwise, we need to check whether the geometric multiplicity of each eigenvalue is equal to its algebraic multiplicity. Recall that the geometric multiplicity of an eigenvalue λ, written by $m_g(\lambda)$, is defined as the dimension of $\ker(\lambda\mathbf{I} - \mathbf{M})$. Suppose $\mathbf{M}$ has r distinct eigenvalues $\lambda_1, \lambda_2, \ldots, \lambda_r$ for some $1 \le r \le n$. The matrix $\mathbf{M}$ is diagonalizable if and only if $m_a(\lambda_i) = m_g(\lambda_i)$ for every $1 \le i \le r$.

If the matrix $\mathbf{M}$ is diagonalizable, the next step is to determine n linearly independent eigenvectors associated with the eigenvalues of $\mathbf{M}$. To make our analysis easier, suppose the eigenvalues of $\mathbf{M}$, counted with multiplicity, are $\lambda_1, \lambda_2, \ldots, \lambda_n$. For each λ_i, we can find a nonzero vector $\mathbf{p}_i \in \ker(\lambda_i\mathbf{I} - \mathbf{M})$. Furthermore, the diagonalizability property of $\mathbf{M}$ ensures that there are n linearly independent vectors $\mathbf{p}_1, \mathbf{p}_2, \ldots, \mathbf{p}_n$ where each $\mathbf{p}_i$ is associated with λ_i.

The diagonalizability of $\mathbf{M}$ also implies that the quantities of algebraic multiplicities and geometric multiplicities of every eigenvalue are equal. Accordingly, we denote the algebraic multiplicity of eigenvalue λ_i by d_i, which is also identical to the exponent of the factor $(x-\lambda_i)$ within the factorization of $p_{\mathbf{M}}(x)$. To avoid working with all eigenvalues of $\mathbf{M}$ at once, we do not construct the matrix $\mathbf{P}$ whose columns are eigenvectors of $\mathbf{M}$. Instead, for every eigenvalue λ_i where $1 \le i \le n$, we work with the extension field

$\mathbb{K}(\lambda_i) = \mathbb{F}_q[x]/(x-\lambda_i)^{d_i} \cong \mathbb{F}_{q^{d_i}}$. Afterward, we consider the known vectors $\mathbf{v}$ and $\mathbf{w}$ as row vectors in $\mathbb{F}^n_{q^{d_i}}$ and the eigenvector $\mathbf{p}_i$ as column vector in $\mathbb{F}^n_{q^{d_i}}$. Then, we compute $\mathbf{v}\mathbf{p}_i$ and $\mathbf{w}\mathbf{p}_i$ as scalar values in $\mathbb{F}_{q^{d_i}}$ and accordingly denote these values by $\hat{v}_i$ and $\hat{w}_i$. Using (13), we calculate $\hat{u}_i = \hat{w}_i/\hat{v}_i$ and obtain the DLP $\lambda_i^t = \hat{u}_i$ in $\mathbb{F}_{q^{d_i}}$. The value t that satisfies $\lambda_i^t = \hat{u}_i$ in $\mathbb{F}_{q^{d_i}}$ can be found using the algorithm for solving the DLP in a finite field (e.g.: [18, 25]). One of the commonly used nontrivial algorithms for solving the DLP in $\mathbb{F}_q$ is the Shanks's baby step – giant step method, which solves the DLP in $\mathbb{F}_{q^{d_i}}$ using $\mathcal{O}\left(d_i \cdot q^{d_i/2}\log q\right)$ scalar operations and $\mathcal{O}\left(q^{d_i/2}\right)$ space in $\mathbb{F}_{q^{d_i}}$ [14].

Observe that with this piecewise approach, for every $1 \le i \le n$ we solve each of the DLP $\lambda_i^t = \hat{u}_i$ in $\mathbb{K}(\lambda_i) \cong \mathbb{F}_{q^{d_i}}$ rather than in $\mathbb{K}(\lambda_1, \lambda_2, \dots, \lambda_n)$. Assuming that the secret exponent t of the MVMP exists, the process of solving the DLP $\lambda_i^t = \hat{u}_i$ in $\mathbb{F}_{q^{d_i}}$ for every $1 \le i \le n$ yields an integer t_i that is unique modulo $|\lambda_i|$. Because there are n instances of the DLP $\lambda_i^t = \hat{u}_i$, the secret exponent t must satisfy $t \equiv t_i \pmod{|\lambda_i|}$ with $1 \le t_i \le |\lambda_i|$ for all $1 \le i \le n$ as in (14). The value of t can be determined using the generalized Chinese Remainder Theorem for non-coprime moduli. Moreover, the solution t exists if and only if $t_i \equiv t_j \pmod{\gcd(|\lambda_i|, |\lambda_j|)}$ for all $1 \le i < j \le n$. Additionally, if t exists, then t is unique modulo $m = \operatorname{lcm}(|\lambda_1|, |\lambda_2|, \dots, |\lambda_n|)$. From [26], the complexity of this generalized Chinese Remainder Theorem is $\mathcal{O}\left(\log^2(m)\right)$.

The pseudocode for the efficient piecewise approach for solving the MVMP is depicted in Algorithm 1. The efficiency of this method is measured in terms of the total number of scalar operations in $\mathbb{F}_q$ and the storage consumption under the assumption that one scalar value in $\mathbb{F}_q$ requires $\mathcal{O}(1)$ space. Elements in $\mathbb{F}_{q^{d_i}}$ are considered as d_i dimensional vectors or polynomials of degree d_i over $\mathbb{F}_q$. Consequently, one scalar operation in $\mathbb{F}_{q^{d_i}}$ is equivalent to d_i finite field operations in $\mathbb{F}_q$ and one scalar value in $\mathbb{F}_{q^{d_i}}$ needs $\mathcal{O}(d_i)$ space. The asymptotic running time and storage complexities of Algorithm 1 is stated in Theorem 1.

Theorem 1 *Assuming that Algorithm 1 uses Hessenberg procedure, von zur Gathen–Shoup factorization technique, and Shanks's baby step – giant step method, the MVMP* $\mathbf{v}\mathbf{M}^t = \mathbf{w}$ *for any diagonalizable matrix* $\mathbf{M} \in GL_n(q)$ *and* $\mathbf{v}, \mathbf{w} \in \mathbb{F}_q^n$ *is solvable using* $\mathcal{O}\left(n^3 \cdot q^{n/2} \cdot \log q\right)$ *scalar operations and* $\mathcal{O}\left(n^2 \cdot q^{n/2}\right)$ *space in* $\mathbb{F}_q$.

Proof: We first analyze the asymptotic running time for Algorithm 1 as follows:

(1) Line 1 is performed using Hessenberg algorithm which requires $\mathcal{O}\left(n^3\right)$ scalar operations in $\mathbb{F}_q$.

(2) Line 2 is the factorization of the characteristic polynomial of $\mathbf{M}$. Using von zur Gathen–Shoup algorithm, this step requires $\mathcal{O}\left(n^2\log^2 n \log\log n\right)$ operations in $\mathbb{F}_q$.

(3) Lines 3 and 5 involve arrays of length n whose entries are integers between 1 and n (inclusive). These lines are respectively performed by simple array scan and array matching procedures, both require at most $\mathcal{O}(n\log n)$ time. This is because if $q \ge n$, then both array scan and array matching procedures require no more than $\mathcal{O}(n)$ time while if $q < n$, then $q^k = n$ for some $k > 0$, and thus both procedures require no more than $\mathcal{O}(n\log n)$ time.

IOP Conf. Series: Journal of Physics: Conf. Series **971** (2018) 012025 doi:10.1088/1742-6596/971/1/012025

Algorithm 1 Piecewise approach for solving the MVMP

Require: Vectors $\mathbf{v}, \mathbf{w} \in \mathbb{F}_q^n$ and a public matrix $\mathbf{M} \in GL_n(q)$.

1: Compute the characteristic polynomial of $\mathbf{M}$, i.e., $p_{\mathbf{M}}(x)$.
2: Factorize $p_{\mathbf{M}}(x)$ into $\prod_{i=1}^{n}(x-\lambda_i)$ using factorization algorithm.
3: For each $1 \le i \le n$, determine the value $d_i = m_a(\lambda_i)$ which is the algebraic multiplicity of λ_i (i.e., the exponent of $(x-\lambda_i)$ within the factorization of $p_{\mathbf{M}}(x)$).
4: For each $1 \le i \le n$, determine the value $m_g(\lambda_i)$ which is the geometric multiplicity of λ_i (i.e., $m_g(\lambda_i) = \dim(\ker(\lambda_i\mathbf{I} - \mathbf{M}))$).
5: For each $1 \le i \le n$, determine whether $m_a(\lambda_i) = m_a(\lambda_i)$. If their quantities are different, then $\mathbf{M}$ is not diagonalizable and the algorithm is terminated. Otherwise, $\mathbf{M}$ is diagonalizable and the algorithm is continued.
6: For all $1 \le i \le n$, find a nonzero vector $\mathbf{p}_i \in \mathbb{F}_{q^{d_i}}^n$ such that $\mathbf{p}_i \in \ker(\lambda_i\mathbf{I} - \mathbf{M})$ and the set $\{\mathbf{p}_1, \mathbf{p}_2, \ldots, \mathbf{p}_n\}$ is linearly independent.
7: **for all** $i \in [1, n]$ **do**
8:     Compute $\hat{v}_i \leftarrow \mathbf{v}\mathbf{p}_i$ as a value in $\mathbb{F}_{q^{d_i}}$.
9:     Compute $\hat{w}_i \leftarrow \mathbf{w}\mathbf{p}_i$ as a value in $\mathbb{F}_{q^{d_i}}$.
10:     Compute $\hat{u}_i \leftarrow \hat{w}_i/\hat{v}_i$ as a value in $\mathbb{F}_{q^{d_i}}$.
11:     Determine the exponent t_i of the DLP $\lambda_i^{t_i} = \hat{u}_i$ in $\mathbb{F}_{q^{d_i}}$.
12: **end for**
13: Determine the solution of the systems of congruences $t \equiv t_i \pmod{|\lambda_i|}$ for $1 \le i \le n$ (as in (14)) using the generalized Chinese Remainder Theorem.
14: If the integer t exists, then t is the solution of $\mathbf{v}\mathbf{M}^t = \mathbf{w}$. Otherwise, the MVMP $\mathbf{v}\mathbf{M}^t = \mathbf{w}$ has no solution for t.

Ensure: If the algorithm is terminated and an integer t exists, then t is the solution of the MVMP $\mathbf{v}\mathbf{M}^t = \mathbf{w}$.

(4) Each of lines 4 and 6 can be done in $\mathcal{O}(Dn^4)$ time where $D = \max\{d_1, d_2, \ldots, d_n\}$. There are n eigenvalues $\lambda_1, \lambda_2, \ldots, \lambda_n$ and each λ_i is an element of $\mathbb{F}_{q^{d_i}}$. The computation $\ker(\lambda_i\mathbf{I} - \mathbf{M})$ for each $1 \le i \le n$ is performed over $\mathbb{F}_{q^{d_i}}$ and it requires $\mathcal{O}(d_i n^3)$ scalar operations in $\mathbb{F}_q$. Since there are n eigenvalues, the upper bound for each of lines 4 and 6 becomes $\sum_{i=1}^{n}\mathcal{O}(d_i n^3) = \sum_{i=1}^{n}\mathcal{O}(Dn^3) = \mathcal{O}(Dn^4)$.

(5) Each of lines 8 and 9 requires $\mathcal{O}(d_i n^2)$ scalar operations in $\mathbb{F}_q$. This is because the vectors $\mathbf{v}$, $\mathbf{w}$, and $\mathbf{p}_i$ are respectively considered as matrices over $\mathbb{F}_{q^{d_i}}$ of size $1 \times n$, $1 \times n$, and $n \times 1$.

(6) Line 10 requires $\mathcal{O}(d_i^2)$ operations in $\mathbb{F}_q$ because both $\hat{v}_i$ and $\hat{w}_i$ are considered as polynomials of degree d_i over $\mathbb{F}_q$.

(7) Line 11 is performed using the Shanks's baby step – giant step algorithm which requires $\mathcal{O}(d_i \cdot q^{d_i/2} \cdot \log q)$ operations in $\mathbb{F}_{q^{d_i}}$ or equivalently $\mathcal{O}(d_i^2 \cdot q^{d_i/2} \cdot \log q)$ operations in $\mathbb{F}_q$.

(8) Line 13 is the generalized Chinese Remainder Theorem for non-coprime moduli which requires $\mathcal{O}(\log^2 D) = \mathcal{O}(\log^2 n)$ time. This upper bound follows from the condition $D = \max\{d_1, d_2, \ldots, d_n\}$ and $d_i \le n$ for all $1 \le i \le n$.

From the previous analysis, the asymptotic time complexity for lines 8–11 is given by $\mathcal{O}\left(d_i n^2\right) + \mathcal{O}\left(d_i n^2\right) + \mathcal{O}\left(d_i^2\right) + \mathcal{O}\left(d_i^2 \cdot q^{d_i/2} \cdot \log q\right)$. From the facts in finite field theory we have $d_i \leq n$, which leads to the following upper bound for this asymptotic expression:

$$\begin{aligned}\mathcal{O}\left(n^3\right) + \mathcal{O}\left(n^3\right) + \mathcal{O}\left(n^2\right) + \mathcal{O}\left(n^2 \cdot q^{n/2} \cdot \log q\right) &= \mathcal{O}\left(n^3\right) + \mathcal{O}\left(n^2 \cdot q^{n/2} \cdot \log q\right) \\ &= \mathcal{O}\left(n^2 \cdot q^{n/2} \cdot \log q\right), \text{ because } q \geq 2.\end{aligned} \tag{15}$$

Consequently, the upper bound for the asymptotic running time of the while loop (lines 7–12) is $n \cdot \mathcal{O}\left(n^2 \cdot q^{n/2} \cdot \log q\right) = \mathcal{O}\left(n^3 \cdot q^{n/2} \cdot \log q\right)$. Therefore, the overall running time complexity of the algorithm can be analyzed as follows:

Table 2: Running time complexity of the algorithm

line	time complexity
1	$\mathcal{O}\left(n^3\right)$
2	$\mathcal{O}\left(n^2 \log^2 n \log\log n\right)$
3, 5	$\mathcal{O}\left(n \log n\right)$
4, 6	$\mathcal{O}\left(Dn^4\right) = \mathcal{O}\left(n^5\right)$, since $D \leq n$
7–12	$\mathcal{O}\left(n^3 \cdot q^{n/2} \cdot \log q\right)$
13	$\mathcal{O}\left(\log^2 n\right)$
overall	$\mathcal{O}\left(n^3 \cdot q^{n/2} \cdot \log q\right)$

since $\mathcal{O}\left(n^3 \cdot q^{n/2} \cdot \log q\right)$ dominates other terms.

For the space complexity analysis of Algorithm 1, observe that an element of $\mathbb{F}_{q^{d_i}}$ can be viewed as a polynomial (or a vector) of degree (or dimension) d_i over $\mathbb{F}_q$. To ease our analysis in determining the upper bound for the memory complexity, we may assume that the storage requirement of an element in $\mathbb{F}_{q^{d_i}}$ is equivalent to the storage requirement of n elements in $\mathbb{F}_q$ (because $d_i \leq n$). The asymptotic space complexity is analyzed as follows:

(1) Line 1 requires $\mathcal{O}\left(n^2\right)$ space for storing a constant number of $n \times n$ matrices over $\mathbb{F}_q$.
(2) Line 2 requires $\mathcal{O}\left(d_i n^{3/2}\right) = \mathcal{O}\left(n^{5/2}\right)$ space for the factorization of a polynomial of degree n where each coefficient takes no more than n possible values.
(3) Each of lines 3 and 5 uses an array of length n whose entries are integers between 1 and n (inclusive). If $q \geq n$, then both arrays require no more than $\mathcal{O}(n)$ space; otherwise $q^k = n$ for some $k > 0$ and thus the space requirement becomes $\mathcal{O}(n \log n)$. Therefore, each of these lines requires at most $\mathcal{O}(n \log n)$ space.
(4) In each of lines 4 and 6, the algorithm uses n matrices of size $n \times n$ over $\mathbb{F}_{q^{d_i}}$ where each $d_i \leq n$. Thus, each of these lines needs $\mathcal{O}\left(n^4\right)$ space.
(5) Each of lines 8 and 9 utilizes an n dimensional vector over $\mathbb{F}_{q^{d_i}}$. Since $d_i \leq n$, each of these lines requires $\mathcal{O}\left(n^2\right)$ space.

(6) Line 10 involves the value $\hat{u}_i$, $\hat{v}_i$, and $\hat{w}_i$. The space complexity for this line is $\mathcal{O}(n)$ because $\hat{u}_i$, $\hat{v}_i$, and $\hat{w}_i$ are elements of $\mathbb{F}_{q^{d_i}}$ and $d_i \leq n$.

(7) Line 11 involves the computation of Shanks's baby step – giant step algorithm in $\mathbb{F}_{q^{d_i}}$ which requires $\mathcal{O}\left(n \cdot q^{n/2}\right)$ space. This is because the Shanks's baby step – giant step procedure needs $\mathcal{O}\left(q^{d_i/2}\right)$ space in $\mathbb{F}_{q^{d_i}}$ or equivalently $\mathcal{O}\left(d_i \cdot q^{d_i/2}\right)$ space in $\mathbb{F}_q$.

(8) Line 13 requires $\mathcal{O}\left(n^2\right)$ space for storing an array of integers between 1 and $q^n - 1$ (inclusive) of length n. There are n entries where each entry requires at most $\mathcal{O}(n)$ space.

Based on the aforementioned analysis, the asymptotic space complexity of lines 8–11 is $\mathcal{O}\left(n^2\right) + \mathcal{O}\left(n^2\right) + \mathcal{O}(n) + \mathcal{O}\left(n \cdot q^{n/2}\right) = \mathcal{O}\left(n \cdot q^{n/2}\right)$ since $q \geq 2$. Consequently, the upper bound for storage requirement of the while loop (line 7–12) becomes $\mathcal{O}\left(n^2 \cdot q^{n/2}\right)$. Thus, the overall space complexity of the algorithm can be evaluated as follows:

Table 3: Space complexity of the algorithm

line	space complexity
1	$\mathcal{O}\left(n^2\right)$
2	$\mathcal{O}\left(n^{5/2}\right)$
3, 5	$\mathcal{O}(n \log n)$
4, 6	$\mathcal{O}\left(n^4\right)$
7–12	$\mathcal{O}\left(n^2 \cdot q^{n/2}\right)$
13	$\mathcal{O}\left(n^2\right)$
overall	$\mathcal{O}\left(n^2 \cdot q^{n/2}\right)$

because $\mathcal{O}\left(n^2 \cdot q^{n/2}\right)$ dominates other terms. □

The result in Theorem 1 states that our proposed diagonalization method can solve the MVMP in $\mathbb{F}_q^n$ using $\mathcal{O}\left(n^3 \cdot q^{n/2} \cdot \log q\right)$ scalar operations, which means that Algorithm 1 solves the MVMP for diagonalizable matrix faster than the collision algorithm in [9] by a factor of $\mathcal{O}(n)$. Nevertheless, the diagonalization technique requires $\mathcal{O}\left(n^2 \cdot q^{n/2}\right)$ space in its execution, which is larger than that of the collision algorithm by a factor of $\mathcal{O}(n)$. This theoretical result indicates a theoretical time-memory trade-off of Algorithm 1.

4.3. Optimized diagonalization method for primitive matrices

Superficially, it is naturally desirable if the public matrix $\mathbf{M} \in GL_n(q)$ is constructed using primitive matrix as in [8, Algorithm 2]. This is because if $\mathbf{M}$ is a primitive matrix, then the value of $\mathbf{vM}^t$ are all distinct for any nonzero $\mathbf{v} \in \mathbb{F}_q^n$ and any integer $0 \leq t < q^n - 1$, which implies that $\mathbf{vM}^t$ can be any nonzero vector in $\mathbb{F}_q^n$ [8, Theorem 3]. Consequently, the utilization of primitive matrix for the formation of $\mathbf{M}$ undoubtedly thwarts exhaustive search attack for unraveling the secret exponent. However, in this section we show that primitive matrices are diagonalizable and thus are vulnerable to the attack method described in Section 4.2. Moreover, we devise a computationally faster version of Algorithm 1 for solving the MVMP which involves primitive matrices.

If $\mathbf{M} \in GL_n(q)$ is a primitive matrix, then its characteristic polynomial, $p_{\mathbf{M}}(x)$, is a primitive polynomial of degree n over $\mathbb{F}_q$ [12]. Moreover, $p_{\mathbf{M}}(x)$ can be factored as $p_{\mathbf{M}}(x) = \prod_{i=0}^{n-1}\left(x - \xi^{q^i}\right)$ where ξ is a primitive root of $\mathbb{F}_{q^n}$ [11, 12]. In addition, the values of ξ^{q^i} are different for all $0 \le i \le n-1$ and are also primitive elements of $\mathbb{F}_{q^n}$. This implies that $\mathbf{M}$ is diagonalizable and the diagonal matrix $\mathbf{\Lambda}$ in (9) has the form $\mathrm{diag}\left(\xi, \xi^q, \ldots, \xi^{q^{n-1}}\right)$. Assuming that $\lambda_i = \xi^{q^{i-1}}$ for $1 \le i \le n$, the expressions (12) and (13) respectively become

$$\hat{v}_i\left(\xi^{q^{i-1}}\right)^t = \hat{w}_i \text{ for all } 1 \le i \le n \tag{16}$$

$$\left(\xi^{q^{i-1}}\right)^t = \hat{w}_i \cdot (\hat{v}_i)^{-1} = \hat{w}_i/\hat{v}_i = \hat{u}_i \text{ for all } 1 \le i \le n. \tag{17}$$

For each $1 \le i \le n$, the integer t that satisfies (17) can be found using the Shanks's baby step – giant step algorithm.

Suppose t_i where $1 \le i \le n$ is the solution of (17), then the solution of the MVMP exists if for all $1 \le i \le n$ the congruences $t \equiv t_i \left(\mathrm{mod}\left|\xi^{q^{i-1}}\right|\right)$ have a unique solution. Using the fact in [21], the value of t exists if and only if $t_i \equiv t_j \left(\mathrm{mod}\,\gcd\left(\left|\xi^{q^{i-1}}\right|, \left|\xi^{q^{j-1}}\right|\right)\right)$ for all $1 \le i < j \le n$. Since each $\xi^{q^{i-1}}$ for every $1 \le i \le n$ is a primitive element in $\mathbb{F}_{q^n}$, then we have $\left|\xi^{q^{i-1}}\right| = q^n - 1$ for all $1 \le i \le n$. Hence, the solution of the MVMP exists if and only if $t_i \equiv t_j \,(\mathrm{mod}\, q^n - 1)$ for all $1 \le i < j \le n$. Accordingly, we have the following corollary.

Corollary 1 *Suppose* $\mathbf{M} \in GL_n(q)$ *is a primitive matrix,* $\mathbf{v}, \mathbf{w} \in \mathbb{F}_q^n$*, and* ξ *is an eigenvalue of* $\mathbf{M}$*. The MVMP* $\mathbf{v}\mathbf{M}^t = \mathbf{w}$ *is solvable for* t *if and only if the DLP* $\xi^t = \hat{u}_1$ *in* $\mathbb{F}_{q^n} \cong \mathbb{F}_q/(p_{\mathbf{M}}(x))$ *has a solution for* $t \in [0, q^n - 2]$*.*

Proof: The condition $t_i \equiv t_j \,(\mathrm{mod}\, q^n - 1)$ for all $1 \le i < j \le n$ implies $t_i \equiv t_1 \,(\mathrm{mod}\, q^n - 1)$ for all $1 \le i \le n$. Hence, t must satisfy $\xi^t = \hat{u}_1$. □

Corollary 1 stipulates that instead of solving n instances of the DLP as in Section 4.2, the MVMP in $\mathbb{F}_q^n$ that involves primitive matrix can be solved by solely solving a specific instance of the DLP in $\mathbb{F}_{q^n}$. This condition also implies that we only need to determine a pair of eigenvalue and eigenvector of $\mathbf{M}$ instead of n pair of eigenvalues and eigenvector as in Section 4.2. The pseudocode for this efficient method is described in Algorithm 2. The asymptotic running time and memory complexities of this algorithm is briefly discussed in Theorem 2.

Theorem 2 *Assuming that Algorithm 2 uses Hessenberg procedure and Shanks's baby step – giant step method, the MVMP* $\mathbf{v}\mathbf{M}^t = \mathbf{w}$ *for any primitive matrix* $\mathbf{M} \in GL_n(q)$ *and* $\mathbf{v}, \mathbf{w} \in \mathbb{F}_q^n$ *is solvable using* $\mathcal{O}\left(n^2 \cdot q^{n/2} \cdot \log q\right)$ *scalar operations and* $\mathcal{O}\left(n \cdot q^{n/2}\right)$ *space in* $\mathbb{F}_q$*.*

Algorithm 2 Optimized diagonalization approach to solve the MVMP for primitive matrices

Require: Vectors $\mathbf{v}, \mathbf{w} \in \mathbb{F}_q^n$ and a public matrix $\mathbf{M} \in GL_n(q)$ which is a primitive matrix.

1: Compute the characteristic polynomial of $\mathbf{M}$, i.e., $p_{\mathbf{M}}(x)$.
2: Set the finite field $\mathbb{F}_{q^n} \cong \mathbb{F}_q / (p_{\mathbf{M}}(x))$.
3: Set ξ where ξ is a root of $p_{\mathbf{M}}(x)$ as the primitive element of $\mathbb{F}_{q^n} \cong \mathbb{F}_q / (p_{\mathbf{M}}(x))$.
4: Find a nonzero vector $\mathbf{p} \in \mathbb{F}_{q^n}^n$ such that $\mathbf{p} \in \ker(\xi \mathbf{I} - \mathbf{M})$.
5: Compute $\hat{v} \leftarrow \mathbf{vp}$ as a value in $\mathbb{F}_{q^n}$.
6: Compute $\hat{w} \leftarrow \mathbf{wp}$ as a value in $\mathbb{F}_{q^n}$.
7: Compute $\hat{u} \leftarrow \hat{w}/\hat{v}$ as a value in $\mathbb{F}_{q^n}$.
8: Determine the exponent t of the DLP $\xi^t = \hat{u}$ in $\mathbb{F}_{q^n}$.

Ensure: If the value t exists, then t is the solution of the MVMP $\mathbf{v}\mathbf{M}^t = \mathbf{w}$.

Proof: The asymptotic analysis of this algorithm is almost similar to that described in the proof of Theorem 1. One substantial difference is that Algorithm 2 only considers an eigenvalue and one of its associated eigenvector instead of n eigenvalues and their corresponding eigenvectors as in Algorithm 1. Hence, both asymptotic time and space complexities of Algorithm 2 are lower by a factor of $\mathcal{O}(n)$ than those for Algorithm 1. Therefore, the running time and space complexities of Algorithm 2 are respectively $\mathcal{O}\left(n^2 \cdot q^{n/2} \cdot \log q\right)$ and $\mathcal{O}\left(n \cdot q^{n/2}\right)$. The detail of the proof of this theorem is left to readers. □

From the result in Theorem 2, we infer that Algorithm 2 is asymptotically faster than the collision algorithms in [9] by a factor of $\mathcal{O}\left(n^2\right)$. Additionally, the space requirement of Algorithm 2 is asymptotically equivalent to those of the collision algorithms in [9]. Hence, from theoretical perspective, Algorithm 2 asymptotically outperforms the collision algorithms in [9].

5. Example and Numerical Experiments

In this section we discuss an example and numerical experiments regarding the formerly explained diagonalization technique for solving the MVMP. We first demonstrate a small numerical implementation of Algorithm 2 in Example 1.

Example 1 Suppose we consider the MVMP $\mathbf{v}\mathbf{M}^t = \mathbf{w}$ in $\mathbb{F}_7^3$ with $\mathbf{v} = (1,1,0)$, $\mathbf{w} = (2,0,0)$, and $\mathbf{M} = \begin{bmatrix} 4 & 4 & 0 \\ 4 & 5 & 5 \\ 0 & 4 & 2 \end{bmatrix}$ as in [9, Example 1]. The characteristic polynomial of $\mathbf{M}$ is $p_{\mathbf{M}}(x) = x^3 + 3x^2 + 2x + 2$ and it is a primitive polynomial of degree 3 over $\mathbb{F}_7$. Thus, we can apply Algorithm 2 to determine the secret exponent t of the MVMP. Suppose ξ is a root of $p_{\mathbf{M}}(x)$ which is also an eigenvalue of $\mathbf{M}$. The eigenvalue ξ is associated with the eigenvector $\mathbf{p} = \left(1, 2\xi + 6, 6\xi^2 + 2\xi + 3\right)^T$. Consequently, we have $\hat{v} = \mathbf{vp} = 2\xi$ and $\hat{w} = \mathbf{wp} = 2$, and thus the solution of the MVMP $\mathbf{v}\mathbf{M}^t = \mathbf{w}$ is identical to the secret exponent t in the following condition:

$$2\xi\left(\xi^{t}\right) = 2, \text{ or equivalently} \tag{18}$$

$$\xi^{t} = 2/2\xi = 3\xi^{2} + 2\xi + 6. \tag{19}$$

By solving the DLP (19) in $\mathbb{F}_{7^3}$ for the value of t, we have $t = 341$.

The numerical experiments are performed to assess the practical effectiveness of the diagonalization method for solving the MVMP. We confine our subject of interest to the MVMP that uses primitive matrix as one of its public parameters. One underlying reason is because the utilization of primitive matrix purportedly makes the protocol more secure against exhaustive search attacks. The experiments were conducted by comparing both running time and memory aspects of the deterministic collision algorithm in [9] and Algorithm 2. The algorithms were implemented using SageMath version 7.1 beta3 and Python 2.7.17 over Mac OS X El Capitan system with 3.5 GHz 6-Core Intel Xeon E5 processor and 16 gigabytes of memory.

The experiments considered several instances of the MVMP in $\mathbb{F}_q^n$ for $q = 2, 3, 5, 7$ and various values of n. For each of the scenarios, the public matrix $\mathbf{M} \in GL_n(q)$ is a primitive matrix, the public vector $\mathbf{v} \in \mathbb{F}_q^n$ is a random nonzero vector, and the value of the secret exponent t is equal to $\lfloor (q^n - 1)/2 \rfloor$. One objective of these experiments is to find the crossover points at which Algorithm 2 practically works faster than the deterministic collision algorithm. The graphs of the running time and storage requirement of both algorithms for solving the MVMP in $\mathbb{F}_2^n$ where $3 \leq n \leq 21$ are correspondingly depicted in Figure 1 and Figure 2 . For the finite field $\mathbb{F}_2$, our experiments infer a crossover point of $n = 17$. This result shows that Algorithm 2 is practically faster than the collision algorithm in [9]. However, it requires more memory than the collision algorithm. We can see the memory usage comparison between Algorithm 2 and the collision algorithm in Figure 2. Their memory consumptions have constant difference around 9 megabytes. This disparity is constant because both algorithms have equivalent space complexity. Nevertheless, in our experiments, our proposed algorithm requires more memory than the collision algorithm. This is probably due to the utilization of extension fields in SageMath—which costs more space than the prime fields.

Table 4 presents the running time and memory consumption of the collision algorithm and Algorithm 2 for $\mathbb{F}_q^n$, where $q = 3, 5, 7$ and various values of n. Table 4 shows a similar property as in Figure 1 that Algorithm 2 is faster than the collision algorithm in [9]. The crossover points for these experiments are as low as $n = 17$ for $q = 2$, $n = 11$ for $q = 3$, $n = 7$ for $q = 5$, and $n = 7$ for $q = 7$. In addition, Table 4 also indicates similar characteristics as in Figure 2 that Algorithm 2 requires more memory than the collision algorithm. However, in our experiments, there are several anomalies. The memory consumption of our proposed algorithm when the matrix size is 14, 15, and 16 are higher than those other values in $\{3, 4, ..., 21\}$. Despite our best efforts, we do not know yet the explanation of these anomalies. We are currently planning to resolve this issue by conducting further experiments with more comprehensive test cases. Nevertheless, one should note that our experimental results are supplementary. Theoretically, our proposed algorithm uses asymptotically equivalent amount of memory to that of the collision algorithm as described in Theorem 2.

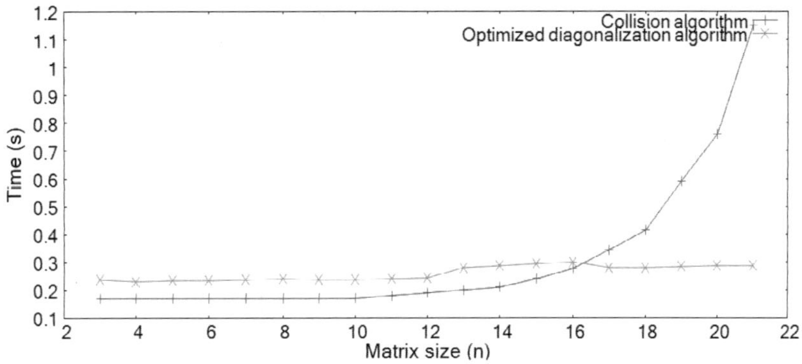

Figure 1: Average running time for collision algorithm in [9] and optimized diagonalization algorithm (Algorithm 2) for solving the MVMP in $\mathbb{F}_q^n$ for $q = 2$ and $n = 3, ..., 21$.

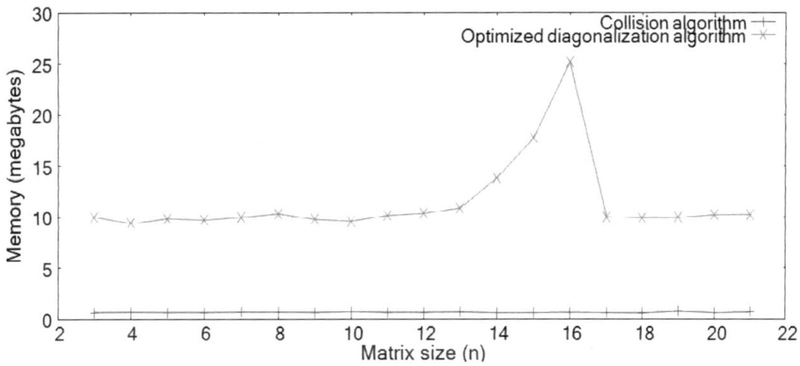

Figure 2: Average memory usage for collision algorithm in [9] and optimized diagonalization algorithm (Algorithm 2) for solving the MVMP in $\mathbb{F}_q^n$ for $q = 2$ and $n = 3, ..., 21$.

6. Conclusion and Future Works

We study the Megrelishvili vector-matrix problem (MVMP) as a linear algebra problem that strongly correlates with the security of Megrelishvili protocol. In particular, we formulate an approach for solving the MVMP using matrix diagonalization method. The theoretical result in Theorem 1 asserts that the MVMP $\mathbf{v}\mathbf{M}^t = \mathbf{w}$ in $\mathbb{F}_q^n$ for any diagonalizable $\mathbf{M} \in GL_n(q)$ is solvable using $\mathcal{O}\left(n^3 \cdot q^{n/2} \cdot \log q\right)$ finite field operations and $\mathcal{O}\left(n^2 \cdot q^{n/2}\right)$ storage. Moreover, our investigation also yields an important counterintuitive result: the utilization of a primitive matrix for the construction of the public matrix $\mathbf{M}$ makes the protocol more vulnerable to attacks. In Theorem 2 we show that the MVMP in $\mathbb{F}_q^n$ which uses a primitive matrix is solvable in $\mathcal{O}\left(n^2 \cdot q^{n/2} \cdot \log q\right)$ time and $\mathcal{O}\left(n \cdot q^{n/2}\right)$ space. Therefore, our diagonalization methods are asymptotically faster than the collision algorithms in [9] whenever the public matrix is diagonalizable.

Table 4: Average running time and memory consumption for collision algorithm in [9] and optimized diagonalization method (Algorithm 2) for solving the MVMP in $\mathbb{F}_q^n$ for $q = 3, 5, 7$ and several n.

q	n	Collision Algorithm in [9]		Algorithm 2	
		Running Time (seconds)	Memory Usage (megabytes)	Running Time (seconds)	Memory Usage (megabytes)
3	11	0.54	1.1	0.46	10.53
	12	0.81	1.17	0.51	10.53
	13	1.35	1.47	0.64	10.5
5	7	0.40	1	0.39	10.2
	8	0.63	1.13	0.48	10.23
	9	1.25	1.13	0.63	10.73
7	6	0.46	0.93	0.47	10.2
	7	0.9	1.23	0.54	10.4
	8	2.23	2.13	0.87	10.8

Additionally, if the public matrix is also a primitive matrix, then diagonalization method asymptotically outperforms the collision algorithm in [9].

We also conducted numerical experiments to support the theoretical asymptotic analysis in Section 4. The experiments were restricted to the MVMP that uses primitive matrix as its public matrix. The results of the experiments show that the diagonalization method practically works faster than the deterministic collision algorithm for solving the MVMP in $\mathbb{F}_q^n$, with crossover points as low as $n = 17$ for $q = 2$, $n = 11$ for $q = 3$, $n = 7$ for $q = 5$, and $n = 7$ for $q = 7$. Nevertheless, experimental results indicate that the diagonalization method requires more memory than the collision algorithm—albeit the theoretical analysis shows that the space complexities of both techniques are asymptotically equivalent whenever the public matrix is also a primitive matrix.

In Theorem 1 and Theorem 2 we provide asymptotic running time upper bounds for solving the MVMP. These upper bounds improve the previously discovered bounds in [9]. However, these new upper bounds are confined to the condition where the public matrix is diagonalizable. We conjecture that a similar and more general upper bound for any public matrix can be obtained by applying the Jordan decomposition technique for solving the MVMP. We believe that this approach is attainable by modifying methods in [19, 20]. Finally, we do not know yet whether the upper bound for solving the MVMP that involves primitive matrices in Theorem 2 is optimal or not. The research of the optimal running time upper bound for solving the MVMP is essential to quantify the security of Megrelishvili protocol.

References

[1] Diffie W and Hellman M E 1976 *New directions in cryptography, Information Theory, IEEE Transactions on* **22** 644–654

[2] Megrelishvili R, Chelidze M and Chelidze K 2006 *New matrix sets generation and the cryptosystems, Applied Mathematics, Informatics and Mechanics (AMIM)* **11** 29–36

[3] Megrelishvili R and Sikharulidze A 2009 *New matrix sets generation and the cryptosystems, Proceedings of the European Computing Conference and the Third International Conference on Computational Intelligence, Tbilisi, Georgia* pp 253–255

[4] Megrelishvili R, Chelidze M and Besiashvili G 2010 *Investigation of new matrix-key function for the public cryptosystems, Proceedings of The Third International Conference, Problems of Cybernetics and Information* vol 1 pp 6–8

[5] Megrelishvili R 2011 *On the original one-way function and the new digital signature scheme, Applied Mathematics, Informatics and Mechanics (AMIM)* **16** 43–48

[6] Beletsky A, Beletsky A and Kandyba R 2013 *Matrix analogues of the Diffie-Hellman protocol, ICTERI* pp 352–359

[7] Arzaki M 2016 *Elementary algorithms analysis of Megrelishvili protocol, Indonesian Journal on Computing (Indo-JC)* **1** 11–23

[8] Arzaki M and Wahyudi B A 2016 *On the construction of secure public parameters for Megrelishvili protocol, 2016 4th International Conference on Information and Communication Technology (ICoICT 2016)* (IEEE) pp 1–5

[9] Arzaki M and Wahyudi B A 2016 *Collision algorithms for breaking Megrelishvili protocol: theory and numerical experiments, 2016 8th International Conference on Advanced Computer Science and Information Systems (ICACSIS 2016)* (IEEE) pp 53–60

[10] Arzaki M and Wahyudi B A 2017 *Extending Megrelishvili protocol for multi-party key agreement, 2017 5th International Conference on Information and Communication Technology (ICoICT)* (IEEE) pp 290–297

[11] McEliece R J 1987 *Finite fields for computer scientists and engineers* (Springer Science & Business Media)

[12] Lidl R and Niederreiter H 1994 *Introduction to finite fields and their applications* (Cambridge university press)

[13] Durbin J R 2008 *Modern algebra: An introduction* (John Wiley & Sons)

[14] Hoffstein J, Pipher J C and Silverman J H 2014 *An introduction to mathematical cryptography* 2nd ed (Springer)

[15] Niven I 1948 *Fermat's theorem for matrices, Duke Mathematical Journal* **15** 823–826

[16] Horn R A and Johnson C R 2012 *Matrix analysis* (Cambridge university press)

[17] Cormen T H, Leiserson C E, Rivest R E and Stein C 2009 *Introduction to algorithms* (MIT press)

[18] Shanks D 1971 *Class number, a theory of factorization, and genera, Proc. Symp. Pure Math* vol 20 pp 415–440

[19] Menezes A J and Wu Y H 1997 *The discrete logarithm problem in* $GL(n, q)$, *Ars Combinatoria* **47** 23–32

[20] Freeman D M 2004 *The discrete logarithm problem in matrix groups* URL `http://theory.stanford.edu/~dfreeman/papers/discretelogs.pdf`

[21] Cohen H 2013 *A course in computational algebraic number theory* vol 138 (Springer Science & Business Media)

[22] Abdeljaoued J and Malaschonok G I 2001 *Efficient algorithms for computing the characteristic polynomial in a domain, Journal of Pure and Applied Algebra* **156** 127–145

[23] Pernet C and Storjohann A 2007 *Faster algorithms for the characteristic polynomial, Proceedings of the 2007 international symposium on Symbolic and algebraic computation* (ACM) pp 307–314

[24] Shoup V 1993 *Factoring polynomials over finite fields: asymptotic complexity vs. reality, Proc. IMACS Symposium, Lille, France*

[25] Pollard J M 1978 *Monte Carlo methods for index computation* (mod$p$), *Mathematics of Computation* **32** 918–924

[26] Shoup V 2009 *A computational introduction to number theory and algebra* (A computational introduction to number theory and algebra, Cambridge university press)

Multicore runup simulation by under water avalanche using two-layer 1D shallow water equations

B A R H Bagustara[1,*], C A Simanjuntak[2] and P H Gunawan[3]

School of Computing, Telkom University, Jalan Telekomunikasi No. 1 Terusan Buah Batu, Bandung 40257, Indonesia.

E-mail: [1]bajoerajahusein@student.telkomuniversity.ac.id;
[2]cassrio@student.telkomuniversity.ac.id;
[3]phgunawan@telkomuniversity.ac.id
*Corresponding author

Abstract. The increasing of layers in shallow water equations (SWE) produces more dynamic model than the one-layer SWE model. The two-layer 1D SWE model has different density for each layer. This model becomes more dynamic and natural, for instance in the ocean, the density of water will decreasing from the bottom to the surface. Here, the source-centered hydro-static reconstruction (SCHR) numerical scheme will be used to approximate the solution of two-layer 1D SWE model, since this scheme is proved to satisfy the mathematical properties for shallow water equation. Additionally in this paper, the algorithm of SCHR is adapted to the multicore architecture. The simulation of runup by under water avalanche is elaborated here. The results show that the runup is depend on the ratio of density of each layers. Moreover by using grid sizes $N_x = 8000$, the speedup and efficiency by 2 threads are obtained 1.74779 times and 87.3896 % respectively. Nevertheless, by 4 threads the speedup and efficiency are obtained 2.93132 times and 73.2830 % respectively by similar number of grid sizes $N_x = 8000$.

1. Introduction

Simulating fluid flow in estuaries, lake, river, coastal area, etc, can be done by shallow water equations (SWE). The type of this system equations is a hyperbolic type of partial differential equation. The equation is also known as Saint-Venant equations which is derived from the Navier-stokes equations for describing fluid motion [1]. Commonly, the original shallow water equations is described in one layer fluid surface. Moreover, recently the two- and multi-layer shallow water equations are developed for giving the model more dynamic and natural. Here the model consists of various density for each multiple layers.

In two-layer form of SWE, for $i \in \mathcal{L} = \{1, 2\}$, the equations can be rewritten in compact form as follows

$$\partial_t \mathbb{U}_i + \partial_x \mathbf{F}(\mathbb{U}_i) = -\mathbf{Q}(\mathbb{U})\partial_x \mathbb{S}_i, \qquad (1)$$

where,

$$\mathbb{U}_i = \begin{pmatrix} h_i \\ h_i u_i \end{pmatrix}, \qquad \mathbf{F}(\mathbb{U}_i) = \begin{pmatrix} h_i u_i \\ h_i u_i^2 + \frac{g}{2} h_i^2 \end{pmatrix}, \tag{2}$$

$$\mathbf{Q}(\mathbb{U}_i) = \begin{pmatrix} 0 \\ g h_i \end{pmatrix}, \quad \mathbb{S}_i = \begin{pmatrix} 0 \\ \sum_{k<i}^{2} \frac{\rho_k}{\rho_i} h_k + \sum_{k>i}^{2} h_k + Z + f_{fric} \end{pmatrix}, \tag{3}$$

where $h(x,t)$ denotes the water height, $u(x,t)$ the average velocity, g the gravitational force, and $Z(x)$ the bottom elevation. The subscripts 1 and 2 are used to describe the first and second layers of water respectively. Moreover, the space and time are shown as x and t respectively. Figure 1 shows the configuration of primitive variables of (1-3).

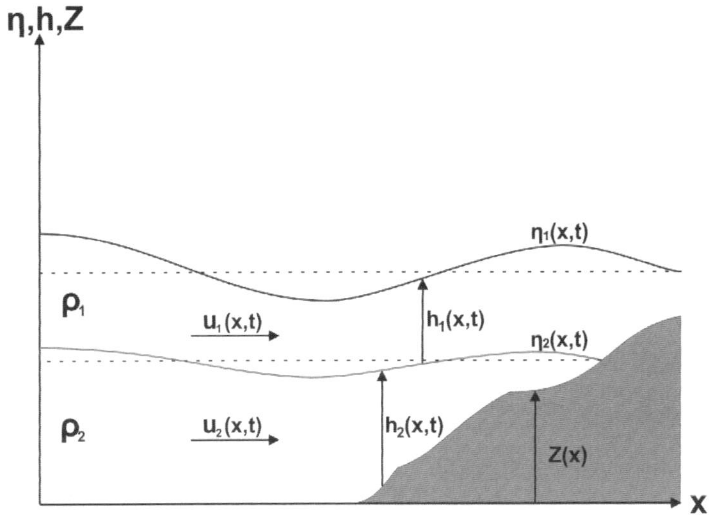

Figure 1. The configuration of two-layer shallow water equations. The surface of first layer is denoted as $\eta_1(x,t)$ with depth $h_1(x,t)$. Meanwhile the surface and depth of second layer are given as $\eta_2(x,t)$ and $h_2(x,t)$.

Generally, the numerical methods for approximating one-layer SWE should have following properties[2, 3, 4, 5]:

- preserving the positivity of water height ($h > 0$),
- gratifying well-balanced,
- satisfying discrete entropy balance, and
- satiating the dry/vacuum condition ($h = 0$).

In other side, the numerical scheme of two-layer SWE also should be satisfied previous mathematical properties for each layers. Bouchut, et. al.,[2], introduce the source-centered hydrostatic reconstruction (SCHR) scheme which is a robust scheme for approximating the mass and momentum conservation of two-layer SWE. Moreover, the SCHR also is proved to satisfied the previous basic properties of SWE (see [2] for more detail).

The aims of this paper are given as follows:

- to elaborate two-layer model with SCHR scheme for describing runup by under water avalanche phenomena and,
- to analyze the computational time of SCHR scheme for multiple cores architecture.

Here, the original algorithm of SCHR scheme will be modified such that the algorithm in accordance with the parallel architecture.

2. Source-centered Hydrostatic Reconstruction (SCHR) Scheme

Follow the paper of Bouchut [2], the source-centered hydrostatic reconstruction (SCHR) scheme is a modified original hydrostatic reconstruction (HR) scheme [3, 4, 6] for two-layer model of SWE due to the problem of momentum conservation. In [7], two schemes using original HR are called splitting and sum scheme. The problem using splitting method with original HR produces wrong solutions due to the total momentum did not conserve. Even, this scheme is developed into a variant of splitting which is called sum scheme still produces wrong solutions.

The detail of numerical scheme of SCHR can be found in [2], however the brief summary of this scheme will be written here. Let start with the discrete domain of one-dimensional space $\Omega = [0 : L]$, $L \in \mathbb{R}$ and time $T = [0, \infty)$ as follows

$$\Delta x = \frac{L}{N_x}, \quad N_x \in \mathbb{Z}^+, \qquad x_j = j \times \Delta x, \quad j \in \mathcal{M} = \{0, 1, \cdots, N_x\},$$
$$t^n = n \times \Delta t, \quad n \in \mathcal{T} = \{0, 1, \cdots\},$$

where here the uniform of discrete space Δx is considered. The time step Δt will be given depend on the stability condition (15). Therefore, the variable $\mathbb{U}_j^n$ denotes the the value of variable $\mathbb{U}$ at point x_j and current time t^n.

The finite volume numerical scheme using numerical fluxes SCHR is given as

$$\frac{\mathbb{U}_{i,j}^{n+1} - \mathbb{U}_{i,j}^n}{\Delta t} + \frac{\mathcal{F}_l(\mathbb{U}_{i,j}^n, \mathbb{U}_{i,j+1}^n, b_{i,j}^n, b_{i,j+1}^n) - \mathcal{F}_r(\mathbb{U}_{i,j-1}^n, \mathbb{U}_{i,j}^n, b_{i,j-1}^n, b_{i,j}^n)}{\Delta x} = 0,$$
$$i \in \mathcal{L}, \quad j \in \mathcal{M}, \quad n \in \mathcal{T}, \tag{4}$$

where

$$b_i = \sum_{k<i}^{2} \frac{\rho_k}{\rho_i} h_k + \sum_{k>i}^{2} h_k + Z. \tag{5}$$

Moreover the numerical fluxes $\mathcal{F}_{l/r}(u_l, u_r, b_l, b_r)$ are defined as

$$\mathcal{F}_l = \mathcal{F}_l^{HR} + \mathcal{P}_l, \tag{6}$$
$$\mathcal{F}_r = \mathcal{F}_r^{HR} + \mathcal{P}_r, \tag{7}$$

where $\mathcal{F}_{l/r}^{HR}$ denote the numerical fluxes of the original hydrostatic reconstruction scheme which can be found in [4] and $\mathcal{P}_{l/r} = (\mathcal{P}^0, \mathcal{P}_{l/r}^1)$.

Under the Proposition 1 in the paper of Bouchut [2], in order to satisfy the semi-discrete entropy condition, thus $\mathcal{P}^0$ and $\mathcal{P}_{l/r}^1$ with $(\cdot)_+ = \max(\cdot, 0)$ and $(\cdot)_- = \min(\cdot, 0)$ are defined as follows

$$\mathcal{P}^0 = \frac{\kappa\left((1+\theta)u_l + (1-\theta)u_r\right)}{2} \tag{8}$$
$$\mathcal{P}_l^1 = \frac{\kappa(1+\theta)(h_r - h_l + \Delta b)g}{2} + u_l(\mathcal{P}^0)_+ + u_r(\mathcal{P}^0)_- \tag{9}$$
$$\mathcal{P}_r^1 = -\left(\frac{\kappa(1-\theta)(h_r - h_l + \Delta b)g}{2}\right) + u_l(\mathcal{P}^0)_+ + u_r(\mathcal{P}^0)_- \tag{10}$$

with θ and κ are the parameters for hydrostatic solver which are given as follows

$$\theta = \min\left(1, \frac{(u_l)_+}{\sqrt{gh_l}}\right) - \max\left(1, \frac{(-u_r)_+}{\sqrt{gh_r}}\right), \tag{11}$$

$$\kappa = \min(\bar{\kappa}, 2.5 \times \min(h_l, h_r)), \tag{12}$$

$$\bar{\kappa} = \frac{|\overline{\Delta b}|}{2}\begin{cases} \dfrac{(\Delta b - h_l)h_r}{2(hr + \Delta b - h_l)} & \text{if } \Delta b > h_l, \\ 0 & \text{if } -h_l \leq \Delta b \leq h_l, \\ \dfrac{-(h_r + \Delta b)h_l}{2h_l - 2(h_r + \Delta b)} & \text{if } \Delta b < -h_r, \end{cases} \tag{13}$$

$$\overline{\Delta b} = \begin{cases} \min(\Delta b, h_l) & \text{if } \Delta b \geq 0, \\ \max(\Delta b, -h_r) & \text{otherwise.} \end{cases} \tag{14}$$

The stability condition of the SCHR scheme is presented as

$$\Delta t \leq \frac{\Delta x}{\lambda(\mathbb{U}^n_{i,j}, \mathbb{U}^n_{i,j+1}, b^n_{i,j+1} - b^n_{i,j})} \tag{15}$$

where λ is the wave speed which is defined in [2] in detail.

3. Numerical Algorithms using Multicore Architecture

In this paper, the multicore architecture algorithm for computing the numerical scheme of two-layer will be given. Figure 2 shows the serial and parallel parts of main algorithm. In terms of memory, parallel architecture is divided into two parts, shared memory and distributed memory [8, 9, 10]. Shared memory is used in this simulation to run the parallel algorithm using OpenMP (Open-Multi-Processing) platform. Here, OpenMP is used because it is straightforward and simple to implement to the parallel algorithm [11, 12, 13, 14, 15].

Here is the explanation of parallelization algorithm process :

- **Initialization process**: all variables at $t = 0$ are given.
- **Compute mass**: when time less than final time, mass equation is computed using (4). It will obtain new water depth h.
- **Compute momentum**: after mass equation is computed, momentum equation will be computeed to deliver new discharge hu using (4).
- **Determine boundary condition**: in this case, two treatments are used to determine the boundary. On left boundary, Dirichlet is used for the velocity and Von-Neumann is used for the water depth. Meanwhile, Von-Neumann is used for both velocity and water depth on the right boundary [16, 17].
- **Update mass and momentum**: all variables are updated.

4. Numerical Simulation and OpenMP Performance

In this section, the numerical simulation of runup due to under water avalanche will be elaborated. Moreover, the parallel performance using OpenMP also will be discussed in this section.

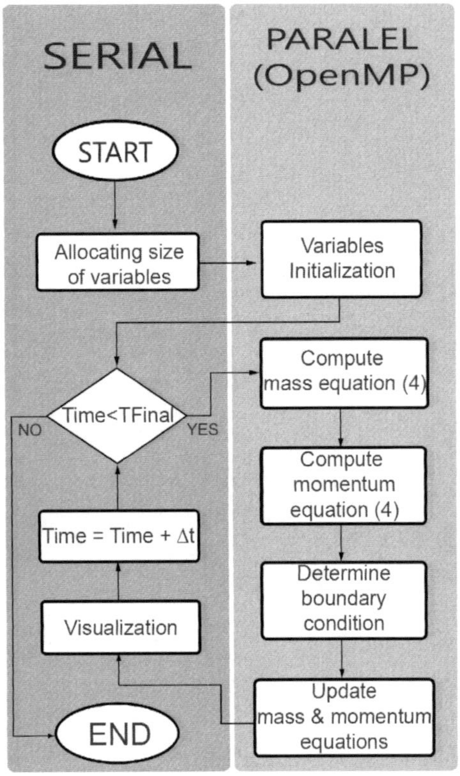

Figure 2. The original algorithm will be divided into two parts. The first part is devoted to the serial computing. Here, the light computation is taken place such as the initialization of parameters. The second part is used to multicore computing since the long time simulation will be needed.

4.1. Simulation of runup by under water avalanche

Here, the configuration of avalanche simulation is illustrated in Figure 3. The detail of this configuration is given as follows

$$Z(x) = \begin{cases} 0 & \text{if } x < 70, \\ \alpha x + c & \text{otherwise,} \end{cases}, \tag{16}$$

$$\eta_1(x,0) = \max(30, Z(x)), \tag{17}$$

$$\eta_2(x,0) = \begin{cases} 20 & \text{if } 30 \le x \le 60, \\ 0 & \text{otherwise,} \end{cases}, \tag{18}$$

$$u_1(x,0) = u_2(x,0) = 0, \tag{19}$$

where $\alpha = 1$ and $c = -70$.

In this simulation, the domain is given in $\Omega = [0 : 500]$ and various ratio of density are given. According to (19), the initial velocity for first and second layer is set to be zero. Thus, the steady condition at initial time is described in Figure 3.

The results by various ratio of density are given in Figure 4. Figure 4(a) is shown the runup profile in three various final time using ratio $\rho_1/\rho_2 = 0.5$. Runup profile at $t = 5.40$, $t = 10.4$ and $t = 14.2$ are described by blue straight line, red line points, and green dash line respectively. It shown that the first runup reached $x = 108.5$ and wetting the bottom (Z) with water depth $h = 0.3592$ at $t = 5.40$. The second runup is observed only reach $x = 102$ with water depth $h = 0.1676$. Moreover, the third runup is observed reach $x = 104.5$ with depth $h = 0.4423$.

Here, the first runup is shown has the highest value, meanwhile the second runup is shown has the lowest value. Indeed the impact of first sediment slides (layer 2) on bottom causes the

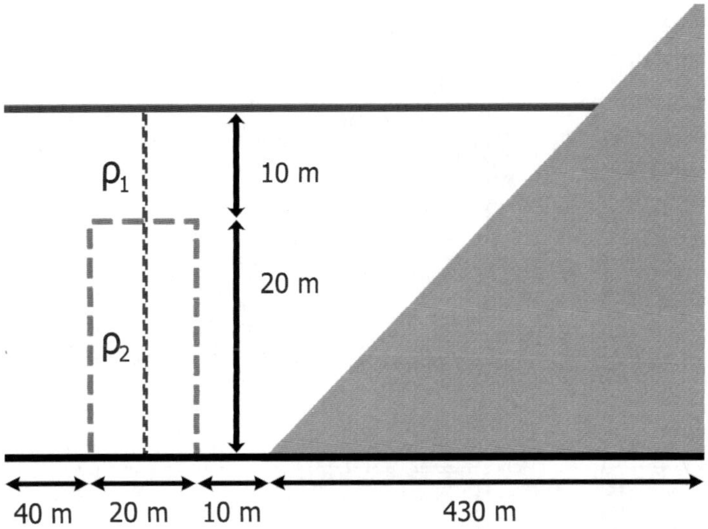

Figure 3. The scenario of wave propagation simulation in enclosed bay. A slope bottom is given with 45 degrees from $x = 70$ until $x = 500$. The spatial domain of this simulation is $\Omega = [0 : 500]$ m. The avalanche will be simulated using second layer (dash line), thus it will generating the wave which is described by first layer (straight line).

energy and velocity of first layer increasing (see Figure 4(b)). When the sediment avalanche moves to the left side, the velocity becomes negative then it makes the second runup loss of energy. However, due to the moves of sediment back again to the right side, then the third runup is pushed the water higher than the second runup.

Similar profiles are observed for ratio of density $\rho_1/\rho_2 = 0.25$ in Figure 4(c) and $\rho_1/\rho_2 = 0.125$ in Figure 4(e). However using small ratio of density, the runup for three final time are obtained increasing. For instance, the first runup using ratio of density $\rho_1/\rho_2 = 0.25$ and $\rho_1/\rho_2 = 0.125$ are conducted at $x = 113.5$ and $x = 116$ respectively. The distance of first and second runup for ratio of density $\rho_1/\rho_2 = 0.5$ is 2.00208 meters. Meanwhile, The distance of first and second runup for ratio of density $\rho_1/\rho_2 = 0.125$ is 6.0077 meters.

4.2. Parallel performance

Here, the parallel performance using OpenMP platform for simulating the runup due to avalanche is given using the following specifications of computer (Table 1).

Table 1. The specifications of computer which is used for measuring the OpenMP performances.

Computer Specifications	
Processor	Intel(R) Core(TM) i5-2500
Num of cores (threads)	4
Clock (MHz)	2600.00
Memory DDR3 (GB)	8
Operating System	Ubuntu 16.04.1 LTS

In Table 2, the parallel performances for using two and four threads are conducted from the average of 10 times experiments. It is shown that the CPU time for each of simulations is accelerated well by OpenMP. By two threads in parallel algorithm, serial time is obtained around 1.65 times the parallel time. Meanwhile, the parallel time is 2.81 times faster than the

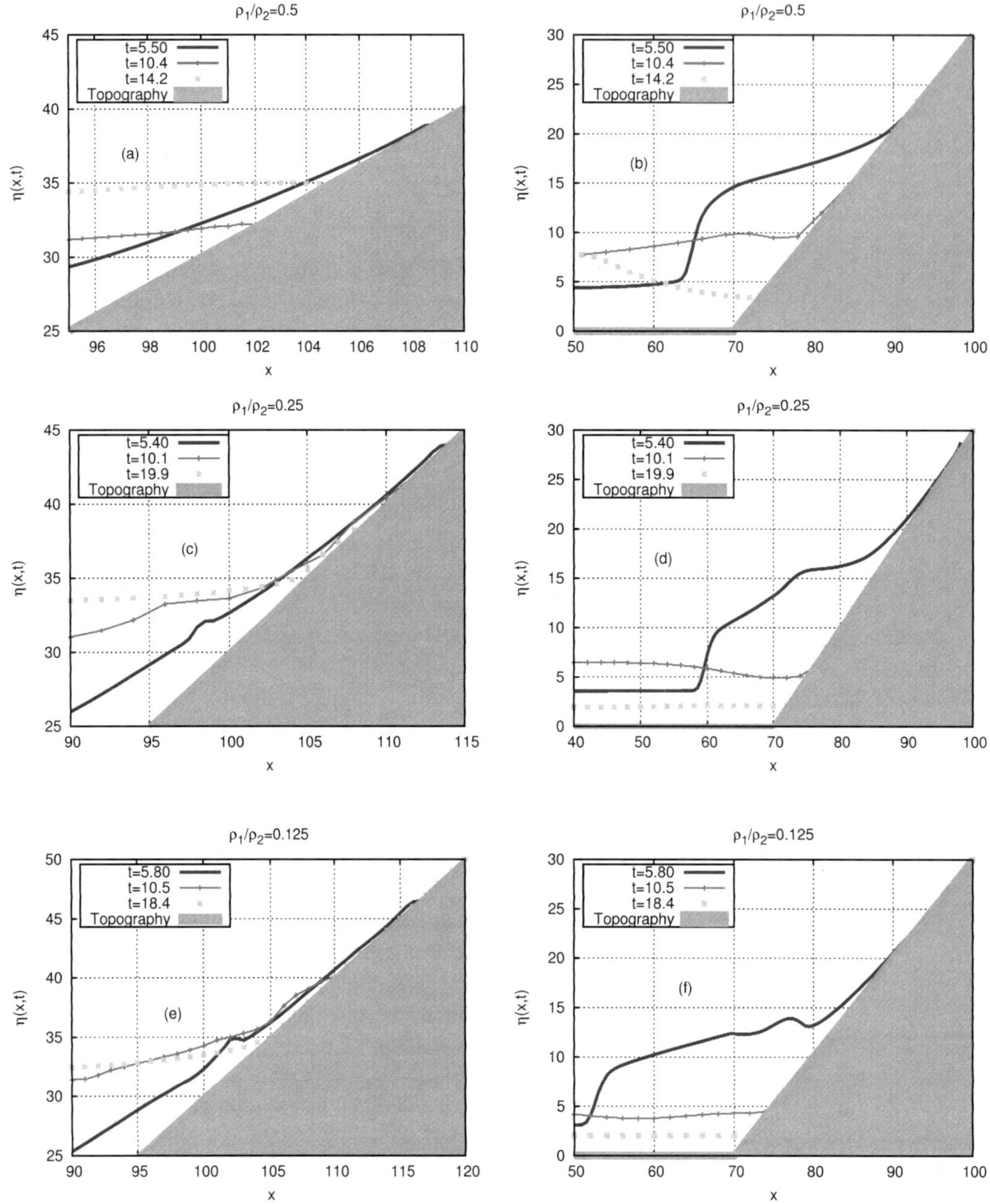

Figure 4. The runup profile of the first layer (on the left side) and sediment profile of the second layer (on the right side). (a) The runup profile of first layer using ratio $\rho_1/\rho_2 = 0.5$. (b) The sediment profile of second layer using ratio $\rho_1/\rho_2 = 0.5$. (c) The runup profile of first layer using ratio $\rho_1/\rho_2 = 0.25$. (d) The sediment profile of second layer using ratio $\rho_1/\rho_2 = 0.25$. (e) The runup profile of first layer using ratio $\rho_1/\rho_2 = 0.125$. (f) The sediment profile of second layer using ratio $\rho_1/\rho_2 = 0.125$.

Table 2. The CPU time, speedup and efficiency of the program.

		2 threads			**4 threads**		
N_x	**Serial time**	**Parallel time**	**Speedup**	**Efficiency (%)**	**Parallel time**	**Speedup**	**Efficiency (%)**
500	17.9467	10.9958	1.63214	81.6071	7.00126	2.56335	64.0838
1000	35.3949	21.0549	1.68108	84.0538	12.4394	2.84539	71.1347
2000	70.1040	41.6684	1.68243	84.1213	24.3071	2.88410	72.1024
4000	139.959	83.0917	1.68439	84.2196	48.1345	2.90767	72.6916
8000	278.771	159.499	1.74779	87.3896	95.1009	2.93132	73.2830

serial time using four threads. This means that the time acceleration using four threads is better than two threads.

In contrast with the CPU time, the performance of efficiency using two threads is observed better than using four threads. For instance in Table 2, using $Nx = 2000$, the efficiency for two and four threads are conducted 84.1213% and 72.1024% respectively.

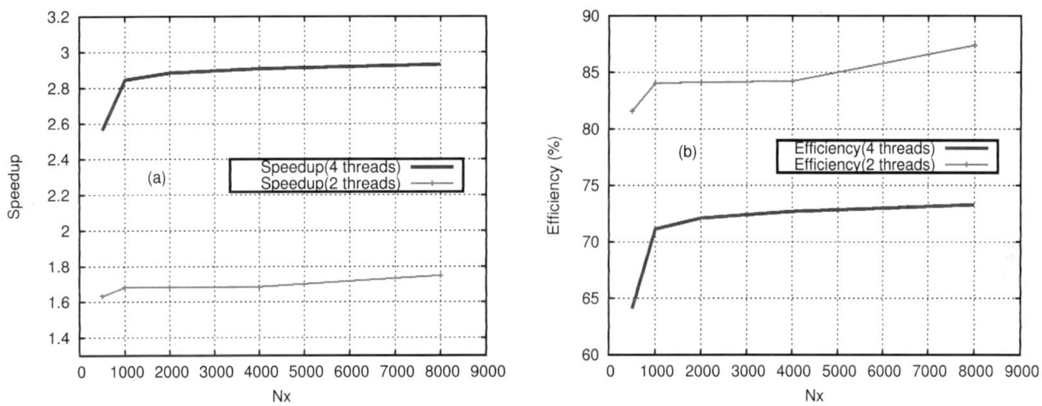

Figure 5. The comparison of parallel performance in speedup (a) and efficiency (b) using 2 and 4 threads.

Profile of speedup and efficiency can be seen in Figure 5. Figure 5(a) is shown the speedup profile for both experiments, two and four threads. Overall, the simulations using four threads have higher speedup than using two threads. Meanwhile in Figure 5(b), the efficiency profile is clearly shown in contrast with the speedup profile. The simulations using four threads have lower efficiency percentage than using two threads.

5. Conclusion

Runup simulation by under water avalanche using two-layer 1D SWE and SCHR method is successfully simulated. The SCHR method is used to approximate the solution of two-layer 1D SWE model. Three simulations have been done with various ratio of density ρ_1/ρ_2 to investigate the runup and sediment movement. From the results, the runup and sediment movement are

IOP Conf. Series: Journal of Physics: Conf. Series **971** (2018) 012026 doi:10.1088/1742-6596/971/1/012026

observed depend on the ratio of density ρ_1/ρ_2. By lighter ratio, the highest runup is obtained and vice versa. Moreover, the experiments of parallel computing in order to analyze the parallel performances using OpenMP are given. Two experiments are elaborated, using 2 and 4 threads. The results of parallel performance are shown satisfying by the observation of acceleration in CPU time for computing. Using a large number of grid sizes $N_x = 8000$, the speedup by 2 and 4 threads are obtained 1.74779 and 2.93132 times respectively. However, the efficiency 87.3896 % and 73.2830 % are observed by using 2 and 4 threads respectively in similar number of grid sizes.

References

[1] Tan W Y 1992 *Shallow water hydrodynamics: Mathematical theory and numerical solution for a two-dimensional system of shallow-water equations* vol 55 (Elsevier)
[2] Bouchut F and Zeitlin V 2010 *Discrete and Continuous Dynamical Systems-Series B* **13** 739–758
[3] Bouchut F 2004 *Nonlinear stability of finite Volume Methods for hyperbolic conservation laws: And Well-Balanced schemes for sources* (Springer Science & Business Media)
[4] Audusse E, Bouchut F, Bristeau M O, Klein R and Perthame B t 2004 *SIAM Journal on Scientific Computing* **25** 2050–2065
[5] Doyen D and Gunawan P H 2014 *Finite Volumes for Complex Applications VII-Methods and Theoretical Aspects* (Springer) pp 227–235
[6] Bouchut F 2007 *Edited Series on Advances in Nonlinear Science and Complexity* **2** 189–256
[7] Bouchut F and de Luna T M 2008 *ESAIM: Mathematical Modelling and Numerical Analysis* **42** 683–698
[8] Gunawan P H 2016 *Information and Communication Technology (ICoICT), 2016 4th International Conference on* (IEEE) pp 1–5
[9] Neal J, Fewtrell T and Trigg M 2009 *Environmental Modelling & Software* **24** 872–877
[10] Mulyani, Putri N D and Gunawan P H 2017 *Information and Communication Technology (ICoIC7), 2017 5th International Conference on* (IEEE) pp 1–5
[11] Pan X M, Pi W C and Sheng X Q 2011 *Progress In Electromagnetics Research* **112** 199–213
[12] Juliati S and Gunawan P H 2017 *Information and Communication Technology (ICoIC7), 2017 5th International Conference on* (IEEE) pp 1–5
[13] Pahlevi M R and Gunawan P H 2017 *Information and Communication Technology (ICoIC7), 2017 5th International Conference on* (IEEE) pp 1–4
[14] Alamsyah M N A, Simanjuntak C A, Bagustara B A R H, Pradana W A and Gunawan P H 2017 *Information and Communication Technology (ICoIC7), 2017 5th International Conference on* (IEEE) pp 1–6
[15] Iryanto and Gunawan P H 2017 *Information and Communication Technology (ICoIC7), 2017 5th International Conference on* (IEEE) pp 1–5
[16] David L, Nacer B, Pascal B and Gérard J 2006 *Heat and mass transfer* **42** 322–337
[17] Gersborg-Hansen A, Bendsøe M P and Sigmund O 2006 *Structural and multidisciplinary optimization* **31** 251–259

IOP Conf. Series: Journal of Physics: Conf. Series **971** (2018) 012027 doi:10.1088/1742-6596/971/1/012027

Simulation of shoreline development in a groyne system, with a case study Sanur Bali beach

P H Gunawan[1,*] and S R Pudjaprasetya[2]

[1]School of Computing, Telkom University, Jalan Telekomunikasi No. 1 Terusan Buah Batu, Bandung 40257, Indonesia.

[2]Department of Mathematics, Faculty of Mathematics and Natural Sciences, Institut Teknologi Bandung, Jalan, Ganesha No. 10, Bandung 40132, Indonesia

E-mail: [1]phgunawan@telkomuniversity.ac.id; [2]sr_pudjap@math.itb.ac.id
*Corresponding author

Abstract. The process of shoreline changes due to transport of sediment by littoral drift is studied in this paper. Pelnard-Considère is the commonly adopted model. This model is based on the principle of sediment conservation, without diffraction. In this research, we adopt the Pelnard-Considère equation with diffraction, and a numerical scheme based on the finite volume method is implemented. Shoreline development in a groyne system is then simulated. For a case study, the Sanur Bali Beach, Indonesia is considered, in which from Google Earth photos, the beach experiences changes of coastline caused by sediment trapped in a groyne system.

1. Introduction

Groyne is a hydraulic structure built along a seashore to control erosion. Commonly, a groyne is installed perpendicular or slightly oblique to the shoreline. If waves approach the coastline obliquely, fluid flow propagating along the coast will be generated. On a sandy coast, this flow together with the stirring action of breaking waves causes a longshore sand transport, also known as *littoral drift*.

Top view of the Sanur Bali beach in Indonesia is shown in Figure 1. The coastline changes form time to time in response to the installation of a breakwater groyne system. Although groyne system serves as a breakwater, the presence of groyne interrupts the water flow and reduces sand movement, thus reducing erosion. So that the presence of groyne also affects the changing of the coastline, as a result of the accumulation of sand sediments in the up current part of the groyne, as well as the erosion in the down current area of the groyne, see Figure 1 (top).

Shoreline development due to the newly constructed breakwater groyne will be studied here. The mathematical model of shoreline change is derived from mass conservation of sand. Without diffraction, the equation is quite simple, in literature it is known as the Pelnard-Considère equation [1, 2, 3, 4, 5]. In this article, shoreline change in groyne system is considered, and diffraction effect is incorporated. Further, we develop a finite volume numerical scheme for simulating shoreline change due to groyne installment. As recorded in [6, 7], finite volume is a suitable method for solving problems based on conservation equation.

This paper is organized as follows, in Section 2 we derive a diffusion model based on the conservation of sediment, to describe coastline development in a groyne system. In the model,

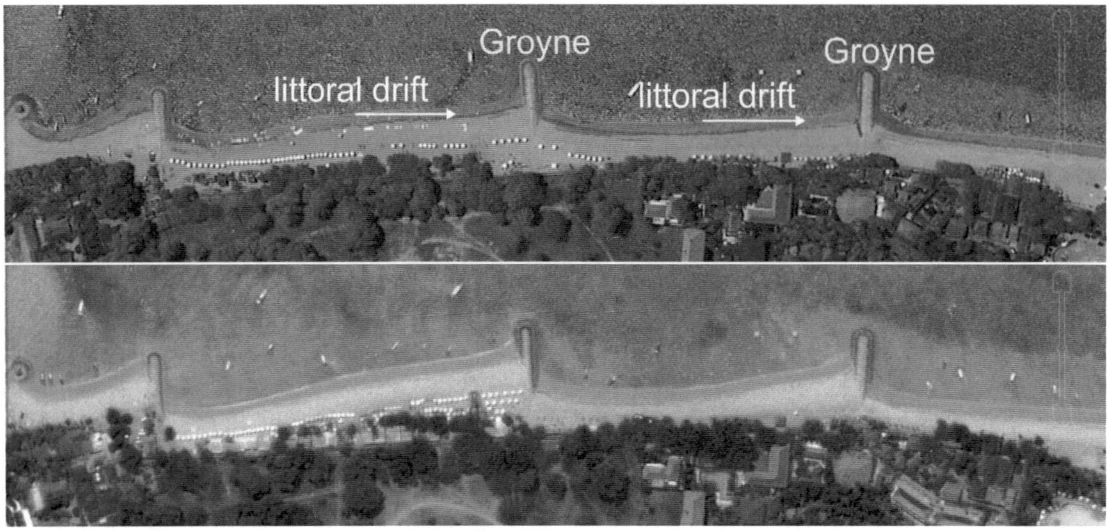

Figure 1. Google Earth photos of Sanur Bali Beach, in Indonesia, which shows changes of the coastline due to breakwater groyne. The top photo is from 2009, and below photo is from 2014.

the groyne is formulated as boundary condition. In Section 3, the finite volume method is formulated. Several numerical experiments are given in Section 4. Finally, some conclusions and discussion are drawn in the last section.

2. The Shoreline Model with Diffraction Effect

Without going deeply into the physics, we can have insight of the consequences of constructing a groyne. Here, we adopt the approach by [2, 8] which we resumed here for the sake of clarity.

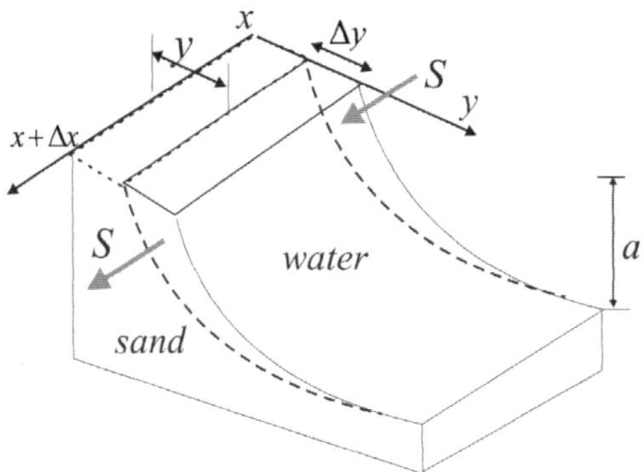

Figure 2. Change of coastline due to sand transport by littoral drift.

Let $y(x,t)$ denotes the shoreline at time t, whereas S (m^3/sec) denotes the sediment discharge brought by the littoral drift, see Figure 2. Transport of sand is governed by the following balance of sand in the control interval $[x, x+\Delta x]$ as follows

$$S|_x\Delta t - S|_{x+\Delta x}\Delta t = (y|_{t+\Delta t} - y|_t)\, a\Delta x,$$

with a denotes the water depth. Divide both sides with $\Delta x\Delta t$, followed by taking the limit $\Delta x \to 0$, $\Delta t \to 0$, will result the conservation of mass in terms of differential equations as follows

$$\frac{\partial S}{\partial x} + a\frac{\partial y}{\partial t} = 0. \tag{1}$$

Further, sand discharge is given by

$$S = S_0 - q\frac{\partial y}{\partial x}. \tag{2}$$

An estimate formula from the CERC (Coastal Engineering Research Center) give us

$$S_0 = Eh^2\phi_x, \quad \text{and} \quad q = Eh^2, \tag{3}$$

with E (m^3/s) a coefficient proportional to incident wave energy flux, ϕ_x the angle of wave crest with respect to the $x-$axis, and h is the ratio of wave height at x to wave height at $x = \infty$. Using (2), equation (1) becomes

$$\frac{\mathrm{dS_0}}{\mathrm{dx}} - \frac{\partial}{\partial x}\left(q\frac{\partial y}{\partial x}\right) + a\frac{\partial y}{\partial t} = 0.$$

Next step, the physical shoreline is split into two parts; y_0 as the stationary effect from diffraction, and $y'(x,t)$ as the non-stationary effect, to be precise

$$y(x,t) = y'(x,t) + y_0(x). \tag{4}$$

Substituting (3) into (1) gives us the following equation

$$a\frac{\partial y'}{\partial t} - q\frac{\partial^2 y'}{\partial x^2} - \frac{\mathrm{dq}}{\mathrm{dx}}\frac{\partial y'}{\partial x} = -\frac{\mathrm{dS_0}}{\mathrm{dx}} + q\frac{\partial^2 y_0}{\partial x^2} + \frac{\mathrm{dq}}{\mathrm{dx}}\frac{\mathrm{dy_0}}{\mathrm{dx}}. \tag{5}$$

Further, we choose $y_0(x)$ such that it satisfies an ordinary differential equation

$$\frac{\mathrm{dS_0}}{\mathrm{dx}} - q\frac{\mathrm{d^2y_0}}{\mathrm{dx^2}} - \frac{\mathrm{dq}}{\mathrm{dx}}\frac{\mathrm{dy_0}}{\mathrm{dx}} = 0, \tag{6}$$

whereas $y'(x,t)$ satisfies the following partial differential equation

$$a\frac{\partial y'}{\partial t} - q\frac{\partial^2 y'}{\partial x^2} - \frac{\mathrm{dq}}{\mathrm{dx}}\frac{\partial y'}{\partial x} = 0. \tag{7}$$

In this new formulation, the stationary effect can be obtained by solving the ordinary differential equation (6), whereas the non-stationary effect from littoral sand drift is governed by a diffusion equation (7). Next, we formulate the boundary conditions. Far away at $x \to \infty$ the breakwater groyne has no effect, hence

$$\lim_{x\to\infty} y = 0, \lim_{x\to\infty} y' = 0, \quad \text{and} \quad \lim_{x\to\infty} y_0 = 0, \lim_{x\to\infty} y_0' = 0. \tag{8}$$

At the breakwater $x = 0$, there is no discharge or $S(0) = 0$, then

$$\phi_x(0) = \frac{\partial y}{\partial x}(0). \tag{9}$$

Moreover, we can rewrite the Equation (6) as

$$\frac{\mathrm{d}}{\mathrm{dx}}\left(S_0 - q\frac{\mathrm{dy_0}}{\mathrm{dx}}\right) = 0. \tag{10}$$

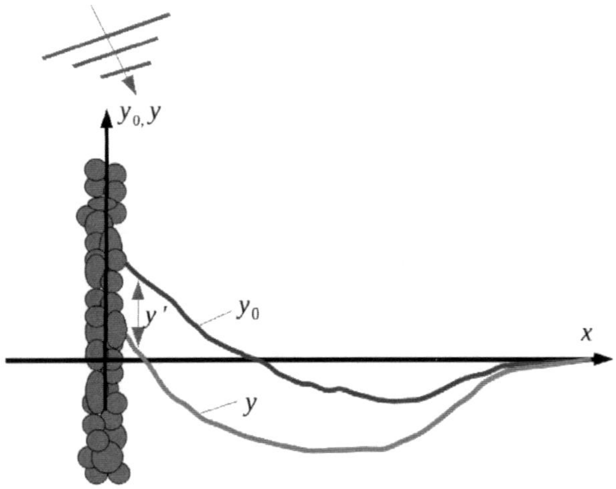

Figure 3. Sketch and notations of shoreline development in the lee side of breakwater.

Direct integration will give us

$$S_0 - q\frac{\mathrm{d}y_0}{\mathrm{dx}} = Eh_\infty^2\phi_\infty - Eh_\infty^2\frac{\mathrm{d}y_0}{\mathrm{dx}}|_\infty.$$

Further, as h be the ratio of wave height to wave height as infinity, obviously $h_\infty = 1$. By adopting boundary conditions (8) and relation (3) the equation above can be written as

$$\frac{\mathrm{d}y_0}{\mathrm{dx}} = \phi_x - \frac{\phi_\infty}{h^2}. \tag{11}$$

We note here that relation (11) holds in the general case, but in the case of no diffraction, wave height is constant which make the wave height ratio $h = 1$, and the equation (3) reduces to $\frac{\mathrm{d}y_0}{\mathrm{dx}} = \phi_x - \phi_\infty$, which is exactly the Pelnard-Considère approximation.

Following Bakker [2], and under the assumption that ϕ_x and ϕ_∞ are small, the angle of wave crest ϕ_x can be approximated as

$$\phi_x = \phi_\infty - \frac{\lambda}{2\pi}\frac{\mathrm{d}\theta}{\mathrm{dx}}, \tag{12}$$

where λ is the wave length and θ the phase difference. Substituting (12) into Equation (11) yields

$$\frac{\mathrm{d}y_0}{\mathrm{dx}} = -\frac{\lambda}{2\pi}\frac{\mathrm{d}\theta}{\mathrm{dx}} + \phi_\infty\left(\frac{h^2-1}{h^2}\right). \tag{13}$$

Integrating (13) give us

$$y_0(x/\lambda) = -\frac{\lambda}{2\pi}\theta(x/\lambda) + \phi_\infty\int_0^{x/\lambda}\frac{h^2-1}{h^2}\,\mathrm{d}x, \tag{14}$$

after using $y_0(0) = 0$ and $\theta(0) = 0$. The first and second terms on the right hand side of (14) are due to diffraction and changing wave height, respectively. For computation, (14) is rewritten in normalized 'length' variables x/λ, y_0/λ as

$$y_0/\lambda = -\frac{1}{2\pi}\theta(x/\lambda) + \phi_\infty\int_0^{x/\lambda}\frac{h^2-1}{h^2}\,\mathrm{d}(x/\lambda). \tag{15}$$

The first term on the right hand side of (15) shows the influence of wave diffraction, whereas the second term is the influence of the changing in wave height.

2.1. Computation of the stationary shoreline y_0

Assuming the angle of wave incidence ϕ_∞ is known, and the length of the groyne is l, we first compute the phase difference w as a function of x/λ

$$w = \frac{r-\eta}{\lambda} = \sqrt{(x/\lambda)^2 + (l/\lambda)^2} - l/\lambda \cos\phi_\infty - x/\lambda \sin\phi_\infty \tag{16}$$

The relative wave height h and the change in phase by diffraction θ will be computed with diffraction theory. In order to calculate the integral in (15) the simplified theory by Putnam and Arthur [9] will be used. Let $u = 2\sqrt{w}$, and define

$$f(u) = -\frac{1}{\sqrt{2}} e^{i\pi/4} \int_{-\infty}^{u} e^{-i\pi v^2/2} \mathrm{d}v. \tag{17}$$

The wave height ratio h and phase difference θ as functions of x/λ are obtained from the modulus and argument of $f(u)$, respectively. With the help of Maple software, the curves of $h(x/\lambda)$ and $\theta(x/\lambda)$ can be obtained, and the results for $\phi_\infty = 0.1$ and $l = 10\lambda$ are plotted in Figure 4.

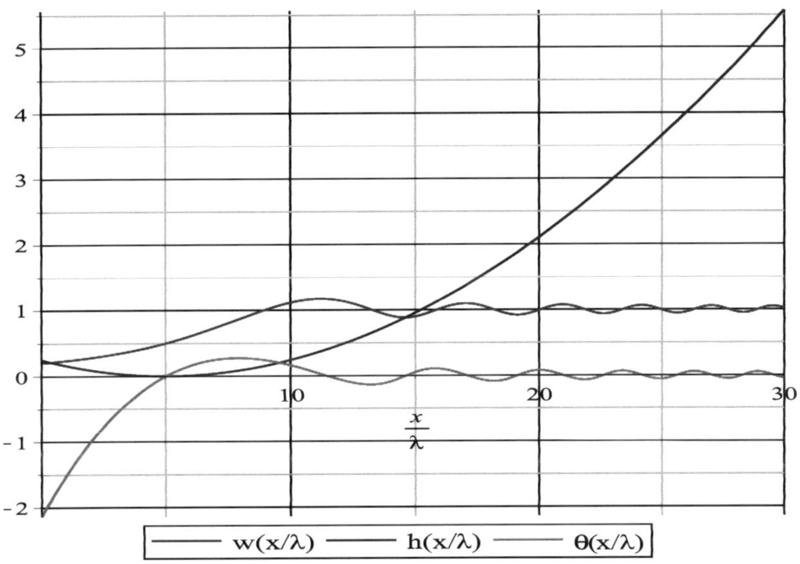

Figure 4. The curves of $w(x/\lambda)$ (red), $h(x/\lambda)$ (blue) and $\theta(x/\lambda)$ (green) plotted using $l = 10\lambda$ and $\phi_\infty = 0.1$.

2.2. Computation of the non-stationary y'

Further we consider Equation (7) for the non-stationary shoreline $y'(x,t)$ subject to boundary conditions $\lim_{x\to\infty} y = 0$ and $\lim_{x\to\infty} y' = 0$. Remember that $y' = y - y_0$, and since (9) we thus have

$$\left(\frac{\partial y'}{\partial x}\right)_{x=0} = (\phi_x)_{x=0} - \left(\frac{\mathrm{d}y_0}{\mathrm{d}x}\right)_{x=0},$$
$$\left(\frac{\partial y'}{\partial x}\right)_{x=0} = \frac{\phi_\infty}{(h)^2_{x=0}}. \tag{18}$$

Note that the last formulation is obtained after applying (11) for $\left(\frac{\mathrm{dy}_0}{\mathrm{dx}}\right)_{x=0}$.

The following algorithm summarizes the procedure of calculating shoreline change at the lee side of the breakwater

Algorithm 1 The algorithm for calculating shoreline change.

Step 0. Start.
Step 1. Given an angle of wave incidence ϕ_∞ and breakwater length l.
Step 2. Compute the phase difference w as a function of x/λ using (16).
Step 3. Compute θ and h as functions of x/λ using (17).
Step 4. Compute the stationary condition $y_0(x)$ using (15) and given the initial condition of $y'(x,0) = -y_0(x)$.
Step 5. While (time $\leq$ final time) do Steps **6-7**
  Step 6. Compute $y'(x,t)$ using (7).
  Step 7. Update the real shoreline as $y = y' + y_0$.
Step 8. (The procedure was successful.) STOP
Step 9. End.

Note: in the computation, diffraction effect is only considered in the lee side of breakwater.

3. Finite Volume Method

In order to solve (7) numerically, here we present the finite volume method for one-dimensional unsteady follows as given in of Versteeg et. al. [10]. First stepof this method is to define the domain into sub-domains in discrete control volumes. Let us dividing our domain $\Omega = [0, L]$ into a mesh $\mathcal{M} = \{1, 2, \ldots, N_x\}$ nodal of points. Here, we denote $i \in \mathcal{M}$ is the general nodal point in our discretization. Moreover, the interfaces of control volume C_i of nodal i are defined by $i \pm \frac{1}{2}$ (see Figure 5). In this case, we assume that the distance between i-th node to the its neighbours nodal points is in uniform which is Δx.

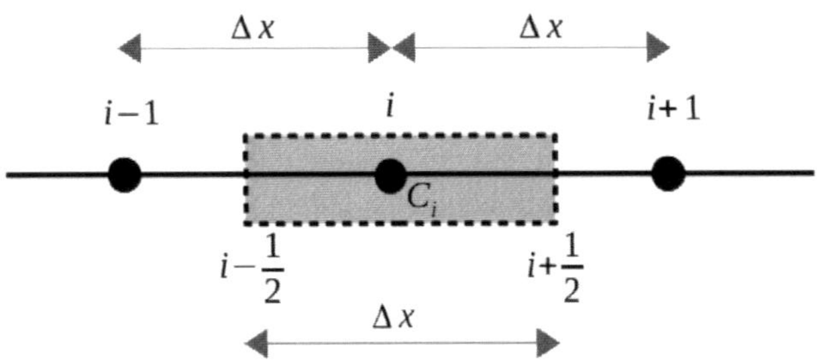

Figure 5. The control volume C_i of nodal point i with uniform discrete space distance Δx. The interfaces of C_i are defined at points $i - \frac{1}{2}$ and $i + \frac{1}{2}$.

The sense of the finite volume method is the integration of the governing equation (7) over a control volume C_i and over a time interval t to Δt in nodal point i. Then for all $i \in \mathcal{M}$ we have,

$$\int_t^{t+\Delta t}\int_{C_i}\frac{\partial y}{\partial t}\,\mathrm{d}C_i\,\mathrm{d}t=\int_t^{t+\Delta t}\int_{C_i}D\frac{\partial^2 y}{\partial x^2}\,\mathrm{d}C_i\,\mathrm{d}t, \tag{19}$$

$$=\int_t^{t+\Delta t}\int_{C_i}\frac{\partial}{\partial x}\left(D\frac{\partial y}{\partial x}\right)\,\mathrm{d}C_i\,\mathrm{d}t. \tag{20}$$

This can be written as

$$\int_{i-\frac{1}{2}}^{i+\frac{1}{2}}\left[\int_t^{t+\Delta t}\frac{\partial y}{\partial t}\,\mathrm{d}t\right]\,\mathrm{d}C_i=\int_t^{t+\Delta t}\left[\left(DA\frac{\partial y}{\partial x}\right)_{i+\frac{1}{2}}-\left(DA\frac{\partial y}{\partial x}\right)_{i-\frac{1}{2}}\right]\,\mathrm{d}t. \tag{21}$$

In Equation (21), A is the face area of control volume C_i, which is equal to $\Delta V/\Delta x$, where ΔV denotes the volume. Further, we denote y_i^{n+1} and y_i^n are the distance of shoreline in ith node at time $t+\Delta t$ and t respectively. Thus, the left hand side of equation (21) can be written as

$$\int_{i-\frac{1}{2}}^{i+\frac{1}{2}}\left[\int_t^{t+\Delta t}\frac{\partial y}{\partial t}\,\mathrm{d}t\right]\,\mathrm{d}C_i=\left(y_i^{n+1}-y_i^n\right)\Delta V. \tag{22}$$

If we approximate the diffusion terms on right hand side of Equation (21) by central difference, thus we have

$$\left(y_i^{n+1}-y_i^n\right)\Delta V=\int_t^{t+\Delta t}\left[\left(D_{i+\frac{1}{2}}\frac{y_{i+1}-y_i}{\Delta x}\right)-\left(D_{i-\frac{1}{2}}\frac{y_i-y_{i-1}}{\Delta x}\right)\right]A\mathrm{d}t. \tag{23}$$

where $D_{i+\frac{1}{2}}=(D_{i+1}+D_i)/2$ and $D_{i-\frac{1}{2}}=(D_i+D_{i-1})/2$.

By generalizing the integral of shoreline distance with respect to time with weighted parameter $\theta\in[0,1]$,

$$\int_t^{t+\Delta t}y_i\,\mathrm{d}t=[\theta y_i^{n+1}+(1-\theta)y_i^n]\Delta t, \tag{24}$$

hence we get

$$\left(\frac{y_i^{n+1}-y_i^n}{\Delta t}\right)\Delta x=\theta\left[\frac{D_{i+\frac{1}{2}}(y_{i+1}^{n+1}-y_i^{n+1})}{\Delta x}-\frac{D_{i-\frac{1}{2}}(y_i^{n+1}-y_{i-1}^{n+1})}{\Delta x}\right],$$

$$+(1-\theta)\left[\frac{D_{i+\frac{1}{2}}(y_{i+1}^n-y_i^n)}{\Delta x}-\frac{D_{i-\frac{1}{2}}(y_i^n-y_{i-1}^n)}{\Delta x}\right]. \tag{25}$$

Note that in Equation (25), if $\theta=0$ is called fully explicit and $\theta=0.5$ is called the Crank-Nicolson scheme. As explained, in Chapter 5 and 8 in [10], in the case of explicit scheme, the scheme are conditionally stable. The numerical scheme should satisfy the positiveness of coefficients which are defined as

$$\Delta t\le 0.5\frac{(\Delta x)^2}{D}. \tag{26}$$

However, in the case of fully implicit ($\theta=1$), this condition is not necessary since implicit scheme is unconditionally stable.

4. The Results of Numerical Simulation

In this section, three scenarios of numerical simulation will be given. The simulation of simple coastline development model without diffraction effect will be elaborated in first simulation. In the second simulation, the coastline model which takes into account the diffraction effect will be given. Finally, simulation in study case of Sanur Bali beach will be elaborated in the third simulation.

4.1. Simple coastline development model

Here, we implement the numerical scheme to simulate shoreline change due to the installation of one groyne without diffraction. Omitting the diffraction effect, $y_0(x) = 0$, the discharge q and the angle of wave crest ϕ_x in (7) is assumed to be constant. Thus the shoreline development is governed by the diffusion equation below

$$a\frac{\partial y'}{\partial t} - q\frac{\partial^2 y'}{\partial x^2} = 0.$$

The finite volume scheme (25) as described in the previous section is adopted, with $D = q/a = S_0\left(a\phi_x\right)^{-1}$ after using the relation (3).

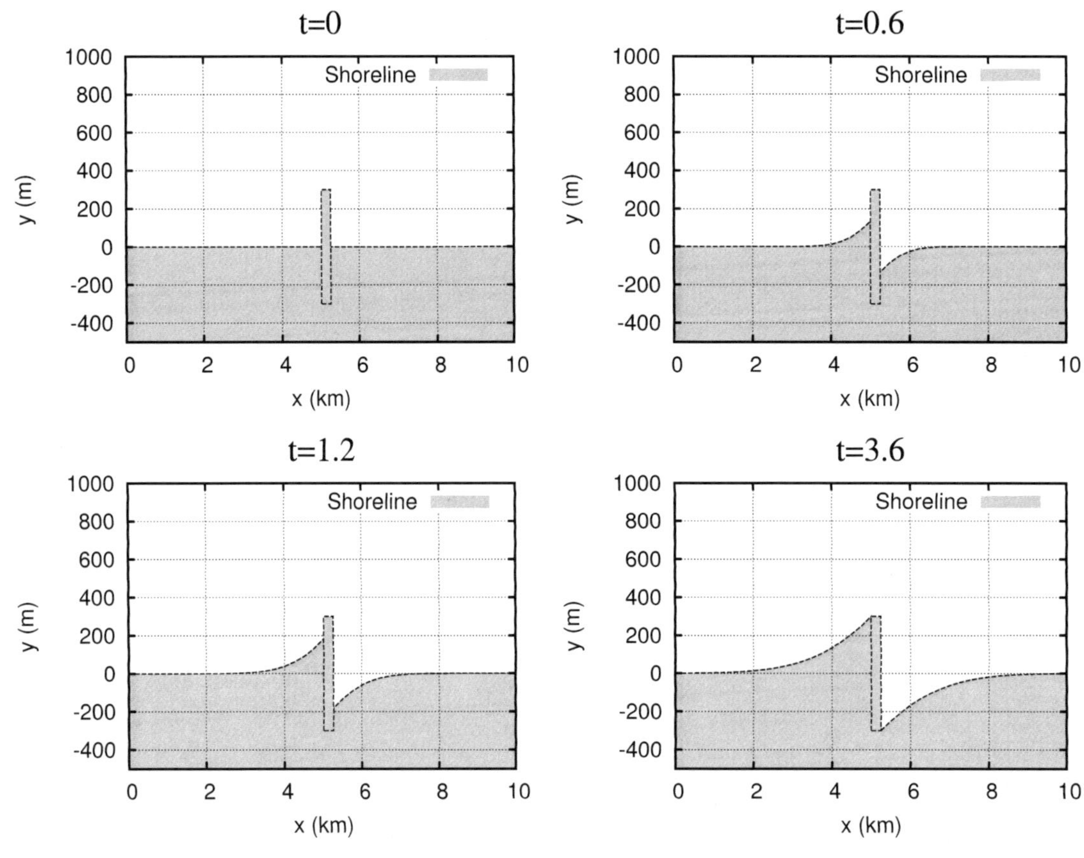

Figure 6. Shoreline change development around a groyne as functions of time $t = 0, 0.6, 1.2,$ and 3.6 years.

Taking the computational domain as $\Omega = [0, 10]$ km with $\Delta x = 0.25$ km. A groyne is located at the middle of domain, whereas the length of the groyne is assumed to be infinite. The coefficient D is computed using parameters $S_0 = 1.5 \times 10^6$ m^3/year, water depth $a = 18$ m, and $\phi_x = 0.2$ rad. In this simulation, we use the explicit finite volume scheme (25) with $\theta = 0$, and we did not take into account the diffraction effect.

Simulation results at subsequent times $t = 0, 0.6, 1.2$ and 3.6 years are shown in Figure 6. We can see clearly that on the up current part of the domain there is accumulation of sand, whereas on the downstream part erosion occurs. Here, we can observe that the accumulation and erosion of sand is in equilibrium. Indeed this equilibrium occurs since the diffraction effect is not considered.

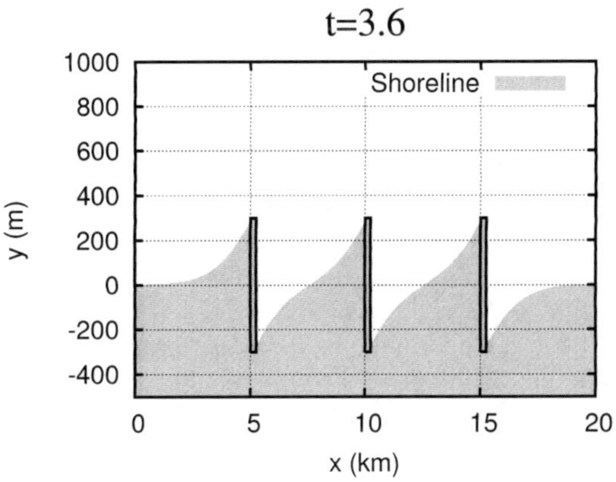

Figure 7. Shoreline change in a system of three groynes after $t = 3.6$ year.

Further simulation, we present the shoreline development due to the installation of a groyne system consisting of three groynes. Domain of the simulation is $\Omega = [0, 20]$ km. The first groyne is installed at 5 km in domain of simulation. Moreover, the second and third groyne are installed at 10 km and 15 km of the domain respectively. The other parameters are the same with the previous simulation. Result of this simulation is given in Figure 7. Similar to the result using one groyne, the accumulation of sand is observed in the left side of groynes. Meanwhile, erosion of sand is occurred on the right side of groynes. Without diffraction, the simulation results of shoreline change were not physically relevant. Therefore in the next subsections, the simulation which takes into account the diffraction effect will be elaborated and validated by a real shoreline from Google Earth photo.

4.2. Shoreline change due to diffraction effect

Here, two numerical simulations with the diffraction effect will be conducted. In the first simulation, only the right part of the groyne is considered. Parameters used in this simulation are as as follows

- The length of groyne is 10λ.
- The angle of wave incidence is $\phi_\infty = 5^o$.

Computational domain is taken to be $\Omega = [0, 10]/\lambda$ and it is discretized uniformly with a discrete space $\Delta x = 0.5$ and discrete time $\Delta t = \Delta x^2/8$. Using Algorithm 1, the results of this simulation can be found in Figure 8. We can see that simulation result is now physically relevant, at the groyne $x/\lambda = 0$ the transported sand is not as much as those at $x/\lambda = 0.87$. This is because of the wave incident is set to be $\phi_\infty = 5^o$, thus the point where $w = 0$ is calculated by $x/\lambda = l/\lambda \tan \phi_\infty = 0.87$. Moreover, at $x \to \infty$, no sand is transported. This result is in a good agreement with the numerical simulation by Bakker, 1970 [2].

The second simulation is the case where the groyne is installed in the middle. The same computational domain and parameters as the previous simulation in Subsection 4.1 is used.

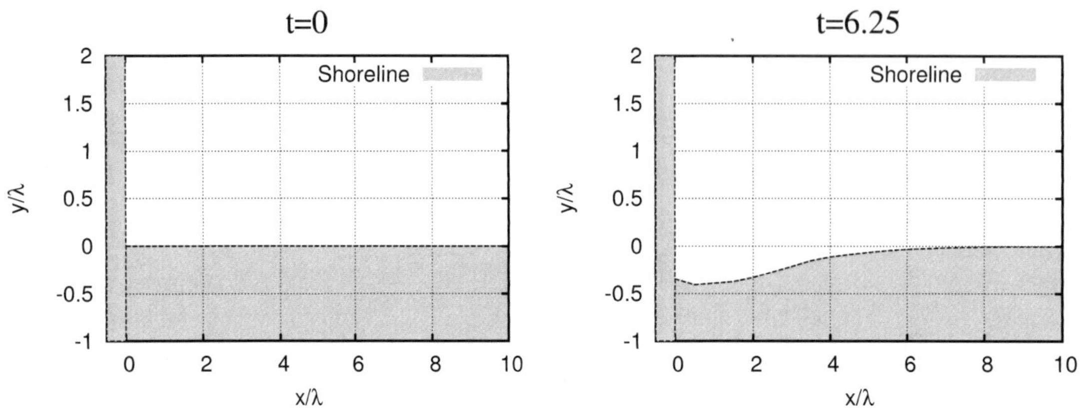

Figure 8. The shoreline change with diffraction effect at initial $t = 0$ (left) and $t = 200 \times \Delta t$ (right).

However, the only difference is, here, the wave crest is set to be $\phi_\infty = 5^o$. Moreover, here the step time space is set to be $\Delta t = \Delta x^2/(8D)$ to keep the stability of numerical scheme.

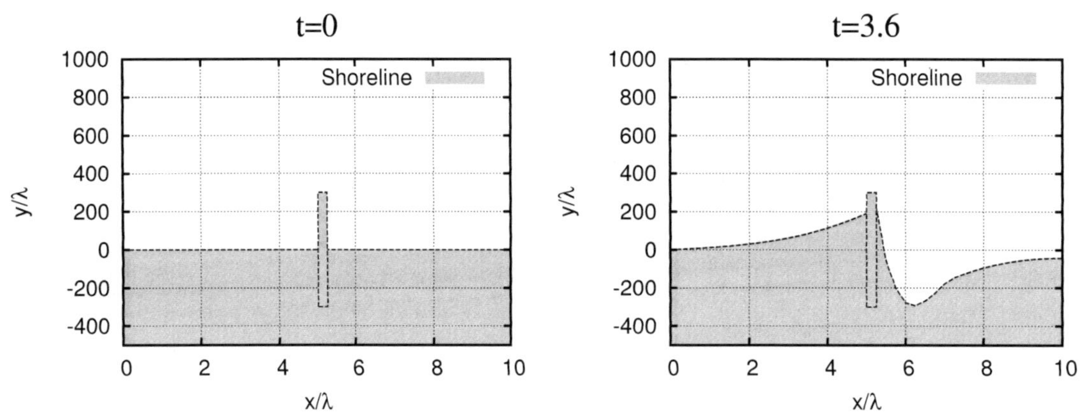

Figure 9. The shoreline changed around a middle installed groyne with diffraction effect at time $t = 0$ (left) and final time $t = 3.6$ (right).

Simulation result is given in Figure 9. In the Figure 9 (right), we can see clearly that diffraction has an effect only on the right part of the groyne, whereas on the left part diffraction seems to have no effect. This is because here we take the angle of incident wave to be $\phi_\infty = 5^o$.

5. Simulation of Shoreline Change in Sanur Bali Beach

In this research, Sanur Bali beach, Indonesia has been chosen for validating the solution of purposed model. Sanur Bali beach is located at southeast of Denpasar city (capital city of Bali island). This location is one of the most famous touristic place for surfing water sport in Bali. Moreover, along behind the coastal of Sanur Bali beach, some big and luxury hotels are existed. The box area in Figure 10 shows the location of Sanur Bali beach and the coastal area where some groynes are constructed (see Figure 1).

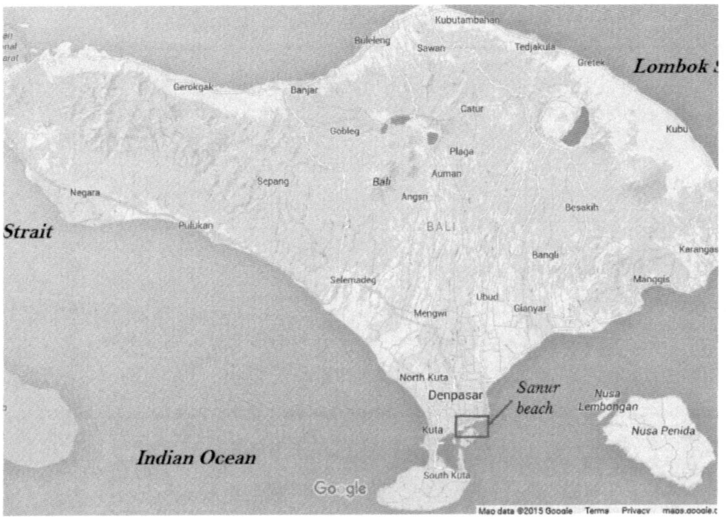

Figure 10. Location of the case study taken from Google Earth photo. The location is in Sanur Bali beach which some groynes are build up.

The initial configurations of this simulation are given as follows

$$a = 1, \qquad \phi_\infty = 0.1 \text{ rad}, \qquad S_0 = 1.5 \times 10^2.$$

Further, the water depth is assumed to be 1 meter on average. The wave incident and water depth can be obtained from the websites which provide the ocean waves and weather information for surfing sport. For our computations, here we use the the parameters which are obtained from magicseaweed.com.

The spatial domain of this computation is $\Omega = [0, 500\ km]$. This domain is discretized with the spatial grid size $\Delta x = 5$ km. From Google Earth photo in 2009, we can get shoreline data, which is used as the initial condition. By using procedure in Algorithm 1, the results of our simulation in three years time simulation are given in Figure 11. Numerical results is compared with shoreline data from Google Earth photo in 2012, for the case of (top) without diffraction, (bottom) with diffraction. From Figure 11 (top) discrepancy are clearly observed in the simulation result without diffraction, especially on the right side of the groyne.

Nevertheless, by accommodating the diffraction term, the numerical result is shown to be in a good agreement with shoreline data, Figure 11 (bottom). After three years simulation time, the shape of shoreline on the left part of the groyne is shown to be in a good agreement with the shoreline data.

6. Conclusion

Shoreline development due to the installation of breakwater groyne have been studied in this paper. Our study adopted the Pelnard-Considère equation with diffraction effect. The finite volume method was adopted here, and physically relevant results were obtained when diffraction effect was incorporated. For the case of Sanur Bali beach, simulation result of the shoreline changes due to a groyne system presents a good agreement with Google Earth photos.

Acknowledgement

The second author would like to thank Institut Teknologi Bandung for the financial support from Riset KK with contract 007/I1.C01/PL/2016.

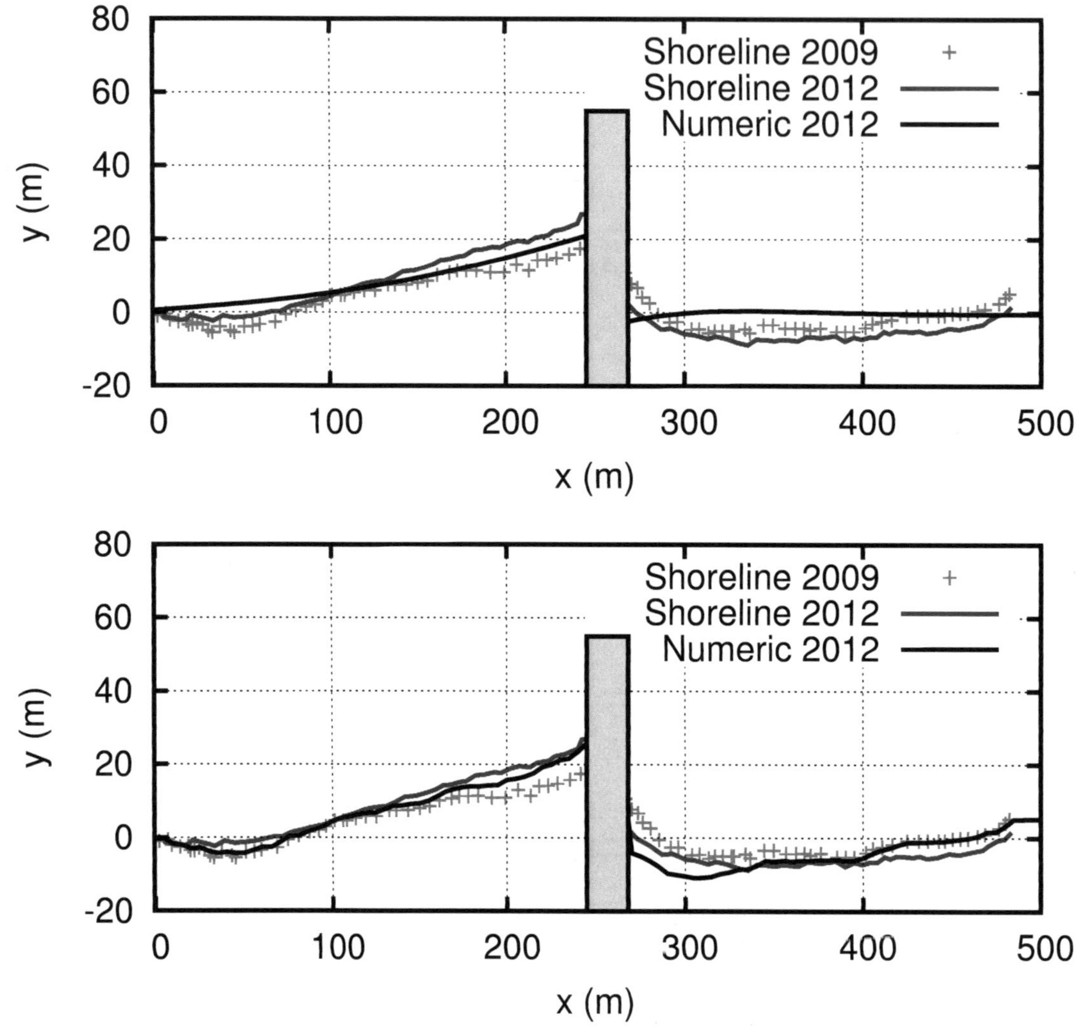

Figure 11. Numerical simulations after three years computation, in comparison with shoreline data extracted from Google Earth photo, (top) without diffraction, (bottom) with diffraction.

References

[1] Pelnard-Considère R 1956 *4th Journees de l'Hydraulique, Les energies de la Mer, Paris* **3** 289–298
[2] Bakker W 1970 *Studierapport WWK* 70–2
[3] Bakker W 1969 *Coastal Engineering 1968* pp 492–517
[4] Perlin M and Dean R 1978 *Coastal Engineering 1978* pp 1818–1838
[5] Valsamidis A and Reeve D E 2017 *Continental Shelf Research* **132** 49–57
[6] LeVeque R J 2002 *Finite volume methods for hyperbolic problems* vol 31 (Cambridge university press)
[7] Toro E F 2009 *Riemann solvers and numerical methods for fluid dynamics: a practical introduction* (Springer Science & Business Media)
[8] LeMehaute B and Soldate M 1977 Mathematical modeling of shoreline evolution. Tech. rep. DTIC Document
[9] Putnam J A and Arthur R S 1948 *Eos, Transactions American Geophysical Union* **29** 481–490
[10] Versteeg H K and Malalasekera W 2007 *An introduction to computational fluid dynamics: the finite volume method* (Pearson Education)

IOP Conf. Series: Journal of Physics: Conf. Series **971** (2018) 012028 doi:10.1088/1742-6596/971/1/012028

OpenACC performance for simulating 2D radial dambreak using FVM HLLE flux

P H Gunawan and M R Pahlevi

School of Computing, Telkom University, Jalan Telekomunikasi No. 1 Terusan Buah Batu, Bandung 40257, Indonesia.

E-mail: phgunawan@telkomuniversity.ac.id; rezhapzz@student.telkomuniversity.ac.id

Abstract. The aim of this paper is to investigate the performances of openACC platform for computing 2D radial dambreak. Here, the shallow water equation will be used to describe and simulate 2D radial dambreak with finite volume method (FVM) using HLLE flux. OpenACC is a parallel computing platform based on GPU cores. Indeed, from this research this platform is used to minimize computational time on the numerical scheme performance. The results show the using OpenACC, the computational time is reduced. For the dry and wet radial dambreak simulations using 2048 grids, the computational time of parallel is obtained 575.984 s and 584.830 s respectively for both simulations. These results show the successful of OpenACC when they are compared with the serial time of dry and wet radial dambreak simulations which are collected 28047.500 s and 29269.40 s respectively.

1. Introduction

Shallow water equations (SWE) are the set of hyperbolic equations which have two parts, called mass and momentum conservation.This set of equation is also known as Saint-Venant equations. Several applications for describing the fluid flows using shallow water model are available. For instance, the model is used to describe the water flow of rivers, ocean, channels, and etc [1]. In two-dimensional, the shallow water equations are written as

$$U_t + F(U)_x + G(U)_y = 0 \tag{1}$$

where

$$U = (h, hu, hv)^T, \tag{2}$$

$$F(U) = \left(hu, hu^2 + \frac{1}{2}gh^2, huv\right)^T, \tag{3}$$

$$G(U) = \left(hv, huv, hv^2 + \frac{1}{2}gh^2\right)^T. \tag{4}$$

The notation $h(x, y, t)$ describes the water depth, g the gravitational force, $u(x, y, t)$ and $v(x, y, t)$ the velocity in x and y-direction respectively. Moreover, the variables x and y are space, and t is time (see Figure 1 for more detail about the situation).

IOP Conf. Series: Journal of Physics: Conf. Series **971** (2018) 012028 doi:10.1088/1742-6596/971/1/012028

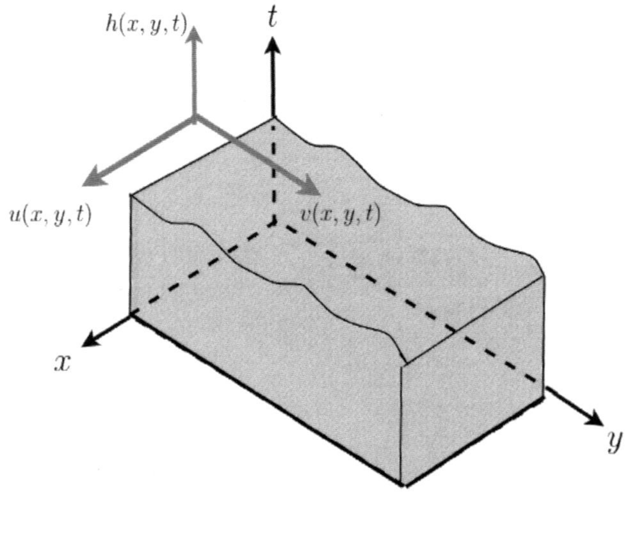

Figure 1. The configuration of 2D shallow Water Equations. Here, two velocities are acting on x and y-direction of domain.

The robust scheme for approximating (1 - 4) is known as collocated scheme of Finite Volume Method (FVM) [2, 3, 4, 5] and [6]. This scheme has widely used to simulate the fluid flow and has been proved to satisfy the entropy condition.

The discretization of two-dimensional of SWE using FVM produces a large discrete points (grids). Thus, this will lead to the long computational time for solving all discrete variables in all grids. Several ideas to reduce the computation in time is available. One of the examples can be used is to the parallel computing techniques. for instance, in the references [7, 8, 9, 10, 11, 12, 13, 14, 15] and [16] a good performance of parallel computing in accelerate the numerical computation is shown satisfied. Therefore, the aim of this paper is to show the performance of parallel computing for simulating dry-wet radial dambreak cases in two-dimensional. In this case, the parallel platform OpenACC (Open Accelerator) will be used.

2. Finite Volume Method for Shallow Water Equations

Given the control volume $V_{i,j}$ over domain $\Omega := [0, L_x] \times [0, L_y]$, in the discrete space point i, j as shown in Figure 2. Since this control volume is on 2D spatial domain, thus $V_{i,j}$ can be defined on $(x_{i-1/2}, x_{i+1/2}) \times (y_{j-1/2}, j_{j+1/2})$ at time t^n to t^{n+1}. Using the uniform discrete space, the space is discretized by $N_x \times N_y \in \mathbb{N}^2$ such that $x_i = i \times \Delta x$, $y_j = j \times \Delta y$ with $i, j \in \mathcal{M} = \{0, 1 \cdots, N_x\} \times \{0, 1 \cdots, N_y\}$. Therefore the space step for x and y-direction are given as $\Delta x = L_x/N_x$ and $\Delta y = L_y/N_y$ respectively.

In another hand, the time is discretized with time step Δt which is defined by the CFL (Courant-Frederich-Levy) condition (11). Thus, the discrete space t^n can be defined as $t^n = n \times \Delta t$ with $n \in \mathcal{T} = \{0, 1, \cdots\}$.

In order to approximate the shallow water equations (1), the discrete notation of $U^n_{i,j}$ is defined as the average approximation over control volume as

$$U^n_{i,j} :\approx \frac{\displaystyle\int_{V_{i,j}} U(x, y, t^n) \quad d\Omega}{|V_{i,j}|}, \quad \forall i, j \in \mathcal{M}, n \in \mathcal{T}. \tag{5}$$

Finally, the dicretization of shallow water equations in finite volume method can be written

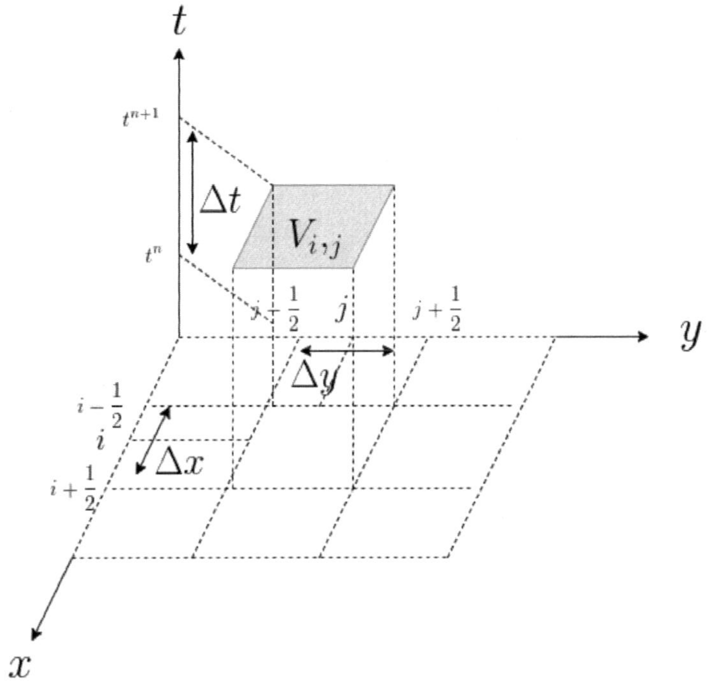

Figure 2. The control volume of discretization space and time. The space steps Δx and Δy, and time step Δt are set to be uniform.

as follows

$$U_{i,j}^{n+1} = U_{i,j}^{n} - \Delta t \left(\frac{F_{i+\frac{1}{2},j}^{n} - F_{i-\frac{1}{2},j}^{n}}{\Delta x} + \frac{G_{i,j+\frac{1}{2}}^{n} - G_{i,j-\frac{1}{2}}^{n}}{\Delta y} \right), \quad \forall i,j \in \mathcal{M}, n \in \mathcal{T} \tag{6}$$

Indeed in the finite volume method, the interesting part is to approximate the numerical fluxes $F_{(i\pm\frac{1}{2},j)}$ and $G_{(i,j\pm\frac{1}{2})}$. Several numerical fluxes can be found in some references [6, 2]. In this paper, the HLLE (Harten, Lax, van Leer and Einfeld) numerical flux will be used for simplicity and robustness. The numerical fluxes are given as

$$F_{i+\frac{1}{2},j}^{n} = \mathfrak{F}(U_{i,j}^{n}, U_{i+1,j}^{n}) = a_1 F(U_{i+1,j}^{n}) + a_2 F(U_{i,j}^{n}) - a_3(U_{i+1,j}^{n} - U_{i,j}^{n}), \tag{7}$$

for the flux in x-direction, and

$$G_{i,j+\frac{1}{2}}^{n} = \mathfrak{G}(U_{i,j}^{n}, U_{i,j+1}^{n}) = a_1 G(U_{i,j+1}^{n}) + a_2 G(U_{i,j}^{n}) - a_3(U_{i,j+1}^{n} - U_{i,j}^{n}), \tag{8}$$

for the flux in y-direction.

The numerical fluxes $F(U_{i,j}^{n})$ and $G(U_{i,j}^{n})$ are the flux function as defined in (3) and (4) with the parameter U at point i,j and current time n. Another parameters are defined as

$$a_1 := \frac{\min(\lambda_2, 0) - \min(\lambda_1, 0)}{\lambda_2 - \lambda_1}, \quad a_2 := 1 - a_1, \quad a_3 := \frac{\lambda_2|\lambda_1| - \lambda_1|\lambda_2|}{2(\lambda_2 - \lambda_1)}. \tag{9}$$

Moreover, the coefficients λ_1 and λ_2 can be obtained in some references such as in [2, 17, 18] for more detail. Thus the final discretization of (6) with numerical fluxes HLEE can be given as follows

$$U_{i,j}^{n+1} = U_{i,j}^{n} - \frac{\Delta t}{\Delta x} \left(\mathfrak{F}(U_{i,j}^{n}, U_{i+1,j}^{n}) - \mathfrak{F}(U_{i-1,j}^{n}, U_{i,j}^{n}) \right) - \frac{\Delta t}{\Delta y} \left(\mathfrak{G}(U_{i,j}^{n}, U_{i,j+1}^{n}) - \mathfrak{G}(U_{i,j-1}^{n}, U_{i,j}^{n}) \right),$$
$$\forall (i,j) \in \mathcal{M}, n \in \mathcal{T}. \tag{10}$$

The numerical scheme (10) has the stability condition. It stability is controlled by the following CFL (Courant-Frederich-Levy) condition

$$\Delta t \leq \nu \frac{\Delta\Psi}{\max_{i,j}(|u_{i,j}| \pm \sqrt{gh_{i,j}}, |v_{i,j}| \pm \sqrt{gh_{i,j}})} \tag{11}$$

Here, grid cells is given in uniform for x and y-direction, hence $\Delta\Psi = \Delta x = \Delta y$, with $0 < \nu \leq 1$ which is called the Courant number.

3. Parallel Algorithm using OpenACC Platform

OpenACC is known as a programming platform for parallel computing. This programming platform is developed by several enterprises such as Cray, CAPS, Nvidia, and PGI [19, 20]. OpenACC is designed to porting the codes with wide variety of heterogeneous HPC hardware platform and architectures with less programming effort than required with a low-level model. Moreover, OpenACC supports C, C++, and Fortran programming languages and multiple hardware architectures. In order to accelerate the application with GPU or accelerator, OpenACC has porting cycle is used.

The preparation steps for using OpenACC is given as follows:

(i) Before start to use OpenACC in an application, the observation of routines and loops which are the most spending the bulk of time should be done. The advantage of this observation indeed to maximize the acceleration of computational time in the application.

(ii) After the important parts of the programming code have been identified, the OpenACC directives should be used to accelerate these regions on the accelerator. Parallel loops within the code should be decorated with `#pragma acc loop` directives to provide OpenACC compiler information to parallelize the code (see Figure 3).

(iii) In the programmers side, the additional instruction to compiler should be done. For instance, how to manage the local memory to the accelerator and to keep the information moved between two memories if absolutely necessary.

(iv) Another important step is when the information coming from the CPU architecture to GPU architecture, restructuring the loops should be done in order to reduce the frequency of data movement

Previous steps have been successful in many applications since its simplicity and applicable such that the programmer can quickly and productively achieve the acceleration of time. Therefore, in this paper OpenACC will be used to accelerate the simulation time.

In Figure 3, the parallel programming can be started in initialization value of $U^0_{i,j}$. Then the initialized value will be sent to the accelerator only once with Peripheral Component Interconnect express (PCIe), it is because transfer value between CPU and GPU is very expensive. It has been explained in [21], the value from the host for each routines and loops will be mapped automatically inside gangs, where a gang consists of several workers. The first parallel part is given to compute Δt for CFL condition, then the second part is given to compute primitive variables at new time step $h^{n+1}_{i,j}$,$u^{n+1}_{i,j}$, and $v^{n+1}_{i,j}$. The third part of parallel is given for boundary condition. Once those parts are done, the results will be sent back to the CPU for the next iteration.

4. Numerical Results and Parallel Performance

All the following numerical results are compiled from HPC lab work server School of Computing, Telkom University with the specifications as shown in Table 1.

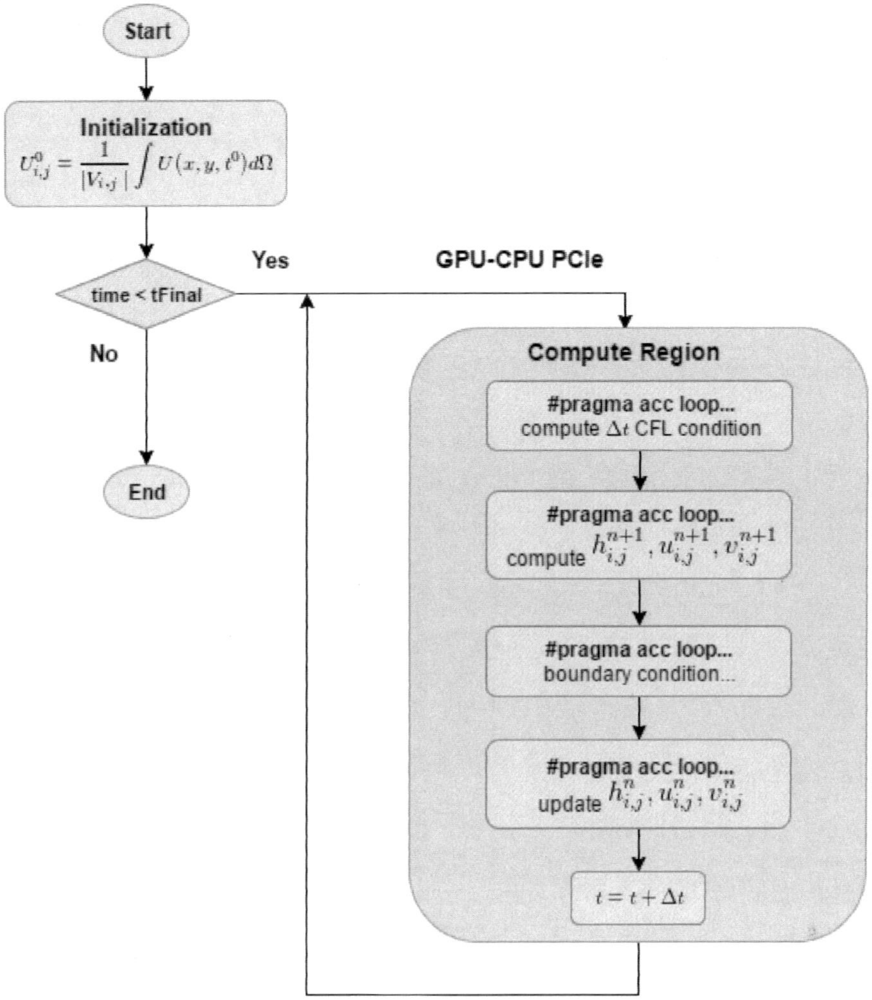

Figure 3. The flowchart of parallel computation.

Table 1. The computer specifications for running the numerical simulations.

Name	Type
Operating System	Ubuntu 16.10
Processors	Intel(R) Xeon(R) 2 socket @16core
RAM	6GB
GPU	Nvidia GTX670
CUDA cores	1024

4.1. Dry and wet radial dambreak simulations

In order to see the performance of parallel computing, two numerical simulations of radial dambreak problem are given. Here, two numerical simulations of radial dambreak which are dry and wet cases are given. Following are the initial configuration of two radial dambreak problems on domain $\Omega = [0,1] \times [0,1]$ and at a middle point $(x_0, y_0) = (0.5, 0.5)$.

IOP Conf. Series: Journal of Physics: Conf. Series **971** (2018) 012028 doi:10.1088/1742-6596/971/1/012028

The initial configurations of dry radial dambreak (water height and fluxes) are given as

$$h(x,y,0) = \begin{cases} 1, \text{ if } \sqrt{(x-x_0)^2+(y-y_0)^2} \leq \dfrac{3}{10}, \\ 0, \text{ otherwise}, \end{cases} , \tag{12}$$

$$q(x,y,0) = p(x,y,0) = 0, \tag{13}$$

where $q(x,y,t) = h(x,y,t).u(x,y,t)$ and $p(x,y,t) = h(x,y,t).v(x,y,t)$.

The results of water surface and velocity in x and y-direction profiles of dry radial dambreak simulation are shown in Figures 4 and 5 respectively. From Figure 4, the numerical method is shown able to handle the vacuum problem. The water surface smoothly flow over dry bed. In this simulation, the final time of simulation is $t = 0.03$ s.

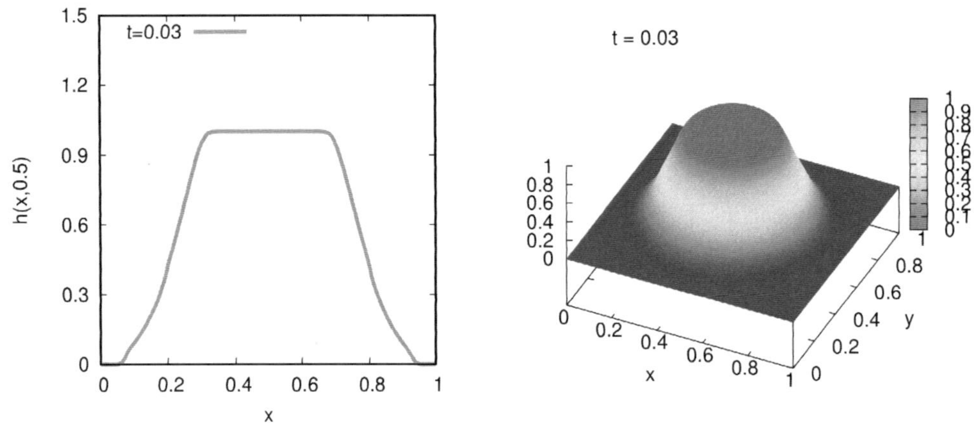

Figure 4. Simulation results of dry radial dambreak problem. (Left) The slice profile of dry radial dambreak at $y = 0.5$. (Right) the 3D view of water height profile dry bed.

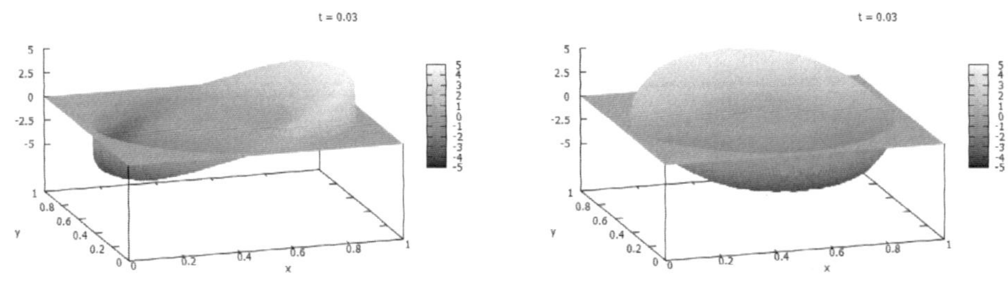

Figure 5. The velocity profile of dry radial dambreak simulation. (Left) velocity profile in x-direction (u). (Right) velocity profile in y-direction (v).

In other side, the initial configurations of wet radial dambreak are given as follow

$$h(x,0) = \begin{cases} 1, \text{ if } \sqrt{(x-x_0)^2+(y-y_0)^2} \leq \dfrac{3}{10}, \\ 0.2, \text{ otherwise}, \end{cases} , \tag{14}$$

$$q(x,y,0) = p(x,y,0) = 0. \tag{15}$$

This initial conditions of wet radial dambreak problem are for water depth and velocity in x and y-direction profiles. The results of this simulation are shown in Figures 6 and 7.

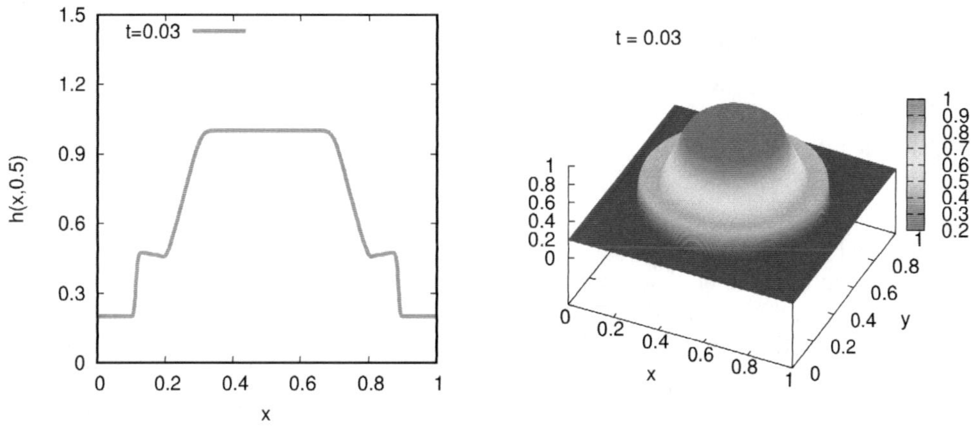

Figure 6. Simulation results of wet radial dambreak problem. (Left) The slice profile of wet radial dambreak at $y = 0.5$. (Right) the 3D view of water height profile wet bed.

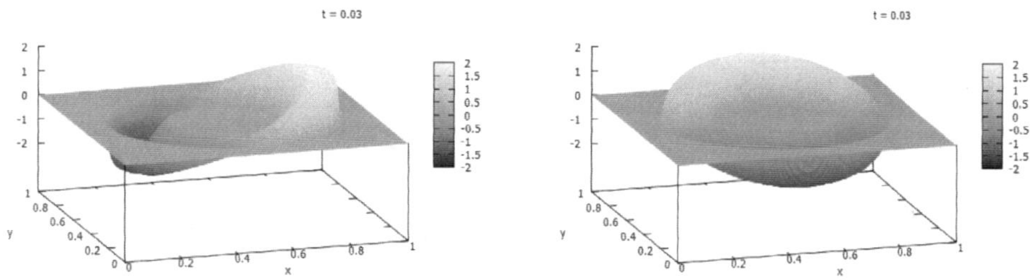

Figure 7. The velocity profile of wet radial dambreak simulation. (Left) velocity profile in x-direction (u). (Right) velocity profile in y-direction (v).

Here, the water surface profile of wet radial dambreak is shown in Figure 6. Moreover, the velocity in x and y-direction profiles are shown in Figure 7(Left) and Figure 7(Right) respectively.

It can be seen in Figure 6(Left), the wet radial dambreak produces the shock waves in around 0.1 and $x = 0.9$. Indeed this shock is also shown in one-dimensional problem (see [1] for more detail). Our focus in this paper is to reduce the computational time of this simulation by using large discrete points. Next section will present the result of parallel computing using OpenACC with GPU architecture model.

4.2. Parallel performance

The computational time in serial and parallel for dry and wet radial dambreak simulation can be found in Table 2. Here, the results are obtained from uniform discrete points $Nx = Ny$ (follow the reference [22]), hence the simulations use several numbers of discrete points $(Nx \times N_y) = (512 \times 512), (1024 \times 1024), (2048 \times 2048)$, and (4096×2048) with final time $T = 1$ s. Moreover, in this simulation, the number of threads for parallel computation is used is 128 threads in GPU.

Table 2. Computational time in serial and parallel for dry and wet radial dambreak simulation.

Grids	Serial time for dry bed (s)	Parallel time for dry bed (s)	Serial time for wet bed (s)	Parallel time for wet bed (s)
512 × 512	359.245	8.495	404.898	8.956
1024 × 1024	3272.480	70.760	3414.270	71.460
2048 × 2048	28047.500	575.984	29269.400	584.830
4096 × 4096	Nan	4823.369	Nan	4881.460

As shown in Table 2, unfortunately, the computer could not able to get the exact time in serial using grid number 4096 due to lack of computer resource to compute such large grid. Here, serial experiment using 4096 grids took more than 3 days. Thus, this computational time is denoted by Nan. However, by using OpenACC, the computational time for dry and wet simulation are obtained 4823.369 s and 4881.460 s respectively.

From the Table 2, the computational time of parallel is shown significantly increasing along the increasing of discrete points. However, compared with the serial, indeed the parallel programming by OpenACC is successfully reduce the computational cost. For instance using grid sizes 2048, the serial time for dry and wet radial dambreak simulation are obtained 28047.500 s and 29269.40 s respectively. Meanwhile, the parallel time for both simulations are collected 575.984 s and 584.830 s respectively. The discrepancy of CPU time of dry and wet dambreak is not to large. This discrepancy can be obtained due to the factor of floating point arithmetic.

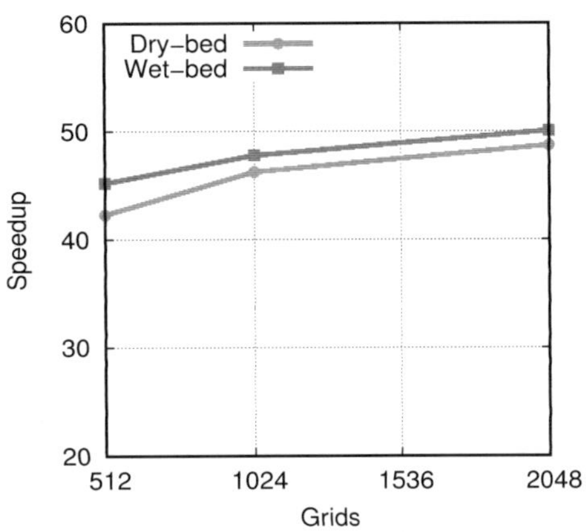

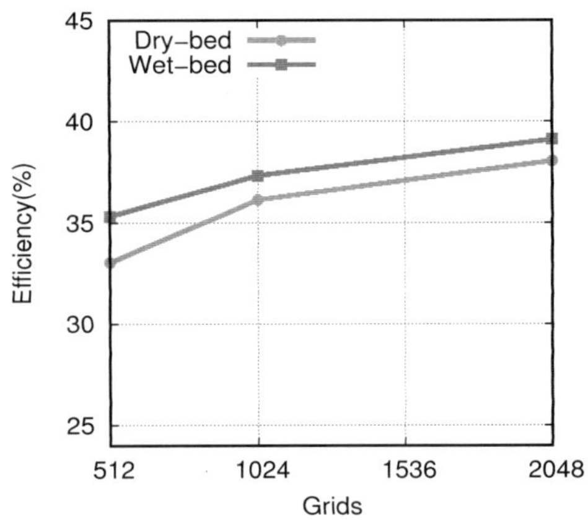

Figure 8. The speedup (left) and efficiency (right) in parallel programming for both simulations.

Figure 8 shows the performances of parallel computing using speedup and efficiency measurements. In Figure 8 (left) the speedup is shown increasing for both simulations. Using 2046 number of grids, the speedup for dry and wet radial dambreak is obtained around 48.69 and 50.04 times respectively. Meanwhile, the efficiency for dry and wet radial dambreak is 38.04% and 39.10% respectively. The detail of speedup and efficiency measurement formula can be found in references [12, 23].

5. Conclusion

The application of OpenACC platform to reduce the computational time of 2D dry and wet radial dambreak simulations are given. The modified parallel algorithm of numerical scheme for approximating shallow water equations are also given. The results show that the computational time of dry and wet cases using the OpenACC are satisfactory. Here, OpenACC platform can minimize the time execution for large discrete points. For the dry and wet radial dambreak simulations, using 2048 grids, the computational time for both simulations are 575.984 s and 584.830 s respectively. Those results show that the OpenACC can reduce the computational time where the serial time for both simulations are obtained 28047.500 s and 29269.40 s respectively using 2048 grid sizes. Moreover, the maximum speedup 50.04 times and efficiency 39.10% are observed at wet radial dambreak simulation by similar number of grids.

References

[1] Doyen D and Gunawan P H 2014 *Finite Volumes for Complex Applications VII-Methods and Theoretical Aspects* (Springer) pp 227–235
[2] Delestre O, Cordier S, James F and Darboux F 2008 *12th International Conference on Hyperbolic Problems* vol 67 (American Mathematical Society) pp 537–546
[3] Delestre O and Marche F 2011 *Journal of Scientific Computing* **48** 41–51
[4] Audusse E, Bouchut F, Bristeau M O, Klein R and Perthame B 2004 *SIAM Journal on Scientific Computing* **25** 2050–2065
[5] Gunawan P H and Lhébrard X 2015 *Computers & Fluids* **121** 44–50
[6] Bouchut F 2004 *Nonlinear stability of finite Volume Methods for hyperbolic conservation laws: And Well-Balanced schemes for sources* (Frontiers in Mathematics. Birkhäuser Verlag, Basel)
[7] Gunawan P H 2016 *Information and Communication Technology (ICoICT), 2016 4th International Conference on* (IEEE) pp 1–5
[8] de la Asunción M, Castro M, Mantas J and Ortega S 2016 *Advances in Engineering Software* **99** 59–72
[9] De La Asunción M, Mantas J M and Castro M J 2011 *The Journal of Supercomputing* **58** 206–214
[10] Brodtkorb A R, Sætra M L and Altinakar M 2012 *Computers & Fluids* **55** 1–12
[11] Mulyani, Putri N D and Gunawan P H 2017 *Information and Communication Technology (ICoIC7), 2017 5th International Conference on* (IEEE) pp 1–5
[12] Pahlevi M R and Gunawan P H 2017 *Information and Communication Technology (ICoIC7), 2017 5th International Conference on* (IEEE) pp 1–4
[13] Juliati S and Gunawan P H 2017 *Information and Communication Technology (ICoIC7), 2017 5th International Conference on* (IEEE) pp 1–5
[14] Alamsyah M N A, Simanjuntak C A, Bagustara B A R H, Pradana W A and Gunawan P H 2017 *Information and Communication Technology (ICoIC7), 2017 5th International Conference on* (IEEE) pp 1–6
[15] Iryanto and Gunawan P H 2017 *Information and Communication Technology (ICoIC7), 2017 5th International Conference on* (IEEE) pp 1–5
[16] Castillo D, Ferreiro A, García-Rodríguez J A and Vázquez C 2013 *Applied Mathematics and Computation* **219** 11233–11257
[17] LeVeque R J 2002 *Finite volume methods for hyperbolic problems* vol 31 (Cambridge university press)
[18] Toro E F 2013 *Riemann solvers and numerical methods for fluid dynamics: a practical introduction* (Springer Science & Business Media)
[19] Wienke S, Terboven C, Beyer J C and Müller M S 2014 *European Conference on Parallel Processing* (Springer) pp 812–823
[20] Wienke S, Springer P, Terboven C and an Mey D 2012 *Euro-Par 2012 Parallel Processing* 859–870
[21] Kirk D B and Wen-Mei W H 2013 *Programming massively parallel processors: a hands-on approach* (Morgan kaufmann)
[22] Cheng J, Grossman M and McKercher T 2014 *Professional Cuda C Programming* (John Wiley & Sons)
[23] Rauber T and Rünger G 2013 *Parallel programming: For multicore and cluster systems* (Springer Science & Business Media)

IOP Conf. Series: Journal of Physics: Conf. Series **971** (2018) 012029 doi:10.1088/1742-6596/971/1/012029

GraDit: graph-based data repair algorithm for multiple data edits rule violations

Wa Ode Zuhayeni Madjida and I Gusti Bagus Baskara Nugraha

School of Electrical Engineering and Informatics, Institut Teknologi Bandung, Bandung, Indonesia

E-mail: zuhayeni@students.itb.ac.id

Abstract. Constraint-based data cleaning captures data violation to a set of rule called data quality rules. The rules consist of integrity constraint and data edits. Structurally, they are similar, where the rule contain left hand side and right hand side. Previous research proposed a data repair algorithm for integrity constraint violation. The algorithm uses undirected hypergraph as rule violation representation. Nevertheless, this algorithm can not be applied for data edits because of different rule characteristics. This study proposed GraDit, a repair algorithm for data edits rule. First, we use bipartite-directed hypergraph as model representation of overall defined rules. These representation is used for getting interaction between violation rules and clean rules. On the other hand, we proposed undirected graph as violation representation. Our experimental study showed that algorithm with undirected graph as violation representation model gave better data quality than algorithm with undirected hypergraph as representation model.

1. Introduction

Data is important for an organization to make a decision, so that the organization should keep their data quality. The better the data quality, the data can be used as a basis for decision making. In fact, existing data is always dirty or contain some glitches because of data collecting and recording process. This leads to inconsistent data and could make organization falls on erroneous decisions. One way to eliminate these inconsistencies and ensure the data quality is by doing data repair or data cleaning [1].

Data cleaning is a process to detect and correct errors in data. Data can be relational database tables and records on file. In data cleaning process, we want to make sure that each record in data is clean or not contain faulty value, so that we can use the data for analysis and describe valid conditions. Constraint-based is one of the data cleaning approach.

In constraint-based data cleaning, data is said to be dirty if there are inconsistencies or violations against data quality rules [2]. The rules describe relationship between attributes in data. There are two kind of data quality rules, they are integrity constraint and data edits [8].

Integrity constraint is used to express the data quality rules for checking inconsistencies in relational database. Several types of integrity constraints are functional dependency (FD) and inclusion functional dependency [2], conditional functional dependency (CFD) [3] and conditional inclusion dependency [2] [4], denial constraint [5], matching dependencies [6], and editing-constraints [12].

Furthermore, data edits is used for data where the structure is corresponding to questionnaires schema. National statistics institution uses data edits to keep their data quality by checking inconsistencies in data that come from census questionnaires. Every set of answer in questionnaires must satisfy defined data edits rules, so it can be referred as the correct answer. If there are inconsistent answer, the data must be repaired.

Various algorithm has been proposed for repairing data inconsistencies in constraint-based data cleaning. [9] and [10] introduce a repairing algorithm that gives some cleaning strategies based on cost. [11] and [12] utilize master data for identifying wrong value in an attribute. [13] uses statistical methods for improving accuracy of inconsistent data. [14] takes advantage of user engagement in providing repair updates and using machine learning to identify and apply the correct updates directly to database without user engagement anymore. Another study was conducted by [15] in proposing a data cleaning algorithm for matching and data repair. However, these researches are focus on how repair inconsistencies based on violation of a rule without consider another defined rules.

In [16] and [17], they proposed an algorithm that deal with interaction among defined rules. They represent rules violation into undirected hypergraph to get conflict attribute in data that must be repaired. Nevertheless, the algorithm is designed for integrity constraint FD and CFD, not for data edits rules. Although they have similar structure, the algorithm can not be applied for data edits rules.

Structurally, data edits rules is similar with constant conditional functional dependency (constant-CFD), which incorporate bindings of related values on attribute. Constant value in data edits rules is identical with *pattern tableau* in CFD. Other related structure is element of the rule that contain left hand side (LHS) and right hand side (RHS). However, we can not represent data edits rule using undirected hypergraph and only for rule violation, due to differences of the rule characteristics.

If there are a set of data edits rules:

$\varphi1$: [A = 1] $\wedge$ [B = 2] $\rightarrow$ [C != null]
$\varphi2$: [C = 1] $\rightarrow$ [D = 2]
$\varphi3$: [C = 2] $\rightarrow$ [D = null]
$\varphi4$: [D = 2] $\vee$ [E = 2] $\rightarrow$ [F != null]
$\varphi5$: [F != null] $\wedge$ [[G = 1] $\vee$ [G = 2]] $\rightarrow$ [H = null]

Problem 1. LHS of data edits rules contains one or more operands, AND ($\wedge$) and OR ($\vee$). In FD or CFD, operand in LHS is only AND. For example, data edits rules $\varphi5$ contains two operands. Undirected hypergraph can not handle these operands, so that conflict attribute in repairing process falls into wrong attribute.

Problem 2. A set of data edits rules contain rules that organize various attribute values based on attribute domain. Consider attribute C has domain value $dom(C) = \{ 1, 2 \}$, $\varphi2$ and $\varphi3$ are rules which determine D value based on all possibility of C value. If we just represent rule violations, we can not get repair expression of cleaning process because it is at any other rule.

Problem 3. One rule and another rule in data edits rules are linked each other. RHS in one rule can be LHS in another rule, i.e RHS [F != null] in $\varphi4$ is LHS in $\varphi5$. By using undirected hypergraph and representing only for rule violation, we can not get interaction between violation rule and clean rule, so that repair one attribute can create other new violation.

These problems motivate the study of novel repair algorithm for data edits rule. We proposed another kind of violation representation and designing the repair algorithm.

2. Related Work

Constraint-based data cleaning is one of data cleaning approach. Several algorithm are proposed for repairing data inconsistencies in constraint-based data cleaning. [9] proposed a data cleaning method, sampling against space of possible repairs. [10] uses cost in comparing data and constraint repair. [11] and [12] utilize master data for identifying wrong value in an attribute. [13] improves accuracy of inconsistent data by using statistical methods. [14] takes advantage of user engagement in providing repair updates. The update is stored in database and will be used by machine learning algorithm to identify and apply the correct updates without user engagement anymore. Another study was conducted by [15] proposed data cleaning algorithm for matching and data repair. However, these research focus on repairing one violation without considering interaction among rules. This can make the repair is not based on other attribute value in different rule, so that it causes new violation of other rule.

In [16], they proposed an algorithm that consider the interaction between violation rules. The violation is represented in undirected hypergraph and using minimum vertex cover (MVC) to define a conflict attribute, namely attribute that must be repaired. The proposed algorithm repair one attribute in every iteration until there is no violation in data. To do the repair holistically, [17] continued this study. The research takes advantage of the undirected hypergraph for getting *context* of conflict attribute. The context are the attribute involvement in other violation rule and its change effect for the other attribute in same violation rule. They used this context to build repair expression and doing repair for several attributes in every iteration (holistic repair). To determine the value of attribute improvements, researchers used Value Frequency Map (VFM) and Quadratic Programming (QP). However these algorithms are designed for integrity constraint FD and CFD, and can not be applied for data edits rules, so that we proposed a repair algorithm for data edits rules.

3. Proposed Algorithm

3.1. Data Edits Rule

National statistics using data edits for checking inconsistencies in data that come form census questionnaires. An example of using data edits for checking data inconsistencies is as follows.

An inconsistent answer in a questionnaire can be declared as:

```
marital status = ''married'', age = ''5 years old''
```

To detect this violations, rule can be made as:

```
if marital status is married, age must not be less than 16 years old
```

Edits express the inconsistency condition, namely:

```
marital status = ''married'' ∧ age < 16
```

Based on the rule, we can define a pattern that must be satisfied by data, as follow:

```
if marital status is married, then age must greater or equal than 16
years old
```

Married is constant for marital status, and *greater than 16 years old* is constant for age. Denoted that φ is *data edits rule*, which refer to rules that must be satisfied by data, then the rule mathematically can be noted as:

$$\varphi : [marital = "married"] \rightarrow [age \geq 16].$$

Based on the rule above, the LHS of the rule is [$marital$ = "$married$"] and the RHS is [$age \geq 16$]. LHS and RHS of the rule can contain more than one expression and implicate AND ($\wedge$) and OR ($\vee$) operand.

A record in data is said violating the rule, if the record meets the condition at LHS, but does not meet the condition on the RHS.

3.2. Proposed Algorithm

In data that come from questionnaire schema, every record must satisfy defined data edits rules. Consequently, violation detection and data repairing are processed per record. Figure 1 shows proposed data cleaning algorithm based on data edits rules. We called it GraDit. Here is step by step of the algorithm:

Step 1: Parsing Rules. This process would generalize all defined rules. Expression in a rule that hold != (not equal), $<$, $\leq$, $>$, $\geq$ is elaborated into equality expression based on domain attribute value. Additionally, if a rule contains two operands AND and OR, the rule is converted to disjunctive normal form (DNF) by using distributive law.

Step 2: Generate Bipartite-Directed Hypergraph. This process represents overall defined rules and attributes in bipartite-directed hypergraph. Node of graph consist of two set, that is attribute nodes and rule nodes. Attribute nodes represent attribute in a rule and rule nodes represent rule it self. This representation would be used for getting interaction between violation rule and clean rule in repairing process.

For every record in data, do step 3 until step 7. This process will continue until number of current violation is greater than previous violation.

Step 3: Hit data. Set flag in bipartite-directed hypergraph based on value of each attribute in data.

Step 4: Identify violation. This process would detect rule violation on data by using edits.

Step 5: Generate conflict graph. Generating conflict graph, namely graph that represent the violations. We use undirected graph as violation representation model. Node in the graph is attribute on violation rule, and edge connects each node based on relationship characteristics between attribute in the same rule.

Step 6: Identify MVC. Identify MVC of the conflict graph by using vertex cover approximation algorithm. Node that is selected as MVC would become repaired attribute candidate in repairing process.

Step 7: Repair data. We proposed algorithm 1 for repairing data. This process would get attribute nodes in bipartite-directed hypergraph, namely attributes that were selected as repair candidate. The process would find attribute nodes which have value besides value attribute of data. These nodes is evaluated based on some parameters and the best nodes would become repair expression. This study proposed some parameters in selecting repair expression, they are:

(i) Number of ingoing degree that come from rule node that has error flag in bipartite-directed hypergraph (IER).

(ii) Number of ingoing degree that come from flagged attribute node to connected error rule node (IHV).

(iii) Number of outgoing degree to rule node that has outgoing edge to flagged attribute node (OHR).

(iv) Number of outgoing degree to flagged attribute node from connected rule node (OHV).

Table 1 shows algorithm comparison between previous research and this study.

IOP Conf. Series: Journal of Physics: Conf. Series **971** (2018) 012029 doi:10.1088/1742-6596/971/1/012029

Algorithm 1 Data Repairing

Input: vertex list from mvc (conf), data, bipartite-directed hypergraph (graph)
Output: Repair data

```
repairCandidate ← []
IER ← []
IHV ← []
OHR ← []
OHV ← []
for all vertex ∈ conf do
    candidate ← get variable nodes of vertex in graph besides hitdata
    repairCandidate ← repairCandidate ∪ candidate
    for all rc ∈ repairCandidate do
        er ← compute number of inbound error rule node to rc
        IER.add(er)
        hv ← compute number of inbound hit variable node to error rule node of rc
        IHV.add(hv)
        hr ← compute number of outbound rule node of rc which has outbound hit variable node
        OHR.add(hr)
        hv ← compute number of outbound hit variable node from outbound rule node of rc
        OHV.add(hv)
    end for
end for
repair ← get repairCandidate with the highest number in IER
if repair.size() = 1 then
    data.update(repair)    ▷ repair data with repairCandidate with the highest number in IER
else
    repair ← get repair with the highest number in IHV
    if repair.size() = 1 then
        data.update(repair)
    else
        repair ← get repair with the highest number in OHR
        if repair.size() = 1 then
            data.update(repair)
        else
            repair ← get repair with the highest number in OHV
            data.update(repair)
        end if
    end if
end if
return data
```

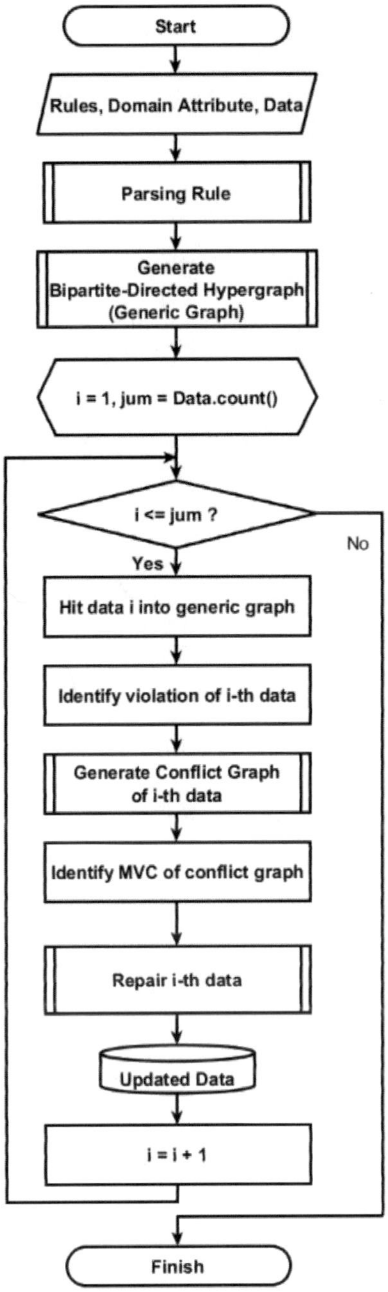

Figure 1. GraDit Algorithm.

4. Experiments

4.1. Experimental Settings

Datasets and rules. We use national socio-economic survey datasets which is collected by Statistics of Indonesia. Datasets consist of 100 record and 209 attributes where 149 attributes of them are set in rules. The datasets is divided into three instances:

(i) Clean Data I_c. This data is cleaned data that has satisfy all defined rules. We use this data as ground truth.

Table 1. Algorithm Comparison of Previous Research and Proposed.

	FindVRepair [16]	Holistic Repair [17]	GraDit
Data quality rule	FD and CFD	FD and CFD	Data edits rule
Representation for overall rules	NO	NO	YES
Violation representation model	undirected hypergraph	undirected hypergraph	undirected graph
Repair value of conflict attribute	using other attribute value in same rule	using Value Frequency Map (VFM) and Quadratic Programming (QP)	using proposed parameters

(ii) Dirty Data I_d, data that has some rule violations. This data comes from I_c where some attribute value are changed by a certain error rate, that is, ratio of the number of dirty attribute to the total number of attribute which is set in the rule. Error rate which would be generated is 10 to 40 percent. Error is generated randomly on the attributes of the datasets based on domain value of each attribute.

(iii) Repaired Data I_r. This data is repair of I_c, that is output of repair algorithm.

There are 483 data edits rule that manage relationship between attribute of dataset. These rules consist of 286 rules that manage relationship between two attributes (no operand in rule), 131 rules with one operand AND or OR in a rule, and 66 rules that consist of two operand AND and OR in a rule. One rule and other rules are linked each other.

Algorithms. We implement two algorithms, they were proposed algorithm in figure 1 and proposed algorithm where the violation representation is changed into undirected hypergraph. Both of them have been implemented in Java. Based on these algorithm, we would show how violation representation takes effect to data repairing result.

Metric. We use F-measure to measure repair quality. The formula of F-measure is $(2x(PxR)/(P+R))$, where P is precision and R is recall. This metric is adopted from [9].

4.2. Experimental Results

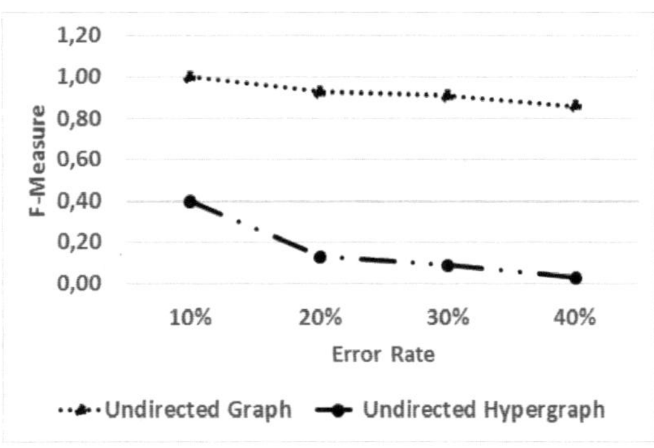

Figure 2. F-Measure Maximal Value of Data Repairing Based on Error Rate.

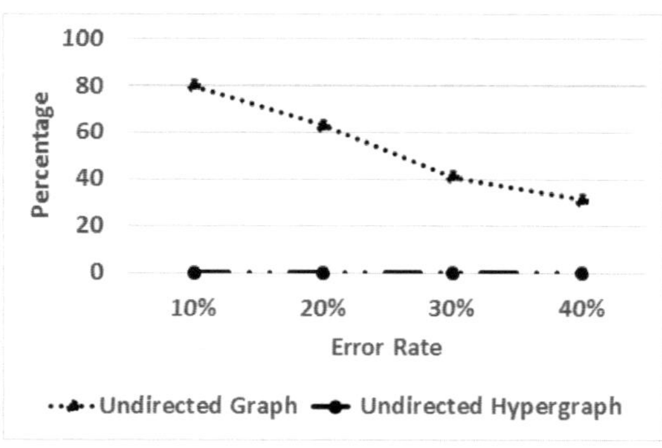

Figure 3. Percentage of Record that Have F-Measure >0.5 Based on Error Rate.

4.2.1. F-Measure Maximal Value of Data Repairing Based on Error Rate Figure 2 shows maximal value of f-measure that is obtained from impemented algorithm. Out of 100 tested record on the error rate of 10%, f-measure can reach maximum value 1 for the algorithm by using undirected graph as a violation representation model. In other words, the algorithm can fix all the errors properly. This is because the generated error is an attribute on RHS of rule, so that violations can be captured and repaired overall by the algorithm. However, although the algorithm can provide maximum value, algorithm by using undirected hypergraph as violation representation can not deliver maximum value 1.

Violations that are represented using undirected hypergraph lead to mistake in identifying attribute candidates to be repaired. In violation rule, there is an LHS attribute that exist in almost every rule that were violated. This makes the attribute become vertex cover when calculating MVC. In fact, the attribute is not error attribute.

In data with error rate 20 to 40%, f-measure of algorithm with undirected graph violation representation is close to 1. In contrast, the algorithm by using undirected hypergraph, f-measure decline close to 0 by increasing the error rate in the data.

F-measure value of each repaired record is greatly affected by violation representation model. This representation model determine attribute candidates to be repaired. Inaccuracy in defining repaired candidate attribute, it would affect repaired data quality. Undirected graph can give better results than undirected hypergraph because undirected graph representation consider relationship characteristic of attributes in a rule, which is based on the operand in the rule, either AND nor OR. Otherwise, undirected hypergraph can not describe attribute relationship characteristic because all attributes in rule that become node on the hypergraph were covered by one edge.

Other factor that influences the value of f-measure is generated error attribute. The attribute can be involved in LHS of a rule, RHS, or LHS in one rule and RHS in another rule. If the error attribute is only in LHS and error value of the attribute is not a LHS conditions that must be met by the rule, then the error can not be captured by the violations. The other case, there were conditions that are satisfied by LHS attributes, then led the RHS attributes captured as a violation. The RHS attributes have possibility to be repaired by algorithm. These improvements is not a proper repair based on ground truth, but in terms of cost, which is the number of attribute change in data repairing process, it is an improvement with the lowest cost.

4.2.2. Percentage of Record that Have F-Measure >0.5 Based on Error Rate Based on percentage of record that have f-measure value above 0.5, namely data that can be corrected more than 50% of the overall error attribute, undirected graph representation could give significant result than algorithm by using undirected hypergraph. Figure 3 shows the results. Algorithms with undirected hypergraph does not produce any of record that have f-measure above 0.5, either on the data with an error rate of 10% to 40%. In contrast, undirected graph algorithms can provide 80% of record that have f-measure above 0.5 in data with 10% of error rate. This percentage continues to decrease with higher error rate in the data.

5. Conclusion

In this paper, we proposed an algorithm for constraint-based data cleaning of *data edits rule*. We introduce bipartite-directed hypergraph as model representation of overall *data edits rule* for getting interaction between violation rule and clean rule. For representing rules violation, we use undirected graph that could describe attribute relationship characteristic in a rule. Based on experintal study, algorithm that using undirected graph as violation representation could give higher f-measure than algorithm which undirected hypergraph. Out of 100 tested record by using undirected graph algorithm, 80% records give f-measure above 0.5 in data with 10% error rate, 63% records in data with 20% error rate, 41% records in data with 30% error rate, and

31% records in data with 40% error rate. This result shows that by representing violation in undirected graph, the proposed algorithm can give better repair quality than using undirected hypergraph.

References

[1] T. Dasu and T. Johnson, *Exploratory Data Mining and Data Cleaning.* New York, NY, USA: John Wiley & Sons, Inc., 1 ed., 2003.
[2] W. Fan and F. Geerts, *Foundations of Data Quality Management.* Morgan & Claypool Publishers, 2012.
[3] W. Fan, F. Geerts, X. Jia, and A. Kementsietsidis, "Conditional functional dependencies for capturing data inconsistencies," *ACM Trans. Database Syst.*, vol. 33, pp. 6:1–6:48, June 2008.
[4] P. Z. Yeh and C. A. Puri, "Discovering conditional functional dependencies to detect data inconsistencies," in *Proceedings of the Fifth International Workshop on Quality in Databases at VLDB2010*, vol. 176, Citeseer, 2010.
[5] X. Chu, I. F. Ilyas, and P. Papotti, "Discovering denial constraints," *Proc. VLDB Endow.*, vol. 6, pp. 1498–1509, Aug. 2013.
[6] W. Fan, X. Jia, J. Li, and S. Ma, "Reasoning about record matching rules," *Proc. VLDB Endow.*, vol. 2, pp. 407–418, Aug. 2009.
[7] W. Fan, J. Li, S. Ma, N. Tang, and W. Yu, "Towards certain fixes with editing rules and master data," *The VLDB Journal*, vol. 21, no. 2, pp. 213–238, 2012.
[8] M. Carey, S. Ceri, P. Bernstein, U. Dayal, C. Faloutsos, J. Freytag, G. Gardarin, W. Jonker, V. Krishnamurthy, M. Neimat, *et al.*, *Data-Centric Systems and Applications.* Springer, 2006.
[9] G. Beskales, I. F. Ilyas, and L. Golab, "Sampling the repairs of functional dependency violations under hard constraints," *Proc. VLDB Endow.*, vol. 3, pp. 197–207, Sept. 2010.
[10] F. Chiang and R. J. Miller, "A unified model for data and constraint repair," in *2011 IEEE 27th International Conference on Data Engineering*, pp. 446–457, April 2011.
[11] T. Diallo, J.-M. Petit, and S. Servigne, "Discovering editing rules for data cleaning," in *Proceedings of AQB conference*, p. 40, 2012.
[12] W. Fan, J. Li, S. Ma, N. Tang, and W. Yu, "Towards certain fixes with editing rules and master data," *The VLDB Journal*, vol. 21, no. 2, pp. 213–238, 2012.
[13] G. Cong, W. Fan, F. Geerts, X. Jia, and S. Ma, "Improving data quality: Consistency and accuracy," in *Proceedings of the 33rd International Conference on Very Large Data Bases*, VLDB '07, pp. 315–326, VLDB Endowment, 2007.
[14] M. Yakout, A. K. Elmagarmid, J. Neville, M. Ouzzani, and I. F. Ilyas, "Guided data repair," *Proc. VLDB Endow.*, vol. 4, pp. 279–289, Feb. 2011.
[15] W. Fan, J. Li, S. Ma, N. Tang, and W. Yu, "Interaction between record matching and data repairing," in *Proceedings of the 2011 ACM SIGMOD International Conference on Management of Data*, SIGMOD '11, (New York, NY, USA), pp. 469–480, ACM, 2011.
[16] S. Kolahi and L. V. S. Lakshmanan, "On approximating optimum repairs for functional dependency violations," in *Proceedings of the 12th International Conference on Database Theory*, ICDT '09, (New York, NY, USA), pp. 53–62, ACM, 2009.
[17] X. Chu, I. F. Ilyas, and P. Papotti, "Holistic data cleaning: Putting violations into context," in *2013 IEEE 29th International Conference on Data Engineering (ICDE)*, pp. 458–469, April 2013.
[18] A. S. Asratian, T. M. Denley, and R. Häggkvist, *Bipartite graphs and their applications*, vol. 131. Cambridge University Press, 1998.
[19] G. Gallo, G. Longo, S. Pallottino, and S. Nguyen, "Directed hypergraphs and applications," *Discrete Appl. Math.*, vol. 42, pp. 177–201, Apr. 1993.
[20] G. Ausiello, P. G. Franciosa, and D. Frigioni, "Directed hypergraphs: Problems, algorithmic results, and a novel decremental approach," in *Proceedings of the 7th Italian Conference on Theoretical Computer Science*, ICTCS '01, (London, UK, UK), pp. 312–327, Springer-Verlag, 2001.
[21] N. J. Nilsson, *Problem-Solving Methods in Artificial Intelligence.* McGraw-Hill Pub. Co., 1971.
[22] C. zturan, "On finding hypercycles in chemical reaction networks," *Applied Mathematics Letters*, vol. 21, no. 9, pp. 881 – 884, 2008.
[23] E. Mendelson, *Introduction to Mathematical Logic.* CRC Press, 2015.
[24] G. Ausiello, A. D'Atri, and D. Saccà, "Graph algorithms for functional dependency manipulation," *J. ACM*, vol. 30, pp. 752–766, Oct. 1983.
[25] P. Cohn, *Basic Algebra: Groups, Rings and Fields.* Springer London, 2004.

Survey data and metadata modelling using document-oriented NoSQL

Lutfi Rahmatuti Maghfiroh, I Gusti Bagus Baskara Nugraha

School of Electrical Engineering and Informatics, Institut Teknologi Bandung, Bandung, Indonesia

lutfirm@stis.ac.id, baskara@stei.itb.ac.id

Abstract. Survey data that are collected from year to year have metadata change. However it need to be stored integratedly to get statistical data faster and easier. Data warehouse (DW) can be used to solve this limitation. However there is a change of variables in every period that can not be accommodated by DW. Traditional DW can not handle variable change via Slowly Changing Dimension (SCD). Previous research handle the change of variables in DW to manage metadata by using multiversion DW (MVDW). MVDW is designed using relational model. Some researches also found that developing nonrelational model in NoSQL database has reading time faster than the relational model. Therefore, we propose changes to metadata management by using NoSQL. This study proposes a model DW to manage change and algorithms to retrieve data with metadata changes. Evaluation of the proposed models and algorithms result in that database with the proposed design can retrieve data with metadata changes properly. This paper has contribution in comprehensive data analysis with metadata changes (especially data survey) in integrated storage.

1. Introduction

Data and information are important to an organization. Organization can use data analysis to make a better decision. Statistical data is usually collected through census and survey using questionnaire. To get up-to-date analysis, organization needs to integrate data storage to get statistical data faster and easier. However, data collected from year to year may have metadata change. One of ways to store the data is by using data warehouse.

Data warehouse (DW) is a subject-oriented, integrated, time-varying and nonvolatile data collection [1]. We can use data warehouse to store and organize the data so that we can use it for analyzing easily. DW can integrate all metadata and make it easy to maintain.

To manage the metadata change in DW, Kimball [2] introduce slowly changing dimension (SCD) concept to track the changes. However, it can only handle change of member attribute structure in multidimensional model. It cannot handle change of non-member attribute structure, for example, adding new attribute, removing attribute, etc. Body et al. in [3] introduce a temporal multidimensional model to handle evolution of DW structure. However, this approach only use one fact table to store all data version. This cause limitation in schema change and only support structural change in dimension table.

Bebel et al. [4] handle DW evolution using multiversion DW (MVDW). They provided format model to create MVDW. The study was continued in [5], which gave definitions to manage and perform queries in MVDW. Other researchers proposed query in MVDW [6] and use MVDW to manage metadata [7] [8]. The proposed model was implemented using a relational model of Oracle PL/SQL.

However, relational model has assumption that data are structured and stored in rows and columns of a table. The table has same column number and same column data type [9]. This model is not suitable for survey data with variable change. NoSQL (Not Only SQL) can be used to address those needs because NoSQL model is flexible. NoSQL database also provide a mechanism that is easier and simpler to store and retrieve data than relational database [10]. In [11] [12] [13] [14] are they proposed implementation of the DW and multidimensional models using NoSQL. However, these studies did not handle how to manage metadata changes or evolutions. In addition, there is no way of how to handle changes in concept and definition of a variable.

The remainder of this paper is organized as follows. Relevant work on data warehouses, metadata management and NoSQL is discussed in section 2. Section 3 presents motivational example and proposed design. Section 4 discuss about experiment and results to evaluate the propose model and algorithm to get data. Section 5 presents the conclusions and suggests directions for further research.

2. Related Research

Researchers in [4] handled changes in evolution of external data sources using Multiversion DW (MVDW) model. They provide a model format in making MVDW include the components of the model, operators to change scheme and operators to change structure of instances dimension. The model is basic for building a prototype system of MVDW. The study was continued in [5]. In [5], researchers gave definitions to manage and perform queries in MVDW. Based on previous research, the researchers then used MVDW to manage metadata [7]. In [7], researchers also propose multiversion data warehouse (MVDW) as a framework for dealing with schema and content of external data sources changes so that they could perform simulations and manage alternative business scenarios. MVDW can handle the limitations of SCD in a dimension instance change. DW schema version describes the structure of a specific time period. Version instance represents a collection of data that is described by the version of the scheme. When the name or definition of several attributes on the version level is changed, then the new level version will be created in the new version of the DW. Similarly, when the name or structure of a version of the facts changed, the new version of the facts will be made in the new version of the DW. Each version is only valid on a particular time frame. To perform a query of this proposal, researchers developed a traditional SQL select command. To apply metamodel that was proposed in [7], researchers built a prototype system using Java MVDW and Oracle PL/SQL, where the data and metadata stored in an Oracle database.

Researchers in [12] investigated use of NoSQL models for decision-making system (decision support system or DSS). In addition, researchers also proposed a rule mapping to transform multidimensional data model to logical document oriented models of NoSQL. In more detail researchers use multidimensional model implementation in column-based [13] and documents oriented [14] NoSQL database. Researchers also create an automatic rule mapping of multidimensional models to NoSQL [15]. However, there is no research about managing metadata change or evolution of data.

3. Design

3.1. Motivational Example

To collect data, we usually use a questionnaire which is designed especially for a certain survey. There are some approach in collecting data. Those are household, member of household, establishment, etc.

- Every survey has different characteristics caused by the difference of characteristics of data being collected. Even in the same survey, it may has different characteristics in the different periods.
- Another difference is same question may have different concepts and definitions. Figure 1 column 14 and column 15 and figure 2 column 412 and column 413 show these illustrations.

- The survey has a hierarchy of data, the data for households and facts for household member.
- Facts of households and household member consists of numeric data and data nonnumeric.
- The question on a survey period may not appear again in the next survey period. The question on a survey of certain period may not exist in the previous survey period. These changes can lead to inconsistencies, especially if it convert to the code of a question in a survey period that is used in another question of another survey period. Figure 1 column 10 and column 11 show these illustrations.
- The same question can have different variable name. One question may have a different code of the same variable. An example is sex variable for household member. In a survey, the sex variable is stored with the name "V" while in other surveys, it was stored as "VAR_V". Another example is the variable of ownership status of residence. In a survey, the variable is stored with the name "VAR_RES" whereas in other surveys saved as "STATUS_OF_RESIDENCE". Figure 1 column 5 and figure 2 column 407 show these illustrations.
- The same answer choice can have a different code. Different answer choice can have the same code. An example in a survey, a variable that may have the answer of "Yes" and "No" is coded with codes 1 and 2 whiles in other surveys, this answer has a code 1 and 5. Figure 1 column 8 and figure 2 column 408 show these illustrations.

Data storage should be able to accommodate these historical changes. If the problems or changes are not accommodated, then it can lead to inconsistency of data taken from DW. Thus to solve these problem, we propose DW NoSQL model to manage the changes and an algorithm to get data.

3.2. Proposed Design

To handle metadata changes, we propose DW NoSQL model to manage the changes and an algorithm to get data. The proposed model design use document-oriented NoSQL nonrelational model using method proposed in [12], [13], [14] and [11]. Model in Listing 1 and Listing 2 show model for data and metadata. Questions or variables are stored as columns and the answers are stored as column values.

Collection for store data is composed of the household data, the data of household members, the data module id associated with the survey, households' identity, and survey responses. Each model for data and metadata store in separate structure. Collection for store metadata contains id surveys, period of the survey, survey overview, code questions, information and id code of the code which becomes the parent id. Every new survey data, we will add new metadata to the parent id, which contains a code that correspond to code in metadata of previous year. Data and metadata, each of them is stored in a single document. Data is not normalized, because it will slow down time to read the data [11].

To get data with the metadata change from database, we proposed an algorithm. The algorithm consult the metadata model in order to track metadata historical changes. In the proposed algorithm, we only need consult metadata to track columns changes. Because NoSQL model is flexible, we only need one collection to store metadata with different columns name. New codes from metadata then are used in executed query. Figure 5(b) shows the proposed algorithm. Figure 5(a) shows comparison of algorithm from [8].

IOP Conf. Series: Journal of Physics: Conf. Series **971** (2018) 012030 doi:10.1088/1742-6596/971/1/012030

VI.A. HOUSEHOLD MEMBER INFORMATION																
No	Household member (HM) name (People who live and eat in the household, including adult, children and toddler)	Relation with the head of household (code)	Gender 1. M 2. F	Age (year)	Marital status (code)	Have you become crime victim in the past one year? (code)	If Column 7 = 1 to 6, is it reported to police? 1. Yes 2. No	Travel frequency *) in the past 3 calendar months? If no travel, use code "00"	If traveling (Column 9 ≠ 0)		Household member age 0-17 years		Household member age 0-6 years		Household member age 3-6 years or Column 14 = 1 or 2	
									Main purpose of the last trip (code)	Main destination province of the last trip (code)	Have birth certificate from the Registrar Office? Can be shown? (code)	If Column 12 = 3 or 4, what is the main reason? (Do not read out the answer!) (code)	Ever/ on pre-school education? 1. Yes, once 2. Yes, currently attending 3. No	If Column 14 = 1 or 2, type of pre-school education (code)	Attended pre-school education in the past 3 months? 1. Yes 2. No	If Column 16 = 1, transportation facility used to go to school (code)
(1)	(2)	(3)	(4)	(5)	(6)	(7)	(8)	(9)	(10)	(11)	(12)	(13)	(14)	(15)	(16)	(17)

Figure 1. Examples of variable in questionnaire in survey #1.

BLOCK IV.A DEMOGRAPHY, BIRTH CERTIFICATE, AND EARLY CHILD EDUCATION (ECD)												
No.	**NAME OF HOUSEHOLD MEMBERS** *(Write down the names of all members who usually live and eat in this household starting from the head of household, spouse, unmarried children, married children, children in law, grandchildren, parents/parents in law, other family, servant, etc.)* ***Make sure that all members are registered and no one is overlooked***	WHAT IS RELATIONSHIP OF *(name)* WITH THE HEAD OF HOUSEHOLD **(Code)**	WHAT IS MARITAL STATUS OF *(name)* 1. Single 2. Married 3. Divorced 4. Widowed	IS *(name)* MALE OR FEMALE 1. Male 2. Female	WHAT IS *(name)* DATE OF BIRTH? Month year	HOW OLD IS *(name)* *Record in completed year. Age should be written, if ≥ 97 year, write 97*	*If marital status is married* ***(404 = 2)*** DOES SPOUSE OF *[name]* USUALLY LIVE IN THIS HOUSEHOLD? 1. Yes 5. No	*If ever married* ***(404 = 2, or 3, or 4)*** AT WHAT AGE *(name)* DID THE FIRST MARRIED?	**For household member age 0 - 17 years** DOES *(name)* HAVE A BIRTH CERTIFICATE? FROM THE CIVIL REGISTRATION OFFICE? MAY I SEE IT? **(Code)**	*If does not have a birth certificate* ***(410 = 5)*** WHAT IS THE MAIN REASON **(Code)**	**For household member age 0 - 10 years** HAS *(name)* EVER ATTENDED PRE-SCHOOL EDUCATION? **(Code)**	*If yes/still* ***(412 = 1, 2 or 3)*** WHAT TYPE OF PRESCHOOL EDUCATION **(Code)**
401	402	403	404	405	406	407	408	409	410	411	412	413

Figure 2. Examples of variable in questionnaire in survey #2.

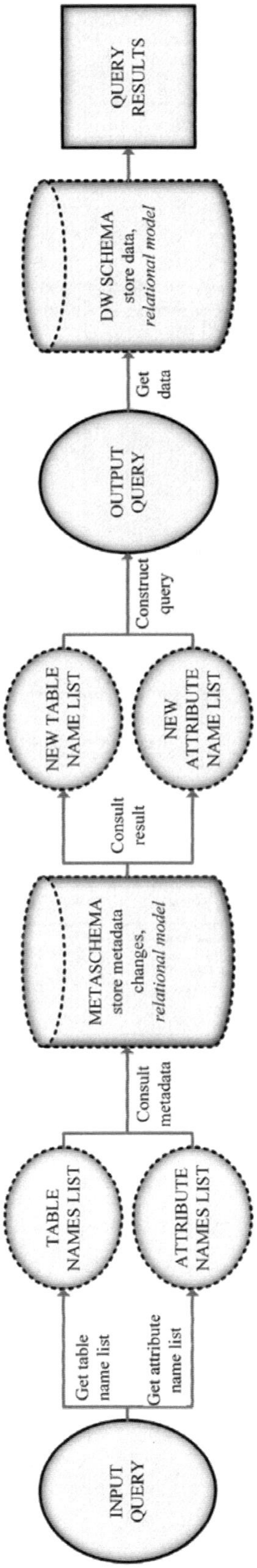

Figure 3. MVDW Model design.

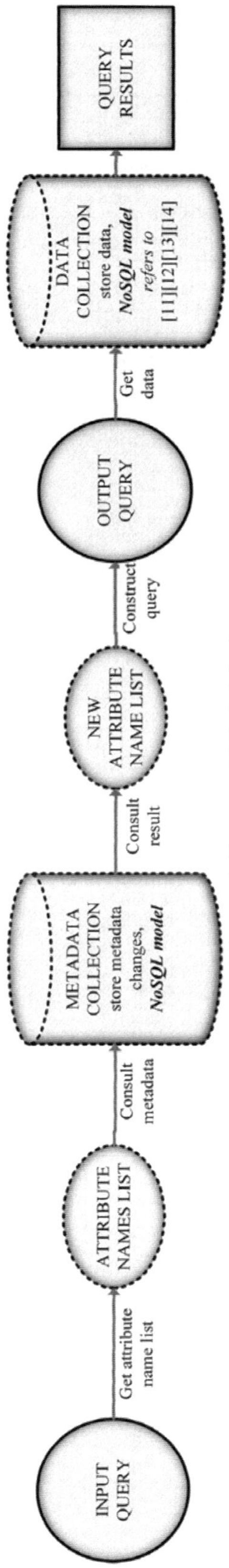

Figure 4. Proposed Model design.

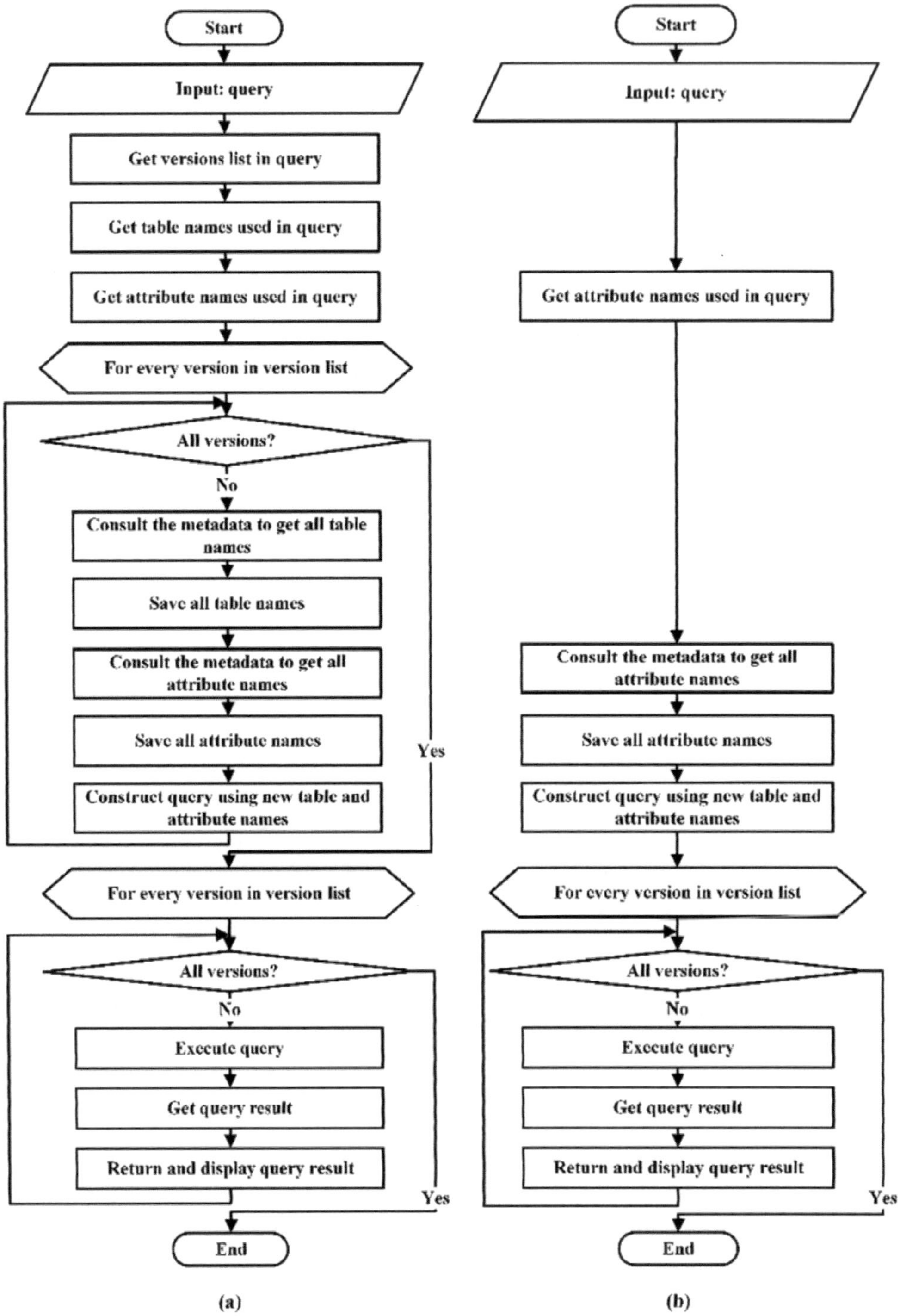

Figure 5. (a) Flowchart of algorithm from [8] (b) Flowchart of proposed algorithm.

Listing 1. NoSQL data model

```
{
"_id" :...,
"province_code" : ...,
"province_name" : ...,
"district_code" : ...,
"district_name" : ...,
....
"subdistrict_code" : ...,
"subdistrict_name" : ...,
"village_code" : ...,
"village_name" : ...,
"HH_code" : ...,
"HH_member_code" : ...,
"VAR_1" : ...,
"VAR_2" : ...,
"VAR_3" : ...,
"VAR_4" : ...,
...,
"VAR_N" : ...,
"survey_id" : ...
}
```

Listing 2. NoSQL metadata model

```
{
"_id" : ...,
"id" : ...,
"survey_name" : : ...,
"month" : ...,
"year" : ...,
"code" : ...,
"definition" : ...,
"parent_id" : ...,
"survey_id" : ...
}
```

4. Experiments

4.1. Experiments Setup

NoSQL models is implemented by using document-oriented NoSQL because it is faster in reading data (query) [12] [16] [17] [18] and we use MongoDB is popular document-oriented system [11]. Data and metadata are stored in a document (database), where each scheme is stored in a single collection. For comparison, we use relational model in Figure 6 to be implemented in MVDW.

For evaluation, we use MongoDB version 3.2 and PostgreSQL 9.5. Hardware that we used is Windows 10 64bit, 4GB RAM, 500GB, Intel® Core™ i3-3217U CPU @1.80GHz. To running the queries, the code is implemented in the Java programming language and Netbeans IDE 8.1. Evaluation is conducted using data Labor Force Survey (Sakernas) [19] 2010, 2012, and 2015 taken randomly.

We compare the result of executed query in 9 scenario to evaluate the proposed model and algorithm. We use 100 rows of data, which consists of 34 rows of data in 2010, 33 rows of the data in 2012, and 33 rows of the data in 2015. Each period has some variable metadata changes.

- Query 1: show the code of the district or city, the name of the district or city, the relationship of household member to head of household (KRT) and gender
- Query 2: show the code of the district or city, the name of the district or city, a relationship of household member to KRT and gender
- Query 3: show the code of the district or city, a relationship of household member to KRT and age of male population
- Query 4: show the code of the district or city, a relationship of household member to KRT and age of the male population, sorted by district or city code
- Query 5: show the code of the district or city, the name of the district or city, a relationship of household member to KRT and last educational level
- Query 6: show the code of the district or city, a relationship of household member to KRT and the last education level of males population
- Query 7: show the code of the district or city, the name of the district or city, a relationship of household member to KRT and the last education level of males population, sorted by code of the district or city

- Query 8: show the number of people grouped by district or city code and name
- Query 9: show the number of people grouped by district or city code and sex

To evaluate that proposed model and algorithm design work properly, we do some evaluation. First we do functionality evaluation to evaluate that proposed model and algorithm design work properly. In this evaluation, we will compare query results running in the proposed system and the query running without metadata management. Then we evaluate query running on proposed model and algorithm. This evaluation is expected to have result that the proposed model has faster query running than others. However, in this paper we only present the first evaluation.

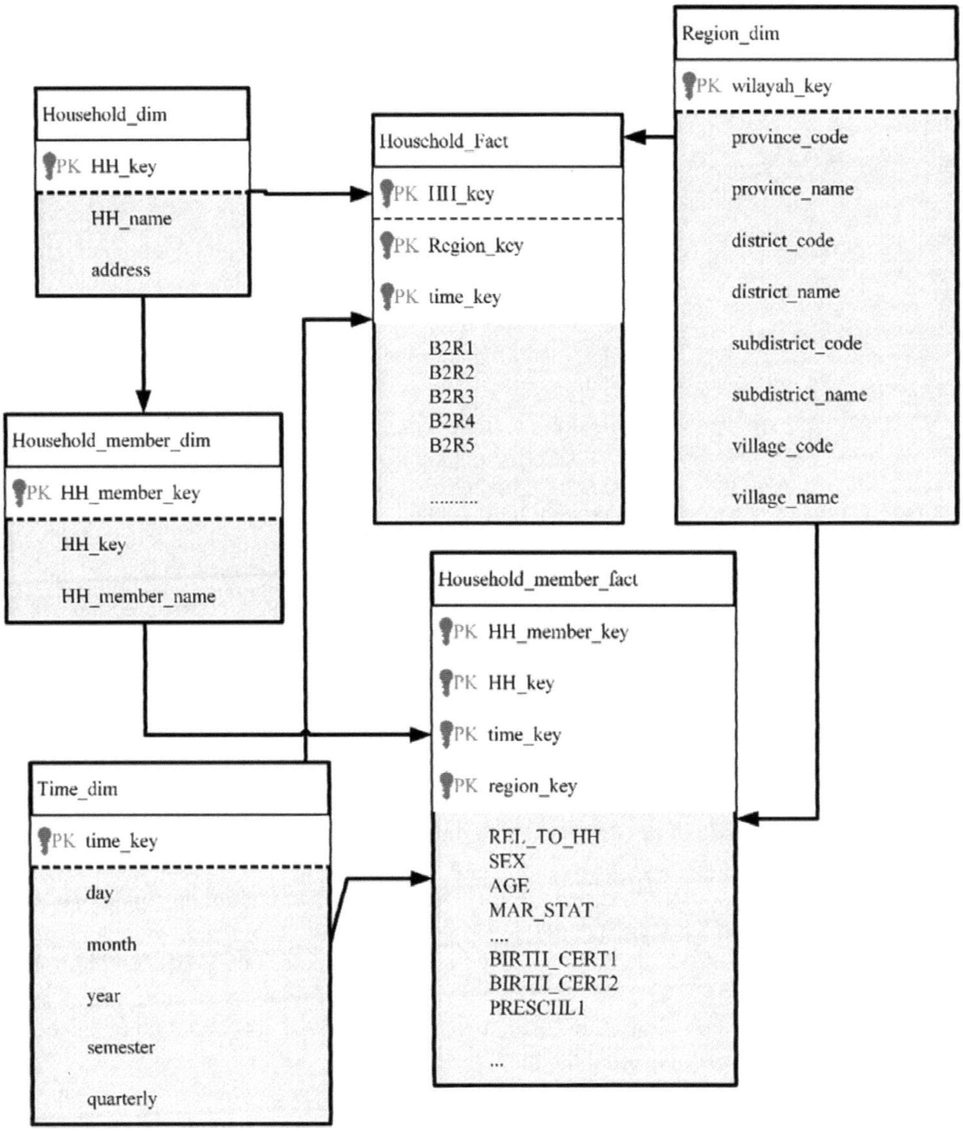

Figure 6. DW schema to compare.

IOP Conf. Series: Journal of Physics: Conf. Series **971** (2018) 012030 doi:10.1088/1742-6596/971/1/012030

Table 1. Functionality evaluation results.

Query	Expected result			Query result of proposed design			(1)			(2)			(3)			Query work properly?			
																Proposed design	(1)	(2)	(3)
1	100 rows			100 rows			34 rows			33 rows			33 rows			Yes	No	No	No
2	100 rows			100 rows			34 rows			33 rows			33 rows			Yes	No	No	No
3	52 rows			52 rows			***36 rows***			***36 rows***			16 rows			Yes	No	No	No
4	52 rows			52 rows			***36 rows***			***36 rows***			16 rows			Yes	No	No	No
5	100 rows			100 rows			34 rows			33 rows			33 rows			Yes	No	No	No
6	52 rows			52 rows			***36 rows***			***36 rows***			16 rows			Yes	No	No	No
7	100 rows			100 rows			34 rows			33 rows			33 rows			Yes	No	No	No
8	district code	district name	Count	district code	district name	Count	district code	district name	Count	district code	district name	Count	district code	district name	Count	Yes	No	No	No
	01	MTB	100	01	MTB	100	01	MTB	34	01	MTB	33	01	MTB	33				
9	Sex	district code	Count	Sex	district code	Count	***Sex***	***district code***	***Count***	***Sex***	***district code***	***Count***	Sex	district code	Count	Yes	No	No	No
	Male	1	52	Male	1	52	***Male***	***1***	***36***	***Male***	***1***	***36***	Male	1	16				
	Female	1	48	Female	1	48	***Female***	***1***	***31***	***Female***	***1***	***31***	Female	1	17				

Note:

(1) Query results using variables of 2010 survey and without metadata management

(2) Query results using variables of 2012 survey and without metadata management

(3) Query results using variables of 2015 survey and without metadata management

Query	Expected result	Query result of proposed design	(1)	(2)	(3)
1	100 rows	100 rows	34 rows	33 rows	33 rows

Figure 7. Result of Query 1 from Table 1.

Query	Expected result	Query result of proposed design	(1)	(2)	(3)
3	52 rows	52 rows	***36 rows***	***36 rows***	16 rows

Figure 8. Result of Query 3 from Table 1.

4.2. Experiments Results on Functionality Evaluation

To evaluate that proposed model and algorithm design work properly, we do functionality evaluation. The proposed design works properly if running query results expected data either there is metadata change or not. Table I shows functionality evaluation results. Columns in the table are number of query, expected results, result of a query with metadata management, the results with the variables in 2010 without metadata management, the results with the variables in 2012 without metadata management, and the results with the variables in 2015 without metadata management.

We can see from the table that the results of a query with metadata and without metadata have different results. All running queries with proposed design result data as expected, while the queries that run without metadata management have wrong results. With metadata management, query not only get data with same variable with same code but also data of the same variables with different code but without metadata management, running query will retrieve data that corresponds to the code of mentioned variables only. For example, the queries that mentions gender variable (coded as "V"), with metadata management, data that have a variable name corresponding to the variable "V" (e.g. variable VAR_V) can be also obtained but without metadata management, queries only retrieve data that have variable name "V". This is reason that explain why query result of column (1), (2) and (3) have not result as expected. The table shows that the query 1, 2, 5, 7, and 8 have different results because there are no similar variable across the three data and there is no metadata management. This is contrast to the results of the query 3, 4, 6, and 9. The queries generate same result of the data in 2010 and 2012 because the data in 2010 and 2012 have the same variable codes used in the queries, although query run without metadata management.

Query 1 (Figure 7) uses three variables, that is district code (coded as X in 2010, X in 2012, and VAR_X in 2015), district name (coded as Y in 2010, VAR_Y in 2012, and VAR_Y1 in 2015) and gender (coded as V in 2010, V in 2012 and VAR_V in 2015). Without metadata management, running query that use variable code of 2010 only get data that have all same variable of variable code X for district code, Y for district name, and V for gender, as well as running query that use variable code of 2012 and 2015. This query results 34 rows of data with variable code in 2010, 33 rows of data with variable code in 2012 and 33 rows of data with variable code in 2015 from 100 rows data, not as expected result. With metadata management, running query can get data with all variable code correspond to district code (X in 2010, X in 2012, and VAR_X in 2015), district name (Y in 2010, VAR_Y in 2012, and VAR_Y1 in 2015) and V (V in 2010, V in 2012 and VAR_V in 2015). This query results 100 rows out of 100 rows data, as expected result (Figure 7). This explanation is similar for query 2, 5, 7 and 8.

Query 3 (Figure 8) uses four variables, that is district code (coded as X in 2010, X in 2012, and VAR_X in 2015), KRT (coded as A in 2010, A in 2012, and VAR_A in 2015), age (coded as C in 2010, C in 2012 and VAR_C in 2015) and gender (coded as V in 2010, V in 2012 and VAR_V in 2015). Without metadata management, running query that use variable code of 2010 only get data that have all same variable of variable code X for district code, A for KRT, C for age and V for gender, as well as running query that use variable code of 2012 and 2015. Running query with variable of data in 2010 and 2012 have same result because all variable have same variable codes. This query results 36 rows of data with variable code in 2010, 36 rows of data with variable code in 2012 and 16 rows of data with variable code in 2015 from 52 rows data, not as expected result. Result from query that use variable code in 2010 is data from 2010 and 2012, as well as query that use variable code in 2012. With metadata management, running query can get data with all variable code correspond to variable of district code (X in 2010, X in 2012, and VAR_X in 2015), KRT (A in 2010, A in 2012, and VAR_A in 2015), age (C in 2010, C in 2012 and VAR_C in 2015) and V (V in 2010, V in 2012 and VAR_V in 2015). This query results 52 rows out of 100 rows of data, as expected result (Figure 8). This explanation is similar for query 4, 6, and 9.

Based on the functionality evaluation, we can conclude that proposed method has expected results. This means that the proposed method is better than other method that does not use metadata management in retrieve data. This means also that the proposed method can solve the problems.

5. Conclusion

To handle metadata change in data warehouse (DW), we propose model and an algorithm. The proposed model is designed using NoSQL models. The proposed model consists of a data model and metadata models without normalization. To get the data with metadata changes, we propose an algorithm. The algorithm checks the corresponding metadata changes to the metadata models.

The proposed model is evaluated by comparing models without and with metadata models. The evaluation results show that the model with metadata generate data corresponding properly. This is what cannot be done by the model without the metadata. Next we will evaluate comparison of query running time of proposed model and algorithm to MVDW model.

References

[1] W. H. Inmon, *Building the Data Warehouse*, 4th ed. New York, NY, USA.: John Wiley & Sons, Inc, 2005.

[2] R. Kimball and M. Ross, *The Data Warehouse Toolkit: The Definitive Guide to Dimensional Modeling*, 3rd ed. Indianapolis: Wiley, Jul. 2013.

[3] M. Body et al., "A Multidimensional and Multiversion Structure for OLAP Applications," in *Proceedings of the 5th ACM International Workshop on Data Warehousing and OLAP*, ser. DOLAP '02. New York, NY, USA: ACM, 2002, pp. 1–6.

[4] B. Be̜bel, Z. Królikowski, and R. Wrembel, "Formal approach to modelling a multiversion data warehouse," in *Bulletin of the Polish Academy of Sciences, Technical Sciences*, vol. **54**, 2006, pp. 51–62.

[5] R. Wrembel and T. Morzy, "Managing and Querying Versions of Multiversion Data Warehouse," in *Advances in Database Technology EDBT 2006*, ser. Lecture Notes in Computer Science, Y. Ioannidis et al., Eds. Springer Berlin Heidelberg, Mar. 2006, no. 3896, pp. 1121–1124.

[6] T. Morzy and R. Wrembel, "On Querying Versions of Multiversion Data Warehouse," in *Proceedings of the 7th ACM International Workshop on Data Warehousing and OLAP*, ser. DOLAP '04. New York, NY, USA: ACM, 2004, pp. 92–101.

[7] R. Wrembel and B. Be̜bel, "Metadata Management in a Multiversion Data Warehouse," in *Proceedings of the 2005 OTM Confederated International Conference on On the Move to Meaningful Internet Systems: CoopIS, COA, and ODBASE* - Volume **Part II**, ser. OTM'05. Berlin, Heidelberg: Springer-Verlag, 2005, pp. 1347–1364.

[8] ——, "Metadata Management in a Multiversion Data Warehouse," in *Journal on Data Semantics VIII*, ser. Lecture Notes in Computer Science, S. Spaccapietra et al., Eds. Springer Berlin Heidelberg, 2007, no. 4380, pp. 118–157.

[9] Z. Parker, S. Poe, and S. V. Vrbsky, "Comparing NoSQL MongoDB to an SQL DB," in *Proceedings of the 51st ACM Southeast Conference*, ser. ACMSE '13. New York, NY, USA: ACM, 2013, pp. 5:1–5:6.

[10] M. O. Fitri, "Trend Penggunaan Nosql Untuk Basis Data Non Relasional," Fakultas Sains dan Teknologi UIN ALauddin, vol. **7**, No 1 (2013): Januari, 2013.

[11] M. Chevalier et al., "Document-oriented Models for Data Warehouses - NoSQL Document-oriented for Data Warehouses," in *Proceedings of the 18th International Conference on Enterprise Information Systems*. Rome, Italy: SCITEPRESS - Science and and Technology Publications, 2016, pp. 142–149.

[12] ——, "Implementing Multidimensional Data Warehouses into NoSQL," in *Proceedings of the 17th International Conference on Enterprise Information Systems*. Barcelona, Spain: SCITEPRESS - Science and and Technology Publications, 2015, pp. 172–183.

[13] ——, "Implementation of Multidimensional Databases with DocumentOriented NoSQL," in *Big Data Analytics and Knowledge Discovery*, ser. Lecture Notes in Computer Science, S. Madria

and T. Hara, Eds. Springer International Publishing, Sep. 2015, no. 9263, pp. 379–390.

[14] ——, "Implementation of Multidimensional Databases in ColumnOriented NoSQL Systems," in *Advances in Databases and Information Systems*, ser. Lecture Notes in Computer Science, M. Tadeusz, P. Valduriez, and L. Bellatreche, Eds. Springer International Publishing, Sep. 2015, pp. 79–91.

[15] ——, "How Can We Implement a Multidimensional Data Warehouse Using NoSQL?" in *Enterprise Information Systems*, ser. Lecture Notes in Business Information Processing, S. Hammoudi et al., Eds. Springer International Publishing, Apr. 2015, no. 241, pp. 108–130.

[16] R. Yangui, A. Nabli, and F. Gargouri, "Automatic Transformation of Data Warehouse Schema to NoSQL Data Base: Comparative Study," *Procedia Computer Science*, vol. **96**, pp. 255–264, 2016.

[17] L. Bonnet et al., "Reduce, You Say: What NoSQL Can Do for Data Aggregation and BI in Large Repositories," in *2011 22nd International Workshop on Database and Expert Systems Applications*, Aug. 2011, pp. 483–488.

[18] J. Duda, "Business Intelligence And Nosql Databases," *Information Systems in Management Katedra Informatyki Szkoła Główna Gospodarstwa Wiejskiego w Warszawie*, vol. **1**, no. 1, pp. 25–37, 2012.

[19] Subdirektorat Rujukan Statistik Badan Pusat Statistik, "Survei Angkatan Kerja Nasional," Feb. 2017. [Online]. Available: https://sirusa.bps.go.id/index.php?r=sd/view&kd=5&th=2011

IOP Conf. Series: Journal of Physics: Conf. Series **971** (2018) 012031 doi:10.1088/1742-6596/971/1/012031

Ontology to relational database transformation for web application development and maintenance

Kamal Mahmudi, M M Inggriani Liem, Saiful Akbar

School of Electrical Engineering and Informatics, Institut Teknologi Bandung
Jl. Ganesha 10 Bandung 40132, Indonesia
23514050@std.stei.itb.ac.id, inge@informatika.org, saiful@informatika.org

Abstract. Ontology is used as knowledge representation while database is used as facts recorder in a KMS (Knowledge Management System). In most applications, data are managed in a database system and updated through the application and then they are transformed to knowledge as needed. Once a domain conceptor defines the knowledge in the ontology, application and database can be generated from the ontology. Most existing frameworks generate application from its database. In this research, ontology is used for generating the application. As the data are updated through the application, a mechanism is designed to trigger an update to the ontology so that the application can be rebuilt based on the newest ontology. By this approach, a knowledge engineer has a full flexibility to renew the application based on the latest ontology without dependency to a software developer. In many cases, the concept needs to be updated when the data changed. The framework is built and tested in a spring java environment. A case study was conducted to proof the concepts. **Keywords**—ontology to database transformation, converter, application generator.

1. Introduction

The multilayered architecture approach is one of best practices in application development. Separating domain, infrastructure and user interface layers give us a cleaner and easier application to be maintained. Each layer has its responsibility and responds to change with different need and speed. Separating domain layer from other layers also helps a reengineering process in the future development [1].

Ontology as explicit specification of conceptualization [2] can be used as domain layer representation. Ontology has been used to solve many problems with incomplete, complex or evolved knowledge representation. Some researches mapped knowledge domain represented in an ontology to an object model represented as a database. But, methods, procedures and tools for application developers to develop and maintain ontology-based application are still limited [3]. Ontology as a basis of a domain model is important in the application development process, and it can be combined with relational database [4]. As another component of application, ontology is also prone to evolution. Especially as the backbone of the application, change in the conceptual model will lead to change in the requirement, and then to change the model domain [5].

This research harvests methods and tools for data transformation to ontology and code generators in order to create an updated system that shall generate a web application from the ontology. The system shall be able to adapt to a new conceptual model as well as the last updated data.

2. Problem Statement and Proposed Solution

Database to ontology transformation and ontology to database transformation have been reported in many researches. Some of them used a migration rule and mapping [4][8][9][10][11]. However, the concept evolves and the development of ontology can still occur even after it's used to build the application. Hence, there is a need to implement a system that is not only able to transform ontology to relational database, but also be able to generate the application from the ontology. And when a change happened to the database or ontology, the system shall synchronize the change and regenerate the application.

This research proposes a set of tools to transform ontology component for database and vice versa, generate application code, build and run the application. All processes are being done automatically within the framework, by pushing a button. The main function of the generated application is to manage the data that ontology has. But since the generated source code is editable, developer can modify it to add his sophisticated functional that he needs.

3. Related Works

3.1. Ontology

Ontology can be defined as explicit specification of conceptualization [2]. Ontology in a program can be illustrated as the definition of all representation of that program including classes and relationships. Ontology has no standard format, it can have different structures depend on the language it used.

Generally, an ontology consists of classes, attributes, individual/instances, relations, rules, and axioms [6]. These components can be coded in XML based ontology language, such as DARPA Agent Mark-up Language + Ontology Interchange Language (DAML+OIL), Resource Description Framework (RDF), or Web Ontology Language (OWL). Classes can be defined and organized in a hierarchical structure using OWL. OWL consists of individuals, properties, and classes [7].

The individual represents object from the domain of discourse. The individual is also known as instance, or instance of a class. The properties represent the relationship that an individual has. Properties can be object or datatype properties. Object properties represent the relationship between individuals while datatype properties represent the relationship between individual and XML datatype (integer, boolean, string, etc.)

The properties can be either functional, transitive, or symmetric. Functional property means that each object has only one value. An example of this property is HasBirthMother which every human only has one birth mother.

A property can be a transitive property, that is if property P connects individual A and B, and connects individual B and C, then P also connects individual A and C. HasAncestor is one of many examples of this type of property. A symmetric property connects individual A to B, the same way it connects individual B to A. A perfect example of this relationship is HasSibling relationship.

Some researches have proposed algorithms, mappings, and rules for transforming an ontology to a relational database [4][8] and from relational databases to ontology [9][10][11]. Some studies combine both technologies, for example, using a relational database for storing ontology [12], using a relational database as data source in reasoning [13].

3.2. Relational Database

A database is a set of symbols which organized and machine readable [14]. A relational database organizes the symbols in a collection of relations. Its consists of relation, attribute, tuple, and key [15].

A relational database uses a two-dimensional table to represent entities of application domain, such as Customer or Order, or to represent the relationship between entities. This two-dimensional table is called a relation. A tuple in a relational database represents an instance of an entity. Each row (tuple) of the Customer table represents a customer.

An entity has properties representing the attributes or fields of the table. For example, a Customer table may have Name and Address attributes. The domain of an attribute is a range of possible values that the attribute may have. Each table in a relational database has one or more fields to identify each of rows on that table called key. The key may be used on another table as a foreign key.

To be a self-described entity, database must contain metadata, a description of the data held in the database. This description is also called schema. With this schema, a program can determine the format, the size and the type of the data. For example, the schema would show that the first 20 bytes of the stored records are the name of the customer, and the next 30 bytes of the records are his address.

3.3. Web Based Application

A web based application is a client-server computer program in which the client runs in a web browser. The multilayered architecture approach appropriates to be used in a web application development. By separating the domain layer from the infrastructure layer and the user interface layer, the application development is easier to maintain and the codes are cleaner. Each layer responds to change independently from others. Separating domain layer from other layers also helps the re-engineering process in the future development [1].

A three-tier architecture separates presentation layers, business layer and data layer [16]. This three-tier architecture, then evolved to various architectures depends on the needs of the application. Figure 1 is an example of a multilayered architecture where the business layer is split into an application layer and a domain layer. The application layer handles tasks which triggered by the presentation layer. The domain layer handles business logic which can be used in different presentation layers or even different applications.

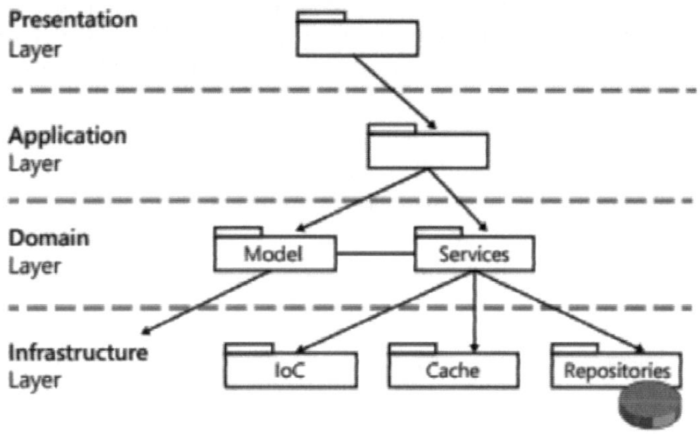

Figure 1. An example of multilayered architecture [16].

Ontology can be used in the application development phase or during the application usage [17]. Zviedries used ontology in the development phase to generate an application for information system [18]. In that research, he created generic component for information system, then the information system will load the ontology and adjust the information system to the ontology specification. There is no editable source code for a developer to edit, because all specifications of his application were coded in the ontology, so a developer needs to have knowledge in ontology to modify the way information interacts.

On the other hand, Kremen used ontology as the persistence layer in his multilayered application [3]. He builds all application components to process data based on Open World Assumption where the data and the knowledge may not complete and will be considered as unknown information rather than invalid information.

In our previous research [19], we were able to generate a knowledge repository (an application) from a given ontology. In that research, the ontology was used once to generate repository components, such as relational database, data editor, and data API. Every tool was used one-by-one manually to perform certain function such as repository component generation, database transformation, etc. The repository was then composed of those components, and operated based on its database. After the repository was being built, it was completely independent from the origin ontology. Changes in the repository will not be synchronized to the ontology and vice versa.

3.4. Application Generator

A large-scale application may consist of a set of similar modules. Developing an application with similar modules may take time and prone to error since the programmer may reckless in copy-paste certain codes. With an application generator, a programmer only needs to specify the definition of each module.

An application generator at least does one of these processes [20]: (a.) Receives DSL code, parse, and generate code in the target language. (b.) Receives the templates and the content of each template, and then generate the code by filling the template using the given content. (c.) Generates the code algorithmically.

3.5. Spring Framework

Spring Framework is a Java platform with some ready-to use, comprehensive infrastructures to develop a Java application [21]. With Spring Framework infrastructure, a developer can focus on the development of the core of the application. Spring Framework can be combined with several extensions to build a web based application on the Java EE Platform.

Spring MVC Web Framework consists of dispatcher servlet, controller, model, and view [22]. An HTTP request is processed by a dispatcher servlet to determine which tasks will be executed by determining the URL pattern, parameters, HTTP headers, and other factors. Servlet maintains lifecycle of HTTP request, determines and executes controller class, receives and returns the model and renders the suitable view.

Spring Roo is created to increase the productivity of programmers in a Spring environment. The Spring Roo generates the code based on the given commands from a terminal. It can also generate a model and the controller classes based on class diagram that have been designed on Spring Roo notation. Spring Roo is also able to reverse-engineer an existed database, to generate codes and modules. The results are Java code and not limited to Spring (Spring Security and Spring Web Flow). The codes are also compatible with other Java technology like Java Persistence API (JPA), Java Server Pages, Apache Maven, and AspectJ.

Spring Roo uses a "Convention over Configuration" paradigm: every class in the class diagram will be associated with a model class, a controller class, and a table in a database [23]. However, Spring Roo only generates the code and the database scheme. It cannot be used to populate data in the database

4. Analysis and Solution

The following section starts with a brief discussion on database to ontology migration and vise-versa, and the generation of web based application

4.1. Metadata and Data

In this research, a mapping rule between the ontology components and the relational database components is defined as shown in table 1. This mapping is used for the ontology to relational database transformation. We combine the Vysnauskas's rules [4] and the Spring Framework characteristic.Vysnauskas's rules are used to generate the tables from the ontology. The Spring Roo generates the models and controller classes for the Spring application.

In Description Logics (DL) community, ontology components are categorized into Terminological Box (TBox)and Assertional Box (ABox). TBox contains assertions about individuals, i.e. OWL facts such as type, property-value, equality or inequality assertions. ABox contains axioms about classes, i.e. OWL axioms such as subclass, equivalent class or disjointness axioms [24]. TBox component is used to generate the tables in the database and generating the model and controller classes from ontology.

Table 1. Mapping between ontology and relational database

No	Ontology	Relational Database	Type
1	Class	Table	TBox
2	Subclass	Table	TBox
3	Object properties	Relationship table	TBox+ABox
4	Data type properties	Field/column	TBox+ABox
5	Constraint	Datatype, constraint	TBox
6	Individual	Row	ABox

The classes and the subclass in an ontology are similar to tables in a database. Object property in the ontology connects an individual to another individual which is similar to the relationship table in a relational database. A functional object property is transformed into a field in the relational database (TBox) and defined as a foreign key. The value of functional object properties is transformed as a row of data in relationship table (ABox). The following equation is used to check the result of the transformation of classes in the ontology to tables in the database:

$$\mathrm{OKL} = \mathrm{BTB} - \mathrm{BTR} \tag{1}$$

where:
OKL : the number of classes in the ontology
BTB : the number of tables in the relational database
BTR : the number of relationship tables in the relational database

An individual in ontology becomes a row of data in a relational database. But a relationship table in a relational database may also have row(s) of data. The following equation is used to check the transformation of individuals in the ontology to rows of data in the relational database:

$$\mathrm{OIN} = \mathrm{BBD} - \mathrm{BBR} \tag{2}$$

where:
OIN : the number of individuals in the ontology
BBD : the number of rows of data in the relational database
BBR : the number of rows of data in the relationship tables in the relational database

Each knowledge engineer may design a different ontology for the same concept. This may happen because of different requirement in reasoning or other specific requirement that engineer faces. For example, in the pizza ontology [25], names of countries are being represented as individuals while names of pizzas are being represented as subclasses of Pizza class. The system should support a variation from default rules. In our tools, a developer can configure the transformation by ignoring some classes, considering a subclass as a row of a table, or transforms the subclasses to enumerations or constants.

A modified version of the class transformation algorithm from Vysnauskas's algorithm can be seen in figure 2. In our algorithm, each class, is queued, flagged according to the configuration (if there's any) while the transformation is done.

IOP Conf. Series: Journal of Physics: Conf. Series **971** (2018) 012031 doi:10.1088/1742-6596/971/1/012031

The subclasses of that class have the same flag as their superclass's flag. This modified algorithm changes the equation (1) and the equation (2). Since some classes in an ontology maybe ignored, or may be transformed as row(s) of data in a relational database, the equation (1) is then modified into

$$\mathrm{OKL} - \mathrm{OIG} - \mathrm{ODA} - \mathrm{OKE} - \mathrm{OSE} = \mathrm{BTB} - \mathrm{BTR} \tag{3}$$

where:

OKL : the number of classes in the ontology

OIG : the number of classes in the ontology which will be ignored during transformation

ODA : the number of subclasses in the ontology which will be considered as rows of data in the relational database

OKE : the number of classes in the ontology which will be considered as enumeration during transformation

OKT : the number of classes in the ontology that are not part of OIG, ODA, and OKE

OSE : the number of subclasses of OKT

and the equation (2) is modified into:

$$\mathrm{ODA} + \mathrm{OIN} = \mathrm{BBD} - \mathrm{BBR} \tag{4}$$

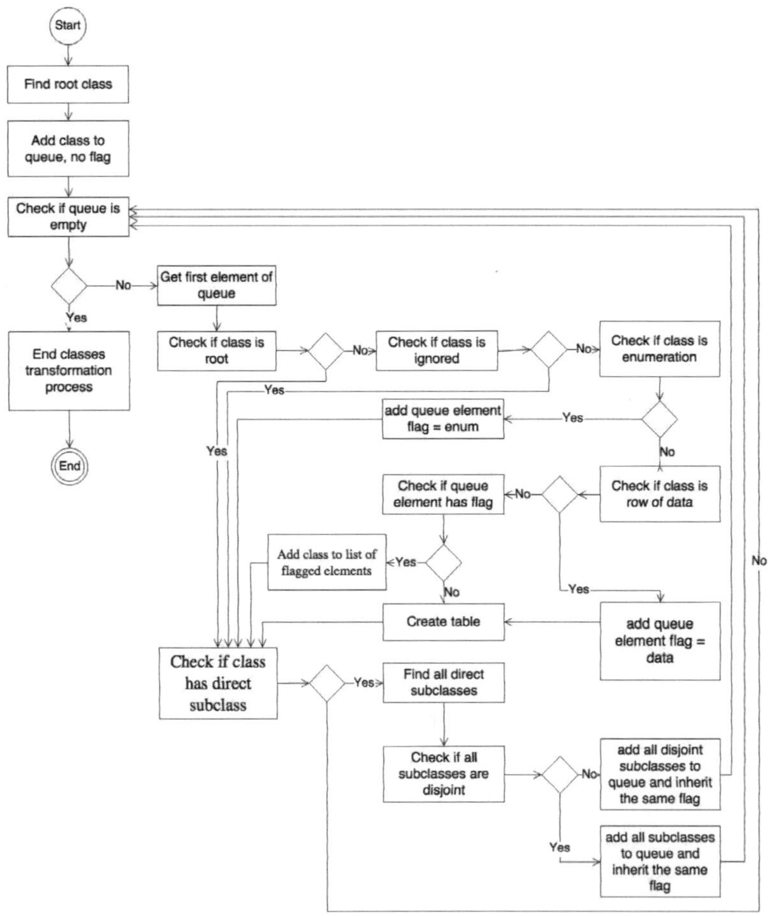

Figure 2. Class Transformation Algorithm.

IOP Conf. Series: Journal of Physics: Conf. Series **971** (2018) 012031 doi:10.1088/1742-6596/971/1/012031

4.2. Change's Impact

Change can happen in the rows of data or the domain model itself. The usage of web application leads to relational database's change. In this research, we restrict to handle the change in the data, not in the database schema.

In the beginning of the life cycle of the application, the knowledge engineer represents the domain concept in the ontology. The MVC components and relational database will be generated by the system from this ontology considering the configuration rule. The application is ready to be used.

When the knowledge engineer needs to update the ontology, the application and the database must be changed accordingly. On the other side, the data in the database is changed while the application is used. This change must be detected and added to ontology after the data are sent to the ontology. TBox component will be adjusted according to the change, as ABox component from old application will be deleted from the database and repopulated.

4.3. Code Generation

The codes (MVC components) are generated by Spring Roo upon accepting domain model's description in DSL. The forms and data viewers have similar themes and interfaces. The displays of tables' fields are generated conformed to the description in the DSL. This DSL is generated by TBox and ABox extractor using transformation mapping (table 1) and algorithm (figure 2).

4.4. System Automation

The automation mechanism generates a web application following a workflow as illustrated in figure 3. At first, a knowledge engineer creates or provides an ontology. Then this ontology with an appropriate configuration file are loaded into the system. At the next step, TBox components from the ontology are extracted. Then the extracted components will be used to generate the code and to build the web application. After the code and database schema generation, ABox components from the ontology are extracted and imported to the database.

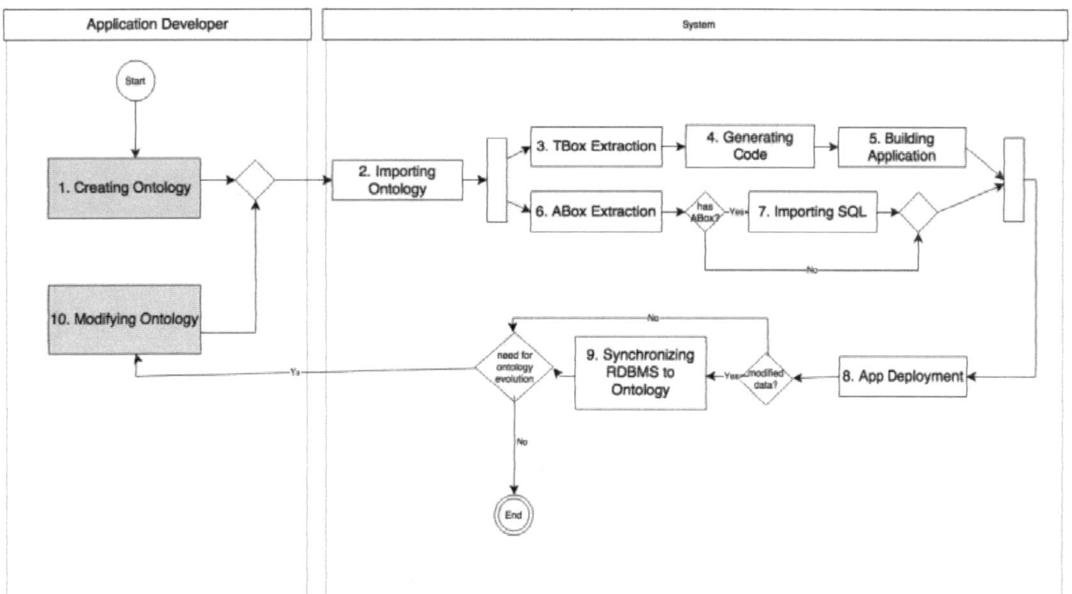

Figure 3. Application Generation Workflow

At this point, the system is a runnable web application with data in RDBMS. As the application is running and being used, the data in the RDBMS are updated. Our framework checks the RDBMS periodically, and synchronize automatically the ontology from the latest data if changes is detected. In this case, the application does not need to re-generated.

At the last step, when the knowledge engineer changes the ontology, he triggers the process for importing the new ontology that will trigger the code and database re-generation.

The framework has two constraints in the recent version. The domain engineer can't modify the ontology, while web application is running. The system does not consider the scalability, such as data volume and transaction volume will not be considered in generating this ontology-based application. However, the system does not prevent a developer to modify the generated code.

4.5. Implementation

This section described tools harvested through the open source repositories or implemented by authors. The framework architecture is illustrated in figure 4. The tools are:

1. Apache Maven, a comprehensive tool which is used to make software development project building process easier.
2. Spring Framework, a Java platform that provides comprehensive infrastructure support to develop Java application.
3. SpringKR, a Spring Framework extension to accommodate the Spring application to have knowledge repository functionalities.
4. Spring Roo, a code generator using the infrastructures provided by Spring Framework.
5. RooKR, an add-on for Spring Roo to automatically add SpringKR characteristic into the MVC code.
6. OWL2Roo, an ontology converter from OWL representation to Spring Roo notation which is used in generating web based application.
7. OWLUpdater, a tool to detect change(s) in relational database and add it into the ontology.

Apache Maven, Spring Framework, and Spring Roo are harvested through the open source repositories. SpringKR, RooKR, OWL2Roo, and OWLUpdater are implemented by authors.

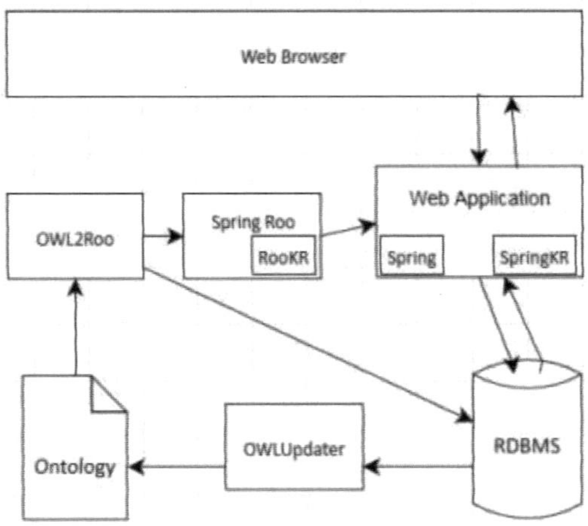

Figure 4. Framework Architecture

5. Experiment and Proof of Concept

To test our system, pizza ontology [25] and a configuration file is loaded to the framework to generate MVC component and pizza database. The configuration file in this testing specifies that: DomainConcept, ValuePartition, Food and IceCream are ignored, subclasses of Pizza, PizzaBase, and PizzaTopping are flagged as data, and Spiciness is flagged as enumeration.

Equation (3) is used to check the result of class transformation, and equation (4) is used to check the result of data transformation. Both equations use the number of component in pizza ontology and pizza database (table 2). Generated MVC code is shown at figure 5.

Table 2. Number of pizza components

No	Code	Component	Location	Total
1	OKL	The number of classes	Ontology	100
2	OIG	The number of classes that will be ignored (classIgnored)	Ontology	4
3	ODA	The number of subclasses of classes which flag as data (classAsData)	Ontology	88
4	OKE	The number of classes which flag as data enumeration (classAsEnum)	Ontology	1
5	OSE	The number of subclasses of classes which flag as data enumeration (classAsEnum)	Ontology	3
6	OIN	The number of individuals	Ontology	5
7	BTB	The number of tables	Database	6
8	BTR	The number of relational tables	Database	2
9	BBD	The number of rows of data	Database	259
10	BBR	The number of rows of data in the relationship tables	Database	166

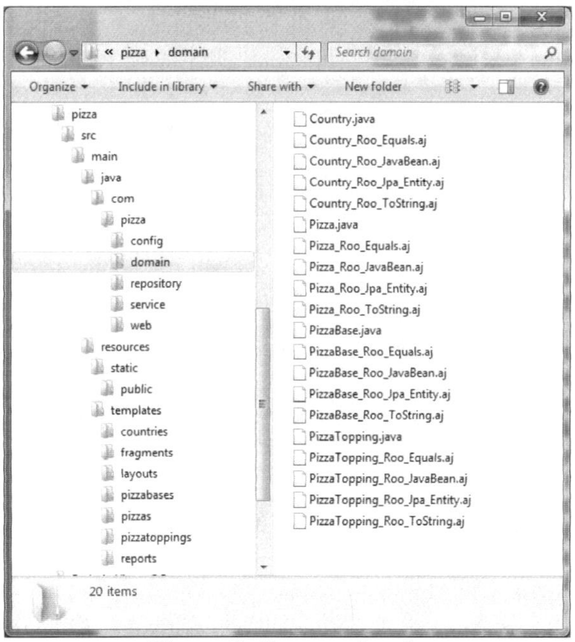

Figure 5. Generated code

6. Result and Discussion

In this research, we successfully mapped and transformed an ontology into a relational database by a default transformation rules and configuration file. This transformation can be used to generate a web application from the ontology, detect changes in the databases, and synchronize the changes to the ontology. A knowledge engineer can use this as a cycle to maintain ontology and relational database used by the application.

The transformation can be configured as needed by the application developer. Because of the different viewpoint and style among domain engineer in defining a domain, not all ontology elements can be mapped into a certain element in relational database. For example, domain engineer can define a concept is a subclass or an entity from an existing class.

The change in the ontology doesn't require application developer to change the code manually. The developer can restart the application generator, and the framework adjusts the code and the database accordingly. Also, the change in the database doesn't require the developer to adjust the change to ontology, because data synchronization can be scheduled by the framework automatically. At this time of writing, synchronization tool is still being developed.

In this research, ontology is chosen as base for generating web based application. We hope in the future, the framework can maximize the role of ontology as knowledge base which can be used in Knowledge Management System.

Our framework generates web application to maintain data (CRUD operations). Axioms in the ontology are transformed into database constraint only. Further development of the framework is planned for generating more complex business rules.

References

[1] E. Evans, "Domain-Driven Design: Tackling Complexity in the Heart of Software", 2003.
[2] T. R. Gruber, "Towards Principles for the Design of Ontologies Used for Knowledge Sharing," International Journal of Human-Computer Studies, 43, pp. 907-928, 1995.
[3] P. Kremen, "Building Ontology-Based Information System," Doctoral Thesis Faculty of Electrical Engineering Department of Cybernetics, 2012.
[4] E. Vysnauskas and L. Nemuraite, "Transforming Ontology Representation from OWL to Relational Database," InformationTechnology and Control, Vol. 35, No. 3 A, pp. 333-343, 2006.
[5] N. Stojanovic, et al., "Evolution in The Ontology-Based Knowledge Management System", 2002.
[6] J. Bormejo, "A Simplified Guide to Create an Ontology," 2007.
[7] M. Horridge, "A Practical Guide to Building OWL Ontologies Using Protégé 4 and CO-ODE Tools, Edition 1.3", The University of Manchester, 2011.
[8] M. M. Roldan & J. A. Montes, "A Protege Plugin for Storing OWL ontologies in Relational Databases", 2015.
[9] K. J. Lee & T. K. Whangbo, "Semantic Mapping between RDBMS and Domain Ontology," Computers and Communications, 2007. ISCC 2007.
[10] G. Bumans, "Mapping between Relational Databases and OWL Ontologies: an example", Scientific Papers, University Of Latvia, 2010.
[11] N. Cullot, et al., "DB2OWL: A tool for automatic database-to-ontology mapping", Proceedings of the Fifteenth Italian Symposium on Advanced Database Systems, SEBD 2007.
[12] M. D. M. Roldan & J. A. Montes, "A Protege Plugin for Storing OWL ontologies in Relational Database," 2015.
[13] Y. Liu & P. McBrien, "SQOWL2: Transactional Type Inference for OWL 2 DL in an RDBMS", British International Conference on Databases 2015, pp. 206-219.
[14] H. Darwen, "An Introduction to Relational Database Theory", Ventus Publishing ApS, 2010.
[15] C. Ritchie, "Database Principles and Design", 3rd edition, Delmar Learning, 2008.
[16] D. Esposito, "Modern Web Development: Understanding Domains, Technologies, and User Experiences", Microsoft Press, 2016.
[17] N. Guarino, "Formal Ontology and Information System," Proceedings of FOIS'98, Trento, Italy, 6-8 June 1998. Amsterdam, IOS Press, pp. 3-15, 1998.
[18] M. Zviedris, et al., "Ontology-Based Information System," JIST 2013, pp.33-47.
[19] K. Mahmudi & I. Liem, "Ontology-Based Knowledge Repository: using Java Spring Framework," KNIF 2015.

[20] I. Liem & Y. Nugroho, "An Application Generator Framelet," Software Engineering, Artificial Intelligence, Networking, and Parallel/DistributedComputing, 2008. DOI: 10.1109/SNPD.2008.113
[21] Pivotal Software, "Spring Framework Reference Documentation" ,4.2.1.RELEASE.
[22] K. Rimple & S. Penchikala, "Spring Roo in Action", Manning Publications Co, 2012.
[23] Pivotal Software, "Spring Roo – Reference Documentation",1.3.1.RELEASE. https://github.com/rabarro/spring-roo-1.3.1.RELEASE/blob/master/docs/pdf/spring-roo-docs.pdf
[24] E. Sirin, et al., "Pellet: A Practical OWL-DL Reasoner", Web Semantics:Science, Services and Agents on the World Wide Web, Volume 5, Issue2, pp. 51-53, 2007.
[25] Stanford University, "Pizza Ontology", last access at October 2nd, 2015,http://protege.stanford.edu/ontologies/pizza/pizza.owl

IOP Conf. Series: Journal of Physics: Conf. Series **971** (2018) 012032 doi:10.1088/1742-6596/971/1/012032

Computational parallel for shallow water-sediment concentration coupled model

I B P A Pranidhana[1*] **and P H Gunawan**[2]

School of Computing, Telkom University, Jalan Telekomunikasi No. 1 Terusan Buah Batu, Bandung 40257, Indonesia.

E-mail: [1]`gusangga@student.telkomuniversity.ac.id`;
[2]`phgunawan@telkomuniversity.ac.id`
`*Corresponding author`

Abstract. The goal of this paper is to analyze the parallel performance using OpenMP platform on shallow water-sediment concentration coupled model. The sediment model is coupled with the shallow water model for generating the water flow. In this paper, convection-diffusion equation is used to describe the sediment movement. The numerical results using OpenMP platform is shown satisfying to reduce the computational time. Indeed, the parallel time is faster than the serial time even along the increasing of the discrete points. The result using number of grid sizes 1600, the speedup is obtained 1.583 times. Meanwhile the efficiency is observed 39.57%. Moreover, the average efficiency from 5 times experiments is found 38.9%.

1. Introduction

Suspended sediment model is a mathematical model of sediment transport over the water flow. For instance, the sandy bottom in coastal area will be driven by the water flow and will eventually fall down until it is suspended. This can produce the propagation of concentration of sediment from high to low regions. Generally, the high concentration of sediment is located at the bottom, meanwhile the low concentration is located near the water surface. This concentration profile can be modeled by the steady suspended sediment model. The steady model of suspended sediment can be found for instance in [1].

The main focus of sediment model in [1] is two-dimensional (2D) vertical steady state of sediment. The formula of this model is given as follows

$$\frac{\partial \Phi}{\partial x} + \frac{\omega_s}{u(z)}\frac{\partial \Phi}{\partial z} = \frac{D}{u(z)}\frac{\partial^2 \Phi}{\partial z^2}, \tag{1}$$

where Φ is the concentration of sand sediment, ω_s the settling velocity of sand particles, $u(z)$ the velocity function depend on z and D a constant diffusion coefficient of turbulent flow. Moreover, x and z are the horizontal and vertical coordinates respectively.

The numerical result of this steady model is shown in Figure 1. The detail of this numerical simulation will be given in Section 4.1. Here, the result by a constant water velocity depend on the vertical height $u(z)$ is shown that the water flow disturbs the concentration of sand at surface. Additionally, Figure 1 shows the increasing of sand concentration along horizontal direction at steady state.

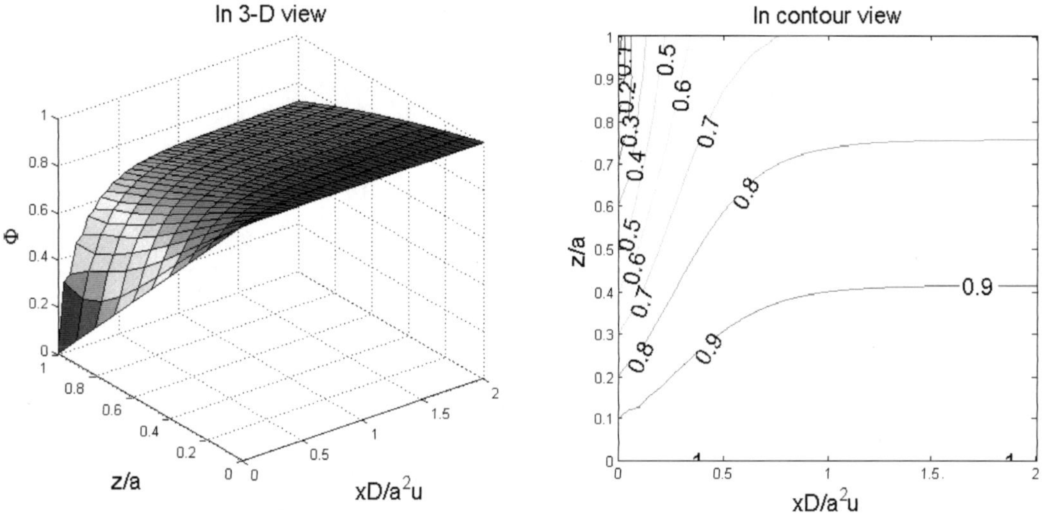

Figure 1: Suspended sediment concentration (Φ/Φ_0) in steady state case with 3D view (left) and 2D contour view (right).

In this paper, the goal is to investigate suspended sediment change (1) where the water velocity is changing in time. This means the velocity function in (1) should be depend on spatial vertical coordinate z and time t. In order to transform velocity function of sediment, the interaction of sediment and water dynamic should be considered since the velocity of water is depend on spatial and time variables. Thus in this paper, the steady state suspended sediment model will be coupled with shallow water model which can be connected trough velocity function.

Another contribution of this paper is to analyze the computational time of parallel algorithm for water-sediment interaction. The increasing of number of grid sizes in horizontal and vertical directions can produce the increasing of computational cost. Thus, the parallel algorithm is required to minimize this cost. Several references such as in [2, 3, 4, 5, 6] are shown the parallel computing using OpenMP can accelerating the numerical computation. Therefore here, architecture shared parallel programming by OpenMP platform will be used for simplicity.

2. Coupled model of water-sediment interaction and its numerical scheme

2.1. Dynamical model of water flow

The well known water flow model for describing the shallow area is called shallow water equations [7]. This model consists of two equations, mass (2) and momentum (3) equations, which can be written with zero bottom as

$$\frac{\partial h}{\partial t} + \frac{\partial (hv)}{\partial x} = 0, \tag{2}$$

$$\frac{\partial v}{\partial t} + v\frac{\partial v}{\partial x} + g\frac{\partial h}{\partial x} = -S_f, \tag{3}$$

where $h(x,t)$ denotes the water depth, $v(x,t)$ horizontal velocity, g gravitational force, and S_f a friction term.

The approximation solutions (2 - 3) by some numerical methods can be found in some references. The one well known method is called the finite volume method (FVM), for instance see [8, 7, 9, 10, 11, 12, 13, 14] and [15]. This method is a robust and shock capturing method which is still widely used for approximating shallow water equations.

Here, the staggered grid scheme FVM [7] will be used. This scheme is shown straight forward and simple to implement. Moreover, in [8] the scheme is proved satisfying the entropy balance. Following the notation in [7], let the discrete properties of shallow water equations on domain $\Omega = [-L, L]$ are given as follows

$$\Delta x = \frac{2L}{N_x}, \qquad x_k = \Delta x \times k, \qquad k \in \mathcal{M} = \{0, 1, 2, \cdots, N_x\},$$
$$t^n = \Delta t \times n, \qquad n \in \mathcal{T} = \{0, 1, 2, \cdots\}.$$

The discretization of mass balance (2) at full spatial discrete point $k \in \mathcal{M}$ is given as

$$\frac{h_k^{n+1} - h_k^n}{\Delta t} + \frac{q_{k+\frac{1}{2}}^n - q_{k-\frac{1}{2}}^n}{\Delta x} = 0, \tag{4}$$
$$q_{k+\frac{1}{2}}^n = h_{k+\frac{1}{2}}^n v_{k+\frac{1}{2}}^n, \qquad h_{k+\frac{1}{2}}^n = \begin{cases} h_k^n & \text{if } v_{k+\frac{1}{2}}^n > 0, \\ h_{k+1}^n & \text{otherwise} . \end{cases}.$$

Meanwhile, the discrete form of momentum balance (3) at half spatial discrete point $k + \frac{1}{2} \in \mathcal{M} - \{0, N_x\}$ is given as

$$\frac{v_{k+\frac{1}{2}}^{n+1} - v_{k+\frac{1}{2}}^n}{\Delta t} + \frac{1}{h_{k+\frac{1}{2}}^{n+1}} \left(\frac{q_{k+1}^n v_{k+1}^n - q_k^n v_k^n}{\Delta x} - v_{k+\frac{1}{2}}^n \frac{q_{k+1}^n - q_k^n}{\Delta x} \right) + g \frac{h_{k+1}^{n+1} - h_k^{n+1}}{\Delta x} = -S_f|_k^n, \tag{5}$$
$$q_k^n = \frac{q_{k+\frac{1}{2}}^n + q_{k-\frac{1}{2}}^n}{2}, \qquad v_k^n = \begin{cases} v_{k-\frac{1}{2}}^n & \text{if } q_k^n > 0, \\ v_{k+\frac{1}{2}}^n & \text{otherwise} . \end{cases}.$$

Note that, Equations (4-5) are used full and half grid notations for mass and momentum variables respectively. Authors encourage the readers to see references [7, 8] for more detail about staggered grid scheme.

2.2. Suspended sediment model

Suppose Φ denotes concentration of sand sediment. When sand tends to move downwards due to gravity force, then sand particle has fall velocity $\omega_s > 0$. Directly, this gravitational effect produces the influx $-\omega_s\Phi$. Moreover, diffusion mechanism for sand particles occurs such that the concentration profile is flat (outflux). Thus, the relation of two fluxes can be given as

$$Q = -\omega_s \Phi - D\frac{\partial \Phi}{\partial z}. \tag{6}$$

In steady state situation the mass balance for suspended sand is given by the following equation:

$$u(z)\frac{\partial \Phi}{\partial x} + \frac{\partial Q}{\partial z} = 0. \tag{7}$$

Finally, substituting (6) to (7), the suspended sediment model in Equation (1) can be achieved. Here, the velocity is depend on the vertical direction and in parabolic shape

$$u(z) = u_0 \frac{(-z^2 + 2az)}{a^2}. \tag{8}$$

This velocity function describes the velocity of water at bottom will be less than the surface in sediment area.

In this paper, a coupled model of SWE (2-3) and suspended sediment (1) will be elaborated. The idea is to give a small modification of (8) such that the average velocity from SWE can be transformed to the velocity with parabolic shape depend on z. The illustration of this coupled model can be seen in Figure 2. Then here, velocity of suspended sediment model will becomes

$$u(z,t) = v(x_0,t)\frac{(-z^2+2az)}{a^2}. \tag{9}$$

This velocity will be acting on the interface between SWE model and sediment model.

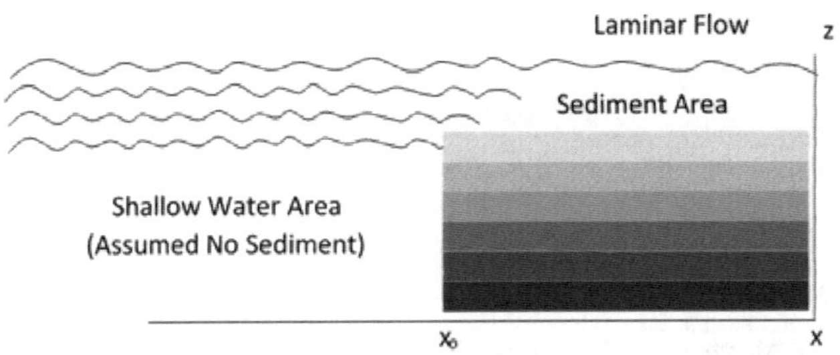

Figure 2: The illustration of coupled model using SWE and suspended sediment model. The influx of water velocity for sediment change at x_0 will be given from SWE.

In order to approximate the solution of sediment equation (1), Crank-Nicolson method can be used. The discrete properties of z direction on domain $\Omega_z[0, L_z]$ are given as follows

$$\Delta z = \frac{L_z}{N_z}, \qquad z_i = \Delta z \times i, \qquad i \in \{0, 1, \cdots, N_z\}.$$

Discrete form of (1) by Crank-Nicolson method can be seen as the following

$$\frac{\Phi_i^{k+1}-\Phi_i^k}{\Delta x} + \Theta\left\{\frac{\omega_s}{u}\left(\frac{\Phi_{i+1}^{k+1}-\Phi_{i-1}^{k+1}}{2\Delta z}\right) - \frac{D}{u}\left(\frac{\Phi_{i+1}^{k+1}-2\Phi_i^{k+1}+\Phi_{i-1}^{k+1}}{\Delta z^2}\right)\right\}$$
$$+(1-\Theta)\left\{\frac{\omega_s}{u}\left(\frac{\Phi_{i+1}^{k}-\Phi_{i-1}^{k}}{2\Delta z}\right) - \frac{D}{u}\left(\frac{\Phi_{i+1}^{k}-2\Phi_i^{k}+\Phi_{i-1}^{k}}{\Delta z^2}\right)\right\} = 0. \tag{10}$$

where $\Theta \in [0, 1]$. Here, by using $\Theta = 0$ then the numerical scheme (10) is called explicit method. Meanwhile, it called implicit method if $\Theta = 1$. Generally, Crank-Nicolson use $\Theta = 0.5$ to gained the second order approximation which can produces system of equations. Further, this system of equations can be solved numerically by Thomas algorithm [1].

3. Parallel architecture using OpenMP

The application of parallel programming in this paper uses OpenMp platform. According to [6, 16, 17, 18, 19, 20], OpenMp is an application programming interface (API) for parallel programming which is designed for shared memory multiprocessors architecture. In computing process of coupled model SWE and suspended sediment, there are 9 stages of the processes. Where, 4 processes can be done in parallel and other processes can done without parallel (serial). The flow of this stages can be seen in Figure 3 for more detail.

The computation process start from serial part which are allocating size of variables. The goal of this stage is to create the type and structure of data and to initialize the variables. Next, program will compile the parallel section trough several stages. First step is to compute

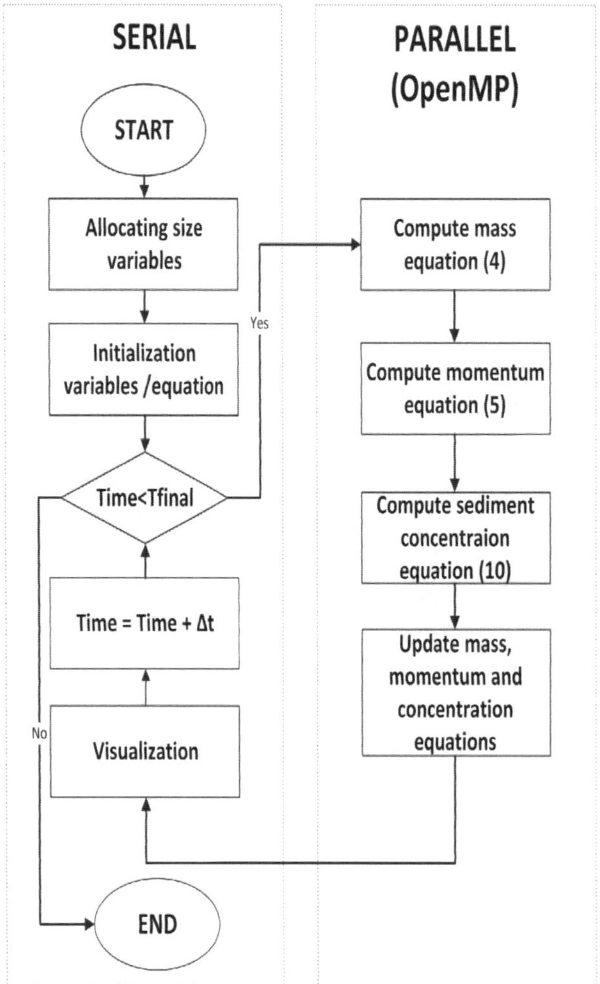

Figure 3: The serial and parallel computation processes. Several tasks and procedures which can be parallelized are grouped in parallel part. Meanwhile, the rest of them are grouped in serial part.

the mass equation to obtain the solution of water height h by (4). Second step is to compute momentum equation to obtain the average velocity v by (5). Third step is to compute sediment concentration in order to obtain the solution of sediment profile Φ/Φ_0 by (10). And the forth step is to update the variables for next iterations. Finally, the code leads back to serial section for entering the visualization stage.

4. Numerical results

4.1. Simulation of steady sediment change

In this section, the simulation of steady flow of sediment change is elaborated. The simulation only use Equation (1) with (8) and unassisted SWE. In this simulation, the discrete properties are given as:

- The discrete domain: $\Delta x = 0.05$, $\Delta z = 0.1$, and $\Delta t = 0.0001$.
- The coefficients: $\Phi_0 = 1$, $u_0 = 1$, $D = 1$, $\omega_s = 0.3$ and $\Theta = 0.5$.
- At boundaries:

$$\begin{cases} \Psi = \Psi_0, & \text{if } z = 0, \ \{\text{Equilibrium concentration at bottom}\}, \\ \omega_s \Phi - D\dfrac{\partial \Phi}{\partial z} = 0, & \text{if } z = a, \ \{\text{No sediment enters and leaves at water surface}\}, \end{cases}$$

IOP Conf. Series: Journal of Physics: Conf. Series **971** (2018) 012032 doi:10.1088/1742-6596/971/1/012032

- At initial condition:

$$\Phi(z) = \Phi_0 \exp\left(\frac{-\omega_s z}{D}\right).$$

The result of this simulation can be seen in Figure 1. The initial condition gives the linear profile of concentration to the z direction at $xD/a^2u = 0$. Afterward, form the result, it can seen clearly that the horizontal current $u(z)$ shift the contour plot to the right. The profile of sediment concentration (Φ/Φ_0) is in a good agreement with the adaptation length of this simulation. The profile is shown in parabolic shape at domain $[0 : 1]$ which agrees with the velocity profile (8).

4.2. Unsteady sediment

Here, the unsteady sediment change due to the change of water velocity will be elaborated. The dynamical system of water is governed by SWE (2-3). Meanwhile, the suspended sediment model is governed by (1) with velocity function (9).

The properties of initial sediment can be found in Section 4.1. However, several additional coefficients are given such as g and μ are 9.8 and 40 respectively. Moreover note that here, the velocity function (9) is used, thus u_0 is no longer a coefficient.

Here, the domain in z- and x-direction of simulation are given as $0 \leq z/a \leq 1$ and $0 \leq xD/a^2u \leq 2$ respectively. Similar to the steady case, Crank-Nicolson method ($\Theta = 0.5$) is used to approximate (1). The initial condition of SWE is given as follows

$$h(xD/a^2u, 0) = \begin{cases} 15 & \text{if } -4.5 \leq xD/a^2u \leq -2.5 \\ 1 & \text{otherwise} \end{cases}$$

$$v(xD/a^2u, 0) = 0$$

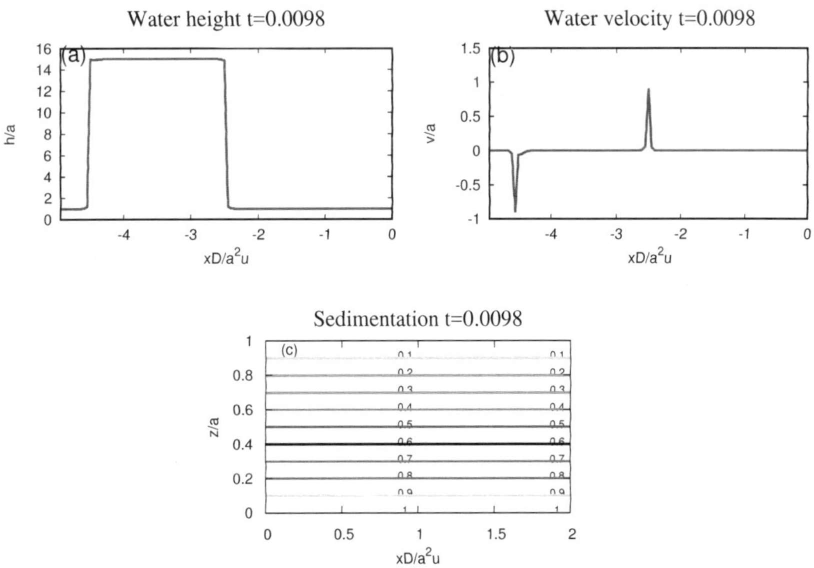

Figure 4: Simulation results at time 0.0098 s. (a) Water height profile.(b) Water velocity profile. (c). Suspended sediment concentration profile.

The results of this simulation are shown in Figures 4-7. Several sequential of final time are recorded at $t = 98 \times 10^{-4}$ s, $t = 19 \times 10^{-1}$ s, $t = 99 \times 10^{-1}$ s and $t = 11.9$ s. At time $t = 98 \times 10^{-4}$ s, the initial dam break propagates the water level to the left and right side 4(a), thus increases the velocity at the discontinue points 4(b). Since at the interface ($xD/a^2u = 0$), the velocity is zero, then no sediment movement is detected in Figure 4(c).

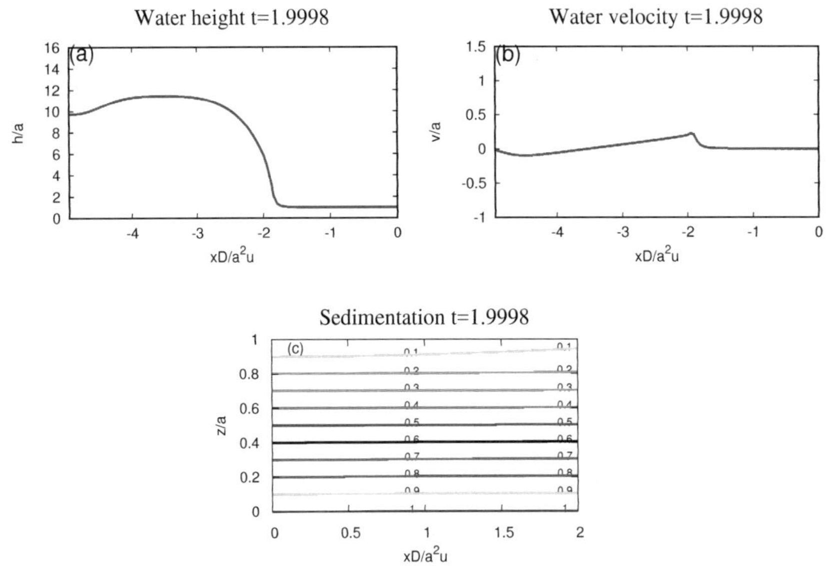

Figure 5: Simulation results at time 1.9998 s. (a) Water height profile .(b) Water velocity profile. (c). Suspended sediment concentration profile.

Figure 5 shows the propagation of water height (a), velocity (b) and sediment movement (c) at final time $t = 19 \times 10^{-1}$ s. Since the velocity at $xD/a^2u = 0$ slightly increasing, sediment movement in the top layer can be detected.

Meanwhile, in Figures 6 and 7, sediment movement (c) can be easily observed since the propagation of water height and velocity is increasing at interface. This means velocity u_0 increasing and sediment profile will be close to the solution of steady (Figure 1).

Indeed from this results, the coupled model SWE and suspended sediment model is successfully implemented. The simulations show that the increasing velocity of shallow water at the interface of sediment domain, then sediment change can be observed. Here, since the velocity of SWE in interface close to 1, then the solution of sediment is observed in a good agreement with the steady solution.

4.3. Parallel performance

In order to analyze the parallel computing performance using OpenMP, several simulations in unsteady sediment model are given. The specifications of computer in this experiment is shown in Table 1.

The time result of simulation for several number of discrete sizes N_x in shallow water-sediment concentration coupled model is shown in Table 2. The resulting time is obtained from the simulation with different discrete points N_x, which are $100, 200, 400, 800$ and 160. Table 2 shows the CPU time in parallel code is faster than the serial code along the increasing of the discrete points. By using $N_x = 1600$ points, the execution time in parallel and serial are obtained 1431.259 s and 2265.401 s respectively. This means, by using 4 processors, the speedup of

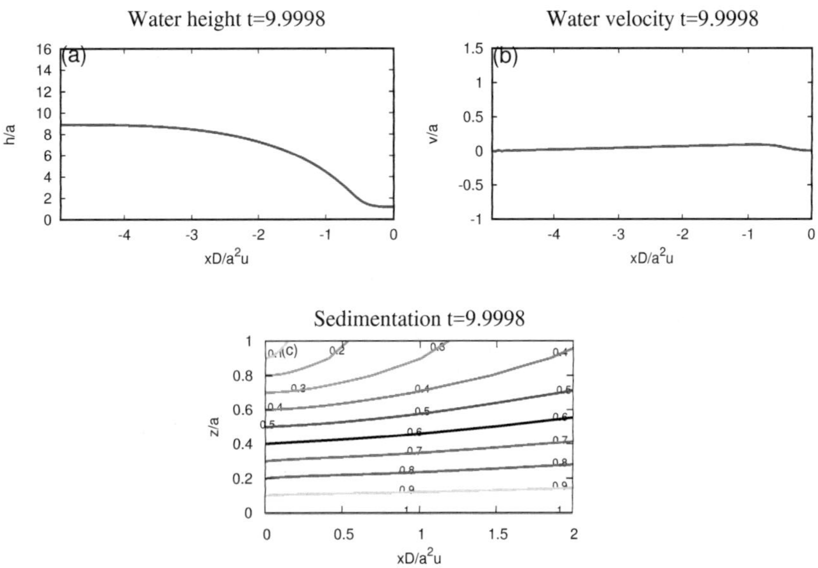

Figure 6: Simulation results at time 9.9998 s. (a) Water height profile.(b) Water velocity profile. (c). Suspended sediment concentration profile.

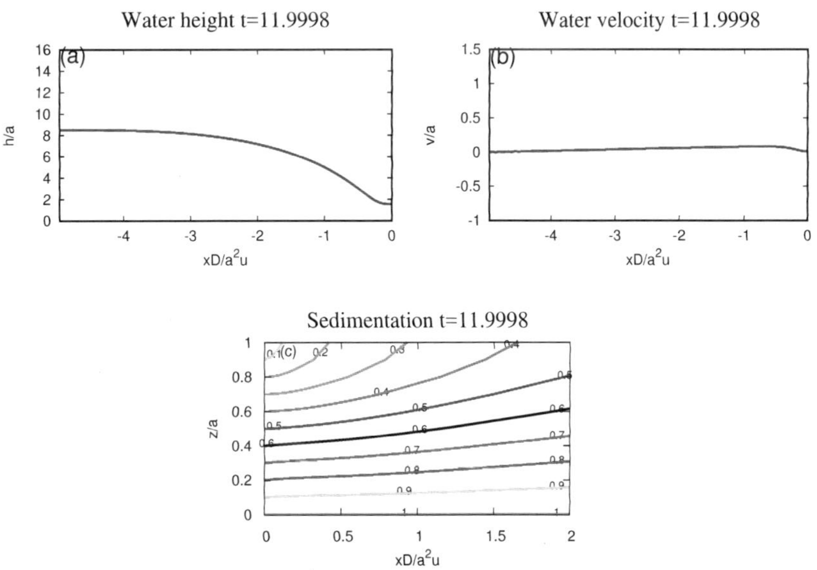

Figure 7: Simulation results at time 11.9998 s. (a) Water height profile.(b) Water velocity profile. (c). Suspended sediment concentration profile.

Table 1: The specifications of computer in experiment of parallel computing.

Name	Type
Processors	Intel (R) Core (TM) i5-2500
Number Of Processors	4
RAM Size	8 GB
Operating System	Ubuntu 12.04.3

parallel performance is obtained 1.583 times and the efficiency is given 39.57 %. Overall, the average of efficiency for all experiments is observed 38.9 %. Indeed here, parallel programming can reduce the time computation successfully.

Table 2: The results of CPU time, speedup and efficiency in parallel computing.

Discrete Points	Execution Time (s)		Speedup	Efficiency(%)	Average Efficiency(%)
	Serial	Parallel			
100	186.159	120.896	1.539	38.49	38.9
200	331.429	214.156	1.548	38.69	
400	681.832	438.632	1.554	38.86	
800	1132.156	725.012	1.562	39.05	
1600	2265.401	1431.259	1.583	39.57	

5. Conclusion

Here, the simulations of shallow water-sediments concentration coupled model has been successfully implemented. The solution of unsteady sediment model is in a good agreement with the steady solution when the velocity at the interface water- sediment tends to 1. The results of parallel programming with OpenMP is shown satisfying. The execution time in serial can be reduced by parallel programming. For simulation with 1600 girds, the computational time of simulation with parallel is 1431.259 s and serial is 2265.401 s. Those results show that OpenMP is successfully reduced the time computation. The maximum speedup and efficiency are observed 1.583 and 39.57% are respectively. Moreover, the average of efficiency in overall experiments is obtained 38.9%.

Acknowledgment

Second author would like to say thank you very much for Prof. Sri Redjeki P. for the great discussion in the beginning of sediment modeling.

References

[1] Vreugdenhil C B 2012 *Computational hydraulics: an introduction* (Springer Science & Business Media)
[2] Gunawan P H 2016 *Information and Communication Technology (ICoICT), 2016 4th International Conference on* (IEEE) pp 1–5
[3] De La Asunción M, Mantas J M and Castro M J 2011 *The Journal of Supercomputing* **58** 206–214
[4] Brodtkorb A R, Sætra M L and Altinakar M 2012 *Computers & Fluids* **55** 1–12
[5] Castillo D, Ferreiro A, García-Rodríguez J A and Vázquez C 2013 *Applied Mathematics and Computation* **219** 11233–11257
[6] Chandra R 2001 *Parallel programming in OpenMP* (Morgan kaufmann)
[7] Stelling G S and Duinmeijer S A 2003 *International Journal for Numerical Methods in Fluids* **43** 1329–1354
[8] Doyen D and Gunawan P H 2014 *Finite Volumes for Complex Applications VII-Methods and Theoretical Aspects* (Springer) pp 227–235

[9] Delestre O, Cordier S, James F and Darboux F 2008 *12th International Conference on Hyperbolic Problems* vol 67 (American Mathematical Society) pp 537–546
[10] Delestre O and Marche F 2011 *Journal of Scientific Computing* **48** 41–51
[11] Audusse E, Bouchut F, Bristeau M O, Klein R and Perthame B 2004 *SIAM Journal on Scientific Computing* **25** 2050–2065
[12] Gunawan P H and Lhébrard X 2015 *Computers & Fluids* **121** 44–50
[13] Crowhurst P and Li Z 2013 *Computer Modelling and Simulation (UKSim), 2013 UKSim 15th International Conference on* (IEEE) pp 55–60
[14] Olsen N R B 2007 *The Norvegian University of Science and Technology, Trondheim*
[15] Bouchut F 2004 *Nonlinear stability of finite Volume Methods for hyperbolic conservation laws: And Well-Balanced schemes for sources* (Frontiers in Mathematics. Birkhäuser Verlag, Basel)
[16] Mulyani, Putri N D and Gunawan P H 2017 *Information and Communication Technology (ICoIC7), 2017 5th International Conference on* (IEEE) pp 1–5
[17] Pahlevi M R and Gunawan P H 2017 *Information and Communication Technology (ICoIC7), 2017 5th International Conference on* (IEEE) pp 1–4
[18] Juliati S and Gunawan P H 2017 *Information and Communication Technology (ICoIC7), 2017 5th International Conference on* (IEEE) pp 1–5
[19] Alamsyah M N A, Simanjuntak C A, Bagustara B A R H, Pradana W A and Gunawan P H 2017 *Information and Communication Technology (ICoIC7), 2017 5th International Conference on* (IEEE) pp 1–6
[20] Iryanto and Gunawan P H 2017 *Information and Communication Technology (ICoIC7), 2017 5th International Conference on* (IEEE) pp 1–5

IOP Conf. Series: Journal of Physics: Conf. Series **971** (2018) 012033 doi:10.1088/1742-6596/971/1/012033

OpenMP performance for benchmark 2D shallow water equations using LBM

Khairul Sabri[1,*], Hasbi Rabbani[2] and Putu Harry Gunawan[3]

School of Computing, Telkom University, Jalan Telekomunikasi No. 1 Terusan Buah Batu, Bandung 40257, Indonesia.

E-mail: [1]irulsabri@student.telkomuniversity.ac.id;
[2]hasbirabb@student.telkomuniversity.ac.id;
[3]phgunawan@telkomuniversity.ac.id
*Corresponding author

Abstract. Shallow water equations or commonly referred as Saint-Venant equations are used to model fluid phenomena. These equations can be solved numerically using several methods, like Lattice Boltzmann method (LBM), SIMPLE-like Method, Finite Difference Method, Godunov-type Method, and Finite Volume Method. In this paper, the shallow water equation will be approximated using LBM or known as LABSWE and will be simulated in performance of parallel programming using OpenMP. To evaluate the performance between 2 and 4 threads parallel algorithm, ten various number of grids Lx and Ly are elaborated. The results show that using OpenMP platform, the computational time for solving LABSWE can be decreased. For instance using grid sizes 1000×500, the speedup of 2 and 4 threads is observed 93.54 s and 333.243 s respectively.

1. Introduction

The fluid flow phenomena such as flow in the rivers, channels, and lake can be described by a known mathematical model which is called shallow water equations (SWE)[1]. Consisting of mass (1) and momentum (2 - 3) equations, SWE in 2-dimension can be given as follows

$$\partial_t h = -\left(\partial_x(hu) + \partial_y(hv)\right), \tag{1}$$

$$\partial_t(hu) = -\left(\partial_x(hu^2 + \frac{1}{2}gh^2) + \partial_y(huv)\right), \tag{2}$$

$$\partial_t(hv) = -\left(\partial_x(huv) + \partial_y(hu^2 + \frac{1}{2}gh^2)\right), \tag{3}$$

where h denotes the water depth/height, u the lateral velocity in x direction, v the lateral velocity in y direction, and g a constant gravitational force. The variables x, y and t denote the space and time respectively.

The approximation solution of Equations (1 - 2) can be obtain using several methods. The most popular method is grid method known as the finite volume method (FVM). This method is developed into several numerical schemes which are widely applied and are shown robust scheme, for instance see the references [2, 3, 4, 5] and [6]. However, the deficiency of this method is to solve the general water flow with a complex structure. Here, the LBM is the numerical method based on particle method. LBM is already applied in many applications of fluid flows [7]. The advantage of LBM is the complexity of boundary condition can be omitted [8]. However, LBM can gain higher computational cost than the other methods.

In order to accelerate the computation in time, the parallel computing techniques will be elaborated. In references [9, 10, 11, 12] and [13], parallel computing is shown as a good idea to accelerate the numerical computation. Thus the goal of this paper is to obtain the performance of parallel computing for simulating LABSWE. Additionally, here the shared parallel architecture using OpenMP platform will be used.

2. Approximation Solution of SWE using LBM

Lattice Boltzmann method (LBM) is the lattice gas automata (LGA) evolved method for solving a discrete computational equation [14]. The characteristics of LBM is composed of three important parts: lattice Boltzmann equation (LBE), lattice pattern which formed by arranged particles, and equilibrium distribution function.

2.1. Lattice boltzmann equation

In the LBM, there are two important processes: streaming step and collision step. On the streaming process, the particles move to another nearby lattice pattern particles in any directions using their velocities which is written by

$$f_\alpha(x + e_\alpha \Delta t, t + \Delta t) = f'_\alpha(x,t) + \frac{\Delta t}{N_\alpha e^2} e_{\alpha i} F_i(x,t), \tag{4}$$

where f_α is a function for distributing particles, f'_α a distribution function before process of streaming, F_i the force function in i-direction, Δx the uniform distance of particles over the lattice, and Δt the time step. The velocity is given by $e = \Delta x/\Delta t$, e_α and a constant N_α is the constant can be determined by the pattern of following lattice

$$N_\alpha = \frac{1}{e^2} \sum_\alpha e_{\alpha i} e_{\alpha i}. \tag{5}$$

Particles move because the influence of the velocity by another particles in collision step. This step can be described by the following equation

$$f'_\alpha(x,t) - f_\alpha(x,t) = \Omega_\alpha[f(x,t)]. \tag{6}$$

where Ω_α is called the collision operator which can control the acceleration of the change of f_α. This operator is determined from the microscopic dynamics in generally. Higuera and Jiménez [15] introduce the linearization of collision operator by the equilibrium condition. Therefore, the collision operator can be simplified becomes

$$\Omega_\alpha(f) = -\frac{1}{\tau}(f_\alpha - f_\alpha^{eq}) \tag{7}$$

where τ is defined as relaxation time and f_α^{eq} is known as distribution function of the local equilibrium.

Finally the combination of two steps (streaming and collision) which consists of Equations (4), (6) and (7) becomes following popular formula known as Lattice Boltzmann Equation (LBE),

$$f_\alpha(x + e_\alpha \Delta t, t + \Delta t) = f_\alpha(x,t) - \frac{1}{\tau}(f_\alpha(x,t) - f_\alpha^{eq}(x,t)) + \frac{\Delta t}{N_\alpha e^2} e_{\alpha i} F_i(x,t). \tag{8}$$

2.2. Lattice pattern

Lattice pattern is used for representing grid points and determining particle's motions in LBM. From the lattice pattern, we could easily know that the LBM doing a 1D, 2D, or 3D model. In addition, the constant N_α from Equation (8) is determined by the lattice pattern.

In 1D model, the lattice pattern is depend on the amount of particle speed in lattice points. Generally it is known as the line lattice with two or three speed model. Meanwhile, there are commonly two kinds of lattice patterns in 2D models: the hexagonal and the square lattices. Hexagonal lattice can have 6 or 7-speed models, while square lattice can have 4, 5, 8, or 9-speed models.

According to recent study, the most accurate lattice pattern of all 2D models is square lattice with 9-speed (D2Q9) models [16]. Hence this paper will use D2Q9 to simulate 2D SWE which the model of lattice and its system are shown in Figure 1 below

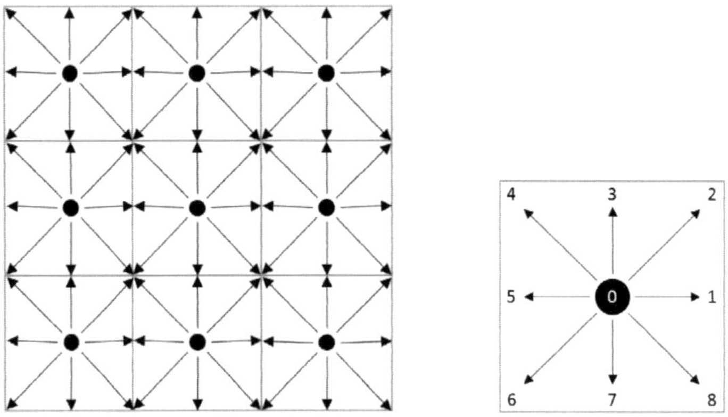

(a) (b)

Figure 1. (a) System of interaction over lattices. (b) The D2Q9 lattice.

In D2Q9 lattice (Figure 1 (b)), each particle moves to its neighbor with certain velocity and direction along the 1-8 links which 0 is indicated as the link with zero speed in the particle. This vector of particle velocity is defined by

$$e_\alpha = \begin{cases} (0,0), & \alpha = 0, \\ e\cos(0.25\pi(\alpha-1)), e\sin(0.25\pi(\alpha-1)), & \alpha = 1,3,5,7, \\ \sqrt{2}e\cos(0.25\pi(\alpha-1)), \sqrt{2}e\sin(0.25\pi(\alpha-1)), & \alpha = 2,4,6,8. \end{cases} \tag{9}$$

and the D2Q9 lattice has basic features as follows [14]

$$\sum_\alpha e_{\alpha i} = \sum_\alpha e_{\alpha i}e_{\alpha j}e_{\alpha k} = 0, \tag{10}$$

$$\sum_\alpha e_{\alpha i}e_{\alpha j} - 6e^2\delta_{ij} = 0, \tag{11}$$

$$\sum_\alpha e_{\alpha i}e_{\alpha j}e_{\alpha k}e_{\alpha l} - 4e^4(\delta_{ij}\delta_{kl} + \delta_{ik}\delta_{jl} + \delta_{il}\delta_{jk}) + 6e^4\Delta_{ijkl} = 0, \tag{12}$$

where $\Delta_{ijkl} = 1$ in terms of $i = j = k = l$, and $\Delta_{ijkl} = 0$ besides.

Using Equation (9), the value of N_α can be found by evaluating Equation (5), yields

$$N_\alpha = \frac{1}{e^2}\sum_\alpha e_{\alpha i}e_{\alpha i} = \frac{1}{e^2}\sum_\alpha e_{\alpha y}e_{\alpha y} = 6. \tag{13}$$

Substituting Equation (13) to the Equation (8), leads to

$$f_\alpha(x + e_\alpha \Delta t, t + \Delta t) = f_\alpha(x,t) - \frac{1}{\tau}(f_\alpha - f_\alpha^{eq}) + \frac{\Delta t}{6e^2} e_{\alpha i} F_i(x,t). \tag{14}$$

Finally, Equation (14) is in the form of LBM which is used for simulating fluid flows.

2.3. Local equilibrium distribution function

Based to the LGA, an equilibrium function is originally obtained from Maxwell-Boltzmann equilibrium distribution function [17, 18]. Here the equilibrium function is expressed as,

$$f_\alpha^{eq} = P_\alpha + Q_\alpha e_{\alpha i} u_i + R_\alpha e_{\alpha i} e_{\alpha j} u_i u_j + S_\alpha u_i u_j, \tag{15}$$

where u_i and u_j are the components of flow velocity, or well known as $u_i = u$ and $u_j = v$.

Equation (15) becomes a popular equation since this equation is successfully applied for approximating the solution of many flow models [19, 20]. Since the lattice on Figure 2.2 (b) has the same shape as the equilibrium function, thus

$$P_1 = P_3 = P_5 = P_7 = \bar{P}, \tag{16}$$

$$P_2 = P_4 = P_6 = P_8 = \tilde{P}, \tag{17}$$

likewise the expressions for Q_α, R_α and S_α. According to Equation (12), the local equilibrium function D2Q9 lattice (Figure 2.2 (b)) can be written as

$$f_\alpha^{eq} = \begin{cases} P_0 + S_0 u_i u_i, & \alpha = 0, \\ \bar{P} + \bar{Q} e_{\alpha i} u_i + \bar{R} e_{\alpha i} e_{\alpha j} u_i u_j + \bar{S} u_i u_i, & \alpha = 1,3,5,7, \\ \tilde{P} + \tilde{Q} e_{\alpha i} u_i + \tilde{R} e_{\alpha i} e_{\alpha j} u_i u_j + \tilde{S} u_i u_i, & \alpha = 2,4,6,8. \end{cases} \tag{18}$$

Basically, the coefficients P_0, $\bar{P}$ and $\tilde{P}$ are defined by a constraint of equilibrium distribution function and are based on the law of conservation such as mass and momentum conservation. Moreover, to get the local equilibrium distribution function in SWE, three conditions should be satisfied,

$$\sum_\alpha f_\alpha^{eq}(x,t) - h(x,t) = 0, \tag{19}$$

$$\sum_\alpha e_{\alpha i} f_\alpha^{eq}(x,t) - h(x,t) u_i(x,t) = 0, \tag{20}$$

$$\sum_\alpha e_{\alpha i} e_{\alpha j} f_\alpha^{eq}(x,t) = \frac{1}{2} g h^2(x,t) \delta_{ij} + h(x,t) u_i(x,t) u_j(x,t). \tag{21}$$

The detail of previous equations can be found in book of Guo Zhou [20] and known as Lattice Boltzmann Method for The Shallow Water Equations (LABSWE).

3. Parallel Architecture

For accelerating the computational time of 2D SWE using LBM, the process will be elaborated by parallel algorithm. Since the existence of patterned particles in the system of LBM, the computation of functions from each grid can be separated and computed in different processor. In this paper, the parallel algorithm will be implemented using an application programming interface (API) OpenMP.

OpenMP is an API that provide multithreading and shared memory [21, 22, 23, 24, 25, 26]. Multithreading is the capability of CPU for running some processes in the same time. This capability may reduce the runtime of process and increase the efficiency of an algorithm. Moreover, shared memory is the architecture of parallel which allows each running process to gain memory source proportionally. With this feature, access time to the storage can be minimum and accelerate the runtime of process can be increased.

In LABSWE algorithm, the parallelization can be implemented on streaming and collision step for each grids. More detail about the parallel algorithm in LABSWE is shown in this flowchart Figure 2.

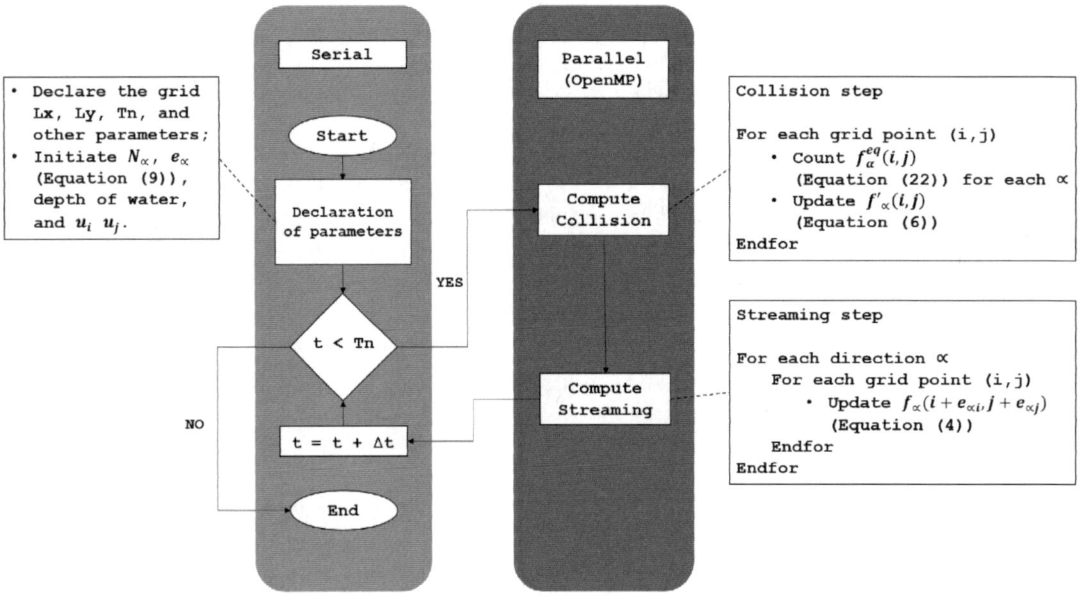

Figure 2. The configuration of parallel algorithm for approximating the solution of LABSWE in accelerating the computational time.

In serial process, all of the parameters are declared and stated the initial condition of the water flows. Meanwhile in parallel process, the procedures to compute and update the collision and streaming step of all lattice particles are elaborated.

4. Numerical Results and Parallel Performances

Here, the initial conditions of 2D benchmark LABSWE are given as follows

$$h(x, y, 0) = 0.05$$
$$u(x, y, 0) = v(x, y, 0) = 0$$

The wall boundary condition is given in this simulation. The wave is generated at the boundary $x = 0$ which is

$$h(x, y, t) = \sum_\alpha f_\alpha(x, y, t).$$

The results of water height $h(x, y, t)$ in two final time observations can be seen in Figure 3. It can be seen clearly that the water wave are generated on the boundary $x = 0$ and produces the highest wave amplitude 0.053 m. The waves propagate to the right side of domain. Figure 3 at top show the three dimensions view of water surface, meanwhile the slice in two dimensions view is shown in Figure 3 at bottom.

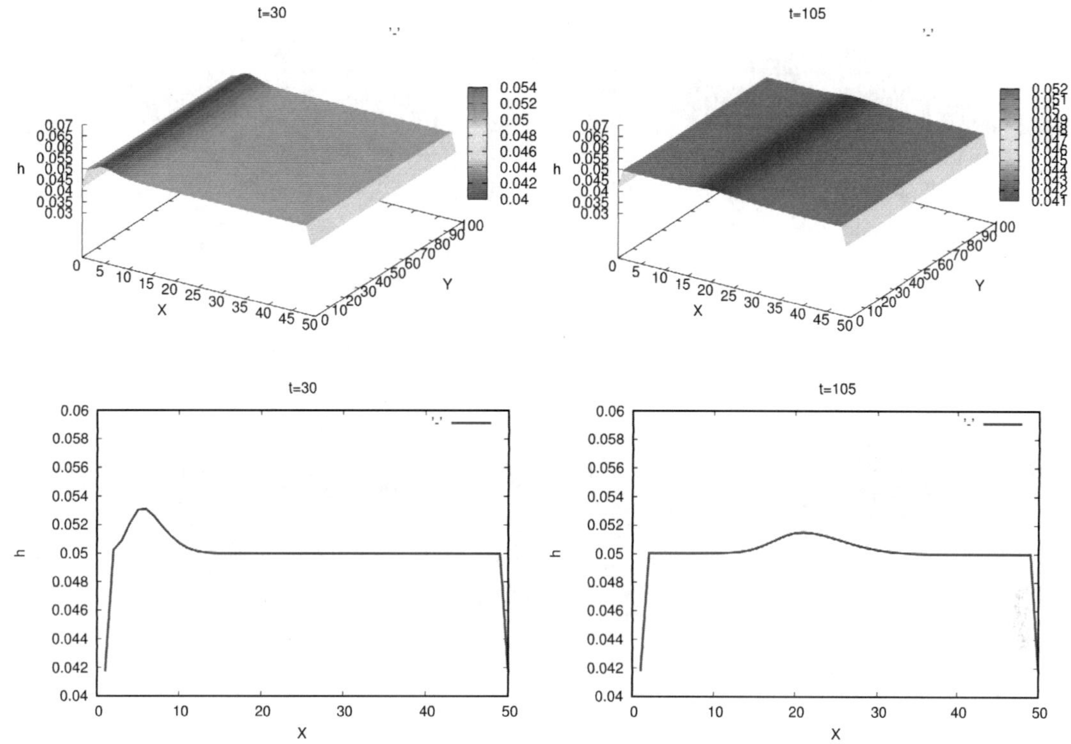

Figure 3. The results of water waves which are generated on the boundary $x = 0$ when $t = 30$ (left) and $t = 105$ (right).

Moreover, the velocity profiles in x and y-direction at final time of simulation $t = 30$ and $t = 105$ seconds are given in Figure 4. From Figure 4, there are discrepancy between the velocities v and u in two final time of simulation. The velocity u is shown more fluctuated than the velocity v since the water waves propagate along the x-direction.

This paper is focus on the parallel performances of previous simulation. The results of the parallel performance can be found in Table 1. Here, two experiments are given, using 2 and 4 threads. The ten various number of L_x and L_y are given in order to evaluate the parallel performance.

From the Table 1, the OpenMP platform is shown able to accelerate the computational time for various number of grid sizes. In using the large number $L_x = 1000$ and $L_y = 500$, the CPU time of serial code is obtained 463.624 s. Meanwhile the CPU time in parallel code using 2 and 4 threads are observed 393.54 s and 333.243 s respectively. This result produces speedup 1.178 and 1.391 times using 2 and 4 threads respectively. Moreover, the efficiency using 2 threads is obtained 58.9% and using 4 threads is calculated 34.8%.

In this paper, by the experiment using 2 and 4 threads, the speedup for all simulations using 4 threads is observed get the best speedup than using 2 threads. However, in contrast with the speedup, the efficiency using 2 threads is obtained better than using 4 threads.

5. Conclusion

According to the results, the parallel algorithm for solving LABSWE can decrease the execution time for serial algorithm. It means parallel algorithm is successfully applied in this problem. Furthermore, parallel code with 2 threads has lower average speed up than using 4 threads.

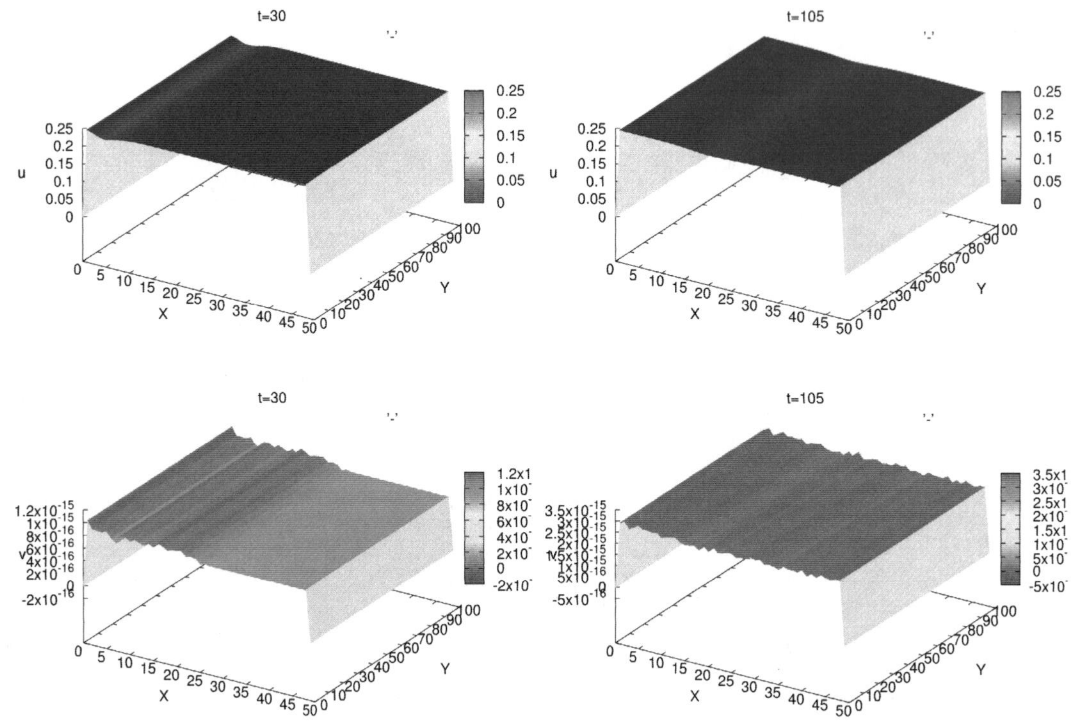

Figure 4. Visualization of $u(x, y, t)$ (top) and $v(x, y, t)$ (bottom).

Table 1. The speedup and efficiency of parallel LABSWE algorithm performances using OpenMP.

Lx	Ly	Execution Time (s)			Speed Up		Efficiency (%)	
		Serial	2 Threads	4 Threads	2 Threads	4 Threads	2 Threads	4 Threads
100	50	2.782	2.85	2.218	0.976	1.254	48.8	31.4
200	100	11.581	9.658	7.099	1.199	1.631	59.9	40.8
300	150	28.677	23.482	18.705	1.221	1.533	61.1	38.3
400	200	53.363	42.515	33.882	1.255	1.575	62.8	39.4
500	250	84.749	67.358	53.064	1.258	1.597	62.9	39.9
600	300	123.766	98.564	77.579	1.256	1.595	62.8	39.9
700	350	170.399	135.97	107.787	1.253	1.581	62.7	39.5
800	400	225.605	180.17	142.236	1.252	1.586	62.6	39.7
900	450	291.857	234.385	183.981	1.245	1.586	62.3	39.7
1000	500	463.624	393.54	333.243	1.178	1.391	58.9	34.8

However, the average efficiency of parallel code using 2 threads is higher than using 4 threads. For instance using grid sizes 1000×500, the speedup of 2 and 4 threads is obtained 393.54 s and 333.243 s respectively. Moreover the efficiency of 2 and 4 threads is conducted 58.9% and 34.8 % respectively.

IOP Conf. Series: Journal of Physics: Conf. Series **971** (2018) 012033 doi:10.1088/1742-6596/971/1/012033

References

[1] Doyen D and Gunawan P H 2014 *Finite Volumes for Complex Applications VII-Methods and Theoretical Aspects* (Springer) pp 227–235
[2] Delestre O, Cordier S, James F and Darboux F 2008 *12th International Conference on Hyperbolic Problems* vol 67 (American Mathematical Society) pp 537–546
[3] Delestre O and Marche F 2011 *Journal of Scientific Computing* **48** 41–51
[4] Audusse E, Bouchut F, Bristeau M O, Klein R and Perthame B 2004 *SIAM Journal on Scientific Computing* **25** 2050–2065
[5] Gunawan P H and Lhébrard X 2015 *Computers & Fluids* **121** 44–50
[6] Bouchut F 2004 *Nonlinear stability of finite Volume Methods for hyperbolic conservation laws: And Well-Balanced schemes for sources* (Frontiers in Mathematics. Birkhäuser Verlag, Basel)
[7] Mohamad A A 2011 *Lattice Boltzmann method: fundamentals and engineering applications with computer codes* (Springer Science & Business Media)
[8] Mei R, Luo L S and Shyy W 1999 *Journal of computational physics* **155** 307–330
[9] Gunawan P H 2016 *Information and Communication Technology (ICoICT), 2016 4th International Conference on* (IEEE) pp 1–5
[10] de la Asunción M, Castro M, Mantas J and Ortega S 2016 *Advances in Engineering Software* **99** 59–72
[11] De La Asunción M, Mantas J M and Castro M J 2011 *The Journal of Supercomputing* **58** 206–214
[12] Brodtkorb A R, Sætra M L and Altinakar M 2012 *Computers & Fluids* **55** 1–12
[13] Castillo D, Ferreiro A, García-Rodríguez J A and Vázquez C 2013 *Applied Mathematics and Computation* **219** 11233–11257
[14] Zhou J G 2004 *Lattice Boltzmann methods for shallow water flows* vol 4 (Springer)
[15] Higuera F and Jimenez J 1989 *EPL (Europhysics Letters)* **9** 663
[16] Skordos P 1993 *Physical Review E* **48** 4823
[17] Chen S and Doolen G D 1998 *Annual review of fluid mechanics* **30** 329–364
[18] He X and Luo L S 1997 *Physical Review E* **55** R6333
[19] Chen H, Chen S and Matthaeus W H 1992 *Physical Review A* **45** R5339
[20] Guangwu Y 2000 *Journal of Computational Physics* **161** 61–69
[21] Dagum L and Menon R 1998 *IEEE computational science and engineering* **5** 46–55
[22] Mulyani, Putri N D and Gunawan P H 2017 *Information and Communication Technology (ICoIC7), 2017 5th International Conference on* (IEEE) pp 1–5
[23] Pahlevi M R and Gunawan P H 2017 *Information and Communication Technology (ICoIC7), 2017 5th International Conference on* (IEEE) pp 1–4
[24] Juliati S and Gunawan P H 2017 *Information and Communication Technology (ICoIC7), 2017 5th International Conference on* (IEEE) pp 1–5
[25] Alamsyah M N A, Simanjuntak C A, Bagustara B A R H, Pradana W A and Gunawan P H 2017 *Information and Communication Technology (ICoIC7), 2017 5th International Conference on* (IEEE) pp 1–6
[26] Iryanto and Gunawan P H 2017 *Information and Communication Technology (ICoIC7), 2017 5th International Conference on* (IEEE) pp 1–5

IOP Conf. Series: Journal of Physics: Conf. Series **971** (2018) 012034 doi:10.1088/1742-6596/971/1/012034

Computational multicore on two-layer 1D shallow water equations for erodible dambreak

C A Simanjuntak[1,*], B A R H Bagustara[2] and P H Gunawan[3]

School of Computing, Telkom University, Jalan Telekomunikasi No. 1 Terusan Buah Batu, Bandung 40257, Indonesia.

E-mail: [1]cassrio@student.telkomuniversity.ac.id;
[2]bajoerajahusein@student.telkomuniversity.ac.id;
[3]phgunawan@telkomuniversity.ac.id
*Corresponding author

Abstract. The simulation of erodible dambreak using two-layer shallow water equations and SCHR scheme are elaborated in this paper. The results show that the two-layer SWE model in a good agreement with the data experiment which is performed by Louvain-la-Neuve Université Catholique de Louvain. Moreover, the parallel algorithm with multicore architecture are given in the results. The results show that Computer I with processor Intel(R) Core(TM) i5-2500 CPU Quad-Core has the best performance to accelerate the computational time. Moreover, Computer III with processor AMD A6-5200 APU Quad-Core is observed has higher speedup and efficiency. The speedup and efficiency of Computer III with number of grids 3200 are 3.716050530 times and 92.9% respectively.

1. Introduction

Dambreak problem is widely used as a benchmark in fluid flow problems. Recently, this problem can be categorized in two types which are dambreak in hard [1] and sediment bed [2]. In sediment bed, the sediment can be moved by the interaction between the water and sediment itself. Thus, the mathematical model of this interaction problem will be needed.

Several mathematical models for bed morphological change due to water flow are available in some references [2, 3, 4]. For instance, the coupled model of shallow water-Exner equations is given [2, 5]. In [2, 5], the one-layer shallow water equations is used for describing the water flow. Meanwhile for the sediment, the Exner equation is elaborated.

In this paper, the two-layer shallow water equations (SWE) will be used to describe the fluid-sediment interaction phenomena. This model is chosen since it consists of various density for each layers, such that can be used to describe the sediment.

The two-layer SWE model is given as follows:

$$\frac{\partial h_1}{\partial t} + \frac{\partial (h_1 u_1)}{\partial x} = 0, \tag{1}$$

$$\frac{\partial (h_1 u_1)}{\partial t} + \frac{\partial \left(h_1 u_1^2 + \frac{g}{2} h_1^2\right)}{\partial x} = -g h_1 \frac{\partial (h_2 + Z)}{\partial x}, \tag{2}$$

$$\frac{\partial h_2}{\partial t} + \frac{\partial (h_2 u_2)}{\partial x} = 0, \tag{3}$$

$$\frac{\partial (h_2 u_2)}{\partial t} + \frac{\partial \left(h_2 u_2^2 + \frac{g}{2} h_2^2\right)}{\partial x} = -g h_2 \left(\frac{\rho_1}{\rho_2}\frac{\partial h_1}{\partial x} + \frac{\partial Z}{\partial x}\right), \tag{4}$$

where the water height is denoted by $h(x,t)$, the lateral velocity by $u(x,t)$, the gravitational acceleration by g, and the bottom elevation by $Z(x)$. The systems (1-2) and (3-4) describe the mass and momentum equations of first and second layers respectively. Moreover, the space and time are denoted by x and t respectively. The detail of primitive variables from (1-4) can be seen in Figure 1. In Figure 1, the water level of second layer with density ρ_2 is denoted by $\eta_2(x,t) = h_2(x,t) + Z(x)$. Meanwhile, $\eta_1(x,t) = h_1(x,t) + h_2(x,t) + Z(x)$ denotes the water level of first layer with density ρ_1.

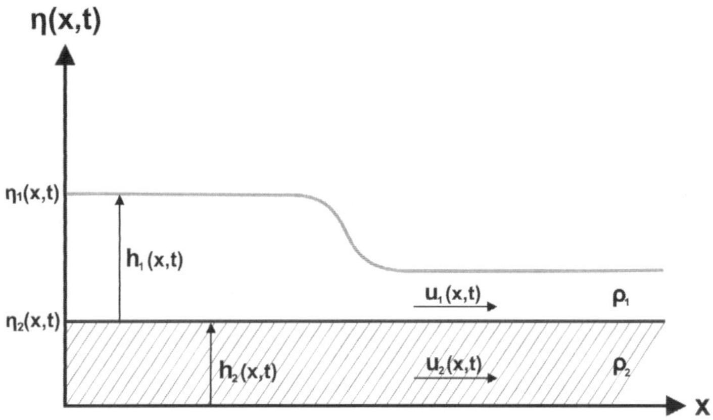

Figure 1. The configuration of primitive variables from two-layer shallow water equations.

Here, the source-centered hydrostatic reconstruction (SCHR) scheme will be used to approximate the (1-4). This numerical method is satisfied the mathematical properties for SWE model [6, 7, 8, 9]. The properties are such as, preserving the non-negative of water height ($h > 0$), satisfying well-balanced, discrete entropy balance, and the dry or vacuum condition ($h = 0$).

The goal of this paper is to perform the two-layer SWE for erodible dambreak problem using SCHR scheme and to analyze the performance of its parallel computing. Here, the numerical simulation will be compared with the data experiment given in [2, 10]. Originally, the experiment were performed in Louvain-la-Neuve Université Catholique de Louvain, which is reported by Fraccarollo and Capart 2002 [11]. Moreover, in this paper, the parallel computing in multicore architecture with OpenMP platform will be elaborated [12].

2. Source-centered Hydrostatic Reconstruction Scheme

In this paper, the source-centered hydrostatic reconstruction (SCHR) scheme will be used to discretize (1-4). This scheme is a modified original hydrostatic reconstruction (HR) scheme [7, 8, 13] from one-layer SWE. The modification is made due to the momentum conservation problem occurs when the original HR is used to discretized two-layer SWE. In [14], two numerical schemes which are called splitting and sum scheme use original HR for approximating two-layer SWE. However, both of them produce wrong solutions because of the total momentum did not conserve. Therefore in [6], the SCHR is introduced to approximate the two-layer SWE which is a robust scheme.

Here, the brief explanations about SCHR will be given, the more detail about this scheme can be found in [6]. Given the discrete domain of 1D spatial space $\Omega = [-L : L]$, $L \in \mathbb{R}$ and time $T = [0, \infty)$ as follows

$$\Delta x = \frac{2L}{N_x}, \quad N_x \in \mathbb{Z}^+, \qquad x_j = j \times \Delta x - L, \quad j \in \mathcal{M} = \{0, 1, \cdots, N_x\},$$

and

$$t^n = n \times \Delta t, \quad n \in \mathcal{T} = \{0, 1, \cdots\}.$$

Here, the discrete space Δx is considered as the uniform step and the time step Δt is depend on the stability condition (16). Let the primitive variables in (1-4) in a vector form,

$$\mathbb{U} = \begin{pmatrix} h_i \\ h_i u_i \end{pmatrix}, \qquad i \in \{1, 2\}. \tag{5}$$

Therefore, the finite volume method for approximating (1-4) is given as

$$\mathbb{U}_{i,j}^{n+1} = \mathbb{U}_{i,j}^n - \frac{\Delta t}{\Delta x}\left(\mathcal{F}_l(\mathbb{U}_{i,j}^n, \mathbb{U}_{i,j+1}^n, b_{i,j}^n, b_{i,j+1}^n) - \mathcal{F}_r(\mathbb{U}_{i,j-1}^n, \mathbb{U}_{i,j}^n, b_{i,j-1}^n, b_{i,j}^n)\right),$$
$$i \in \{1, 2\}, \quad j \in \mathcal{M}, \quad n \in \mathcal{T}. \tag{6}$$

where

$$b_i = \sum_{k<i}^{2} \frac{\rho_k}{\rho_i} h_k + \sum_{k>i}^{2} h_k + Z_i, \qquad i \in \{1, 2\}. \tag{7}$$

Here, the variable $\mathbb{U}_j^n$ denotes the the value of variable $\mathbb{U}$ at point x_j and current time t^n. The important part of SCHR is to define the numerical fluxes such that satisfy the total momentum conservation. The numerical scheme SCHR defines the numerical fluxes $\mathcal{F}_{l/r}(u_l, u_r, b_l, b_r)$ as follows

$$\mathcal{F}_l = \mathcal{F}_l^{HR} + \mathcal{P}_l, \qquad \mathcal{F}_r = \mathcal{F}_r^{HR} + \mathcal{P}_r, \tag{8}$$

where $\mathcal{F}_{l/r}^{HR}$ is the numerical fluxes of the original hydrostatic reconstruction scheme for one-layer SWE which can be found in [8] for more detail and the additional terms $\mathcal{P}_{l/r} = (\mathcal{P}^0, \mathcal{P}_{l/r}^1)$ for the balancing in momentum conservation.

Proposition 1. *In order to satisfy the semi-discrete entropy condition, thus $\mathcal{P}^0$ and $\mathcal{P}_{l/r}^1$ are defined as follows*

$$\mathcal{P}^0 = \frac{\kappa\left((1+\theta)u_l + (1-\theta)u_r\right)}{2} \tag{9}$$

$$\mathcal{P}_l^1 = \frac{\kappa(1+\theta)(h_r - h_l + \Delta b)g}{2} + u_l \max(\mathcal{P}^0, 0) + u_r \min(\mathcal{P}^0, 0) \tag{10}$$

$$\mathcal{P}_r^1 = -\left(\frac{\kappa(1-\theta)(h_r - h_l + \Delta b)g}{2}\right) + u_l \max(\mathcal{P}^0, 0) + u_r \min(\mathcal{P}^0, 0) \tag{11}$$

with θ and κ are the parameters for hydrostatic solver which are given as follows

$$\theta = \min\left(1, \frac{\max(u_l, 0)}{\sqrt{gh_l}}\right) - \max\left(1, \frac{\max(-u_r, 0)}{\sqrt{gh_r}}\right), \tag{12}$$

$$\kappa = \min(\bar{\kappa}, 2.5 \times \min(h_l, h_r)), \tag{13}$$

$$\bar{\kappa} = \frac{|\overline{\Delta b}|}{2} \begin{cases} \dfrac{(\Delta b - h_l)h_r}{2(h_r + \Delta b - h_l)} & \textit{if } \Delta b > h_l, \\ 0 & \textit{if } -h_l \leq \Delta b \leq h_l, \\ \dfrac{-(h_r + \Delta b)h_l}{2h_l - 2(h_r + \Delta b)} & \textit{if } \Delta b < -h_r, \end{cases} \tag{14}$$

$$\overline{\Delta b} = \begin{cases} \min(\Delta b, h_l) & \textit{if } \Delta b \geq 0, \\ \max(\Delta b, -h_r) & \textit{otherwise.} \end{cases} \tag{15}$$

Proof. The proof of this proposition can be found in the paper of Bouchut [6] in detail. □

The numerical scheme (6) with the numerical fluxes of SCHR (8 - 15) has the stability condition in order to keep the positivity of water height for each layer. The stability condition of the SCHR scheme is given as follows

$$\frac{\Delta t \times \lambda(\mathbb{U}^n_{i,j}, \mathbb{U}^n_{i,j+1}, b^n_{i,j+1} - b^n_{i,j})}{\Delta x} \leq 1, \tag{16}$$

where the function $\lambda(\mathbb{U}^n_{i,j}, \mathbb{U}^n_{i,j+1}, b^n_{i,j+1} - b^n_{i,j})$ is the wave speed which can found in [6] in detail.

3. Parallel Algorithm for SCHR Numerical Scheme

OpenMP platform is one of shared memory architecture in parallel computing which are the threads allowed to access main memory for executing algorithm at the same time. In this paper shared parallel by using OpenMP platform will be used for computing the numerical scheme of two-layer because this platform is simple to implement and straightforward [12, 15, 16, 17, 18, 19]. Moreover, the performance of parallel implementation can be evaluated by measuring speedup and efficiency [20]. Figure 2 shows the parallel implementation for computing the numerical scheme.

The parallel algorithm of erodible dambreak simulation is described as follow :

- **Initializing all parameter**: here the parallel computing will be started. Initialize the values of water elevation, density, and velocity for each layer and spatial grid.
- **Computing the mass conservation**: the unknown values $h^{n+1}_{1,j}$ and $h^{n+1}_{2,j}$ will be computed using (6).
- **Computing the momentum conservation**: the momentum equation (6) will be used for computing the unknown variables of velocity u.
- **Update the previous values**: all unknowns variables should be updated for the next time iteration.
- **Output of Simulation**: the data of unknown variables and CPU time are collected.

In this algorithm, each processors will execute the same process on parallel part. The number of tasks in loop process is depend on the number of processors. For instance, using number of grids N_x and processor p, the maximum number of tasks for a processor is defined as $\lceil N_x/p \rceil$.

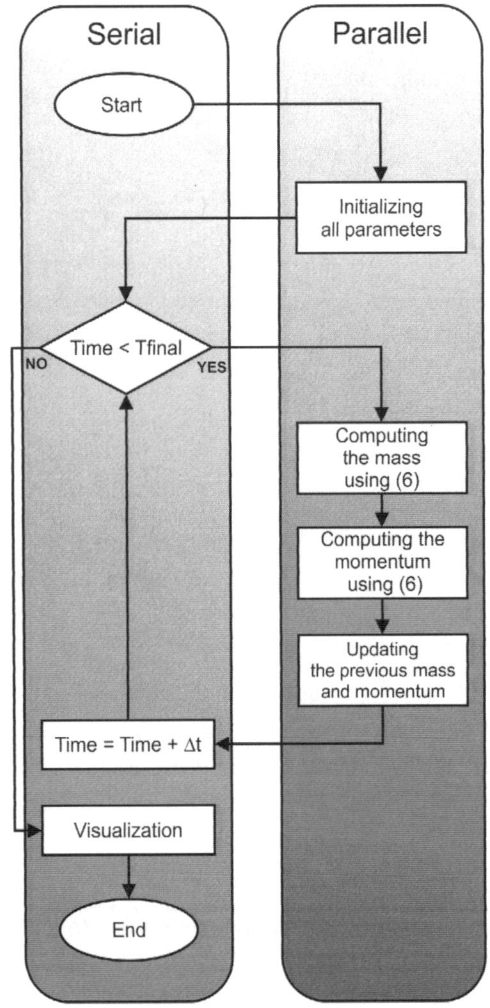

Figure 2. The algorithm consists of serial and parallel part. These two parts are given depend on the subroutines or loops which can be parallelized or not. In this rough algorithm, the parallel code is implemented in the core of computing numerical scheme and time consuming. In this case, computing the mass, momentum and updating the primitive variables are a time-consuming subroutine since they are repeated during simulation.

4. Numerical Simulations

In this section, the numerical simulation of erodible dambreak using SCHR will be given. The comparison numerical result of two-layer SWE using SCHR with data experiment obtained in [2, 10, 11] also will be elaborated in Subsection 4.1. There are two experiments of erodible dambreak flow can be found in the paper of Capart and Young 1998 [21] and [11] which are performed in Taipe Univeristy of Taiwan and Louvain-la-Neuve Université Catholique de Louvain respectively. In this paper, the numerical results will be compared only with the data experiment in Louvain. Moreover, the parallel performance with OpenMP platform to accelerate the computational time will be given in Subsection 4.2.

4.1. Simulation of erodible dambreak

In this simulation, the comparison of results two-layer SWE using SCHR scheme and data experiment in Louvain will be elaborated. The initial condition of erodible dambreak is given

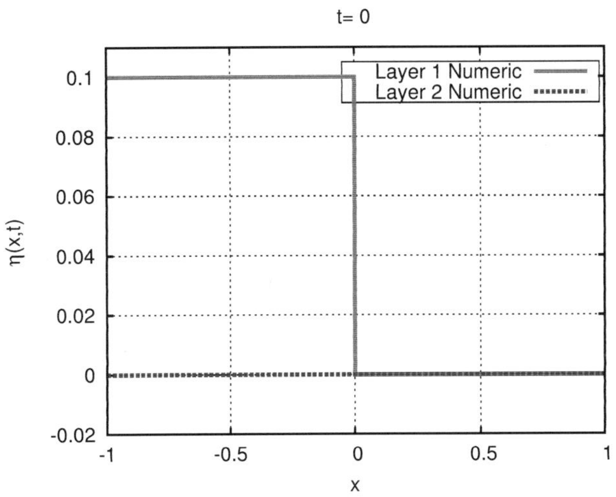

Figure 3. The initial condition of erodible dambreak simulation with spatial domain of this simulation is $\Omega = [-1 : 1]$. Here, first and second layer describe the water surface and erodible bottom or sediment respectively. The density of each layers will be defined in a ratio in the simulation.

in Figure 3 and described as the following equations,

$$\eta_1(x,0) = \begin{cases} 0.1 & \text{if } x < 0, \\ 0 & \text{otherwise,} \end{cases}, \tag{17}$$

$$\eta_2(x,0) = 0, \tag{18}$$

$$u_1(x,0) = u_2(x,0) = 0. \tag{19}$$

First layer acts as the water surface and second layer is used to describe the sediment or erodible bottom. At the initial condition, velocity of water and sediment is set to be zero and the sediment is given in a flat profile.

The results using the ratio of $\rho_1/\rho_2 = 10^{-3}$, 6×10^{-2}, 10^{-1}, 4×10^{-1}, 6×10^{-1} and 8×10^{-1} are given in Figure 4. It can be seen clearly that, the numerical results are depend on the ratio of density of each layer. If the ratio of density ρ_1/ρ_2 small enough, thus the influence of first layer to the second layer is small and less erodible of sediment can be obtained. Meanwhile, when the ratio is high, the relation between water and sediment is more strong. Therefore the water flow causes the erodible on bottom of simulation.

From the results (Figure 4), the lines in the upper and bottom are shown the numerical solutions by SCHR of first and second layer respectively. Meanwhile, the dots '+' and '*' are described the data experiment of water surface and sediment respectively. Figures 4(a), (b) and (c) show the numerical solutions of the water surface and sediment are slightly far from the experiment data. These results are using the ratio of density ≤ 0.1. In other side using the ratio of density > 0.1, Figures 4(d), (e) and (f) show the numerical solutions is in a good agreement with the data experiment for both water surface and sediment profile.

The experiment of erodible dambreak in Louvain uses the sediment porosity $\phi = 0.3$, diameter of sediment particle 3.2 mm, and sediment density $1,540$ kg/m^3. If the water density is given $1,000$ kg/m^3, thus the ratio of water and sediment is obtain around 0.64. Therefore, without consider the porosity and diameter of sediment particle, the numerical result with density ratio $\rho_1/\rho_2 = 0.6$ can be seen close enough to the data experiment (see Figure 4(e)).

4.2. Parallel performances using OpenMP platform

In this paper, three different computers are used for obtaining the performance of parallel algorithm. The specifications of each computer are shown on Table 1. In order to obtain the

IOP Conf. Series: Journal of Physics: Conf. Series **971** (2018) 012034 doi:10.1088/1742-6596/971/1/012034

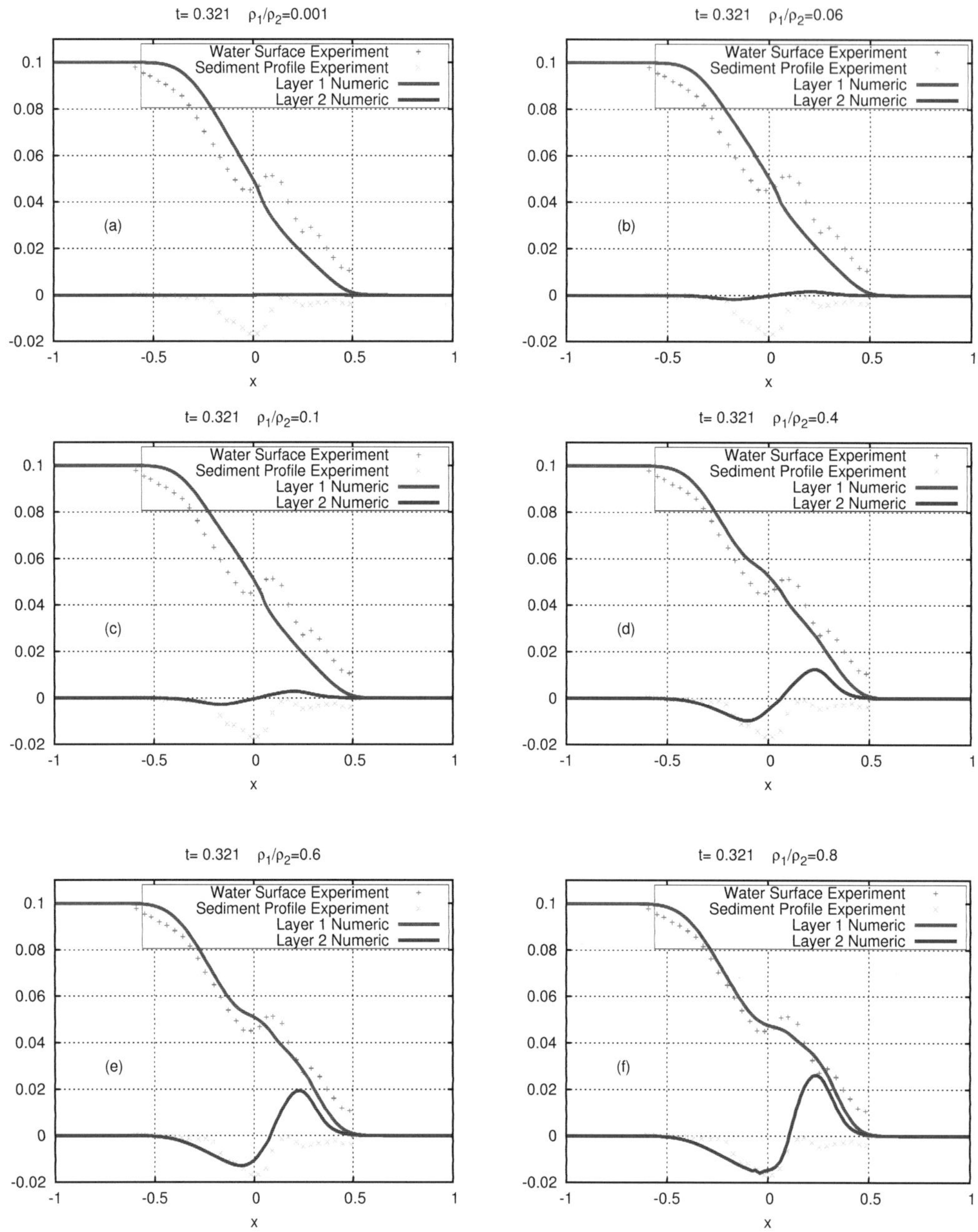

Figure 4. The comparison of two-layer SWE and SWE-Exner model at final time of simulation $t = 0.36$ seconds. (a) The comparison using ratio $\rho_1/\rho_2 = 10^{-3}$. (b) The comparison using ratio $\rho_1/\rho_2 = 6 \times 10^{-2}$. (c) The comparison using ratio $\rho_1/\rho_2 = 10^{-1}$. (d) The comparison using ratio $\rho_1/\rho_2 = 4 \times 10^{-1}$. (e) The comparison using ratio $\rho_1/\rho_2 = 6 \times 10^{-1}$. (f) The comparison using ratio $\rho_1/\rho_2 = 8 \times 10^{-1}$.

performance of each computers, here, the parallel computing use 4 (four) cores/threads for each computers.

Table 1. The specifications of three computers for measuring the performance of parallel computing in the simulation.

Computer	Operating System	Processor	Memory
I	Ubuntu 12.04.3 LTS	Intel(R) Core(TM) i5-2500 CPU Quad-Core	8 GB
II	Ubuntu 14.04.3 LTS	AMD Opteron 63xx class CPU Quad-Core	8 GB
III	Ubuntu 14.04.3 LTS	AMD A6-5200 APU Quad-Core	6 GB

The serial and parallel CPU time using Computer I is given in Table 2. Using processor Intel(R) Core(TM) i5-2500 CPU Quad-Core, the CPU time of serial and parallel is obtained 2.406260 s and 0.763619 s respectively using grid sizes 3200. The speedup is shown increasing along the increasing of grid sizes and reached 3.15 times when the number of grid sizes is 3200.

Table 2. The CPU time and speedup performance of parallel implementation of Computer I.

N_x	Serial time	Parallel time	Speedup
200	0.168599	0.056266	2.996463228
400	0.328057	0.108305	3.029010664
800	0.622869	0.202799	3.071361299
1600	1.178440	0.379376	3.106258698
3200	2.406260	0.763619	3.151126413

The parallel performance using Computer II with processor AMD Opteron 63xx class CPU Quad-Core can be seen in Table 3. By various numbers of grid sizes, the CPU time of serial and parallel is shown increasing along the N_x is doubled. When the $N_x = 200$, the CPU time in serial code is 0.400121 s, which is approximately 2.9 times the parallel time. In other side, by $N_x = 3200$ the CPU time in parallel code is obtained 2.008820 s which is almost 1/3.15 times of serial time.

Table 3. The CPU time and speedup performance of parallel implementation of Computer II.

N_x	Serial time	Parallel time	Speedup
200	0.400121	0.136395	2.933545951
400	0.777113	0.258098	3.010922208
800	1.525450	0.500182	3.049789876
1600	3.063830	0.988666	3.098953539
3200	6.298810	2.008820	3.135577105

Computer III with processor AMD A6-5200 APU Quad-Core gives the parallel performance as shown in Table 4. Using this computer, the CPU time of serial and parallel using $N_x = 3200$ is 8.798530 s and 2.367710 s respectively. Thus the speedup of parallel computing is reached 3.71 times than the serial time.

Table 4. The CPU time and speedup performance of parallel implementation of computer III.

N_x	Serial time	Parallel time	Speedup
200	0.589111	0.173978	3.386123533
400	1.161250	0.330499	3.513626365
800	2.334270	0.648217	3.601062607
1600	4.605950	1.264240	3.643256027
3200	8.798530	2.367710	3.716050530

From Tables 2, 3 and 4, it can be seen that Computer I with processor Intel(R) Core(TM) i5-2500 CPU Quad-Core has the best performance of CPU time for both serial and parallel computing. However, according the speedup measurement, Computer III with processor AMD A6-5200 APU Quad-Core has the best performance for speedup. The speedup is obtained more than 3.3 times along the increasing of the grid size N_x. Comparing with the Computer I and II, their speedup (around 3.15 and 3.13 times respectively) is observed less than the speedup of Computer III which is achieved 3.71 times using $N_x = 3200$.

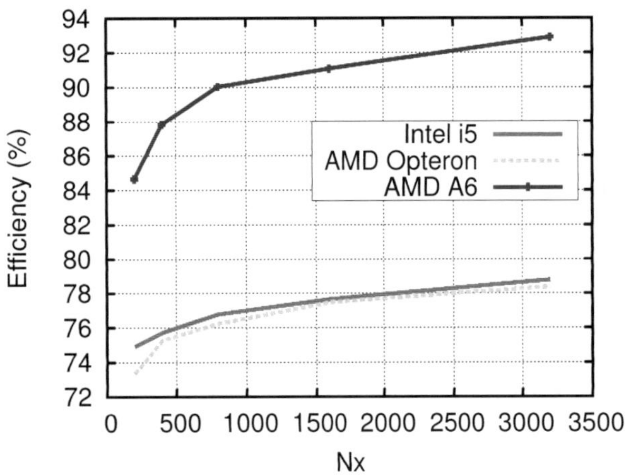

Figure 5. The efficiency profile for measuring the parallel performance using three computers. The straight line shows the efficiency of Computer I. The efficiency of Computer II is shown with a dash line. Moreover, the line-points is described the efficiency of Computer III.

Additionally, Computer III also has a good performance of efficiency using 4 cores/threads which is shown in Figure 5. The gap between efficiency of Computer III and the rest of computers is shown slightly far. For instance, using $N_x = 200$ the efficiency for Computer I, II and II are observed 74.9%, 73.3% and 84.6% respectively. Moreover, by $N_x = 3200$, the efficiency using Computer I, II and III are obtained 78.7%, 78.3% and 92.9% respectively.

Remark 1. *Computers I and II are the server computers, whereas computer III is the personal computer (PC). The performance of CPU time for server computers is observed better than PC. In terms of computational time, the server computer will obviously be superior to the PC due to its high specifications such as memory and processor type. However, the speedup and efficiency by using computer server are obtained lower than PC when using number of grids* $N_x = 3200$. *This can happen because the high specification computer is not efficient enough to compute small number of grids. Nevertheless, it would be different if the large number of grids is used.*

5. Conclusion

The comparison of numerical solution of two-layer SWE with the data experiment in erodible dambreak simulation is elaborated. Without considering the characteristics of sediment, the numerical solutions for ratio of density greater than 0.1 are shown in a good agreement with the data experiment in Louvain case. In parallel performance during erodible dambreak simulation, Computer I with processor Intel(R) Core(TM) i5-2500 CPU Quad-Core is shown the best computer in order to accelerate the computational time. However, the best measurement in speedup and efficiency is observed for Computer III with processor AMD A6-5200 APU Quad-Core. Using the maximum grid size $Nx = 3200$, the speedup is obtained 3.716050530 times and the efficiency is calculated 92.9%.

References

[1] Song L, Zhou J, Li Q, Yang X and Zhang Y 2011 *International Journal for Numerical Methods in Fluids* **67** 960–980
[2] Gunawan P H and Lhébrard X 2015 *Computers & Fluids* **121** 44–50
[3] Simpson G and Castelltort S 2006 *Computers & Geosciences* **32** 1600–1614
[4] Cao Z, Pender G, Wallis S and Carling P 2004 *Journal of hydraulic engineering* **130** 689–703
[5] Gunawan P, Eymard R and Pudjaprasetya S 2015 *Computational Geosciences* **19** 1197–1206
[6] Bouchut F and Zeitlin V 2010 *Discrete and Continuous Dynamical Systems-Series B* **13** 739–758
[7] Bouchut F 2004 *Nonlinear stability of finite Volume Methods for hyperbolic conservation laws: And Well-Balanced schemes for sources* (Springer Science & Business Media)
[8] Audusse E, Bouchut F, Bristeau M O, Klein R and Perthame B t 2004 *SIAM Journal on Scientific Computing* **25** 2050–2065
[9] Doyen D and Gunawan P H 2014 *Finite Volumes for Complex Applications VII-Methods and Theoretical Aspects* (Springer) pp 227–235
[10] Wu W and Wang S S 2007 *Journal of hydraulic engineering* **133** 48–58
[11] Fraccarollo L and Capart H 2002 *Journal of Fluid Mechanics* **461** 183–228
[12] Gunawan P H 2016 *Information and Communication Technology (ICoICT), 2016 4th International Conference on* (IEEE) pp 1–5
[13] Bouchut F 2007 *Edited Series on Advances in Nonlinear Science and Complexity* **2** 189–256
[14] Bouchut F and de Luna T M 2008 *ESAIM: Mathematical Modelling and Numerical Analysis* **42** 683–698
[15] Iryanto and Gunawan P H 2017 *Information and Communication Technology (ICoIC7), 2017 5th International Conference on* (IEEE) pp 1–5
[16] Alamsyah M N A, Simanjuntak C A, Bagustara B A R H, Pradana W A and Gunawan P H 2017 *Information and Communication Technology (ICoIC7), 2017 5th International Conference on* (IEEE) pp 1–6
[17] Pahlevi M R and Gunawan P H 2017 *Information and Communication Technology (ICoIC7), 2017 5th International Conference on* (IEEE) pp 1–4
[18] Juliati S and Gunawan P H 2017 *Information and Communication Technology (ICoIC7), 2017 5th International Conference on* (IEEE) pp 1–5
[19] Mulyani, Putri N D and Gunawan P H 2017 *Information and Communication Technology (ICoIC7), 2017 5th International Conference on* (IEEE) pp 1–5
[20] Rauber T and Rünger G 2013 *Parallel programming: For multicore and cluster systems* (Springer Science & Business Media)
[21] Capart H and Young D 1998 *Journal of Fluid Mechanics* **372** 165–187

Smoothed particle hydrodynamics method for simulating waterfall flow

M G Suwardi, Jondri, and D Tarwidi

School of Computing, Telkom University, Jalan Telekomunikasi No. 1 Terusan Buah Batu, Bandung 40257, Indonesia

E-mail: ghazali.suwardi@gmail.com

Abstract. The existence of waterfall in many nations, such as Indonesia has a potential to develop and to fulfill the electricity demand in the nation. By utilizing mechanical flow energy of the waterfall, it would be able to generate electricity. The study of mechanical energy could be done by simulating waterfall flow using 2-D smoothed particle hydrodynamics (SPH) method. The SPH method is suitable to simulate the flow of the waterfall, because it has an advantage which could form particles movement that mimic the characteristics of fluid. In this paper, the SPH method is used to solve Navier-Stokes and continuity equation which are the main cores of fluid motion. The governing equations of fluid flow are used to obtain the acceleration, velocity, density, and position of the SPH particles as well as the completion of Leapfrog time-stepping method. With these equations, simulating a waterfall flow would be more attractive and able to complete the analysis of mechanical energy as desired. The mechanical energy that generated from the waterfall flow is calculated and analyzed based on the mass, height, and velocity of each SPH particle.

1. Introduction

Indonesia is rich in diversity of natural ecosystem, including waterfall which spreads in the majority of regions. Besides as tourism destination, a waterfall is also as natural resource which is renewable. Energy which come from waterfall is an opportunity and chance to use as sustainable power plants, while the Indonesia is lack of innovation in utilization of renewable resource.

Principle of hydroelectric power plants of waterfall is by changing water flow to mechanical energy (using water turbine) and from the mechanical energy to electrical power (using generator). Waterfall and its mechanical energy surely could help fulfill electrical needed in Indonesia. Amount of the energy resulted could be analyzed with simulating its water flow. The problem could be simulated with several methods. The waterfall simulation had been studied by Uchiyama et al. [1] which was using two-dimensional moving particle semi-implicit (MPS) method. However, another method that is used frequently and implemented in many fields of study is smoothed particle hydrodynamics (SPH) method. While there are numerous alternative methods, such as finite difference and finite element but it could not resolved the discontinuity of fluid motion.

Kellomaki [2] had studied the large-scale water simulation in games in which he compared some methods performance. It was concluded that SPH-method could provide the possibility of almost abundant water topology changes which could not be provided by most other methods.

SPH method is easy to implement and work best when applied to shallow water equation. Moreover, Onderik [3] build an animation of particle based miscible and immiscible fluids with multiple interfaces by using SPH method.

SPH method is an approximation of Lagrangian for computation in field of fluid dynamics. It model fluid flow as group of particles that move in real-time by using law of Hydrodynamics and gravitational force. The advantages of SPH method in fluid simulation can be seen more obvious in [4]. SPH could simulate phenomenon which has domain of complex geometry while other methods were not easy to implement. This paper focuses on SPH method in simulating waterfall flow for analyzing influence of height to mechanical energy resulted.

2. SPH Theory

In this section, SPH theory to simulate water flow will be briefly discussed. The mechanical energy is calculated by summing up the contribution of mechanical energy of each particles.

2.1. Waterfall's mechanical energy

The first Thermodynamics Law explains on conservation of energy, which means that amount of energy at the before and the end of process is the same. Energy could not be demolished as well as be created, but energy could evolve from one form to other form of energy [5]. Furthermore, mechanical energy is accumulation between kinetic and potential energy. Waterfalls mechanical energy could be implemented in hydroelectric power plant.

Process of hydroelectric power plant is by converting waters potential energy become mechanical energy by turbine and then the energy is converted to electrical energy by generator. Waters height and velocity also support the process. Electrical power is obtained by using waterfalls height. Mechanical energy is given by

$$\text{Mechanical energy} = \frac{1}{2}mv^2 + mgh, \tag{1}$$

where m is waters mass, v is waters velocity, g is gravitational acceleration, and h is waterfalls height from ground. In SPH method, the amount of energy can be noted as discrete system (ignoring external parameter) [6]. It is written as

$$E = \sum_i m_i \left(\frac{1}{2}v_i + u_i\right). \tag{2}$$

The subscript i represents the parameter is belongs to particle i. Moreover, u denotes particles heat energy. Because this simulation just explain mechanical energy, thus heat energy should be ignored. Hence, we could use equation (1) for calculating particles mechanical energy. Therefore, in SPH method formulation, the mechanical energy can be written as

$$\text{Mechanical energy} \approx \sum_{i=1}^{n} \frac{m_i\, v_i^2}{2} + m_i\, g\, y_i \tag{3}$$

where, y_i is position of particle i in y axis, which means as particles height.

2.2. SPH Method

In this simulation, we implement SPH method by doing Lagrangian approximation for computing fluid dynamics. In the method, the fluid is represented as a collection of particles which have fluid properties, such as mass, density, velocity, acceleration, and pressure [7]. In each particle, they have velocity, acceleration and position. They have spatial distance, which then be meshed by kernel function that could evaluate accuracy of physical quantity of each particle.

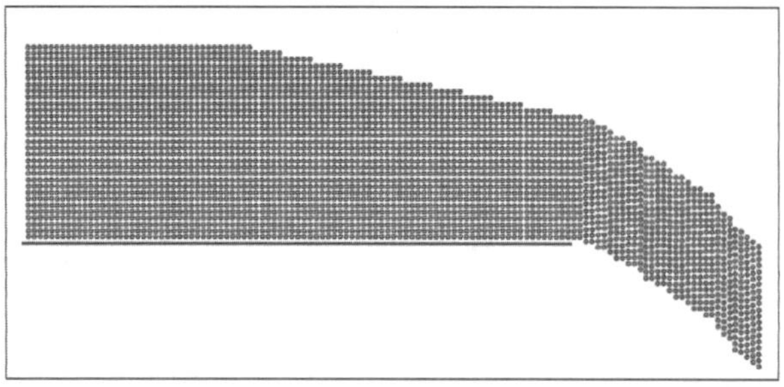

Figure 1. Illustration of SPH particle in waterfall flow simulation with SPH method.

One disadvantage of the SPH method is that its simulation is slower than the grid method. The reason is because particles move freely between one another, so it takes more time to find neighbor particles. In addition, it is necessary to calculate the distance between each pair of neighbor particles involving the calculation of square roots [8]. However, the SPH method results is easier when it is visualized compared to the grid method. Figure 1 shows illustration of particles motion in waterfall flow with SPH method.

There are at least four steps to be considered in the SPH approximation:

- The continuum domain is represented by a series of randomly distributed particles.
- Functions such as density, velocity, and energy are formulated by integral formulas.
- The integral formula of the function is converted into a finite sum of the neighbor particles.
- The kernel function used to satisfy the numerical accuracy of the results.

SPH method is started by giving integral formula for continue equation $f(x)$:

$$f(\mathbf{x}) = \int_{\Omega} f(\acute{\mathbf{x}})\delta(\mathbf{x} - \acute{\mathbf{x}})d\acute{\mathbf{x}}, \tag{4}$$

where x is position vector and Ω is the domain of integral which consist of x. And also, $\delta(\mathbf{x} - \acute{\mathbf{x}})$ is function of Dirac delta which can be described as follows

$$\delta(\mathbf{x} - \acute{\mathbf{x}}) = \begin{cases} \mathbf{x} = 1, & \text{if } \mathbf{x} = \acute{\mathbf{x}} \\ \mathbf{x} = 0, & \text{if } \mathbf{x} \neq \acute{\mathbf{x}}. \end{cases} \tag{5}$$

If Dirac delta function is replaced by a smoothing function, W, it yields the kernel approximation of $f(x)$:

$$< f(\mathbf{x}) > = \int_{\Omega} f(\acute{\mathbf{x}})W(\mathbf{x} - \acute{\mathbf{x}}, h)d\acute{\mathbf{x}}, \tag{6}$$

where h represents the influence area of smoothing function W which is called smoothing length. Illustration of smoothing length of kernel function is shown by Figure 2. Liu and Liu [9] derived kernel approximation for $\nabla \cdot f(x)$ by implementing Divergence Theorem. It is given by

$$< \nabla \cdot f(\mathbf{x}) > = -\int_{\Omega} f(\acute{\mathbf{x}})\nabla \cdot W(\mathbf{x} - \acute{\mathbf{x}}, h)d\acute{\mathbf{x}}. \tag{7}$$

The next step of the SPH method is to discretize a continuous function into a set of particles. The integral starts with the number of neighbor particles of particle i in position x_i in the domain of influence Ω_i. The object divided into small parts with mass m_1, m_2, ..., m_N, then the volume of each part is $\Delta V_j = m_j/\rho_j$. Therefore, the particle approaches of equations (6) and (7) are

$$< f(\mathbf{x}) > = \sum_{j=1}^{N} \frac{m_j}{\rho_j} f(\mathbf{x}_j) W(\mathbf{x} - \mathbf{x}_j, h), \tag{8}$$

$$< \nabla \cdot f(\mathbf{x}) > = - \sum_{j=1}^{N} \frac{m_j}{\rho_j} f(\mathbf{x}_j) \cdot \nabla W(\mathbf{x} - \mathbf{x}_j, h), \tag{9}$$

where N represents the number of neighbor particles, m_j and ρ_j are mass and density of particle j.

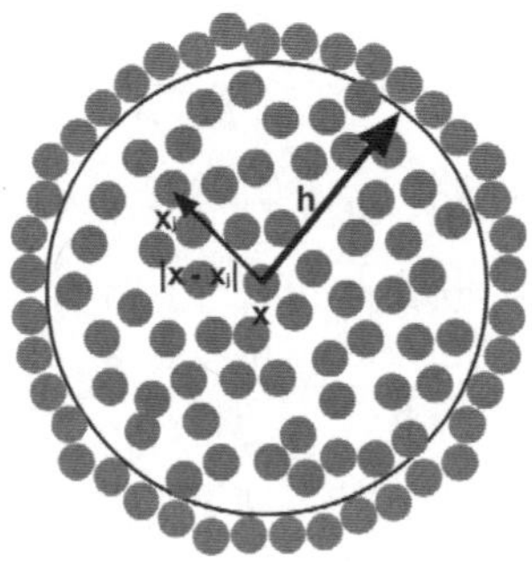

Figure 2. Smoothing length of kernel function.

2.3. Kernel function

The accuracy of SPH method depends on kernel function. In order to implement kernel function, it should fulfill several conditions as follow [6]:

- Normalization condition: $\int_\Omega W(\mathbf{x} - \acute{\mathbf{x}}, h)\, d\Omega = 1$
- Delta function property: $\lim_{h \to 0} W(\mathbf{x} - \acute{\mathbf{x}}, h) = \delta(\mathbf{x} - \acute{\mathbf{x}})$
- Compact support: $W(\mathbf{x} - \acute{\mathbf{x}}, h) = 0$ outside Ω domain.

The are many types of kernel function. One of them is cubic spline. This simulation uses kernel function of cubic spline which is frequently used in SPH simulation:

$$W(r,h) = \frac{10/7\pi}{h^2} \begin{cases} 1 - \frac{3}{2}(\frac{r}{h})^2 + \frac{3}{4}(\frac{r}{h})^3, & 0 \leqslant \frac{r}{h} < 1 \\ \frac{1}{4}(2 - \frac{r}{h})^3, & 1 \leqslant \frac{r}{h} < 2 \\ 0, & \frac{r}{h} \geqslant 2, \end{cases} \tag{10}$$

where r is distance between particle and its neighbor particle. Moreover, the derivative of $W(r,h)$ can be written as

$$\frac{\partial W(r,h)}{\partial r} = \frac{10/7\pi}{h^2} \begin{cases} -3(\frac{r}{h}) + \frac{9}{4}(\frac{r}{h})^2, & 0 \leqslant \frac{r}{h} < 1 \\ -\frac{3}{4}(2 - \frac{r}{h})^2, & 1 \leqslant \frac{r}{h} < 2 \\ 0, & \frac{r}{h} \geqslant 2. \end{cases} \tag{11}$$

2.4. Governing equation

Governing equation for the fluid dynamics is momentum equation (Navier-Stokes equation), continuity equation, and equation of state [10].

2.4.1. Momentum equation Navier-Stokes or momentum equation in continuous domain can be writen as

$$\frac{D\mathbf{v}}{Dt} = -\frac{1}{\rho}\nabla p + \mathbf{F} \tag{12}$$

where $\mathbf{v}$, ρ, and $\mathbf{F}$ are velocity, density, and external force per mass of fluid, respectively. In this case, surface tension is assumed that do not significantly happen, thus it could be ignored. In SPH formulation, momentum equation could be written as

$$\frac{\mathrm{d}\mathbf{v}_i}{\mathrm{d}t} = -\sum_{j=1}^{N} m_j \left(\frac{P_j}{\rho_j^2} + \frac{P_i}{\rho_i^2}\right)\nabla_i W(\mathbf{x}_i - \mathbf{x}_j, h) + \mathbf{F}_i \tag{13}$$

2.4.2. Continuity equation It is assumed that the fluid is weakly compressible. It means that the density of fluid change with time, but the change of density is not significant. The continuity equation can be written as

$$\frac{D\rho}{Dt} = -\rho\nabla \cdot \mathbf{v} \tag{14}$$

Equation (14) above could be written in SPH notation as

$$\frac{\mathrm{d}\rho_i}{\mathrm{d}t} = \sum_{j=1}^{N} m_j(\mathbf{v}_i - \mathbf{v}_j) \cdot \nabla_i W(\mathbf{x}_i - \mathbf{x}_j, h). \tag{15}$$

2.4.3. Equation of state Equation of state represents relationship between density and pressure. In simple form, the equation of state is given by

$$P = c^2(\rho - \rho_0) + P_{atm} \tag{16}$$

where c, ρ_0, and P_{atm} are speed of sound, density reference, atmospheric pressure, respectively. The choice of the speed of sound is very important in the simulation because it influences the change of density. The speed of sound can be selected as $c = \sqrt{gH}$, in order to keep fluctuation of density less than 1% [10].

2.5. Equation of particle movement

Particle movement can be calculated by solving the equation as follow [11]

$$\frac{d\mathbf{r}_a}{dt} = \mathbf{v}_a \tag{17}$$

or XSPH variant

$$\frac{d\mathbf{r}_a}{dt} = \hat{\mathbf{v}}_a = \mathbf{v}_a + \epsilon \sum_b m_b \left(\frac{v_{ba}}{\bar{\rho}_{ab}}\right) W_{ab} \tag{18}$$

where $\bar{\rho} = (\rho_a + \rho_b)/2$ and $\epsilon\,(0 \leq \epsilon \leq 1)$ is a constant. The XSPH variant moves a particle with velocity which is close to average of its neighborhood [12].

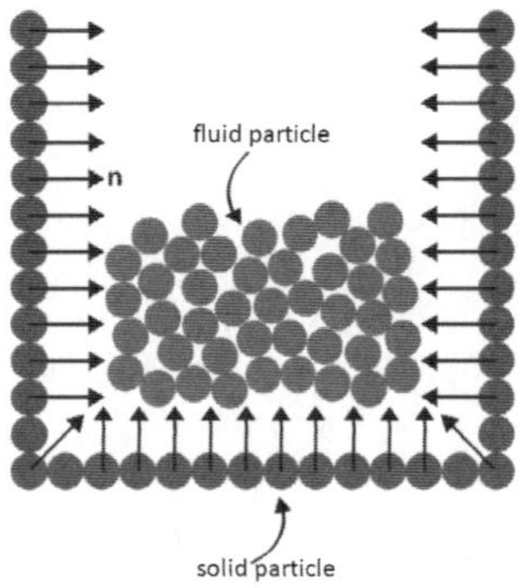

Figure 3. Solid particles boundary with its normal vector.

2.6. Time stepping

Time stepping is differential function which would run the simulation movement (plot of particles position) each step. In this simulation, SPH method is solved by Leapfrog method of time stepping. This method uses general formula in Euler, that is [13]

$$x = x + \frac{h}{2}v, \tag{19}$$

where h is interval of time stepping (time difference). Depend on numerical integration (midpoint method) which is used in this Leapfrog method, we could change v with midpoint of interval [14]:

$$x_1 = x_0 + hv_{1/2}. \tag{20}$$

Assuming that we can calculate $v_{1/2}$ by several ways. Then, we can implement rule of midpoint for calculating value of v at the next step.

$$v_{3/2} = v_{1/2} + hF(x_1) \tag{21}$$

Then we receive the next step x by $x_2 = x_1 + hv_{3/2}$. After we start with x_0 and $v_{1/2}$, we can continue x and v by using Leapfrog method and relating one and each other:

$$x_{n+1} = x_n + hv_{n+1/2}, \tag{22}$$

where

$$v_{n+3/2} = v_{n+1/2} + hF(x_{n+1}). \tag{23}$$

2.7. Boundary treatment

In SPH approximation, particle position should be paid attention when it close to boundary wall. Boundary walls are assumed as solid particle. It is needed for preventing fluid particle to permeate boundary wall. Fluid particles, which close to wall boundary would face repulsive force from solid particle. The repulsive force which is faced by fluid particle j is normal to solid particle i is given by

$$\mathbf{f}_{ji} = \mathbf{n}_i R(y) P(x) \tag{24}$$

By using vector of normal from solid particle boundary, then

$$R(y) = \frac{1}{\sqrt{q}}(1-q) \quad \text{and} \quad P(x) = \frac{1}{2}(1 + cos(\frac{\pi x}{\Delta p}))$$

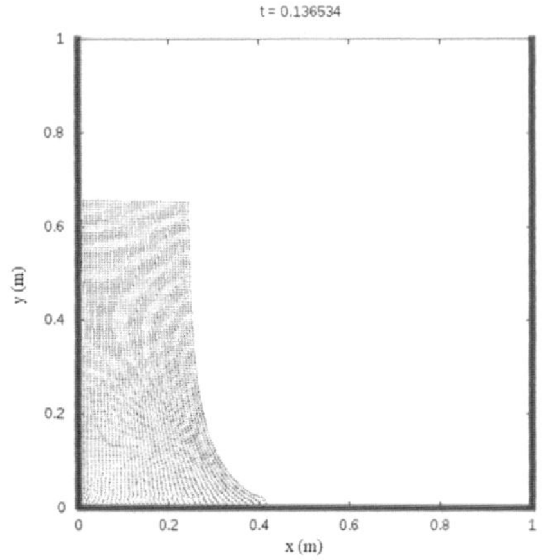

Figure 4. Example of dam-break simulation using SPH method.

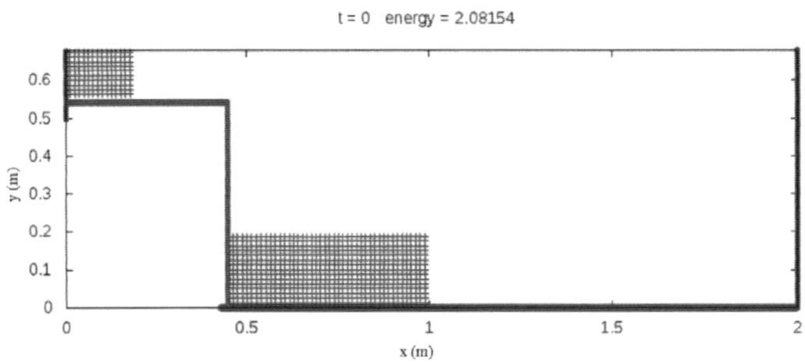

Figure 5. Initial set up of waterfall simulation using SPH method.

where $q = y/(2\Delta p)$, x is projection distance of fluid particle j at vector tangent from solid particle boundary i, y is perpendicular distance of fluid particle j from solid partcle boundary i, Δp is distance of early particle. Distance of solid particle boundary depend on early fluid particle. If Δp is too small, fluid particle will face rejection force before close to wall boundary. On the other hand, if Δp is too big, numerous fluid particles will permeate boundary wall before face repulsive force [10].

3. Experimental Setup

The SPH simulation is build by using programming language C/C++. Gnuplot is also used for visualization of SPH particles. Figure 4 shows the dam-break simulation by using SPH method which is implemented by C/C++ program. Overall, the area of the simulation is established by 2×1 meters. In the plot of the particle position, physical properties such as velocity, density, pressure, and mass are always changing in each time.

There are two types of SPH particles, the first of which is fluid particle which contain physical properties, and solid particle that each particle has no velocity value. Fluid particles are formed into water to be streamed into waterfalls, whereas solid particles are formed into the system's boundary walls. All parameters for the fluid and solid SPH particles are summarized in Table 1 and Table 2, respectively. The initial set up of waterfall simulation using SPH method is depicted by Figure 5. After establishing system, the simulation is run by using Leapfrog time-stepping. The final time of this simulation is 2.2 seconds.

Table 1. Parameter of fluid particle.

Parameter	symbol	value	unit
Gravitational acceleration	g	9.8	m/s^2
Initial density of fluid	ρ_0	1000	kg/m^3
Diameter of a particle	-	0.00625	m
Number of fluid particles	-	10216	-

Table 2. Parameters of solid particle.

Parameter	symbol	value	unit
Gravitational acceleration	g	9.8	m/s^2
Density of solid wall	ρ	1200	kg/m^3
Diameter of a particle	-	0.00625	m
Number of solid particles	-	2563	-

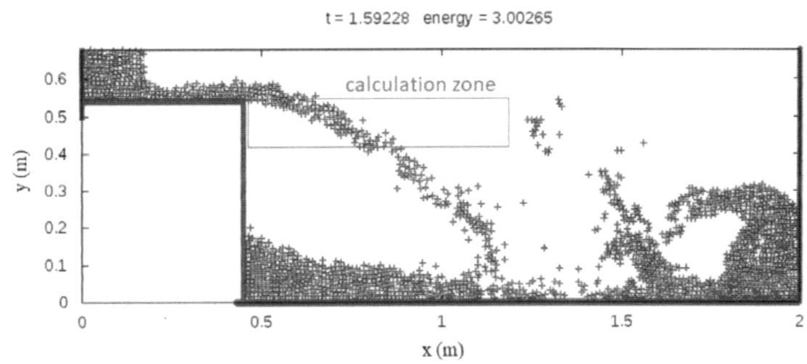

Figure 6. Waterfall simulation by using SPH method with waterfall height of 0.55 meters.

4. Results and Discussion

There are three scenarios in simulating the waterfall flow by using SPH method. The three scenarios are executed by varying the height of waterfall above ground. In each scenario, the mechanical energy resulted will be calculated and analyzed.

4.1. Waterfall height of 0.55 meters

Figure 6 shows the waterfall simulation by using SPH method with waterfall height of 0.55 meters. By this height, the amount of mechanical energy is calculated inside the calculated zone (x-axis from 0.46 to 1.2 meters and y-axis from 0.45 to 0.55 meters). Mechanical energy average in each time inside calculation zone with waterfall height of 0.55 meters is displayed by Figure 7. The figure reveals the amount of mechanical energy in Joule against time. From the figure, it can be seen that the average of mechanical energy for whole time is 23.83 Joule.

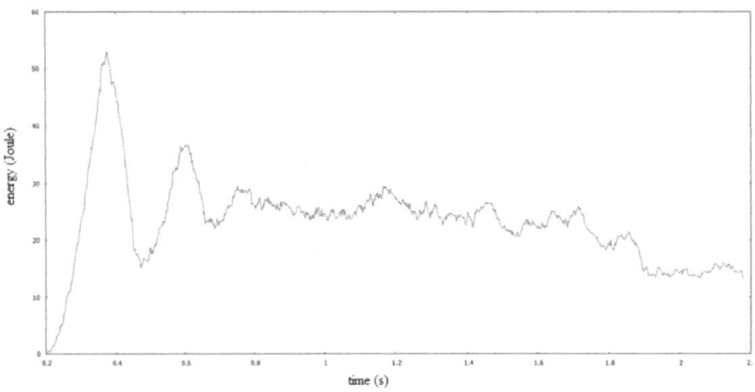

Figure 7. Mechanical energy average (in Joule) in each time inside calculation zone with waterfall height of 0.55 meters.

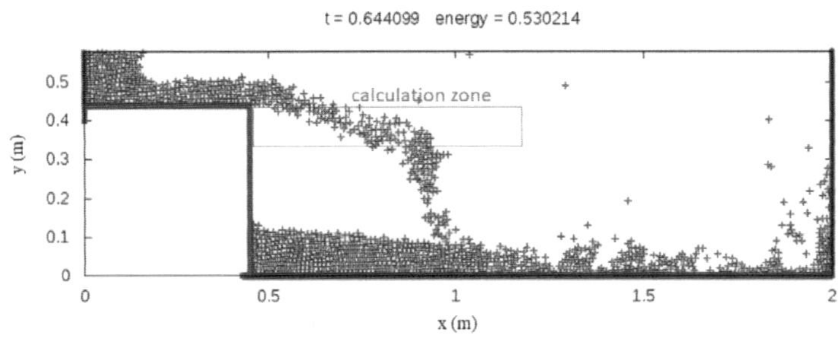

Figure 8. Waterfall simulation by using SPH method with waterfall height of 0.45 meters.

4.2. Waterfall height of 0.45 meters

The waterfall flow simulation using SPH method with height of 0.45 meters is presented in Figure 8. The calculation zone for this scenario is x-axis from 0.46 to 1.2 meters and y-axis from 0.35 to 0.45 meters. Here, the average of mechanical energy is calculated inside the calculated zone. Figure 9 shows mechanical energy average in each time inside calculation zone with waterfall height of 0.45 meters. The figure reveals the amount of mechanical energy in Joule against time. For this scenario, it can be seen that the average of mechanical energy for whole time is 21.17 Joule.

4.3. Waterfall height of 0.35 meters

The calculation zone for third scenario is x-axis from 0.46 to 1.2 meters and y-axis from 0.25 to 0.35 meters. Here, the average of mechanical energy is calculated inside the calculated zone. The waterfall flow simulation using SPH method with height of 0.35 meters is presented in Figure 10. The mechanical energy average in each time inside calculation zone with waterfall height of 0.35 meters is shown by Figure 11. The figure reveals the amount of mechanical energy in Joule against time. It can be seen that the average of mechanical energy for whole time is 18.95 Joule.

4.4. Influence of waterfall height to mechanical energy

The mechanical energy average in each time inside calculation zone with waterfall height of 0.55, 0.45, and 0.35 meters is displayed in Figure 12. It can be seen that the higher the waterfall position, the greater the mechanical energy. Moreover, the average of mechanical energy for

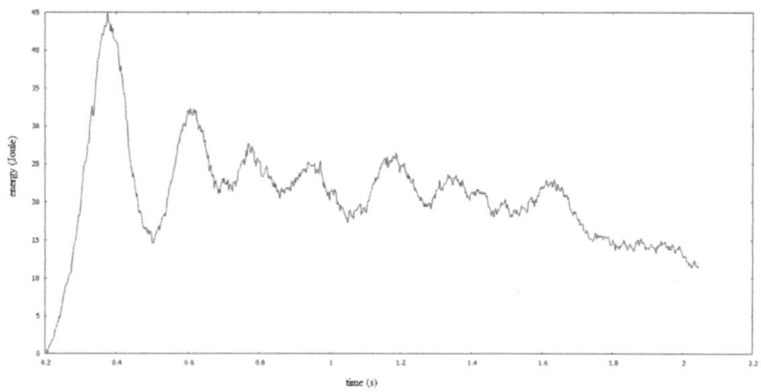

Figure 9. Mechanical energy average (in Joule) in each time inside calculation zone with waterfall height of 0.45 meters.

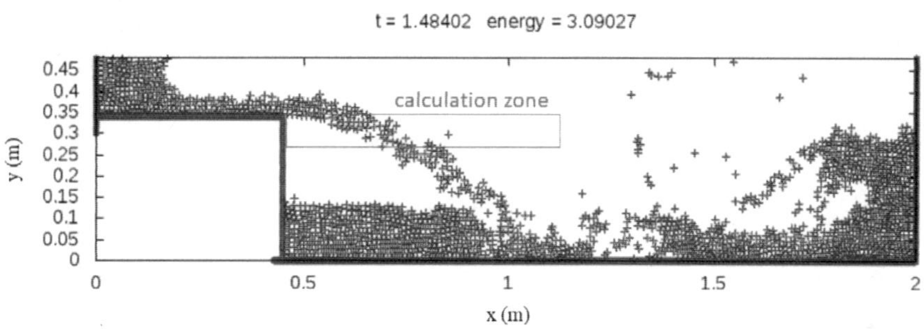

Figure 10. Waterfall simulation by using SPH method with waterfall height of 0.35 meters.

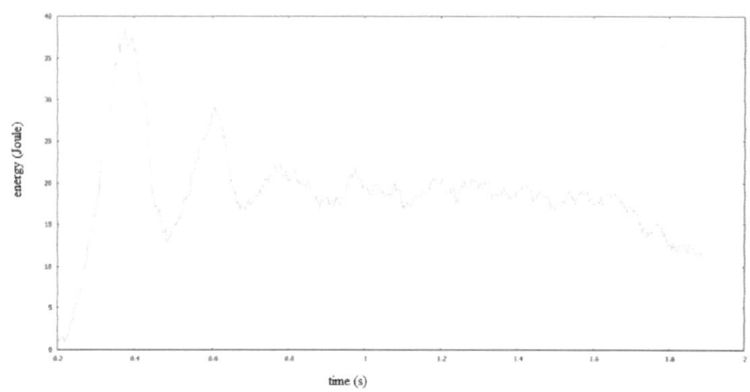

Figure 11. Mechanical energy average (in Joule) in each time inside calculation zone with waterfall height of 0.35 meters.

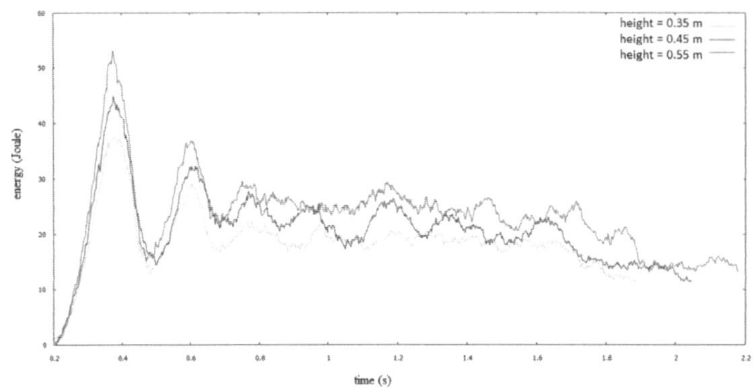

Figure 12. Mechanical energy average (in Joule) in each time inside calculation zone with waterfall height of 0.55, 0.45, and 0.35 meters.

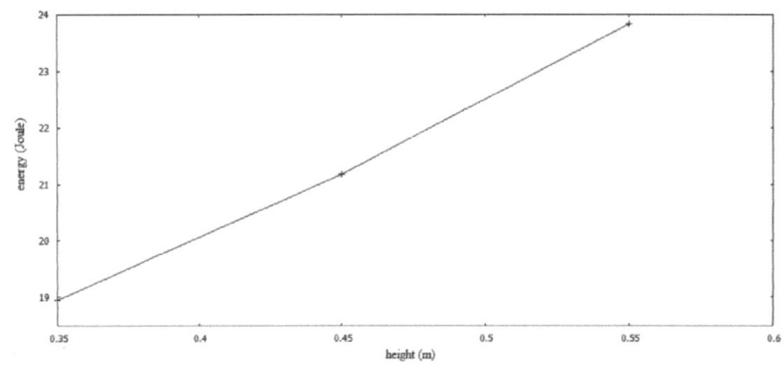

Figure 13. Average of mechanical energy for whole time against waterfall height.

whole time against waterfall height is shown by Figure 13. In the simulation, the waterfall with height of 0.55 meters has mechanical energy of 23.83 Joule. This amount of energy can result 11.9 Watt which is capable of turning on an LED floodlight with a voltage of 220 Volts. Moreover, the numerical results show that by rising the waterfall hight by 0.1 m, the mechanical energy produced is increasing by 2.4 Joule. Therefore, it can be interpreted that if there are two waterfalls with a difference of 0.1 meters of height may result 2.4 Joule range of mechanical energy generated.

5. Conclusion

From the numerical results, it has been shown that the mechanical energy of waterfall is highly influenced by its height. In this case, the higher the waterfall, the greater the mechanical energy. By the height of 0.55 meters, the waterfall generates mechanical energy of 23.82 Joule while the height is 0.45 meters, its mechanical energy is 21.17 Joule. Moreover, the waterfall fliw produces 18.95 Joule. It can be concluded that by increasing the waterfall height of 0.1 meters, the amount of mechanical energy is rising by 2.4 Joule.

References

[1] Uchiyama T, Fukuhara H, Iio S and Ikeda T 2013 *International Journal of Rotating Machinery* **2013**

[2] Kellomäki T 2015 *Tampereen teknillinen yliopisto. Julkaisu-Tampere University of Technology. Publication; 1354*

[3] Onderik J 2011 *Animation of Particle Based Miscible and Immiscible Fluids with Multiple Interfaces* Ph.D. thesis Comenius University
[4] Macia F, Souto-Iglesias A, Gonzalez L and Cercos-Pitam J 2013 Mps = sph *8th International Smoothed Particle Hydrodynamics European Research Interest Community (SPHERIC) Workshop*
[5] Berata W 2003 *Jurnal Teknik Mesin* **3** 46–54
[6] Michael Owen J 2014 *International Journal for Numerical Methods in Fluids* **75** 749–774
[7] Kipfer P and Westermann R 2006 Realistic and interactive simulation of rivers *Proceedings of Graphics Interface 2006* (Canadian Information Processing Society) pp 41–48
[8] Krog T 2003 *Visualization of Large-Scale Fluid* Master's thesis Technical University of Denmark, DTU, DK-2800 Kgs. Lyngby, Denmark
[9] Liu M and Liu G 2010 *Archives of computational methods in engineering* **17** 25–76
[10] Tarwidi D 2012 The smoothed particle hydrodynamics method for two-dimensional stefan problem *Proceedings of The 5th International Symposium on Computational Sciences (ISCS) 2012*
[11] Monaghan J J 1992 *Annual review of astronomy and astrophysics* **30** 543–574
[12] Monaghan J 1989 *Journal of Computational physics* **82** 1–15
[13] Omelyan I, Mryglod I and Folk R 2002 *Computer Physics Communications* **146** 188–202
[14] Young P 2014 *Lecture notes in University of california, santa cruz*

Parallelization of elliptic solver for solving 1D Boussinesq model

D Tarwidi and D Adytia

School of Computing, Telkom University, Jalan Telekomunikasi No. 1 Terusan Buah Batu, Bandung 40257, Indonesia

E-mail: dedetarwidi@telkomuniversity.ac.id

Abstract. In this paper, a parallel implementation of an elliptic solver in solving 1D Boussinesq model is presented. Numerical solution of Boussinesq model is obtained by implementing a staggered grid scheme to continuity, momentum, and elliptic equation of Boussinesq model. Tridiagonal system emerging from numerical scheme of elliptic equation is solved by cyclic reduction algorithm. The parallel implementation of cyclic reduction is executed on multicore processors with shared memory architectures using OpenMP. To measure the performance of parallel program, large number of grids is varied from 2^8 to 2^{14}. Two test cases of numerical experiment, i.e. propagation of solitary and standing wave, are proposed to evaluate the parallel program. The numerical results are verified with analytical solution of solitary and standing wave. The best speedup of solitary and standing wave test cases is about 2.07 with 2^{14} of grids and 1.86 with 2^{13} of grids, respectively, which are executed by using 8 threads. Moreover, the best efficiency of parallel program is 76.2% and 73.5% for solitary and standing wave test cases, respectively.

1. Introduction

Boussinesq type of models have been widely used to solve various coastal engineering problems such as for simulating wave propagation against breakwater for harbor design, tsunami generation and propagation, wave-structure interaction, floating oscillating water column, and ship mooring system. The Boussinesq model was first introduced by J.V. Boussinesq in 1872 [1]. Unlike the Shallow Water Equations (SWE), the Boussinesq model can simulate dispersion effect in wave propagation. Some of numerical methods that have been developed to solve Boussinesq models are finite difference, finite element, finite volume, and spectral method; see for example [2, 3, 4, 5] for these numerical implementations. The model in [2] has been used for simulating long wave such as tsunami propagation (see [5, 6]), as well as short wave such as wind wave (see [2, 7, 8]).

In modeling and simulation of fluid dynamics, number of complex calculations and large data have conduced very long computational time. On the other hand, numerical simulation results are required quickly and precisely, as example in weather and ocean wave forecasting. In last decades, the development of multicore computers to get fast simulation results is growing rapidly. However, the existing code of computer programs only utilized single core even it was executed in multicore computer architecture. Hence, it is required to develop a parallel algorithm that can be implemented in distributed or shared memory architecture. OpenMP (Open Multi-Processing) is an API (application program interface) that supports shared memory multiprocessing. OpenMP has been developed widely in scientific simulation. Marrone et al. [9] studied simulation of ship wave breaking pattern using OpenMP. Wen et al. [10] developed OpenMP parallel computing to simulate wave-structure interactions. Moreover, Michailidis et al. [11] solved tridiagonal system with pipelining technique in OpenMP.

Tridiagonal systems often appear in various scientific computing problems. Tridiagonal system arises from numerical solution of partial differential equation such as an elliptic equation which is solved by finite difference, finite volume, and finite element method. The most known of the tridiagonal system solver is Thomas algorithm [12]. The algorithm is simple and fast to solve tridiagonal system. Unfortunately, for large number of linear equations, Thomas algorithm is not efficient. Besides, Thomas algorithm is not suitable for parallelization. In last few decades, some parallel algorithms are developed to solve tridiagonal system especially those involving large size of linear equations. Allmann et al. [13] developed cyclic reduction algorithm to solve tridiagonal system in shared memory machine. Further, Stone et al. [14] used recursive doubling to large number of linear equation in tridiagonal system. In addition, Qeusada-Barriuso [15] implemented Bondeli's algorithm in OpenMP to solve tridiagonal system.

This paper focuses on parallelization of a tridiagonal system of an elliptic solution which is part of numerical solution of Boussinesq model in [2, 7, 8]. Cyclic reduction is applied to solve tridiagonal system and is implemented on OpenMP parallel architecture. More discussion of cyclic reduction can be found in [16]. Speedup and efficiency of the parallel program is analyzed to evaluate the performance of parallel solver of elliptic equation. This paper is organized as follows. In Section 2, 1D Boussinesq model and its numerical solution based staggered grid is briefly discussed. Parallelization of elliptic solution is presented in Section 3. Further, performance of parallel program is discussed in Section 4. The conclusion of the numerical experiment is given in Section 5.

2. Boussinesq Model & Numerical Solution

There are some mathematical models to simulate water wave propagation. The simplest one is the Shallow Water Equations (SWE). The SWE consists of continuity and momentum equation. However, the drawback of the SWE is the effect of dispersion on wave propagation that is not taken into account. In this study, a Boussinesq model is used to simulate water wave. In the Boussinesq type of model, the dispersion effect is taken into account when deriving the model. In this paper, we use the Boussinesq model that is introduced in [2, 7, 8]. In this Boussinesq model, there is an additional term in continuity equation which represents dispersion effect. Asides from continuity and momentum equation, in the Boussinesq model of [8], there is an auxiliary equation, i.e. an elliptic equation, that has to be solved for correcting dispersion effect.

Let $\eta(x,t)$ and $u(x,t)$ represent elevation of water surface and velocity of water wave in $x-$ direction, respectively. Let $H(x,t) = \eta(x,t) + h(x)$ be the total depth of the fluid where $h(x)$ is depth from mean water level. Moreover, $\psi(x,t)$ is an auxiliary variable which depends on horizontal variable. The 1D Boussinesq model that is proposed by [2, 8] is given by

$$\partial_t \eta = -\partial_x(Hu) - \partial_x(\beta \partial_x \psi) \tag{1}$$

$$\partial_t u = -g\partial_x \eta - u\partial_x u - c_f \frac{|u|u}{H} \tag{2}$$

$$-\partial_x(\alpha \partial_x \psi) + \gamma \psi = \partial_x(\beta u), \tag{3}$$

where g and c_f denote gravitational acceleration and bottom friction coefficient, respectively. Here, $\alpha(z), \beta(z), \gamma(z)$ are coefficients which depend on vertical profile and given by

$$\alpha = \frac{2}{15}H^3, \quad \beta = -\frac{H^2}{3}, \quad \gamma = \frac{H}{3}. \tag{4}$$

Note that, as denoted in shallow water equation, equation (1) and (2) are also called continuity and momentum equation with additional term $-\partial_x(\beta\partial_x\psi)$. Furthermore, equation (3) is called the elliptic equation. In next section, numerical solution of the elliptic equation will be parallelized by using cyclic reduction algorithm.

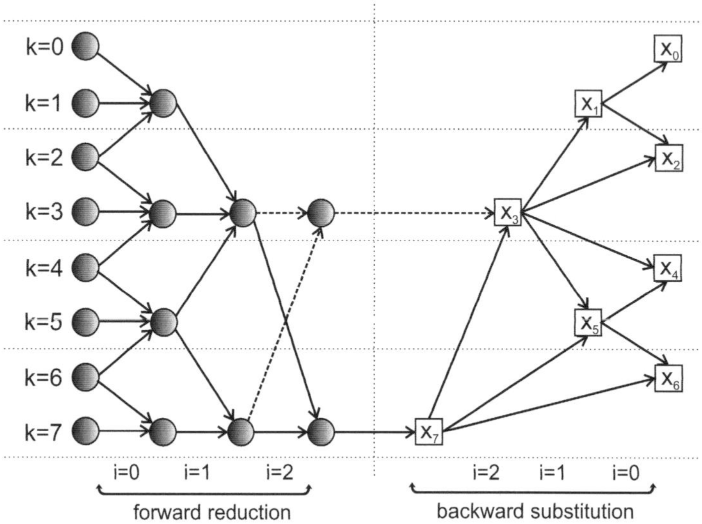

Figure 1. Illustration of parallelization of cyclic reduction algorithm for 8 linear equations and 4 processors.

Numerical solution of Boussinesq model is solved by staggered grid discretization which is proposed in [17]. More review about staggered grid can be found in [18]. Implementation of staggered grid to (1) - (3) can be described as follows. Suppose $\Omega = [L_x^-, L_x^+]$ be the computational domain of fluid. The region of interest is partitioned into N_x subregions or subintervals. Here, N_x is also called number of grids. Moreover, the length of subinterval is $\Delta x = (L_x^+ - L_x^-)/N_x$. Let subscript $i + 1/2$ be the interface between node i and $i + 1$. Here, $x_{i+1/2}$ is a node between x_i and x_{i+1} and is called a half grid. Moreover, x_i denotes a full grid. In staggered grid scheme, the discretization of $\eta(x, t)$ and $u(x, t)$ is assigned in difference position of grid. $\eta(x, t)$ is assigned in full grid meanwhile $u(x, t)$ is assigned in half grid. Moreover, $\psi(x, t)$ is calculated in full grid. Discretization of (1) - (3) based on staggered grid is given by

$$\frac{\eta_i^{n+1} - \eta_i^n}{\Delta t} = -\frac{{}^*H_{i+\frac{1}{2}}^n u_{i+\frac{1}{2}}^n - {}^*H_{i-\frac{1}{2}}^n u_{i-\frac{1}{2}}^n}{\Delta x} - \frac{{}^*\beta_{i+\frac{1}{2}}^n \left(\psi_{i+1}^n - \psi_i^n\right) - {}^*\beta_{i-\frac{1}{2}}^n \left(\psi_i^n - \psi_{i-1}^n\right)}{(\Delta x)^2} \tag{5}$$

$$\frac{u_{i+\frac{1}{2}}^{n+1} - u_{i+\frac{1}{2}}^n}{\Delta t} = -g\frac{\eta_{i+1}^{n+1} - \eta_i^{n+1}}{\Delta x} - (u\partial_x u)_{i+\frac{1}{2}}^n - c_f \left.\frac{|u|u}{H}\right|_{i+\frac{1}{2}}^n \tag{6}$$

$$-\frac{{}^*\alpha_{i+\frac{1}{2}}^n \left(\psi_{i+1}^n - \psi_i^n\right) - {}^*\alpha_{i-\frac{1}{2}}^n \left(\psi_i^n - \psi_{i-1}^n\right)}{(\Delta x)^2} + {}^*\gamma_i\psi_i = \frac{{}^*\beta_{i+\frac{1}{2}}^n u_{i+\frac{1}{2}}^n - {}^*\beta_{i-\frac{1}{2}}^n u_{i-\frac{1}{2}}^n}{\Delta x} \tag{7}$$

The superscript * denotes the variable is still unknown yet. Here, ${}^*H$ is approximated by first-order upwind method:

$${}^*H_{i+\frac{1}{2}} = \begin{cases} H_i, & \text{if } u_{i+\frac{1}{2}} \geq 0 \\ H_{i+1}, & \text{if } u_{i+\frac{1}{2}} < 0. \end{cases} \tag{8}$$

IOP Conf. Series: Journal of Physics: Conf. Series **971** (2018) 012036 doi:10.1088/1742-6596/971/1/012036

The first-order upwind method is also applied to $^*\alpha$, $^*\beta$, and $^*\gamma$.

Finally, the computational procedure in one time step to solve 1D Boussinesq model based on staggered grid is summarized as follows.

(i) Compute $^*\alpha$, $^*\beta$, and $^*\gamma$ using first-order upwind method and equation (4).
(ii) Solve elliptic equation (7) to obtain ψ_i^n .
(iii) Update surface elevation η_i^{n+1} base on equation (5).
(iv) Update velocity $u_{i+\frac{1}{2}}^{n+1}$ based on equation (6).

Algorithm 1 Solve tridiagonal system $A\mathbf{x} = \mathbf{f}$ using Cyclic Reduction

Input: matrix size: n; tridiagonal matrix: $A[i][j]$, $i = 0, 1, 2, ..., n-1, j = 0, 1, 2, ..., n-1$ with $A[i][i] \neq 0$; right hand side: $f[i]$, $i = 0, 1, 2, ..., n-1$

Output: $x[i], i = 0, 1, 2, ..., n-1$

```
{Forward Reduction}
for i = 0 to (log2 n) − 1 do
    for j = 2^(i+1) − 1 to n − 1, j = j + 2^(i+1) do
        idx1 ← j − 2^i
        idx2 ← j + 2^i
        α ← A[j][idx1]/A[idx1][idx1]
        γ ← A[j][idx2]/A[idx2][idx2]
        for k = 0 to n − 1 do
            A[j][k] ← A[j][k] − (αA[idx1][k] + γA[idx2][k])
        end for
        f[j] ← f[j] − (αf[idx1] + γF[idx2])
    end for
end for
{Backward Substitution}
idx2 ← n − 1
idx1 ← n/2 − 1
x[idx2] = f[idx2]/A[idx2][idx2]
x[idx1] = (f[idx1] − A[idx1][idx2] · x[idx2])/A[idx1][idx1]
for i = (log2 n) − 2 to 0 do
    for j = 2^(i+1) − 1 to n − 1, j = j + 2^(i+1) do
        idx1 ← j − 2^i
        idx2 ← j + 2^i
        sum1 = f[idx1]
        sum2 = f[idx2]
        for k = 0 to n − 1 do
            sum1 ← sum1 − A[idx1][k] · x[k]
            sum2 ← sum2 − A[idx2][k] · x[k]
        end for
        x[idx1] ← (sum1 + A[idx1][idx1] · x[idx1])/A[idx1][idx1]
        x[idx2] ← (sum2 + A[idx2][idx2] · x[idx2])/A[idx2][idx2]
    end for
end for
```

3. Parallelization of Elliptic Solver

Discretization of the elliptic equation using staggered grid is given by (7). In solving (7), there is an expensive computational cost due to large number of grids. By rearranging (7) in term of ψ, the linear tridiagonal system is obtained and is given by

$$\begin{bmatrix} b_1 & c_1 & & & 0 \\ a_2 & b_2 & c_2 & & \\ & \ddots & \ddots & \ddots & \\ & & a_{N_x-1} & b_{N_x-1} & c_{N_x-1} \\ 0 & & & a_{N_x} & b_{N_x} \end{bmatrix} \begin{bmatrix} \psi_1 \\ \psi_2 \\ \vdots \\ \psi_{N_x-1} \\ \psi_{N_x} \end{bmatrix} = \begin{bmatrix} f_1 \\ f_2 \\ \vdots \\ f_{N_x-1} \\ f_{N_x} \end{bmatrix} \tag{9}$$

where

$$\begin{aligned} a_i &= -\frac{{}^*\alpha_{i-\frac{1}{2}}}{(\Delta x)^2}, \quad i = 2, \cdots, N_x \\ b_1 &= \frac{{}^*\alpha_{\frac{3}{2}}}{(\Delta x)^2} + {}^*\gamma_1 \\ b_i &= \frac{{}^*\alpha_{i-\frac{1}{2}} + {}^*\alpha_{i+\frac{1}{2}}}{(\Delta x)^2} + {}^*\gamma_i, \quad i = 2, \cdots, N_x - 1 \\ b_{N_x} &= \frac{{}^*\alpha_{N_x-\frac{1}{2}}}{(\Delta x)^2} + {}^*\gamma_{N_x} \\ c_i &= -\frac{{}^*\alpha_{i+\frac{1}{2}}}{(\Delta x)^2}, \quad i = 1, \cdots, N_x - 1 \\ f_i &= \frac{{}^*\beta_{i+\frac{1}{2}} u_{i+\frac{1}{2}} - {}^*\beta_{i-\frac{1}{2}} u_{i-\frac{1}{2}}}{\Delta x} \end{aligned}$$

Cyclic reduction algorithm is used to solve tridiagonal system in (9). This method uses elimination step which is similar to Gaussian elimination. However, cyclic reduction can be implemented into parallel program. In general, cyclic reduction works by eliminating variable x_{i-1} and x_{i+1} from equation i by utilizing equations $i-1$ and $i+1$. By this elimination, the number of equation is reduced to a half number of equations. In the new linear system, the same elimination is also assigned.

There are two phases to solve tridiagonal system by using cyclic reduction algorithm. The first phase is forward reduction. In each step of forward reduction, half of the variables in the tridiagonal system is eliminated. The process is repeated until two equations with two unknown variables are left. Here, two unknown variables with two linear equations is then solved by simple elimination. Cyclic reduction algorithm to solve tridiagonal system $Ax = f$ is given by Algorithm 1. The detail of the algorithm can be found in [19]. Illustration to solve 8-element of tridiagonal system using 4 threads is shown by Figure 1.

Implementation of cyclic reduction in OpenMP is described in Figure 2. At first, N_x/p equations are distributed among the threads. Here, p denotes the number of threads used in parallel program. Tridiagonal matrix A and right hand side f is assigned as shared variables in forward reduction phase. Furthermore, in backward substitution, x, A, f, and N_x is considered as shared variables. Note that, in the simulation, the tridiagonal system solver function is called in each time step.

4. Results and Discussion

The computational time in solving the Boussinesq model numerically has become an important part in the simulation of water waves. One of the numerical solution of the Boussinesq model

```
/*forward substitution*/
for(i=0; i<log2(Nx); i++){
n = pow(2,i+1);
# pragma omp parallel shared ( A, f, Nx ) \
private ( j, k, index1, index2, alpha, gamma )
{
# pragma omp for
for(j=pow(2,i+1)-1;j<Nx;j=j+n){
...
}
}
}
/*backward substitution*/
...
for(i=log2(Nx)-2;i>=0;i--){
n = pow(2,i+1);
# pragma omp parallel shared ( x, A, f, Nx ) \
private ( j, k, index1, index2, sum1, sum2 )
{
# pragma omp for
for(j=pow(2,i+1)-1;j<Nx;j=j+n){
...
# pragma omp critical
for(k=0;k<Nx;k++){
...
}
...
}
}
}
```

Figure 2. Parallel implementation of cyclic reduction algorithm in OpenMP.

that can be optimized its computational time is elliptic solver. As described in Section 3, the numerical solution of elliptic equation is in the form of tridiagonal system. In this case, the cyclic reduction algorithm is used to solve the tridiagonal system. The numerical experiment is conducted in personal computer with specification: Intel Xeon CPU E3-1230 v5 @3.4 GHz, 16 GB of RAM, and 64-bit OS type.

Performance of cyclic reduction algorithm implemented in OpenMP parallel architecture is evaluated. Two test cases, i.e. solitary and standing wave, are employed to measure the performance of parallel program. The benefit of parallel program is measured by the quantity of speedup and efficiency. The speedup represents the comparison of execution time between parallel program (multiprocessor) relative to sequential program (single processor). The speedup $S_p(n)$ is defined as

$$S_p(n) = \frac{T^*(n)}{T_p(n)},$$

where $T^*(n)$ and $T_p(n)$ are execution time of sequential and parallel program with size n, respectively. Here, p is the number of threads used to solve the problem. Moreover, the efficiency

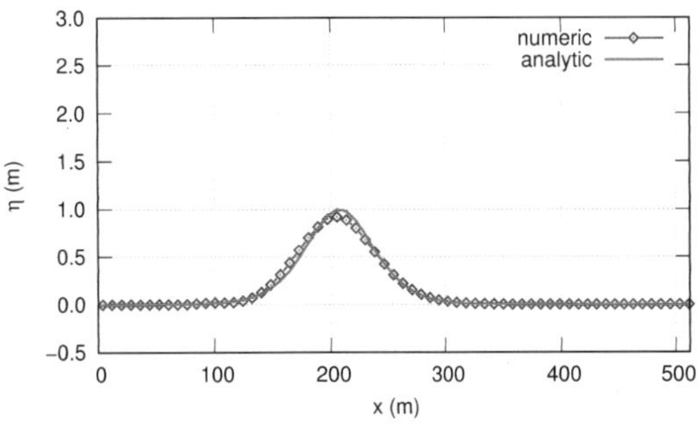

Figure 3. Numerical vs analytical solution for solitary wave.

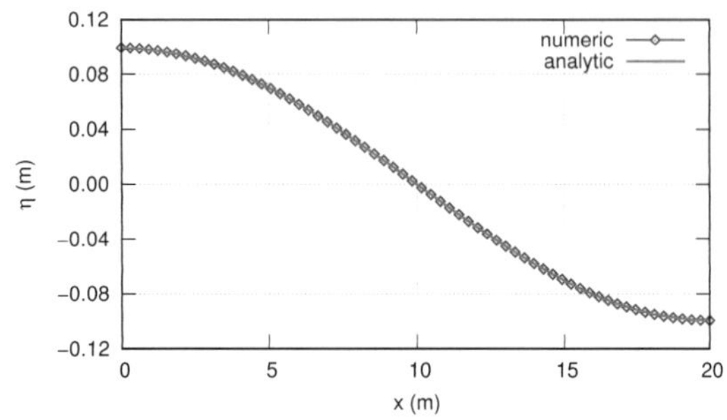

Figure 4. Numerical vs analytical solution for standing wave.

Table 1. Execution time of sequential program (1 thread) and parallel program (2, 4, and 8 threads) for Test Case 1.

# threads	Execution time (in seconds)						
	$N_x = 2^8$	$N_x = 2^9$	$N_x = 2^{10}$	$N_x = 2^{11}$	$N_x = 2^{12}$	$N_x = 2^{13}$	$N_x = 2^{14}$
1	0.57	5.10	17.71	74.29	273.30	1032.00	4201.60
2	0.89	4.55	13.93	49.00	185.11	694.00	2755.75
4	1.42	5.04	16.25	43.76	150.29	538.08	2141.17
8	1.38	6.21	16.12	47.13	137.23	504.02	2026.88

of a parallel program can be expressed as

$$E_p(n) = \frac{S_p(n)}{p}.$$

The efficiency expresses processor utilization of parallel program. In other words, it measures the fraction of time for which a processor is usefully exploited.

4.1. Test Case 1: Solitary wave

In this test case, a solitary wave is considered as initial condition of the Boussinesq model. Let $\Omega = [-512, 512]m$ be the domain of the simulation. The domain is discretized in to finite number of grids. To measure performance of the parallel program, the number of grids is varied from 2^8 to 2^{14}. Moreover, time step used in this simulation is $\Delta t = 0.01s$ with final time is $10s$. The initial condition for solitary wave is given by

$$\eta(x,0) = 2A_0 \operatorname{sech}^2\left(\sqrt{\frac{3A_0}{4h_0^2(h_0 + A_0)}}\,x\right),$$

where $A_0 = 1$ and $h_0 = 10$.

Numerical solution of 1D Boussinesq model with solitary wave as initial condition is depicted in Figure 3. From the figure, it can be seen that the numerical solution shows a good agreement with the analytical solution. Table 1 displays the execution time of sequential and parallel program of Test Case 1 for various number of grids. Table 1 was obtained by executing the parallel program for 1, 2, 4, and 8 threads. When executing the parallel program, there was no

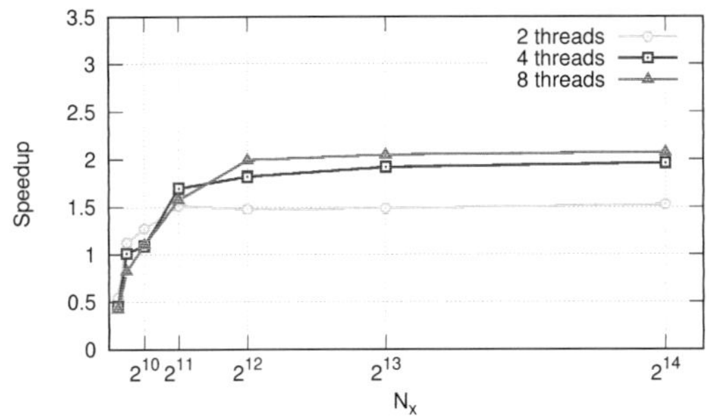

Figure 5. Speedup of parallel program for Test Case 1.

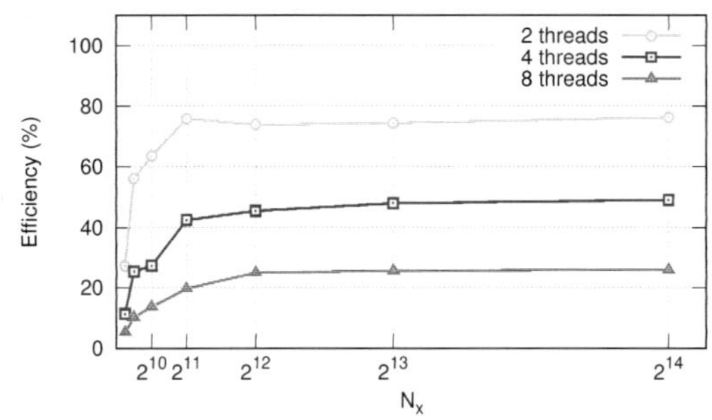

Figure 6. Efficiency of parallel program for Test Case 1.

Table 2. Execution time of sequential program (1 thread) and parallel program (2, 4, and 8 threads) for Test Case 2.

# threads	Execution time (in seconds)						
	$N_x = 2^8$	$N_x = 2^9$	$N_x = 2^{10}$	$N_x = 2^{11}$	$N_x = 2^{12}$	$N_x = 2^{13}$	$N_x = 2^{14}$
1	1.28	10.13	79.41	600.26	3991.28	22671.6	175409
2	1.42	9.33	59.17	421.15	2968.81	15616.4	119344
4	1.77	9.78	54.66	349.93	2569.05	12339.6	109587
8	1.85	9.75	53.79	351.55	2528.66	12181.0	104278

any program that runs in the computer so that each scenario has the same treatment. It can be seen that the execution time of the sequential program is faster than the parallel program for $N_x = 2^8$ and $N_x = 2^9$. Furthermore, The execution time of parallel program is faster than the sequential program if the number of grids is more than 2^{10}.

The speedup of parallel program for Test Case 1 by using 2, 4, 8 threads can be seen in Figure 5. It can be seen that the speedup of parallel program grows sublinearly with respect to number of grids used. From the figure, it can be revealed that the speedup more than 1 if the number of grids is more than 2^{10}. It means that the parallel program is efficient if $N_x \geq 2^{10}$. However, the speedup for N_x greater than 2^{12} does not change significantly. The speedup by using 2, 4, and 8 threads are around 1.5, 1.9, and 2.0, respectively. However, the best speedup for solitary wave test case is 2.07 which is executed in 8 threads and using 2^{14} of grids.

Figure 6 shows efficiency of parallel program for Test Case 1. It can be seen that parallel program by using 2 threads is more efficient than 4 and 8 threads. The highest efficiency is about 76.2% and is achieved when the parallel program is using 2 threads.

4.2. Test Case 2: Standing Wave

Initial condition for the second test case is considered as standing wave. The initial condition of standing wave is given by

$$\eta(x,0) = 2\,A_0 \cos\left(\frac{\pi}{20}\,x\right),$$

where $A_0 = 0.05$. The numerical domain of standing wave simulation is given by $\Omega = [0, 20]m$. Similar to Test Case 1, the standing wave simulation also uses various of number of grids from $N_x = 2^8$ to $N_x = 2^{14}$. The final time of the simulation is also 10 seconds. However, in the standing wave simulation, the time step Δt follows the number of grids that used in

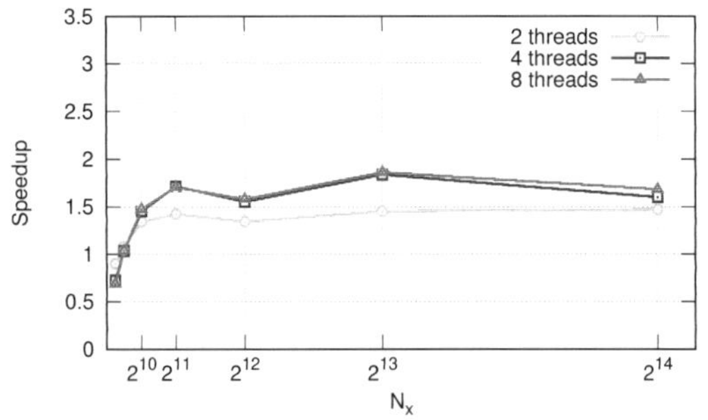

Figure 7. Speedup of parallel program for Test Case 2.

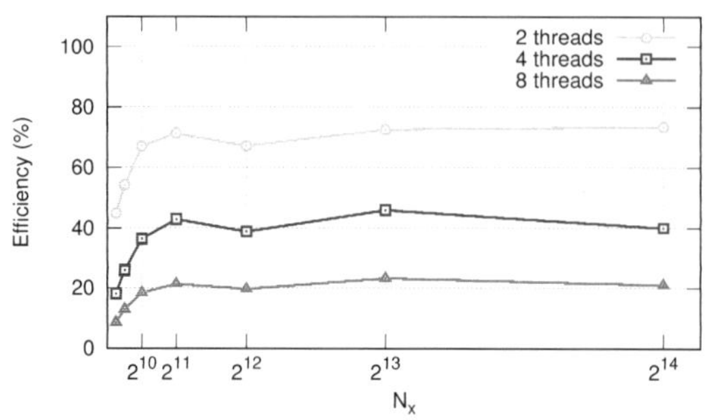

Figure 8. Efficiency of parallel program for Test Case 2.

the simulation. The more number of grids, the smaller time step used in the simulation. As consequence, the number of time iteration becomes difference in each number of grids used.

Numerical result of 1D Boussinesq model with standing wave against its analytical solution is shown by Figure 4. It can be revealed that numerical solution of standing wave shows good agreement with the analytical solution. Table 2 shows execution time of sequential and parallel program for $N_x = 2^8$ to $N_x = 2^{14}$. From this table, it can be seen that the execution time of parallel program is faster than the sequential program if $N_x \geq 2^9$. Speedup and efficiency of parallel program for Test Case 2 are presented in Figure 7 and Figure 8, respectively. It can be seen that the best speedup is 1.86 which is achieved for 8 threads and $N_x = 2^{13}$. The highest of efficiency is achieved for 2 threads and is about 73.5%. Moreover, the efficiency for 4 and 8 threads with $N_x = 2^{13}$ are 45.9% and 23.2%, respectively.

5. Conclusion

Parallelization of elliptic solver for solving 1D Boussinesq model has been successfully conducted. Cyclic reduction algorithm has been implemented in parallel program and executed on multicore processors with shared memory architectures using OpenMP. The best speedup of solitary and standing wave test cases is about 2.07 with 2^{14} of grids and 1.86 with 2^{13} of grids, respectively, which is executed by using 8 threads. By the numerical experiment, it can be concluded that the speedup does not change significantly if the number of grids is greater than 2^{12}. Moreover, the efficiency of parallel program is 76.2% and 73.5% for solitary and standing wave test cases, respectively.

References

[1] Boussinesq J 1872 *J. Math.* **17** 55–108
[2] Adytia D 2014 *The Twenty-fourth International Ocean and Polar Engineering Conference* (International Society of Offshore and Polar Engineers)
[3] Madsen P A and Sørensen O R 1992 *Coastal engineering* **18** 183–204
[4] Nwogu O 1993 *Journal of waterway, port, coastal, and ocean engineering* **119** 618–638
[5] van Groesen E, Adytia D and Andonowati 2008 *Natural Hazards and Earth System Sciences* **8** 175–185
[6] Adytia D and van Groesen E 2009 *Asian and Pacific Coasts 2009, Proceedings of the 5th International Conference on APAC 2009* (World Scientific) pp 122–128
[7] Adytia D 2012 *Coastal zone simulation with variational Boussinesq modelling* (PhD Thesis, University of Twente, The Netherlands)
[8] Adytia D and Lawrence 2016 *ASME 2016 35th International Conference on Ocean, Offshore and Arctic Engineering* (American Society of Mechanical Engineers)
[9] Marrone S, Bouscasse B, Colagrossi A and Antuono M 2012 *Computers & Fluids* **69** 54–66
[10] Wen H, Ren B, Dong P and Wang Y 2016 *Applied Ocean Research* **59** 366–377
[11] Michailidis P D and Margaritis K G 2011 *Journal of Computational and Applied Mathematics* **236** 326–341

[12] Thomas L H 1949 *Watson Sci. Comput. Lab. Rept., Columbia University, New York* **1**
[13] Allmann S, Rauber T and Runger G 2001 *Parallel and Distributed Processing, 2001. Proceedings. Ninth Euromicro Workshop on* (IEEE) pp 290–297
[14] Stone H S 1973 *Journal of the ACM (JACM)* **20** 27–38
[15] Quesada-Barriuso P, Lamas-Rodríguez J, Heras D B, Bóo M and Argüello F 2011 *Int. Conf. on Parallel and Distributed Processing Techniques and Applications (as part of WorldComp2011 Conference)*
[16] Rauber T and Rünger G 2013 *Parallel programming: For multicore and cluster systems* (Springer Science & Business Media)
[17] Stelling G S and Duinmeijer S A 2003 *International journal for numerical methods in fluids* **43** 1329–1354
[18] Pudjaprasetya S, Magdalena I and Tjandra S 2017 *Journal of Earthquake and Tsunami* **11** 1740002
[19] Karniadakis G E and Kirby II R M 2003 *Parallel scientific computing in C++ and MPI: a seamless approach to parallel algorithms and their implementation* (Cambridge University Press)

Text Categorization on Hadith Sahih Al-Bukhari using Random Forest

Muhammad Fauzan Afianto[1], Adiwijaya[2], Said Al-Faraby[3]

Telkom University

E-mail: fauzanafianto@student.telkomuniversity.ac.id[1], adiwijaya@telkomuniversity.ac.id[2], saidalfaraby@telkomuniversity.ac.id[3]

Abstract. Al-Hadith is a collection of words, deeds, provisions, and approvals of Rasulullah Shallallahu Alaihi wa Salam that becomes the second fundamental laws of Islam after Al-Qur'an. As a fundamental of Islam, Muslims must learn, memorize, and practice Al-Qur'an and Al-Hadith. One of venerable Imam which was also the narrator of Al-Hadith is Imam Bukhari. He spent over 16 years to compile about 2602 Hadith (without repetition) and over 7000 Hadith with repetition. Automatic text categorization is a task of developing software tools that able to classify text of hypertext document under pre-defined categories or subject code[1]. The algorithm that would be used is Random Forest, which is a development from Decision Tree. In this final project research, the author decided to make a system that able to categorize text document that contains Hadith that narrated by Imam Bukhari under several categories such as suggestion, prohibition, and information. As for the evaluation method, K-fold cross validation with F1-Score will be used and the result is 90%.

1. Introduction

Islam based on 2 fundamental laws : Al-Qur'ana as the set of words of Allah and Al-Hadith that documenting words, deeds, provisions, and approvals of Rasulullah as the prophet of Allah. Hadith was compiled and categorized by several Imam such as Imam Bukhari, Imam Muslim, and Imam Tirmidzi, etc. All of them based on one source Rasulullah Muhammad PBUH. Imam Bukhari is one of the known Imam that according to Ulama, Hadith that he narrated proven hold onto the most sahih among the other. Imam Bukhari spent 16 years on compiling about 2602 hadith (without repetition) and more than 7000 hadith (with repetition)[1]. Referring to[2], Figure 1 is the component of Hadith.

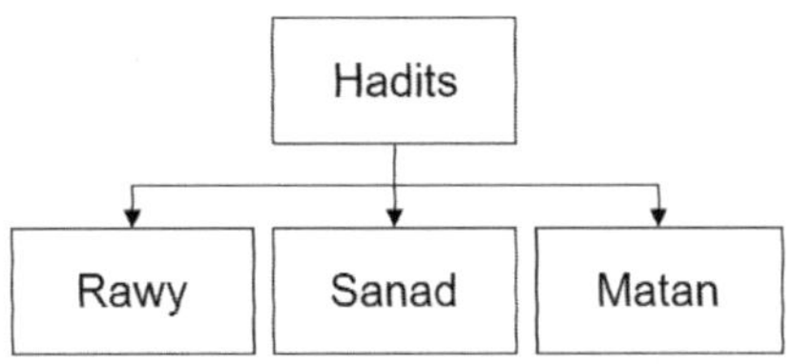

Figure 1. Hadith Components.

Sanad is the chain of conveyor of each Hadith, this component present on the beginning of

Hadith. Matan is the content of Hadith, present after the Sanad, and finally Rawy, this is the person or Imam that compile Hadith such as Imam Bukhari.

By the exponential growth of digitalized document, emerge the need of a system that able to extract high quality information, that is why automatic text categorization become popular.

Automatic text categorization is a task of developing software tools that able to classify text of hypertext document under pre-defined categories or subject code[1].

2. Literature Study

In this reseach, we only used 1650 Hadith from 7008 documented Hadith from "Hadits 9 Imam" from Lembaga Ilmu dan Dakwah serta Publikasi Sarana Keagamaan (LIDWA) in Bahasa Indonesia. Then we divide that into 3 category suggestion, prohibition, and information with 550 record each. As for the label we used hand labelling.

as for the related study, there are several paper that used hadith as their documents such as:

Table 1. Related Works

No	*Author*	*Method*	*Accuracy*
1	Harrag et al. (2011b)[4]	Dictionary-lookup Stemming, and ANN	94%
2	Harrag et al. (2009)[3]	ID3 Decission Tree	81%
3	Najeeb (2014)[5]	Associative Classication	72%

Each of them using different dataset in English Language. In the research by Harrag et al. (2011b)[4]. They use various stamming method in preprocessing. As for the best method they use dictionary lookup-stemming, it is proven to reduce 4055 variance of words into 739. as for the class, they divide the hadith based on 14 class and use ANN classifier, the result is 94% Accuracy.

As for the research by Harrag et al. (2009)[3]. ID3 classifier used with the highest accuracy of 81% while Najeeb (2014)[5] decided to classify Hadith by their authentic, so the class is binary either the hadith is not authentic (daif) or authentic (sahih). They used Association Rule Mining (ARM) and yield 72% as their highest accuracy.

The machine learning method that used in this research is Random Forest. This method developed from Decision Tree that known able to overcome the weakness of Decision Tree, Overfitting. The Decision Tree that used is Classification and Regression Tree (CART) that build By using Gini Index[6]. Basically, this method build lots of CART then doing voting to decide the which category that has the most amount from them. The creation of CART itself use bootstrapping method, in this research we call the bootstrap tree.

to train the Random Forest, we have to select which feature that will be extracted from the documents hence, Term Frequency-Inverse Document Frequency (TF-IDF) used[7].. By using this Feature Extraction, a sparse matrix will be built. the topmost of that matrix will be filled by bag of words and the leftmost will be filled with list of document name, the rest will be filled by the value of TF-IDF that gained by Equation 1

$$Word_{xy} = TermFrequency_{xy} * log\frac{SumDocument}{TermFrequencyDocument_{xy}} \quad (1)$$

Before we extract the feature form Hadith, we have to reduce the variance of each word there, by using the following preprocessing, stemming, case folding, punctuation removal, and stopwords removal.

IOP Conf. Series: Journal of Physics: Conf. Series **971** (2018) 012037 doi:10.1088/1742-6596/971/1/012037

To evaluate, we decided to use K-Fold Cross Validation to observe the overall F1-Score. This mechanism divide the data training to by the amount of K that depicted in Figure 2

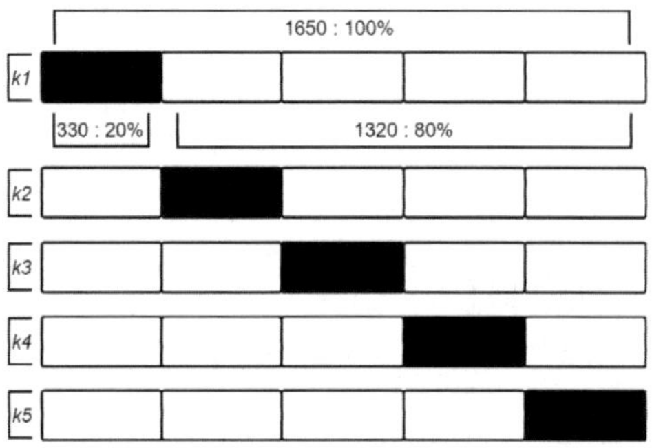

Figure 2. K-Fold Cross Validation.

So, each selected K part would be data testing while the rest become data training. Then the average of F!-Score would be calculated to observe the final scoring.

3. Development and Analyze

In this research, we build the system that able to categorize hadith into 3 category suggestion, prohibition, and information. Here is the figure t represent every step that would be done by the system.

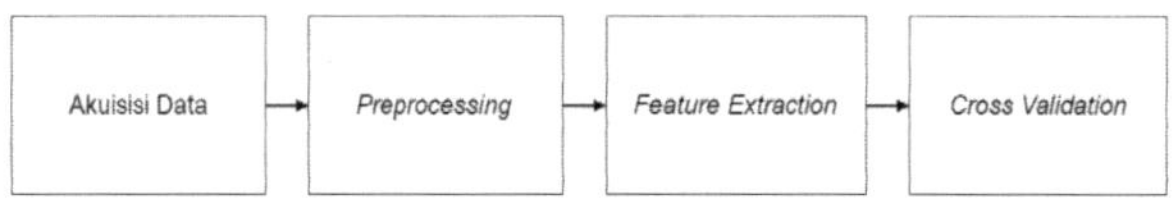

Figure 3. Block Diagram.

According to the Figure 3 The data would go to the preprocessing mechanism to reduce the variance of words, in preprocessing there will be several sub mechanism such as case folding, stemming, punctuation removal, and stopwords removal that proven good enough to reducing the variance of data in Bahasa Indonesia.

after that, the preprocessed data needs to represented by the feature that able to used as the input for classifier. In this research we represent the preprocessed data as sparse matrix with the topmost row filled with bag of word, and the leftmost column filled with the document number, the rest would be filled with the value of words that calculated using TF-IDF formula.

Then, the extracted features ready to be learned by the Random Forest classifier. the amount bootstrap tree would be generated. The evaluation using K-Fold Cross Validation also happened in conjunction with the creation of Random Forest classifier.

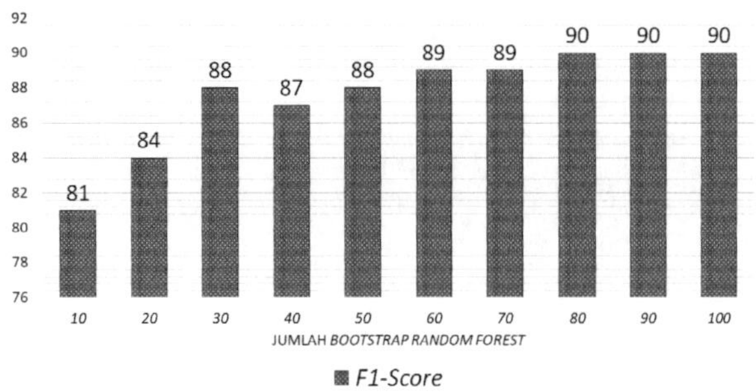

Figure 4. Amount of Bootstrap Tree

Based on Figure 4 The purpose of this scenario is to observe the influence of bootstrap tree towards the score of F1-Score. we divide each test with the addition of 10 bootstrap tree. by the time the amount of bootstrap tree reach 80 there are no improvement of F1-Score, from this we can see that the amount of bootstrap tree able to yield significant effect.

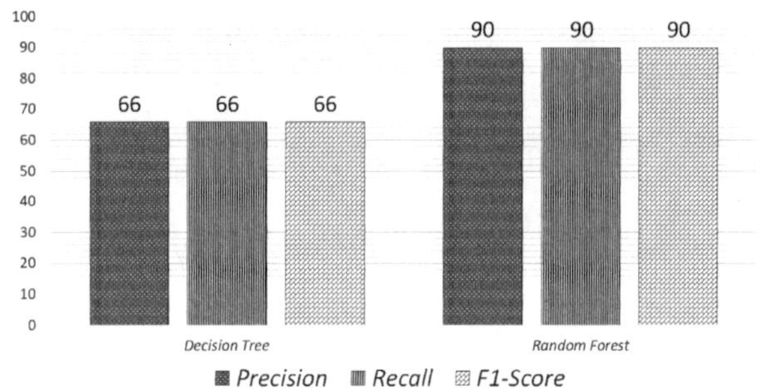

Figure 5. Decision Tree and Random Forest Comparison

The purpose of this scenario is to observe the potency of Random Forest towards Decision Tree as depicted on Figure 5 . As we can see the difference of F1-score while using K-Fold Cross Validation with 5 as the Amount of K in respect of 2:8 of testing and training ratio, yield 66% : 90% of the F1-Score. This is prove that in categorizing Hadith Sahih Al-Bukhari Random Forest able to overcome the weakness of Decision Tree.

4. Conclusion

In conclusion automatic text categorization on Hadith Sahih Al-Bukhari using Random Forest with sample data of 1650 hadith with 550 hadith of each category using TF-IDF as Feature Extraction and 100 as the amount of bootstrap tree evaluated by K-Fold Cross Validation with 5 as the amount of Fold, yield 90% of F1-Score.

The most influential mechanism is the amount of bootstrap tree on the Random Forest, while the rest does not really significant. Also in this case Random Forest also proven able to overcome the weakness of Decision tree by the ratio of 90% : 66% of F1-Score using K-Fold Cross Validation.

As for the further research it would be good if all of the hadith used, other than that combining hadith from the other Imam such as Imam Muslim, Imam Tirmidzi, etc. would be a good idea since the variance would be more.

5. References

[1] Naji Al-Kabi, M., Kanaan, G., Al-Shalabi, R., Al-Sinjilawi, S.I. and Al-Mustafa, R.S., 2005. Al-Hadith text classifier. Journal of Applied Sciences, 5, pp.584-587.

[2] Saloot, M.A., Idris, N., Mahmud, R., Jaafar, S., Thorleuchter, D. and Gani, A., 2016. Hadith data mining and classification: a comparative analysis. Artificial Intelligence Review, 46(1), pp.113-128.

[3] Harrag, F., El-Qawasmeh, E. and Pichappan, P., 2009, July. Improving Arabic text categorization using decision trees. In Networked Digital Technologies, 2009. NDT'09. First International Conference on (pp. 110-115). IEEE.

[4] Harrag, F., El-Qawasmah, E. and Al-Salman, A.M.S., 2011, April. Stemming as a feature reduction technique for Arabic text categorization. In Programming and Systems (ISPS), 2011 10th International Symposium on (pp. 128-133). IEEE.

[5] Najeeb, M.M., 2014. Towards innovative system for Hadith Isnad processing. Int J Comput Trends Technol, 18(6), pp.257-259.

[6] Arifin, A.H.R.Z., Mubarok, M.S. and Adiwijaya, A., 2016, September. Learning Struktur Bayesian Networks menggunakan Novel Modified Binary Differential Evolution pada Klasifikasi Data. In Indonesia Symposium on Computing (IndoSC) 2016.

[7] Aziz, R.A., Mubarok, M.S. and Adiwijaya, A., 2016, September. Klasifikasi Topik pada Lirik Lagu dengan Metode Multinomial Naive Bayes. In Indonesia Symposium on Computing (IndoSC) 2016.

[8] Mubarok, M.S., Adiwijaya and Aldhi, M.D., 2017, August. Aspect-based sentiment analysis to review products using Nave Bayes. In AIP Conference Proceedings (Vol. 1867, No. 1, p. 020060). AIP Publishing.

Indonesian name matching using machine learning supervised approach

1. **Mohamad Alifikri**
School of Computing, Telkom University, Indonesia
alifikri@gmail.com
2. **Moch. Arif Bijaksana,**
School of Computing, Telkom University, Indonesia
arifbijaksana@telkomuniversity.ac.id

Abstract. Most existing name matching methods are developed for English language and so they cover the characteristics of this language. Up to this moment, there is no specific one has been designed and implemented for Indonesian names. The purpose of this thesis is to develop Indonesian name matching dataset as a contribution to academic research and to propose suitable feature set by utilizing combination of context of name strings and its permute-winkler score. Machine learning classification algorithms is taken as the method for performing name matching. Based on the experiments, by using tuned Random Forest algorithm and proposed features, there is an improvement of matching performance by approximately 1.7% and it is able to reduce until 70% misclassification result of the state of the arts methods. This improving performance makes the matching system more effective and reduces the risk of misclassified matches.

1. Introduction

Name matching methods can determine the degree of relatedness between person names and provide score for the strength of the match. The higher the match score, the more likely it is that the records relate to the same individual. However, many existing name matching methods tend to be generic and language or cultural independent. As a result, one method could be best applied in certain country/culture but in other country, it does not always show the best performance.

The mainstream approaches for name matching problem are using phonetic encoding and pattern matching. Further techniques have been developed for both approaches beyond the original ones, even combination of the two with the aim to improve the matching quality. The techniques use similarity measure rather than identity, such as edit distance, n-gram matching and soundex.

Despite extensive exploration of this task, most works have focused on languages, particularly in English. Phonetic matching method like Soundex, Metaphone, Double Metaphones developed using phonetic for English names primarily [1]. Pattern Matching methods such as Edit Distance is language independent, but too general; it does not capture the majority of possible variations in specific culture or country. For example: existing methods could not capture Indonesia names: Cecep and Tjetjep, as a matching name pair. This presents a new challenge for name matching particularly in specific country like Indonesia.

One of the main obstacles to perform name matching task for Indonesia names is that there is no available dataset of Indonesia names which are naturally formed into pairs exist and ready for the research. This research beside propose a solution which include selection of machine learning classification algorithm and suitable features that have best performance on matching Indonesia names, also develop representative Indonesia name pair dataset as a basis of current and future name matching research.

2. Review of Literature

2.1. Name Matching Variations

Most works with respect to name matching focus on cross language domain. This is justifiable due to the most difficult issue when dealing with distinctive languages which have diverse characteristics, and require alignment using transliteration, even custom one, to do name matching task. In any case, the origin of name matching issue is from the same fundamental issue, that is name variations. One work about name variations was reported by [2] in her report describing all variations commonly appeared in Indonesia names. Another reference [3] divide name variations into 2 group of categories which are linguistic (transcription variants, homophones, name alias) and non-linguistic (the use of initials, change order of name parts, spelling error).

2.2. Name Matching Methods

Many variations for approximate string matching have been developed, there are two main categories [1][4]; pattern matching and phonetic matching techniques. Some of most commonly used techniques are pattern and phonetic matching.

Name matching is also considered as a classification task [5][6]. Each string pair is an instance: a positive classification means that two strings can refer to the same name. When the compared candidate name pairs are only classified into matches and non- matches (but not potential matches), then this classification is known as a **binary classification** problem. Further, if preparing training set in the form of name pairs with their label match or non-match, at that point **a supervised classification approach** can be utilized to train a classification model. The prepared model is then utilized to distinguish record pairs with an unknown status into matches or non-matches. Three popular supervised classification techniques that have been employed in the area of data matching and deduplication are Linear SVM, Logistic Regression and Random Forest. They are used to evaluate name matching performance during this thesis research.

3. Research Methodology

There are four stages use in this research as depicted in **Figure *1***:

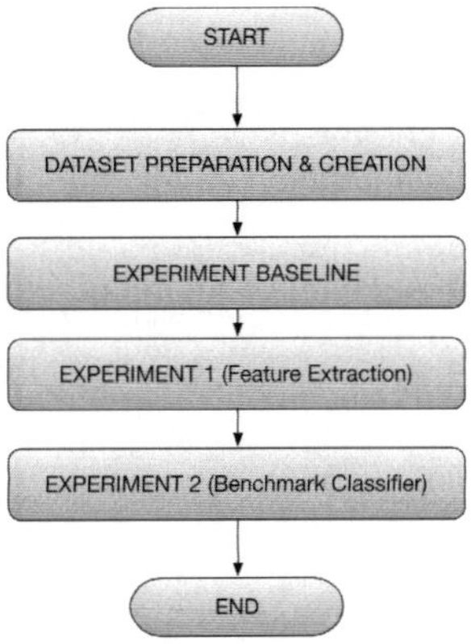

Figure 1. Experiment Scenario

3.1. Data Preparation & Dataset Creation

This research started with collecting Indonesian names from two online sources by performing web crawling to obtain the person names data which are list of high school student names (taken from Data Pokok Pendidikan SMA-SMK website) and constituent data of Pilkada 2017 (Simultaneous Regional Election) organized by KPU. Total number of Indonesia person names collected are 2,125,210 names. This stage was conducted by running series of data cleaning and standardization process which are stripping whitespace, removing honorific/titles, duplication/missing values handling.

Concerning pairwise format, in binary classification problem, the dataset was commonly categorized into two types. They were positive and negative examples (instances). Positive examples are pair of names that considered as matching/similar names, on the contrary negative are dissimilar pair. Data set generation stage was performed with goal to compose dataset containing both positive and negative examples as shown in **Table *1***.

Table 1. Data preparation result

	Jarowinkler_score	Name1	Name2	Pairname	Variant_type	label
1706	0.576923	Dhian Prianto	Sofian	(Dhian Prianto, Sofian)		0
3170	0.887500	Syahrul Syahrozi	Syahrul	(Syahrul Syahrozi, Syahrul)	Word addition/removal	1
1961	0.595238	Mustofa	Lusi	(Mustofa, Lusi)		0
4360	0.891429	Siti Arini	Siti A.	(Siti Arini, Siti A.)	Short form (initial)	1
897	0.758333	Rudiy Supriyatna	Supriyanta	(Rudiy Supriyatna, Supriyanta)		0
919	0.529365	Grace Nathascha	Thascha	(Grace Nathascha, Thascha)		0
3025	0.940308	Intan Purnama Sari Ertanus	Intanpurnama Sari Ertanus	(Intan Purnama Sari Ertanus, Intanpurnama Sari...	Typo random space deletion	1
3781	0.969474	Raihan Fathurrahman	Raihan Fathoerrahman	(Raihan Fathurrahman, Raihan Fathoerrahman)	Transcription variants	1
3298	0.849351	Yohan Hendrik Manibor	Yohan H. M.	(Yohan Hendrik Manibor, Yohan H. M.)	Short form (initial)	1
3111	0.890909	Yuyun Jufri	Yuyun	(Yuyun Jufri, Yuyun)	Word addition/removal	1

3.2. Experiment 1 Baseline

This experiment utilized pattern and phonetic matching methods as the baseline of experiment. The goal is to setup the baseline performance compared to our proposed method. There are 30 different pattern matching method, from winkler, permute-winkler, bag distance, edit distance, until qgram, etc.

And for phonetic matching methods, there are 15 different methods (soundex, phonex, phonix, caverphone, until koelner).

Name matching was performed by comparing the score of each method with their best threshold. The score below the threshold was non-matching pair, and otherwise was matching pair. This classification was then compared with actual label of each pair and count how many pair were correctly or incorrectly classified. Using F1 measure metric, this experiment then determined how strong the predictive power for each tested method.

The best performing overall matching algorithm for Indonesia pair names dataset was achieved by **permwinkler** with F1 score 0.972. However, if name matching requires both accuracy and speed, then the method best suited this need is **qgram2P.** In general, pattern matching algorithm beat phonetic matching algorithm in term of performance (F1 measure score).

3.3. Experiment 2 (Feature Extraction)

This experiment focused on finding the way to improve the performance of the baseline methods using Machine Learning classification methods, considering that the opportunity area could be improved from those identified on previous experiment. This experiment selected three classifiers to conduct classification, which were Linear SVM (SVM), Logistic Regression (LR) and Random Forest (RF).

The experiment 1 test through some of pattern matching methods, phonetic matching methods, and each of them evaluated one-by-one as single feature by machine learning algorithm. List of features can be observed in **Table** *2*. Beside using existing pattern and phonetic methods, the research also introduced and proposed new feature group named context_feature.

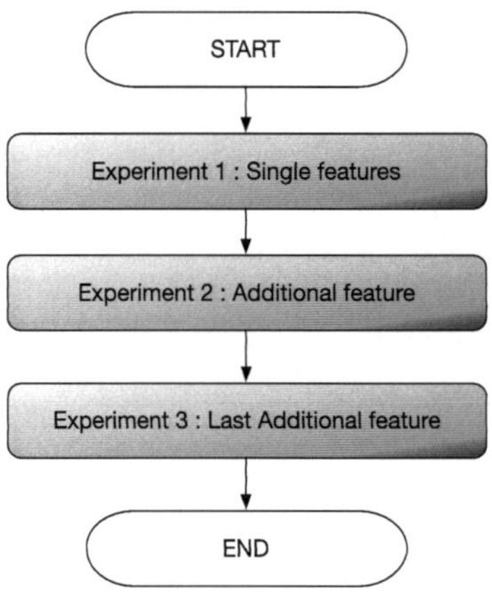

Figure 2. Feature extraction experiment

As shown in **Figure** *2*, after experiment 1 ran, the best single feature which had the best F1 score was selected. It was then passing through to experiment 2. This experiment 2 testing used combination of 2 features, consisted of one feature obtained from experiment 1, and another one was a new additional feature. The combination of 2 features that gave the best F1 score was then used as baseline for experiment 3. In experiment 3, an additional feature was added to result of experiment 2 to form 3 features. The combination of 3 features that gave the maximum F1 score were then selected as the final and the best feature set of this experiment.

The goal of this experiment was to find what feature group contribute the most significant to the F1-score performance. It was indicated from the result (**Table** *3*) that the best feature group is achieved by using **context_features+permwinkler+ontolcs3avrg** with F1-score 0.993. However, this score is the same with previous experiment that using only 2 features **context_features+permwinkler** or **context_features+lcs3avrg**. By this result it can be concluded that the increasing number of feature is not always making the performance better anymore. Another important result is among 3 classifiers, the best classifier was achieved by **Random Forest.**

Table 2. List of features for feature extraction experiment

No.	Feature group	Feature type	Permitted value	Description
1	permwinkler	numerical	0.0 – 1.0	Winkler permutation score
2	lcs2	numerical	0.0 – 1.0	Longest common substring with minimum length of substrings 2
3	lcs3	numerical	0.0 – 1.0	Longest common substring with minimum length of substrings 3
4	ontolcs3	numerical	0.0 – 1.0	Ontology longest common substring with minimum length of substrings 3
5	editex	numerical	0.0 – 1.0	Phonetic aware edit-distance
6	qgram2	numerical	0.0 – 1.0	q-grams of length 2
7	qgram3	numerical	0.0 – 1.0	q-grams of length 3
8	qgram2P	numerical	0.0 – 1.0	Padded q-grams of length 2
9	qgram3P	numerical	0.0 – 1.0	Padded q-grams of length 3
10	tfidf	numerical	0.0 – 1.0	Cosine similarity of TF-IDF character 1,2,3,4-gram
11	nysiis4	numerical	0.0 – 1.0	NYSIIS limited/padded to length 4, use 5 sets of rules to convert name into code
12	phonex4	numerical	0.0 – 1.0	Phonex limited/padded to length 4, combination of Soundex and Metaphone
13	phonix4	numerical	0.0 – 1.0	Phonix limited/padded to length 4, applies string standardisation rules before a nine numerical code transformation
14	soundex4	numerical	0.0 – 1.0	Soundex limited/padded to length 4
15	context features	categorical	string and T/F (True or False)	The number of features in this group is variable, depend on how many word on string name. The features are extracted per word. 1. Extracting prefix/suffix features • 1,2,3 char prefix and suffix (char) • has_initial (T/F) • has_tj (T/F) • has_oe (T/F) • has_sj (T/F) • has_dj (T/F) • has_ie (T/F) 2. Extracting aligned transformation features As many as 12 features extracted by utilizing sequence matching difflib library, algorithm based on Ratcliff and Obershelp pattern matching. Names are aligned at first, and then the features extracted per word: • insert_char_position (start/mid/end) • insert_is_char_a_consonant (T/F) • insert_is_with_rightchar_ejaanlama (T/F) • insert_is_with_leftchar_ejaanlama (T/F) • delete_char_position (start/mid/end) • delete_is_char_a_consonant (T/F) • delete_is_with_rightchar_ejaanlama (T/F) • delete_is_with_leftchar_ejaanlama (T/F) • replace_char_position (start/mid/end)

				• equal_samewith_oneofstrings (T/F) • replace_in_rules (T/F) 3. Extracting context features Six features are extracted following the idea from what kind of misclassification occured : • context_same_last_char (T/F) • context_singleword (T/F) • context_string1_in_string2 (T/F) • context_difference_stringlength (T/F) • context_differ_only_last_char (T/F) • context_same_truncated_words_remove_space_ between_words (T/F)

Table 3. Result of feature extraction experiment using 3 features

Feature Group	FN			FP			Misclassification			# features	F1 on testing set		
	SVM	LR	RF	SVM	LR	RF	SVM	LR	RF		SVM	LR	RF
context_features|permwinkler|ontolcs3avrg	9	8	8	14	13	5	23	21	13	6841	0.987	0.988	0.993
context_features|lcs3avrg|lcs2avrg	12	7	11	10	9	3	22	16	14	6841	0.988	0.991	0.992
context_features|lcs3avrg|tfidf	13	10	9	9	9	5	22	19	14	6841	0.988	0.989	0.992
context_features|permwinkler|qgram2Pavrg	9	11	10	12	10	4	21	21	14	6841	0.988	0.988	0.992
context_features|permwinkler|lcs2avrg	9	9	12	12	10	3	21	19	15	6841	0.988	0.989	0.991
context_features|permwinkler|tfidf	10	9	10	11	9	5	21	18	15	6841	0.988	0.990	0.991
context_features|lcs3avrg|permwinkler	10	9	11	13	11	5	23	20	16	6841	0.987	0.989	0.991
context_features|lcs3avrg|qgram2Pavrg	13	11	12	11	10	4	24	21	16	6841	0.986	0.988	0.991
context_features|lcs3avrg|nysiis4	13	11	12	9	10	5	22	21	17	6841	0.988	0.988	0.990
context_features|lcs3avrg|ontolcs3avrg	11	10	10	12	10	7	23	20	17	6841	0.987	0.989	0.990
context_features|permwinkler|nysiis4	11	12	13	14	12	5	25	24	18	6841	0.986	0.986	0.990

3.4. Experiment 3 Benchmark Classifier

The purpose of this experiment was to provide brief snapshot on the performance of multiple classifiers using their default parameters to get the idea on what was the best performing classifier to solve problem of name matching.

Table 4. Experiment 3 result applying multiple classifiers and using context + permwinkler features

	f1_score	**processtime**
Logistic Regression	0.990009	0.439947
AdaBoost	0.989262	5.564991
NeuralNet	0.988628	17.073669
Random Forest	0.988261	0.663623
Perceptron	0.987778	0.865366
Linear SVM	0.987624	8.571252
Decision Tree	0.985388	0.676483
Stochastic Gradient Descent	0.984654	0.181748
Bernoulli Naive Bayes	0.977275	0.273834
KNN	0.972321	6.108816

4. Result and Discussion

From all experiments conducted on each method either using baseline or machine learning classification, the result could be compared by looking on **Table 5**.

Table 5. Comparison of experiment results

Experiment	Methods	TP	TN	FP	FN	F1	#Samples	Error	Ratio error
Baseline	permwinkler	863	810	23	20	0.976	1,716	43	2.51%
Classification machine learning	context features	863	826	7	20	0.985	1,716	27	1.57%
Classification machine learning	context features+permwinkler	875	828	5	8	0.993	1,716	13	0.76%

- Based on comparison of the result, it is revealed that machine learning classification have shown better performance for name matching task compare to what can be best achieved by baseline or conventional method (using pattern matching techniques).
- Machine learning classification algorithms have wide range of classifier algorithms and parameters that could be tuned in order to find what is the best one. Experimental evaluation in this thesis shows that by using proposed features (context features) and selecting the best classifier (Random Forest), shows that the performance could be improved by 0,9% (using context features only) and 1,7% (by using context and permwinkler features), compared to baseline. This small improvement on F1 measure is due to baseline already have very good number of F1 (F1 = 0,976).
- Context features and combination of context features and permwinkler can reduce many misclassifications significantly by 70% reduction that present on baseline permwinkler.

5. Conclusion and Future Works

5.1. Conclusion

Indonesia name dataset by this thesis is developed. This dataset hopefully will open more research for improving Indonesian name matching system. From the experiments performed for name matching of Indonesian names using this dataset, it is concluded that machine learning supervised classification techniques have good chance to outperformed the state of the art using conventional pattern matching techniques. The key importance aspect for achiving superior performance is on how feature constructed and hyperparameter tuning of machine learning classification algorithms.

Although conventional method had achieved near perfect performance (high F1 score), this research had shown that by using context features combined with other selected feature (permwinkler), it could surpass the best achieved performance of conventional pattern matching techniques (permute-winkler).

5.2. Recommendation

It is advisable to provide more comprehensive solution, that future research can be covered the other source of name variation which is due to name alias.

Thorough analysis on Indonesia names variation could be improved by focusing on majority of Indonesia population which is Java tribe or others. By performing deep analysis, hopefully one can identify better predictive and discriminative features to improve the performance of name matching task.

It has also the possibility for improvement if the misclassification result can be analyzed and combined further using heuristic rule to make additional classification.

IOP Conf. Series: Journal of Physics: Conf. Series **971** (2018) 012038 doi:10.1088/1742-6596/971/1/012038

References

[1] P. Christen, "A Comparison of Personal Name Matching: Techniques and Practical Issues," *Data Min. Work. 2006. ICDM Work. 2006. Sixth IEEE Int. Conf.*, no. September, pp. 290–294, 2006.

[2] D. Isyanti, "Changes in the Rules for Indonesian Name Personal Headings," 2006.

[3] B. Lisbach and V. Meyer, *Linguistic Identity Matching*. Springer Science & Business Media, 2013.

[4] C. Varol and C. Bayrak, "Hybrid Matching Algorithm for Personal Names," *ACM J. Data Inf. Qual.*, vol. 3, no. 4, pp. 1–18, 2012.

[5] W. W. Cohen, P. D. Ravikumar, S. E. Fienberg, P. D. Ravikumar, and S. E. Fienberg, "A Comparison of String Distance Metrics for Name-Matching Tasks," *Proc. IJCAI-03 Work. Inf. Integr. Web*, pp. 73–78, 2003.

[6] P. Christen, *Data Matching Concepts and Techniques for Record Linkage, Entity Resolution, and Duplicate Detection*. Springer Berlin Heidelberg, 2012.

Negation handling in sentiment classification using rule- based adapted from Indonesian language syntactic for Indonesian text in Twitter

Rizkiana Amalia, Moch Arif Bijaksana, Dhinta Darmantoro
School Of Computing, Telkom University, Indonesia

munief.rizkiana@gmail.com, arifbijaksana@telkomuniversity.ac.id, dhinta@telkom.co.id

Abstract. The presence of the word negation is able to change the polarity of the text if it is not handled properly it will affect the performance of the sentiment classification. Negation words in Indonesian are '*tidak*', '*bukan*', '*belum*' and '*jangan*'. Also, there is a conjunction word that able to reverse the actual values, as the word '*tetapi*', or '*tapi*'. Unigram has shortcomings in dealing with the existence of negation because it treats negation word and the negated words as separate words. A general approach for negation handling in English text gives the tag '*NEG_*' for following words after negation until the first punctuation. But this may gives the tag to un-negated, and this approach does not handle negation and conjunction in one sentences. The rule-based method to determine what words negated by adapting the rules of Indonesian language syntactic of negation to determine the scope of negation was proposed in this study. With adapting syntactic rules and tagging "*NEG_*" using SVM classifier with RBF kernel has better performance results than the other experiments. Considering the average F1-score value, the performance of this proposed method can be improved against baseline equal to 1.79% (baseline without negation handling) and 5% (baseline with existing negation handling) for a dataset that all tweets contain negation words. And also for the second dataset that has the various number of negation words in document tweet. It can be improved against baseline at 2.69% (without negation handling) and 3.17% (with existing negation handling).

1. Introduction

Social media such as blogs, Facebook, discussion forums, Twitter, Instagram, has large users. Twitter as a social media states in their website that the users of Twitter until 30 June 2016 are approximately 1 billion unique visits monthly to the sites with embedded Tweets, 313 million active users each month, and 82% are active users on mobile [1]. However, the structure of the text on Twitter is unstructured texts, in the sense that words are not formal, abbreviations, slang, and misspellings.

Sentiment Analysis (SA) can be framed as a text classification task where the categories are *polarities* such as *positive* and *negative*. Many companies use sentiment analysis approach to finding out what is happening with the market. For example the feedback of customers on products released, problems or disorders are often experienced by customers. They give a potential comparison to competitors of the same product and others. Preceded by Pang Lee et.al. [2] perform classification on movie review into two classes, the positive reviews and negative reviews, the results of experiments showing that the use unigram based (a bag of individual word) as a classification feature goes well with classifiers such as Multinomial Naive Bayes or Support Vector Machine.

Besides the unstructured text, there are several challenges in sentiment analysis fields, one of those is a negation. The important part when handling negation is to determine correctly what part of the meaning expressed is modified by the presence of the negation, this is called the scope of negation. The negations can appear in various forms in opinion reviewed by customers. For example from previous work in Indonesian Twitter Supriyono, Budhi [3], from 20844 tweets there are 4672 tweets or 22,41% from data containing at least one negation term in a tweet. It can be seen that the negation is often used in the tweet opinion, therefore, important to note the effect on the polarity of the tweet.

The most common way is modeling document into a vector space of numerical linear algebra and perform operations on these vectors, such representation is known as the *"Bag-Of-Word"* or BOW. It uses frequencies of words or terms ("word occurrences") on a document. So it is closely related to the number of occurrences of words in a document. With traditional methods, unigram features, it can not deal the existence of negation, because unigram treats negation words with words that negated as a separate word.

Motivated by this problems, this research is to investigate the influence of negation handling for sentiment classification in Indonesian Twitter, also to handle the existence of transitional words in a sentence, and to enhance the sentiment analysis with the proposed negation handling method at this study. With adapting Indonesian syntactic rules as negation representation capable to handling negation as negated representation in a sentence and can enhance the sentiment classification process. This study is conducted based on the Twitter dataset in telecommunication product domain, specifically, they are the product of PT. Telkom Indonesia (*Speedy, Indihome*, and *UseeTV*) and used two classifiers, Multinomial Naive Bayes and SVM.

2. Literature Review

2.1. Related Works

Preceded by Pang Lee et. al [2], they classify documents not by topics but by sentiments, e.g determining whether the review is positive or negative. For negation handling, if a word *x* follows the negation word then a new feature 'NOT_*x*' created tag every word from *x* until first punctuation mark. But this method cannot model the scope of negation, because it is heuristically tagging all word until it finds the mark, without concerning with negation words or not. Addition in preprocessing task, mostly the punctuation marks are removed; this is for simplification in preprocessing stage.

For the negation in sentiment, there are some of the researchers that focus on the impact of the negation in sentiment sentences. A survey conducted by Wieghan et. al [4], they survey for negation role in sentiment analysis. They state that effective negation model for sentiment analysis usually requires the knowledge of polar expression. Jia et. al. [5] studied the impact of each occurrence of a negation term in a sentence on its polarity and introduced the concept of scope of the negation term *t*. As research in Indonesia, Bojar [6] who conducted research about the resources of the lexicon for Indonesian sentiment also did the negation handling. By adapting the technique from Das and Chen [7] handled the negation of sentiment caused by a negation word. Bojar uses negation words such as *'tidak', 'tak', 'tanpa', 'belum',* and *'kurang'*. The words that occur between the negation words and the first punctuation after the negation word are tagged with 'NOT_'. Example, there is a sentence: *'kameranya kurang bagus gambarnya'* became *'kameranya kurang NOT_bagus NOT_gambarnya'*.

Several studies also concern with the scope of negation, Moral Dadvar et. al.[8], conduct a study dealing with different negation scopes to investigate how it affects the polarity identification of the sentences and assume that opinions are mostly expressed by the use of adjective and adverbs. Hogenboom et. al. [9] they state that for English review sentences, the best performing method is considering 2 words following the negation to be negated. Not only in the English language, in Chinese [10] they have constructed a Chinese corpus and a new syntactic structure-based framework to identify the linguistic scope of a cue, instead of the traditional chunking-based framework. In Hindi [11], initially annotated dataset to test the proposed algorithm was conducted then the rules are devised for handling negation with static windows (3 to 5 words). And in Arabic [12], they stated that sentiment is related to culture and language morphology, so they also defined a set of rules that capture the morphology of negations in Arabic. These rules are then used to detect sentiment taking care of negated words. It can be seen that linguistic features that define the structure or syntactic from languages may influence to determine the polarity of sentiment.

2.2. *Bag-of-Words and TFIDF Weight*

Model representation of a document or text most frequently and widely used is the VSM (Vector Space Model) or known as BOW (Bag Of Words). BOW ignores the order of the words in a document and disregards all grammar. The exact ordering of the terms in a document is ignored but the number of occurrences of each term is material [13]. However, it performs generally very well in text-based classification. Another approach besides calculates term frequency, there are Term Frequency-Inverse Document Frequency (TF-IDF). Since each word has a different degree of importance in the document, for each word it is given an indicator, i.e the term weight. It can be used for reduction of the term dominance that often appears in various documents. This is necessary because the term that appears in many documents, can be considered as a common term so that it does not matter its value. Conversely, the term scarcity factor in the document collection should be considered in weighting.

$$Weight(i) = tf.idf(i) = tf_i * \log\frac{N}{df_i(w)}$$

Where *Tf* is the term frequency or a number of a term *(w)* in a document, *IDF* is the frequency of the document or the number of documents that contained the term *(w)* in it, *N* is total documents, and *Df* is the number of the document that contained the term *(w)*.

2.3. *Multinomial Naive Bayes*

Naive Bayes classification is one of classification methods based on applications of simple possibility Bayes theorem. It is assumed that the presence or absence of a particular occurrence of a group is not associated with the presence or absence of other events. Bayes theorem is used to calculate the probability of occurrence of an event based on the effect obtained from the observation. Bayesian theory can also be used as a decision-making tool to renew the confidence level of the information. They can be formulated as,

$$p(C_k|x) = \frac{p(x|C_k)\,p(C_k)}{p(x)}$$

Where x is a vector that representing some n features (independent variables), it assigns to this instance probability $p(C_k|x_1, \dots x_n)$ for each of k possible outcomes or classes C_k. In text classification, our goal is to find the best class for the document. The best class in Multinomial Naive Bayes classification is the most likely or *maximum a posteriori* (*MAP*) decision rule. A Bayes classifier is the functions that assign a class label $\hat{y} = C_k$ for some k as follow,

$$\hat{y}(predicted\ class) = \underset{k\in\{1,\dots,K\}}{argmax}\, p(C_k)\prod_{i=1}^{n} p(x_i|C_k)$$

Naive Bayes Classifier is a general term which refers to the conditional independence of each of the features in the model, while Multinomial Naive Bayes classifier is a specific instance of a Naive Bayes classifier which uses a multinomial distribution for each of the features. The term Multinomial Naive

Bayes simply lets us know that each $p(x|C_k)$ is a multinomial distribution, rather than some other distribution. This works well for data which can easily be turned into counts, such as word counts in the text. Multinomial Naive Bayes can be used for various purposes including for the classification of documents, spam detection or filtering spam, and sentiment classification [14].

2.4. Support Vector Machine

The aim of SVM is to find the best classification function to distinguish between classes in the training data and to produce a model based on training data which predicts the target values of the test data given only the test data attributes [15]. In linear data set, where the data can be separated linearly, example with training data set of the instance label pair $(x_i, y_i), i = 1, \dots, l$ where $x_i \in R^n$ and $y \in \{1, -1\}^l$ is the label class from x_i. Because there are many such linear hyperplanes, SVM additionally guarantees that the best function is found by maximizing the margin between the two classes. Mathematically, the SVM optimization problem formulation for the case of linear classification in primal space is:

$$\min \frac{1}{2} \|w\|^2$$

$$subject\ to$$

$$y_i(wx_i + b) \geq 1, i = 1, \dots, n$$

To solve nonlinearity problems that often occur in real cases, the kernel method is applied. The kernel method provides an alternative approach by mapping data x from the input space to feature space F through a function φ then $\varphi: x \rightarrow \varphi(x)$. The kernels in this are used linear and RBF. Much like research on text classification using SVM before [16]. And also in [15] stated that RBF kernel function has the same performance with linear SVM on certain parameters.

2.5. Negation in Indonesian

The negations words (*tidak, bukan, belum, jangan*) not only used to express the denial or inverse the sentences but also in conjunction with the yes or no questions. Besides the direct negation marker, there are also another words that indirectly can invert the meaning of sentences even it is not there negation marker, such *tetapi* or *tapi.*

2.6. Negation Handling

From literature survey, many studies have been performed with respect to sentiment classification problem at the sentence or phrase level approach, including handling the negation in subtasks of sentiment analysis. It becomes important because, in traditional methods such unigram, two sentences can end up with same VSM representation, for example, sentences *"responnya cepat sekali"* dengan *"responnya gak cepat sama sekali"*. This two sentences may be represented with same feature vector, and it can be leading to misclassified. Some ways in the previous work to handle negation in English are by determining the scope of negation first that have been done, in [8], [7] Das and Chen proposed using tagging "NOT_", or using static windows [2], [9]. For scope of negation using fixed static window has been tested in various number, start with simplest way to invert the polarity of the sentiment word directly following the negation word [17], 5 next words to the negation are inverted [18], or from three to six words following after negation [19]. Not only in English language, research about negation effect in sentiment classification also has been done for another language [10],[11],[12].

3. Baseline & Proposed Methods

3.1. Baseline System

There was two baselines system used as a comparison to see how far the handling effect of negation. First the baseline without negation handling. It was the experiment where there was not a treatment for any negation in the text, and it was called ***B-NEG*** for the experiment. The second baseline with existing negation handling method, it was the experiment to determine the scope of negation in order to negation handling using tag *'NEG_'* for words between negation and first punctuation mark (period, comma, question marks). And it was called ***B+NEG*** for the experiment.

3.2. Dataset

There were two datasets crawled and labeled manual from Twitter using the keyword. First dataset (*all_negation*), were all tweets containing negation or conjunction at least one word. A total number of negation tweets 300 tweets, consists of 128 negative tweets, 100 non-opini tweets, and 72 positive tweets. The second dataset (*training_negation*) contained tweets that did not all contain negations or conjunctions. Consists of 100 negative tweets, 102 non-opini tweets, and 98 positive tweets. With 30% (~90 tweets) contains negation words. The domain was in telecommunication Telkom product, Speedy, and Indihome. For testing data, we applied split in a dataset with 80:20 composition. 80% for the training set and 20% for testing data for each dataset. Negation handling applied after preprocessing phase.

3.3. Proposed Methods

The proposed method experiments were determined the scope of negation using linguistic approach, by defining the set of rules based on Indonesian syntactic (see table 1). In Indonesian grammars referring to [20], there are four negative words: *'tidak', 'bukan', 'belum'* and *'jangan'*. And also from a research about comparison negation for English and Indonesia, [21] and [22] stated that Indonesian negation markers come after various syntactic categories such as noun, adjective, adverb, prepositional phrase. In addition, heuristic rules based on Part-of-Speech from [23] (JJ_NN and VB_NN forms) were used. The syntactic grammars were adapted from these references. We used this rules with 2 differents experiments way to determine and treat negations in the text (see Figure 1).

- Proposed rules as negation term, called it with *Neg_Term* for the experiment, a pair or set of words that obtained from rules were inserted into the text as a presentation from negation.
- Proposed rules as the scope of negation, called it with *Neg_Term+NEG* for the experiment, a pair or set of words that obtained from rules were given the tag *'NEG_'* as a presentation from negation.

Example raw text : *'@TelkomCare Bukan salah Telkom Speedy tapi PLN yang padam kebetulan UPS dan Genset ku rusak.'*
Preprocessing : *'bukan salah telkom speedy tapi padam genset rusak'*
Negation handling (*Neg_Term*) : *'**bukan_salah** telkom speedy **tapi_padam** genset rusak'*
Negation handling (*Neg_Term+NEG*) : *'bukan **NEG_salah** telkom speedy tapi **NEG_padam** genset rusak'*

Table 1. Syntactic Rules for Negation and Conjunction

Negation Words	Adjacent	Examples
Tidak	Adjective – JJ	('tidak', u'NEG'), ('adil', u'JJ')
	Verb – VB	('tidak', u'NEG'), ('beri', u'VB')
	Noun – NN	('tidak', u'NEG'), ('masalah', u'NN')
	Indefinite numbers – CD	('tidak', u'NEG'), ('banyak', u'CD')
	Adjective followed by Noun – JJ_NN	('tidak', u'NEG'), ('jelas', u'JJ'), ('terimakasih', u'NN')
	Verb followed by Noun – VB_NN	('tidak', u'NEG'), ('paham', u'VB'), ('sistem', u'NN')
	Modals (bisa,dapat,	('tidak', u'NEG'), ('dapat', u'MD'),

	pernah,boleh,perlu,harus, mungkin) followed by one word after – MD_word	('bantu', u'NN') ('tidak', u'NEG'), ('mau', u'RB')
	Adverb (mau, ingin) – RB	
Bukan	Noun – NN	('bukan', u'NEG'), ('speedy', u'NN')
	Adjective – JJ	('bukan', u'NEG'), ('salah', u'JJ')
	Quantity clause – CD	('bukan', u'NEG'), ('sedikit', u'CD')
Belum	Verb – VB	('belum', u'NEG'), ('selesai', u'VB')
	Adjective – JJ	('belum', u'NEG'), ('lama', u'JJ')
	Modals – MD (bisa, dapat, pernah, boleh, perlu,harus mungkin)	('belum', u'NEG'), ('bisa', u'MD') ('belum', u'NEG'), ('mau', u'RB')
	Adverb (mau, ingin) – RB	
Jangan	Verb – VB	('jangan', u'NEG'), ('mati', u'VB')
	Adjective –JJ	('jangan', u'NEG'), ('sungkan', u'JJ')
	Modals (pernah) – MD	('jangan', u'NEG'), ('pernah', u'MD')
	Adverb (mau) – RB	('jangan', u'NEG'), ('mau', u'RB')
Tetapi / Tapi	Adjective – JJ	('tapi', u'CC'), ('lambat', u'JJ')
	Noun – NN	('tapi', u'CC'), ('hubung', u'NN')
	Adverb – RB	('tapi', u'CC'), ('pasti', u'RB')
	Noun followed by Adjective – NN_JJ	('tapi', u'CC'), ('indihome', u'NN'), ('cepat', u'JJ')
	Negation – NEG, for this forms, applied negation rules for each negator.	('tapi', u'CC'), ('belum', u'NEG'), ('datang', u'VB')

4. Result

The final result of the experiment was the classification process done by machine learning approach, using Multinomial Naive Bayes and Support Vector Machine. Performance measurements performed measuring the average of F1-score (percentage) and accuracy (percentage).

Table 2. Performance based on average F1-score

Experiments	*All_negation*		*Training_negation*	
	M-NB	SVM	M-NB	SVM
B - Neg	**85,03%**	84,95%	83,08%	83,73%
B + Neg	80,01%	81,74%	82,22%	83,25%
Neg_Term	81,67%	81,83%	84,93%	85,23%
Neg_Term + Neg	84,96%	**86,74%**	**86,26%**	**86,42%**

Table 3. Performance Based on Accuracy

Experiments	*All_negation*		*Training_negation*	
	M-NB	SVM	M-NB	SVM
B - Neg	**85,00%**	85,00%	83,33%	83,33%
B + Neg	80,00%	81,67%	83,33%	83,33%
Neg_Term	81,67%	81,67%	85,00%	85,00%
Neg_Term + Neg	85,00%	**86,67%**	**86,67%**	**86,67%**

4.1. Classification Performance

In Table 2, it can be seen in *all_negation* dataset for our proposed methods there are not increasements in average F1-score when compared to the baseline without negation handling *B-NEG* with Naive Bayes classifier, but compared to baseline negation handling *B+NEG* our proposed method there is an increase of 1,66% (*Neg_Term*) and 4,95% (*Neg_Term+NEG*). While with SVM classifier our proposed method *Neg_Term+NEG* seen there was a better improvement when compared with both baseline, *B-NEG* (1.79%) and *B+NEG* (5%).

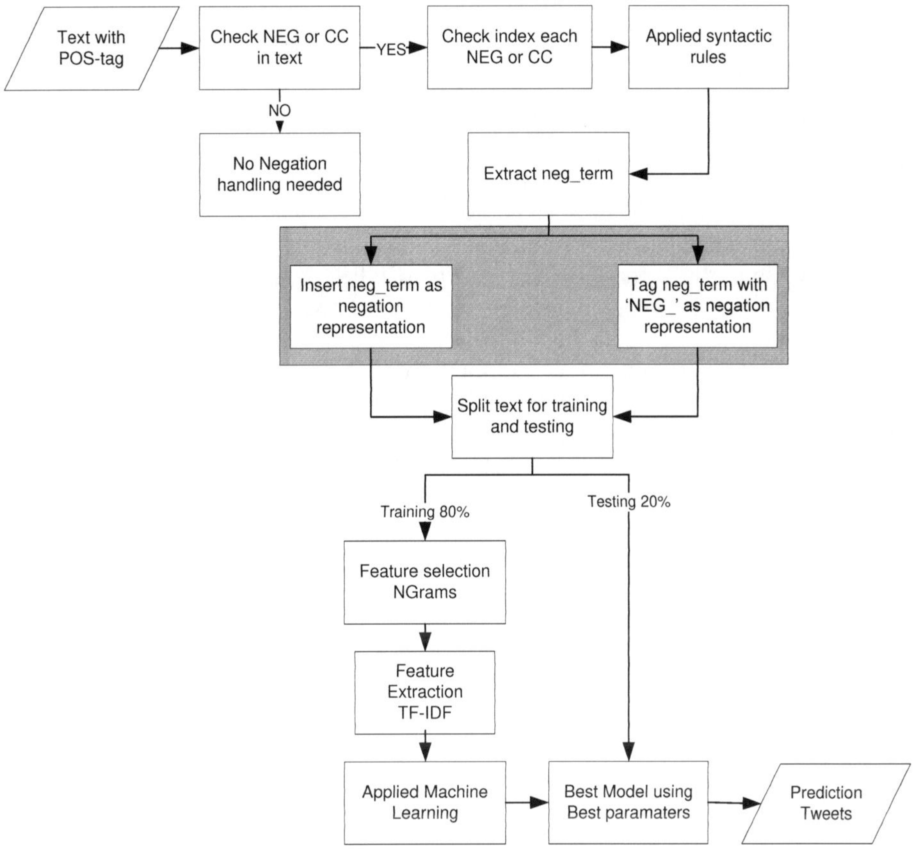

Figure 1. Negation Handling Process for Proposed Methods

And for *training_negation*, in this type of data, both of our proposed methods can increase average F1-score performance using Naive Bayes classifier at 1.85% (*Neg_Term*) and 3.18% (*Neg_Term+NEG*) against *B-NEG* and 2.71% (*Neg_Term*) and 4.04% (*Neg_Term+NEG*) against *B+NEG*. The same result for both our proposed method and SVM classifier compared with baseline *B-NEG* was 1.5% (*Neg_Term*) and 2.69% (*Neg_Term + NEG*), and compared against with baseline *B+NEG* was 1.98% (*Neg_Term*) and 3.17% (*Neg_Term + NEG*). From experiment results based on F1-score and accuracy, our proposed methods have no significant effect on Naive Bayes classifier in *all_negati*on data. But has improved with using SVM classifier, it also for another dataset.

5. Discussion

In the first dataset, *all_negation*, each tweet document contained a component of negation word, so negation had an almost be spread evenly distributed across the document tweets. It is possible that the negation words forming a new form (tagging 'NEG_' or *negation_term*) will be present in multiple document tweets and distributed to in 3 classes. For Multinomial Naive Bayes, it has to deal with the count of the word for each class and the total count off of all words in the class, which is a calculation of conditional probability or likelihood events for each term. TFIDF weighting is closely related to the number of terms in a document belonging to a particular class (TF) and the number of terms in the entire document (IDF). Like the word *'bisa'*, there are 98 words *'bisa'* in 300 document tweets in *all_negation* dataset. For example, in *B+NEG* experiment, there were 55 *'bisa'* and *Neg_Term+NEG* had 49 *'bisa'* which was negated into *'NEG_bisa'* which could be in positive, negative, and non_opini tweets. The calculation of the total TFIDF weight of the words *'bisa'* and *'NEG_bisa'* in the entire training document would be not much different. So when there was data testing that contained the word *'NEG_bisa'*, the term weight value in conditional probability for each given class had no significant effect. As if the process without negation handling with negation handling result was not significant.

While in SVM, although experiment with negation handling (*B + NEG* and *Neg_Term*) the F1-score average score was lower than experiment without negation handling, but the F1-score average on *Neg_Term + NEG* was better than the other three. Through TF-IDF the important words which have used to train SVM model were obtained. The accuracy of the model generated from the training process with SVM was highly dependent on the kernel function and the parameters used. It was different from Multinomial Naive Bayes which was treated the features as independent, whereas SVM looks at the interactions between them to a certain degree. But in the second dataset, *training_negation,* the more varied data the number of negations in the tweets, negation handling more visible influence. Since the negation representation word distribution had a smaller DF if it was compared to *all_negation* data. So the difference in the weight of the word representation negation of the document looks more significant for a particular class.

In addition, when reanalyzed on the implementation of syntactic rules for proposed methods on both datasets, there were some tweets that despite the word negation but unsuccessful syntactic rules apply. These were 32 tweets (10.67%) in the *all_negation* dataset and 8 tweets (2.67%) in the *training_negation* dataset. The tweets that were not applied by syntactic rules were caused by the following:

a. The position of the word negation was at the end of the sentences.
b. The word after the negation had a POS-Tag which was not included in the syntactic rules.
c. The unigram POS-tag had default tag *'NN'* for unrecognized tokens through tagging training.

6. Conclusions & Future Work

6.1. Conclusions

The traditional method of negation handling using the *'NEG_'* tag for words after negation until the first punctuation, there will be irrelevant words that were negated. In this study, two ways to handle negation were proposed namely by representing negation and word negated to determine the scope of negation. Based on experiment our proposed methods, did not work well with Multinomial Naive Bayes in the data that had a negation word distribution almost across the tweet document, but it had increased compared to the baselines (without negation handling and negation handling existing) in the data whose amount of negation did not dominate the overall tweets (in this study, percentage of the word negation is 30% tweet of the document tweet). For overall, the second proposed methods with adapting syntactic rules and tagging *"NEG_"* using SVM classifier with RBF kernel had better performance results than the other experiments. The RBF kernel hyperparameters from 10-fold cross validation auto-tuning process were obtained (C = 100 and gamma = 0,001) for both datasets. Considering the best average F1-score value was 86,74% for the first dataset that all tweets contain negation words and 85,42% for the second dataset that had the various number of negation words in document tweet. The performance of proposed method can be improved against the baseline, equal to 1.79% (without negation handling) and 5% (with existing negation handling) for the first dataset. And also for the second dataset, it can be improved against baseline at 2.69% (without negation handling) and 3.17% (with existing negation handling).

6.2. Future Work
Based on the findings and the conclusion of the study, there are some recommendations. In the future, the handling of negations may be done on another domain for the Indonesian language. But with resource adjustments such as dictionaries and datasets used. And also by using a complete dictionary or automatic translator for deep preprocessing. Because a good preprocessing process can improve the performance of negation handling better. Meanwhile, POS-tagging is an important support task in the implementation of syntactic rules, in the future it can be tested POS-tagging method other than unigram tagger for Bahasa Indonesia.

References

[1] "Twitter Corp.," [Online]. Available: https://about.twitter.com/company.

[2] B. Pang, L. Lee and S. Vaithyanathan, "Thumbs Up? Sentiment Classification Using Machine Learning Techniques," *Proceedings of the ACL-02 conference on Empirical methods in natural language processing - EMNLP '02,* vol. 10, pp. 79-86, 2002.

[3] B. Supriyono, "WEB DATA MINING FOR CUSTOMER'S SENTIMENT CLASSIFICATION FOR TELKOM SPEEDY USING TWITTER IN INDONESIAN," no. August 2015.

[4] M. Wiegand, A. Balahur, B. Roth, D. Klakow and A. Montoyo, "A Survey on the Role of Negation in Sentiment Analysis," *Proceedings of the Workshop on Negation and Speculation in Natural Language Processing,* no. July, pp. 60-68, 2010.

[5] L. Jia, C. Yu and W. Meng, "The effect of negation on sentiment analysis and retrieval effectiveness," *Proceeding of the 18th ACM conference,* no. c, pp. 1827-1830, 2009.

[6] Franky, O. Bojar and K. Veselovská, "Resources for Indonesian Sentiment Analysis," *The Prague Bulletin of Mathematical Linguistics,* vol. 103, no. 1, pp. 21-41, 2015.

[7] S. Das and M. Chen, "Yahoo! for Amazon: Sentiment Extraction from Small Talk on the Web," *Management science,* vol. 53 (9), 2004.

[8] M. Dadvar, C. Hauff and F. D. Jong, "Scope of Negation Detection in Sentiment Analysis," *Proceedings of the Dutch-Belgian Information Retrieval Workshop (DIR 2011),* pp. 16-20, 2011.

[9] A. Hogenboom, P. Van Iterson, B. Heerschop, F. Frasincar and U. Kaymak, "Determining negation scope and strength in sentiment analysis," *Conference Proceedings - IEEE International Conference on Systems, Man, and Cybernetics,* pp. 2589-2594, 2011.

[10] B. Zou, Q. Zhu and G. Zhou, "Negation and Speculation Identification in Chinese Language," *Proceedings of the 53rd Annual Meeting of the Association for Computational Linguistics and the 7th International Joint Conference on Natural Language Processing,* pp. 656-665, 2015.

[11] N. Mittal and B. Agarwal, "Sentiment Analysis of Hindi Review based on Negation and Discourse Relation," *Sixth International Joint Conference on Natural Language Processing,* pp. 57-62, 2013.

[12] R. M. Duwairi and M. A. Alshboul, "Negation-Aware Framework for Sentiment Analysis in Arabic Reviews," *Proceedings - 2015 International Conference on Future Internet of Things and Cloud, FiCloud 2015 and 2015 International Conference on Open and Big Data, OBD 2015,* pp. 731-735, 2015.

[13] C. D. Manning, P. Raghavan and H. Schutze, "Chapter 6: Scoring, term weighting, and the vector space model," in *An Introduction to Information Retrieval*, Cambridge, England, Cambridge Press, 2009, pp. 109 - 134.

[14] S. Raschka, "Naive Bayes and Text Classification: Introduction and Theory," 2014.

[15] W. H. Chih, C. C. Chih and J. L. Chih, "A Practical Guide to Support Vector Classification," *BJU International,* vol. 101, no. 1, pp. 1396-400, 2008.

[16] T. Joachims, "Text Categorization with Support Vector Machines: Learning with Many Relevant Features," *Machine Learning,* vol. 1398, no. LS-8 Report 23, p. 137–142, 1998.

[17] P. V. Iterson, B. Heerschop, A. Hogenboom, F. Frasincar and U. Kaymak, "Accounting for Negation in Sentiment Analysis," *DIR,* no. February 2011.

[18] G. Grefenstette, Y. Qu, J. Shanahan and D. Evans, "Coupling Niche Browsers and Affect Analysis for an Opinion Mining Application.," *In Proceedings of the 12th International Conference Recherche d'Information Assistee par Ordinateur,* pp. 186-194, 2004.

[19] R. Narayanan, B. Liu and A. Choudhary, "Sentiment analysis of conditional sentences," *Proceedings of the 2009 Conference on Empirical Methods in Natural Language Processing Volume 1 EMNLP 09,* no. August, p. 180, 2009.

[20] J. N. Sneddon, Indonesian A Comprehensive Grammar, 1996.

[21] D. N. Syafar, "NEGASI DALAM BAHASA INDONESIA DAN BAHASA INGGRIS," *Jurnal Arbitrer,* vol. 3, no. 1, April 2016.

[22] R. Rakhmania, "A COMPARISON BETWEEN ENGLISH AND INDONESIAN NEGATION MARKERS: A SYNTACTIC STUDY.," 2012.

[23] U. Farooq, H. Mansoor, A. Nongaillard, Y. Ouzrout and M. Abdul, "Negation Handling in Sentiment Analysis at Sentence Level," *Journal of Computers,* vol. 12, no. 5, pp. 470-478, 2016.

IOP Conf. Series: Journal of Physics: Conf. Series **971** (2018) 012040 doi:10.1088/1742-6596/971/1/012040

Social media insights for sustainable development and humanitarian action in Indonesia

Imaduddin Amin, Zakiya Pramestri, George Hodge and Jong Gun Lee

Pulse Lab Jakarta - United Nations Global Pulse

E-mail: {imaduddin.amin, zakiya.pramestri, george.hodge, jonggun.lee}@un.or.id

Abstract.
Tracking human development and humanitarian action has been enhanced by the growth of social media. Twitter is a data source with potential, when used alongside data from surveys, especially the national census, to understand the situation on the ground and track changes. In Indonesia, a country with one of the highest Twitter penetration rates, we seize this opportunity by using Twitter data to produce more timely insights and to enhance evidence-based decision-making. Despite social media's limitations, namely representativeness and validity, we are able to show its potential by looking at case studies on five different topics; (a) food and agriculture, (b) public health (c) economic well-being (d) urban resilience and (e) humanitarian action. We observe that the insights gained by using Twitter data were derived not only from the content of posts such as understanding public opinion or sentiment, but also from activities related to it, for instance the location and time-stamp of the post, which furthers our real-time understanding of the situation and user behavior changes. In this paper, we also briefly explain "social listener", a social media monitoring tool that used by Government of Indonesia to understand citizen opinions in social media related to government priorities.

1. Introduction

Traditional statistics, surveys and census data are used to understand public policy issues and to track development. These traditional data, however, are less effective at capturing a real-time snapshot of the situation on the ground in order to inform timely action. New data sources hold the opportunity to complement and supplement traditional data to inform and enhance decision-making in the public sector.

As a core component of the data revolution, social media generate useful data on a variety of issues. Social media platform enables public discourse on a variety of topics, including development and public sector issues. With over 20 million of users, Indonesia ranks third in the world in the number of active Twitter users. This wealth of data has huge potential to provide real-time insights on public interests and issues, including for the Government to understand citizens' priorities, perceptions and behavior in aggregate.

On Twitter, a user can post content such as text, images, and short videos on any topic. Alongside the content of the post, data is captured on the time, time zone, location, device, and other variables. Others also include gender information, users language and users demographics such as age and gender to extend that basic information. The user can choose whether to attach certain data to the post, such as the location where the tweet was posted.

We group the attributes available from Twitter into two classes, namely "what people say", and "what people do". "What people say" contains text-based information shared by a user, such as a tweet text or profile narrative, while "What people do" concerns the information associated with the post or Twitter profile that is related to user activities such as location information, time information, or device used.

In this paper, we would like to address the usage of Twitter data, specifically in the development and humanitarian sectors. We will explore five case studies including (a) food and agriculture, (b) public health (c) economic well-being (d) urban resilience and (e) humanitarian action. In addition, we briefly explain "social listener", a tool developed by Pulse Lab Jakarta and used by Government of Indonesia (GoI) to monitor public discourse on Twitter.

This paper is organized as follows: Section 2 describes related work on the application of Twitter data to issues of international development and public policy in general. In Section 3, we explain how Pulse Lab Jakarta and GoI use Twitter for sustainable development and humanitarian action, followed by a description of social listener in section 4. In Section 5 we conclude this work.

2. Related Work

The online activities in social media have been subject to a number of studies. Various methodologies have been examined including text mining, sentiment analysis, spatiotemporal analysis, network analysis, among others, to be applied in different fields such as health, humanitarian actions, economy and urban dynamics.

A set of works has demonstrated the potential of using Twitter for investigating **public health**. A topic model was used to track illness symptoms and awareness, with the quantitative correlations with public health data [1, 2]. Real-time content analysis were explored to provide more real-time insights on public concerns [3, 4]. Forecasting and "nowcasting" model were also developed using surveillance signals from Twitter to predict and track level of diseases, such as influenza [5, 6, 7, 8, 9, 10, 11], cholera outbreak [12], and mental illness [13]. Another work by Sadilek *et al.* explored interaction of social activity, human mobility inferred from Twitter, and the spread of infectious disease in a large real-world population [14]. Besides, Twitter streams also used to investigate public concerns related to efforts in improving disease surveillance, for instance, on vaccination [15, 16, 17, 4, 18]. Public sentiments on vaccination were assessed, to investigate the implication on likelihood of disease outbreak [16]. The Twitter pattern behavior were also found to be highly correlated with vaccination behavior based on official government data [17], and the Twitter narrative related to particular vaccination were analyzed, to explore the structure of discussions, connections, and its correlation to the geographical spaces [18].

Prior works also investigated the potential of social media to provide insights during **critical situation, such as in natural and man-made disasters**. A machine learning methods was proposed to extract relevant information for disaster response [19, 20, 21, 22, 23]. A study proposed framework for using Twitter on event detection over crisis situation [24, 25, 26, 27]. Aside from the data mining methods, other study suggested visual analytic systems based on spatiotemporal analysis to support disaster management [28]. Geographical approached also proposed to incorporate social media analysis for gaining better insight including the spatial patterns of the disaster, particularly in flood [29], and wildfire [30]. A set of works on data-mining and machine-learning techniques that has been applied for disaster management were also summarized in this study [31, 32].

A number of research works attempted to utilize social media data for understanding **urban dynamics** and social media has been studied for capturing the social behavior in real world events. Wu *et al.* analyzed social media data to cluster urban hot spots, and found the relationship between the hot spots and housing prices in Shenzhen, China [33]. Another work attempted to analyze geo-tagged tweets for understanding the social-inequality in the city [34]. A

IOP Conf. Series: Journal of Physics: Conf. Series **971** (2018) 012040 doi:10.1088/1742-6596/971/1/012040

probabilistic topic model to investigate a set of topics related to citizens activities [35]. Gallegos *et al.* measured public sentiment in tweets to gain insights on land use and services in specific areas [36]. Related to environment, an interesting work by Jiang *et al.* suggested geo-targeted social media analytic method to monitor air quality dynamics [37]. Other studies analyzed the spatio-temporal dynamics for inferring human mobility and traffic insights [38, 39, 40, 41, 42, 43]. Origin-destination of people' movement was inferred from the social media mobility network in global-level [44], and local-level [45, 41, 46, 47]. Furthermore, Chan *et al.* proposed a unified framework combining linguistic model with Hinge-loss Markov Random Fields to identify traffic congestion locations [48].

In Indonesian context, a set of works utilizing social media data also has been conducted. A study extracted information from authority' Twitter account on reporting traffic condition, to be visualized in a map [49]. The sentiment of tweets using Bahasa Indonesia was classified by applying pre-processing noisy text and machine learning for classification [50]. Further, a sentiment analysis method has been implemented to detect sarcasm messages [51]. Text mining also used to investigate user personality [52] and cyber-bullying pattern in Twitter [53],

3. Twitter Data for Development in Indonesia

In this section, we explain how Pulse Lab Jakarta and GoI use Twitter data for development with five case studies in five sectors, (a) food and agriculture, (b) public health (c) economic well-being (d) urban resilience and (e) humanitarian action.

3.1. Humanitarian actions

Forest and peatland fires continue to affect not only Indonesia but also in regional of Southeast Asia. This situation leads to extensive environmental destruction, economic loss, health problem and others. At present, Indonesia disaster management authorities manage the haze situation by utilizing hot spot data from satellite, sensors deployed in the affected area and official statistics such as population density and distribution. All of those information will inform about the peat fire condition but do not capture a ground truth situation from citizen perspectives such as citizen needs, opinion or citizen behaviour. In order to complement the information gap as well as having more frequent data update, Pulse Lab Jakarta and GoI test the potential of using social media, specifically twitter data.

With the help of domain expert, we defined a set of initial keyword or taxonomy that related to haze situation. For instance 'kabut asap' for a general topic on haze, 'masker', 'sesak napas' for a health-related issue. These keywords are then used to capture relevant message from twitter space to generate insights. From our experiment, we show that these messages are strongly correlated with hotspot dynamics [54].

To understand user mobility, we use location information that available in a geo-stamp tweet. Each user will have set of places that he/she already visit and mobility characteristics inferred from these data. We then compare the user typical behaviour during the normal (non-haze) situation and haze situation. Our result shows that residents make more short-distance movements on normal situation compare to the weeks with haze. Furthermore, we also able to show that some people increase significantly their mobility by hundreds of kilometers during the evacuation week on haze disaster [55].

In addition, we then combine this insights with other dataset used by authority such as hotspot information, place of interest, demographic information and others into one platform[1]. This platform is now used by authority and opened for public access.

[1] http://hazegazer.org

3.2. Urban Dynamics
In 2014, the Indonesian Central Bureau Statistics (BPS) conducted their first commuting survey for Greater Jakarta area that covers 13,120 households. The aims of the survey are to understand the commuting behaviour of citizens in 10 cities and 3 regencies in Greater Jakarta area. This survey result filled the initial data gap needed by the government but the process takes one year from design to delivery.

In order to have similar statistics that relatively easy to frequently reproduce, we then test the possibility of using social media data. To this end, we use more than 38 millions of tweets from 1.4 million of users. Per-user, we inferred two most important locations: origin and destination in sub-district level. Origin location was inferred as the most tweeted sub-district location between 9 pm and 7 am destination location was determined as the most tweeted sub-district location during weekdays, excluding the origin location. Using this approach, we found 305,761 users who meet the criteria. This number of users represents 2.8 percent of the whole population or 14 percent of the total commuting population in Greater Jakarta.

This simple approach produces good results without significant differences. A cross-correlation score between official statistics result and our approach shows 0.97 similarity. We also show that only 3 out of 25 origin-destination pair that has the rank difference between official statistics and our approach [56]. Even though three origin cities have different rank pairs, it is worth to mention that the score of those cities is close each other from official statistics.

3.3. Food and Agriculture
Understanding price dynamics in near real-time is an important aspect concerning the government as a basis to have good policy. Currently, GoI through Ministry of Trade rely on people that deployed in a number of local markets to daily report about price dynamics, including food price as basic commodities. This initiative has good implication to fill the data gaps and reduce the time delay, but has a scalability issue. Driven by this need, we then explore the usability of Twitter data to nowcast food price in national level.

For this purpose, we collect two different types of dataset. First, twitter data as a primary source for nowcasting and official statistics data as a secondary or ground-truth dataset. Twitter data was collected using a set of keyword that related to food commodities such as price and commodity name. For this research, four commodities (beef, chicken, onion and chili) were selected. These commodities were chosen considering data availability and the country-level priorities for food security monitoring in consultation with the Ministry of National Development Planning.

From all relevant tweet messages, we extract two main information, a price mentioned in tweet messages and the volume of relevant messages over time. The numerical filter was applied to extract the first information while the latter was calculated based on tweets captured by taxonomy. We then develop a model to nowcast each commodity based on price information extracted in tweet messages, total tweets today and total tweets yesterday. Once we have the price for a commodity we then compare the result with official data.

The result shows that a price built by the model has a high correlation with the official price such as 0.76 for chili and more than 0.8 for other commodities[2] [57, 58]. This method is later improved by incorporating Google trends data,it and shows higher correlation compared to the previous model.

3.4. Public Health
There are numerous challenges in implementing immunization program including mis-perceptions about vaccines and dissemination of incorrect information. These factors lead to

[2] http://nowcasting.unglobalpulse.org/

low vaccine coverage and high dropout rates, which means that children do not receive full immunization. Hence, understanding the situation in real-time is important to identify the misinformation disseminated on the ground and taking rapid response accordingly.

Pulse Lab Jakarta together with UNICEF and Government of Indonesia, conduct a project to understand public perception on immunization. We capture all conversation related to immunization and identify the influencer of a topic discussed by the citizen.

To refine the focus of the project, four main sub-topics were chosen based on relevance to immunization program; religious concerns on immunization, conversations around disease outbreaks, symptoms or health conditions discussed as vaccine side effects, and the launch of the new vaccine product. All sub-topics has related keyword or taxonomy that defined by our domain expert. By using the keywords, 88,368 relevant public tweets were collected from January 2012 to December 2013 for Bahasa Indonesia language. From these relevant messages, some analysis was conducted such as understanding the conversation dynamics, word clustering to identify the top keyword for each sub-topic and which content is shared the most by the citizen. We also identify the network of influencers, which we could leverage for rapid response to public concerns and misunderstood information about vaccines and immunization [59].

3.5. Economic Well-being

Public sentiment towards policy changes is one of the topic that can be identified through social media, particularly to the policies which is directly has impact on citizen' economic well-being. For example, a hashtag #OccupyNigeria was triggered, when government cut fuel subsidy and cause young people mobilize themselves to protest via social media[3]. Another case was in Yemen, where a number of citizens turned to social media to express their anger on fuel hikes[4]. Fuel pricing has been one of sensitive issues for citizens, as it may shake political stability as well as hurt the economy of a country.

In order to understand the public sentiments on fuel subsidy in Indonesia, we analyze tweets in 10 - 30 November 2014 using social media monitoring tools namely Crimson Hexagon[5]. The result shows that the relevant tweets' volume has significant spike in 17 November 2014, the day when the GoI started to cut the fuel subsidy, with more than 600,000 tweets per day. It is also interesting to see that public sentiment was changed over time toward this issue. The negative sentiments was raised significantly from one week before announcement, then decreased after two weeks as shown in Figure 1

4. Social Listener

Pulse Lab Jakarta developed "Social listener", a platform that collects, analyze and present information from text data such as social media data, citizen feedback and others. It allows a user to launch a dashboard that monitor conversation from selected data set. To launch the dashboard, user need to provide a list of topic and sub-topics of interest. Each sub-topics is represented by keywords or taxonomy in Boolean search format. This keywords later used as a filter to capture relevant messages from data firehouse, which the insights will be presented in an easy way.

GoI use this platform to monitor citizen opinion related to the government national priorities. For this purpose, two datasets are selected; a citizen complain mechanism, Lapor![6] and Twitter. Lapor! is considered as active participation from public as they need to submit to specific platform, while Twitter is considered as passive participation from public. The dashboard

[3] http://www.bbc.com/news/world-africa-16591389
[4] http://gulfnews.com/news/gulf/yemen/yemenis-protest-against-fuel-price-hike-1.1367338
[5] https://www.crimsonhexagon.com/
[6] https://lapor.go.id/

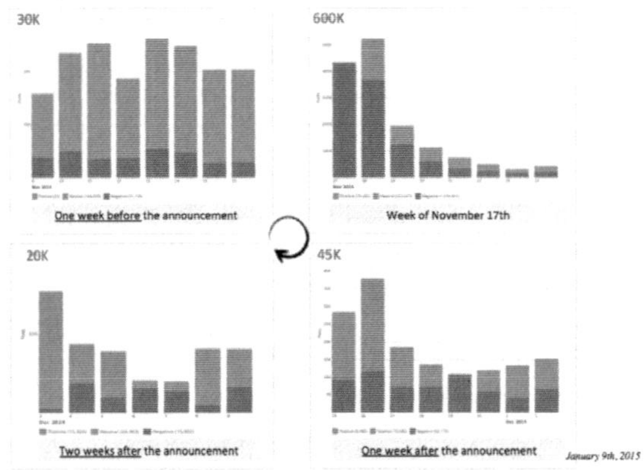

Figure 1. Sentiment classification on fuel subsidy cut related tweets

launched by Government of Indonesia is an extended version of citizen dashboard that we built for mining citizen feedback in city level [60]. Currently, the platform is able to provide messages distribution and its dynamics, topic cluster and trending topics.

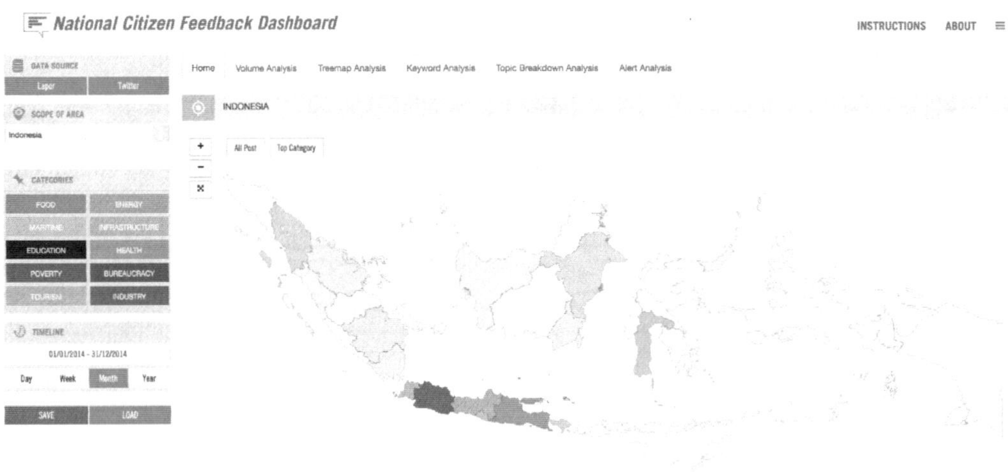

Figure 2. Geographic Distribution at Social Listener Platform

Figure 2 shows the snapshot of Social Listener for National Citizen Feedback Dashboard from Twitter data for a year. From the image, we are able to see that the monitor has 10 sub-topics of interest. The map shows geographic distribution of relevant messages. A user then able to go deeper in city level to filter only relevant messages from selected city. The darker color means the area has more relevant messages. In this case, Java island has most relevant messages compare to other main islands.

In social listener, a topic of interest can be group into two levels such as topic and sub-topics. In Figure 3 it is shown the proportion of relevant messages and the keyword associate with that. Figure 2 shows that poverty category dominate the conversation and keyword such as "raskin" and "subsidy" are the most common word mentioned in messages. It is also possible that a message will have two or more categories. The relation between a category and other is then presented as shown in Figure 4 .

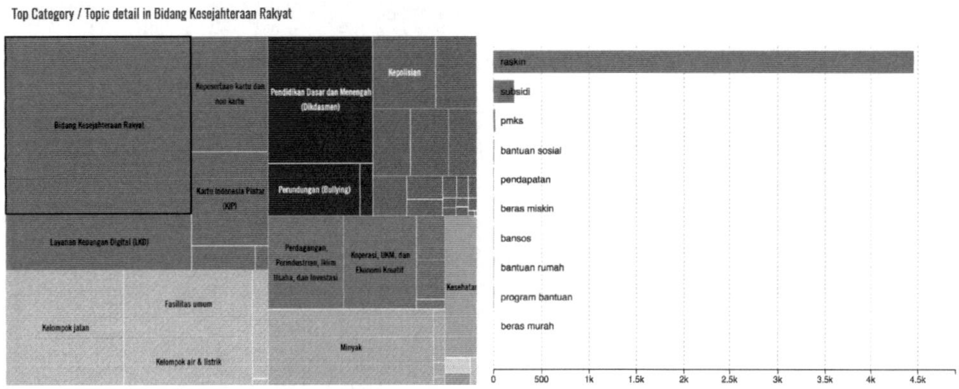

Figure 3. Topic Breakdown and related keywords

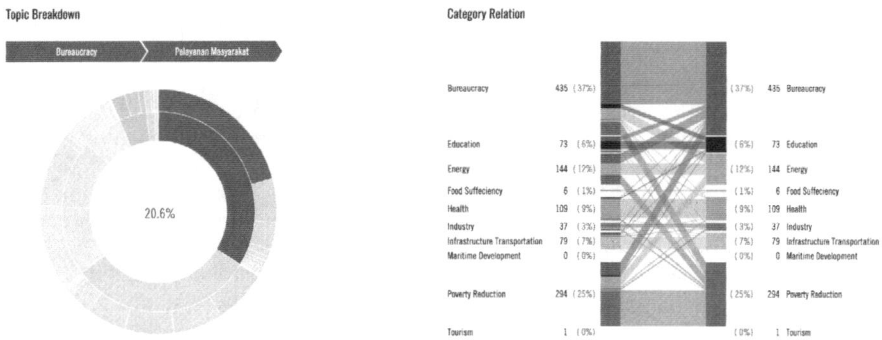

Figure 4. Topic Relations

5. Conclusions

Social media have great potential as sources of aggregated citizen perspectives in real-time. However, there are some limitations and challenges that we need to consider before using social media as primary sources, such as representativeness and validity [61]. Despite its limitations, the case studies described in this paper show that Twitter can be used to complement existing traditional data such as surveys and the national census. In addition, it is shown that the potential of Twitter data is not limited to only textual information or "what people say", but also "what people do" such as location, time and other variables.

References

[1] Paul M J and Dredze M 2011 You are what you tweet: Analyzing twitter for public health. *ICWSM*
[2] Lamb A, Paul M J and Dredze M 2013 Separating fact from fear: Tracking flu infections on twitter *In NAACL*
[3] Chew C and Eysenbach G 2010 Pandemics in the age of twitter: Content analysis of tweets during the 2009 h1n1 outbreak vol 5 (Public Library of Science) pp 1–13
[4] Mollema L, Harmsen I A, Broekhuizen E, Clijnk R, Melker II D, Paulussen T, Kok G, Ruiter R and Das E 2015 Disease detection or public opinion reflection? content analysis of tweets, other social media, and online newspapers during the measles outbreak in the netherlands in 2013 vol 17 p e128
[5] Lampos V and Cristianini N 2012 Nowcasting events from the social web with statistical learning vol 3 pp 1–22
[6] Paul M J, Dredze M and Broniatowski D 2014 Twitter improves influenza forecasting (Public Library of Science (PLoS))

[7] Gesualdo F, Stilo G, Agricola E, Gonfiantini M V, Pandolfi E, Velardi P and Tozzi A E 2013 Influenza-like illness surveillance on twitter through automated learning of naive language vol 8
[8] Broniatowski D A, Paul M J and Dredze M 2013 National and local influenza surveillance through twitter: An analysis of the 2012-2013 influenza epidemic vol 8
[9] Signorini A, Segre A M and Polgreen P M 2011 The use of twitter to track levels of disease activity and public concern in the u.s. during the influenza a h1n1 pandemic vol 6 pp 1–10
[10] Nagar R, Yuan Q, Freifeld C C, Santillana M, Nojima A, Chunara R and Brownstein J S 2014 A case study of the new york city 2012-2013 influenza season with daily geocoded twitter data from temporal and spatiotemporal perspectives vol 16 p e236
[11] Lee K, Agrawal A and Choudhary A 2017 Forecasting influenza levels using real-time social media streams *2017 IEEE International Conference on Healthcare Informatics (ICHI)*
[12] Chunara R, Andrews J R and Brownstein J S 2012 Social and news media enable estimation of epidemiological patterns early in the 2010 haitian cholera outbreak vol 86 pp 39–45
[13] Choudhury M D, Gamon M, Counts S and Horvitz E 2013 Predicting depression via social media *ICWSM* ed Kiciman E, Ellison N B, Hogan B, Resnick P and Soboroff I
[14] Sadilek A, Kautz H and Silenzio V 2012 Modeling spread of disease from social interactions *In Sixth AAAI International Conference on Weblogs and Social Media (ICWSM*
[15] Dunn A G, Leask J, Zhou X, Mandl K D and Coiera E 2015 Associations between exposure to and expression of negative opinions about human papillomavirus vaccines on social media: An observational study vol 17 p e144
[16] Salath M and Khandelwal S 2011 Assessing vaccination sentiments with online social media: Implications for infectious disease dynamics and control vol 7 pp 1–7
[17] Huang X, Smith M C, Paul M, Ryzhkov D, Quinn S, Broniatowski D and Dredze M 2017 Examining patterns of influenza vaccination in social media *AAAI Joint Workshop on Health Intelligence (W3PHIAI)*
[18] Radzikowski J, Stefanidis A, Jacobsen K H, Croitoru A, Crooks A and Delamater P L 2016 The measles vaccination narrative in twitter: A quantitative analysis vol 2 p e1
[19] Mendoza M, Poblete B and Castillo C 2010 Twitter under crisis: Can we trust what we rt? *Proceedings of the First Workshop on Social Media Analytics* SOMA '10 (New York, NY, USA) pp 71–79
[20] Qu Y, Huang C, Zhang P and Zhang J 2011 Microblogging after a major disaster in china *Proceedings of the ACM 2011 conference on Computer supported cooperative work - CSCW 11* (ACM Press)
[21] Imran M, Elbassuoni S, Castillo C, Diaz F and Meier P 2013 Extracting information nuggets from disaster-related messages in social media *ISCRAM* (ISCRAM Association)
[22] Imran M, Elbassuoni S, Castillo C, Diaz F and Meier P 2013 Practical extraction of disaster-relevant information from social media *Proceedings of the 22nd International Conference on World Wide Web - WWW'13 Companion* (ACM Press)
[23] Ashktorab Z, Brown C, Nandi M and Culotta A 2014 Tweedr: Mining twitter to inform disaster response *11th Proceedings of the International Conference on Information Systems for Crisis Response and Management* ed Hiltz S R, Plotnick L, Pfaf M and Shih P C
[24] Terpstra T, de Vries A, Stronkman R and Paradies G 2012 Towards a realtime twitter analysis during crises for operational crisis management *Proceedings of the 9th International ISCRAM ConferencProceedings of the 9th International ISCRAM Conference*
[25] Zhou X and Chen L 2013 Event detection over twitter social media streams vol 23 pp 381–400
[26] Sakaki T, Okazaki M and Matsuo Y 2010 Earthquake shakes twitter users: real-time event detection by social sensors *Proceedings of the 19th international conference on World wide web* (ACM) pp 851–860
[27] Gupta A, Kumaraguru P, Castillo C and Meier P 2014 Tweetcred: A real-time web-based system for assessing credibility of content on twitter vol abs/1405.5490
[28] Chae J, Thom D, Jang Y, Kim S, Ertl T and Ebert D S 2014 Public behavior response analysis in disaster events utilizing visual analytics of microblog data vol 38 pp 51–60
[29] de Albuquerque J P, Herfort B, Brenning A and Zipf A 2015 A geographic approach for combining social media and authoritative data towards identifying useful information for disaster management vol 29 pp 667–689
[30] Wang Z, Ye X and Tsou M H 2016 Spatial, temporal, and content analysis of twitter for wildfire hazards vol 83 pp 523–540
[31] Houston J B, Hawthorne J, Perreault M F, Park E H, Hode M G, Halliwell M R, McGowen S E T, Davis R, Vaid S, McElderry J A and Griffith S A 2014 Social media and disasters: a functional framework for social media use in disaster planning, response, and research vol 39 pp 1–22
[32] Li T, Huang Y, Chen S C, Navlakha J, Iyengar S S, Xie N, Zeng C, Zhou W, Zheng L, Jiang Y, Yang Y, Ha H Y and Xue W 2017 Data-driven techniques in disaster information management vol 50 pp 1–45
[33] Wu C, Ye X, Ren F, Wan Y, Ning P and Du Q 2016 Spatial and social media data analytics of housing

prices in shenzhen, china vol 11 ed Choo K K R

[34] Shelton T, Poorthuis A and Zook M 2015 Social media and the city: Rethinking urban socio-spatial inequality using user-generated geographic information

[35] Kling F and Pozdnoukhov A 2012 When a city tells a story: Urban topic analysis *Proceedings of the 20th International Conference on Advances in Geographic Information Systems* SIGSPATIAL '12 pp 482–485

[36] Gallegos L, Lerman K, Huang A and Garcia D 2015 Geography of emotion: Where in a city are people happier? vol abs/1507.07632

[37] Jiang W, Wang Y, Tsou M H and Fu X 2015 Using social media to detect outdoor air pollution and monitor air quality index (aqi): A geo-targeted spatiotemporal analysis framework with sina weibo (chinese twitter) vol 10 pp 1–18

[38] Jurdak R, Zhao K, Liu J, AbouJaoude M, Cameron M and Newth D 2015 Understanding human mobility from twitter vol 10 ed Wu Y

[39] Pan B, Zheng Y, Wilkie D and Shahabi C 2013 Crowd sensing of traffic anomalies based on human mobility and social media *Proceedings of the 21st ACM SIGSPATIAL International Conference on Advances in Geographic Information Systems - SIGSPATIAL 13*

[40] Tejaswin P, Kumar R and Gupta S 2015 Tweeting traffic *Proceedings of the 2nd IKDD Conference on Data Sciences - CODS-IKDD 15*

[41] Wu L, Zhi Y, Sui Z and Liu Y 2014 Intra-urban human mobility and activity transition: Evidence from social media check-in data vol 9 pp 1–13

[42] Liu Y, Sui Z, Kang C and Gao Y 2014 Uncovering patterns of inter-urban trip and spatial interaction from social media check-in data vol 9 pp 1–11

[43] Chua A, Marcheggiani E, Servillo L and Moere A V 2015 FlowSampler: Visual analysis of urban flows in geolocated social media data *Lecture Notes in Computer Science* pp 5–17

[44] Hawelka B, Sitko I, Beinat E, Sobolevsky S, Kazakopoulos P and Ratti C 2014 Geo-located twitter as proxy for global mobility patterns vol 41 pp 260–271

[45] Gabrielli L, Rinzivillo S, Ronzano F and Villatoro D 2013 From tweets to semantic trajectories: Mining anomalous urban mobility patterns *Lecture Notes in Computer Science* pp 26–35

[46] Yang F, Jin P J, Cheng Y, Zhang J and Ran B 2014 Origin-destination estimation for non-commuting trips using location-based social networking data vol 9 pp 551–564

[47] Gao S, Hu Y and Janowicz K 2014 Detecting origin-destination mobility flows from geotagged tweets in greater los angeles area

[48] Chen P T, Chen F and Qian Z 2014 Road traffic congestion monitoring in social media with hinge-loss markov random fields *2014 IEEE International Conference on Data Mining*

[49] Endarnoto S K, Pradipta S, Nugroho A S and Purnama J 2011 Traffic condition information extraction visualization from social media twitter for android mobile application *Proceedings of the 2011 International Conference on Electrical Engineering and Informatics*

[50] Naradhipa A R and Purwarianti A 2011 Sentiment classification for indonesian message in social media *Proceedings of the 2011 International Conference on Electrical Engineering and Informatics*

[51] Lunando E and Purwarianti A 2015 Indonesian social media sentiment analysis with sarcasm detection vol abs/1505.03085

[52] Pratama B Y and Sarno R 2015 Personality classification based on twitter text using naive bayes, KNN and SVM *2015 International Conference on Data and Software Engineering*

[53] Margono H, Yi X and Raikundalia G K 2014 Mining indonesian cyber bullying patterns in social networks *Proceedings of the Thirty-Seventh Australasian Computer Science Conference - Volume 147* pp 115–124

[54] Kibanov M, Amin I and Lee J G 2015 Supporting peat fire management using social media

[55] Kibanov M, Stumme G, Amin I and Lee J G 2017 Mining social media to inform peatland fire and haze disaster management

[56] Amin I, PP N L P S, Riyadi Y, Lee J G, Pramana S and Kurniawan R 2017 Inferring commuting statistics in greater jakarta from social media locational information from mobile devices *Netmob Book of Abstracts* pp 24–26

[57] 2014 Nowcasting food prices in indonesia using social media signals *Global Pulse Project Series no. 1* (UN Global Pulse)

[58] Kim J, Cha M and Lee J G 2017 Nowcasting commodity prices using social media vol 3 (PeerJ) p e126

[59] 2015 Understanding immunisation awareness and sentiment through social and mainstream media *Global Pulse Project Series no. 19* (UN Global Pulse)

[60] 2017 *Fostering Innovation in the Public Sector* (OECD Publishing)

[61] Tufekci Z 2014 Big questions for social media big data: Representativeness, validity and other methodological pitfalls

IOP Conf. Series: Journal of Physics: Conf. Series **971** (2018) 012041 doi:10.1088/1742-6596/971/1/012041

Classifying emotion in Twitter using Bayesian network

Muhammad Surya Asriadie, Mohamad Syahrul Mubarok [1] and Adiwijaya

Fakultas Informatika, Telkom University, Jalan Telekomunikasi No. 1, Bandung, ID

E-mail: `msyahrulmubarok@telkomuniversity.ac.id`, `suryaa@student.telkomuniversity.ac.id`, `adiwijaya@telkomuniversity.ac.id`

Abstract. Language is used to express not only facts, but also emotions. Emotions are noticeable from behavior up to the social media statuses written by a person. Analysis of emotions in a text is done in a variety of media such as Twitter. This paper studies classification of emotions on twitter using Bayesian network because of its ability to model uncertainty and relationships between features. The result is two models based on Bayesian network which are Full Bayesian Network (FBN) and Bayesian Network with Mood Indicator (BNM). FBN is a massive Bayesian network where each word is treated as a node. The study shows the method used to train FBN is not very effective to create the best model and performs worse compared to Naive Bayes. F1-score for FBN is 53.71%, while for Naive Bayes is 54.07%. BNM is proposed as an alternative method which is based on the improvement of Multinomial Naive Bayes and has much lower computational complexity compared to FBN. Even though it's not better compared to FBN, the resulting model successfully improves the performance of Multinomial Naive Bayes. F1-Score for Multinomial Naive Bayes model is 51.49%, while for BNM is 52.14%.

1. Introduction

Language is used to express not only facts, but also emotions. Emotions are noticeable from behavior up to the social media statuses written by person. Classification of emotions in a text is done in a variety of media such as Twitter. Twitter is a popular social networking service where users can post and interact with messages called tweets. There are two types of users on twitter, the user who tweets about themselves and the user who tweets to share information. In both cases, the tweets can provide information about emotion [4] [15]. Therefore, Twitter is considered to have suitable data for experiments related to emotion classification.

Classification of emotion has a wide range of applications. One of the example is for measuring the happiness index of people which is useful for political consideration for public well-being. In Indonesia, happiness index is measured through Happiness Index Measurement Survey held by Central Bureau of Statistics which is done only once every couple of years [2]. By applying emotion detection, the happiness index can be measured at anytime using any popular social media as the medium. Another example is to apply emotions detection in messaging application to check the message before sending the message to avoid misunderstanding. Emotions detection is also applicable to measure the popularity of a product or a brand [14].

[1] Corresponding author.

Naive Bayes is one of the most commonly used methods for sentiment classification. Even though it usually has a good performance, it has a weakness to assume independencies between words [10]. Bayesian network is a probabilistic graphical model that can represents dependency between variables, but there's not so much research about sentiment classification using this method. The previous study of sentiment classification using Bayesian network is done by Wan and Gao [21] and Ortigosa-Hernndez et. al [16], but both of the studies did not focus on how Bayesian network can be used to classify sentiment. In this paper, we aims study emotion classification on twitter using Bayesian network because of the ability to model uncertainty and relationships between features.

2. Related works

Sentiment analysis aims to extract subjective information from a text. In general, it can detect the writer's attitude towards a topic. An example problem is to detect the polarity of sentiment in a text, whether it's positive, negative or neutral. Sentiment analysis can also detect another type of aspect like emotional states.

Balabantaray, Mohammad, and Sharma studies emotion classification of twitter data acquired from the web which are manually annotated by five judges [3]. The tweets are consists of seven classes, six Ekman emotion (joy, sadness, anger, disgust, surprise, and fear) and an addition of a neutral class. SVM then used to classify features that were constructed from the combination of Unigrams, Bigrams, Personal-pronouns, adjectives, Word-net Affect emotion lexicon, Word-net Affect emotion lexicon with left/right context, Word-net Affect emotion POS, POS, POS-Bigrams, Dependency-Parsing Feature, and Emoticons. The experiment was done using Leave-one-out cross-validation. The highest accuracy achieved is 73.24% that is accomplished using combination of all available features.

Roberts et. al studies emotion classification in twitter into seven classes which are anger, disgust, fear, joy, love, sadness, and surprise [20]. His dataset is acquired by selecting 14 specific subjects that are often believed to contain emotions. Seven Binary SVM are used to build the system; each of the Binary SVM is used to classify one class. Unigrams, Bigrams, Trigrams, Contains !, Contains ?, WordNet synsets, WordNet hypernyms, Topic scores, and Significant words are used as the features. The best feature combination of each class is selected using greedy additive feature selection. The experiment is done using 10-fold cross-validation with micro-averaged F1-Score of 66.8%.

Wan and Gao studies sentiment analysis of Twitter Data for Airline Services using ensemble classification strategy [21]. Majority vote principle is used to determine the result of multiple classification methods including Naive Bayes, SVM, Bayesian Network, C4.5 Decision Tree, and Random Forest. The study shows this method could improve the overall accuracy of the system. Bayesian Network can compete or perform even better than the other methods.

3. System design

In this study, we propose two models based on Bayesian Network called Full Bayesian Network and Bayesian Network with Mood Indicator. The following are the steps needed to build our models:

(i) Data
Data is prepared using k-fold cross-validation scheme with five folds. The data will be shuffled before divided using k-fold cross-validation.

(ii) Pre-processing
Pre-processing step is required to clean data and reduce the number of words used. The details of this step will be explained in the next section.

(iii) Binary Term Frequencies
Feature are represented using Binary Term Frequencies which treats every occurrence of words in a document as 1 and 0.

(iv) Feature selection
This step aims to reduce the number of features using chi-squared feature selection algorithm.

(v) Structure learning
Structure Learning aims to find the best structure of a Bayesian Network based on data. The structure of Full Bayesian Network is optimized using simulated annealing search and $\widehat{f}$CLL as the scoring function. Bayesian Network with Mood Indicator doesn't require structure learning because the number of the possible structure is small.

(vi) Parameter learning
The probability distribution of each node is learned on train data using MAP-estimation.

(vii) Evaluation
The performance of the model is evaluated using F1-Score.

4. Dataset
In this experiment, we use #Emotional Tweets datasets by Saif Mohammad that is acquired by making use of hashtag feature on Twitter [13]. It consists of six classes with 21051 total tweets (Figure 1).

Figure 1: 21051 tweets categorized into six Ekkman emotion: anger, fear, disgust, joy, sadness, and surprise.

4.1. Pre-processing
Data gathered from the web is usually not clean and contains a lot of unnecessary information. Therefore, pre-processing steps are required to clean the data. The pre-processing steps are as follow:

(i) *HTML escape*
Data gathered from the web usually contain escaped character as a form of HTML code (`<`, `>` dan `&`). This process aims to convert HTML code back to their original character.

Table 1: *#Emotional Tweets*'s dataset sample for each emotion classes.

Tweet	Emosi
Soooo dooowwwn!! Move on, get some sleep... Me deserve better.#forgetit #yawning	*anger*
Making art and viewing art are different at their core!	*fear*
Can parents teach kids how to cover their mouths when theycough? Ewah.	*disgust*
the moment when you get another follower and you cheer.	*joy*
I said no to the same boy like 5 times last night!	*sadness*
Tomorrow I'm going to get @HerBraceFaceAss nd Jaela frmschool early nd takin my grls to lunch. L0L	*surprise*

(ii) *Apostrophe lookup*
In this step, every word that contains apostrophes is converted into standard lexicons. For example *I've* become *I have.*

(iii) *Username lookup*
In twitter, every user can mention their friends in their tweets by prepending `@` symbol before their friend's username. This step aims to group usernames into one tag which is {{`username`}}.

(iv) *URL lookup*
It's not rare to see that Twitter is used as a medium to share information or links to other websites. This step transforms links into one tag which is {{`url`}}.

(v) *Money lookup*
In this step, every word that begins with money symbol (example: *$*) is transformed into one tag which is {{*money*}}.

(vi) *Remove number*
This step converts every numbers into one tag which is {{`number`}}.

(vii) *Remove repeats*
This step remove repeated characters of the text.

(viii) *Stemming*
This step aims to transform a word into its base or root form. Porter Stemmer algorithm is used in this step [18].

(ix) *Remove punctuations*
In this step, punctuations are removed of the text.

(x) *Hashtag removal*
Hashtag is a feature where users can tag tweets into a specific topic. It's written by prepending "#" into a word which usually a topic. To reduce the number of features while retaining the information, any hashtag that rarely appears is grouped into a specific tag {{`hashtag`}}, while any hashtag that often appears remains.

(xi) *Stop words removal*
In this step, every word that doesn't contain significant meaning is removed.

5. Bayesian Network

Bayesian Network is a graphical representation of probabilistic relationships between several random variables [19]. It's very suitable for representing knowledge about an uncertain domain.

Each node in a Bayesian Network represents a random variable, while each edge represents the dependency between the random variables.

5.1. Parameter learning

Parameter learning aims to estimate the probability distribution of every node in Bayesian Network based on data. Some of the methods to learn parameters using complete data are Maximum Likelihood Estimate (ML-estimation) and Maximum A Posteriori Estimation (MAP-estimation) [19]. MAP-estimation can be seen as the general form of ML-estimation, where it adds information about prior knowledge to the estimation. MAP-estimation with equivalent sample size is formulated as follow:

$$\theta_{ijk} = \frac{\frac{\alpha}{r_i \cdot q_i} + N_{ijk}}{\sum_k \frac{\alpha}{r_i \cdot q_i} + N_{ijk}} \quad (1)$$

The explanation of each symbol is in the next section.

5.2. Structure learning

Structure learning is used to learn a Bayesian Network structure from data. There are two kinds of structure learning algorithm, constrained-based method, and heuristic search. Hill climbing, tabu search, and simulated annealing are some of the methods based on heuristic search. Simulated Annealing is a heuristic algorithm inspired by the process of annealing in metallurgy. It's often used when the search space is discrete (e.g., Traveling Salesman Problem).

Algorithm 1 Simulated Annealing for Bayesian Network structure learning [5]

function SIMULATEDANNEALING(τ_0, T_{ni}, ni)
  for $i = 1, ..., n$ **do**
    $\pi_i \leftarrow \emptyset$
    $\pi_{best,i} \leftarrow \emptyset$
  end for
  $T \leftarrow \tau_0 \cdot N$
  $\alpha \leftarrow (T_{ni}/T_0)^{1/ni}$
  $k \leftarrow 0$
  repeat
    repeat
      select two indices i, j randomly
    until $v_j \in \pi_j$ or adding v_j to π_i does not introduce cycle
    if $v_j \in \pi_i$ **then**
      if $\exp(m(v_i, \pi_i \setminus v_j) \Leftrightarrow m(v_i, \pi_i)) > random[0..1]$ **then**
        $\pi_i \leftarrow \pi_i \setminus v_j$
      end if
    else
      if $\exp(m(v_i, \pi_i v_j) \Leftrightarrow m(v_i, \pi_i)) > random[0..1]$ **then**
        $\pi_i \leftarrow \pi_i v_j$
      end if
    end if
    if $\sum_{i=1}^{n} m(v_i, \pi_{best,i}) < \sum_{i=1}^{n} m(v_i, \pi_i)$ **then**
      $\forall_{i \in \{1,...,n\}} \pi_{best,i} \leftarrow \pi_i$
    end if
  until $k = ni$
  return $\pi_{best,i}, ... \pi_{best,n}$
  $T \leftarrow T * \alpha$
  $k \leftarrow k + 1$
end function

Where π_i is the parent node of v_i and $\pi_{best,i}$ as its best parent so far. T is temperature and τ_0 is the initial factor for initial temperature T_0 based on total nodes N. α is temperature change factor based on final temperature T_{ni}, initial temperature T and maximum iteration ni. m is optimization function used to calculate fitness score.

Optimization function is used to find the best structure on each iteration based on their score. Factorized Conditional Log Likelihood ($\widehat{f}CLL$) is a discriminative and decomposable optimization function based on an approximation of conditional log likelihood [8].

$$\begin{aligned}\widehat{f}CLL(G|D) = (\alpha + \beta)\widehat{LL}(B|D) - \beta\lambda \sum_{i=1}^{n} \sum_{j=1}^{q_i^*} \sum_{k=1}^{r_i} \sum_{c=0}^{1} N_{ijk}(\log(\frac{N_{ijck}}{N_{ij*k}}) - \log(\frac{N_{ijc}}{N_{ij*}})) \\ - \beta\lambda \sum_{c=0}^{1} N_c \log(\frac{N_c}{N}) - \beta N \rho\end{aligned} \quad (2)$$

Where n is the number of nodes, r_i is the number of possible random variable values, q_i is the number of possible parent nodes configuration, q_i^* is q_i without class node as its parent and C is the total number of classes.

α, β, dan Λ is a constant that minimalize the mean squared error (MSE) of $\widehat{f}(U_t, V_t) = \alpha \log(U_t) + \beta \log(V_t) + \lambda$ to approximate $f = \log(\frac{U_t}{U_t+V_t})$. Where $\widehat{f}$ is an approximation of f.

$$\alpha = \frac{\pi^2 + 6}{24} \tag{3}$$

$$\beta = \frac{\pi^2 - 18}{24} \tag{4}$$

$$\lambda = \frac{\pi^2}{12 \ln 2} - (2 + \frac{(\pi^2 - 6) \log p}{12}) \tag{5}$$

$\widehat{LL}$ is Log-Likelihood defined as follow [7]:

$$\widehat{LL}(G|D) = \sum_{i=1}^{n} \sum_{j=1}^{q_i} \sum_{k=1}^{r_i} N_{ijk} \log(\frac{N_{ijk}}{N_{ij}}) \tag{6}$$

The observed parameters are defined as follow:

$$N_{ij} = \sum_{k=1}^{r_i} N_{ijk} \tag{7}$$

$$N_{ij*k} = \sum_{c=1}^{s} N_{ijck} \tag{8}$$

$$N_{ijc} = \sum_{k=1}^{r_i} N_{ijck} \tag{9}$$

$$N_{ij*} = \sum_{k=1}^{r_i} \sum_{c=1}^{s} N_{ijck} \tag{10}$$

$$N_c = \frac{1}{n} \sum_{i=1}^{n} \sum_{j=1}^{q_i^*} \sum_{k=1}^{r_i} N_{ijck} \tag{11}$$

Where N is the documents in the datasets, X_i is observed variable, PA_i is parent node of X_i and C_i is the label of X_i. N_{ijk} is the number of appearances when $X_i = k$ and $PA_i = j$. N_{ijck} is the number of appearances when $X_i = k$, $PA_i = j$ and $C_i = c$.

6. Bayesian Network models

This section describes Full Bayesian Network and Bayesian Network with Mood Indicator model in more detail.

6.1. Full Bayesian Network (FBN)

Full Bayesian Network uses words as nodes with the addition of one class node. The result is a massive Bayesian Network structure consist of more than one thousand nodes. Chi-squared feature selections method is used to reduce the number of nodes that are independent towards the document's class.

A metaheuristic algorithm, simulated annealing is used to approximate the best Bayesian Network structure in the solution space. Fitness score is calculated using $\widehat{f}$CLL because it's discriminative and decomposable. The initial structure of Bayesian Network is set to Naive Bayes to allow a faster optimization. To avoid overfitting because of $\widehat{f}$CLL, the number of maximum parent nodes is limited.

6.2. Bayesian Network with Mood Indicator (BNM)

Bayesian Network with Mood Indicator is an improvement of Multinomial Naive Bayes by connecting indicator nodes to the class node. It affects the class prior probability, resulting in a better inference performance. There are two indicator nodes used by the model, a node for positive mood indicator and a node for negative mood indicator. We use emoticons in tweets to define the tweet's mood, it's positive when any emoticons that show a happy expression exist and negative mood otherwise.

This model doesn't require a structure learning algorithm because the maximum number of nodes are three. We obtain the best structure by iterating all possible structures.

7. Experiment and analysis

This section describes the result and analysis of the experiment for Full Bayesian Network and Bayesian Network with Mood Indicator.

7.1. Full Bayesian Network

Performance analysis on test and train Data

Table 2: The performance of Naive Bayes (NB) and Full Bayesian Network (FBN) using maximum two and three parent nodes on train data for every significant level.

Model	Significant Level	Precision	Recall	F1-Score
NB	0.001	62.76%	61.45%	60.28%
NB	0.01	65.09%	63.88%	62.96%
NB	0.1	72.27%	70.92%	70.37%
FBN 2 Parents	0.001	63.02%	61.69%	60.5%
FBN 2 Parents	0.01	65.5%	64.28%	63.35%
FBN 2 Parents	0.1	72.67%	71.29%	70.73%
FBN 3 Parents	0.001	63.2%	61.85%	60.66%
FBN 3 Parents	0.01	65.69%	64.45%	63.52%
FBN 3 Parents	0.1	**72.89%**	**71.48%**	**70.92%**

Table 2 shows the performance of Naive Bayes and Full Bayesian Network on train data. Full Bayesian Network has better performance over Naive Bayes for every significant level. This shows that optimizing conditional log-likelihood (using $\widehat{f}CLL$) could affect the performance of Bayesian Network for classification.

Table 3: The performance of Naive Bayes (NB) and Full Bayesian Network (FBN) using maximum two and three parent nodes on test data for every significant level.

Model	Significant Level	Precision	Recall	F1-Score
NB	0.001	54.74%	54.56%	52.75%
NB	0.01	**55.58%**	55.46%	53.89%
NB	0.1	55.46%	**55.74%**	**54.07%**
FBN 2 Parents	0.001	54.60%	54.50%	52.68%
FBN 2 Parents	0.01	55.32%	55.30%	53.71%
FBN 2 Parents	0.1	54.37%	54.89%	53.24%
FBN 3 Parents	0.001	54.48%	54.42%	52.60%
FBN 3 Parents	0.01	55.00%	55.09%	53.49%
FBN 3 Parents	0.1	54.11%	54.65%	53.01%

Table 3 shows the performance of Naive Bayes and Full Bayesian Network on test data. Full Bayesian Network with maximum two parent nodes is better compared to Full Bayesian Network with maximum three parent nodes. However, Full Bayesian Network has worse performance compared to Naive Bayes. This shows the improvement of the conditional log-likelihood score doesn't affect the performance of model directly.

Performance analysis on each Iteration

To analyze the cause of poor performance of Full Bayesian Network compared to Naive Bayes, we compare the performance of the best FBN model on each iteration towards the train and test data.

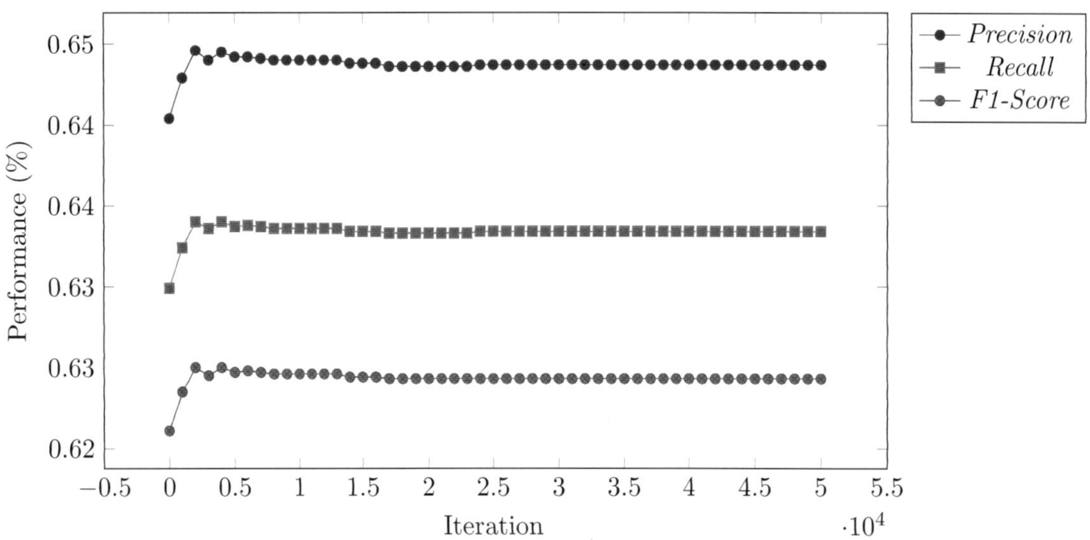

Figure 2: The performance of first fold's FBN 2 parents with significant level 0.01 on each iteration towards train data on the first cross-validation's fold.

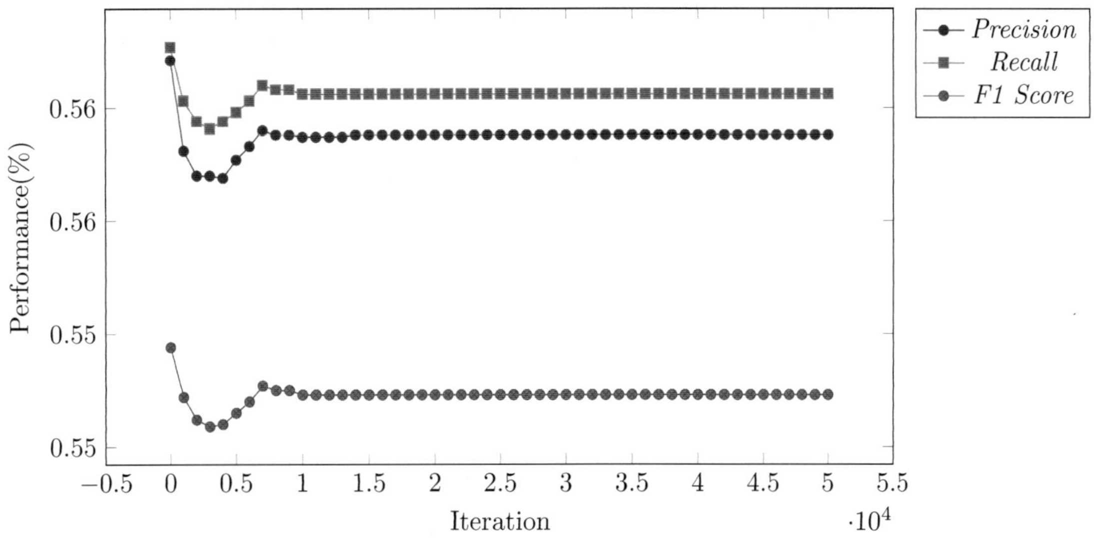

Figure 3: The performance of first fold's FBN 2 parents with significant level 0.01 on each iteration towards test data.

Figure 2 and 3 shows the performance of Bayesian Network on each iteration. The performance on train data is drastically improved in the beginning of the iteration before declined and became stable in the later iteration. The performance on test data declined drastically in the beginning of the iteration before it improves and becomes stable in the later iteration. This shows the improvement of $\widehat{fCLL}$ doesn't directly affect the performance of the Bayesian Network which makes it possible for the model to performs worse when the $\widehat{fCLL}$ score is improved. It's caused by the characteristics of $\widehat{fCLL}$ which is an approximation of conditional log-likelihood. The improvement of conditional log-likelihood not only able to improve the performance but can also decline the performance. The performance declined when conditional log-likelihood is improved by improving the ability to recognize a class while sacrificing the ability to recognize the other class.

Performance analysis for each emotion

To analyze the cause of poor performance of Full Bayesian Network compared to Naive Bayes, we analyze the performance of the best FBN model on each iteration towards the train and test data.

Table 4: The performance of first fold's FBN using maximum two parent nodes with significant level of 0.01 for each emotion.

Emotion	Train			Test		
	Precision	Recall	F1-Score	Precision	Recall	F1-Score
anger	62.25%	48.93%	54.79%	44.64%	34.90%	39.17%
disgust	86.74%	72.71%	79.11%	29.79%	18.79%	23.05%
fear	79.16%	55.82%	65.47%	64.92%	44.93%	53.11%
joy	63.28%	83.98%	72.18%	60.54%	79.11%	68.59%
sadness	51.99%	52.44%	52.21%	43.86%	45.40%	44.61%
surprise	67.11%	42.68%	52.17%	63.62%	41.56%	50.27%

Table 4 shows the performance of Full Bayesian Network for each emotion. The overall performance of the model on train data is better compared to test data with each emotion scored at least 50% F1-score, meanwhile the performance on test data is worse compared the train data, especially for anger, disgust, and fear emotions which have lesser number of samples compared to other emotions. It's likely caused by the class imbalance of the data which led to miss classification.

Inference complexity analysis

There are only two possible values (*true* when the word exist in a document and *false* otherwise) for each word node because the data is represented using Binary Term Frequency. Because of that, the value of each node become known and marginalization can be avoided. Therefore the inference of complexity of Full Bayesian Network is the same as Naive Bayes, which is $O(CN)$. Where C is the number of possible class and N is the number of nodes.

7.2. Bayesian Network with Mood Indicator

This section analyzes the performance of Bayesian Network with Mood Indicator. Figure 4 and table 5 shows the structure and the performance respectively.

Table 5: The performance of Bayesian Network with Mood Indicator for every possible structures (Figure 4).

Struktur	*Precision*	*Recall*	*F1-Score*
1	54.97%	54.98%	51.49%
2	55.08%	55.0%	51.64%
3	55.56%	55.44%	52.13%
4	55.56%	55.44%	52.13%
5	**55.57%**	**55.44%**	**52.14%**
6	**55.57%**	**55.44%**	**52.14%**
7	**55.57%**	**55.44%**	**52.14%**
8	55.08%	55.0%	51.64%
9	**55.57%**	**55.44%**	**52.14%**
10	55.56%	55.44%	52.13%
11	**55.57%**	**55.44%**	**52.14%**
12	**55.57%**	**55.44%**	**52.14%**
13	**55.57%**	**55.44%**	**52.14%**
14	55.08%	55.0%	51.64%
15	55.08%	55.0%	51.64%
16	**55.57%**	**55.44%**	**52.14%**
17	**55.57%**	**55.44%**	**52.14%**
18	**55.57%**	**55.44%**	**52.14%**
19	55.08%	55.0%	51.64%
20	55.08%	55.0%	51.64%
21	55.49%	55.3%	51.97%
22	55.49%	55.3%	51.97%
23	55.49%	55.3%	51.97%
24	55.49%	55.3%	51.97%
25	55.5%	55.32%	51.99%
26	55.49	55.3%	51.97%

Table 5 shows structure 5, 6, 7, 9, 11, 12, 13, 16, 17 and 18 has the better performance compared to the other structures. The addition of indicator node also improves the performance

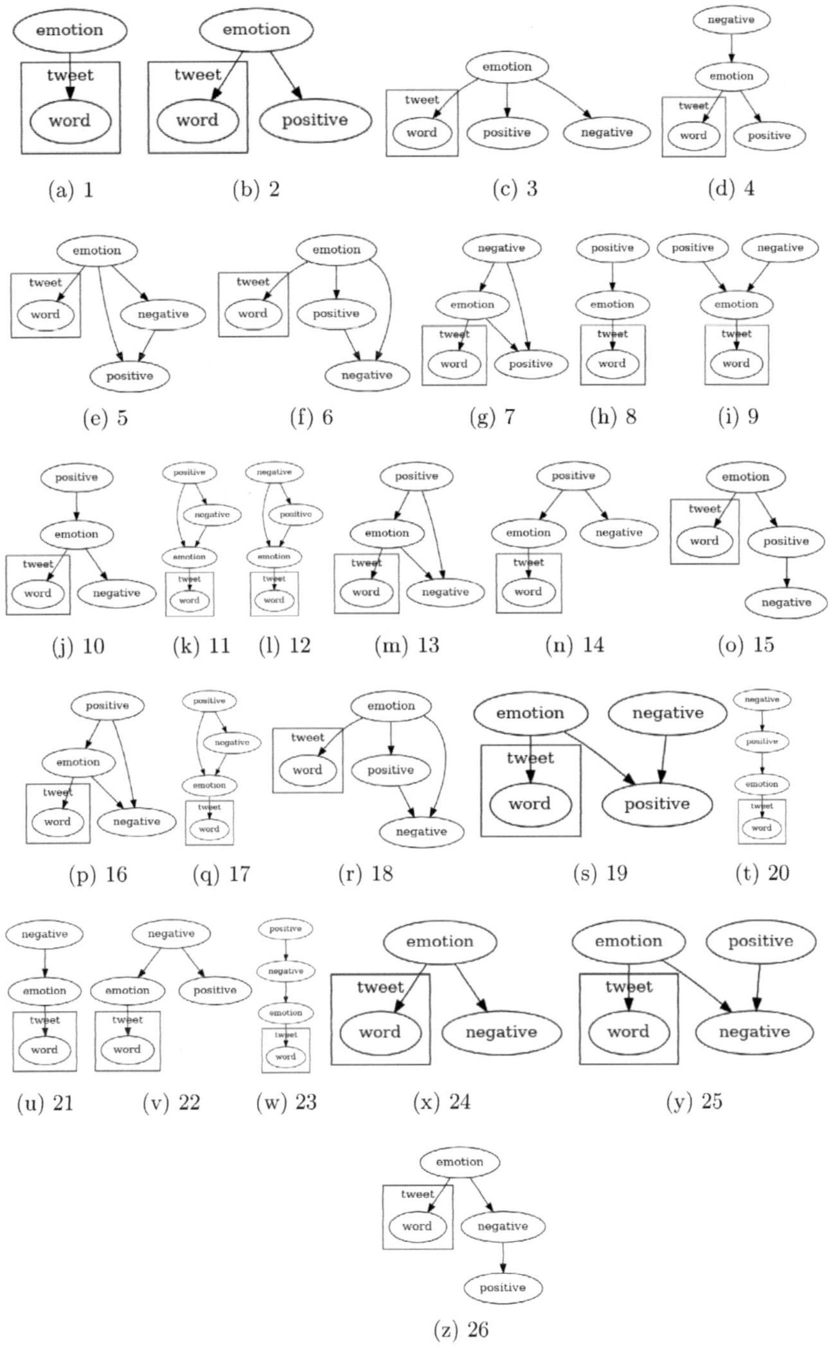

Figure 4: All possible structures of Bayesian Network with Mood Indicator. Words node is shown using plate notation which illustrates a Multinomial Naive Bayes relationship. The emotion's node is the class node while the positive and the negative node is the indicator nodes. Only nodes within the Markov Blanket of the class node are displayed.

of the model compared to Multinomial Naive Bayes without indicator node. Additional information provided by indicator node affects the prior probability of the class, resulting in better posterior probability towards a class.

Inference complexity analysis

The inference complexity of Multinomial Naive Bayes depends on the input data. The complexity is $O(CW)$, where C is the number of classes and W is the number of terms (input data). Bayesian Network with Mood Indicator has the similar complexity because of the modification on prior calculation. The complexity is $O((2+C)W)$, where 2 comes from the number of nodes that affects class node, which is positive and negative nodes.

8. Conclusion

We have presented the experiment result of Full Bayesian Network and Bayesian Network with Mood Indicator for emotion classification on twitter data. Based on the result of the experiment, it shows that the performance of both model is not good enough for classifying emotion from twitter data due to the simplicity for the model.

Full Bayesian Network trained using Simulated Annealing and $\widehat{f}CLL$ performs better compared to Naive Bayes on train data. However, Full Bayesian Network has worse performance compared to Naive Bayes on test data. It's caused by the characteristics of $\widehat{f}CLL$ which is an approximation of conditional log-likelihood. In future work, we suggest trying different search algorithm and the score function that directly affects the performance (e.g., classification rate).

Even though it's not better compared to Full Bayesian Network, Bayesian Network with Mood Indicator successfully improves the performance of Multinomial Naive Bayes with much lower computational complexity. The addition of indicator node on Multinomial Naive Bayes could improve the performance of Multinomial Naive Bayes. For future work, we suggest trying different indicator nodes such as the result of K-means clustering on the text and experimenting with n-grams for the multinomial naive bayes features.

References

[1] Grimace. http://www.grimace-project.net/. Accessed: 2016-11-10.

[2] Indeks kebahagiaan indonesia tahun 2017. Badan Pusat Statistik Indonesia, August 2017.

[3] Balabantaray, R. C., Mohammad, M., and Sharma, N. Multi-class twitter emotion classification: A new approach. *International Journal of Applied Information Systems 4*, 1 (September 2012), 48–53. Published by Foundation of Computer Science, New York, USA.

[4] Bollen, J., Mao, H., and Pepe, A. Modeling public mood and emotion: Twitter sentiment and socio-economic phenomena, 2011.

[5] Bouckaert, R. *Bayesian Belief Networks: from Construction to Inference.* PhD thesis, Utrecht, Netherlands, 1995.

[6] Bougie, R., Pieters, R., and Zeelenberg, M. Angry customers don't come back, they get back: The experience and behavioral implications of anger and dissatisfaction in services. *Journal of the Academy of Marketing Science 31*, 4 (2003), 377–393.

[7] Carvalho, A. M. Scoring functions for learning bayesian networks, 2009.

[8] Carvalho, A. M., Roos, T., Oliveira, A. L., and Myllymäki, P. Discriminative learning of bayesian networks via factorized conditional log-likelihood. *J. Mach. Learn. Res. 12* (July 2011), 2181–2210.

[9] Ekman, P. Facial expression and emotion. *American Psychologist 48*, 4 (1993), 384–392.

[10] Gamallo, P., and Garcia, M. Citius: A naive-bayes strategy for sentiment analysis on english tweets. pp. 171–175.

[11] Korb, K. B., and Nicholson, A. E. *Bayesian Artificial Intelligence, Second Edition*, 2nd ed. CRC Press, Inc., Boca Raton, FL, USA, 2010.

[12] Manning, C. D., Raghavan, P., and Schütze, H. *Introduction to Information Retrieval.* Cambridge University Press, New York, NY, USA, 2008.

[13] Mohammad, S. M. #emotional tweets. In *Proceedings of the First Joint Conference on Lexical and Computational Semantics - Volume 1: Proceedings of the Main Conference and the Shared Task, and Volume 2: Proceedings of the Sixth International Workshop on Semantic Evaluation* (Stroudsburg, PA, USA, 2012), SemEval '12, Association for Computational Linguistics, pp. 246–255.

[14] Mohammad, S. M., and Yang, T. W. Tracking sentiment in mail: How genders differ on emotional axes. In *Proceedings of the 2Nd Workshop on Computational Approaches to Subjectivity and Sentiment Analysis* (Stroudsburg, PA, USA, 2011), WASSA '11, Association for Computational Linguistics, pp. 70–79.

[15] Naaman, M., Boase, J., and Lai, C.-H. Is it really about me?: Message content in social awareness streams. In *Proceedings of the 2010 ACM Conference on Computer Supported Cooperative Work* (New York, NY, USA, 2010), CSCW '10, ACM, pp. 189–192.

[16] Ortigosa-Hernndez, J., Rodrguez, J. D., Alzate, L., Lucania, M., Inza, I., and Lozano, J. A. Approaching sentiment analysis by using semi-supervised learning of multi-dimensional classifiers. *Neurocomputing 92* (2012), 98 – 115. Data Mining Applications and Case Study.

[17] Pang, B., and Lee, L. Opinion mining and sentiment analysis. *Foundations and Trends in Information Retrieval 2*, 12 (2008), 1–135.

[18] Porter, M. F. Readings in information retrieval. Morgan Kaufmann Publishers Inc., San Francisco, CA, USA, 1997, ch. An Algorithm for Suffix Stripping, pp. 313–316.

[19] Rish, I. A tutorial on inference and learning in bayesian networks. IBM T.J.Watson Research Center, 2000.

[20] Roberts, K., Roach, M. A., Johnson, J., Guthrie, J., and Harabagiu, S. M. Empatweet: Annotating and detecting emotions on twitter. In *Proceedings of the Eight International Conference on Language Resources and Evaluation (LREC'12)* (Istanbul, Turkey, may 2012), N. C. C. Chair), K. Choukri, T. Declerck, M. U. Doan, B. Maegaard, J. Mariani, A. Moreno, J. Odijk, and S. Piperidis, Eds., European Language Resources Association (ELRA).

[21] Wan, Y., and Gao, Q. An ensemble sentiment classification system of twitter data for airline services analysis. In *2015 IEEE International Conference on Data Mining Workshop (ICDMW)* (Nov 2015), pp. 1318–1325.

[22] Adiwijaya. Aplikasi Matriks dan Ruang Vektor. In *Graha Ilmu* (Yogyakarta, 2014).

[23] Adiwijaya. Matematika Diskrit dan Aplikasinya. In *Alfabeta* (Bandung, 2014).

[24] MS Mubarok, Adiwijaya, MD Aldhi. Aspect-based sentiment analysis to review products using Nave Bayes. In *AIP Conference Proceedings 1867, 020060* (2017).

[25] Aziz, R.A., Mubarok, M.S. and Adiwijaya. Klasifikasi Topik pada Lirik Lagu dengan Metode Multinomial Naive Bayes. In *Indonesia Symposium on Computing (IndoSC)* (2016).

[26] Arifin, A.H.R.Z., Mubarok, M.S. and Adiwijaya. Learning Struktur Bayesian Networks menggunakan Novel Modified Binary Differential Evolution pada Klasifikasi Data. In *In Indonesia Symposium on Computing (IndoSC)* (2016).

[27] Yulita, I.N., Houw Liong The and Adiwijaya. Fuzzy Hidden Markov Models for Indonesian Speech Classification. In *JACIII*, 16(3), pp.381-387 (2012).

[28] Pratiwi, A.I., Adiwijaya. On The Feature Selection and Classification Based on Information Gain for Document Sentiment Analysis. In *Applied Computational Intelligence and Soft Computing*, (2017).

IOP Conf. Series: Journal of Physics: Conf. Series **971** (2018) 012042 doi:10.1088/1742-6596/971/1/012042

Analysis and implementation of cross lingual short message service spam filtering using graph-based k-nearest neighbor

Dyah Ayu Cyntya Dewi, Shaufiah, Ibnu Asror

Data Mining Center Laboratory, School Of Computing. Telkom University, Bandung 40257, Indonesia

E-mail: cintyadewi.ring@gmail.com, shaufiah@telkomuniversity.ac.id, iasror@telkomuniversity.ac.id

Abstract. SMS (Short Message Service) is on e of the communication services that still be the main choice, although now the phone grow with various applications. Along with the development of various other communication media, some countries lowered SMS rates to keep the interest of mobile users. It resulted in increased spam SMS that used by several parties, one of them for advertisement. Given the kind of multi-lingual documents in a message SMS, the Web, and others, necessary for effective multilingual or cross-lingual processing techniques is becoming increasingly important. The steps that performed in this research is data / messages first preprocessing then represented into a graph model. Then calculated using GKNN method. From this research we get the maximum accuracy is 98.86 with training data in Indonesian language and testing data in indonesian language with K 10 and threshold 0.001.

1. Introduction

SMS (Short Message Service) is one of the communication services that are still the main choice in various parties. Along with the development of other communication media, the country lowered the SMS rate to remain the top choice for mobile users [1]. This results in increased spamming of spam that some irresponsible people use, such as advertising and fraud. So with the number of SMS fraud or advertising action is causing discomfort for most SMS recipients. SMS spam is an SMS that contains information unwanted by the recipient of the message. Spam SMS is sent from one sender to many numbers obtained randomly (randomly).

Given the multi-lingual nature of documents in a message, web, etc., the need for effective multilingual or cross-lingual processing techniques is becoming increasingly important. Therefore, the hope of this research system can recognize a message with multi-language exactly that is Indonesian or English language and classify it into class of spam or ham (not spam)[2][3].

In this research, a message will be classified in the category of spam or not spam (ham). The dataset in this research using English language and Indonesian language. For the classification process, the data already obtained needs to be preprocessed in advance to clear the data without changing any contained information [4].

There are classification methods that have been used in previous research, such as Nave bayes [5][6][7], KNN, GKNN with accuracy maximum 95,5. From the research mentioned that Graph-based K-nearest Neighbor can perform SMS classification with accuracy level reaches 98,9 [8].

Based on this research, GKNN method was chosen as algorithm used to perform the process of classification on SMS filtering in this research. From this research the system will classify data in spam or ham class (not spam).

2. Related Works

Palmieri et all says that The mobile networks are indeed very popular in our lives but raises serious security concerns their increasing vogue and widespread coverage [9]. It means that smartphones may now represent an most ideal target for malicious authors. Kim et all says SMS is not only used for texting among people but it is also used as security like authentication method, token (mobile banking, double authentication, one-time password delivery, etc)[8].

When we embarked on this line of research, we did not nd any publications addressing the area of Cross-lingual SMS spam filtering. On the other hand, there is a rich literature addressing the related problem of Cross-Lingual Information Retrieval (CLIR) or Cross-Lingual Text Categorization(CLTC). olsson et al says that Both CLIR and CLTC are based on some computation of the similarity between texts, comparing documents with queries or class proles[10]. The main contrasting between CLIR and CLIC is that CLIR is queries based, consisting of a few words only, whereas in CLTC have class and each class is describe by an major profile (may be seen as a adequate documents collection)[10]. In this research we used CLTC to solve our problem.

3. Proposed Schema

System development will be divided in four phase, these are: documents translation as needed, preprocessing, build graph, classification using graph-based K-Nearest Neighbor. In figure 3-1, represent system overview.

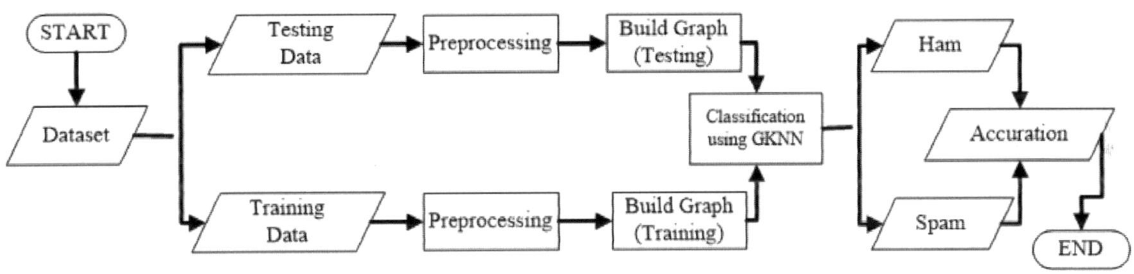

Figure 1. System Overview.

3.1. Dataset

A dataset is an object that represents data. The dataset that used are contain of unstructured text which later data will be transformed into a form of data that is structured to facilitate the process of storing, calling, and classification on the system. The Indonesian dataset is the result of manual collection from previous researchers[6][5], while the English dataset is obtained from SMS Corpus: SMS Spam Collection v.1.

3.2. Document Translation

The data used will be adjusted to the data partition scenario. For a cross-lingual approach, data will be translated into other languages with the help of machine translator i.e. Google Translate.

Documents translation is only done for test scenario 3, 4 and 5. Before preprocessing, data will be translated. For example, English data will be translated into Indonesian language, figure 2 illustrated this example.

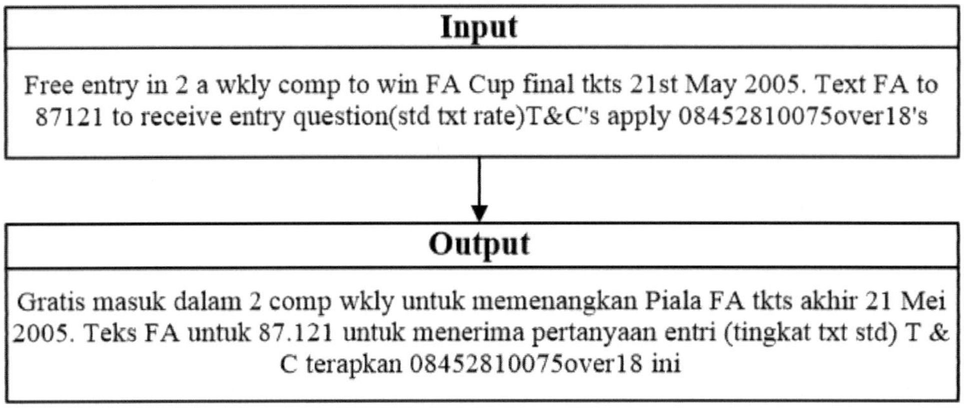

Figure 2. Data Translation.

3.3. Preprocessing Data

At this phase, training data and testing data will be processed to produce better data to be processed in the next phase. In the preprocessing phase, there are 5 step of the process that must be done. Here is an explanation of each step done on preprocessing:

3.3.1. Case Folding Case Folding and remove punctuation: is the process of converting capital letters contained in the dataset into lowercase for all datasets and also remove characters and punctuation.

3.3.2. Tokenizing Tokenizing is the process to separate the sentence contained in the review into a snippet of words, this piece of words that will be input of the next step after preprocessing.

3.3.3. Slang Handling In dataset, there are many informal words, it called slang word. To solve these cases, so we created a dictionary containing the words slang and equipped with the meaning of the word. Glossary for the English slang dictionary from " Slang Dictionary-Text Slang Internet Slang Words" on the website http://www.noslang.com/dictionary/, while for Bahasa Indonesia obtained from http: //en.wikipedia .org /Wiki/Indonesian$_{s}lang$.

3.3.4. Stopword Removal Eliminate Common words that have lesser significance. To remove the words included in the stopword list, then use a word matching technique with a dictionary that lists the word stopwords. For the list of word stopwords, in this study used dictionary word stopword Indonesian sourced from the site https://github.com/masdevid/ID-Stopwords, and dictionary word stopword English sourced from the site http://www.ranks.nl/Stopwords.

3.3.5. Stemming In the dataset there are many words that have affixes, so it takes stemming that aims to change back words that have additional affixes or changes in the form of words. The stemmer for Indonesian language is Sastrawi Indonesian stemmer is sourced from https://pypi.python.org/pypi/Sastrawi/1.0.1. And for English stemmer use Porter Stemmer sourced from NLTK library: https://pypi.python.org/pypi/stemming/1.0.

3.4. Build Graph
After preprocessing, it will generate data that ready to be processed in the next step. Data will be represented in graph. Formation of graph in data training is done by dividing data into 5 documents / messages into 1 (one) graph. So suppose, there are 100 documents/messages, then split into 20 graphs. While for data testing representation, each message/document will be 1 (one) graph.

Each node that forms in a graph shows a token selected at the preprocessing step. Edge is formed based on the order of occurrence between 2 words. The weights of the edge are shown with the Feature Weight Matrix. Graph formed consists of 2 graphs, namely training graph and testing graph. Training graph will group the data by category, spam and ham. While testing graph is a graph formed from data testing.

3.5. Classification Using G-KNN
Graph-based K-Nearest Neighbors (GKNN) is a evolution of K-Nearest Neighbors, which in GKNN data will be represented in the form of graph models. From that model will be calculated the similarity between documents classified by the large number of documents [8]. In a graph consists of nodes, edges, and weights of edges. To measure the similarity between 2 graphs (spam graph and ham graph), then the classification is measure the similarity. There is a Feature Weight (FW) which defines the similarity between 2 graphs based on the weight of the nodes and edges on the graph. The Node Fit Percent (NFP) shows how many nodes in the sample graph with their weights bigger than zero also appear in the test graph NFP can be formulated as follows:

$$Nfp = \frac{|\{node \mid node \in tg \; node \in sg\}|}{|\text{Number of Feature terms}|} \tag{1}$$

Calculate the value of NFP from testing graph and training graph by calculating the frequency of each feature on the graph that is the value of W (i, i). If the Nfp value is greater than the threshold, it will calculate the FW value of 2 pieces of graph. Otherwise if the Nfp value is smaller than the threshold, then the 2 calculated graphs are not in one category, so there is no need to calculate the FW value. For the calculation of NFP and Feature Weight can be seen in Figure 3 and Figure 4.

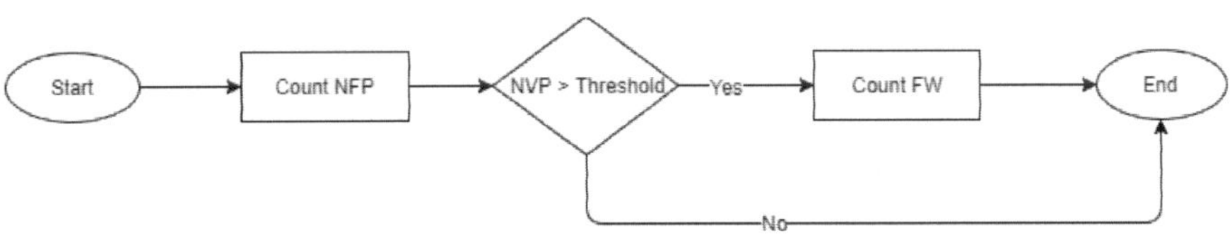

Figure 3. Calculation of NFP.

The process of comparing the values of NFP and FW is done until all training and testing graph have been compared. In the end, the initially empty RL list will be filled with the FW value and the category of the training graph. The most categories that appear on the list will be a new category of testing graphs tested that represent category of SMS testing.

4. Analysis and Experiment
For the classification in the Graph K-Nearest Neighbor method is used K value that will determine the length of list / array. The K value used in this test is 10, as well as by varying the threshold value to see the effect of the threshold value in find the most optimal threshold value in this research.

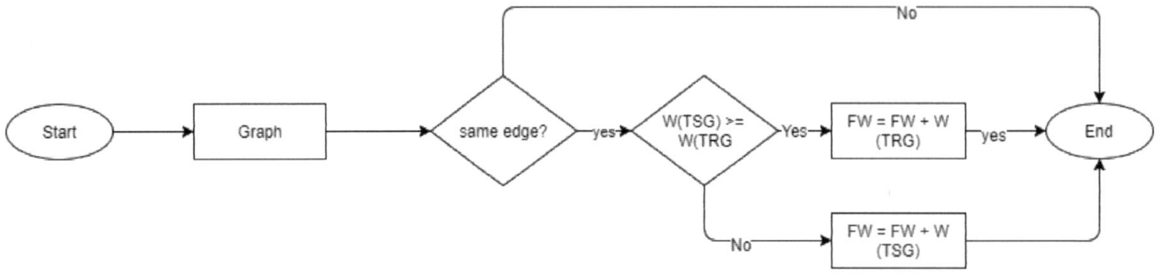

Figure 4. Calculation of Feature Weight.

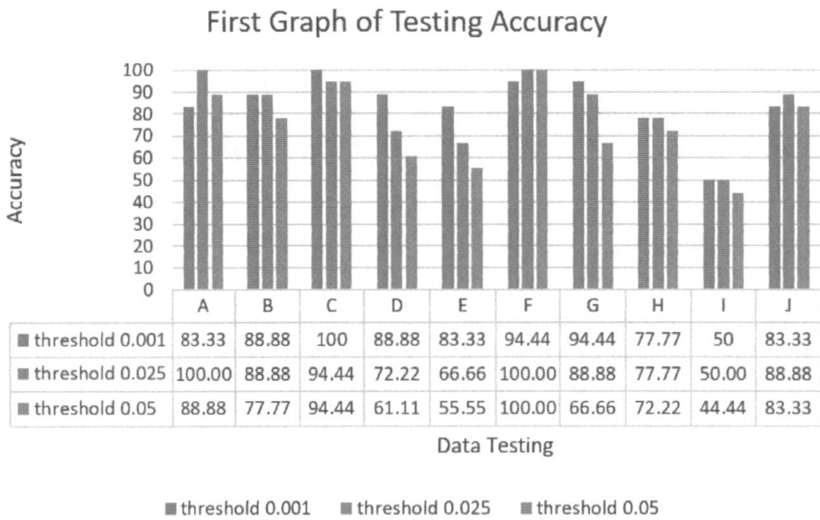

	A	B	C	D	E	F	G	H	I	J
threshold 0.001	83.33	88.88	100	88.88	83.33	94.44	94.44	77.77	50	83.33
threshold 0.025	100.00	88.88	94.44	72.22	66.66	100.00	88.88	77.77	50.00	88.88
threshold 0.05	88.88	77.77	94.44	61.11	55.55	100.00	66.66	72.22	44.44	83.33

Figure 5. First Graph of Testing Accuracy.

4.1. First Scenario Testing Analysis

The first test was conducted using Indonesian training data, and data testing using Indonesian Language. The threshold values used in this test are 0.001, 0.025 and 0.05. Based on the first test (Figure 5), with 10-fold by changing the threshold value obtained the highest accuracy of the highest average accuracy with the threshold of 0.001 is 84.44 with the highest accuracy value of 100 available in the sample data C. In the Indonesian language dataset especially the spam class has a pattern Almost similar, it creates a weighted value on each edge that is getting higher, so when searching for FW value, then having a higher weight value will have a higher chance of being in that category.

4.2. Second Scenario Testing Analysis

The second test result for 10-fold cross validation that consist of training data using Indonesian, and data testing using English (without translated) with a combination of threshold value 0.0005 and 0.0001.

Based on the results of the second test (Figure 6), by changing the threshold value obtained average accuracy of 86.4 and the average value of F1 of 2.84. It can be seen that the accuracy values have the same and low result of each sample of data testing. This low F1-measure result can be influenced by the language of different training data and data testing, so in the

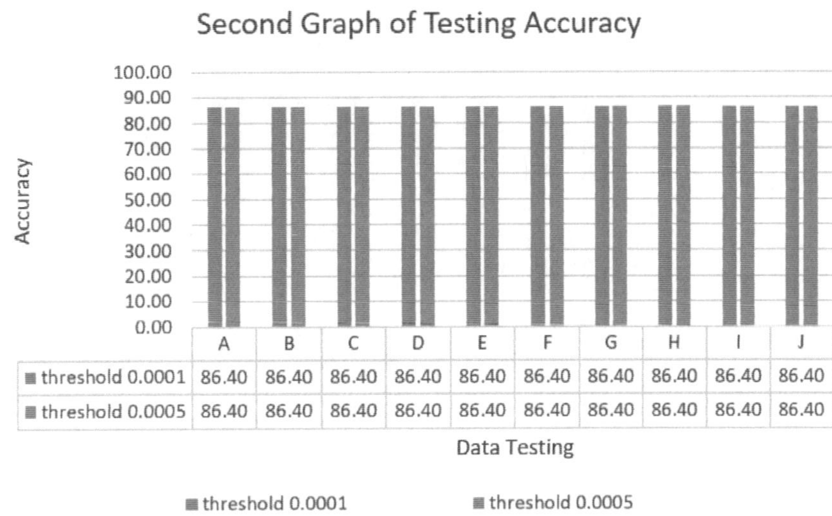

	A	B	C	D	E	F	G	H	I	J
threshold 0.0001	86.40	86.40	86.40	86.40	86.40	86.40	86.40	86.40	86.40	86.40
threshold 0.0005	86.40	86.40	86.40	86.40	86.40	86.40	86.40	86.40	86.40	86.40

Figure 6. Second Graph of Testing Accuracy.

classification process few find the same edge on the training graph.

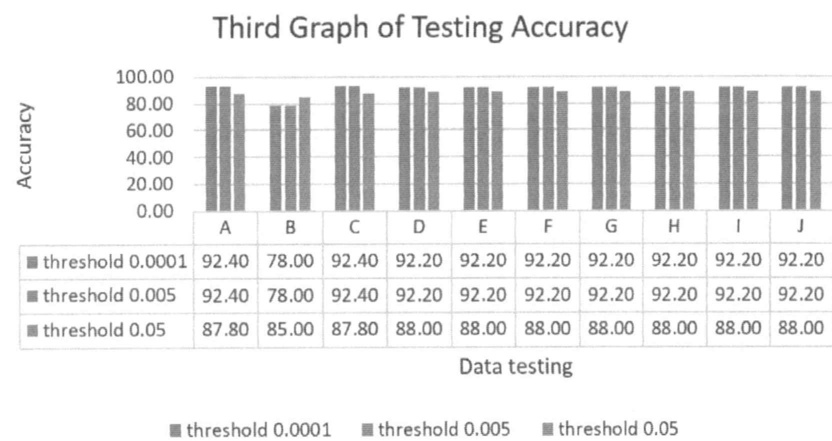

	A	B	C	D	E	F	G	H	I	J
threshold 0.0001	92.40	78.00	92.40	92.20	92.20	92.20	92.20	92.20	92.20	92.20
threshold 0.005	92.40	78.00	92.40	92.20	92.20	92.20	92.20	92.20	92.20	92.20
threshold 0.05	87.80	85.00	87.80	88.00	88.00	88.00	88.00	88.00	88.00	88.00

Figure 7. Third Graph of Testing Accuracy.

4.3. Thrid Scenario Testing Analysis

The third test result for 10-fold validation that consist of data training using Indonesian and English (without translation), and data testing using English (translated to Indonesian) with the combination of threshold value 0.0001, 0.005 and 0.05.

Based on the third test (Figure 7), by changing the threshold value obtained the highest accuracy with the highest threshold 0.0001 is 90.82, with the highest accuracy of 92.4 in sample data A and C. It can be seen also that the value of F1-measure is low, it is also caused by the use of different language in training data and data testing.

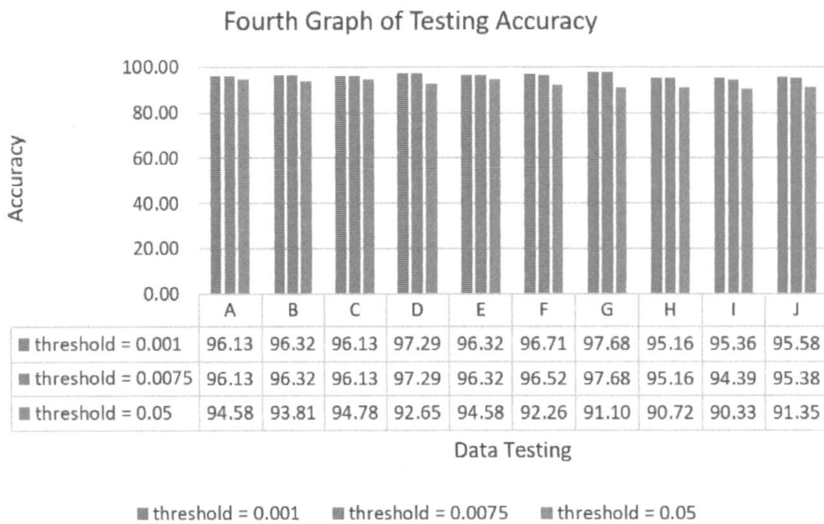

Figure 8. Fourth Graph of Testing Accuracy.

4.4. Fourth Scenario Testing Analysis

The fourth test was performed using 10-fold validation using Indonesian and English training data (translated into Indonesian), and testing data using Indonesian and English (translated to Indonesian). The threshold values used in this test are 0.001, 0.0075 and 0.05.

Based on the fourth test (Figure 8), by changing the threshold value obtained the highest accuracy on the threshold 0.001 of 96.27, with the highest accuracy of 97.68 in a sample data G. From the results of this test is considered the most optimal for cross-lingual handling, because in this scenario has a good combination of data sharing and data translation. The best results of the overall experimental results are from this fourth scenario.

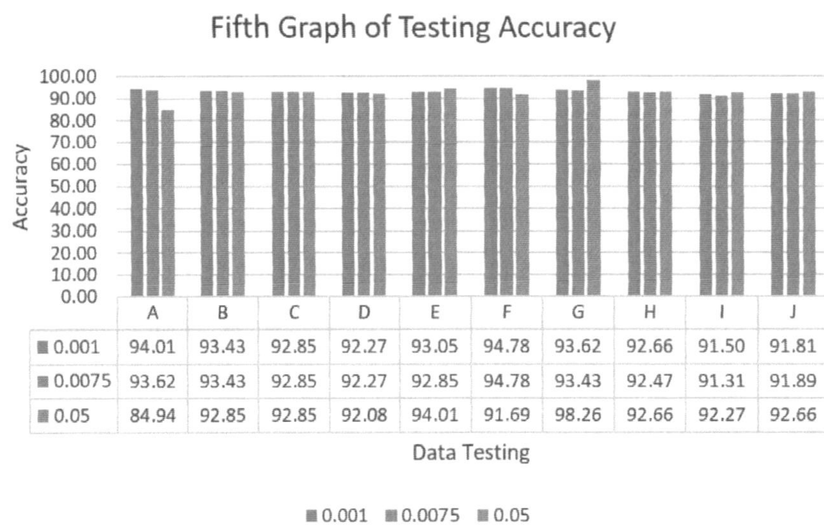

Figure 9. Fifth Graph of Testing Accuracy.

4.5. Fifth Scenario Testing Analysis
The fifth test was performed using 10-fold validation using Indonesian (translated to English) and English training data and testing data using Indonesian (translated to English) and English. The threshold values used in this test are 0.001, 0.075 and 0.05.

Based on the fifth test (Figure 9), by changing the threshold value obtained the highest accuracy on the threshold 0.001 of 93, with the highest accuracy of 94.78 in a sample data F. From the results of this test is considered the most optimal for cross-lingual handling, because in this scenario has a good combination of data sharing and data translation, but this scenario is not the best result.

5. Conclusion
Based on experiment result we found that the fourth scenario is the optimum one. with the accuracy 97.86. with based model in indonesia language(Bahasa), data training translated into bahasa and data testing translated into bahasa. That can be conclude with this dataset is the best using Indonesia language/Bahasa as primary language for analysis.

References
[1] A. Garz, Cross-Lingual Web Spam Classification, pp. 11491156, 2013
[2] M. Xiao and Y. Guo, Semi-Supervised Representation Learning for Cross-Lingual Text Classification, no. October, pp. 14651475, 2013.
[3] M. Xiao and Y. Guo, A Novel Two-Step Method for Cross Language Representation Learning, Adv. Neural Inf. , pp. 19, 2013.
[4] Sunardi, M., Listiyono, H. (2009). Aplikasi SMS Gateway. Jurnal Teknologi Informasi DINAMIK, XIV(1).
[5] T. Akhir, Sistem SMS Spam Filtering Berbahasa Indonesia Pada Smartphone Android Menggunakan Algoritma Multinomial Nave Bayes SMS Spam Filtering System For Indonesian SMS On Android Smartphone Using Multinomial Naive Bayes Algorithm Program Studi Sarjana Teknik Inf, 2016.
[6] , Shaufiah, Imanudin and Asror, Ibnu, ANDROID SHORT MESSAGES FILTERING FOR BAHASA USING MULTINOMIAL NAIVE BAYES, 2016.
[7] Gomez Hidalgo, J.M., Bringas, G.C., Sanz, E.P., Garca, F.C.: Content based SMS spam filtering; Proceedings of the 2006 ACM sympo-sium on Document engineering, pp. 107-114, 2006.
[8] S. Kim, Graph-based KNN Algorithm for Spam SMS Detection, vol. 19, no. 16, pp. 24042419, 2013.
[9] Palmieri, Michela and Impey, Soren and Kang, Hyojin and di Ronza, Alberto and Pelz, Carl and Sardiello, Marco and Ballabio, Andrea, "Human molecular genetics", vol 2011, number 19, pages 3852-3866, year 2011, publisher Oxford University Press
[10] J. S. Olsson, D. W. Oard, and J. Haji, Cross-Language Text Classification, pp. 645646.

On feature augmentation for semantic argument classification of the Quran English translation using support vector machine

Dina Khaira Batubara[1], Moch Arif Bijaksana[2], Adiwijaya[3]

School Of Computing, Telkom University, Indonesia[1,2,3]

khairadina@gmail.com[1], arifbijaksana@telkomuniversity.ac.id[2], adiwijaya@telkomuniversity.ac.id[3]

Abstract. Research on the semantic argument classification requires semantically labeled data in large numbers, called corpus. Because building a corpus is costly and time-consuming, recently many studies have used existing corpus as the training data to conduct semantic argument classification research on new domain. But previous studies have proven that there is a significant decrease in performance when classifying semantic arguments on different domain between the training and the testing data. The main problem is when there is a new argument that found in the testing data but it is not found in the training data. This research carries on semantic argument classification on a new domain that is Quran English Translation by utilizing Propbank corpus as the training data. To recognize the new argument in the training data, this research proposes four new features for extending the argument features in the training data. By using SVM Linear, the experiment has proven that augmenting the proposed features to the baseline system with some combinations option improve the performance of semantic argument classification on Quran data using Propbank Corpus as training data.

1. Introduction

1.1. Semantic Argument Classification

Semantic Argument Classification is the process of analyzing the sentence to investigate the pattern of WHO did WHAT to WHOM, WHEN, WHERE, WHY, HOW, from a structure text data. Semantic argument classification is also referred to as semantic role labeling (SRL) process. Originally building a good performance SRL system for one domain is conducted by manually semantic labeling process on a large data, called corpus. In the preliminary research, two types of corpus have been built, namely FrameNet [1] and Propbank [2], both are from news domain or news genre. Because building a corpus is costly and time-consuming, recently many studies have used FrameNet and Propbank corpus as training data to conduct semantic argument classification research on new domains such as domain literature, biomedical abstract, spoken of data, social media data, etc without the need to build a new corpus for those new domains [3] [4]

However, conducting the semantic argument classification on different domains is not the same as on the same domain. Previous studies have proven that there is a significant decrease in performance when classifying semantic arguments on different domain between the training and the testing data. Despite the existing SRL system has been tested and worked well on a sentence from the same domain, but it showed a sharp decline in performance when tested on a different domain [5]. The main problem is when there is a new argument that found in the testing data but it is not found in the training data. To recognize the new argument in the training data, extending the argument features in the training data to accommodate the new features of the new argument becomes one of the solutions.

This study performs a research related to semantic argument classification on a new domain that is Quran English Translation by utilizing Propbank corpus as training data. The Quran English translation is a translation of the original Arabic Quran. Hence the composition of grammar and the sentence structure, English-Quran is still affected by its original languages, Arabic. In Arabic, the holy Quran has a significant difference from the newswire domain, being closer to poetic language, more creative linguistic expression, and has many variations of vocabulary and sentence structure.

For a case study, we choose the verses with predicate “say” as the samples. Basic of choosing this keyword because “say” is one of the many emerging predicates in Quran that called as much as 1722 times. In addition, compared to the other predicate, a sentence with predicate phrase “say” has a variety of more complex patterns. So it would be better if it is used as a reference.

By using SVM Linear Classifier, the ultimate goal is to improve the performance of semantic argument classification on Quran domain by taking the advantage of existing corpora and significantly reduce system development costs for Quran domain. This research focuses on the improvement of ARG0, ARG1, ARG2, and ARGM-TMP by augmenting a set of feature to deal with a new argument on Quran domain.

1.2. Semantic Annotation and Propbank Corpus

The corpora or corpus is a collection of data in the form of text documents used in the case study text mining. Today, it has two kinds of the corpus that can be used for development of semantic annotation argument data, that FrameNet and PropBank (Proposition Bank). This research used Propbank corpus as training data. It is used because the annotation process between one label with another has an independent labeling. It is able to further facilitate the classification process.

Based on PropBank annotation rule, the argument of the verb labeled sequentially from ARG0 to ARG4, where ARG0 is the proto-agent (usually the subject of a transitive verb) and ARG1 are proto-patient (usually a direct object), etc. In addition to the core argument (ARG1-ARG4), there is some additional arguments marked as ARGMs. Examples of ARGMs are ARGM-LOC (indicate location), ARGM-DIR (indicates direction), ARGM-PRP (implying the destination), and so on [2]. Table 1 shows the argument label related to predicate say.

Table 1. Argument Related With Predicate "say"

Tag	Description
ARG0	Agent, The people who say
ARG1	Thing, Utterance
ARG2	Patient, To whom said
ARG3-5	Explicit Argument
ARGM	Modifier (Additional Argument)

2. Related Work

The annotated corpuses FrameNet and Propbank of the initial research has greatly driven the development of Semantic Role Labeling. [6]was first introduced SRL system automatically based on statistical classifier trained on hand-annotated corpora FrameNet. In their initial work, gold or auto parse syntax tree was used as input and then extracting various lexical and syntactic features to identify the semantic role for a given predicate.

After [6], a lot of progress produced on automatic semantic role labeling. Progress can be linked to better modeling techniques, more relevant features and in small sizes, clean annotations and machine learning models. For features and machine learning models, [6] as initial research used basic features depicted on Table 2 and using the maximum entropy classifier. [7] resulted in two systems using decision tree classifier. The first system uses the same features with [6]. Then they showed performance improvement of another system which used some additional features. The state-of-the-art reported by Pradhan et al [8] where a wide range of novel features, including features extracted from named entities, verb clusters, verb sense, temporal cue word, dynamic context, etc were tested with an SVM classifier. And thereafter, most studies in SRL use SVM and maximum entropy classifier [9] [5]

In features engineering, the terms of finding the proper syntactic and semantic knowledge, the SRL researchers investigated features based on two formalisms, namely constituency grammar and dependency grammar [10] [11]. The initial basic features introduced by [6], and a number of additional features were later introduced by [12] [8] [13] [14]. It has been a long history that SRL systems have tried to use the dependence among semantic arguments.

In the Quran domain, there were some efforts to build the Quran corpus. [15] present preliminary work on the creation of a unique Arabic proposition repository for Quranic Arabic. They annotated the semantic roles for the 50 most frequent verbs in the Quranic Arabic Dependency Treebank (QATB) [16]. [17] present an initial research task for building a lexical database of the verb valences in the Arabic Quran using FrameNet frames. They studied the context of verbs in the Arabic Quran and compared them with matching frames and frame evoking verbs in the English FrameNet. These two studies focused on building corpus using the rules of Propbank and FrameNet. Both did not report any experiments to test the performance of the built corpus. While in the same domain that is Bible, the author has not found a similar research, either related to the preparation of corpus for semantic labeling as well as research related to the classification of a semantic argument. Research related to semantic research on Bible generally discusses the linguistic context and meaning of words.

So far, authors has not found a similar study using the Qur'an domain as test data. Hence, for analytical material, this research compare the performance of the baseline system that using the existing semantic features with the system after augmented with the new proposed features. In the sentence patterns, there are some similarities and differences in Propbank and Quran's translation. This supports the idea that the Propbank data is reusable for Quran domain, without having to build a corpus Quran thoroughly. By augmented some appropriate features to minimize the mismatch between Propbank and Quran domain data will improve the performance of semantic argument classification on Quran domain.

3. Baseline & Proposed Scheme

3.1. Baseline System

Previous research shows that different features are required for different subtasks [12]. As baseline features, there are two kinds of features, the basic features, and additional features. Basic features are a set of features commonly used for semantic argument classification research, while additional features are a set of features added to the system to increase the performance. Basic features and additional features are adopted from previous research.

Table 2. Baseline Features

Basic Features [6]	Additional Features
Predicate, Phrase Type, Path, Position, Voice, Headword, Sub-Categorization	Noun Head of PP, First/Last Word In Constituent, First/Last POS In Constituent [8]
	Syntactic Frame [12]
	Constituent Order [18]
	Argument Order [19]

3.2. Baseline System Performance

Table 3 shows the performance of classification using baseline features when tested on auto labeled and hand-labeled data. The performance is indicated by the value of accuracy, precision, recall, and F_1. The table shows that the performance of the system increased significantly when additional features were added to the baseline system.

Table 3. Baseline System Performance

Features	Auto Labeled Data				Hand Labeled Data			
	A	P	R	F1	A	P	R	F1
Basic	77.12	79.10	77.10	77.80	76.47	79.90	76.50	77.70
Basic + Additional	81.92	83.20	81.90	82.10	87.40	89.30	87.40	88.10

Table 4 shows the accuracy of the baseline system for the ARG0, ARG1, ARG2, and ARGM-TMP which become the focus of this research. As well as the performance of all arguments, the accuracy of these four arguments also increases when additional features are added to the system. From the results of this initial experiment, it can be concluded that the performance of the system can be improved by adding some features corresponding to the type of data.

Table 4. Accuracy of ARG0, ARG1, ARG2 and ARGM-TMP

Arguments	Auto Labeled Data		Hand Labeled Data	
	Basic	Basic + Additional	Basic	Basic + Additional
ARG0	91.37	94.06	88.40	92.17
ARG1	77.64	78.79	79.92	86.41
ARG2	51.85	85.19	45.16	78.71
ARGM-TMP	27.66	77.66	14.05	77.27
ALL	77.12	81.92	76.47	87.40

3.3. Dataset

Propbank frameset data version 1.7[1] in XML format is used as training data. This data originally consists of 27.629 arguments. Due to resource limitations in running the experiment, resample technique using weka to filter data is conducted and only 90% data that consist of 24,865 arguments with 7261 predicates are taken.

For testing data, the Quran English translation by Ministry of Religious Affairs downloaded through the website Tanzil – Quran Navigator[2], translator by Shahih International is used. The Qur'an translation is unlabeled data. Because the supervised learning method is applied, the unlabeled Quran's translation data must first be labeled. There are two scenarios for the labeling process. The first scenario uses the existing automatic semantic role labeler, in this research using practnlptools 1.0[3]. The result of this scenario is called Quran Auto Labeled Data. The second scenario uses hand-labeled data namely manually labeling the data by the author refer to Propbank annotation rule [2]. The result is called Quran Manual Labeled or Hand Labeled Data. This process generates labeled Qur'an translation data stored in XML format, according to the frameset of Propbank data.

3.4. Proposed Scheme

The steps undertaken generally consisted of feature extraction process and classification of a semantic argument. The system was able to perform the feature extraction process of semantic argument

[1] http://propbank.github.io/
[2] www.tanzil.net
[3] https://pypi.python.org/pypi/practNLPTools/1.0

classification of an English sentence. This process played a role in producing quality features in the classification process. The general process of the system is shown in Figure 1.

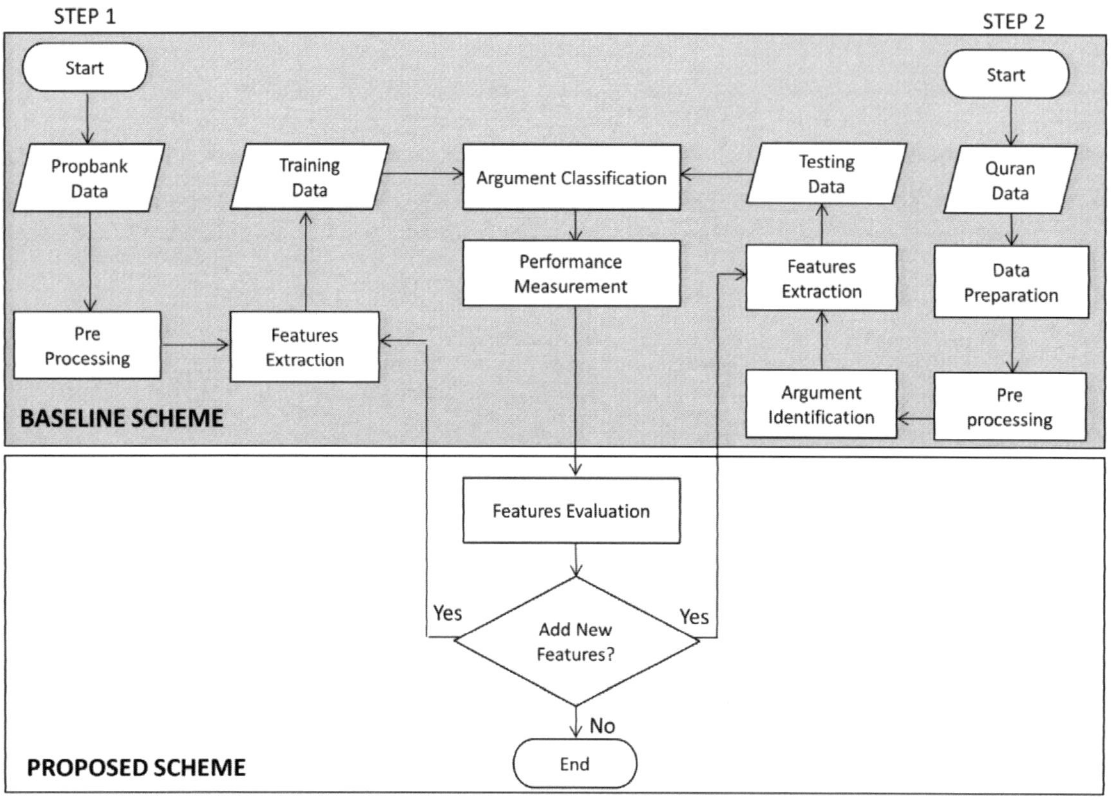

Figure 1. Experiment Scenario

3.4.1. Data Preparation. In data preparation process, the XML file for the English translation of the Qur'an in according to the rules of Propbank's frameset XML file is constructed [4].

3.4.2. Argument Identification. The focus of this research is to improve the argument classification process. Therefore, this research is limited on the process of argument identification which is assumed to have been done or using the data that has been conducted through the argument identification step.

3.4.3. Preprocessing. In preprocessing, the input were an XML file of PropBank data and Al-Quran translation data that has been structured and labeled. The necessary argument information was taken from this labeled data. This was the stage of extracting information from sentence where sentences, predicates, and arguments were taken.

3.4.4. Feature Extraction. Feature extraction is the process of extracting features from the data. The input was data generated form the preprocessing, where in one record data consisted of sentence, predicate, and argument. These three components were extracted and the features are shown in Table 2. The results of this process were the features extracted data, consisting of features for each record or data.

3.4.5. Feature Evaluation. Features evaluation is a process of selecting or evaluating the most relevant attribute on the entire data with predictive modeling problems are being worked on.

From features extracted data, then the features evaluation process was performed. This process was performed by Weka, using Gain Ratio Attribute Evaluator and search method Ranker. The process is by finding out the attributes or features from Quran domain that have a high value of correlation with the class. Gain Ratio Attribute Evaluator evaluate the worth of an attribute by measuring the gain ratio with respect to the class. The selected features will be developed into a new feature that will be augmented to the training and testing data sets. The result obtained was the features rank based on the value of the gain ratio. Based on the results obtained in Figure 2, the proposed features were constructed by developing of Position, Phrase Type, Path, 1st POS and Last POS features. Noun Head of PP feature was not included because this feature did not work thoroughly, it was only for preposition type phrases.

```
=== Attribute Selection on all input data ===

Search Method:
        Attribute ranking.

Attribute Evaluator (supervised, Class (nominal): 16 Arg):
        Gain Ratio feature evaluator

Ranked attributes:
 0.6845    4 Position
 0.4826    3 Phrase Type
 0.4561    8 NH_PP
 0.3436    2 Path
 0.3098   11 1St POS
 0.3033   12 Last POS
 0.2723    9 1St Word
 0.2531    6 Head Word
 0.2428   13 Syntactic Frame
 0.2364   10 Last Word
 0.2336   14 Cons Order
 0.2245   15 Arg Order
 0.0827    7 Subcategorization
 0.0553    1 Predicate
 0.0234    5 Voice

Selected attributes: 4,3,8,2,11,12,9,6,13,10,14,15,7,1,5 : 15
```

Figure 2. Features Evaluation Process

3.4.6. Proposed Features. The proposed features was constructed from the most important features resulting from features evaluation process from Quran domain. The following are the proposed features developed from baseline features based on features evaluation process :

Table 5. The Proposed Features

Feature	Description
Position Order (PO)	PO was built from position and path features. PO represented the position of an argument from the predicate, but in this feature, the distance was also mentioned, for example, one position on the left, formulated into "left_1".
Phrase Type Order (PTO)	PTO was built from phrase type and position features. PTO represented the phrase type and position of an argument in one package, for example, "NP_left".
Second POS in Constituent (SP)	SP as well as the first/last POS in constituent, but they took the POS of the second word of the constituent.
Second Word in Constituent (SW)	SW as well as the first/last word in constituent, but they took the second word of the constituent, as a pair of SP features.

3.4.7. Argument Classification. This research will use SVM as a classifier *[8] [9]*. This research use libSVM *[20]* with linear kernel, degree 3, tolerance of the termination criterion, e = 0.001 and cost per unit violation of the margin is C = 1.0. The type of SVM that will be used is a standard algorithm (C-SVC) for classification.

4. Result

Table 6 shows that the four new features improved the performance of the system with some combination option. Table 7 shows the enhancement in accuracy and F1 score for each ARG0, ARG1, ARG2, ARGM-TMP and overall arguments.

Table 6. Performance of Proposed Features Separately (%)

Features	Auto Labeled		Hand Labeled	
	Acc	F_1	Acc	F_1
Baseline (B)	81.92	82.10	86.72	87.60
B+SP	83.13	83.40	87.46	88.20
B+SW	81.96	82.10	86.93	87.80
B+PO	82.23	82.50	87.20	88.00
B+PTO	82.54	82.70	87.40	88.10

Table 7. The Enhancement of Accuracy and F-Measure With Proposed Features (%)

Argument	Auto Labeled Data				Hand Labeled Data			
	Acc	Features	F_1	Features	Acc	Features	F_1	Features
ARG0	0.67	B+SW	1.30	B+PO+SP	1.10	B+PTO+SW	0.00	-
ARG1	3.28	B+PO+SP	2.20	B+PO+SP	0.27	B+PO	1.10	B+PO+PTO+SP
ARG2	0.00	-	1.20	B+PTO	5.81	B+PTO+SP	1.90	B+PO+PTO
ARGM-TMP	6.38	B+PO+PTO	7.00	B+PTO+SP	7.44	B+PTO	4.50	B+PTO
ALL	1.48	B+PO+SP	1.60	B+PO+SP	0.47	B+PO+PTO	0.40	B+PO+PTO

4.1. Argument Performance

The SW feature only contributed positively to ARG0 when tested on both types of data, this was due to the Quranic data for there were quite a few verses that had similar patterns in ARG0. The SW feature contributed negatively to ARG2, this was due to the considerable variation of the word that had a large number of possible values with the same Part-Of-Speech but did not present in Propbank, e.g. "to Allah", "to Moses", "to Satan", etc.

The SP feature was capable of contributing positively to ARG0 and ARG1, both in terms of accuracy and F_1. The PO feature also contributed positively to ARG0 and ARG1, both in terms of accuracy and F_1. The PTO feature contributed positively to ARG2 and ARGM-TMP, both in terms of accuracy and F_1. The PO+SP combination greatly affected the performance gains of accuracy and F_1 in ARG1 when tested on auto labeled data. The combination PTO+SP had greatly affected the performance gains of accuracy in ARG2 when tested on hand-labeled data and the F_1 in ARGM-TMP when tested on auto labeled data. And the combination of PO+PTO had greatly affected the performance gains of accuracy in ARGM-TMP when tested on auto labeled data.

4.2. Feature Performance

The following are the features performance of the proposed system results for all arguments. Figure 3 show the features performance when tested on auto labeled data. For all arguments, the highest accuracy and F_1 are obtained by adding PO+SP features, that is 83.40%, increase of 1.48% from the baseline features and 83,70%, increase by 1,60% from the baseline features. Figure 4 shows the features performance when tested on hand-labeled data. For all arguments, the highest accuracy and F_1 are obtained by adding PO+PTO features. Accuracy is 87,88%, increase 0.47% from the baseline system and F_1 is 88,50%, increase 0,40% from the baseline features.

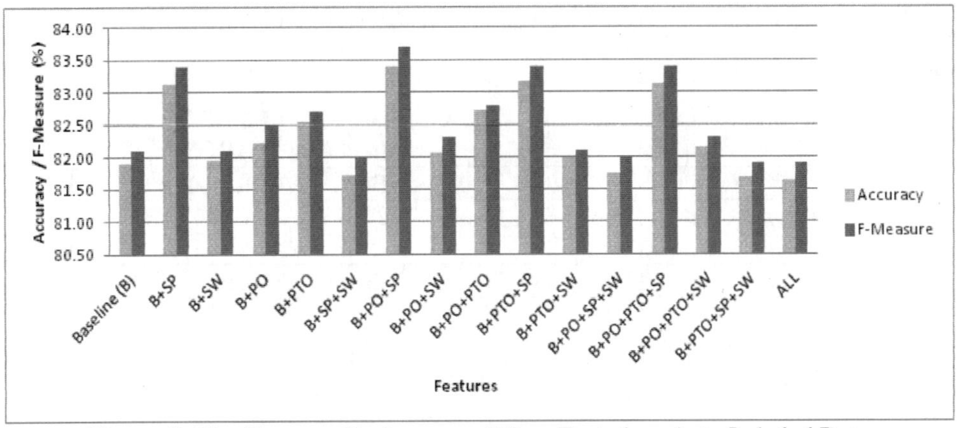

Figure 3. The Features Performance When Tested on Auto Labeled Data

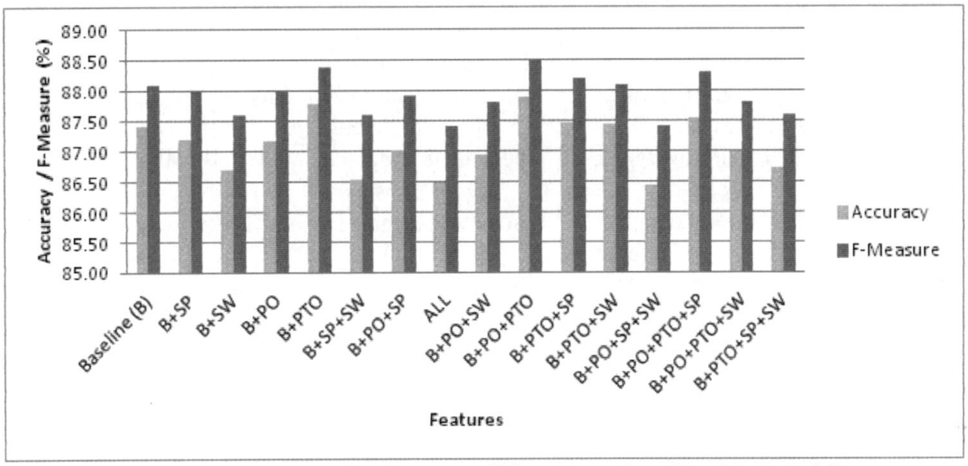

Figure 4. The Features Performance When Tested on Hand Labeled Data

5. Discussion

The objectives of this research is to consider the best feature combination which produce the best performance, this section analyzes against the performance of accuracy and F1 scores for overall arguments, taking into account the average accuracy and F_1 scores of ARG0, ARG1, ARG2, and ARGM-TMP.

When tested on auto labeled data, the highest accuracy and F-1 for all arguments was obtained by adding a combination of PO+SP features. While the highest average accuracy for the above four arguments was obtained by adding a combination of PO+PTO features, and for the highest average F-1 values was obtained by adding the PTO+SP features. This section analyzes the three combinations of features above.

Table 8 shows that the highest accuracy and F-1 for overall arguments was obtained by adding a combination of PO+SP features, namely 83.40% and 83.70%. But in more detail on the performance of ARG0, ARG1, ARG2, and ARGM-TMP, the average accuracy for these four arguments was only 84.24%. For this features combination, the ARGM-TMP accuracy was not good only 75.53%, lower than the baseline features.

The addition of this feature combination to the system improved accuracy and F_1 for overall arguments but did not result in stable average performance for all arguments. One possible cause was as described in Table 6 that the increase in ARGM-TMP correlated to the addition of the PTO feature.

Table 8. Performance of Proposed Features When Tested on Auto Labeled Data

Features	Performance	All Arguments	4 Arguments	Enhancement	Rank
PO+SP	Accuracy	83.40	84.24	Best Acc (All ARG) Low Acc (4 ARG)	
	F_1	83.70	76.05	Best F_1 (All ARG)	
PTO+SP	Accuracy	83.16	85.52	2nd Best Acc (All ARG) 2nd Best Acc (4 ARG)	**1**
	F_1	83.40	76.45	Best F_1 (4 ARG)	
PO+PTO	Accuracy	82.75	85.82	Good Acc (All ARG) Best Acc (4 ARG)	**2**
	F_1	82.80	75.78		

The second highest overall accuracy performance of the argument was obtained by adding the PTO and SP features of 83.16% with F_1 of 83.40%. In more detail on the four arguments above, these features combination achieved the second highest for the average value of accuracy that equaled to 85.52%. For the average of the F_1 score on these four arguments achieved the highest score, which was 76.50%. Therefore, the addition of PTO+SP features to the baseline system was considered very good to improve the overall performance both all arguments and against the four arguments above.

While the highest average accuracy for the above four arguments was obtained by adding a combination of PO+PTO features, but the accuracy for overall arguments was not really good. Therefore, the addition of these features to the baseline system was not really good for overall arguments.

When tested on hand-labeled data, the highest accuracy and F_1 for all arguments was obtained by adding a combination of PO+PTO features. While the highest average accuracy for the above four arguments was obtained by adding PTO features, and for the highest average F_1 values was obtained the same value between addition the combination of PO+PTO features and addition of PTO features. This section analyzes these two combinations of features.

Table 9. Performance of Proposed Features When Tested on Hand Labeled Data

Features	Performance	All Arguments	4 Arguments	Enhancement	Rank
PO+PTO	Accuracy	87.88	85.65	Best Acc (All ARG) 2nd Best Acc (4 ARG)	**1**
	F_1	88.50	85.53	Best F_1 (All ARG) Best F_1 (4 ARG)	
PTO+SP	Accuracy	87.79	85.71	2nd Best Acc (All ARG) Best Acc (4 ARG)	**2**
	F_1	88.40	76.45	2nd Best F_1 (All ARG) 2nd Best F_1 (4 ARG)	

Table 9 shows that the addition of the proposed features improved the performance for ARG0, ARG1, ARG2 and ARGM-TMP and overall arguments. From the combination of proposed features, the highest accuracy performance of overall arguments was obtained by adding PO and PTO features of 87.88%. In more detail on the performance of the ARG0, ARG1, ARG2, and ARGM-TMP, the average accuracy for the four arguments above achieved the second highest value of 85.65%, this shows that the accuracy distribution for the four arguments above with the addition of these two features was quite stable. The average F_1 for the four arguments above also reached the highest value of 85.50%. Then the addition of these two features on the baseline system improved the accuracy and F_1 score as for all arguments and for the above four arguments very well and stable.

The second highest overall accuracy performance of the argument was obtained by adding the PTO feature of 87.79%. In more detail on the performance of the ARG0, ARG1, ARG2, and ARGM-TMP, the average accuracy for the four arguments above achieve the highest value of 85.71%, meaning that the accuracy distribution for the four arguments above with the addition of these two features is fairly stable. The average F_1 for the four arguments above reached the second highest score of 85.30%. Therefore the addition of this feature on the system improved the accuracy and F_1 for all and the four arguments above well and stable.

Generally, the proposed features improved the system performance. Adding some combination options of proposed features had increased the performance of ARG0, ARG1, ARG2, ARGM-TMP and all arguments. However, the counter-use of all proposed features simultaneously decreased the performance (lower than baseline). The performance declined in the use of all these features was likely due to the use of a combination of SW and SP features. From a series of experiments showed that the combination of SP and SW on one package did not produce a good performance, even in accuracy and F_1 score.

6. Conclusions & Future Work

The semantic argument classification on Quran data can be constructed by using the training data from the Propbank corpus. The decreasing in the performance when the training data of the propbank were tested on the Quranic data was because of the new argument, the argument contained in the Quran but was not in the Propbank. To address this problem, this research proposes four new features for extending the argument features in the training data. The four new features was constructed from the most important features resulting from features evaluation process from Quran domain. From the experiment in accordance with the hypothesis, it concluded that augmentation of four new features separately or with some combination options to the baseline system improved the performance of the system. Based on the conclusion, there are some recommendations for feature work; replacing the manual process on argument identification to become integrated in one system, experimenting on verses of Quran using other predicates, using other features combinations and using domain adaptation method.

References

[1] C. F. Baker, C. J. Fillmore and J. B. Lowe., "The Berkeley Framenet Project," *Proceedings of the COLING-ACL, Montreal, Canada,* 1998.

[2] M. Palmer and P. Kingsbury, "From Treebank to PropBank," *Proceedings of the 3rd International Conference on Language Resources and Evaluation (LREC-2002), Las Palmas, Spain,* 2002.

[3] R. Tzong and H. T. e. al, "BIOSMILE: Adapting Semantic Role Labeling for Biomedical Verbs: An Exponential Model Coupled with Automatically Generated Template Features," *Proceedings of the BioNLP Workshop on Linking Natural Language Processing and Biology at HLT-NAACL, vol. VI,* pp. 57-64, 2006.

[4] Ng, D. Dahlmeier and H. Tou, "Domain Adaptation for Semantic Role Labeling in the Biomedical Domain," *Bioinformatics Advance Access,* 2010.

[5] S. Pradhan, W. Ward and J. Martin, "Towards Robust Semantic Role Labeling," *Computational Linguistics 34(2),* pp. 289-310, 2008.

[6] D. Gildea and D. Jurafsky, "Automatic Labeling of Semantic Roles," *Association for Computational Linguistics, vol. 28, no. 3,* pp. 245-288, 2002.

[7] M. Surdeanu, S. Harabagiu, J. Williams and P. Aarseth, "Using predicate-argument structures for information extraction," *Proceedings of ACL-2003,* 2003.

[8] S. Pradhan, W. Ward, K. Hacioglu, J. H. Martin and D. Jurafsky, "Shallow Semantic Parsing using Support Vector Machines," *HLT-NAACL,* pp. 233-240, 2004.

[9] N. E. Ozgenzil and N. McCracken, "Semantic Role Labeling Using libSVM," *Preceedings of CONLL-2005,* 2005.

[10] M. A. a. A. M. Mubarok, "Aspect-based sentiment analysis to review products using Naïve Bayes," in *AIP Conference Proceedings*, 2017.

[11] A. A. Pratiwi, "On The Feature Selection and Classification Based on Information Gain for Document Sentiment Analysis," *Applied Computational Intelligence and Soft Computing,* 2017.

[12] N. Xue and M. Palmer, "Calibrating Features For Semantic Role Labeling," *In Proceedings of the Conference on Empirical Methods in Natural Language Processing (EMNLP'04),* 2004.

[13] L. Yang, "Generalizable Features Help Semantic Role Labeling," *23rd Pacific Asia Conference on Language, Information and Computation,* pp. 859-866, 2009.

[14] H. Yang and C. Zong, "Learning Generalized Features for Semantic Role Labeling," *ACM Trans Asian Low-Resour. Lang. Inf. Process, Vol. 15, No. 4, Article 28,* 2016.

[15] Z. Wajdi, H. Abdelati and D. Mona, "A Pilot Propbank Annotation for Quranic Arabic," *Workshop on Computational Linguistics for Literature,* pp. 78-83, 2012.

[16] K. Dukes and T. Buckwalter, "A Dependency Treebank of the Quran using Traditional Grammar," *Proceedings of the 7th International Conference on Informatics and Systems (INFOS),* 2010.

[17] A. B. Sharaf and E. Atwell, "Knowledge Representation pf The Quran Through Frame Semnatic - A Corpus-Based Approach," *Proceedings of the Fifth Corpus Linguistics Conference,* 2009.

[18] S. Pradhan, K. Hacioglu, V. Krugler, W. Ward, J. H. Martin and D. Jurafsky, "Support Vector Learning for Semantic Argument Classification," *Machine Learning, vol. 60, no. 1-3,* pp. 11-39, 2005.

[19] M. Fleischman and N. Kw, "Maximum Entropy Models for FrameNet Classification," *Proceedings of the 2003 Conference on Empirical Methods in Natural Language Processing (EMNLP-03),* pp. 49-56, 2003.

[20] C. Chang, L. A and J. Chin, "LibSVM : A Library For Support Vector Machines," *ACM Transaction on Intelligent Systems and Technology; vol, 2; no. 3,* pp. 27:1 - 27:27, 2011.

[21] N. Xue and M. Palmer, "Calibrating Features For Semantic Role Labeling," *In Proceedings of the Conference on Empirical Methods in Natural Language Processing (EMNLP'04),* 2004.

IOP Conf. Series: Journal of Physics: Conf. Series **971** (2018) 012044 doi:10.1088/1742-6596/971/1/012044

Experiment on building Sundanese lexical database based on WordNet

Sari Dewi Budiwati, Novihana Nurani Setiawan

Information System Study Program,
School of Applied Science, Telkom University
Jl. Telekomunikasi, Ters. Buah Batu – Bandung

saridewi@tass.telkomuniversity.ac.id

Abstract. Sundanese language is the second biggest local language used in Indonesia. Currently, Sundanese language is rarely used since we have the Indonesian language in everyday conversation and as the national language. We built a Sundanese lexical database based on WordNet and Indonesian WordNet as an alternative way to preserve the language as one of local culture. WordNet was chosen because of Sundanese language has three levels of word delivery, called language code of conduct. Web user participant involved in this research for specifying Sundanese semantic relations, and an expert linguistic for validating the relations. The merge methodology was implemented in this experiment. Some words are equivalent with WordNet while another does not have its equivalence since some words are not exist in another culture.

1. Introduction

Indonesia has 719 individual languages. Of these, 706 are living and 13 are extinct [1]. Sundanese language is the second biggest language used in Indonesia. It is used by 34.000.000 population, most of the language is used in West Java and Banten province and consist of different dialect: Sumedang, Cianjur, Ciamis, Serang, Bogor, Subang, Tasikmalaya, Purwakarta and Cirebon dialect [2]. Sundanese language is considered as stable but threatened language [2] as because of some reason:

- Sundanese language rarely used by youth
  Regional languages are considered as an old-fashioned language by Indonesian youth, this is also happening with Sundanese. They are prouder using Indonesian or English language instead Sundanese as it is one of the effects of globalization. Now, there is only 40% Sundanese youth who know and understand Sundanese language in West Java Province [3].
- Competition between regional language, national language, and foreign language
  After the Indonesian Youth pledge on October 28th, 1928, a new generation used the Indonesian language in their daily activities: school, books, news, office work, etc. Additionally, English as global language affecting the youth generation also. Due to these, many Indonesian youths tend to use Indonesian or English language than regional language causing the decreasing in regional language.

In order not to be categorized as an endangered language, there is need some efforts to preserve the language [2]:

- Using Sundanese language in the family or social environment.

This effort had been proclaimed by Bandung Mayor in 2013. Every Wednesday, Bandung citizens are encouraged to use Sundanese for communication, it is known as Sundanese Wednesday *(Rebo nyunda)*. But this culture is done in Bandung city only, not in another area of West Java and Banten province, therefore the chance for the decreasing still can be found.

- The inclusion of Sundanese language in school curriculum
  Indonesian school curriculum was encouraged to have the local subject as a distinction with other region and to preserve regional wisdom. This also applied in West Java and Banten province. Elementary to high school had been introduced Sundanese languages in their curriculum.
- The planning of corpus can be pursued by accelerating the alignment of Sundanese language through the absorption of Indonesian vocabulary, or other languages to express concepts, especially science and another modern life.

In this paper, we described our effort to maintain and preserve Sundanese language using information technology approach. We built a Sundanese lexical database based on Princeton WordNet (2) and Indonesian WordNet (3), called Sunda WordNet. It is developed by merge approach as Sundanese synset is specified first the define its semantic relations and then mapped into WordNet synsets. The selection method was chosen based on Sundanese language characteristics that it has rich of same meaning for one word, called Code of Conduct *(undak-usuk basa)*.

2. Sundanese language code of conduct

Sundanese language has four [5] speech level or language code of conduct [4], but almost all of Sundanese literature makes into three level:

- Ribaldry level. It is used for friends whose age is below, or same age or for a best friend.
- Standard level. It is used for his/herself
- Polite level. It is used for upper level, someone whose doesn't know, or for someone who is being respectful.

In Table 1 given examples how different words can be used based on the language level although it has the same meaning. Some words changes are simple as we merely need to replace the words, for instance in no 2, we can replace the word "sick" for each level with "*damang*", "*udur*", and "*gering*", without having to change the previous sentence: "What's the matter, ..." *("Ku naon Jang,...")*. But for other, it will affect the whole sentence. For example, in no 4, the word "eat" for each level are "*tuang*", "*neda*" and "*dahar*", but for ribaldry levels, we have to change "want to" with "*arek*. Therefore, Sundanese words are rich for synonym words. Some example of language code of conduct with WordNet equivalence is shown in Table 2.

Table 1. Sundanese language changes

No	English sentence	Sundanese sentence		
		Polite	Standard	Ribaldry
1	You don't hear it, do you?	*Teu kadangu ku ceupil anjeun?*	*Teu kadenge ku ceuli maneh?*	*Nya teu kadenge ku jejebir sia teh?*
2	What's the matter, are you sick?	*Ku naon Jang, teu damang?*	*Ku naon Jang, udur?*	*Ku naon Jang, gering?*
3	I am sorry, I could not come yesterday	*Punten we kamari teh abdi mah teu tiasa sumping*	*Punten we kamari teh abdi mah teu tiasa dongkap*	*Punten we kamari teh abdi mah teu datang.*

4	Do you want to <u>eat</u>, mom?	*Bade <u>tuang</u>, bu?*	*Bade <u>neda</u>, bu?*	*Arek <u>dahar</u>, bu?*

Table 2. Sundanese language code of conduct example

No	English words	Polite	Standard	Ribaldry
1	Enter, come in, get into, get in, go into, go in, move into	*Lebet*	*Lebet*	*Abus, Asup*
2	Little brother	*Rai, rayi*	*Adi*	*Adi*
3	Compete, contend	*Aben*	*Adu*	*Adu*
4	Learn, larn, acquire	*Wulang*	*Wuruk*	*Ajar*
5	Grandfather, gramps, granddad, grandad, granddaddy, grandpa	*Tuang eyang*	*Pun aki*	*Aki*
6	Admit, acknowledge	*Angken, ngangken*	*Aku, ngaku*	*Aku, ngaku*
7	Hoop, ring	*Cingcin, lelepen*	*Ali*	*Ali*
8	Sick	*Teu damang*	*Udur*	*Gering*
9	Arrive	*Sumping, rawuh*	*Dongkap*	*Cunduk, datang*
10	Eat	*Tuang*	*Neda*	*Dahar*
11	Ear	*Cepil*	*Ceuli*	*Ceuli, Jejebir*

3. Princeton WordNet

We take WordNet definition from [6] because our goal in this research is to collect Sundanese words and classified as lexical and semantic relations. They stated that "Princeton WordNet, or simply PWN, is a lexical resource for the English language containing a large database of open-class words, i.e. nouns, verbs, adjectives, and adverbs. These words are clustered together based on their meaning into synonym sets, or synsets. Each instance of a word in a synset is marked as a separate sense of that word. PWN further specifies semantic relations between synsets and specific word senses, among others antonym, hypernym, and meronym" [6].

Further definition of relations in WordNet are described below:

a. Lexical relations, consist of synonym and antonym
   - Synonym;
     Two expressions are synonymous if the substitution of one for the other never changes the truth value of a sentence in which the substitution is made [7]. For synonym in Sundanese language is explained in table 1.
   - Antonym;
     The antonym of a word x is sometimes not-x, but not always [7]. Like in English, "not" (*henteu/teu*) can be placed before words, for Sundanese generally the same. For example, in the figure below, "sick" for Sundanese is *"teu damang"*, meanwhile "healthy" for Sundanese is *"damang"*. But like [7] said, it is not always "not-x", for example "in" in Sundanese is *"lebet"*, meanwhile "out" for Sundanese is *"kaluar"*.

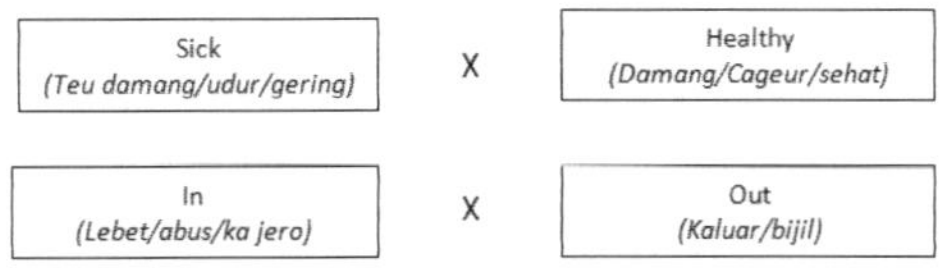

Figure 1. Antonym example

b. Semantic relations, consist of Hyponym/Hypernym and Meronym/Holonym
- Hyponymy/hypernym is a subordination/superordination, subset/superset or IS A relation [7]. Sundanese example for hyponymy/hypernym is shown in figure below

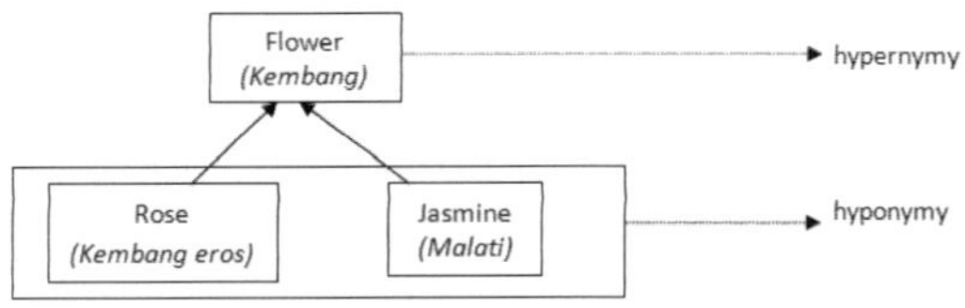

Figure 2. Hyponym/Hypernym example

- Meronym/Holonym is a part-whole (or HAS A) relation [7]. Sundanese example for meronym/holonym is shown in figure below

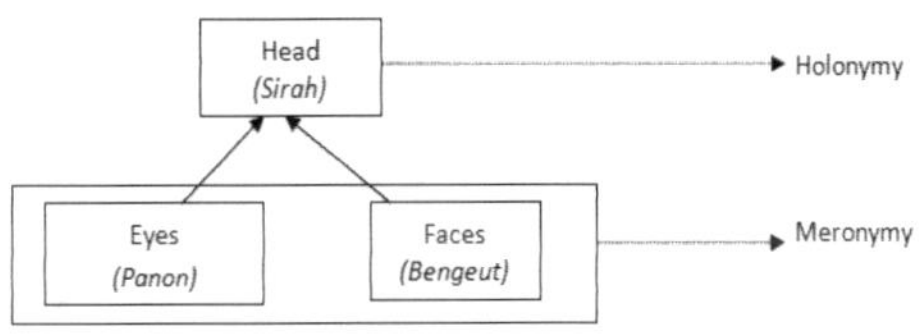

Figure 3. Meronym/Holonym example

4. Related Work
The Sundanese language research in Natural Language Processing field still minor, this is shown as there's still few research publications and basic resources tools in Sundanese for advanced end-user technologies, such as corpora, part of speech, parser, etc. Therefore, Sundanese is considered as under resource language. The first research is The Crubadan Project, where one of its research resources is Sundanese lexicons. The Crubadan project is developed by a web crawler, using simple character trigrams for language recognition. They collect the lexicon from 330 web pages and resulted in 50.000 words, 50.000 words bigram and 9.520 chartigrams [8].

5. Methodology and approach
There are two general approaches to construct a WordNet [6]:
- The expand model
  The synsets in WordNet are translated to the target language, afterward, its semantic relationships holding between synsets, are also transferred over.
- The merged model
  The synsets and semantic relationship are independently specified, afterward, its mapped into WordNet synsets through the definition of an equivalence relation.

Meanwhile, another WordNet research language use a hybrid approach [9].

a. Design and Implementation

Sunda WordNet is developed by merge approach. We choose this method as Sundanese language is rich with synonyms. In this research, first Sundanese synset is specified, further define its semantic relations, and then mapped into WordNet synset (Figure 4). As for implementation design, we choose use-case diagram to describe whole system Figure 5.

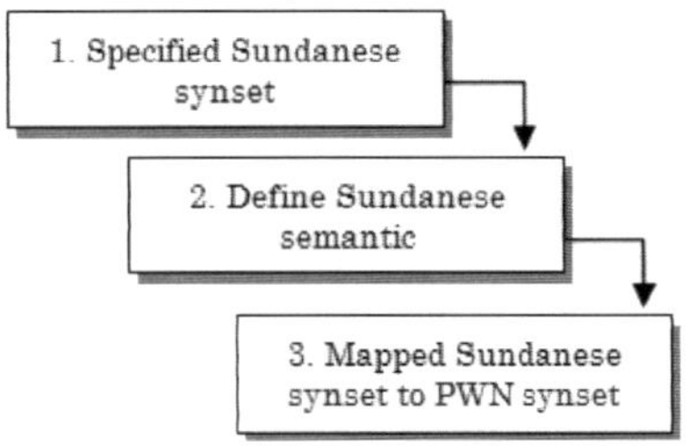

Figure 4. Merge approach for experiment

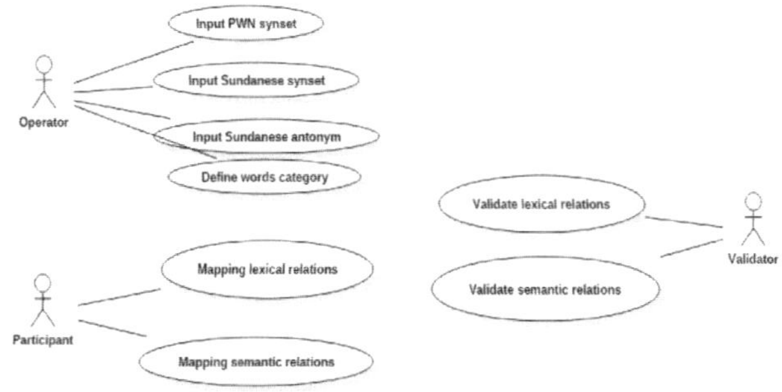

Figure 5. Use-case diagram process

i. *Specified Sundanese synset*. Sundanese synset is easier to be defined since it is based on Sundanese language code of conduct.

In Figure 6, we use a random number for synonym ID and we collect synonym words (synset) from [10] which consist of 2.259 words. Meanwhile, we collect the description for definition column from [11]. Last, example column and idKelasKata is added by participant users. idKelasKata is a field to define whether the synset is noun, verb, and adverb.

idsinonim	sinonimwords	definition	example	idkelaskata
4028818b4ed8133c014ed8172dce0001	Sirah	Kepala	Sirahna tidagor kana tihang	K001
4028818b4ed8133c014ed81b7a3b0002	Mastaka	Kepala (bahasa halus)	Dupi Ã©ta nu dina mastaka bapa sanÃ©s kaca soca ?	K001
4028818b4ed8133c014ed81d8d140003	Hulu	Kepala (bahasa kasar)	Maneh mah gede hulu	K001
4028818b4ed8133c014ed821fdf90004	Babatok	Kepala (bahasa kasar)	Maksud maneh naon neunggeul babatok aing!	K001
4028818b4ed8133c014ed8267a360005	Panon	Mata, sebagai alat penglihatan	Bulu panon	K001
4028818b4ed8133c014ed82b00800006	Soca	Mata, sebagai alat penglihatan	Kaca soca abdi ical	K001
4028818b4ed8aef2014ed8b44bec0001	Kumis	Rambut yang tumbuh di atas bibir, di bawah hidung.	Kumis Pa Camat panjang sabeulah	K001

Figure 6. Example of synonym table

ii. *Define Sundanese semantic*. Lexical antonym was defined from the same source [10] which results as 94 pair synonym-antonym. Since the result is still a very small number, we added from [12] as we show some example in Table 3. We change the synset_ID from a random number in Figure 6 to be less simple as seen in Table 3.

Table 3. Sundanese antonym mapping

Antonym_ID	Synset_ID	English Words	Sundanese Words
A01	S25	Slow	*Alon*
A02	S25	Slow	*Lalaunan*
A03	S26	Man	*Lalaki*
A04	S27	Man	*Pameget*

For semantic relations, we involved web user participant due to no resource or dictionary of Sundanese semantic words. Another objective is to promote and actively participate from Sundanese people. In Figure 7, we design the graphical user interface for the participant to choose the relation between synset, whether it is holonym-meronym or hypernym-hyponym. The example result of the relations is shown in Table 4 and Table 5.

Table 4. Sundanese hypernym-hyponym mapping

HH_ID	Parent_ID	Words	Sinonym_ID	Words
HH_01	S09	Kembang	S10	Kembang Eros
HH_02	S09	Kembang	S11	Malati

Table 5. Sundanese holonym-meronym mapping

HM_ID	Parent_ID	Words	Sinonym_ID	Words
HM_01	S12	Sirah	S18	Panon
HM_02	S12	Sirah	S21	Bengeut

Choosing the synset relation will be difficult for some people because not all Sundanese people understand the synset words. As the result, many of participant gives wrong answer and it will be affected the result. Therefore we provide a validator user who's a Sundanese expert language which will approve the answer (Figure 8).

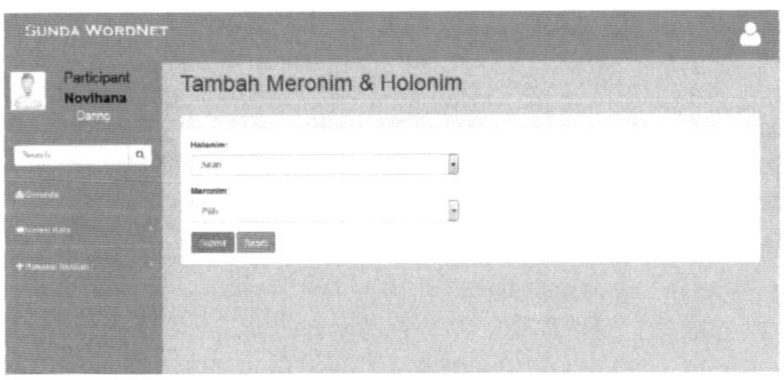

Figure 7. Synonym mapping to meronym and holonym

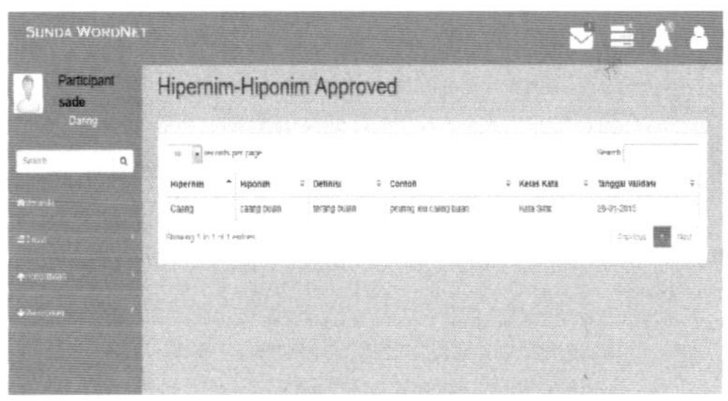

Figure 8. Validation page

iii. *Sundanese synset mapping into WordNet synset.* Next step is mapping the Sundanese synset into WordNet synset. The mapping process is done by sense of understanding from expert to minimize the error. We use WordNet description from Indonesian WordNet [6] and [7] as shown in Figure 9 and then define the Sundanese synset as shown in Figure 10. Some examples of mapping result are shown in Table 6 and Table 7. Meanwhile, the entity relationship diagram for the whole system is shown in Figure 11.

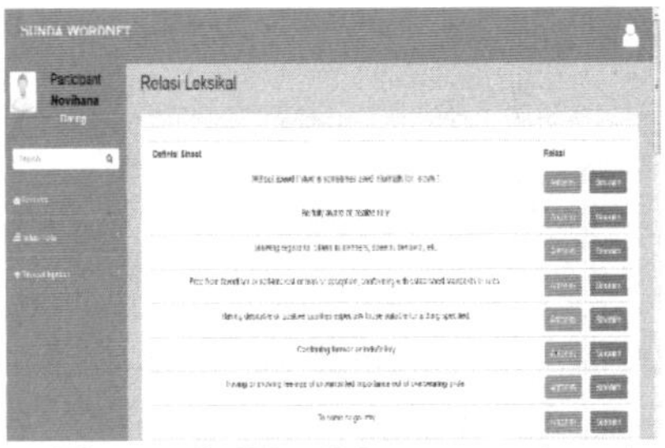

Figure 9. WordNet synset description

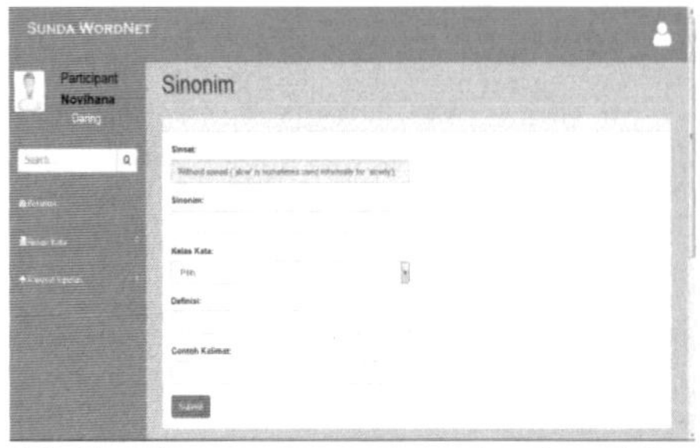

Figure 10. Sundanese synset mapping into WordNet

Table 6. WordNet synset

Synset ID	Word(s)	Definition
01755024	ageless, aeonian, eonian, eternal, everlasting, perpetual, unending, unceasing	continuing forever or indefinitely
11669921	Flower	a plant cultivated for its blooms or blossoms
12620196	rose, rosebush	any of many shrubs of the genus Rosa that bears roses

Table 7. Sundanese synset mapping

Sinonym_ID	Synset_ID	Words
S09	11669921	Kembang
S10	12620196	Kembang eros
S11	12306717	Malati
S12	05538625	Sirah
S13	055386625	Mastaka
S18	05311054	Panon
S23	05320899	Ceuli
S07	01123148	Alus
S08	01123148	Hade

In this step, we found words which still does not have its equivalence in WordNet. For example:

- *Ngaso (aso, leleson)* is a Sundanese word for having rest-activity from works. We found its equivalence as a "break" words, but after examining from WordNet synset, we concluded it is still not best equivalence or best answer.
- *Karembong (kekemben)* is a Sundanese word for clothes which is wrapped around the body, yet we do not find its equivalence in WordNet.

The example above is some words as a result of a culture, as explained by [4]:

"On the other hand, what is seen in a language will be reflected in its culture. For example, the British and other European people, who are not familiar with eating rice, only have one word, rice, to express the concept of gabah (rice in the husk), beras (uncooked rice with the husk removed), and nasi (cooked rice)." (Sudaryat, 2015)

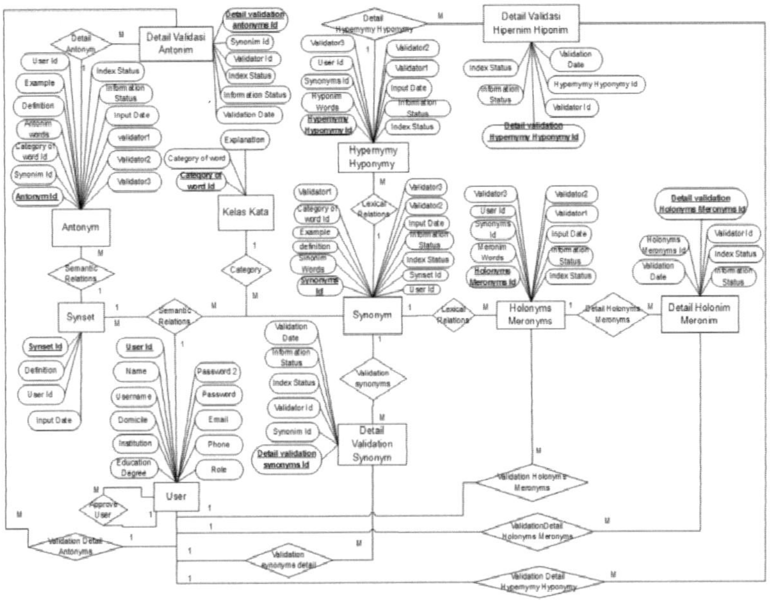

Figure 11. Sundanese entity relation diagram

6. Evaluation and future research

In this experiments the result data still rough due to several reasons

1. Participant involvement for specifying semantic relation is helpful but causing some problems, for example, many participants gives false answer, therefore validator have to check carefully. Another problem rise because it takes a lot of time due to its voluntary participation. In future research, this would might be done with simple Cartesian Product algorithm.
2. Since synset mapping from Sundanese synset to WordNet is done by expert manualy and it needs carefull attention, for future research this step might be done using the Sundanese-English dictionary as a seed to mapping the synset. In addition, web crawling can be used to collect the words or parallel sentences from web page and mapping it to WordNet using dictionary.
3. Altough we already define three words category (noun, verb and adverb) but it still an issue in this experiment since it is use participant decision which caused same problem in (1). Sundanese-English dictionary can be use in this step since most of dictionary already define whether the words is noun, verb or adverb.

7. Conclusion

Merge approach is suitable in this experiments since Sundanese has language code of conduct, which consists of three speech levels of words for same meaning. We collect 2.259 words from language code of conduct and is used as base for synset. Next, specifying semantic relations need more effort because it is still based on user's participation. In this experiment, we found words which are still not available its equivalence in WordNet, due to language is a result of a specific culture.

IOP Conf. Series: Journal of Physics: Conf. Series **971** (2018) 012044 doi:10.1088/1742-6596/971/1/012044

References

[1] Simons GF, Fennig CD. Ethnologue: Languages of the World, Twentieth edition Dallas, Texas: SIL International; 2017.

[2] Sobarna C. Bahasa Sunda Sudah Di Ambang Kematiankah? HUBS Asia. 2007 June; 11: p. 13-17.

[3] Mulyanah A. Republika Online. [Online].; 2013 [cited 2017. Available from: http://nasional.republika.co.id/berita/nasional/jawa-barat-nasional/13/08/26/ms4nkw-bahasa-sunda-terancam-punah.

[4] Sudaryat Y. The Mindset of Sundanese People in Their Language Expressions. Indonesian Journal of Applied Linguistics. 2015 July; 5: p. 126-133.

[5] S A. Tatakrama (Undak-Usuk) Basa Sunda: Bandung: Pustaka Adhigama; 1997.

[6] Putra DD, Arfan A, Manurung R. Building an Indonesian WordNet.. In In Proceedings of the 2nd International MALINDO Workshop; 2008.

[7] Miller A, Beckwith , Fellbaum , Gross , Miller. Introduction to WordNet: An On-Line Lexical Database. International Journal of Lexicography. 1993.

[8] Scannell P. The Crubadan Project: Corpus building for under-resourced languages. In In Proceedings of the 3rd Web as Corpus Workshop; 2007; Louvain-la-Neuve, Belgium.

[9] Prabhu , Desai , Redkar , Prabhugaonkar , Nagvenkar , Karmali R. An Efficient Database Design for IndoWordNet Development. In In COLING 2012 for 3rd Workshop on South and Southeast Asian Natural Language Processing (SANLP); 2012.

[10] Haerudin D. Indonesia University of Education. [Online].; 2012 [cited 2014. Available from: http://file.upi.edu/Direktori/FPBS/JUR._PEND._BAHASA_DAERAH/196408221989031-DINGDING_HAERUDIN/

[11] Danadibrata RA. Kamus Basa Sunda Bandung: PT. Kiblat Buku Utama; 2015.

[12] Tamsyah BR, Purmasih P, Supratman. Kamus Sinonim-Antonim Bahasa Sunda Bandung: Pustaka Setia; 1998.

IOP Conf. Series: Journal of Physics: Conf. Series **971** (2018) 012045 doi:10.1088/1742-6596/971/1/012045

Gold-standard evaluation of a folksonomy-based ontology learning model

E Djuana

Computer System Laboratory, Electrical Engineering Department, Faculty of Industrial Technology, Trisakti University, Grogol, Jakarta 11440, Indonesia

edjuana@trisakti.ac.id

Abstract. Folksonomy, as one result of collaborative tagging process, has been acknowledged for its potential in improving categorization and searching of web resources. However, folksonomy contains ambiguities such as synonymy and polysemy as well as different abstractions or generality problem. To maximize its potential, some methods for associating tags of folksonomy with semantics and structural relationships have been proposed such as using ontology learning method. This paper evaluates our previous work in ontology learning according to gold-standard evaluation approach in comparison to a notable state-of-the-art work and several baselines. The results show that our method is comparable to the state-of the art work which further validate our approach as has been previously validated using task-based evaluation approach.

1. Introduction

Collaborative tagging is a collective process whereby web users annotate web resources with their own keywords (tags) as users' defined metadata for those resources (Golder and Huberman, 2006; Marlow et al., 2006). Because of this process, there is emerging a kind of informal categorization system for searching and browsing of web resources which is defined by users themselves. This categorization system is known as folksonomy or folks' generated taxonomy (Peter, 2009).

Folksonomy has been acknowledged for its potential in improving categorization and searching of web resources (Peter, 2009; Bischoff et al., 2008; Robu et al., 2008). Also, there are studies which discover the potential of folksonomies for building semantic resources such as lightweight ontologies or taxonomies (Heymann and Garcia-Molina, 2006; Schmitz, 2006; Mika, 2007; Garcia-Silva, 2012). However, folksonomy may contain inherent semantic ambiguities e.g. synonymy, polysemy and generality problem (Golder and Huberman, 2006). Besides that, it has no explicit structural and semantic relationships among tags (Djuana et al, 2012). Nonetheless, since tags are contributed by users, there are many personal tags which may only be meaningful to themselves such as "chapter 1", "101", etc (Bischoff et al., 2008). All these challenges may hinder folksonomy's potential for improving search, browsing and other potential applications such as recommendation.

There are many attempts to overcome these challenges, one of which is by associating tags with semantic entities such as lexical resources, dictionary, taxonomy or ontology to make the meaning of tags explicit (Mika, 2007; Garcia-Silva, 2012). Other stream of approaches (Garcia-Silva, 2012) is using clustering based on similarity-based approach (Heymann and Garcia-Molina, 2006) or set-theoretical approach (Schmitz, 2006) to consolidate tags into one structure such as taxonomy or ontology. In this context, how to evaluate the accuracy and effectiveness of the built structures and relationships becomes a crucial issue for these entities to be useful.

This paper presents an attempt to evaluate our ontology learning method from folksonomy which has been conducted and presented in previous papers (Djuana et al, 2012, Djuana et al, 2013, Djuana et al, 2014). Previously, the result of this ontology learning method has been evaluated using task-based evaluation approach or evaluation by application which is in tag recommendation scenario. In this paper, we focus on formal evaluation using gold-standard evaluation approach (Dellschaft and Staab, 2008).

The motivation of this further evaluation is to evaluate coverage of this built ontology as it presents the usefulness of this built ontology for wider application as well as to evaluate intrinsic quality of it. The other motivation is to evaluate our method in comparison to other people's method in ontology learning including state-of-the-art method, how well our method is improving over baseline (classic) method in comparison to state-of-the-art's improvement over the baseline against gold standard universal ontology. We compare this built ontology to one notable state-of-the-art work by Liu et al. (2010) which has been compared to two notable baseline methods: Heymann and Garcia-Molina (2006) which represents similarity-based approach and Schmitz (2006) which represents set-theoretical approach.

This paper is structured as follows. In Section 2 we discuss the key concepts in ontology learning, with specific emphasis in ontology learning from folksonomy, and ontology evaluation. Section 3 discusses related works in terms of ontology learning and ontology evaluation, which discusses the two baselines and the state the art work as the benchmark. Section 4 summarizes our proposed approach from our previous papers. In Section 5 we discuss the evaluation settings and in Section 6 we discuss the evaluation results. Finally, Section 7 concludes this paper with a conclusion and direction for future work.

2. Key Concepts

In this section, we provide brief introduction on ontology and ontology learning, specific class of ontology learning from folksonomy and ontology learning evaluation approaches.

2.1. Ontology and Ontology Learning

Gruber defined that "ontology is formal description and explicit specification of a shared conceptualization" (Gruber, 1992). Depending on the types of stored knowledge, ontology can be differentiated in two types: domain ontology and general ontology (Navigli et al., 2003). General ontology defines concepts that are general for all domains while domain defines specific concepts that forms the core knowledge for one specific domain.

Construction of ontology from Web contents is one important task in Web intelligence according to Zhong and Hayazaki (2002). This work points out that ontologies serves the Semantic Web by providing a controlled vocabulary of concepts, each with explicitly defined and machine-processable semantics.

However, manual ontology construction is time consuming and very costly. Therefore, automatic and semi-automatic ontology constructions have been eagerly studied over the last decade (Maedche and Staab, 2001). One stream of approach relies on machine learning and automated language-processing techniques to extract concepts and ontological relations from structured or unstructured data such as database and text (Navigli et al., 2003).

2.2. Ontology Learning from Folksonomy

Mika stated that folksonomy which is emerging from collaborative tagging has been acknowledged as potential source for constructing ontology. As it captures vocabulary of users which may be aggregated to produce emergent semantics, people may develop lightweight ontologies (Mika, 2007). In this context, folksonomies can be referred to as a new data source for ontology learning that can be analyzed using techniques already used in this area such as clustering, natural language processing, and formal concept analysis (Garcia-Silva et al, 2012).

Construction of ontology from Web contents is one important task in Web intelligence according to Zhong and Hayazaki (2002). This work points out that ontologies serves the Semantic Web by providing a controlled vocabulary of concepts, each with explicitly defined and machine-processable semantics.

In our previous work we have taken Garcia-Silva et al (2012)'s view which describes the most relevant approaches described in the literature whose main objective is either to extract ontologies from tags in folksonomies or to associate tags to external semantic entities to make explicit the meaning of those tags. They have identified three group of approaches which are based on 1) clustering techniques i.e. to cluster tags according to some relations among them (statistical techniques); 2) ontologies i.e. aiming at associating semantic entities e.g. WordNet, Wikipedia, to tags to formally define their meaning; 3) hybrid approach i.e. mixing clustering techniques and ontologies (Djuana et al, 2012).

2.3. Ontology Learning Evaluation

In this section, we summarize the strategies for evaluating ontology learning approaches according to comprehensive evaluation strategies proposed by Dellschaft and Staab (2008).

2.3.1. Preliminaries. Dellschaft and Staab (2008) have introduced two scenarios in ontology learning evaluation which are 1) evaluation of the learning algorithm in the broader context of an automatic or semi-automatic approach to ontology engineering whereby not only the learning algorithm influences the results but also the choice of the correct corpus which must contain information relevant for the task; and 2) evaluation of the quality of the learning algorithm itself.

There are two dimensions of ontology which needs to be evaluated: functional and structural. The functional dimension of an ontology is related to its conceptualization while the structural dimension is related to the representation of an ontology as a graph. While the structural evaluation may pinpoint areas where the problems could exist, it is the functional evaluation that more crucial for evaluating the usefulness or effectiveness of the learned ontology.

Dellschaft and Staab (2008) have argued that for the first scenario, the functional dimension of an ontology should be evaluated by means of an extrinsic, task-based evaluation, i.e. in the running application for which the ontology is engineered while in the second scenario, an intrinsic or task-neutral evaluation by means of a gold-standard based evaluation is usually the better choice.

Task-based approaches (among other approaches such as corpus-based and criteria-based) is "trying to measure in how far an ontology helps to improve the results of a certain task. A task-based evaluation is influenced by many aspects which must be kept constant during all evaluations so that changes in the results can be put down to the changes in the used ontologies" (Dellschaft and Staab, 2008).

Gold-standard based approaches (as supposed to manual evaluation by human experts) "compare the learned ontology with a previously created gold standard which represents an idealized outcome of the learning algorithm. A learning algorithm is better when the learned ontology has a high similarity with the gold standard" (Dellschaft and Staab, 2008).

2.3.2. Gold-standard evaluation approach. The gold-standard based evaluation approach for evaluating ontologies may involve several measures. "They can be distinguished between measures which only evaluate the lexical layer of an ontology, the ones which also take the concept hierarchy or taxonomic layer into account and the ones which evaluate the non-taxonomic relations contained in an ontology" (Dellschaft and Staab, 2008). In this paper, we will concentrate on the measures for evaluating the lexical and the taxonomic layer.

The lexical layer or also known as coverage measure are often used for comparing the terms from the reference and the learned ontology based on an exact match of strings. Examples for this kind of measure are the Term Precision and Term Recall or also known as Lexical Precision and Recall.

The taxonomic layer or also known as relationships measure compares the similarity of the positions of two concepts in the learned and the reference hierarchy as the local measure. The global measure is then computed by averaging the results of the local measure for concept pairs from the reference and the learned ontology. It is usually calculated using local taxonomic over-lap which compares two concepts based on the set of all their super- and sub concepts. These two measures will be described in more details in Section 5.

3. Related Works

In this section, we describe related body of works in ontology learning from folksonomy data for evaluating our proposed method (Djuana et al, 2012, Djuana et al, 2013, Djuana et al, 2014) using gold-standard approach in addition to previous evaluation using task-based approach (tag recommendation).

First, we describe two classic baseline approaches by 1) Heymann and Garcia-Molina (2006) and 2) Schmitz (2006); which are well known for their effectiveness and ease of implementation in Section 3.1. Then, we describe related state-of-the-art approaches and we specifically describe a method by Liu et al. (2010) which we are using as benchmark for our ontology validation in Section 3.2.

3.1. Similarity-based approach

The algorithm published by Heymann and Garcia-Molina (2006) is classified as similarity-based approach according to classification by Liu et al (2010). The algorithm works on tag vectors whose index is equal to the number of times that a tag annotates an object. Then it calculates the similarity between tags using the cosine similarity between tag vectors to build tag similarity graph where each tag is represented by a vertex, and two vertices are connected by an edge if the similarity of the nodes they represent is above some set threshold.

In the explanation provided by Liu et al (2010), for building up the hierarchy of tags or taxonomy, "the algorithm starts with a single node tree whose only node is the "root" node representing the top of the tree. Then, it adds each tag in the tagging system to the tree in decreasing order of how central the tag is to the similarity graph described above. It decides where to put each candidate tag by computing its similarity to every node currently present in the tree, keeping track of the most similar node. The candidate tag is then either added as a child of the most similar node if its similarity to that node is greater than some threshold, or it is added to the root node if there does not currently exist a good parent for that node" (Liu et al, 2010).

3.2. Set-theoretical approach

The algorithm published by Schmitz (2006) is classified as set-theoretical approach according to classification by Liu et al (2010). It is an extension of the algorithm published by Sanderson and Croft (1999). The algorithm is based on subsumption model which partially order the concepts with the pairwise subsumption relations. The subsumption relation of two concepts is usually derived by a set-theoretical method according to the inclusion relation between their specific attribute sets, such as the occurrences of terms in documents.

It is originally defined as follows, for two terms, x and y, x is said to subsume y if the following two conditions hold:

$$P(x|y) = 1, P(y|x) < 1$$

In other words, x subsumes y if the documents which y occurs in are a subset of the documents which x occurs in. Because x subsumes y and because it is more frequent, in the hierarchy, x is the parent of y. Although a respectable number of term pairs were found that adhered to the two subsumption conditions, it was noticed that many were just failing to be included because a few occurrences of the subsumed term, y, did not co-occur with x. Subsequently, the first condition was relaxed and subsumption was redefined as:

$$P(x|y) \geq 0.8, P(y|x) < 1$$

as is described by Liu et al (2010).

Schmitz (2006) adjusting Sanderson and Croft (1999)'s statistical thresholds to reflect the ad hoc usage and adding filters to control for highly idiosyncratic vocabulary as follows:

$$\begin{aligned} P(x|y) \geq t, &\quad P(y|x) < t, \\ Dx \geq Dmin, &\quad Dy \geq Dmin, \\ Ux \geq Umin, &\quad Uy \geq Umin \end{aligned}$$

Where: t is the co-occurrence threshold, Dx is the # of documents in which term x occurs, and must be greater than a minimum value $Dmin$, and Ux is the # of users that use x in at least one annotation, and must be greater than a minimum value $Umin$.

3.3. State of the Art Approaches

According to framework proposed by Garcia-Silva et al (2012), our proposed work falls into the second group which is based on ontologies. It will be discussed in detail in Section 4.

Outside the list from survey conducted by Garcia-Silva et al (2012), we have described previously in Djuana et al (2012) that there are several recent works which tried to extract ontological structures from user tagging systems. Lin, Davis and Zhou (2009) extracted ontological structures by exploiting low support association rule mining supplemented by WordNet. Trabelsi, Jrad and Yahia (2010) focused more on extracting non-taxonomic relationships from folksonomies using triadic concepts with external resources: WordNet, Wikipedia and Google.

Tang et al. (2009) and Liu et al. (2010) represents state-of-the-art work for generating ontology from folksonomy based on generative probabilistic models i.e. tag-topic model and set-theoretical approach i.e. to produce tag subsumption graph respectively.

We chose state-of-the-art work by Liu et al. (2010) as a benchmark to evaluate our proposed approach as it is a major improvement to the subsumption models based on set-theoretical approach, by proposing to reduce noisy subsumptions including irrelevant and inconsistent paths. It is also chosen for its comprehensive evaluation which has been conducted against the classic baselines (Heymann and Garcia-Molina, 2006; Schmitz, 2006) according to evaluation framework proposed by Dellschaft and Staab (2008).

The approach by Liu et al (2010) consists of 3 steps. In the first step, it identifies subsumption tags with a set-theoretical method. Since the subsumption relations discovered in this step may be inconsistent, it then resorts to ranking the tags by generality to settle this problem. Therefore, in the second step, it constructs a tag subsumption graph to compute the generality scores of tags with a random walk based procedure. In the last step, it uses an agglomerative clustering approach, which leverages the result of the tag generality ranking procedure, to generate the concept hierarchy.

4. Proposed Model

To describe our proposed approach, we present several definitions below. This approach has been summarized here as a background for the evaluation purpose and the details has been published in previous papers (Djuana et al, 2012, Djuana et al, 2013, Djuana et al, 2014).

4.1. Definitions

4.1.1. Collaborative Tagging System. A collaborative tagging system contains three entities: users, tags, and items, which are described below:

- Users $U = \{u_1, u_2, .. u_{|U|}\}$ contains all users in an online community who have used tags to annotate their items;

- Tags $T = \{t_1, t_2, .. t_{|T|}\}$ contains all tags used by users in U. Tags are typically arbitrary strings which could be a single word or short phrase. In this respect, a tag is defined as a sequence of terms. For $t \in T$, $t =< term_1, term_2, \ldots\ldots, term_m >$, a function $tagset(t) = \{term_1, term_2, .. term_m\}$ is defined to return the terms in a tag;
- Items $I = \{i_1, i_2, .. i_{|I|}\}$ contains all domain-relevant items or resources. What is considered by an item depends on the type of collaborative tagging system, for instance, in Delicious the items are mainly bookmarks;

Based on these three entities, a collaborative tagging system is formulated as *Folksonomy* which consists of 4-tuple: $F = (U, T, I, Y)$ where U, T, I are finite sets, whose elements are the users, tags and items, respectively. Y is a ternary relation between those elements, i.e. $Y \subseteq U \times T \times I$, whose elements are called the tag assignments, whereby an element $(u, t, i) \in Y$ represents that user u annotated item i using tag t.

4.1.2. General Ontology. The general ontology is defined as a 2-tuple $GeneralONTO = (C, R)$. where $C = \{c_1, c_2, .., c_{|C|}\}$ is a set of concepts; $R = \{r_1, r_2, .., r_{|R|}\}$ is a set of relations representing the relationships between concepts.

A concept c in C is a 3-tuple $c = (id, synset, category)$ where id is a unique identification of concept c; $synset$ is a synonym set containing synonymic terms which represent the meaning of the concept c; and $category$ is a taxonomic category to classify this concept c. A relation r in the relation set R is a 3-tuple $r = (type, x, y)$, where $type \in \{is\_a,\}$; $x, y \in C$ are the concepts that hold the relation r.

Specifically, there are the set of synonyms representing c which represented as $synset(c)$ and the category of c as $category(c)$. For each term w in $synset(c)$, w is represented as a 2-tuple $(w, freq_c(w))$ where w is a synonym term of the concept c; $freq_c(w)$ is the frequency as an indication of how frequently this term has been used to represent the meaning of the concept c based on the accompanying corpus. For a term w, the set of concepts for which w is a synonymic term is defined as $con(w) = \{c | (w, f) \in synset(c)\}$.

4.1.3. Domain Ontology. The domain ontology is defined as 2-tuple $DomainOnto = (TC, TR)$. where $TC = \{tc_1, tc_2, .., tc_{|TC|}\}$ is a set of tag-concepts, i.e., $TC \subseteq C \times 2^T$, and $TR = \{tr_1, tr_2, .., tr_{|TR|}\}$ is a set of tag relations. Each element in TC is a pair of a concept c and a set of tags $\{t_1, t_2, .. t_n\}$, i.e., $tc = (c, \{t_1, t_2, .. t_n\}) \in TC$, which represents that each tag in $\{t_1, t_2, .. t_n\}$ can be mapped to concept c. TR is defined as:

$$TR = \left\{ r = (type, c_1, c_2) \middle| \begin{matrix} r \in R, \\ Concept\_Tag(c_1) \neq \emptyset, \\ Concept\_Tag(c_2) \neq \emptyset \end{matrix} \right\}$$

4.2. Ontology Learning Process

From the backbone ontology, it was expected by conducting ontology learning process; domain ontology which represents a tag collection can be generated. It is expected that this domain ontology will contain sub ontology of the backbone ontology which contextualized to the tag vocabulary and possibly personalized to users' tag usage in that collection. This sub ontology is expected to have tag to concept mapping and the taxonomic relationships between tags which extracted from concept to concept relationships in the backbone ontology.

The lexical knowledge base WordNet (Fellbaum, 1998) was chosen as the backbone ontology as it has wide coverage of concepts (over 200,000) and richness of relationships such as semantic relationships "is-a", "part-of", lexical relationships "synonymy" and "antonymy" as well as availability of accompanying corpus and other facility for disambiguation process.

We have summarized the 3 stages in the domain ontology generation process which are mapping tags to concepts, mapping disambiguation and relationships extraction.

4.2.1. Mapping Tags to Concepts. One tag may contain one or more terms. It is possible that a tag can map directly to one of synonym terms of a concept in the backbone ontology. In other cases, only part of a tag that can map to one of synonym terms.

These cases where handled by three mapping approaches which are (1) whole mapping where by whole tag string can be mapped to a synonym terms in a concept; (2) partial mapping where by partial tag string, after phrase identification stage, can be mapped to a synonym terms in the concept; and (3) term mapping where by each individual term in tag string is mapped to a synonym terms in a concept. This mapping is represented as tag to concept mapping and one tag may map to more than one concepts. Overall, for $\forall t \in T$, the tag to concept mapping is defined as follows:

$$Tag\_Concept(t) = \begin{cases} Tag\_Concept_{whole}(t), & directly\ mapped \\ Tag\_Concept_{partial}(t), & partially\ mapped \\ Tag\_Concept_{term}(t), & term\ mapped \end{cases}$$

4.2.2. Mapping Disambiguation. After all the possible mappings are found, the next stage was mapping disambiguation to choose the most appropriate concept from mapped concepts to represent the meaning of tag for this tag collection.

Two disambiguation strategies were performed which are (1) *disambiguation by frequency* which comes from an expert point of view about general meaning of tags. This mapping strength comes from frequency in a representative corpus of documents which indicate how frequent one synonym terms would be used to represent the meaning of concept that contains these terms; (2) *disambiguation by tag relevance* which comes from users' point of view about a personal meaning in the tags collection. This mapping strength comes from the tag relevance in relation to similar users' understanding and usage of tags. Given a related tag that has been used for an item, this mapping is chosen according to the relevance to other tags.

After mapping disambiguation, each tag t will be mapped to one and only one concept. This can be defined by a one to one disambiguation mapping $M_\gamma: T \rightarrow C$,

$$M_\gamma(t) = \underset{c \in Tag-Concept(t)}{argmax} (T_{C_\gamma}[t,c]) \quad (1)$$

where matrix $T\_C[t_i, c_j]_n$ is defined to represent the strength of the mapping between tags and concepts, where $m=|T|$ and $n=|C|$ and γ is a mapping disambiguation strategy.

On the other hand, multiple tags may also be mapped to one concept. The following function defines the mapping from a concept to tags: $Concept\_Tag: C \rightarrow 2^T$,

$$Concept\_Tag(c) = \{t | \forall t \in \mathrm{T}, M_\gamma(t) == c\}$$

At the end, the confirmed mapping according to two disambiguation strategies were: $M_{frequency}(t)$ and $M_{relevance}(t)$.

4.2.3. Relationships Extraction. Once mapping tags to concepts and mapping disambiguation processes are completed, each tag will map to a concept on the backbone ontology. Based on tag to concept mappings, available relationships ("is-a" relation) among concepts in general ontology were extracted to form the domain ontology.

5. Evaluation

5.1. Evaluation Method

To compare the intrinsic quality of our proposed method we use the gold standard evaluation approach as discussed in Section 2.3.2. For task-based evaluation results please refer to our previous papers (Djuana et al, 2012, Djuana et al, 2013, Djuana et al, 2014).

Following gold-standard ontology chosen by Liu et al. (2010), we use the concept hierarchy from Open Directory Project (ODP)[1]. ODP is a free, user-maintained hierarchical web directory. Each node in the ODP hierarchy has a topic label (e.g. Sports or Arts) and a set of associated URLs. ODP is generated by collaborating users.

In the following the simplified definition of a core ontology will be used. This definition of an ontology only contains the lexical layer and the concept hierarchy. In Dellschaft and Staab (2008) a core ontology is defined as follows: The structure $\mathcal{O} := (\mathcal{C}, \mathrm{root}, \prec \mathrm{c})$ is called a core ontology. $\mathcal{C}$ is a set of concept identifiers and root being a designated root concept for the partial order $\prec$ c on $\mathcal{C}$. This partial order is called concept hierarchy or taxonomy. The equation $\forall \mathrm{c} \in \mathcal{C}: \mathrm{c} \prec \mathrm{root}$ holds for this concept hierarchy.

Given a computed core ontology $\mathcal{O}_C$ and a reference ontology $\mathcal{O}_R$, the lexical precision (LP) and lexical recall (LR) are defined as follows:

$$LP(\mathcal{O}_C, \mathcal{O}_R) = \frac{|\mathcal{C}_C \cap \mathcal{C}_R|}{|\mathcal{C}_C|} \quad (2)$$

$$LR(\mathcal{O}_C, \mathcal{O}_R) = \frac{|\mathcal{C}_C \cap \mathcal{C}_R|}{|\mathcal{C}_R|} \quad (3)$$

To compare between the gold standard and the learned ontology, each sub tree which starts with different ODP topic label will be compared to its corresponding ODP sub tree using lexical precision and lexical recall. The higher the value of precision and recall the more accurate the learned ontology to the gold standard ontology.

5.2. Experiment Setup

Dataset and Experiment Run. One public *folksonomy* dataset was used for the experiment. We use the subset of Delicious dataset provided by Wetzker et al (2008) which contains all public bookmarks of users posted on delicious.com between September 2003 and December 2007. In our experiment, we use the data between September 2003 and July 2005. We also perform a filtering for the dense part of the *folksonomy* using p-core calculation according to Batagelj and Zaversnik (2002). This p-core 15 calculation reduces the folksonomy population to tags, user, and items which appear in at least 15 posts. The overall statistics is presented in Table 1.

Table 1. Delicious dataset statistics

	All	**Filtered by p-core=15**
#users	75,245	24,562
#items	3,158,4	45,793
#tags	456,697	9,718
#posts	7,698,6	1,436,52

We have implemented the two baselines by Heyman and Garcia-Molina (2006) and Schmitz (2006) and ran their algorithms with our dataset and produced two versions of learned ontology structures. We have run our proposed algorithm with the same dataset and produced a version of learned ontology structure. For each of this learned ontology we are calculating lexical precision and recall against the gold standard.

We did not implement Liu's method, but instead we are using results from Liu et al (2010)'s paper directly since they are also comparing against the same two baselines. To compare with Liu's, we are measuring the percentage of improvement from our results in comparison to the two baselines using our dataset with the percentage of improvement in Liu et al (2010)'s paper. By comparing the percentage of improvement, we are indirectly comparing the accuracy of the two algorithms.

[1] http://dmoz.org

Table 2. Improvement Results between the Proposed Method to Baselines

Sub tree	Lexical Precision					Lexical Recall				
	Ours	Heymann	Schmitz	% impr. Ours vs Heymann	% impr. Ours vs Schmitz	Ours	Heymann	Schmitz	% impr. Ours vs Heymann	% impr. Ours vs Schmitz
sport	0.7196	0.7289	0.6794	-0.0128	0.0591	0.0507	0.0227	0.0439	1.2368	0.1543
science	0.5065	0.2012	0.4152	1.5172	0.2199	0.0171	0.0543	0.0872	-0.6854	-0.8041
news	0.2430	0.0478	0.0967	4.0822	1.5141	0.0681	0.0523	0.3020	0.3027	-0.7746
program	0.3738	0.2089	0.3019	0.7895	0.2382	0.3422	0.0739	0.0243	3.6334	13.0933
history	0.3060	0.4241	0.3550	-0.2785	-0.1381	0.0457	0.0526	0.1368	-0.1303	-0.6657
book	0.1312	0.0424	0.0629	2.0961	1.0867	0.0109	0.0280	0.0251	-0.6127	-0.5678
culture	0.2295	0.0051	0.1295	44.4153	0.7729	0.1340	0.0579	0.1868	1.3150	-0.2825
computers	0.4969	0.1600	0.5212	2.1056	-0.0465	0.0174	0.0290	0.0262	-0.4000	-0.3349
game	0.5049	0.1751	0.4695	1.8837	0.0754	0.1474	0.0456	0.1225	2.2346	0.2028
education	0.0612	0.0048	0.0118	11.7031	4.1996	0.0205	0.0531	0.0191	-0.6139	0.0765
resources	0.9180	0.0023	0.3019	392.5994	2.0406	0.0529	0.5025	0.0491	-0.8947	0.0765
media	0.4590	0.0114	0.2588	39.1632	0.7736	0.0943	0.1899	0.0584	-0.5034	0.6155
shop	0.4432	0.2042	0.4063	1.1701	0.0909	0.0943	0.0258	0.0545	2.6584	0.7290
graphic	0.9180	0.0389	0.0476	22.6160	18.2665	0.0209	0.0181	0.0389	0.1584	-0.4631
health	0.4590	0.3563	0.5020	0.2882	-0.0857	0.0461	0.0461	0.1038	0.0013	-0.5553
all	0.4465	0.0503	0.2668	7.8770	0.6738	0.0520	0.0536	0.0491	-0.0310	0.0580
average	0.4513	0.1741	0.3040	34.9426	1.9378	0.0775	0.0834	0.0852	0.5133	0.7000

Table 3. Improvement Results between the State-of-the-Art Method to Baselines

Sub tree	Lexical Precision					Lexical Recall				
	Liu	Heymann	Schmitz	% impr. Liu vs Heymann	% impr. Liu vs Schmitz	Liu	Heymann	Schmitz	% impr. Liu vs Heymann	% impr. Liu vs Schmitz
sport	0.7838	0.9375	0.7500	-0.1639	0.0451	0.0504	0.0261	0.0470	0.9310	0.0723
science	0.5517	0.2588	0.4583	1.1318	0.2038	0.0170	0.0626	0.0934	-0.7284	-0.8180
news	0.2647	0.0615	0.1067	3.3041	1.4808	0.0677	0.0602	0.3233	0.1246	-0.7906
program	0.4162	0.2687	0.3333	0.5489	0.2487	0.3404	0.0851	0.0260	3.0000	12.0923
history	0.3333	0.5455	0.3919	-0.3890	-0.1495	0.0455	0.0606	0.1465	-0.2492	-0.6894
book	0.1429	0.0545	0.0694	1.6220	1.0591	0.0108	0.0323	0.0269	-0.6656	-0.5985
culture	0.2500	0.0065	0.1429	37.4615	0.7495	0.1333	0.0667	0.2000	0.9985	-0.3335
computers	0.5532	0.2058	0.5753	1.6880	-0.0384	0.0173	0.0334	0.0280	-0.4820	-0.3821
game	0.5621	0.2252	0.5183	1.4960	0.0845	0.1466	0.0525	0.1312	1.7924	0.1174
education	0.0667	0.0062	0.0130	9.7581	4.1308	0.0204	0.0612	0.0204	-0.6667	0.0000
resources	1.0000	0.0030	0.3333	332.3333	2.0003	0.0526	0.5789	0.0526	-0.9091	0.0000
media	0.5000	0.0147	0.2857	33.0136	0.7501	0.0938	0.2188	0.0625	-0.5713	0.5008
shop	0.4828	0.2627	0.4485	0.8378	0.0765	0.0938	0.0297	0.0584	2.1582	0.6062
graphic	1.0000	0.0500	0.0526	19.0000	18.0114	0.0208	0.0208	0.0417	0.0000	-0.5012
health	0.5000	0.4583	0.5542	0.0910	-0.0978	0.0459	0.0531	0.1111	-0.1356	-0.5869
all	0.4864	0.0647	0.2945	6.5178	0.6516	0.0517	0.0618	0.0526	-0.1634	-0.0171
average	0.4938	0.2239	0.3356	29.4489	1.9036	0.0771	0.0961	0.0913	0.3065	0.5793

IOP Conf. Series: Journal of Physics: Conf. Series **971** (2018) 012045 doi:10.1088/1742-6596/971/1/012045

6. Results and Discussion

The results for our proposed method are presented in Table 2 while the results from Liu et al (2010)'s paper are copied in Table 3 for easy comparison.

The "average" value is the average over all the sub trees but not include the all root. The "all" root is when all of sub trees are connected which means for all overlap sub trees, the overlapped part will be merged.

Overall results for the proposed method are better against the two baselines with the improvement over Heymann and Garcia algorithm is higher than the improvement over Schmitz algorithm. In this dense part of folksonomy, the proposed method is also following the trend shown by Liu et al's method because the absence of all idiosyncratic and personal tags.

Overall results also show that the proposed methods in average and all situations are better in terms of coverage which are shown by higher value of lexical precision and recall. The improvement in recall value are higher than in precision which shows that the proposed method is better in coverage against Liu et al's method.

However, as we compare the percentage of improvement over Schmitz algorithm, there are several sub trees such as "program" (programming), "computers" and "game" which has lower improvement of precision value than Liu et al.'s method, although comparing to Heymann algorithm the improvement is still better. It is suspected that since the proposed method is partly based on WordNet which is contains more general vocabularies in comparison to ODP, these more technical sub trees were not performed that well.

In this evaluation, we haven't included the taxonomic evaluation in terms of taxonomic precision and recall which evaluates the closeness of relationships structure of the learned ontology to the gold standard. At this stage, we can conclude that based on coverage alone, the proposed method is better and for overall evaluation, the proposed method may be comparable. This is subject to taxonomic evaluation to be conducted as future work.

7. Conclusions

We have presented an attempt to evaluate the proposed ontology learning approach using gold-standard evaluation approach, which previously evaluated using task-based evaluation approach. The lexical evaluation (coverage) show a positive indication that the proposed method is better than notable state-of-the-art method. This is subject to a taxonomic (relationships) evaluation which is to be conducted as a near future work.

References

[1] Batagelj, V. and Zaversnik, M. 2002. Generalized cores, *Arxiv preprint cs/0202039.*

[2] Bischoff, K., Firan, C.S., Nejdl, W. and Paiu, R. 2008. Can all tags be used for search? In *Proceedings of the 17th ACM Conference on Information and Knowledge Management,* ACM, New York, NY, 193-202.

[3] Dellschaft, K. and Staab, S. 2008. Strategies for the evaluation of ontology learning, In *Proceedings of the 2008 conference on Ontology Learning and Population: Bridging the Gap between Text and Knowledge*, IOS Press, Amsterdam, The Netherlands, 253-272.

[4] Djuana, E., Xu, Y. and Li, Y. 2012. Learning personalized tag ontology from user tagging information. In *Proceedings of the Tenth Australasian Data Mining Conference* (Sydney, Australia, December 05 - 07, 2012). AusDM 2012. ACS, Sydney, Aus, 183 - 190.

[5] Djuana, E., Xu, Y., Li, Y and Cox, C. 2012. Personalization in tag ontology learning for recommendation making. In *Proceedings of the 14th International Conference on Information Integration and Web-based Applications & Services* (Denpasar, Bali, December 03 - 05, 2012). iiWAS '12. ACM, New York, NY, 368-377.

[6] Djuana, E., Xu, Y., Li, Y., Josang, A. and Cox, C. 2013. An ontology based method for sparsity problem in tag recommendation. In *Proceedings of the 15th International*

Conference on Enterprise Information Systems (Angers, France, July 04 - 07, 2013). ICEIS 2013. SCITEPRESS, INSTICC, Portugal, 467-474.

[7] Djuana, E., Xu, Y., Li, Y., and Josang, A. 2014. A Combined Method for Mitigating Sparsity Problem in Tag Recommendation. In *Proceedings of the 47th Hawaii International Conference on System Sciences* (Waikoloa, HI, USA, January 6-9, 2014). IEEE Computer Society 2014, 906-915.

[8] Fellbaum, C. (ed.), 1998. *WordNet: An Electronic Lexical Database*, Cambridge, MA: MIT Press.

[9] García-Silva, A., Corcho, O., Alani, H. and Gómez-Pérez, A. 2012. Review of the state of the art: discovering and associating semantics to tags in folksonomies. *The Knowledge Engineering Review*. 27, 1 (Feb. 2012), Cambridge University Press, 57-85.

[10] Golder, S.A. and Huberman, B.A. 2006. Usage patterns of collaborative tagging systems. *J. Inf. Sci.* 32, 2 (Apr. 2006), 198-208.

[11] Gruber, T.R. 1993. A translation approach to portable ontology specifications. *Knowledge Acquisition*, 5, 2, 199-220.

[12] Heymann, P. and Garcia-Molina, H. 2006. *Collaborative creation of communal hierarchical taxonomies in social tagging systems*. Technical Report 2006-10. Stanford University.

[13] Lin, H., Davis, J. and Zhou, Y. 2009. An integrated approach to extracting ontological structures from folksonomies, *The Semantic Web: Research and Applications*, Springer, 654-668.

[14] Liu, K., Fang, B. and Zhang, W. 2010. Ontology emergence from folksonomies, In *Proceedings of the 19th ACM International Conference on Information and Knowledge Management*, ACM, New York, NY, 1109-1118.

[15] Maedche, A. and Staab, S. 2001. Ontology learning for the semantic web. *IEEE Intelligent Systems*, 16, 2, (March 2001), IEEE, NJ, US, 72-79.

[16] Marlow, C., Naaman, M., Boyd, D. and Davis, M. 2006. HT06, tagging paper, taxonomy, Flickr, academic article, to read, In *Proceedings the Seventeenth Conference on Hypertext and Hypermedia*, ACM, New York, NY, 31-40.

[17] Mika, P. 2007. Ontologies are us: A unified model of social networks and semantics. *Web Semantics: Science, Services and Agents on the World Wide Web.* 5, 1 (March. 2007), 5-15.

[18] Navigli, R., Velardi, P. and Gangemi, A. 2003. Ontology learning and its application to automated terminology translation. *IEEE Intelligent Systems*, 18, 1, (Jan 2003), IEEE, NJ, US, 22-31.

[19] Peters, I. 2009. *Folksonomies. Indexing and Retrieval in Web 2.0.* De Gruyter Saur, Berlin, Germany.

[20] Robu, V., Halpin, H., and Shepherd, H. 2009. Emergence of consensus and shared vocabularies in collaborative tagging systems. *ACM Trans. Web.* 3, 4 (Sep. 2009), 14:1-34.

[21] Sanderson, M. and Croft, B. 1999. Deriving concept hierarchies from text. In *Proceedings of the 22nd Annual International ACM SIGIR conference on Research and Development in Information Retrieval*. SIGIR'99, 206-213.

[22] Schmitz, P. 2006. Inducing ontology from flickr tags. In *Proceedings of the Collaborative Web Tagging Workshop at WWW'06,* (Edinburgh, Scotland, May 23 - 26, 2006).

[23] Tang, J., Leung, H., Luo, Q., Chen, D. and Gong, J. 2009. Towards ontology learning from folksonomies, In *Proceedings 21st International Joint Conference on Artificial Intelligence*, 2089-2094.

[24] Trabelsi, C., Jrad, A.B., and Yahia, S.B. 2010. Bridging folksonomies and domain ontologies: Getting out non-taxonomic relations", In *Proceedings IEEE International Conference on Data Mining Workshops*, IEEE, 369-379.

[25] Wetzker, R., Zimmermann, C. and Bauckhage, C. 2008. Analyzing social bookmarking systems: A del.icio.us cookbook, In *Proceedings of European Conference on Artificial Intelligence (ECAI).*

[26] Zhong, N. and Hayazaki, N., 2002. Roles of ontologies for web intelligence. In M.-S. Hacid, Z. Ras, D. Zighed & Y. Kodratoff (Eds.), *Foundations of Intelligent Systems*, 2366, Springer Berlin, Heidelberg, 569-573.

IOP Conf. Series: Journal of Physics: Conf. Series **971** (2018) 012046 doi:10.1088/1742-6596/971/1/012046

Classification of hadith into positive suggestion, negative suggestion, and information

Said Al Faraby, Eliza Riviera Rachmawati Jasin, Andina Kusumaningrum, Adiwijaya

School of Computing, Telkom University

E-mail: saidalfaraby@telkomuniversity.ac.id, eliza@jasin.net, andinakusumaningrum@gmail.com, adiwijaya@telkomuniversity.ac.id

Abstract. As one of the Muslim life guidelines, based on the meaning of its sentence(s), a hadith can be viewed as a suggestion for doing something, or a suggestion for not doing something, or just information without any suggestion. In this paper, we tried to classify the Bahasa translation of hadith into the three categories using machine learning approach. We tried stemming and stopword removal in preprocessing, and TF-IDF of unigram, bigram, and trigram as the extracted features. As the classifier, we compared between SVM and Neural Network. Since the categories are new, so in order to compare the results of the previous pipelines, we created a baseline classifier using simple rule-based string matching technique. The rule-based algorithm conditions on the occurrence of words such as "*janganlah, sholatlah, and so on*" to determine the category. The baseline method achieved F1-Score of 0.69, while the best F1-Score from the machine learning approach was 0.88, and it was produced by SVM model with the linear kernel.

1. Introduction

A hadith is a report that describes words, actions, or habits of Muslim Prophet, Muhammad [1]. There are many collections of hadith compiled by different scholars, some of them are more famous than the others. Hadith is the second reference to Islamic laws for Muslim after Qur'an. Usually, there are two types of content regarding a law, which are positive suggestion and negative suggestion of doing something. However, in hadith, there are also contents that do not fall under the two types, which will be categorized into information. The task of categorizing hadith into aforementioned types will benefit bigger system such as information retrieval or recommender system of hadith, because the system can suggest not only other hadith in the same topic, but also more specifically in the same type of either positive suggestion, negative suggestion, or just general information. Moreover, this kind of categorization is not a special task for hadith, but can be done also for other kinds of documents, for example, agreement documents.

One of the well-known hadith classification tasks is to decide if a hadith is authentic or not. This type of classification requires experts in the area of hadith science [2]. On the other hand, we try to position our hadith classification task into a common natural language text classification task, where the label of a text is determined only by its content without relying too much on the external knowledge. Also, we consider that common people are able to decide the appropriate label of a hadith given that they understand the language used by the hadith.

Our contribution through this paper is providing preliminary results on a new kind of text classification namely positive suggestion, negative suggestion, and information, especially text data from hadith translation in Bahasa.

2. Related Work

2.1. Hadith Authenticity

Usually, classification of hadith done by many Muslim scholars is to decide whether a hadith is authentic or unauthentic. There are common criteria in deciding the authenticity which are the reference to a particular authority, links of Isnad, number of reporters involved in each stage of Isnad, nature of the text and Isnad, and reliability and memory of the reporters [3]. To evaluate all of the criteria requires deep knowledge in the hadith science. Moreover, this is a very sensitive and important task, thus it should be done only by an expert in the hadith science, and it usually does not involve any computational method.

Nevertheless, there was a quite rare research to build an expert system for this type of classification, although the authors already stated that the purposes were to assist an expert in making the decision or to help non-expert to understand how the rules behind a decision [2]. Another similar research tried to focus on handling missing data in the hadith database and experimented with a decision tree and naive bayes as the classifiers [4]. On the other hand, our research goes to a different type of classification and is a bit less sensitive than deciding authenticity. Therefore, we can not relate our work very well to their research.

2.2. Text Classification

Text classification has been studied widely for decades. Joachims showed that SVM was superior to some other classification algorithms including Bayes, C.45, and KNN where tested on Reuters news dataset [5]. Besides SVM, Neural Network is another algorithm that performs good results on text classification [6, 7, 8]. Regarding text classification in Bahasa, there is previous research on lyric classification in Bahasa using Naive Bayes, which shows F1-Score of 0.89 [9]. Naradhipa and Purwarianti experimented with Naive Bayes, Maximum Entropy, and SVM to classify sentiment for product review in Bahasa, and the best accuracy of 81.6% was produced by SVM model [10]. Based on those previous results, we choose to experiment with SVM and Neural Network as the classifiers reported in this paper.

3. Data and Method

3.1. Hadith Dataset

Hadith dataset used in this work is the translation from Arabic to Bahasa, and they were collected from Imam Bukhari Collection, which is part of "Kitab Hadits 9 Imam" provided by Lidwa application[1]. The original dataset has no labels and still contain the *"isnad"* (list of scholars' names who reported the hadith). Data preparation was needed to separate and remove the *isnad* from the actual content of the hadith. In order to create a proper dataset for training and testing, each hadith was labeled manually by authors and colleagues[2]. There is a total of 1650 hadiths were labeled, 550 for each class of positive suggestion, negative suggestion, and information. Samples of hadith from each class are given in table 1.

Note that the keyword in the positive suggestion hadith is *"hendaklah"* (let him), while in negative suggestion is *"janganlah"* (do not). By identifying these keywords, we created a baseline classifier using rule-based string matching which will be elaborated in section 3.

[1] www.lidwa.com

[2] Aknowledgement for Andina, Eliza, Syair, and Fauzan for their contributions in labeling the hadiths

Table 1. Sample of hadith.

Class	Sample Hadith
Positive Suggestion	Jika seorang dari kalian memasuki masjid sedang imam sedang berkhuthbah atau dia telah keluar (kemudian masuk lagi) maka hendaklah dia shalat dua raka'at. English translation: *If one of you entered the mosque while the imam is preaching or he has come out (then come back) then let him pray two rak'ahs*
Negative Suggestion	Janganlah kalian berdusta terhadapku (atas namaku), karena barangsiapa berdusta terhadapku dia akan masuk neraka. English translation: *Do not tell lies about me (on my behalf), for whoever tells lies about me will enter Hell*
Information	Bukanlah disebut miskin orang yang bisa diatasi dengan satu atau dua suap makanan. Akan tetapi yang disebut miskin adalah orang yang tidak memiliki kecukupan namun dia menahan diri (malu) atau orang yang tidak meminta-minta secara mendesak. English translation: *The poor person is not the one who asks a morsel or two (of meals) from the others, but the poor is the one who has nothing and is ashamed to beg from others*

3.2. Method

Methods used in the experiment involve three main stages, preprocessing, feature extraction, and classification.

Preprocessing

As the common practice in text processing, the experiments began with preprocessing the dataset with one or more of these techniques: punctuation removal, case folding, stemming[3], and stopword removal.

Feature Extraction

The default feature extraction method used in our experiments was the combination of unigram and TF-IDF (Term Frequency-Inverse Document Frequency). But we also tried with different feature representations such as normal bag-of-word, bigram, and also trigram [11].

Baseline Classifier

To the best of our knowledge, the classes (i.e positive suggestion, negative suggestion, and information) and the dataset used in this work have not been studied before, so we do not have baseline results to compare with. Instead, we create our own baseline classification model using a simple rule-based and string matching. The rules are derived from a quick observation of the sample dataset, as described by Algorithm 1.

[3] https://github.com/sastrawi/sastrawi

```
if one or more words in Negative Suggestion list appear in hadith then
 | classify as Negative Suggestion;
else if one or more words in Positive Suggestion list appear in hadith then
 | classify as Positive Suggestion
else
 | classify as Information
end
```

Algorithm 1: Rule-based classifier for the baseline

where the examples of words in each category list are given by Table 2.

Table 2. Example of words in Negative Suggestion and Positive Suggestion list.

Negative Suggestion	Positive Suggestion
jangan (do not)	hendaklah (let)
larang (forbid)	sholatlah (do pray)
tidak boleh (should not)	bersedekahlah (do give alm)
	Lakukanlah (do)
	Berpuasalah (do fasting)
	Jagalah (take care)

Support Vector Machine
Originally, Support Vector Machine (SVM) is a non-probabilistic classifier that only handles linear binary classification [12]. Since we have three classes, we have to build a multiclass SVM, in this case using one vs all method. First, we introduce a default SVM model called $SVM_{default}$, which uses Linear kernel, and C (trade-off parameter) value of 1. Later in the experiments, we tried various combinations of C value, Kernel method, and the kernel specific parameter.

Neural Network
As for Neural Network, our $ANN_{default}$ is configured with 10 hidden layers, and 10 neurons at each of the layers [13]. An alternative configuration we tried consisted of 20 hidden layers with 10 neurons at each of the layers.

4. Experiments and Results
Models were evaluated using 5-fold cross validation[14], so that at each fold 80% of the dataset was used for training and the remaining 20% for testing. The evaluation metric used throughout the experiments was macro-average F1-score [15].

4.1. Default models
In this experiment, we compared three classifiers on 5-fold cross validation settings. Preprocessing and feature extraction steps were not needed for baseline model. Meanwhile, for $SVM_{default}$ and $NN_{default}$, both preprocessing and feature extraction were set using combination of unigram and TF-IDF. Macro-average F1-score for the three models are presented in table 3.

Table 3. F1-Score for different classification models.

Model	F1-Score
Baseline	0.69
$\text{SVM}_{\text{default}}$	0.88
$\text{ANN}_{\text{default}}$	0.79

As we can see, our Baseline model just scored 0.69, which is far below the others. It means that the simple rule string matching classifier is not enough to predict the labels, and we need a more sophisticated classifier to do better. One of the examples where simple rules will not work is given in table 4.

Table 4. Missclassified hadith.

Bahasa	Barangsiapa berdusta terhadapku maka **hendaklah** ia persiapkan tempat duduknya di neraka.
English	*Whoever lies against me then* ***let him*** *prepare his seat in hell.*

If only we looked at the keyword, it would be classified as a positive suggestion, but the actual meaning of the hadith is a negative suggestion. This is a similar problem with sentiment analysis when processing sarcastic sentence[16].

4.2. Kernels

$\text{SVM}_{\text{default}}$ used Linear Kernel, so in this experiment, we want to know how different kernel influence the F1-Score of the SVM model. We tried Linear, Polynomial, and RBF kernels with various combination of degree (polynomial) and γ (rbf) values with fix C=1. In table 5, we reported the best result of each kernel.

Table 5. F1-Score for different SVM Kernels.

Kernel	F1-Score
Linear	0.8839
Polynomial	0.8779
RBF	0.8758

The best F1-score was produced by the linear kernel, not the more complex kernel like polynomial and RBF. The reason is probably that because of the number of features is a lot, so mapping the data point to a higher dimension will not help anymore [17].

4.3. Feature Representation
In this experiment, we tried different feature representation, namely unigram, bigram, and trigram which will be combined with TF-IDF weighting score.

Table 6. F1-Score for different feature representations.

Feature	F1-Score
Unigram	0.79
Bigram	0.70
Trigram	0.48

As shown in table 6, bigram and trigram representation did not help to improve what was achieved by unigram model.

4.4. Network Architecture
To investigate the potential improvement of Neural Network model, we construct a more complex architecture that contains 20 hidden layers and 10 neurons at each layers. The comparison with the NN with 10 hidden as default can be seen in table 7.

Table 7. F1-Score for different network architecture.

Architecture	F1-Score
10 hidden layers @ 10 neurons	0.79
20 hidden layers @ 10 neurons	0.85

With changing the architecture of NN, the performance changed significantly. In this case, an improvement of 0.06 point to the F1-Score. The need of more hidden layers can be related to the high input dimension of word features.

5. Conclusion
Classifying text into positive suggestion, negative suggestion, and information can be handled as normal text classification. In our experiments, SVM with linear kernel achieved the best F1-Score of 0.88. The classification problem is not as simple as we thought, as the simple rule-based string matching classifier did not perform well. The future work might be to investigate different feature extraction method, since suffixes (e.g lah in makanlah, sholatlah) play an important role in the meaning of the text, especially in positive suggestion class.

References
[1] Brown JA. Hadith: Muhammad's legacy in the medieval and modern world. Oneworld Publications; 2009 Jun 1.
[2] Bilal K, Mohsin S. Muhadith: A cloud based distributed expert system for classification of ahadith. InFrontiers of Information Technology (FIT), 2012 10th International Conference on 2012 Dec 17 (pp. 73-78). IEEE.
[3] Suhaib H. An introduction to the science of Hadith. London: Al-Quran Society, 1994.
[4] Aldhaln K, Zeki A, Zeki A, Alreshidi H. Improving knowledge extraction of Hadith classifier using decision tree algorithm. InInformation Retrieval & Knowledge Management (CAMP), 2012 International Conference on 2012 Mar 13 (pp. 148-152). IEEE.

[5] Joachims T. Text categorization with support vector machines: Learning with many relevant features. Machine learning: ECML-98. 1998:137-42.
[6] Aggarwal CC, Zhai C. A survey of text classification algorithms. Mining text data. 2012:163-222.
[7] Zhang ML, Zhou ZH. Multilabel neural networks with applications to functional genomics and text categorization. IEEE transactions on Knowledge and Data Engineering. 2006 Oct;18(10):1338-51.
[8] Sebastiani F. Machine learning in automated text categorization. ACM computing surveys (CSUR). 2002 Mar 1;34(1):1-47.
[9] Rifqi AA, Mubarok MS, Adiwijaya. Klasifikasi Topik pada Lirik Lagu dengan Metode Multinomial Naive Bayes. Indonesia Symposium on Computing (IndoSC) 2016. 2016.
[10] Naradhipa AR, Purwarianti A. Sentiment classification for Indonesian message in social media. InCloud Computing and Social Networking (ICCCSN), 2012 International Conference on 2012 Apr 26 (pp. 1-5). IEEE.
[11] Jurafsky D, Martin JH. Speech and language processing. London: Pearson; 2014 Dec 30.
[12] Cortes C, Vapnik V. Support-vector networks. Machine learning. 1995 Sep 1;20(3):273-97.
[13] Demuth HB, Beale MH, De Jess O, Hagan MT. Neural network design. Martin Hagan; 2014 Sep 1.
[14] Kohavi R. "A study of cross-validation and bootstrap for accuracy estimation and model selection." Ijcai. Vol. 14. No. 2. 1995.
[15] Yang Y. An evaluation of statistical approaches to text categorization. Information retrieval. 1999 Apr 1;1(1):69-90.
[16] Maynard D, Mark AG. Who cares about Sarcastic Tweets? Investigating the Impact of Sarcasm on Sentiment Analysis. LREC. 2014.
[17] Hsu CW, Chang CC, Lin CJ. A practical guide to support vector classification.

Semantic text relatedness on Al-Qur'an translation using modified path based method

Yudi Irwanto[1], Moch Arif Bijaksana[2], Adiwijaya[3]

School Of Computing, Telkom University, Indonesia[1,2,3]

yuditube@gmail.com[1], arifbijaksana@telkomuniversity.ac.id[2], adiwijaya@telkomuniversity.ac.id[3]

Abstract. Abdul Baquee Muhammad [1] have built Corpus that contained *AlQur'an* domain, *WordNet* and dictionary. He has did initialisation in the development of knowledges about *AlQur'an* and the knowledges about relatedness between texts in *AlQur'an*. The Path based measurement method that proposed by Liu, Zhou and Zheng [3] has never been used in the *AlQur'an* domain. By using *AlQur'an* translation dataset in this research, the path based measurement method proposed by Liu, Zhou and Zheng [3] will be used to test this method in *AlQur'an* domain to obtain similarity values and to measure its correlation value.
In this study the degree value is proposed to be used in modifying the path based method that proposed in previous research. Degree Value is the number of links that owned by a lcs (lowest common subsumer) node on a taxonomy. The links owned by a node on the taxonomy represent the semantic relationship that a node has in the taxonomy. By using degree value to modify the path-based method that proposed in previous research is expected that the correlation value obtained will increase.
After running some experiment by using proposed method, the *correlation* measurement value can obtain fairly good *correlation* ties with 200 Word Pairs derive from Noun *POS SimLex-999*. The *correlation* value that be obtained is 93.3% which means their bonds are strong and they have very strong *correlation*. Whereas for the *POS* other than Noun *POS* vocabulary that owned by *WordNet* is incomplete therefore many pairs of words that the value of its similarity is zero so the correlation value is low.

1. Introduction

Semantic is the knowledge which learning about meaning of the word in the language. The semantic related with meaning of the word relation, it's like in synonim, antonim and hyponim. Morrist and Hirst 2004 [2] stated that semantic *relatedness* explains the power of semantic relationships between two words or concepts are measured. It includes various relationships among concepts, belonging classical relationships as hypernymy, hyponymy, meronymy, antonymy, synonymy, and any other 'non classical relations'.

Weeds [2] defined that semantic similarity is a specific case of *relatedness*, where the power of relatedness depends on the 'degree of synonyms', that is normally explained by classical relations. In the taxonomy or lexical hierarchy generaly are shaped from hypernym and hyponym words. Hypernym is the words that have general meaning from the other words. Hyponyms are a group of words that are of special significance meaning rather than measured words. Hyponym is a member of group of a hypernym meaning groups, eg hypernym = sport, hyponym = football, basketball, tennis, run.

Semantic similarity between word is often represented by similarity between concept that associated with the word. Concepts can have different meanings but remain semantically interconnected like sports and balls. Other instance that antonyms are respected has a semantic relationship such as smart and stupid, however they are dissimilar. A way that can be used to find out the value of similarity is to calculate the value of shortest path length between two words. The similarity value is also influenced by the depth of subsumer value of the lowest common subsumer (*lcs*) nodes from two concept. The similarity value measured between two concepts will increase if the depth of subsumer value of the two concepts also grows higher in the taxonomy.

2. Literature Review

2.1. Related Work

In the study that conducted by Liu, Zhou and Zheng [3] on a semantic similarity using path-based method managed to get a correlation value of 92.6%. The correlation value still has an opportunity to be improved because of the research by Liu, Zhou and Zheng [3] only optimizing the semantic relationship of hypernym and hyponym. In the Semantic relatedness there are many semantic relationships besides hypernym and hyponym such as synonym, antonym, meronym and holonym. Taking advantage of all semantic relationships is expected to increase the correlations value that have been achieved previously.

Abdul Baquee Muhammad [1] have built Corpus that contained *AlQur'an* domain, *WordNet* and dictionary. He has did initialisation in the development of knowledges about *AlQur'an* and the knowledges about the relatedness between texts in *AlQur'an*. The Path based measurement method that proposed by Liu, Zhou and Zheng [3] has never been used in the *AlQur'an* domain. In this research the path based measurement method proposed by Liu, Zhou and Zheng [3] will be used to test this method in *AlQur'an* domain to obtain similarity values and to measure its correlation value.

To obtain a better correlation value the degree value is proposed to be used in modifying the path based method that proposed by Liu, Zhou and Zheng [3]. Degree Value is the number of links that owned by a lcs (lowest common subsumer) node on a taxonomy. The links owned by a node on the taxonomy represent the semantic relationship that a node has in the taxonomy. By using degree value to modify the path-based method that proposed by Liu, Zhou and Zheng [3] is expected that the correlation value obtained will increase.

2.2. Semantic Similarity

Weeds [2] defined that semantic similarity is a specific case of *relatedness* where the power of relatedness depends on the 'degree of synonyms', that is normally explained by classical relations. Concepts that are semantically related are not necessarily similiar such as car and fuel. Another example is that antonyms are considered to be semantically related such as beautiful and ugly, however they are dissimilar.

The semantic similarity between words has become a widespread topic in natural language processing and information retrieval for many years. It is widely used in a variety of applications such as sense disambiguation, detection and correction of malapropisms, multimodal documents retrieval, and automatic hypertext linking [3].

Similarity between words is often represented by similarity between concepts associated with the words. Many methods are developed to compute word similarity nowadays. Generally, these methods can be classified into two categories, one is edge counting based method that using only semantic link and second is corpus based methods that combining corpus statistics with taxonomic distance. The information sources that commonly be used in similarity measures are the shortest *path* length between words and the depth of subsumer in the taxonomy hierarchy [3].

The key of calculating semantic similarity lies in simulating human thinking behavior. Semantic similarity between words is decided by processing the first hand information sources in the human brain. Therefore the semantic similarity measurement can simulate human judgement by using multiple information sources [3].

Human tends to decide similarity between concepts through comparing their common features and different features. Therefore it be considered that the common and different features of two words to simulate the process of human judgement. It is assumed that human judgement for semantic similarity is a process of comparing equal attribute with aggregate attrbute that is the sum of equal and different attribute [3] . The model of semantic similarity is shown in formula [1].

$$S(w_1,w_2) = \frac{f(d)}{f(d)+f(l)} \quad [1]$$

Where l denotes the shortest *path* length between w_1 and w_2, d denotes the depth of subsumer in the hierarchy semantic net, f is the transfer function for d and l [3].

Table 1. Reference Results

Semantic Similarity Method	Correlation with RG-28	Correlation with MC-28	Correlation with RG-65
WordNet Edges	0.740	0.739	0.787
Hirst-St.Onge	0.671	0.682	0.732
Jiang-Conrath	0.670	0.684	0.731
Leacock-Chodorow	0.801	0.820	0.852
Lin	0.773	0.814	0.834
Resnik	0.706	0.763	0.800
Yang-Powers	0.889	0.921	0.897
Li	0.891	0.883	0.881
Liu-Zhou-Zheng	0.909	0.926	0.891

Tabel 1 shows the performance of semantic similarity measures reported by previous researchers. From table 1 shows that Liu, Zhou and Zheng [3] methods achieves the best corelation 0.926 with the MC's average human value of 28 pairs of words compared with the whole method. This mean that Liu, Zhou and Zheng [3] method's is more accurate than most individual subjects. Therefore human judgement for semantic similarity can be simulated by the ratio of common features to the sum of common and different features between words.

3. Experiment Scenario

Figure 1 is a general stage of the system developed in this research. The system built in this research can perform the relatedness measurement and correlation measurement. The query is used as *WordNet* input to find their *Wordset*. In data preprocesing stage, the *Wordset* and 60 verses of *AlQur'an* are processed to make it being standart form. With Taxonomy and proposed method, the *Wordset* and 60 verses of *AlQur'an* are measured to find their relatedness score. With correlation method, the relatedness score and *SimLex* score are compared and measured to find their correlation score.

The general process of the system as follows

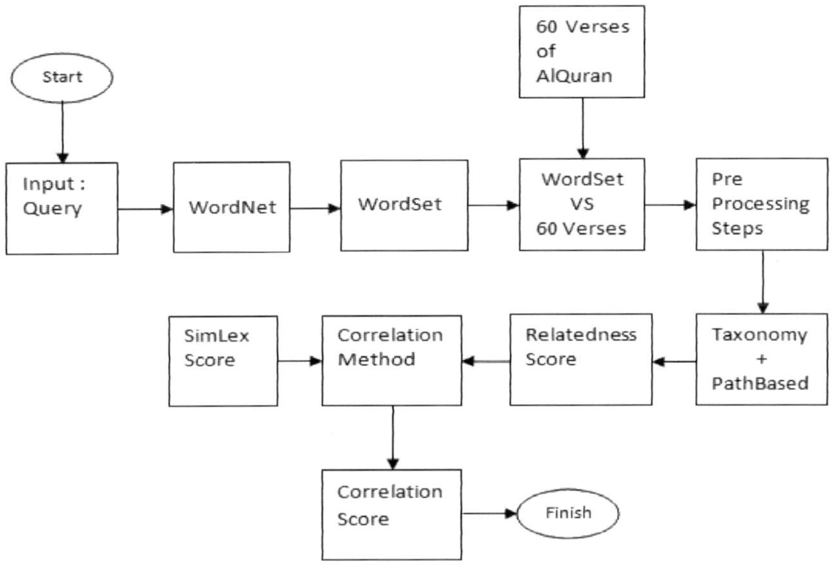

Figure 1. Experiment Scenario

3.1. Presentation of Wordset Dataset

Wordset is output from a query in the semantic sense of the word search process from the WordNet input, where the output shape set of *semantic* words and collected in one set of words. One word of *WordNet* input can produce multiple outputs, so the *WordNet* outputs are said can be collected in a set of words or *Wordset*. The query words that be used in this experiment are taken from the Pillar of Faith in Islam are faith, Allah, god, angel, holy, messenger, prophet, death, disaster, destiny.

Table 2. Query and Wordset

No.	Query	Wordset
1	faith	[u'faith',u'organized_religion',u'religion',u'religious_belief',u'trust']
2	Allah	[u'Allah']
3	god	[u'God',u'Supreme_Being',u'deity',u'divinity,u'god',u'graven_image', u'idol',u'imortal']
4	angel	[u'Angel',u'Angel_Falls',u'angel',u'backer',u'holy_man',u'holy_person', u'saint']
5	holy	[u'holy',u'holy_place',u'sanctum',u'unholy']
6	messenger	[u'courier',u'messenger']
7	prophet	[u'oracle',u'prophesier',u'prophet',u'seer',u'vaticinator']
8	death	[u'Death',u'birth',u'death',u'decease',u'demise',u'destruction',u'dying', u'end',u'expiry',u'last']
9	disaster	[u'calamity',u'cataclysm',u'catastrophe',u'disaster',u'tragedy']
10	destiny	[u'circumstances',u'destiny',u'fate',u'fortune',u'lot',u'luck',u'portion']

3.2. Presentation of 60 verses of AlQuran Dataset

The second *dataset* is a *dataset* that containing 60 verses translations of *AlQur'an*. The second *dataset* must be processed through *preprocessing* stage so the output can be more accurately. There are three steps in the *preprocessing* process, those process namely TSS (*tokenization, stopword and stemming*). 60 verses of *AlQur'an* translation comes from surah *Al-'Asr* verse 1 to the surah of *An-Naas* verse 6.

3.3. Presentation of SimLex-999 Dataset

The *correlation* value calculation need *dataset* that be used to compare data from *similarity* measurement. The *dataset* can be extracted from *SimLex-999* [13], while the *SimLex-999* has 999 words pair which completed with their *similarity* score. The similaity measurement needs 200 words pair that be used in the computation. 200 words pair that taken from *SimLex-999* are completed with their *similarity* score, then 200 words pair are used in *similarity* computation through proposed method.

3.4. The System Improvement

Keijo Ruohonen [16] defined that degree of the vertex is the number of edges with vertex as and end vertex. Number of edges that exist at a vertex are the edge that derived from that vertex or towards to the vertex. The degree value can be used to compute the number of the edge that established at the vertex. Each vertex in the vertex network will have a degree because each vertex will have a relationship with another vertex.

WordNet has a word hierarchy or commonly called taxonomy. In the taxonomy there are relationship between *synsets*, the relatinships are synonym, antonym, hypernym, hyponym, meronym and holonyms. That relationship between *synsets* are connected by several links to form a word or *synsets* network. Each *synset* in taxonomy can have multiple links depending on the relationships that are formed and owned with other *synsets*.

Therefore the degree value can be obtained from the number of links formed from a node connected to another node through the hypernym, hyponym, meronym and holonym relationships. And degree value which is proposed to modify the formula is the number of edge which exists at the lowest common subsumer (*lcs*) vertex. Therefore the degree value can be proposed to modify the formula derive from Liu, Zhou and Zheng [3], and the formula changed like shown below.

$$S(w_1, w_2) = \frac{b*d}{b*d + l} \quad [2]$$

where *b* is the degree value from the lowest common subsumer (*lcs*) node.

In the *path based* method that proposed by Liu, Zhou, and Zheng [3] the *semantic* relationship that be used is the relationship of hypernym and hyponym only. While in *semantic* relatedness there are more existing relationships, not only hypernyms and hyponyms relationships but also synonyms, antonyms, meronyms and holonyms relationships. These relationships have not been optimized in the *path based* method that proposed by Liu, Zhou, and Zheng [3].

In the proposed method it uses the variable degree value, where the variable degree value is the number of links that owned by a node. Those links are representations of *semantic* relationships that owned by a node in *WordNet* taxonomy. So by using the variable degree value all existing semantical relationships can be accommodated well in the proposed method.

4. Result

From the experiments that have been done, in accordance with the hypothesis, with the addition of degree value in *semantic relatedness* method has proven can improve the performance of relatedness value measurement. For analyzis the effectivenes of the proposed method, each *POS* in *WordNet* are measured their *correlation* value. The following are the result of *relatedness* value measurement and *correlation* value measurement.

4.1. Relatedness Value Measurement

In the tables displayed in this research, Path denotes method used by Liu, Zhou and Zheng [3] in doing research that is path based method, Proposed denotes the method that be proposed in this research.

Table 3 Best *relatedness* value based on the query in AlQur'an domain

		Method 1	Method 2	Best Relatednees Value of Verses of AlQur'an		
No	Set of Query	Path	Proposed	Document		Verses of AlQur'an
1	faith	0.664	0.862	d37	109:6	For you is your religion, and for me is my religion
2	Allah	0.577	0.980	d37	109:6	For you is your religion, and for me is my religion
3	god	0.427	0.683	d37	109:6	For you is your religion, and for me is my religion
4	angel	0.309	0.702	d56	114:2	The Sovereign of mankind
5	holy	0.170	0.693	d56	114:2	The Sovereign of mankind
6	messenger	0.560	0.987	d56	114:2	The Sovereign of mankind
7	prophet	0.503	0.984	d56	114:2	The Sovereign of mankind
8	death	0.279	0.731	d1	103:1	By time,
9	disaster	0.326	0.747	d1	103:1	By time,
10	destiny	0.286	0.703	d1	103:1	By time,

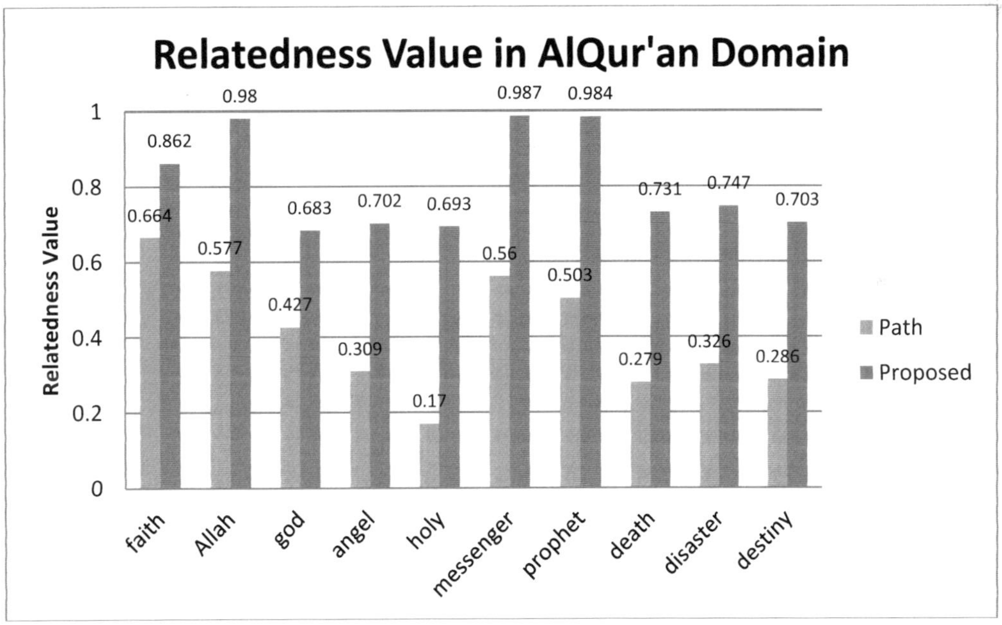

Figure 2. Relatedness Value in AlQur'an Domain

Table 4. Best *relatedness* value based on the query in General domain

		Method 1	Method 2	Best Relatednees Value of Verses of AlQur'an	
No	Set of Query	Path	Proposed	Document	Verses of AlQur'an
1	faith	0.149	0.508	d30	For you is your religion, and for me is my religion
2	Allah	0.127	0.344	d29	For you is your religion, and for me is my religion
3	god	0.172	0.484	d29	For you is your religion, and for me is my religion
4	angel	0.268	0.513	d22	The Sovereign of mankind
5	holy	0.167	0.557	d29	The Sovereign of mankind
6	messenger	0.484	0.733	d22	The Sovereign of mankind
7	prophet	0.437	0.726	d22	The Sovereign of mankind
8	death	0.145	0.513	d30	By time,
9	disaster	0.142	0.489	d30	By time,
10	destiny	0.159	0.519	d30	By time,

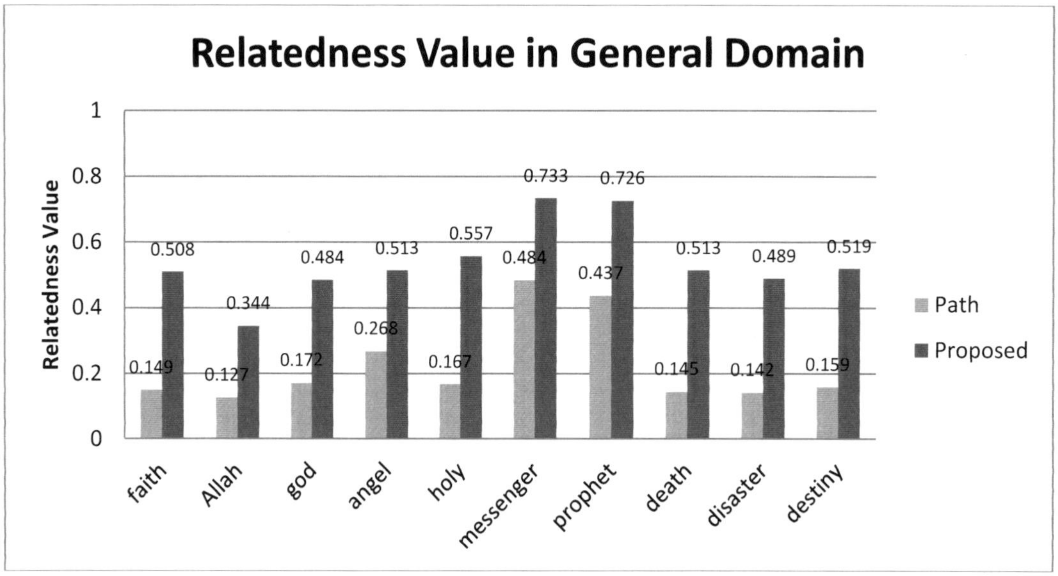

Figure 3. Relatedness Value in General Domain

This experiment is run to find out the highest relatedness value that can be obtained by each method on *AlQur'an* domain or general domain. At Figure 2 shows the result from the experimental measurements in *AlQur'an* domain indicate that the value of *relatedness* measurements by using the proposed method get better result than the Path based method proposed by Liu, Zhou and Zheng [3]. And at Figure 3 shows the result from the experimental measurements in General domain indicate that the value of *relatedness* measurements by using the proposed method get better result than the Path based method proposed by Liu, Zhou and Zheng [3]. From the experiment of *relatedness* measurements value in the both domains obtained results that the value of *relatedness* measurements by using the proposed method get highest result better than the Path based method proposed by Liu, Zhou and Zheng [3].

4.2. Analyzis in Part of Speech of Wordnet

This experiment is running to measure the effectiveness and reliability of the method. The analyzis process that will be performed is comparing the results output from *relatedness* calculation process using the proposed method and with a list of *similarity* score of words pairs that come from *SimLex-999.* Both values are compared to check their *correlation* whether the change in value on the results of the calculation process in line with changes in the *similarity* score of the pair of words from *SimLex-999* or in the reverse line or completely unrelated to the *Pearson* method.

The details about strength of *correlation* has explained by Jonathan Sarwono [15]. The strength of *correlation* according to Jonathan Sarwono [15] are shown at Table 5.

Table 5. Degree of *Correlation* by Jonathan Sarwono [15]

No.	Range	Degree of Correlation
1	0	no correlation
2	0.001 – 0.250	the correlation is very weak
3	0.251 – 0.500	correlation enough
4	0.501 – 0.750	strong correlation
5	0.751 – 0.999	the correlation is very strong
6	0	perfect correlation

After through some *correlation* value computation step which involve some *POS dataset*, the *correlation* value from three *POS dataset* will be displayed and be compared each other. The comparation is performed to find where the *POS* parameter that can make *correlation* value obtain maximal result or to find where the *POS* parameter that cannot make *correlation* value obtain maximal result. The *correlation* value which computed with three *POS* parameter *dataset* are shown at Table 6 and Figure 4.

Table 6. Comparation *correlation* value with three *POS dataset*

	Maximum Correlation Value		
POS	Path	Proposed	Degree of Correlation
Adjective POS	0.101	0.015	the correlation is very weak
Noun POS	0.780	0.933	correlation enough
Verb POS	0.339	0.279	strong correlation

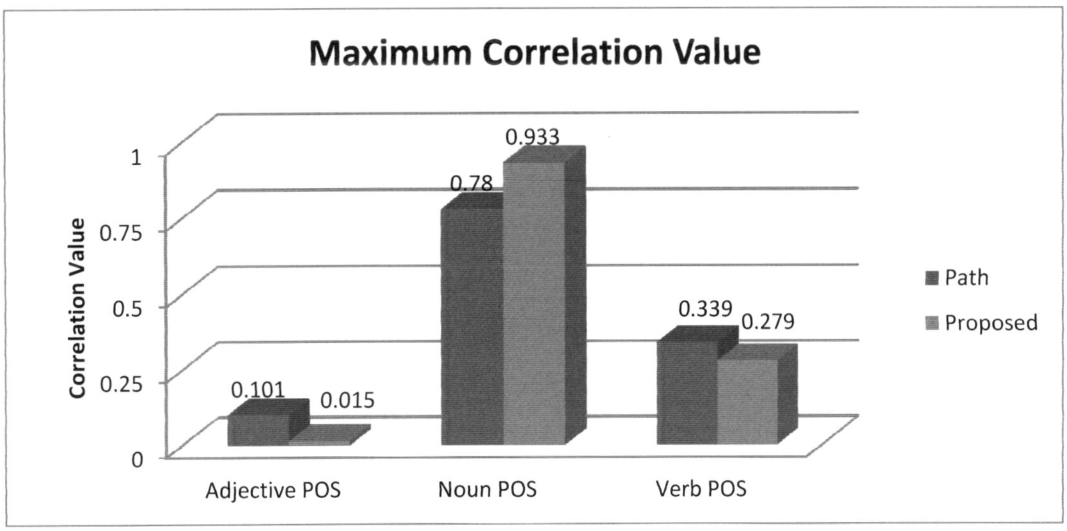

Figure 4. *Correlation* Value Diagram

4.3. Analyzis in Variable of Proposed Method

Proposed methods consist of three variables, those variables are variable *d* denotes depth of subsumer, variable *l* denotes shortest path length between two concept and variable *b* denotes degree of lcs node or number of links in the lcs node. This experiment is running to analysis the variables to find the variable that most influence in the changing of similarity value and correlation value and the variables that are less influential on changes in similarity value and correlation value.

After through some similarity value and *correlation* value computation with involving the changing of varibale *d*, *l* and *b* which each variable has value 1, the similarity value and *correlation* value from three variables will be displayed and be compared each other. The comparation is performed to find where the variable that more influences in similarity value and *correlation* value measurement and to find where the variable that less influences in similarity value and *correlation* value measurement. The comparation of similarity value and correlation value are shown at Table 7 and Figure 5.

Table 7. Comparation of deviation value in relatedness value measurement

Devlation Value	Proposed d = 1	Proposed l = 1	Proposed b = 1
Relatedness Value	0.083	0.128	0.380

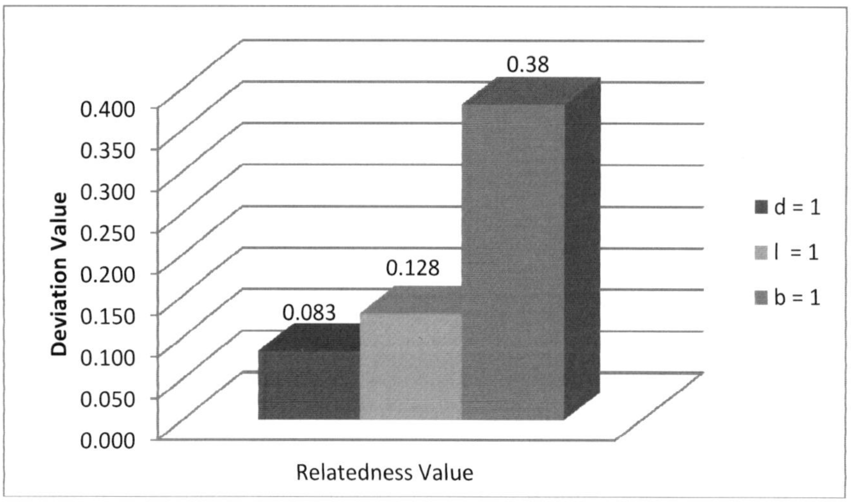

Figure 5. Comparation of deviation value in relatedness value measurement

Table 8. Comparation of deviation value and correlation value measurement

Devlation Value	Proposed d = 1	Proposed l = 1	Proposed b = 1
Relatedness Value	0.027	0.267	0.146

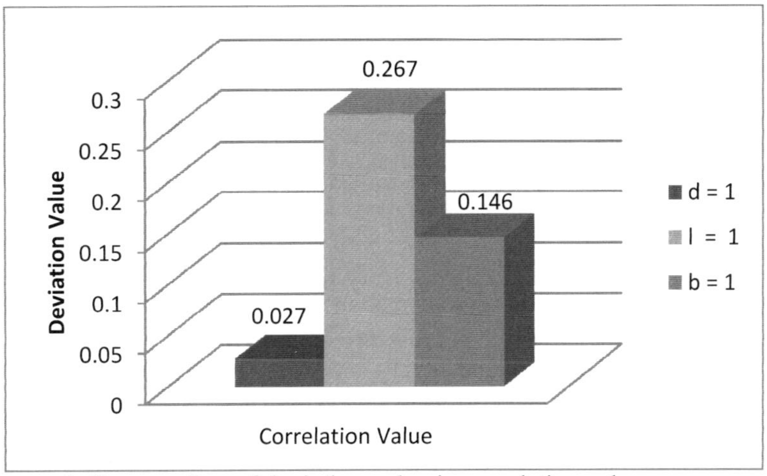

Figure 6. Comparation of deviation value in correlation value measurement

5. Discussion

From the result of the *relatedness* computation experiment revealed that proposed method obtain better *relatedness* value if be compared with *Path based* method that used in Liu, Zhou and Zheng [3] research. Proposed Method obtains better result because its involves degree value in the path based formula so it can increases *relatedness* value and correlation value.

The degree value is used in this proposed method, degree value has means the number of links that owned by a node. Those links are representations of *semantic* relationships that owned by a node in *WordNet* taxonomy. So by using the variable degree value all existing semantical relationships can be accommodated well in the proposed method.

Based on the *correlation* computation experiment indicated that parameter which influences in *correlation* measurement is Noun *POS dataset*. With Noun *POS dataset*, *correlation* value obtains *78%* using path method and *93,3%* using proposed method. The result has mean that *relatedness* computation result has *correlation* with Noun *POS dataset* from *SimLex-999* and the strength of their *correlation* is very strong. The lowest *correlation* with *relatedness* computation result is adjective *POS dataset* from *SimLex-999* because their *correlation* value just *10,1%* using path method and *1,5%* using proposed method.

For Noun *POS* obtains high correlation value because in *WordNet* for Noun *POS* has a more complete vocabulary so rarely there are pairs of words whose value of similarity is zero so that the correlation value becomes high. Whereas for the *POS* other than Noun *POS* vocabulary owned *WordNet* is incomplete so many pairs of words that the value of its similarity is zero so the correlation value is low.

From analysis of the proposed method and based on the experiment of the similarity value measurement, the most influential variable is the variable *b* which has deviation value 0.380 and the less influential is the variable *d* which has deviation value 0.083. In the measurement of correlation value, the most influential variable is the variable *l* which has deviation value 0.267 and the less influential is the variable *d* which has deviation value 0.027.

The variable b becomes the most influential variable in the relatedness value measurement because the variable b is the number of links owned by a node on taxonomy and a node in taxonomy *WordNet* has many semantic links so that when value b is made minimalis (1) it will be very influential to the relatedness value obtained. The variable l becomes the most influential variable on the correlation value measurement because the value of l is the distance or the number of links formed between two words (concept) measured on the taxonomy and the distance between the concept in taxonomy

WordNet far apart so that when the value l is made minimalis (1) will be very influences the correlation value obtained.

6. Conclusions & Future Work

6.1. Conclusions
By involving the degree value in the *path based* formula, proposed method will get better *relatedness* value while in the example query '*messenger*' and the verse is 114:2, it can reaches 98,7% compared by using the path based method that proposed by Liu, Zhou and Zheng [3] its reaches 56%.

With the proposed method, *correlation* value measurement can obtain fairly good *correlation* ties with 200 Word Pairs derive from Noun *POS SimLex-999*. 93.3% is the maximum *correlation* score that can be reached in this research, that value has means their bonds are strong and they have very strong *correlation.*

The *correlation* value measurement in Adjective *POS* and Verb *POS* do not obtain high result as Noun *POS* result because for the *POS* other than Noun *POS* vocabulary that owned by *WordNet* is incomplete, therefore many pairs of words that the value of its similarity is zero so it makes the correlation value is low.

Measurement of correlation value is influenced by relatedness value that obtained from calculation process and relatedness value that obtained from external source such as *Simlex-999* which *Simlex-999* is used as human judgment and comparator in calculation of correlation value.

Based on the proposed method analysis in the similarity value measurement, the most influential variable is degree value which has deviation value 0.380 and the less influential is depth of subsumer which has deviation value 0.083. In the measurement of correlation value, the most influential variable is shortest path length which has deviation value 0.267 and the less influential is depth of subsumer which has deviation value 0.027.

This study has several side that still can be developed to achieve better value for *relatedness* and to assist the development of research on text and *AlQuran* both by the Qur'anic and text researchers or the academic community from all coleges.

6.2. Future Works
The number of existing methods in the measurement of text *relatedness* such vector based method, information content based method, gloss based method can be tried in the experiment to open opportunities for possible improvement that can be found in the text *relatedness* research in the future work later.

By using other knowledge bases besides the *Wordnet*, there is a big opportunity to get an increasing of *relatedness* value, because there are some limitation in this experiment so it has not had enough time to test. This would open the opportunities for future work in the development of research on the *semantic* text *relatedness* in the *AlQuran* domain.

With so many verses in the *AlQuran* which reached 6622 verses which in this research has not been performed, additional number of verses or use other set of verses in the *AlQuran* can be used in the subsequent research to open opportunities the possibility of increasing *relatedness* value that will be get.

Due to several limitations in this study led to the possibilities that be mentioned above have not had enough time to do the research to find the likely possibilities that can be obtained from several variations of the research and experiments that could unlock the barriers that may exist and hinder the increasing relatedeness value.

References

[1] Abdul Baquee Muhammad, " Annotation of Conceptual Co-Reference and Text Mining the Quran ", Submitted in accordance with the requirements for the degree of Doctor of Philosophy, England, September 2012.

[2] Ziqi Zhang; Anna Lisa Gentile; Fabio Ciravegna, "Recent Advances in Methods of Lexical Semantic Relatedness - a Survey", Natural Language Engineering, Volume 19, Issue 04, October 2013.

[3] Xiao-Ying Liu; Yi-Ming Zhou; Rou-Shi Zheng, "Measuring Semantic Similarity in Wordnet", Proceedings of the Sixth International Conference on Machine Learning and Cybernetics, Hongkong, 19-22 August 2007.

[4] Samer Hasan; Rada Mihalcea, "Semantic Relatedness Using Salient Semantic Analysis", Proceedings of the Twenty-Fifth AAAI Conference on Artificial Inteligence, Texas, 2011.

[5] Abdul Baquee Muhammad Sharaf, Eric S.Atwell, "QurSim : A Corpus for Evaluation of Relatedness in Short Texts", Proceedings of the Eighth International Conference on Language Resource and Evaluation (LREC), Istanbul, 21-27 May 2012.

[6] Akip Maulana, Moch. Arif Bijaksana, Ir., M.Tech, PhD, M. Syahrul Mubarok, ST., M.T, "Perancangan Semantic Similarity based on Word Thesaurus Menggunakan Pengukuran Omiotis Untuk Pencarian Aplikasi pada I-GRACIAS", Indonesian Symposium on Computing, Sept 2016.

[7] Lingling Meng, Runqing Huang, Junzhong Gu, "A Review of Semantic Similarity Measure in WordNet", International Journal of Hybrid Information Technology, Vol. 6, No.1, January 2013.

[8] Lingling Meng, Runqing Huang, Junzhong Gu, "Measuring Semantic Similarity of Word Pairs using Path and Information Content", International Journal of Future Generation Communication and Networking (IJFGCN), Vol. 7, No.3, 2014.

[9] Hongzhe Liu, Pengfei Wang, "Assessing Text Semantic Similarity Using Ontology", Journal of Software, Vol. 9, No.2, February 2014.

[10] Issa Atoum, Ahmed Otoom, "Efficient Hybrid Semantic Text Similarity using Wordnet and a Corpus", International Journal of Advanced Computer Sciene and Applications, Vol. 7, No.9, 2016.

[11] Tingting Wei, Huiyou Chang, "Measuring Word Semantic Relatedness using WordNet-based Approach", Journal of Computer, Vol. 10, No.4, July 2015.

[12] Gang Liu, Ruili Wang, Jeremy Buckley, Helen Min Zhou, "A WordNet-based Semantic Similarity Measure Enhanced by Internet - based Knowledge", Proceedings of the Twenty Third International Conference on Software Engineering & Knowledge Engineering (SEKE-2011), Miami, 7-9 Juli 2011.

[13] Felix Hill, Roi Reichart, Anna Korhonen, "SimLex-999 : Evaluating Semantic Models with (Genuine) Similarity Estimation", Journal of Computational Linguistics, Vol. 41, No.4,

December 2015.

[14] Fellbaum, Christine, "About Wordnet, A Lexical Database for English", 2005, [Online], Avalaible at : https://wordnet.princeton.edu/.

[15] Jonathan Sarwono, "Mengenal Apa Itu Analisis Korelasi", 2006, [Online], Avalaible at : http://www.jonathansarwono.info /.

[16] Mubarok, M.S., Adiwijaya and Aldhi, M.D., 2017. Aspect-based sentiment analysis to review products using Naïve Bayes. In AIP Conference Proceedings (Vol. 1867, No. 1, p. 020060). AIP Publishing.

[17] Pratiwi, A.I., Adiwijaya. 2017. On The Feature Selection and Classification Based on Information Gain for Document Sentiment Analysis, Applied Computational Intelligence and Soft Computing

[18] Keijo Ruohonen, "Graph Teory", 2013.

[19] Budi Raharjo, "Mudah Belajar Python untuk Aplikasi Desktop dan Web", 2015.

Identification of four class emotion from Indonesian spoken language using acoustic and lexical features

Fatan Kasyidi and Dessi Puji Lestari

School of Electrical Engineering and Informatics, Bandung Institute of Technology, Bandung, Indonesia

23515025@std.stei.itb.ac.id, dessipuji@informatika.org

Abstract. One of the important aspects in human to human communication is to understand emotion of each party. Recently, interactions between human and computer continues to develop, especially affective interaction where emotion recognition is one of its important components. This paper presents our extended works on emotion recognition of Indonesian spoken language to identify four main class of emotions: Happy, Sad, Angry, and Contentment using combination of acoustic/prosodic features and lexical features. We construct emotion speech corpus from Indonesia television talk show where the situations are as close as possible to the natural situation. After constructing the emotion speech corpus, the acoustic/prosodic and lexical features are extracted to train the emotion model. We employ some machine learning algorithms such as Support Vector Machine (SVM), Naive Bayes, and Random Forest to get the best model. The experiment result of testing data shows that the best model has an F-measure score of 0.447 by using only the acoustic/prosodic feature and F-measure score of 0.488 by using both acoustic/prosodic and lexical features to recognize four class emotion using the SVM RBF Kernel.

1. Introduction

In recent time, research on emotion recognition for several languages has shown significant improvement. This helps to increase the quality of human-computer interaction to have more natural and affective communication. Speech emotion recognitions use many classification techniques such as Support Vector Machine, Naive Bayes, Random Forest, Gaussian Mixture Model (GMM) and Artificial Neural Network (ANN) [1–4]. In [2-3], GMM and ANN showed better performances comparing to the Random Forest and Naive Bayes. For Indonesian Language, research of emotion recognition from speech can be found in [1] where SVM-based model using the acoustic features has been developed. The model they developed shows accuracy of 68,31% for recognizing four class of emotions (angry, happiness, contentment, and sadness). They used only the acoustic/prosodic characteristics. Other research has also been established recently for Indonesian spoken language to recognize the positive and negative emotion using acoustic and lexical features [6].

As for the features, [2,3,6] used not only the acoustic/prosodic features, but also lexical features to improve performance of the system. Although, previous researches show that acoustic features are very dominant to be used for speech emotion recognition, adding lexical features also give promising results to increase performance of the system. Several type of acoustic features have been explored, including the prosodic features (such as pitch, energy, duration and higher order formants), spectral features and cepstral features [1–3]. It includes the derivation of statistical function such as mean, standard deviation, maximum, minimum, range etc.

In this paper, we study four class emotion consist of angry emotion, happiness emotion, contentment, and sadness emotions as in [1]. However other than acoustic feature, we also employ the lexical features. In order to train the emotional model, the speech emotion corpus is collected from Indonesian TV Shows, Indonesia Lawyers Club (ILC), Mata Najwa, Tonight Show and Sarah Sechan. In order to extract the lexical features, we annotate the transcript manually by listening to the audio data. We use the Bag of Words (BoW) and Term Frequency-Inverse Document Frequency (TF-IDF) as has shown an effective performance of recognizing the emotion along with acoustic/prosodic feature [2].

The remainder of this paper is written as follows: the related works is explained in section 2. In section 3, the database collection is discussed. We explain the configuration of the experiment in section 4. In section 5, we describe the result of the experiment, and finally in Section 6 a conclusion and future works are presented.

2. Related Works

In recent years, some research for emotion recognition from spoken language have been established. A study for emotion recognition has been established in English [2]. They used the USC-IEMOCAP as the database that consists of 12 hours of audio-visual data, including speech and text transcription. Every utterance has an emotion label category such as angry, happy, etc and dimensional labels such as valence and activation. There are in total 5531 utterances and 419.4 minutes for the overall duration. The feature extraction process performed to get acoustic and lexical features from those utterances as feature representation. In this study, SVM linear kernel was utilized to model the emotion representation.

Other work on emotion recognition was performed in Chinese [3]. In this study, lexical feature along with acoustic/prosodic feature has been utilized to construct the emotion model. Acoustic/prosodic feature such as MFCC, jitter-related, shimmer related and emotional salient segment was included. For lexical feature, they use semantic labels as the lexical aspect using the HowNet. HowNet is an online common-sense knowledge base unveiling inter-conceptual relations and inter-attribute relations of connoted in the lexicon containing Chinese and English equivalents. The accuracy of this model is 83.55%.

3. Data Collection

The availability of speech corpus containing emotions is unavoidable in order to develop emotion model developed using the machine learning technique. Thus, it is necessary to construct speech emotion corpus for Indonesian language since it is difficult to find publicly available corpus for this task. To construct the corpus, the corpus construction procedures from the previous work was followed [1]. In addition to that, we transcribe manually all utterances in the speech corpus to get their corresponding transcript. There are 3 main steps when collecting the speech corpus: data collection, utterance segmentation, and emotional and transcript labeling as can be seen in Figure 1.

Speech corpus is collected from four Indonesian TV talk show where there are many emotional dialog occur: Indonesian Lawyer club, Tonight Show, Mata Najwa, and Sarah Sechan. After conducting the segmentation process, we get 3652 utterances with total duration of recording is 5 hours 10 minutes and 40 seconds including the neutral emotion. However since the the neutral emotion utterance can bias the emotion recognition, we separate the utterances having neural emotion from the dataset. Thus, after removing the neutral utterances, our data set contains only 1082 utterance with total duration of recording is 1 hour and 7 minutes. Each utterance contains one emotion to ensure the consistency and to avoid the change and transition of emotion for simplicity purpose. Before we annotate the emotion class, we transcribe each utterances manually to get their corresponding transcript. The transcribing process was conducted by 4 annotators that transcribe the speech utterances carefully by listening to each utterances in the database.

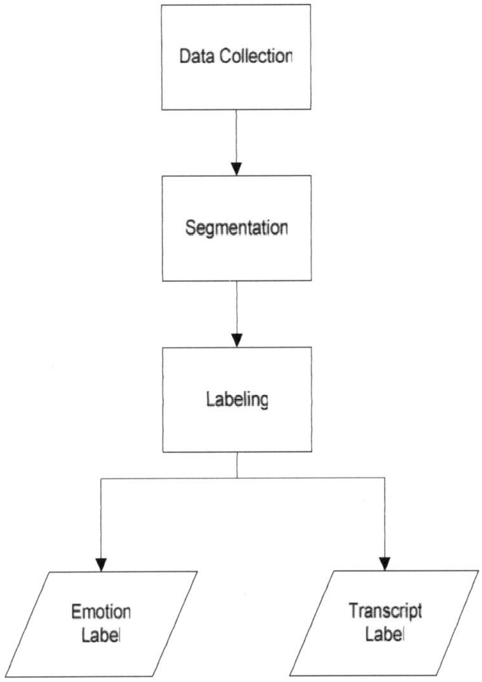

Figure 1. Corpus Construction Steps

After that, every annotator was asked to give their judgment regarding the emotion label of each utterance. To evaluate and decide the emotion in each utterance, majority voting was employed to conclude the emotion class of each utterance. One of the observer is a psychologist that has more weight value than that of another observer. We use an expert decision if there is more than one decision with the same number of votes. In the end, we collect 362 utterances for angry, 262 utterances for happiness, 209 utterances for contentment and 149 utterances for sadness. The data for each class is not balanced since it is difficult to gather such data. Later on, to handle the imbalance data, we apply the resampling technique Synthetic Minority Oversampling Technique (SMOTE). In Figure 2, we can see clearly that the data is not balanced. We have angry as majority class and sadness as minority class.

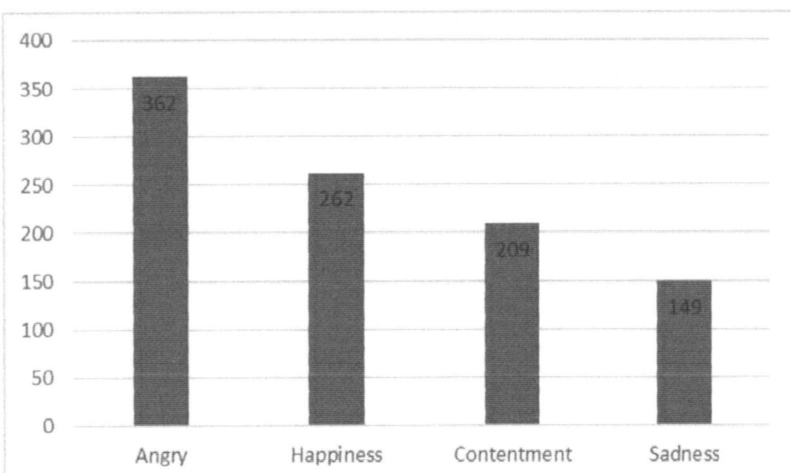

Figure 2. Data for Each Emotion Class

For conducting the experiments, the data that we have collected is divided into two sets: one for development set and one for testing set. The development set contains 982 utterances while the testing set contains 100 utterance with equal distribution between each class, i.e. 25 utterances per emotion class.

4. Experimental Setup

For determining the best features set, we conduct three kind of model development: 1) train the model by using only the acoustic features; 2) train the model by using lexical features; 3) train the model by using both acoustic and lexical features. Those 3 features is employed to train the emotional model using 3 classifiers: SVM, Naive Bayes, and Random Forest. Before we start the experiment, we resampled the data to cope with the data imbalance so there are experiments with and without sampling. We used 5-fold cross-validation and testing schema. The important thing that related to the cross-validation is how to apply the SMOTE in that scheme. As we know that 5-fold cross validation divides the data into two groups in every fold: training and testing set. The SMOTE method only uses the training set in every fold to avoid the duplication and overfitting model in the result. Since the angry class is the majority class, we did not over-sampled the data in this class, while for other three emotion classes, we conduct the oversampling with sampling percentage as shown in Table 1.

Table 1. Percentages of SMOTE for emotion class

Emotion	Percentage (%)
Angry	0
Happiness	30
Contentment	50
Sadness	120

In total, there are 40 experiments that combine features set, experimental schemes, and type of classifier. We compared the performance of every model to find the highest accuracy model.

4.1. Acoustic/prosodic Feature

We extract acoustic/prosodic features for constructing the emotion model. Two acoustic/prosodic features, the emo large feature set, the Low Level Descriptor of Interspeech 2010 set, and the Interspeech 2010 Paralinguistic Challenge feature set are employed and compared as previous work of detecting emotion from Indonesia spoken language used only the emo large feature and the Low Level Descriptor of Interspeech 2010 [1]. The extracted feature for the emo large feature and the Low Level Descriptor of Interspeech 2010 consists of 57 low-level descriptors (LLD) as shown in Table 2 and Table 3.

Table 2. Low Level Descriptor of Emo Large [1]

Ceptral features	Spectral features
MFCC 0-12	Mel-Spectrum bins 0-25
	Zero crossing rate
	25% spectral roll-off points
	50% spectral roll-off points
	75% spectral roll-off points
	90% spectral roll-off points
	Spectral flux
	Centroid
	Relative position of spectral maximum and minimum

Table 3. Low Level Descriptor of Interspeech 2010[1]

Energy features	Voicing-related features
Logarithmic energy	F0
Energy bands 0-250Hz	F0 envelope
Energy bands 0-650Hz	Probability of voicing
Energy bands 250-650Hz	

The Low Level Descriptor of Interspeech 2010 Paralinguistic consists of 31 LLD and 1582 number of features [2]. This feature set has not been used in previous work [1] . However, recently this feature set has been used in Indonesia language emotion recognition to recognize 7 classes of emotions as can be found in [7]. The 31 LLD features set can be seen in Table 4.

Table 4. Low Level Descriptor of Interspeech 2010 Paralinguistic [2]

LLD		
Continuous	Qualitative	Cepstral
Energy: loudness	Local jitter	MFCC 0-14
Pitch: F0 final	JitterDDP	LogMelFreqBand 0-7
F0 final envelope	Local shimmer	
Line Spectral Pair Frequencies	Voicing final unclipped	

4.2. Lexical Feature

We compute Bag-of-Word (BoW) and Term Frequency-Inverse Document Frequency (TF-IDF) to extract the lexical features from each transcript of the speech utterance. The BoW method count the frequency of a word in the document without considering the word order, semantic structure and the grammar [2]. The vector representation of BoW is a normally used feature representation for a text document. The TF-IDF is used to evaluate how important a word is to the certain document in the corpus. The TF-IDF weight of term i in document j as calculated using formula (1). $tf_{i,j}$ is defined as term i in a document j, N as the number of all documents and Df_i as the total number of documents which contains term i [6,7].

$$w_{i,j} = tf_{i,j} \times log\frac{N}{Df_i} \quad (1)$$

In this study, we perform several preprocessing step to the transcription before BoW and TF-IDF are being constructed. We used two preprocessing: formalization and stop-word removal.

5. Result

In this section, the result of all experiments are described. To construct the models, 982 speech utterances of development dataset are being used. We calculate the average of F-measure in every model. As we mention before, there are three models having different feature set: the acoustic features set, lexical features set, and acoustic lexical (AL) feature set. Each features set will be used to train emotional model using the Naive Bayes, Random Forest, and SVM with Radial Basis Function (RBF) kernel and linear kernel. The result of the acoustic model can be seen Table 5.

Table 5. Result of Experiment for Acoustic Model (Two First Column using the Low Level Descriptor of Interspeech 2010 Paralinguistic and the last column using the Emo Large Feature Set)

	Without Resampling	Resampling with SMOTE	Emo Large Feature Set
Model	F-Measure	F-Measure	F-Measure
Naive Bayes	0.377	0.393	-
Random Forest	**0.389**	**0.409**	-
SVM RBF	0.378	0.400	0.411
SVM Linear	0.388	0.340	-

It is shown in Table 5 that when using the Low Level Descriptor of Interspeech 2010 Paralinguistic acoustic feature set, the Random Forest (RF) is the most optimal model for acoustic model with SMOTE and the detail of F-measure for each emotion can be seen in Table 6. It can be seen from Table 6 that it is easier to identify angry and happiness emotions (emotion with more active energy), rather than recognizing sad and contentment emotions.

Table 6. F-measure of Acoustic Model using RF

Emotion	F-Measure
Angry	0.535
Happiness	0.449
Contentment	0.264
Sadness	0.237

To see the effectiveness of using the lexical features, we used the lexical features only and the result of using BoW and TF-IDF can be seen in Table 7.

Table 7. Result of Experiment for Lexical Model

	Without Resampling		Resampling with SMOTE	
Model	BoW	TF-IDF	BoW	TF-IDF
Naive Bayes	0.359	0.339	0.366	0.382
Random Forest	0.399	0.375	0.383	**0.404**
SVM RBF	0.199	0.198	0.199	0.199
SVM Linear	0.394	0.364	0.379	0.341

It is shown in Table 7 that Random Forest is the most optimal model for lexical model TF-IDF with SMOTE and the detail of F-measure for each emotion can be seen in Table 8.

Table 8. F-measure of Lexical Model TF-IDF using RF

Emotion	F-Measure
Angry	0.494
Happiness	0.399
Sadness	0.224
Contentment	0.381

Finally, we combine the acoustic and lexical features and the results can be seen in Table 9.

Table 9. Result of Experiment for Acoustic Lexical Model

	Without Resampling F-measure		Resampling with SMOTE F-measure		Emo Large Feature Set F-Measure	
Model	BoW	TF-IDF	BoW	TF-IDF	BoW	TF-IDF
Naive Bayes	0.380	0.375	0.397	0.400	-	-
Random Forest	0.386	0.380	0.400	0.394	-	-
SVM RBF	0.388	0.389	**0.408**	0.402	0.374	0.380
SVM Linear	0.336	0.335	0.325	0.329	-	-

It is shown in Table 9 that SVM RBF kernel is the most optimal model for acoustic lexical (AL) model BoW with SMOTE and the detail of F-measure for each emotion can be seen in Table 10.

Table 10. F-measure of AL Model BoW using SVM RBF

Emotion	F-Measure
Angry	0.559
Happiness	0.377
Contentment	0.257
Sadness	0.273

The result from the experiment of development data modeling shows that F-measure for each optimal model consecutively are 0.447, 0.404 and 0.408. After that, we construct the testing scenario using testing dataset consist of 100 utterances with 25 utterances for each emotion. Testing scenario constructed only for the most optimal model from 3 global model (acoustic, lexical and AL). The result can be seen in Table 11.

Table 11. F-measure of Testing Scenario

	F-measure				
Emotion	Acoustic RF	Lexical TF-IDF RF	AL BoW SVM RBF	emo_large SVM RBF	emo_large TF-IDF SVM RBF
Angry	0.639	0.493	0.702	0.589	0.627
Happiness	0.471	0.356	0.531	0.553	0.405
Contentment	0.293	0.293	0.318	0.136	0.191
Sadness	0.392	0.356	0.400	0.559	0
Average	0.447	0.374	**0.488**	0.467	0.380

According to Table 11, the emo large as the configuration used in the previous work [1] have a lower accuracy than the acoustic lexical features (AL). This result shows that lexical features have an effect to enhance the accuracy of emotion recognition. Although, the lexical features can improve the recognition result, but it is not always the case. The data with id file P001111 is the example of inconsistency in recognizing the emotion. The AL model could not improved the ability to recognizing the emotion. But there are some examples where the AL model can enhance and fix the recognition such as can be seen in the data with id file P018015. Both of these examples can be seen in Table 12.

Table 12. Result of Prediction

File	Transcipt	Label	Acoustic Model	Lexical Model	AL Model
P001111	selamat berpromo juga semoga di album kedua ini penjualan albumnya meningkat kembali	happiness	angry	happiness	angry
P018015	sebegitu dekatnya yang bikin aku makin jatuh cinta	happiness	sadness	happiness	happiness

Human character and psychology is also affects the recognition. Ethnic in a certain area usually affect the character, especially how the way someone talking. For example, there are some people that talk in a relatively high pitch, even though they are not angry. Thus, it make the recognition is more difficult.

6. Conclusion

In In this paper, we have extended the work on emotion recognition in Indonesian spoken language from our previous works. For this study, we constructed emotional corpus in Indonesian spoken language from Indonesian television talk show which provides the natural emotional data for the corpus. By using this corpus, we construct three emotion recognition model using acoustic: acoustic by using both Emo Large Feature Set that has been used in the previous research and the Interspeech 2010 Feature Set, lexical features using the Bag of Word and TF-IDF, and the lexical and Acoustic Lexical (AL) model. Experimental results show that the best model is achieved when using both acoustic (the interspeech 2010 feature set) and lexical features (BoW) with 0.488 of F-measure score using the SVM RBF Kernel.

References

[1] Lubis N, Lestari D, Purwarianti A, Sakti S, Nakamura S. Emotion recognition on Indonesian television talk shows. In: 2014 IEEE Spoken Language Technology Workshop (SLT). 2014. p. 466–71.

[2] Jin Q, Li C, Chen S, Wu H. Speech emotion recognition with acoustic and lexical features. In: 2015 IEEE International Conference on Acoustics, Speech and Signal Processing (ICASSP). 2015. p. 4749–53.

[3] Wu CH, Liang WB. Emotion recognition of affective speech based on multiple classifiers using acoustic-prosodic information and semantic labels (Extended abstract). In: 2015 International Conference on Affective Computing and Intelligent Interaction (ACII). 2015. p. 477–83.

[4] Noroozi F, Akrami N, Anbarjafari G. Speech-based emotion recognition and next reaction prediction. In: 2017 25th Signal Processing and Communications Applications Conference (SIU). 2017. p. 1–4.

[5] Bhakre SK, Bang A. Emotion recognition on the basis of audio signal using Naive Bayes classifier. In: 2016 International Conference on Advances in Computing, Communications and Informatics (ICACCI). 2016. p. 2363–7.

[6] Kurniawati P, Lestari D. Positive and Negative Emotions Identification from Speech Signal using Acoustic and Lexical Features. In: 2017 International Conference on Advanced Informatics: Concept Theory and Applications (ICAICTA). 2017.

[7] Kurniawati P, Lestari D, Khodra M. Speech Emotion Recognition From Indonesian Spoken Language Using Acoustic and Lexical Features. In: The 20th Conference of the Oriental Chapter of the International Coordinating Committee on Speech Databases and Speech I/O Systems and Assessment. Seoul, Korea; 2017.

[8] Elhadad MK, Badran K, Salama GI. A novel approach for ontology-based dimensionality reduction for web text document classification. In: 2017 IEEE/ACIS 16th International Conference on Computer and Information Science (ICIS). 2017. p. 373–8.

Sentiment analysis: a comparison of deep learning neural network algorithm with SVM and naïve Bayes for Indonesian text

Wahyu Calvin Frans Mariel[1,3], Siti Mariyah[1,2,3], and Setia Pramana[1,2,3]

[1]Sekolah Tinggi Ilmu Statistik (STIS), Jakarta, Indonesia
[2]Center of Computational Statistics Study, STIS, Jakarta, Indonesia
[3]Badan Pusat Statistik (BPS), Indonesia

calvinrevelation@gmail.com, sitimariyah@stis.ac.id, setia.pramana@stis.ac.id

Abstract. Deep learning is a new era of machine learning techniques that essentially imitate the structure and function of the human brain. It is a development of deeper Artificial Neural Network (ANN) that uses more than one hidden layer. Deep Learning Neural Network has a great ability on recognizing patterns from various data types such as picture, audio, text, and many more. In this paper, the authors tries to measure that algorithm's ability by applying it into the text classification. The classification task herein is done by considering the content of sentiment in a text which is also called as sentiment analysis. By using several combinations of text preprocessing and feature extraction techniques, we aim to compare the precise modelling results of Deep Learning Neural Network with the other two commonly used algorithms, the Naïve Bayes and Support Vector Machine (SVM). This algorithm comparison uses Indonesian text data with balanced and unbalanced sentiment composition. Based on the experimental simulation, Deep Learning Neural Network clearly outperforms the Naïve Bayes and SVM and offers a better F-1 Score while for the best feature extraction technique which improves that modelling result is Bigram.

1. Introduction

Text mining is an application of data mining technique that devoted to managing textual data [1]. It refers generally to the process of extracting interesting and non-trivial patterns or knowledge from unstructured data text [2,3]. Basically, text mining is a multidisciplinary field. It involving information retrieval, statistics, mathematics, machine learning, linguistic, and natural language processing. Textual data type is the most natural form on storing information. Based on recent study, it is indicated that 80% of company's information is contained in text documents [2]. The massive use of text-type data demands more intensive use of the knowledge that can be extracted from it. However, text mining is a much more complex task which dealing with unstructured data. One of the most popular domains on text mining is sentiment analysis. This analysis can classify documents based on sentiment contained therein. The sentiment is generally can be classified into two groups (positive and negative classes), or three groups (positive, negative, and neutral classes).

Technically, sentiment analysis can be divided into four types of approach [4]: *Machine learning approach*, it uses a learning algorithm to create a model from training data for supervised classification. (This approach is also used in this paper.). *Lexicon-based approach*, it involves calculation of the polarity of the sentiment by using the semantic orientation of words or sentences in a document. *Rule-based approach*, it searches for opinion words and then classifies them based on the number of positive and negative words. This approach also considers several different rules such as dictionary polarity, negation words, booster words, idioms, emoticons, and mixed opinions. *Statistical model approach*, it makes any opinion as an aspect and a latent rating. This approach represented by using divisions and then clustered into rank form.

For the machine learning approach, sentiment analysis has some commonly used classification algorithms such as Naïve Bayes and Support Vector Machine (SVM). In terms of modelling paradigm, both algorithms are also usually classified as traditional algorithms. But recently, with the emergence of new method that utilize Deep Learning techniques, now sentiment analysis can perform a new era of text classification tasks. As demonstrated by the recent study, this algorithm has several advantages such as it can work on any text genres with minimal restrictions and with no task-specific or data-specific manual feature engineering [5]. In this study, we aim to measure the ability of that new algorithm (Deep Learning Neural Network) by comparing it with the traditional algorithms (Naïve Bayes and SVM) using several combination of preprocessing and feature extraction techniques for Indonesian textual data. That three algorithms tested with balanced and unbalanced sentiment composition documents.

2. Related Work

2.1. Deep Learning Neural Network (DLNN)

Deep learning is a new era of machine learning techniques which imitates the structure and function of the human brain on recognizing something. This algorithm has a unique feature that allows to automatically grasps the relevant features required for. Theoretically, the model used in deep learning is a mathematical function with $f\colon X \rightarrow Y$. Deep learning is the development of a more in-depth Artificial Neural Network (ANN) that uses more than one hidden layer on modelling the given dataset [6]. It consists of three main layers, namely:

1. *Input Layer* : the layer consists of neurons that receive data from variable X.
2. *Hidden Layer* : it consists of neurons that receive data from previous input layer. Each hidden layer is responsible for training the unique set of features. The more number of the hidden layers, the more increasing the complexity and the abstraction.
3. *Output Layer* : the layer consist of neurons that receive data from hidden layer that produces the output value and is the result of the calculation of the variable X to variable Y.

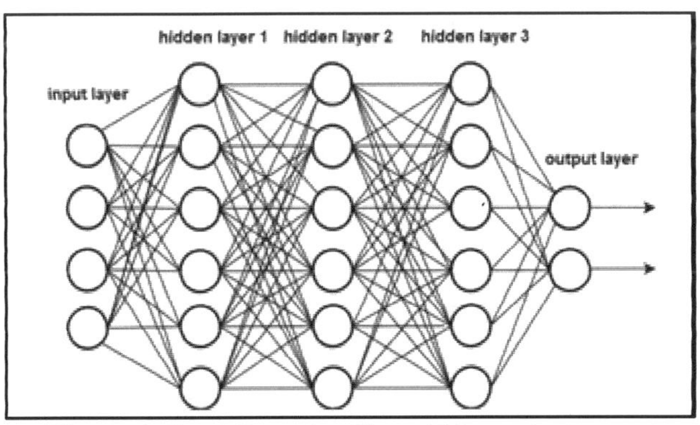

Figure 1. Deep Learning Neural Network structure

2.2. Evaluation Measurement
For the modelling result comparison for each models, the authors used three types of evaluation measurement namely: Precision, Recall and F-measure.

Precision is the relevancy ratio of the result of the prediction answer. That ratio is a comparison between the prediction accuracy and the total number of predictable answers for a particular class (e.g. positive class). That precision can be formulated as follows:

$$Precision = \frac{true\ postives}{true\ positives + false\ positives} \quad (1)$$

Recall is the ratio of how many truly relevant result are returned. The ratio is a comparison between the accuracy of the predicted answer and the true value. The formula used is as follows:

$$Recall = \frac{true\ postives}{true\ positives + false\ negatives} \quad (2)$$

F-measure or so-called F-1 score is a measure of the accuracy of a predicted result by considering precision and recall values. The result value of this measure is the harmonic average between precision and recall which can be formulated as follows:

$$F1\ Score = 2 \times \frac{precision \times recall}{precision + recall} \quad (3)$$

3. Methodology

3.1. Dataset Collection Method
Basically, the dataset used in this research is a primary data which is sourced from Twitter (social media). That text data was gained by utilizing web crawling techniques with two search keywords that is the name of one of the government institution in Indonesia and one of the most prominent Indonesian government figures in 2017. Furthermore, the text data gained from the web crawling is still a raw data that has not been labelled, while to do supervised classification we need a labelled data. Therefore the next step process we did is labeling each data one by one manually. The labelled data is used as the base data (or also called corpus) as the training dataset in creating the model.

From the result of labelling process, the data composition used in this study can be detailed as follows:

Table 1. Dataset composition

Keyword Topic	**Composition Category**	**Sentiment**		
		Positive	**Negative**	**Neutral**
Government Institution	*Imbalanced*	46	30	1.070
	Balanced	46	30	-
Government Figure	*Balanced*	126	132	-

3.2. Text Mining Method
Text data is an unstructured data type. Therefore, the additional text mining steps are required in order to transforming that text data into a a form that can be processed computationally by the machine. This step also usually called text refining [2]. One of the major limitations on text mining is the natural language used. Each language has it own rules. Because of these limitations, the authors restrict the research is only processing text data that use Indonesian language.

To discover the useful patterns from the text data (sentiment pattern), we use the text mining techniques. That techniques used to create the required model which later used to predict any text with topic about it. In this study, the text mining technique is consist of few stages as follows:

3.2.1 Text preprocessing. Text preprocesing is an early stage of semantic analysis (meaning accuracy) and syntactic analysis (arrangement accuracy). The preprocessing for Indonesian language text steps in this study consists of: case folding, word tokenizing, stop word removal, and stemming [7]. At this stage we also carried out the text formalization and word spell improvement (typo).

- *Case Folding,* is the process of uniforming all the characters of the letters that exist in a text into lowercase. For example, the phrase "*Teknologi Informasi dan Komunikasi*" after the case folding will be "*teknologi informasi dan komunikasi*".

- *Word Tokenizing,* is the process of decomposing the whole text into a words form. For example, the phrase "*teknologi informasi dan komunikasi*" after tokenized will be "*teknologi*", "*informasi*", "*dan*", and "*komunikasi*".

- *Stop Word Removal,* is a process of omitting the meaningless words that often appear in a text such as "*yang*", "*di*", "*ke*", "*atau*", etc.

- *Stemming,* is the process of converting a word into its basic form or by other terms it is the process of omitting affixes from a word. This word conversion should be done to make sure that every words that is same but has different affixes can be recognized as a similar value to avoid bias in the transformation stage.

- *Text Formalization,* is the conversion of an informal word into a standard-formal word. Thus, the words that are supposed to be one thing or the same (semantically) can be equalized when the text transformation stage.

- *Word Spell Improvement (typo),* is the alteration of an incorrect word form (syntactically) which may also be due to the typo errors into the proper arrangement. For example, the word "*beruasha*" after the word spell improvement will be "*berusaha*".

3.2.2. Feature selection. Feature selection is the stage of selecting which feature extraction technique will be used later to transform the text. In this research, the authors use four techniques of feature exraction, among others: Bag of Word, Binary Term Frequency (TF), Term Frequency-Inverse Document Frequency (TF-IDF), and Bigram (N-Grams).

3.2.3. Feature extraction (transformation). After selecting the feature at the previous stage, the text data then transformed into a matrix form (or also called feature) to make it able to be processed computationaly.

3.2.4. Pattern discovery. This is the main stage of the series of text mining process. In this stage, the important patterns from text data (training) discovered and formed into a model. As descibed earlier, the authors used three machine learning algorithms to be compared which include: Deep Learning Neural Network, Naïve Bayes, and Support Vector Machine (SVM) on modeling the Indonesian language dataset.

3.2.5. Models evaluation. After the pattern recognition stage, for evaluate the formed models, the authors uses the three measurement: Precision, Recall, and F-1 score.

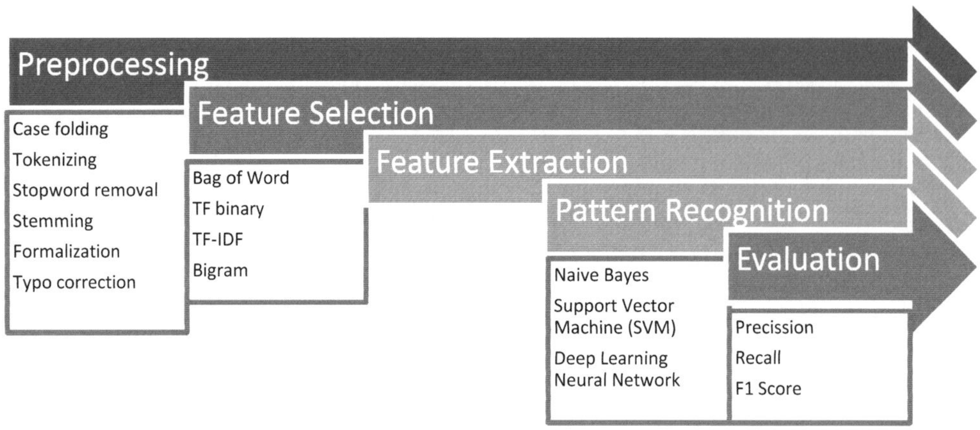

Figure 2. Text mining process used for *Bahasa Indonesia*

4. Experiments

For the implementation of this research, the authors uses Python programming language and its library to build up and process these three algorithms. Furthermore, for the Deep Learning Neural Network, we use TensorFlow and Keras as the development tools in building the Deep Learning models. TensorFlow is an open source library used for rapid numerical computation. It is created and maintained by Google and released under the Apache 2.0 license [6]. TensorFlow is the foundation library in making Deep Learning models either directly build or by using a wrappers library such as Keras.

4.1. Deep Learning Classification Architecture

For the Deep Learning Neural Network classification architecture, the authors divide the modelling algorithm into two main parts of classification:

4.1.1 Three Classes of Sentiment (positive, negative, and neutral). For this three classes classfication, the Keras model used is Sequential with the compile method is Categorical Crossentropy and layers as follows:

- 1 input layer with the number of input dim equal to the length of the vocabulary stored in the transformation stage.
- 2 hidden layer each consisting of 75 neurons and 25 neurons. This hidden layers uses normal init and relu activation.
- 1 output layer with 3 neurons that using normal init and sigmoid activation.

4.1.2 Two Classes of Sentiment (positive or negative). For this two classes classfication, the Keras model used is Sequential with the compile method is Binary Crossentropy and layers as follows:

- 1 input layer with the number of input dim equal to the length of the vocabulary stored in the transformation stage
- 2 hidden layer each consisting of 60 neurons and 20 neurons. This hidden layers uses uniform init and relu activation.

- 1 output layer with 1 neuron that using uniform init and sigmoid activation that can generate output value in range 0 up to 1 where if the resulting value is less than or equal to 0.5 will be categorized as a negative class and vice versa.

4.2. Research Frame of Mind
To simplify the text mining research in this study, the authors described the process on a frame-of-mind form as follows:

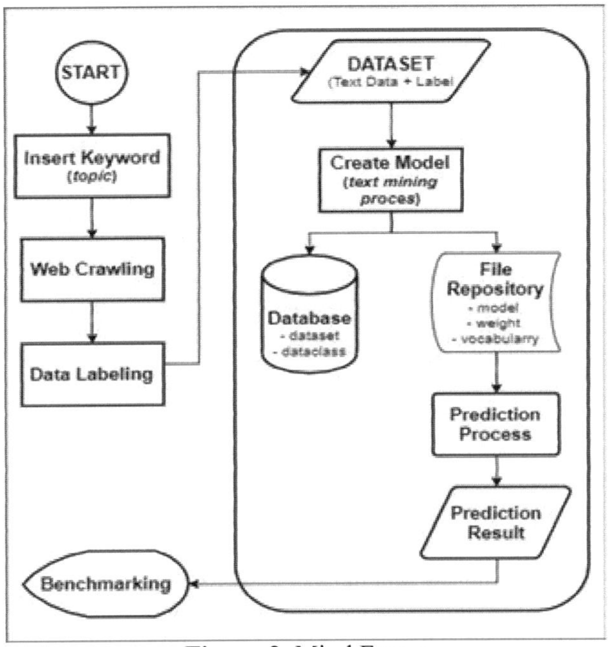

Figure 3. Mind Frame

The first process is starting from determining the topic of the text data to be studied. As described earlier, we use two topics that is about one of the government institution in Indonesia and one of the most prominent Indonesian government figures in 2017. In applying the web crawling technique, the predetermined topic then used as the text search keyword (in this research is on Twitter). After getting the raw text data needed, the next process is labelling that data one by one manually for further supervised classification process. Now we have got the dataset (data & label) needed. Next step is the main process of text mining. As described earlier, that main process is consist of 5 sub-process, they are: Preprocessing, Feature Selection, Feature Extraction, Pattern Recognition, and Evaluation (Figure2). The result of the text mining process is the models. After that, each models then evaluated by the three measurement: precission, recall, and F1 Score using the prediction dataset.

4.3. Experimental Result
Before comparing the three algorithm, each dataset has been splitted up into two group of data: 90% part of data for training set (create model) and 10% more for testing set (prediction). There are three set of data used. The first dataset is consist of three classes, they are positive, negative, and neutral. This first dataset is also categorized as imbalanced composition dataset because its neutral class has a very large size while compared to the other two classes. While the second and third class is consist of two classes, positive and negative. They are categorized as balanced composition dataset. Finally, we show the experimental result of our research on the following benchmark tables:

Table 2. Experimental Result

Algorithm	Feature Extraction Technique	Dataset: Government Institution (3 classes) with Imbalanced Composition			Dataset: Government Institution (2 classes) with Balanced Composition			Dataset: Government Figure (2 classes) with Balanced Composition		
		Precision	Recall	F1 Score	Precision	Recall	F1 Score	Precision	Recall	F1 Score
Naïve Bayes	Bag of Word	0.85	0.88	0.86	0.75	0.75	0.75	0.85	0.85	0.85
	Binary TF	0.85	0.90	0.87	0.75	0.75	0.75	0.85	0.85	0.85
	TF-IDF	0.85	0.87	0.86	0.60	0.75	0.67	0.82	0.81	0.81
	Bigram (nGram)	0.85	0.76	0.81	0.40	1.00	0.57	0.76	0.69	0.72
Support Vector Machine (SVM)	Bag of Word	0.88	0.90	0.89	0.40	0.50	0.44	0.81	0.80	0.80
	Binary TF	0.88	0.92	0.90	0.40	0.50	0.44	0.78	0.76	0.77
	TF-IDF	0.88	0.91	0.87	0.43	0.67	0.52	0.77	0.77	0.77
	Bigram (nGram)	0.88	0.92	0.90	0.40	0.50	0.44	0.83	0.69	0.75
Deep Learning Neural Network	Bag of Word	0.99	0.84	0.91	0.94	0.93	0.93	1.00	0.92	0.96
	Binary TF	0.99	0.74	0.85	0.94	0.93	0.93	1.00	0.92	0.96
	TF-IDF	0.96	0.15	0.26	0.94	0.95	0.95	1.00	0.92	0.96
	Bigram (nGram)	1.00	0.98	0.99	0.97	0.96	0.96	1.00	0.90	0.95

5. Conclusion

The sentiment classification performance of the different combination of algorithms and feature extraction techniques gives quite diverse results. Based on the research above, on all of the dataset given, the performance of Deep Learning Neural Network algorithm is over-whelming. At least, one of the feature extraction technique combinations in Deep Learning algorithm for each dataset always gets the top score. In addition, unlike Naive Bayes and SVM, the modelling results produced by Deep Learning Neural Network are also not significantly affected by either balanced or unbalanced data composition conditions. The resulting score is always high. On average, for the combination of Deep Learning Neural Network algorithms, the best feature extraction technique that can improve the final score of the modelling is Bigram (N-Grams). Thus, based on this research can be concluded that the combination of Deep Learning Neural Network algorithm with bigram technique is very suitable for conducting sentiment analysis for Indonesian text data.

Acknowledgment
The authors also thanks to institute of statistics (STIS) and The Central Bureau of Statistics (Badan Pusat Statistik or BPS) for the opportunity given in carrying out this research.

References

[1] Feldman, R. & Dagan, I. (1995) Knowledge discovery in textual databases (KDT). In proceedings of the First International Conference on Knowledge Discovery and Data Mining (KDD-95), Montreal, Canada, August 20-21, AAAI Press, 112-117.

[2] Tan, A. H. (1999, April). Text Mining: The state of the art and the challenges. In Proceedings of the PAKDD 1999 Workshop on Knowledge Discovery from Advenced Databases (Vol. 8, pp. 65-70). sn.

[3] Han, J., Kamber, M., & Pei, J. (2006). Data preprocessing. Data mining: concepts and techniques. San Fransisco: Morgan Kaufmann, 47-97.

[4] Collomb, A., Costea, C., Joyeux, D., Hasan, O., & Brunie, L. (2014). A study and comparison of sentiment analysis methods for reputation evaluation. Rapport de recherche RR-LIRIS-2014-002.

[5] Singhal, P., & Bhattacharyya, P. Sentiment Analysis and Deep Learning: A Survey.

[6] Brownlee, Jason. (2017).Deep Learning With Python: Develop Deep Learning Models on Theano and TensorFlow using Keras.

[7] Weiss, S. M., Indurkhya, N., Zhang, T., & Damerau, F. (2010). Text mining: predictive methods for analyzing unstructured information. Springer Science & Business Media.

Implementation of support vector machine for classification of speech marked hijaiyah letters based on Mel frequency cepstrum coefficient feature extraction

Wisnu Adhi Pradana, Adiwijaya, Untari Novia Wisesty

School of Computing, Telkom University, Bandung 40257, Indonesia

wa.pradana01@gmail.com, adiwijaya@telkomuniversitiy.ac.id, untarinw@telkomuniversitiy.ac.id

Abstract. Support Vector Machine or commonly called SVM is one method that can be used to process the classification of a data. SVM classifies data from 2 different classes with hyperplane. In this study, the system was built using SVM to develop Arabic Speech Recognition. In the development of the system, there are 2 kinds of speakers that have been tested that is dependent speakers and independent speakers. The results from this system is an accuracy of 85.32% for speaker dependent and 61.16% for independent speakers.

1. Introduction

Arabic is one of the oldest languages in the world, there is little Speech Recognition in Arabic than any other language. In its development, Arabic Speech Recognition has many problems that make it difficult to develop. The most influencing factor in developing the Speech Recognition Arabic language is the large variety of dialect and morphological complexity [2]. Dialect variations can cause meaning differences in spoken word. In addition, in arabic there are some rules on how to read the word (*tajwid*) so it can affect the meaning of a word.

Speech Recognition or abbreviated SR is a technology that enables computers to recognize words spoken by humans on a mic or phone. In its use, SR has been used in a variety of problems, such as the use of computer command with voice or learning a foreign language. In addition, SR also helps interact with foreign people.

There are many methods that have been used in building an SR. However, the method often used to build an SR is the Hidden Markov Model or commonly called HMM. HMM is a model that assumes parameters of unknown parameters and hidden parameters. Therefore, the Speech Recognition Arabic that has evolved today is using the HMM approach [6].

Support Vector Machine or commonly called SVM is one method that can be used to process the classification of a data [4]. The author intends to use the approach in the development of Arabic Speech Recognition. With this system, it is expected to increase the choice in using Arabic Speech Recognition as a classifier of hijaiyah marked letters.

In this paper, the result of implementation Support Vector Machine for Arabic Speech Recognition is given. The proses of Mel Frequency Cepstrum Coefficient will be elaborated in section 2. In the next section, Support Vector Machine proses will be explained. In the section 4, the result has been given to in detail. In the last section, there are the conclusion for all paper.

2. Mel Frequency Cepstrum Coefficient

Mel Frequency Cepstrum Coefficient or MFCC is the most commonly used approach for speech recognition. MFCC is the only acoustic approach that takes the human perception (physiology and behavioral aspects of the sound production organs) as a consideration for building speech recognition which means this approach can not process frequencies above 1KHz. This approach has two types of filters that are linearly at low frequencies below 1000Hz and logarithmic distances above 1000Hz.

There are several steps in the MFCC algorithm to generate a feature vector. These steps are pre-emphasis, frame blocking, windowing, Fast Fourier Transform, Mel Filter Bank Processing, discrete cosine transformation and Delta Energy and Delta Spectrum. The result of MFCC is a feature vector that will be used as a database to determine the classification of voice data.

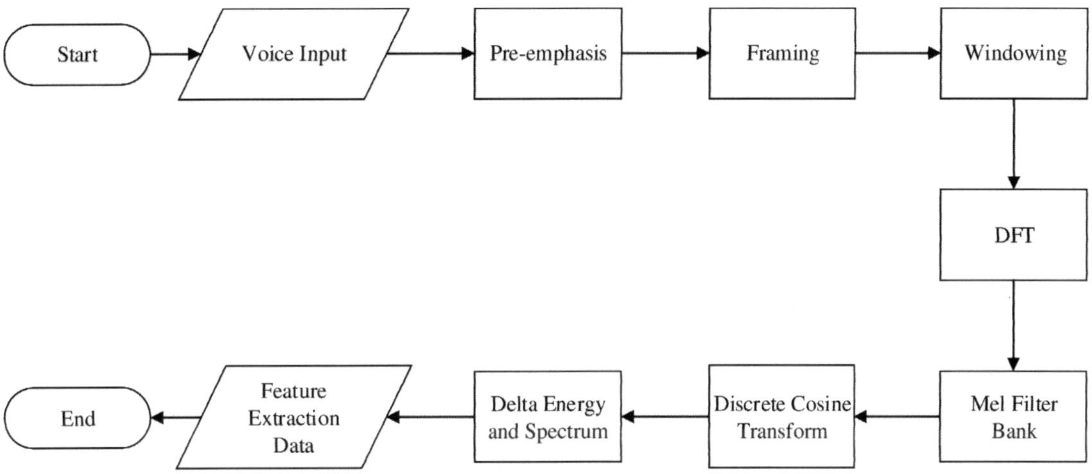

Figure 1. The Step of MFCC[9]

3. Support vector machine

Support vector machine or can be called SVM is one method of supervised learning that can be used for data classification [4]. SVM divides the data into two classes by maximizing margins between hyperplane with the two closest data in both classes [4]. The most optimal hyperplane is called a support vector. To obtain the best hyperplane calculations are needed to determine the distance between data with existing hyperplane. In the first learning process, we will find the minimum distance from each training data on hyperplane using formula 1 and 2.

$$y_i(w^T.x_i + b) \geq 1 \quad ; i = 1,2 \ldots n \tag{1}$$

x_i = data input
y_i = target
b = position of the plane towards the coordinate center
w = normal plane

Equation 2 can be solved using the optimal point of the lagrange function.

$$L_p = L_{(w,b,\alpha)} = \frac{1}{2}\|w\|2 - \sum_{i=1}^{n} \alpha_{i(y_i(w^T x_i+b)-1)} \tag{2}$$

α_i is a lagrange multiplier. To find the optimal point, then w and b must be minimized so that L = 0. Under these conditions, then α can be known.

To make a decision from data testing, it takes a decision function created from the kernel function. There are many kernel functions that can be used for decision making, here are examples of frequently used functions:

$$\textit{Linear kernel} \quad : K(x_i, x_j) = x_i^T x_j \tag{3}$$
$$\textit{Polynomial kernel} : K(x_i, x_j) = (\gamma x_i^T x_j + r)^d \tag{4}$$

d = degree
x_i dan x_j = *vector* data
γ dan r = parameter
Quadratic kernels also include polynomial kernels, but with a second order [16].

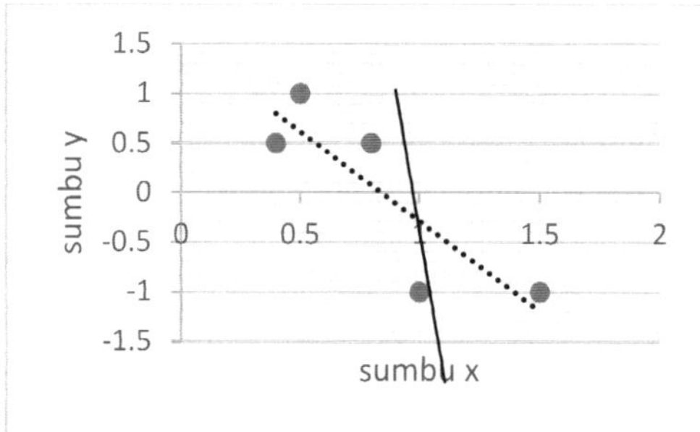

Figure 2. finding proses of best hyperplane

3.1. Multi-class SVM
Initially SVM can only be used to classify 2 classes only. However, in its development the researchers have developed a method that makes SVM can be used for the classification of more than 2 data classes. There is a way to implement multi-class SVM by combining some binary SVM or combining all data into an optimization system.

The "one against all" method is one of the methods that can be used for multi-class SVM problems. This method builds n the binary SVM model (n is the number of classes). Each model is in-training with all the data to look for solution. See table 2.1 and figure 3.2

Table 1. the example of SVM biner one-against-all

$y_i = 1$	$y_i = -1$	Hypothesis
Class 1	Not class 1	$f_1(x) = (w_1)x + b_1$
Class 2	Not class 2	$f_2(x) = (w_2)x + b_2$
Class 3	Not class 3	$f_3(x) = (w_3)x + b_3$
Class 4	Not class 4	$f_4(x) = (w_4)x + b_4$

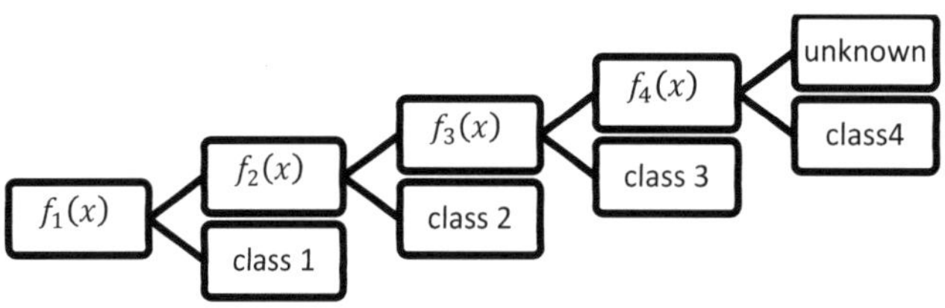

Figure 3. classifying the data

4. Result and Comparison

4.1. Dependent speaker
Accuracy is a measure that approximates the correct size or is received from the quantity amount being measured. Accuracy gained from the number of test results is true divided by the amount of tested data. Table 4.1 is given to represent the results of the types of dependent speakers that have been produced during the study.

Table 4.1 illustrates the accuracy obtained on each letter that has been tested against the model with 2 different kernel functions. The biggest accuracy obtained by using linear and quadratic kernel functions is *ro* (ر) with 97.22% for linear kernel function and 61.11% for quadratic kernel function. The smallest accuracy obtained by using linear kernel function and quadratic kernel function is *tsa* (ث) with 72.22% for linear kernel function and 22.22% for quadratic kernel function. This indicates that the optimized model is used to classify *ro* letters rather than *tsa*. In addition, the average accuracy obtained from systems with linear kernel function is 85.32% and quadratic kernel function is 43.45%. The distribution of letters that run on the system also affects the results of accuracy. With this division, the system does not need to create 168 class models directly. In addition the process of making the model can diroleh more quickly

Table 2. result of testing data with dependent speaker

No	Letter	Hijaiyah	Accuracy		No	Letter	Hijaiyah	Accuracy	
			Linear	Quadratic				Linear	quadratic
1	Alif	ا	88,89%	44,44%	15	Dhod	ض	88,89%	47,22%
2	Ba	ب	75%	33,33%	16	Tho	ط	91,67%	52,78%
3	Ta	ت	77,78%	38,89%	17	Zho	ظ	80,56%	44,44%
4	Tsa	ث	72,22%	22,22%	18	Ain	ع	86,11%	58,33%
5	Jim	ج	86,11%	27,78%	19	Ghoin	غ	77,78%	52,78%
6	Ha	ح	88,89%	52,78%	20	Fa	ف	83,33%	33,33%
7	Kho	خ	94,44%	44,44%	21	Qof	ق	86,11%	44,44%
8	Dal	د	88,89%	38,89%	22	Kaf	ك	86,11%	38,89%
9	Dzal	ذ	86,11%	41,67%	23	Lam	ل	88,89%	47,22%
10	Ro	ر	97,22%	61,11%	24	Mim	م	86,11%	30,56%
11	Zai	ز	88,89%	50%	25	Nun	ن	80,56%	47,22%
12	Sin	س	94,44%	47,22%	26	Wa	و	80,56%	41,67%
13	Syin	ش	80,56%	36,11%	27	Ha	ه	88,89%	44,44%
14	Shat	ص	83,33%	38,89%	28	Ya	ي	80,56%	55,56%

4.2. Independent speaker

Tests on independent speaker types are performed using linear kernel functions. Table 4.2 represents the results of testing using linear kernel functions. There are 672 of testing data that has been tested in this system. This data comes from 1 person who speaks all the letters with different *harakat*.

In table 4.2 can be seen that the biggest accuracy obtained the letters *sin* (س) and *tho* (ط) is 87.5%. While the smallest accuracy obtained from the test results of this model is the letter *fa* (ف) with an accuracy of 25%. This indicates that the model is more optimize to classify the letters *sin* and *tho* rather than the letter *fa*. In addition, the average accuracy obtained from the system is 61.16%.

Table 3. result of testing data with independent speaker

No	Letter	Hijaiyah	Accuracy	No	Letter	Hijaiyah	Accuracy
1	Alif	ا	75%	15	Dhod	ض	75%
2	Ba	ب	54,17%	16	Tho	ط	87,50%
3	Ta	ت	75%	17	Zho	ظ	83,33%
4	Tsa	ث	54,17%	18	Ain	ع	75%
5	Jim	ج	66,67%	19	Ghoin	غ	41,67%
6	Ha	ح	41,67%	20	Fa	ف	25%

7	Kho	خ	70,83%	21	Qof	ق	41,67%
8	Dal	د	83,33%	22	Kaf	ك	66,67%
9	Dzal	ذ	54,17%	23	Lam	ل	62,50%
10	Ro	ر	66,67%	24	Mim	م	29,17%
11	Zai	ز	58,33%	25	Nun	ن	37,50%
12	Sin	س	87,50%	26	Wa	و	50%
13	Syin	ش	58,33%	27	Ha	هـ	62,50%
14	Shat	ص	50%	28	Ya	ي	79,17%
Average				61.16%			

4.3. Comparison

The best accuracy that obtained from support vector machine model with linear kernel function is *ro* (ر) with 97,22% accuracy value for dependent speaker data testing. While for independent speaker test data, the best accuracy result obtained from support vector machine model with linear kernel function is *sin* (س) and *tho* (ط) with 87,5% accuracy. In addition, the average accuracy of both types is 85.32% for dependent speakers and 61.16% for independent speakers.

Meanwhile, support vector machine model with quadratic kernel function is only obtained from dependent speaker type. This happens because the support vector machine model with quadratic kernel function fails to classify data testing from independent speakers. The best accuracy results obtained from the support vector machine model with quadratic kernel function is the letter of *ro* (ر) with an accuracy value of 61.11%. In addition, the average accuracy obtained from this model is 43.45%.

From the results of tests that have been done that the support vector machine model with the type of dependent speaker is better than the independent speaker type. This proves the dependency of a data affecting the results of the classification. In addition, kernel function also affects the result of classification, it is evident that linear function is better than quadratic function.

5. Conclusion

Based on the results of tests that have been implemented, it can be drawn some conclusions as follows, the treatment of training and testing into each letter (each letter has 6 classes) makes the system work faster than all data is processed together. The kernel function affects the accuracy of the system with the use of linear functions resulting in an average accuracy of 85.32% for speaker dependent and 61.16% for independent speakers. Dependency of a data affects the classification results. Models that use dependent speakers have better accuracy than independent speakers

References

[1] Abushariah, M.A.A.M., Ainon, R., Zainuddin, R., Elshafei, M. and Khalifa, O.O., 2012. Arabic speaker-independent continuous automatic speech recognition based on a phonetically rich and balanced speech corpus.International Arab Journal of Information Technology (IAJIT), 9(1), pp.84-93.

[2] AbuZeina, D. and Elshafei, M., 2012. Arabic Speech Recognition Systems. In Cross-Word Modeling for Arabic Speech Recognition (pp. 17-23). Springer US.

[3] Alotaibi, Y.A. and Hussain, A., 2010. Comparative analysis of arabic vowels using formants and an automatic speech recognition system.

[4] Durgesh, K.S. and Lekha, B., 2010. Data classification using support vector machine. Journal of Theoretical and Applied Information Technology, 12(1), pp.1-7.

[5] El Choubassi, M.M., El Khoury, H.E., Alagha, C.J., Skaf, J.A. and Al, M.A., 2003. Arabic speech recognition using recurrent neural networks. In IEEE Intl. Symp. Signal Processing and Information Technology ISSPIT.

[6] Juang, B.H. and Rabiner, L.R., 1991. Hidden Markov models for speech recognition. Technometrics, 33(3), pp.251-272.

[7] Kalinli, A. and Sagiroglu, S., 2006. Short Paper_. Journal of Information Science and

Engineering, 22, pp.1555-1568.

[8] Meng, Y.U.A.N., 2004. Speech recognition on DSP: Algorithm optimization and performance analysis (Doctoral dissertation, The Chinese University of Hong Kong).

[9] Muda, L., Begam, M. and Elamvazuthi, I., 2010. Voice recognition algorithms using mel frequency cepstral coefficient (MFCC) and dynamic time warping (DTW) techniques. arXiv preprint arXiv:1003.4083.

[10] Satori, H., Hiyassat, H., Harti, M. and Chenfour, N., 2009. Investigation arabic speech recognition using CMU sphinx system. Int. Arab J. Inf. Technol., 6(2), pp.186-190.

[11] Singh, S. and Rajan, E.G., 2011. Vector quantization approach for speaker recognition using MFCC and inverted MFCC. International Journal of Computer Applications, 17(1), pp.1-7.

[12] Shady, Y. and Zayed, S.H.H., 2009. Speaker independent Arabic speech recognition using support vector machine.

[13] Susanto, C.P. and Setiawan, E.I., 2015. Algoritma Support Vector Machine Untuk Mendeteksi Sms Spam Berbahasa Indonesia.

[14] U.N. Wisesty, M.S. Mubarok, Adiwijaya, A classification of marked hijaiyah letters' pronunciation using hidden Markov model, AIP Conference Procedings 1867(1), 020036

[15] Zhang, S.X., 2014. Structured support vector machines for speech recognition.

[16] Chang, Y.W., Hsieh, C.J., Chang, K.W., Ringgaard, M. and Lin, C.J., 2010. Training and testing low-degree polynomial data mappings via linear SVM. Journal of Machine Learning Research, 11(Apr), pp.1471-1490.

On the structure of Bayesian network for Indonesian text document paraphrase identification

Ario Harry Prayogo, Mohamad Syahrul Mubarok, Adiwijaya

School of Computing, Telkom University

ariohp@telkomuniversity.ac.id, msyahrulmubarok@telkomuniversity.ac.id, adiwijaya@telkomuniversity.ac.id

Abstract. Paraphrase identification is an important process within natural language processing. The idea is to automatically recognize phrases that have different forms but contain same meanings. For examples if we input query “causing fire hazard”, then the computer has to recognize this query that this query has same meaning as “the cause of fire hazard. Paraphrasing is an activity that reveals the meaning of an expression, writing, or speech using different words or forms, especially to achieve greater clarity. In this research we will focus on classifying two Indonesian sentences whether it is a paraphrase to each other or not. There are four steps in this research, first is preprocessing, second is feature extraction, third is classifier building, and the last is performance evaluation. Preprocessing consists of tokenization, non-alphanumerical removal, and stemming. After preprocessing we will conduct feature extraction in order to build new features from given dataset. There are two kinds of features in the research, syntactic features and semantic features. Syntactic features consist of normalized levenshtein distance feature, term-frequency based cosine similarity feature, and LCS (Longest Common Subsequence) feature. Semantic features consist of Wu and Palmer feature and Shortest Path Feature. We use Bayesian Networks as the method of training the classifier. Parameter estimation that we use is called MAP (Maximum A Posteriori). For structure learning of Bayesian Networks DAG (Directed Acyclic Graph), we use BDeu (Bayesian Dirichlet equivalent uniform) scoring function and for finding DAG with the best BDeu score, we use K2 algorithm. In evaluation step we perform cross-validation. The average result that we get from testing the classifier as follows: Precision 75.2%, Recall 76.5%, F1-Measure 75.8% and Accuracy 75.6%.

1. Introduction

Natural Language Processing (NLP) is a technique that is used to analyze and represent human language automatically. It learns mathematical and computational model from many aspects of the language. NLP is used to take grammatical structure and build an output based on rules that present in the language which is the object of the process [1].

There are examples of NLP implementation such as plagiarism detection, information retrieval, text summarization, question answering, and machine translation. There is one essential process for plagiarism detection called paraphrase identification. Paraphrase is a rephrase of a writing, speech, or an expression of a level in language form to another form without altering its meaning; on another word paraphrase can be translated as a different expression of a text to another form to reveal the implicit meaning [2]. There is research regarding paraphrase identification of a social media (twitter).

There's one thing that makes paraphrase identification is important. If we are doing plagiarism detection, then we need to make an automated system that can determine phrases that are different but having same meaning. For example there are pair of phrase, first phrase is "the source of wildfire" and the second phrase is "cause of wildfire". The system must be able to identify the pair of phrase is a paraphrase. In Indonesian Language structure there are prefix, suffix, infix and circumfix which making NLP more complex than English.

To solve the problem, we try a method that consists several steps, the first is preprocessing, second is feature extraction, third is classifier building, and the last is performance evaluation. Preprocessing step is to improve data quality and remove unnecessary computation by reducing data noise. There are four steps in preprocessing that are tokenization, non-alphanumerical removal, stemming and translation. We use Nazief-Adriani algorithm for stemming because it gives the best result for Indonesian Language corpus [3]. After preprocessing we perform feature extraction process. The idea of using the process is to build new features from dataset. The first kind of extracted features is syntactic features which consist of normalized levenshtein distance feature, term-frequency (tf) based cosine similarity feature, and LCS (Longest Common Subsequence) feature. The second kind of feature is semantic features which calculate similarity between term based on distance in semantic tree (in our case we use WordNet). The semantic distance features consist of Wu and Palmer Features and Shortest Path Features. The next phase of paraphrase identification is classification using Bayesian Networks. Parameter that is used in the classifier is Maximum A Posteriori (MAP). For structure learning of Bayesian Networks DAG (Directed Acyclic Graph), we use BDeu (Bayesian Dirichlet equivalent uniform) scoring function and for finding DAG with the best BDeu score, we use K2 algorithm. In evaluation step we perform n-fold cross-validation with the number of n is 20.

Bayesian Networks is a probabilistic based modeling method that represent a variable set and the conditional dependency through a Directed Acyclic Graph (DAG) [4] [5]. There are four reasons why we choose Bayesian Networks as the classifier, first Bayesian Networks can handle incomplete dataset, second it allows the learning process based on cause and effect, third it is based on Bayesian Statistics which facilitate combination between data and knowledge domain, and fourth it provide a way to avoid overfit data [6].

2. Related Works

There are some researches regarding paraphrase identification such as "PDLK: Plagiarism Detection Using Linguistic Knowledge" [7]. The research proposes a method to identify paraphrase by combining semantic similarity along with word order similarity. The idea is to get a new feature that can handle similarity of a document semantically. For a limit value to separate document whether the content is same or not, they attempt an observation to get a threshold value and alfa weighting. The best alfa weighting value is 0.8 and the best threshold value is 0.6 which produce performance evaluation score of 73.9%, the evaluation uses F1-Measure approach. The research is related to ours in way that the research uses similar method for handling the similarity of a document (in our case is phrase or sentence). The difference of their research with ours is their research takes word order into account whilst our research uses two methods of semantic similarity instead and combined them with syntactic features.

Another research that discuss about paraphrase identification is using translated source where the translated language result is confirmed having same value with original document [8]. The research performs paraphrase identification by translating the phrases into desired language. They later perform an unsupervised learning algorithm for paraphrase extraction. They claim that the approach has the ability to extract multiple kind of paraphrases by doing identification of syntactic paraphrases and extraction of morpho-syntactic paraphrasing. There are other researches such as the research which uses named entity anchors [9]. The research main purpose is to automatically extract paraphrases from a corpus which is from headline news articles. They use named entity and POS Tagging to extract pattern from news article [10]. There is research about text classification using POS Tagging to measure polarity of words [11].

Paraphrase Identification and semantic similarity in twitter with simple features [12] perform syntactic and string similarity between pair of sentences by using logistic regression model with 18 features based on N-Grams. They use Edit Distance between sentences as a feature and normalize the score on the number of token of the shortest sentence. They claim that the feature configuration is best in term of performance score during evaluation on training and development data. In classification process they used several classification methods which Voted Perceptron method returns the highest F1-score of them all with F1-score reaching 74.6%.

Paraphrase identification in Indonesian Corpus [13] tried to identify Indonesian sentence pair whether it's a paraphrase or not. The research uses levenhstein distance and wu palmer method as features (nodes). The total nodes number is three including the class, then the nodes are used in classifier Bayesian Networks Classifier. For Bayesian Networks structure, the research specifies all possible combination of nodes, resulting 25 DAG structures. The research performance evaluation using F1-score which is 71.5%.

3. Research Method

In this section we describe our paraphrase identification method. The general process that is performed in this research is described in Figure 1.

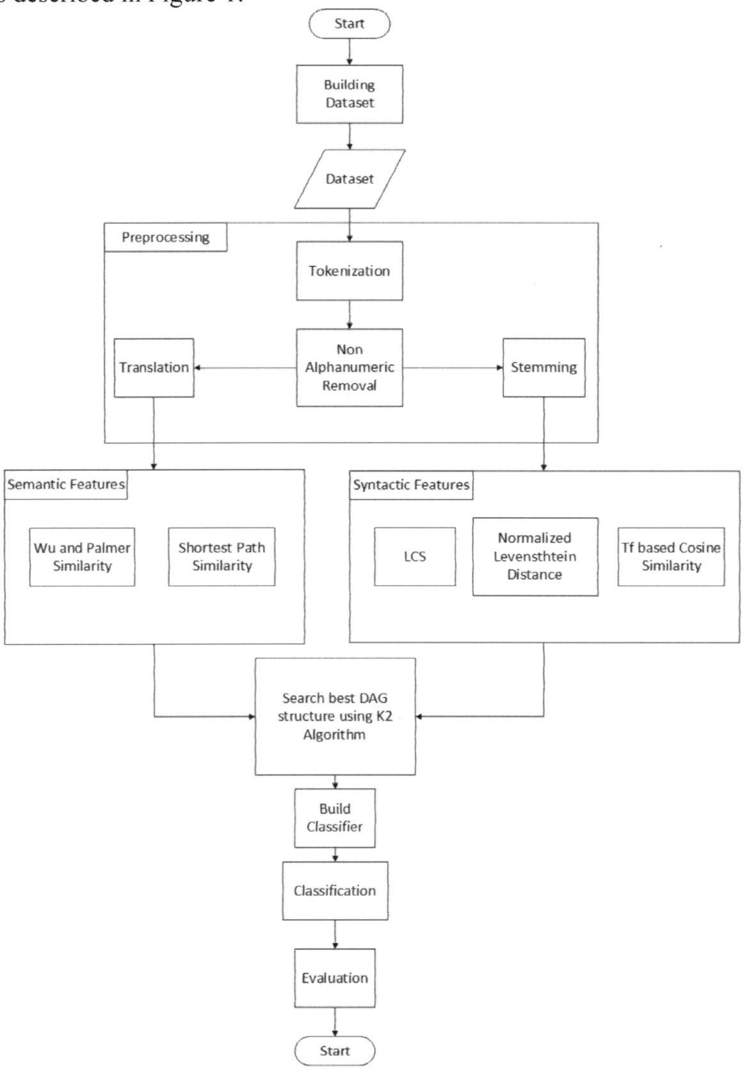

Figure 1 Research Method of Paraphrase Identification

The research method includes four main process. In the first step, preprocessing data with tokenization, nonalphanumeric removal, stemming and translation. The second step is feature extraction which is divided by two types of features, semantic and syntactic. The semantic features consist of Wu and Palmer feature, and Shortest Path feature. The syntactic features consist of normalized levenshtein distance feature, LCS feature, and Tf based cosine similarity feature. All of it means that the number of features used in this research are five features.

Those five features will be used in the third step which is classifier building. Third step is finding the best DAG based on highest number of BDeu score using K2 Algorithm. The fourth step is classifier building using the DAG that is returned from using K2 Algorithm. The detail of each process described in Figure 1 is explained as follows.

3.1 Building Dataset

In this process we are collecting and building paraphrase dataset of Indonesian Language. The data consists of 1004 pair of phrases/sentences along with its label either 1 which is a paraphrase or 0 which is not a paraphrase. This dataset has balance class, it means the number of dataset which is paraphrase is equal with the number of dataset which is not a paraphrase.

When we collect the data, we use several sources which mainly from Indonesian online news articles. We collect all the news from several news websites with similar article. For example, we took news articles from Merdeka[1] and Republika[2]. After we found pair of articles with similar theme, we try to manually find sentences in both articles that is having similar meaning. After that the pair of sentences are compiled as data in the dataset. There are other minor amounts of sources that are used as paraphrase data such as local Indonesian course text book and daily phrase/sentence that usually found in twitter. All of that is just to produce correct paraphrase, for false/incorrect paraphrase we get help from Indonesian Language experts to generate non-paraphrase data. The last phase of building dataset is validation with Indonesian Language experts to identify class/label whether the label is correct or incorrect, if the data is incorrect the experts will change the label to the correct label. Explanation of dataset distribution is explained in Table 1.

Table 1 Dataset Distribution

Class	Number of Data	Percentage
Paraphrase	502	50%
Non-paraphrase	502	50%
Total Dataset	1004	100%

3.2 Preprocessing

This process is the start of processing data before we identify whether a pair phrase is a paraphrase or not. The purpose of this step is to improve data quality and remove unnecessary computation by reducing data noise, all of that is to pursue performance improvement of the classifier. This preprocessing consists of four step which will be explained below.

3.2.1 **Tokenization**, the process that convert phrase or sentences into vector of words. The separation is necessary for stemming process.

3.2.2 **Non-alphanumeric removal**, the process that remove all characters except numbers and letters. By doing this process, the unnecessary information is removed. For example when we perform semantic feature extraction, all characters except numbers and letter will not giving any improvement on the feature quality. That because in this research, semantic feature extraction uses WordNet which only support words. So, if the non-alphanumeric

[1] https://www.merdeka.com/uang/ini-kemampuan-pesawat-n219-buatan-pt-di.html

[2] http://nasional.republika.co.id/berita/nasional/daerah/17/08/16/ouro87377-ini-keunggulan-pesawat-n219-buatan-pt-dirgantara-indonesia

characters will not be useful then we don't need to include those characters into computation in the first place. We need to mention that non-alphanumeric character like "," and "." will not be removed if the character represents decimal number.

3.2.3 **Stemming**, the idea of this process is getting root word from available words in the dataset. The goal of getting root word is to prevent calculation mistake when performing syntactic feature extraction. For example when pair sentences are having same root word for all their words but only different on affix, suffix, infix or circumfix, if we do not perform stemming, the distance score of syntactic feature will not be 0, but if we perform the stemming we will get distance score of 0. If we get score 0 it means the meaning of pair sentence is equal, other than that it means the pair has different meaning whether it's little bit different or explicitly different.

3.2.4 **Translation**, we attempt to translate the vector data into English using Google translator API. The reason why we perform translation because the lack of completed Indonesian thesaurus that is available in the moment. The current Indonesian thesaurus is still on development and not ready to be used as semantic similarity source.

3.3 Feature Extraction

In this process we will extract new features from set of data that is resulted by preprocessing. There are two feature that we will extract which are syntactic feature and semantic feature. Based on [12] that we can measure document similarity using both features. Before we explain each feature in feature extraction we like to explain a vector normalization method called cosine similarity or semantic-vector approach [14] to measure the similarity between sentences. The method is created for each two phrases/sentences, it means each phrase/sentence will have the vector. The pair vectors is containing words from pair phrases/sentences. The equation to calculate sentence similarity as follows:

$$Sim_{cosine}(s_1, s_2) = \frac{\sum_{i=1}^{m}(w_{1i} - w_{2i})}{\sqrt{\sum_{i=1}^{m} w_{1i}^2} \times \sqrt{\sum_{i=1}^{m} w_{2i}^2}} \quad (1)$$

Where $S_1 = (w_{11}, w_{12},..., w_{1n})$ and $S_2 = (w_{21}, w_{22},..., w_{2n})$ are the vectors of sentence S_1 and S_2. The w_{jk} is the weight score of the jth word in vector S_k, n is the number of words in the vector.

3.3.1 Semantic Feature. The semantic feature is a feature that calculate semantic distance of pair phrase/sentence. The goal is to determine similarity of the pair by using WordNet as the semantic tree. Semantic feature is containing two features which are Wu Palmer feature and Shortest Path feature.

a. Wu Palmer

WUP (Wu and Palmer) is our first semantic feature, the method is very simple and having good trade-off between performance and execution time cost [15]. WUP method determines similarity by calculating the depths between two synsets in the Wordnet taxonomies, also with the depth of LCS. The equation of WUP method is as follows [16].

$$Sim_{wup}(s_1, s_2) = \frac{2 * depth(lcs)}{(depth(s_1) + depth(s_2))} \quad (2)$$

For depth(lcs) denotes the depth of the lcs of sentence. LCS itself means the least common subsume.

b. Shortest Path

This method calculates shortest path between pair of term and return minimum distance between them. It based on shortest path that connects the senses in the is-a (hypernym/hyponym) taxonomy. Most of the links in Wordnet come in pairs like hyponym-hyponym, antonym-antonym and meronym-meronym. Most of derivational links are bidirectional which means part of Wordnet is a large undirected graph [17]. A problem of calculating semantic similarity between two words can be translated into shortest path problem of undirected graph G(V,E). It means we can apply basic searching algorithm such as Bidirectional Breadth First Search. Edge length is defined by equation as follows.

$$dist(S,T) \min path\_length(s,t) \tag{3}$$

Where s is source term and t is target term, and $s \in S$ and $t \in T$.

c. Calculating similarity with semantic-vector method

This process is to calculate sentence similarity with semantic-vector approach [7]. The following steps are performed to calculate the semantic similarity.

i. Creating semantic vector
   Each Semantic-vector is created by words from each phrase/sentence. Each value in vector corresponds to a word in phrase/sentence, it means the dimension of the vector equals the number of words that are contained in a phrase/sentence S_1 or S_2. For $S_1 = \{t_{11}, t_{12},...,t_{1n}\}$ and $S_2 = \{t_{11}, t_{12},...,t_{1n}\}$ where t_{ij} is term/word in sentence i and word j.
ii. Adding semantic weight to vector
   Each row of semantic-vector is weighted using semantic similarity value between words from word set of sentences. There some rules for adding weight to the vector:
   1. If the word, w from sentence S_1 appear in the sentence S_2, the weight score for w in the semantic-vector is set to 1. Else, go to next step.
   2. If sentence S_2 does not contain w, then calculate similarity score using either Wu and Palmer or shortest distance method. The result of this calculation is a array represents similarity measure between w and each word in S_2. It means the array dimension is equal with the number of words contained in S_2.
   3. If the array is not empty then we set the weight score for w as maximum value in array result. If array is empty then the weight score for w is set to 0.
iii. After we acquire semantic-vector weight, we compute the similarity between S_1 and S_2 using equation 1.

3.3.2 Syntactic Feature. Syntactic feature is a method which calculates distance of pair phrase/sentence based on lexical characteristics of a sentence. The syntactic feature consists of three features they are normalized levenshtein distance, lcs and tf based cosine similarity.

a. Normalized Levenshtein distance
   The distance calculation is using Normalized Levenshtein Distance method. We use this method because this method able to handle addition, reduction, or changes of a letter in the phrase/sentence. The result of syntactic feature is a number that range between 0 and 1, where 0 means the phrase/sentence is totally equal and 1 means totally different. There is three operations that this method takes into account. The explanation of the operations as follows:
   i. Character addition, this operation basically adds a character into a word. For example in Indonesian, word "asa" becomes "asap" where there's an addition of "a" character p.
   ii. Character removal, this operation removes a character from given word. For example in Indonesian, word "makan" if we remove first character of the word, then the word becomes "akan"

iii. Character substitution, this operation changes a character from given word with another character. For example word "angan" if we change character "a" in digit 4 with character "i" then the word becomes "angin" [18].

The Levenhstein distance [18] gives the result of an edit distance between pair sentences. But the result is not enough because the value returned by the method depends on the length of the sentences. Which means shorter the sentences will likely return smaller value than longer sentences. To handle the problem, we use normalization, which divide the result with the length of the longest sentence between S_1 and S_2.

b. Term frequency based cosine similarity
This feature main point is to consider the number of word appearance in a sentence. The steps performed in the process as follows.
1. Counts term frequency for each word in a sentence, for both S_1 and S_2, the result is in vector form which represents the number of word appearances in a sentence. It means we will have two vectors.
2. Calculate the distance between those vectors using cosine similarity in equation 1.

The result of this process returns a value between 0 and 1, which is a distance in a form of cos value. If the result is 0, it means the pair sentences are completely different. If the result is 1, it means the pair sentences are completely the same.

c. Longest common subsequence (LCS)
To determine similarity with this method we perform comparation between words from S_1 and S_2. Subsequence is any word that is acquired by removing zero or more symbols from w. Sub-word of w means that a subsequence of w that consists of consecutive symbols. For example 'ABCBDAB' and 'BDCABA' give us 'BCBA' as LCS result. A common subsequence of w and w'. We gain words similarity in the sentences by measuring length of LCS [19].
After obtain LCS result for each word, we count the length of LCS for each result and insert all the number into a vector. This means that the vector contains number of LCS length that represent each word in the sentence. We do this for each sentence which means there are two vectors. Next step we compute similarity using cosine similarity in equation 1.

3.4 Building Classifier

In this task we perform two processes, first is finding best DAG structure and the second is building classification. We search the best DAG structure using K2 algorithm with BDeu as scoring function in structure learning process. The BDeu score function is explained in equation 4 [20].

$$BDeu(N : D,\alpha) = \sum_{i}^{n} \left\{ \begin{array}{l} \sum_{j}^{q_i} log\,\Gamma\left(\frac{\alpha}{q_i}\right) - log\,\Gamma\left(\frac{\alpha}{q_i} + n_{ij}\right) + \\ \sum_{k}^{r_i} log\,\Gamma\left(\frac{\alpha}{r_i . q_i} + n_{ijk}\right) - log\,\Gamma\left(\frac{\alpha}{r_i . q_i}\right) \end{array} \right\} \quad (4)$$

Where:
α : an alpha value or hyperparameter
q_i : instantiation number parent X_i
r_i : Value combination of X_i
n_{ij} : The number of instance of $X_i = j$

n_{ijk}: The number of instance of $X_i = k$ and parent from $i = j$

The first process of building classifier is to find best DAG structure using K2 algorithm. The following pseudocode is the algorithm of K2 [21].

```
procedure K2;
{Input: n, o, u, D}
{Output: For each node, a printout of the parents of the node.}
for i := 1 to n do
    πi : = θ;
    Pold := g(i, πi); {This function is computed using equation (4).}
    OKToProceed := true
    while OKToProceed and |πi| < u do
        let z be the node in Pred(xi) - πi that maximizes g(i, πi U {z});
        Pnew := g(i, πi U{z});
        if Pnew > Pold then
            Pold := Pnew;
            πi := πi U{z}
        else OKToProceed := false;
    end {while};
    write('Node:', xi, 'Parents:', πi)
end {for};
end {K2};
```

Where:

n	=	A set of n nodes in graph,
o	=	An ordering on the nodes, in our case the order represents naive bayes form,
u	=	The number of parents a node may have,
D	=	A database containing m cases

After obtain DAG structure, we create joint probability for the graph. The function for joint probability represented in the function 5 as follows.

$$P(U) = \prod_{i}^{n} P(P_k|parents(P_k)) \quad (5)$$

For parameter learning we use Maximum A Posteriori (MAP) that is explained in function 6.

$$\theta_{ijk} = \frac{\alpha_{ijk} + n_{ijk}}{\sum_k (\alpha_{ijk} + n_{ijk})} \quad (6)$$

Where:

θ_{ijk} = Parameter where X_i = k and Parent of i = j
n_{ijk} = The number of instance of X_i = k and parent from i = j
α_{ijk} = Smoothing

The smoothing value can be obtained through function 7.

$$\alpha_{ijk} = \frac{\alpha}{r_i \cdot q_i} \tag{7}$$

Where:

α_{ijk} = Smoothing
α = an alpha value or hyperparameter
r_i = value combination of X_i
q_i = Instantiation parent number of X_i

Alpha (α) value is used to prevent dividing by zero, in our case we use α = 0.1.

4. Result and Evaluation

In this section we will evaluate the result from our work, first we will evaluate the combination of the features built in feature extraction process. Second, we will evaluate the number of parents allowed during DAG structure searching using K2 algorithm. Third we will evaluate the initiation order when using K2 algorithm, the first initiation we will use nodes order that represent naive bayes classifier, and the second initiation we will use random order for the nodes. The evaluation measures that we will be using are accuracy, precision, recall, and F1-Measure. Our performance measures result is the average of n-fold cross-validation results with n = 20.

4.1 Features Evaluation

In this task we will evaluate features quality by creating several combinations of features. We will observe which feature combination returns the best result. The goal is to find out whether the feature is worth to keep during identifying paraphrase or not. As we know the more features means the more nodes exist in the DAG, which means adding time and space complexity. The features combination performance is explained in Table 2.

Table 2 Features combination performance

No	Features	Accuracy	Precision	Recall	F1-Score
1	LEV + COS + LCS + WUP + PATH	0.756	0.752	0.765	0.758
2	LEV + COS + LCS + WUP	0.757	0.753	0.765	0.758
3	LEV + COS + LCS + PATH	0.750	0.745	0.759	0.752
4	LEV + COS + WUP	0.749	0.770	0.709	0.738
5	COS + LCS + PATH	0.729	0.727	0.733	0.730
6	COS + LCS + WUP	0.725	0.719	0.739	0.729
7	COS + WUP + PATH	0.744	0.776	0.685	0.728
8	LEV + WUP + LCS + PATH	0.720	0.713	0.753	0.724
9	LEV + LCS + PATH	0.715	0.710	0.727	0.715
10	LEV + LCS + PATH	0.712	0.705	0.722	0.713

Where:

LEV = Levenshtein distance
COS = Tf based cosine similarity
LCS = Longest common subsequence
WUP = Wu and Palmer
PATH = Shortest path

We can see that that the number of combination displayed in in the Table 2 is ten. The reason of that is we take combination number 10 as performance baseline. Other possible feature combinations which are not displayed in the Table 2 have lower score that the baseline. So, we believe ten combinations are sufficient to represent performance of the model.

Based on features performance results in Table 2 we conclude that the best combination is including all features in this research. The combination returns highest F1-Score measure with 75.8%, although the combination features without PATH feature gives us same F1-Score, higher precision and higher accuracy. We believe that this is not enough reason to remove the PATH feature from the model. Instead of removing the PATH feature to gain very little improvement on performance, we should try to improve or modify the feature.

We also believe that this performance result, especially for PATH feature caused by limited dataset which only has 1004 data. The size of dataset available leaves to choose n-fold cross-validation which means the evaluation performance is averaged through testing process on known environment. On PATH feature function we know that it searches the shortest path between pair of words, but turns out that the shortest path doesn't necessarily represent similarity of the word. The PATH feature may be useful on data test that is properly separated from the development process.

Up until now we only talk about first two feature combination, if we look at feature number three and eight we can see that they have same number of features. The difference between them is the feature number three has COS feature, and the feature number eight has WUP feature. That only difference impacts performance quite significant, feature number three 3.4% better than feature number eight. We call it significant because from ten combination we have, the combination number eight is closer to the base line with only 1.1% higher than the baseline. It means difference between combination number eight and ten is one third of difference between combination number eight and three. Even if we compare the combination number eight with other combination that have COS feature, the performance of combination eight is less than others, even when others have less feature. It goes without saying that COS feature holds better overall performance compared to WUP feature.

We conclude that every feature is important to the model, but the significance is different. The most significant feature is COS feature, it's absence on the model impacts the performance quite a lot. On the other hand, the least significant feature is PATH feature, it's presence has yet gives us any significant impact to the performance. For other features we calculate the order of significance based on performance in Table 2. From highest significance to lower significance we have COS, LEV, WUP, LCS, and PATH. The order of significance is based on the amount of performance drop if we exclude the feature. For example in our case, the absence of COS feature is causing performance drop greater than WUP feature, it means the COS feature significance is higher than WUP feature. Even the significant between four features (COS, LEV, WUP and LCS) is different doesn't mean one feature should be removed. Because the absence of one feature in the list will return lower performance score than combination of the four features.

4.2 Evaluation on number allowed parents on K2 Algorithm

The purpose of this evaluation is to find out the outcome of different u (max allowed parents in a node) value to the DAG structure and its impact to the performance. As we know that K2 algorithm uses a parameter that limits allowed parents in the DAG structure. In the observation we will use 1, 2, and 3 as u number. The scoring function used in the K2 algorithm is BDeu scoring function explained in equation 4. The DAG and the joint probability generated by K2 algorithm as follows.

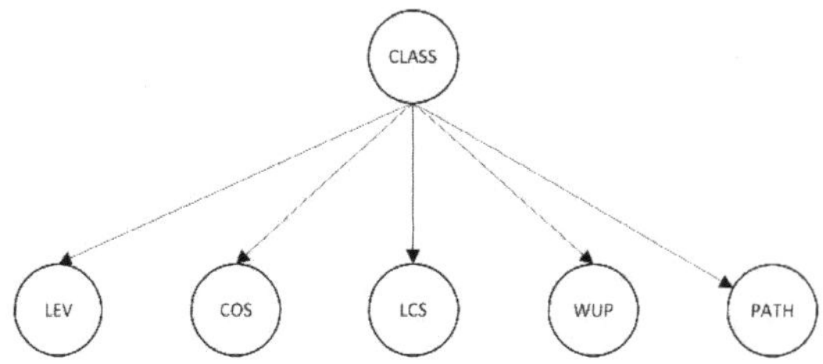

Figure 2 DAG structure where u = 1

The joint probability equation for DAG where u = 1 is:

$$P(CLASS, LEV, COS, LCS, WUP, PATH) \\ = P(LEV|CLASS)P(COS|CLASS)P(LCS|CLASS)P(WUP|CLASS)P(PATH|CLASS)P(CLASS) \quad (8)$$

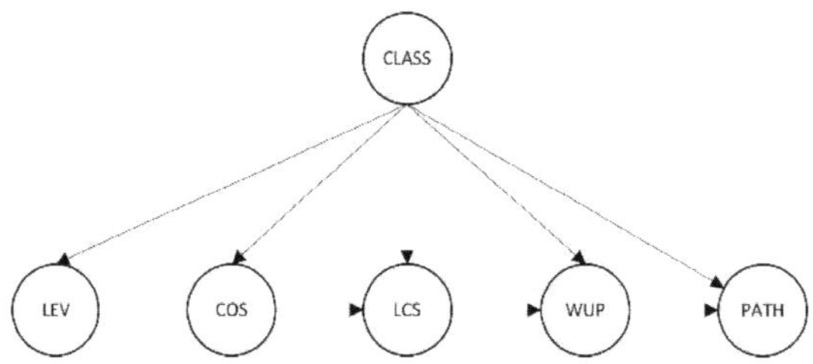

Figure 3 DAG structure where u = 2

The joint probability equation for DAG where u = 2 is:

$$P(CLASS, LEV, COS, LCS, WUP, PATH) \\ = P(LEV|CLASS)P(COS|CLASS)P(LCS|CLASS, COS)P(WUP|CLASS, LCS)P(PATH|CLASS, WUP)P(CLASS) \quad (9)$$

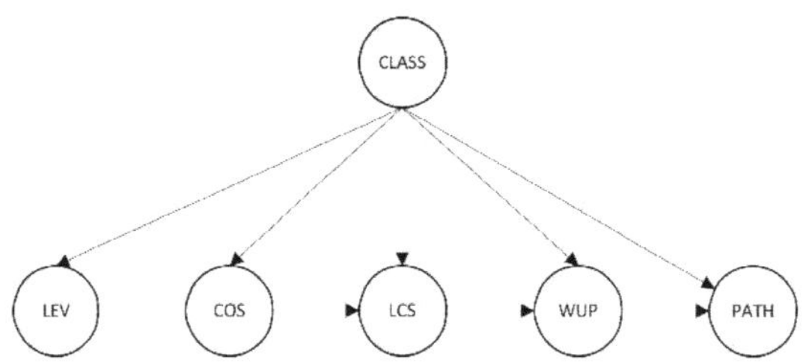

Figure 4 DAG structure where u = 3

The joint probability equation for DAG where u = 3 is:

$$P(CLASS, LEV, COS, LCS, WUP, PATH) \\ = P(LEV|CLASS)P(COS|CLASS)P(LCS|CLASS, COS)P(WUP|CLASS, LCS)P(PATH|CLASS, WUP)P(CLASS) \quad (10)$$

IOP Conf. Series: Journal of Physics: Conf. Series **971** (2018) 012051 doi:10.1088/1742-6596/971/1/012051

After we obtain DAG structures, we attempt to generate two random DAG as comparation to the structure generated by K2 algorithm.

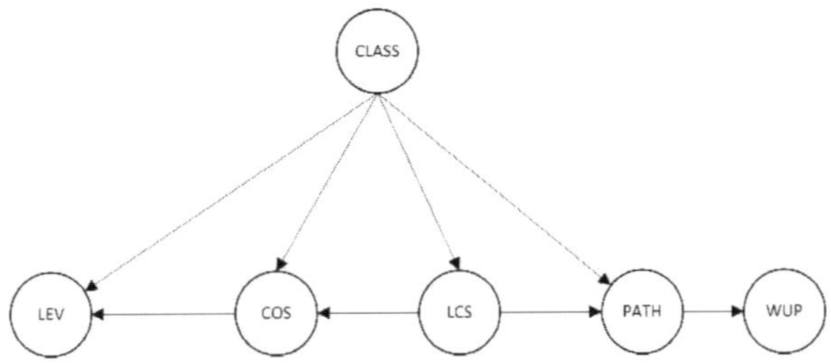

Figure 5 Generated DAG 1

$$P(CLASS, LEV, COS, LCS, WUP, PATH) \\ = P(LEV|CLASS, COS)P(COS|CLASS, LCS)P(LCS|CLASS)P(WUP|PATH)P(PATH|LCS, CLASS)P(CLASS) \quad (11)$$

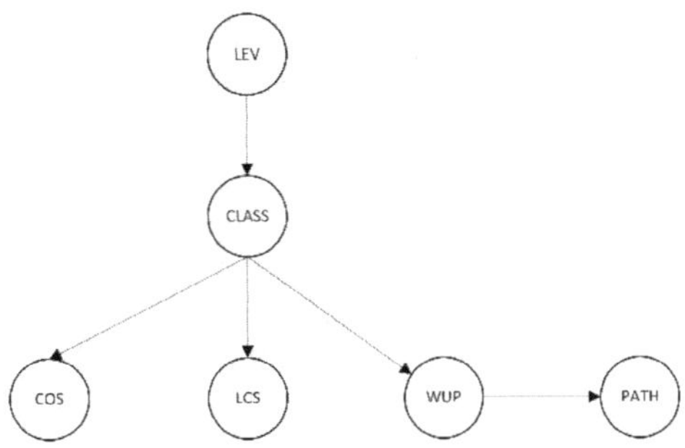

Figure 6 Generated DAG 2

$$P(CLASS, LEV, COS, LCS, WUP, PATH) \\ = P(CLASS|LEV)P(COS|CLASS)P(LCS|CLASS)P(WUP|CLASS)P(PATH|WUP)P(LEV) \quad (12)$$

The BDeu score and average performance of all three DAG from K2 algorithm and two random generated DAG are explained in Table 3 as follows.

Table 3 DAG BDeu score and average performance

No	DAG	BDeu	Accuracy	Precision	Recall	F1-Score
1	DAG where u = 1	-4764.234	0.716	0.727	0.691	0.708
2	DAG where u = 2	-4028.149	0.756	0.752	0.765	0.758
3	DAG where u = 3	-4028.149	0.756	0.752	0.765	0.758
4	Generated DAG 1	-4032.979	0.748	0.740	0.762	0.751
5	Generated DAG 2	-4316.510	0.729	0.740	0.705	0.722

On table 5 we find that the DAG structure generated by K2 algorithm for current features only able to generate two kinds of structure. The first structure is the same with naive bayes structure where the u value is 1. The reason of it because parameter only allows one parent of each node, it means the algorithm will find the best parent for each node. In our case the best parent to our nodes according to K2 algorithm is the class itself, that's why the algorithm returns naive bayes structure.

The second DAG structure generated by K2 is on figure 3 and 4. We notice that there is no difference between DAG on figure 3 and 4. It means the K2 algorithm has found the best DAG it can find. If we use u with value 5, 6, ..., n we will not get any differences. That's because the number of max parent allowed depends on the nodes. Either the number of nodes or the CPT (Conditional Probability Table) of the nodes will affect DAG structure searching. Our total nodes in the graph is 6 which means maximum number of nodes that are assigned as parents is 5 nodes. But with all of that, the result instead gives us DAG in figure 3 and 4 which max parent in its DAG is two parents per node. The cause of it because as mentioned before the generation of the DAG structure depend on CPT in the nodes. The CPT will affect BDeu, which means BDeu score is better on max parents = 2 than max parents > 2.

We use the DAG number 1 in Table 3 as baseline because it takes form of naive bayes classifier, as we know in many cases naive bayes classifier can be used as baseline. From performance point of view we notice that the relation between BDeu score and performance is linier with each other. It means that the higher BDeu score will lead us to higher performance score. The DAG number 2 and 3 perform best, with F1-Score is 75.8%. They performance scores are higher than other two random generated DAG. Even two random generated DAGs have better score than the baseline.

4.3 Evaluation of nodes order in K2 Algorithm

In this evaluation we will observe the model performance when we change nodes order in K2 algorithm. The first case we will use order that represent naive bayes classifier and the second case is random order. For the random nodes order we will perform classification 10 times and take the average score between runs. The result of nodes order observation as follows in Table 4.

Table 4 Performance between random and fixed order during initiation in K2 algorithm

No	DAG	BDeu	Accuracy	Precision	Recall	F1-Score
1	**Fixed Initiation with naive bayes order**	**-4028.149**	**0.756**	**0.752**	**0.765**	**0.758**
2	Random order run 1	-4028.177	0.752	0.747	0.763	0.755
3	Random order run 2	-4028.156	0.753	0.754	0.751	0.752
4	Random order run 3	-4028.428	0.759	0.758	0.761	0.759
5	Random order run 4	-4028.149	0.746	0.747	0.745	0.746
6	Random order run 5	-4028.400	0.751	0.752	0.749	0.750
7	Random order run 6	-4039.326	0.746	0.743	0.753	0.748
8	Random order run 7	-4028.349	0.747	0.745	0.751	0.748
9	Random order run 8	-4028.428	0.747	0.743	0.755	0.749
10	Random order run 9	-4028.149	0.750	0.748	0.755	0.751
11	Random order run 10	-4039.299	0.755	0.758	0.749	0.754
12	**Average between run 1 – 10**	**-4030.49**	**0.750**	**0.749**	**0.753**	**0.751**

Based on results in table 4 we can see that the fixed nodes order gives us better performance result than random order. It's because the not only the performance score but also the consistency of the performance. If we look at random order between run 1 to run 10 we can see constant changes in performance scores. Although there is a run that returns better result than fixed order in run three, it's just slightly better. The chance for us to encounter such result is one in ten runs, it's not justified to use random order in pursue of slightly better performance when the difference actually is ignorable. From average result's point of view, we also find that the overall score of random order is noticeably lower than fixed order.

5 Conclusion and future work

Indonesian paraphrase identification is conducted by performing comparation of pair phrase/sentence whether they are paraphrase or not. During the process we perform preprocessing, feature extraction, classifier building and evaluation. During feature extraction we perform two types of feature, they are syntactic and semantic features. Syntactic features consist of normalized levenshtein distance (LEV), tf cosine similarity (COS), and longest common subsequence (LCS). For semantic feature we have Wu and Palmer feature (WUP) and shortest path feature (PATH). The mentioned features is used in Bayesian Networks classifier.

In this research we can conclude that by performing classification with our method on Indonesian Paraphrase Dataset with our method we obtain average accuracy 75.6%, precision 75.1%, recall 76.5% dan F1-Measure 75.8%. When performing this research, we find difficulties especially in the thesaurus area, where there is no Indonesian thesaurus that is ready to be used with our method. We are choosing alternative by translate the word tokens with Google Translator API, but the problem with that is the ambiguity of translation result. For example if we have a word in Indonesian "tanah" and "air" if we translate the word separately it will mean earth and water, but if we look at another context it can mean homeland. With our method we still encounter problem with hyponym, homonym, polysemy and homograph. We also encounter difficulties with complex sentence that consists of a clause and main clause. In Indonesian Language, the main clause is the main idea of an expression, and the clause is supporting the main clause. With this method we were not able to extract the main idea of complex sentence. Our guess is we need specific rules that able to specify main clause and support clause in a complex sentence to extract its main idea.

From performance point of view, we conclude that the five features (COS, LEV, WUP, LCS, and PATH) should remain in the classifier model. That's because the combination has the highest performance scores along with combination of four features (COS, LEV, WUP, and LCS). We mention that the current PATH feature doesn't have any significant result. It doesn't mean we should remove the feature from the model, because there's possibility that our data doesn't fit quite well with this feature. If we try with another data, there's a chance that this feature (PATH) is having significant role. Speak of feature significance, we conclude that COS feature is the most significant feature, followed by LEV, WUP and LCS. The significance is measured by how much performance drop occurred when a feature is excluded from the model.

For K2 algorithm we conclude that the best number of allowed parents per node are two and the ordering of initiation is better off with the fixed one with naive bayes order. The two conditions are based on performance result during evaluation.

As future work we may try to build a feature model that able to recognize complex sentence. We also have to create feature extraction method that able to extract feature about passive or active sentence, positive or negative sentence (polarity) and significance of each word in sentence/phrase. We also need to develop an feature extraction method that ready to be used if in the future Indonesian Thesaurus is fully completed.

References

[1] A. Reshamwala, D. Mishra and P. Pawar, "Review On Natural Language Processing," *ACST – Engineering Science and Technology: An International Journal (ESTIJ),* vol. 3, no. 1, 2013.

[2] KBBI, "KBBI - Parafrasa," [Online]. Available: kbbi.web.id/parafrasa.

[3] J. Asian, H. E. Williams and S. Tahaghoghi, "Stemming Indonesian," RMIT University, Melbourne.

[4] U. B. Kjaerulff and A. L. Madsen, Bayesian Networks and Influence Diagrams: A Guide to Construction and Analysis, New York: Springer, 2012.

[5] D. Koller and N. Friedman, Probabilistic Graphical Models: Principles and Techniques, Cambridge, Massachusetts: The MIT Press, 2009.

[6] Heckerman and David, A Tutorial on Learning With Bayesian Networks, Redmond: Microsoft Corporation, 1996.

[7] A. Abdi, N. Idris, R. M. Alguliyev and R. M. A., "PDLK: Plagiarism Detection Using Linguistic Knowledge," *Expert Systems With Application,* pp. 1-11, 2015.

[8] R. Barzilay and K. McKeown, "Extracting paraphrases from a parallel corpus," *Annual Meeting of the Association for Computational Linguistics,* pp. 50-57, 2001.

[9] Y. Shinyama, S. Sekine, K. Sudo and R. Grishman, "Automatic paraphrase acquisition from news articles," 2002.

[10] A. A. Pratiwi, "On The Feature Selection and Classification Based on Information Gain for Document Sentiment Analysis," *Applied Computational Intelligence and Soft Computing,* 2017.

[11] M. Mubarok, Adiwijaya and M. Aldhi, "Aspect-based sentiment analysis to review products using Naive Bayes," *In AIP Conference Proceedings,* vol. 1867, no. 1, 2017.

[12] N. P. A. Vo, S. Magnolini and O. Popescu, "Paraphrase Identification and Semantic Similarity in Twitter with Simple Features," *Proceedings of Social NLP,* pp. 10-19, 2015.

[13] Adiwijaya, M. S. Mubarok and A. H. Prayogo, "Identifikasi Parafrasa Pada Dokumen Teks Bahasa Indonesia Menggunakan Bayesian Networks," *Indonesia Symposium On Computing,* 2016.

[14] R. M.Alguliev, R. M.Aliguliyev and C. A.Mehdiyev, "Sentence selection for generic document summarization using an adaptive differential evolution algorithm," *Swarm and Evolutionary Computation,* vol. 1, no. 4, pp. 213-222, 2011.

[15] M. S. K, D. K. C. Shet and D. U. Acharya, "A New Similarity Measure For Taxonomy Based On Edge Counting," *nternational Journal of Web & Semantic Technology (IJWesT),* vol. 3, no. 4, 2012.

[16] M. Palmer and Z. Wu, "Verb Semantics and Lexical Selection," *In Proceedings of the 32nd Annual Meeting of the Association for Computational Linguistics,* 1994.

[17] J. Ferlež and M. Gams, "Shortest-Path Semantic Distance Measure in WordNet v2.0," *Informatica,* vol. 28, pp. 381-386, (2004).

[18] V. Levensthein, "Binary codes capable of correcting deletions, insertions, and reversals," *Soviet Physics Doklady 10,* vol. 8, pp. 707-710, 1966.

[19] B. Bukh and J. Ma, "Longest common subsequence in sets of words".

[20] B. Malone, "Scoring Functions for Learning Bayesian Networks," University of Helsinki, Finland, 2014.

[21] G. F. Cooper and E. Herskovits, "A Bayesian Method for the Induction of Probabilistic Networks from Data," *Machine Learning,* no. 9, pp. 309-347, 1992.

Automatic Semantic Orientation of Adjectives for Indonesian Language Using PMI-IR and Clustering

Dewi Riyanti, M. Arif Bijaksana, and Adiwijaya

School of Computing, Telkom University, Jl. Telekomunikasi No.1, Bandung, 40257, Indonesia

E-mail: dewiriyanti@telkom.co.id, arifbijaksana@telkomuniversity.ac.id, adiwijaya@telkomuniversity.ac.id

Abstract. We present our work in the area of sentiment analysis for Indonesian language. We focus on bulding automatic semantic orientation using available resources in Indonesian. In this research we used Indonesian corpus that contains 9 million words from kompas.txt and tempo.txt that manually tagged and annotated with of part-of-speech tagset. And then we construct a dataset by taking all the adjectives from the corpus, removing the adjective with no orientation. The set contained 923 adjective words. This systems will include several steps such as text pre-processing and clustering. The text pre-processing aims to increase the accuracy. And finally clustering method will classify each word to related sentiment which is positive or negative. With improvements to the text preprocessing, can be achieved 72% of accuracy.

1. Introduction

The need for information encourages the development of research and technology to can answer those needs. The required information is experienced the development of general information to more information specific and specific. Obtain the right and fast information will help the organization to be able to make an answer of the problems. Research in sentiment analysis driven by a thought that information is a sentiment of data which is important and necessary. Sentiment is related to assessment of a context or discourse. Positive sentiment states giving good value to context in text and negative sentiments declared the opposite.

The sentiment analysis which is part of opinion mining [17], is the process of understanding, extracting and processing textual data automatically for information [18]. Conducted to view opinions on a problems, or can also be used to identify trends in the market [1]. Analysis of sentiments in the study this is the process of classifying textual documents into two classes, namely positive and negative sentiment class. The magnitude the effects and benefits of sentimental analysis, leading to research or application of sentiment analysis growing rapidly, even in America approximately 20-30 companies that focus on analytics services sentiment [17]. Basically the sentiment analysis is classified, but the reality is not as easy as the classification process ordinary due to language usage. Where there is ambiguity in the use of words, the absence of inner intonation a text, and the development of the language itself.

As one of the ways to convey, language can be viewed as the tool to express individual subjective valuation that may include the way toward considering, feeling, or a blend of both. The expression can appear as examination, feeling, or assessment, that is communicated towards elements. Sentiment analysis intended as a field of study to analyze the sentiment on the text, or in other words the grouping the polarity of text either in the document level, sentences or features/aspects of level n-gram (noun, verb, adverb, clause, phrase)-whether opinions expressed are positive or negative (there are also neutral). Sentiment analysis aims to determine the attitude of an author of some topics.

The term opinion mining and sentiment analysis are considered interrelated and interchangeable [11] because in the objective opinion mining is gathering public opinions for analysis . Whether opinions expressed in a document, the phrase or entity features/aspects of positive or negative .

One of the techniques of sentiment analysis is semantic orientation or polarity of words. Polarity is binary value of positive or negative sentiment in a word. For identify word polarity can be categorized into positive and negative categories. Two types of techniques have been used in the literature for semantic orientation-based approach for sentiment analysis : corpus based and dictionary or lexicon based. In this study, we explore the corpus-based semantic orientation approach for adjectives. Corpus-based semantic orientation approach requires large dataset to detect the polarity of the words. The main problem with this approach is that it relies on the polarity of the words that have emerged in the training corpus since polarity is computed for the words that are in the corpus.

Research on sentiment analysis continues to grow, each study using data, language, features, different weighting algorithms and calculations. Common corpus data used in sentiment analysis among others, online news web , the wall street journal [3], social media [1], web pages Altavista [5], and movie reviews [14] [15].

Many words communicate the speaker's evaluation of the item that is under discussion as desirable or undesirable. This evaluative character a word is called its semantic orientation. A word with a positive semantic orientation conveys the evaluation that the item is desirable (e.g., nice, good) and a negative orientation conveys the evaluation that the item is undesirable (e.g.,bad, ugly).

Sentiment analysis, also known as opinion mining, refers to the use of natural language processing, text analysis and computational linguistics to identify and extract subjective information in source materials. Sentiment analysis aims to determine the attitude of a writer with respect to some topic. The attitude may be his/her judgement or evaluation based on his/her experience. A basic task in sentiment analysis is classifying the polarity of a given text at the document, sentence, or feature/aspect level whether the expressed opinion in a document, a sentence or an entity feature/aspect is positive, negative, or neutral. In most statistical classification methods, the neutral class is ignored under the assumption that neutral texts lie near the boundary of the binary classifier, several researchers suggest that, as in every polarity problem, three categories must be identified.

Research on sentiment focused on adjectives or adjective phrases as the primary source of subjective content in a document has been gaining a lot of attention in the recent years. The goal is to identified automatically semantic orientation of words and allow a system to further refined the retrieved semantic similarity relationship. Automatic creation of polarity lexicons is a crucial issue to be solved in order to reduce time and efforts in the first steps of sentiment analysis.

For English, the current state-ofthe-art automatic creation of polarity lexicons system achive around 82% accuracy score for adjectives in the semantic orientation identification task [3].

It is said lexicon based because it uses lexicon and phrases (even idioms) that have been stored in the dictionary/lexicon. The use of a dictionary / lexicon is based on the assumption that sentiment orientation in a sentence context can be determined from the sentiment orientation of the vocabulary or the word equivalent in it.

The use of lexicon when conducting sentimental analysis research can be constructed on its own or utilizing dictionaries already built by others [12]. Whether building your own or utilizing an existing dictionary equally has its advantages and disadvantages. By building your own dictionary, it means it will take up a lot of time and energy having to collect from scratch. But the advantages, the contents of the dictionary in accordance with the needs of research. If using an existing dictionary, it is obviously more time-saving and labor-intensive to only modify the language and the dictionary is considered valid because the dictionary is used for academic affairs, as well as the fact that the dictionary is built and used by academics. The drawback, the contents of the dictionary is not necessarily according to the needs of research.

The conceptual framework of sentiment lexicon is described as follow :

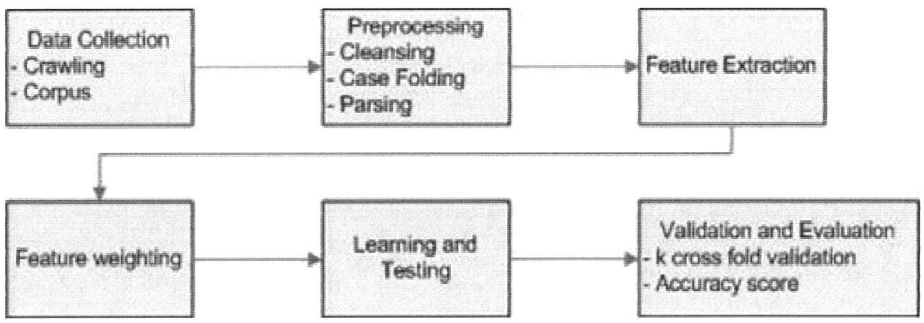

Figure 1. Sentimen Lexicon Process

Dataset is collected, then data preprocessing stage is prepared, this process is include cleansing, case folding, and parsing. Feature extraction and feature weighting selects the related attribute. Next process is learning and testing process, using soft clustering methods with to train a set of base learner to result predictions with each one of them, and then do voting to these predictions. This research is expected to predict the polarity of the lexicon.

2. Theory

This section discusses the literature studies and theories of this study. It is divided into two sections: (1) the related work; and (2) the Existing Method.

2.1. Related work

For the purpose of building such lexicon automatically, a lot of studies have researched unsupervised method of learning polarity of words. In 1997, V. Hatzivassiloglou and K.R. McKeown used 21 million words of Wall Street Journal corpus to predict the semantic orientation of adjective. In this paper, the authors hypothesized that two adjectives joined by and have the same semantic orientation while two adjectives joined by but have the opposite one. They used this idea along with a loglinear regression model and a set of supplementary morphological rules in text preprocessing and clustering. The method detected the word polarity based on the strength of co-occurrence with the seeds to predict whether a pair of adjectives joined by any of these conjunctions has the same or different semantic orientation. Once pairs of adjectives are extracted, they utilized a clustering method to separate all the adjectives conjoined into two clusters, placing as many words of the same orientation as possible into the same subset. The cluster with more elements was labeled as positive adjectives and the other as negative. The strengths of this method are this method fully unsupervised because the algorithm started from nothing and the result was convincing. This work achieved accuracy score between 78% and 92% in the classification of positive and negative adjectives.

P.D. Turney presented a general strategy for learning semantic orientation from semantic association. The unsupervised classification method based on distributional semantics. The method involves issuing queries to a customer reviews and using pointwise mutual information (PMI) to measure an adjective's tendency to appear in positive or negative vicinity. The algorithm is empirically evaluated using a training corpus of approximately 5000 reviews Tested with 410 reviews, the algorithm attains an average accuracy of 74%.

P.D. Turney and M.L. Littman in 2002 showed the accuracy of SO-PMI-IR with a training corpus of 350 million Web pages (at least one hundred billion words). These are the English Web pages that are indexed by AltaVista. Semantic orientation was used as a measure of confidence that the word will be correctly classified. An accuracy of 80% is attained, using an unlabeled corpus of about one hundred billion words.

Y. Matsuo, K. Uchiyama, T. Sakaki and M. Ishizuka proposed an unsupervised method for word clustering based on a word similarity measure by web check. Each pair of words is queried to a search engine, which produced a co-occurrence matrix. By computing the similarity of words, a word cooccurrence graph is obtained. Newman clustering, a new kind of graph clustering algorithm was applied for distinguishing word clusters. Evaluations are made on two arrangements of word groups determined from a open directory and WordNet. In this researched they compared two clustering methods : Newman clustering and word clustering. The word co-occurence graph is created using PMI, Jaccard and Chi-square measures.

S. Vazquez, M. Padro, N. Bel and J. Gonzalo proposed bootstrapping algorithm to automatically extracts all of the polar adjectives joined by y (and) or pero (but) in a given corpus. And they utilized the adjectives that were in the seed polarity lexicon as input for their algorithm to find new adjectives joined and identifying the priority. They propose that polar adjectives and their corresponding polarity values can be automatically identified if they are found in a coordinate construction with the appropriate conjunctions and with other adjectives that were not in the seed lexicon. The process continued until any adjective of lexicon is not found join with any new adjective or untul is no more conjunctive relation of this type.

2.2. Sentiment lexicon

Sentiment lexicon uses the phrases and words (even idioms) that have been stored in the dictionary / lexicon. The use of a dictionary / lexicon is based on the assumption that the orientation of sentiment in a word, phrase or sentence context can be determined from the sentiment orientation of the vocabulary or the word equivalent in it.

The main component of the sentiment lexicon is the list of words and orientation that relate to the word. In this study there are two categories of orientation: positive and negative. The next stage after determining the type of emotion that is determining the words seed. Words seed are a number of words that are directly related to one particular kind of orientation. We constructed this seed by taking all adjectives appearred in Kompas and Tempo corpus 20 times or more, then removed adjectives that have no orientation.

The word list in words seed consists of only a few words, while the sentiment lexicon requires large numbers of words. The word in the lexicon is expanded based on the words seed. After the expansion of words seed, the next step is to determine the weight of each word. This study uses SO-PMI-IR. Then, classify the adjectives in to positive or negative orientation.

2.3. Part-of speech tagging

Part-of-Speech tagging is an NLP task. Most of the activities done in the NLP field such as Information Extraction, Question-Answering, Speech Recognition, Intelligent Tutoring System, Parser, and others require this POS Tagging system for initial processing. Unfortunately, for the Indonesian language the system is few, the publication is lacking, and no one provides for download.

Part-of-speech tagging is a system that automatically labels words to a sentence. In this study used pos tagger method has proposed in [7]. They designed a linguistically motivated

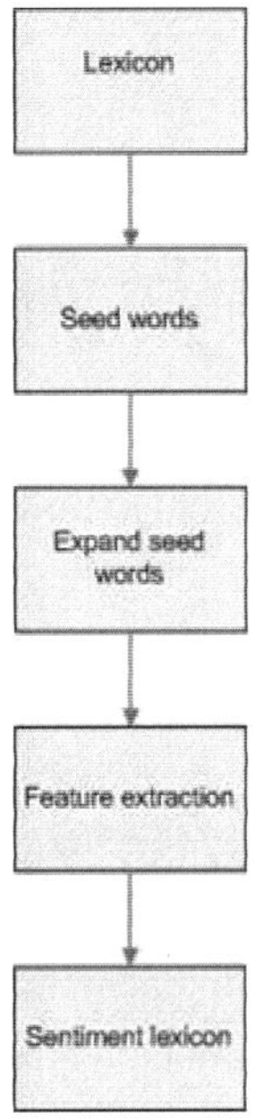

Figure 2. The process of sentiment lexicon

POS tagset for the Indonesian language. The design process is divided into two phases: (1) define initial POS tagset and (2) test and revise POS tagset which involve manually tagging the Indonesian sentences in the IDENTIC corpus. The corpus that manually tagged consists of 10.000 sentences, containing 262.330 lexical tokens.

Suppose, there are sentences "Saya pergi ke sekolah." and there are labels PRP = personal pronouns, VBI = verbs, IN = preposition, NN = noun. The system will accept the input of the sentence, the output is:

Saya/PRP pergi/VBI ke/IN sekolah/NN

The tags used are different depending on the language used in tagging. For Indonesian,below is a table that contains tags with descriptions and examples that has proposed in [7].

2.4. Adjectives

Adjective is a word class that changes a noun or pronoun, by explain and make the word more specific that can explain the quantity, adequacy, order, quality, or emphasis of a word. In Indonesian language, adjectives can be formed due to several processes that occur, as follows:

1. Formed from basic word, for example : tua, muda, tinggi, and rendah.
2. Formed from additional affix (suffix , prefix), for example : tertampan, tercantik.
3. Formed from word repetition, for example : compang-camping, gelap-gulita.
4. Formed from another langunage, for example : kreatif, and legal.
5. Formed from group of words, for example : baik hati, lemah jiwa and keras kepala.

2.4.1. Adjectives pair coordinated by conjuction To achieve coherence or cohesion in a paragraph, it is necessary to connect words or also called conjunctions. It is conjunction that makes a sentence in the paragraph logical, grammatical and concerted. Conjunction itself has a sense of the word task or phrase that connects two equivalent language units, for example

Tag	Description	Example
CC	Coordinating conjunction	dan, tetapi, atau
CD	Cardinal number	dua, juta, enam, 7916, sepertiga, 0,025, 0,525, banyak, kedua, ribuan, 2007, 25
OD	Ordinal number	ketiga, ke-4, pertama
DT	Determiner / article	Para, Sang, Si
FW	Foreign word	climate change, terms and conditions
IN	Preposition	dalam, dengan, di, ke, oleh, pada, untuk
JJ	Adjective	bersih, panjang, hitam, lama, jauh, marah, suram, nasional, bulat
MD	Modal and auxiliary verb	boleh, harus, sudah, mesti, perlu
NEG	Negation	tidak, belum, jangan
NN	Noun	monyet, bawah, sekarang, rupiah
NNP	Proper noun	Boediono, Laut Jawa, Indonesia, India, Malaysia, Bank Mandiri, BBKP, Januari, Senin, Idul Fitri, Piala Dunia, Liga Primer, Lord of the Rings: The Return of the King
NND	Classifier, partitive, and measurement noun	orang, ton, helai, lembar
PR	Demonstrative pronoun	ini, itu, sini, situ
PRP	Personal pronoun	saya, kami, kita, kamu, kalian, dia, mereka
RB	Adverb	sangat, hanya, justru, niscaya, segera
RP	Particle	pun, -lah, -kah
SC	Subordinating conjunction	sejak, jika, seandainya, supaya, meski, seolah-olah, sebab, maka, tanpa, dengan, bahwa, yang, lebih ... daripada ..., semoga
SYM	Symbol	IDR, +, %, @
UH	Interjection	brengsek, oh, ooh, aduh, ayo, mari, hai
VB	Verb	merancang, mengatur, pergi, bekerja, tertidur
WH	Question	siapa, apa, mana, kenapa, kapan, di mana, bagaimana, berapa
X	Unknown	statemen
Z	Punctuation	"...", ?, .

Figure 3. Indonesian Tagset

between words with words, clauses with clauses, phrases with phrases, and sentences with sentences.

Coordinated conjunctions are connecting words that connect two sentences or more in a paragraph having an equal position. Below are the various coordinated conjunctions and their examples

1. Dan : this conjunction is a marker of the relationship between words.
2. Atau : this conjunction is a marker of election relationships in a paragraph or sentence and usually has the opposite meaning.
3. Tetapi or tapi : this conjunction is a marker of inter-sentence resistance in the paragraph.

Adjectives pair coordinated by conjunction is merging or pairing adjectives with conjunction present in the middle. It is part of part-of-speech tagging (POS). POS tags will identify simple adjectives. Below is a an example sentence with a POS tag.

Kita/PRP menginginkan/VBI pembangunan/NN yang/RP damai/JJ dan/CC demokratis/JJ untuk/IN Indonesia/NNP.

The tagged words are then extracted based on the adjective pair pattern, connecting words and adjectives (JJ-CC-JJ). The first word identified is the adjective (JJ), the second word identified is the connector (CC), and the third word identified is the adjective (JJ).

2.5. Pointwise mutual information

Pointwise mutual information (PMI)or point mutual information, is a measure of association used in information theory and statistics [9]. Pointwase Mutual Information (PMI) between two adjectives as follows:

$$PMI(x,y) = log_2(\frac{P(x,y)}{P(x)P(y)}). \tag{1}$$

Where $p(x,y)$ is the probability that $word_1$ and $word_2$ co-occur, $p(x)$ is the probability that $word_1$ co-occur, and $p(y)$ is the probability that $word_2$ co-occur. If there is a true relationship between $word_1$ and $word_2$, then the probabilities that occur in $p(word_1\&word_2)$ will be much larger than $p(word_1)p(word_2)$. And the consequently is $log(word_1 word_2) >> 0$. The log of this ratio is the amount of information obtained about the presence of one word when observing the other.

An adjective can be known the orientation by SO-PMI-IR

$$SO - PMI - IR(word) = PMI(word, positiveparadigms) - PMI(word, negativeparadigms) \tag{2}$$

When the result of PMI is positive then the adjective word has a positive semantic orientation and a negative orientation when the PMI value is negative.

2.6. Clustering

Clustering is a method of data analysis, which is often incorporated as one of the methods of Data Mining, whose purpose is to group data with the same characteristics into an identical 'region' and data with different characteristics to the other 'territories'.

There are several approaches used in developing clustering methods. The two main approaches are clustering with partition and clustering approaches with a hierarchical approach. Clustering with partitioning approach or often called partition-based clustering grouping data by sorting through the data that is analyzed into existing clusters. Clustering with hierarchical approach or often called hierarchical clustering classifies data by creating a hierarchy of dendograms where similar data will be placed in adjacent hierarchies and not in a remote hierarchy. In addition to both approaches, there is also clustering with the approach of automatic mapping (Self-Organizing Map/SOM).

Self-Organizing Map (SOM) is an Artificial Neural Networks type that is training in an unsupervised manner that has been proposed in [13].. SOM generates folders consisting of outputs in low dimensions (2 or 3 dimensions). This folder attempts to locate the properties of the data input. The input and output composition in SOM is similar to the composition of the feature scaling process (multidimensional scaling).

Although the learning process is similar to Artificial Neural Networks, but the process of assigning data input to the map is more similar to K-Means and kNN Algorithm. The procedure taken in doing clustering with SOM is as follows:

1. Determine the weight of the input data randomly. Select one of the data inputs.
2. Calculate the level of similarity (with Eucledian) between the data input and the weight of the input data and select the input data that has similarities with the existing weight (this data is called Best Matching Unit (BMU).
3. Renew weight from input data by bringing the weight to BMU with formula:

   $$Wv(t+1) = Wv(t) + Theta(v,t)xAlpha(t)x(D(t)Wv(t)) \tag{3}$$

   Where Wv(t) is weight at time t. Theta(v, t) is the neighborhood function depends on the Lattice distance between BMU and neuron v. Generally a value of 1 for neurons close enough to BMU, and 0 for the opposite. The use of Gaussian functions is also possible. Alpha(t) is Learning Coefficient is reduced monotonically and D(t) is input data.

2.7. K-Cross fold validation

Cross Validation is one technique to assess or validate the accuracy of a model built on a particular dataset. The model usually aims to make predictions and classifications of new data that may never have appeared in the dataset. The data used in the model development process is called training data, while the data to be used to validate the model is called the test data.

One of the popular cross-validation methods is K-Fold Cross Validation. In this technique the dataset is divided into a number of K partitions at random. Then a number of experimental K-times were conducted, each experiment using K-partition data as data testing and utilizing the rest of the partition as training data. As an illustration, if we do 5-Fold Cross-Validation then design the experimental data as follows:

K1	K2	K3	K4	K5

Figure 4. Dataset

Experiment	Training Data				Testing Data
1	K1	K2	K3	K4	K5
2	K2	K3	K4	K5	K1
3	K3	K4	K5	K1	K2
4	K4	K5	K1	K2	K3
5	K5	K1	K2	K3	K4

Figure 5. Data experiments

To obtain the value of accuracy of the experimental results that we do, can be taken the average value of the entire experiment.

2.8. Method selection
The are some techniques to estimate word co-occurrence frequencies from an available corpus. Based on [18], gave a comparative examination of co-occurence frequency calculation. The research reported that PMI is useful for computing semantic orientation and performed best on average. Another advantages of PMI are fully unsupervised and the result in nominal context makes adjective semantics more interpretable [5]. Then clustering algorithm creates the graph structure to produce two groups of adjectives.

3. Research Methodology
This chapter discusses methodology of the study, it is divided into five sections, namely 1) Research Design; 2) System Implementation; 3) Experiment Scenario; 4) Instrumentation and Data Collection; and 5) Tools for Data Analysis.

3.1. Research design
The design process of this research is describes as follows,

3.1.1. Data preprocessing The dataset we used in this research is 9 million words corpus, collected from Kompas and Tempo articles. The steps performed on the preprocessing data are

1. Cleansing, the process of cleaning documents from words that are not needed to reduce noise. That word omitted are HTML characters, stopwords, and url (http://situs.com)
2. Case folding, change letters as well as deletion of numbers and punctuation. In this case that is used only Latin letters between a to z.
3. Parsing, ie the process of breaking a document into a word. This is in accordance with the feature used ie unigram.

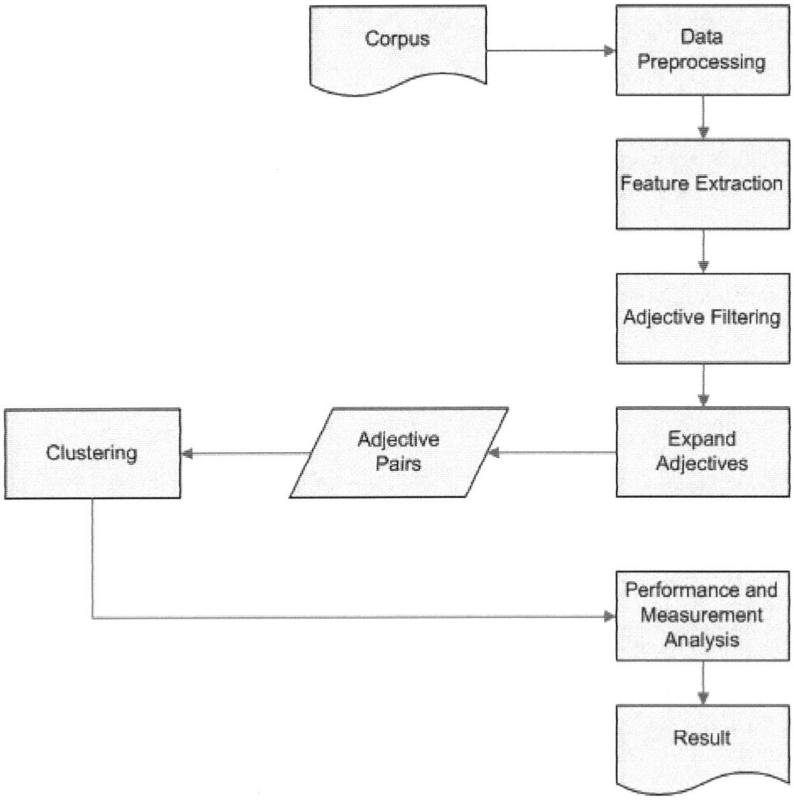

Figure 6. Design Process

3.1.2. Feature extraction Here is a selection process and feature extraction that will be used as the basis of the clustering process.

1. Part of Speech (POS) Tagger, ie the process of giving classes to words. The selected word class is an adjective (adjectives), adverbs, nouns and verbs, in accordance with Dinakaramani [7].
2. Stemming, aimed at reducing variations of words that have the same basic word. In this research, stemming are used in scenario 3.

3.1.3. Adjective filtering Extract from POS TAG corpus (9 million words) adjective words. Adjective words labelled with JJ annotation. From this process, we obtained 923 unique adjective words. The adjective words labbeling by empowering a number of human coders to provide positive and negative orientation. The study used the average result of labeling done by 4 people who had acted as human coders independently. The final set contained 923 adjectives (407 positive and 516 negative terms). From 923 adjective word, we construct a seed words. We constructed this set by taking all adjectives appearing in our corpus 20 times or more. The seed word contain 59 adjective words (39 positive and 20 negative). This seed word will be used in SO-PMI-IR calculation.

3.1.4. Expand adjective Extract from POS tagged corpus (9 million words) and expand adjective seed words coordinated by 'dan', 'tetapi', 'tapi' and 'atau' with another adjective words. We filter from the corpus words 'adjectives seed words (JJ)' + conjuction (CC) + 'adjective words(JJ)'

3.1.5. Clustering This results in graph with same or different links between adjectives. Cluster into two orientations,positive and negative placing as many words of the same orientation as possible into the same subset. Cluster with higher overall frequency is labelled positive.

3.1.6. Performance and measurement analysis Accuracy showed information about how many successful final decisions that the system can analyze against human labeling results.

$$Accuracy = \frac{TP}{N} * 100\% \tag{4}$$

Where TP is true predicted words and N is number of test words.

3.2. System implementation

Here are the hardware and software specifications that were used to implement the prediction system.

3.2.1. Hardware specifications The minimum hardware specification to run this system is the same as with the minimum specifications required by Matlab2012b 64-bit:
a. Processor : Intel Pentium 4 and above, Intel Celeron** , Intel Xeon, Intel Core, AMD64 ** Processor must support SSE2 instruction set.
b. Memory : 2 GB
c. Hard disk : 7 GB* (MATLAB only)
Disk space requirement varies depending on size of partition. The MATLAB installer will inform you of the hard disk space requirement for your particular partition. Installation size will be determined by the installer and can vary for NTFS and FAT formats.

3.2.2. Software specifications The software specifications that were used are:
a. Operating system: Microsoft Windows 8
b. Matlab version R2016b
c. Microsoft Excel 2007

3.3. Experiment scenario

Experiment scenario runs in this study is to split into several sub-systems as shown in the table. Each sub-system will be tested several variations of variables to compare the results. Sasically this study will examine the effect of applying different preprocessing on sentiment classification accuracy results. Three methods of preprocessing will be used and compared, they are:

3.3.1. Scenario 1 The first scenario has propose to testing and investigate the performance of the basic preprocessing. We consider the accuracy of classification which is use clustering from the graph of adjective pairs.

3.3.2. Scenario 2 In this scenario, we add SO-PMI-IR to the clustering process and analyze the effect of SO-PMI-IR to the accuracy of classification which is use clustering from the graph of adjective pairs.

3.3.3. Scenario 3 This scenario has improvement by adding feature extraction stemming . This scenario also use SO-PMI-IR in process.

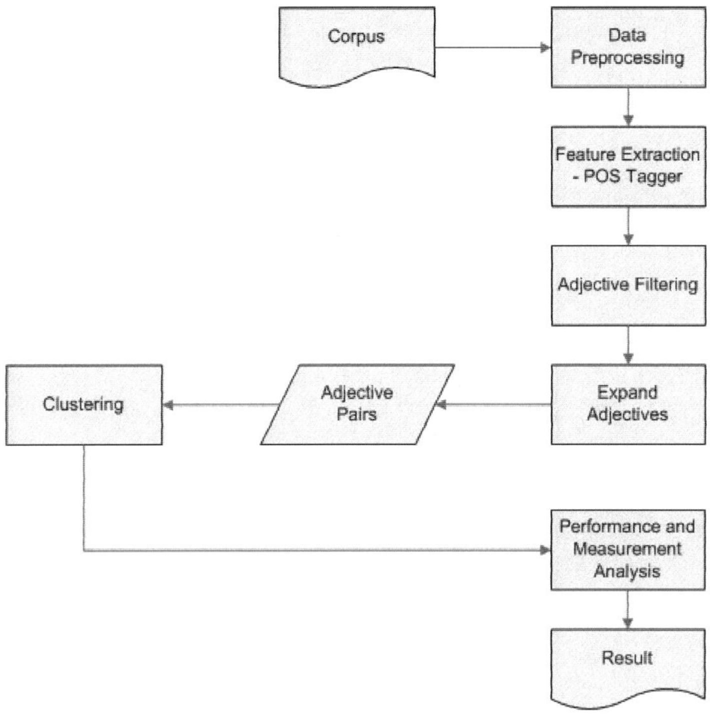

Figure 7. Scenario 1

3.4. Instrument and data collection
The dataset used in this research is 9 million words corpus which was collected from Kompas and Tempo. From Tempo contained 6.022.980 words and Kompas contained 3.697.852 words.

4. Experiment Result
This section show the result of three scenario. The first scenario has propose to testing and investigate the performance of the basic conjunction extraction without preprocessing. In the second scenario, we add PMI to the clustering process and analyze the effect of PMI to the accuracy of classification which is use clustering from the graph of adjective pairs. The last scenario has improvement by adding preprocessing before conjuction extraction. This scenario also use PMI in clustering process.

4.1. Scenario 1
In this scenario, the total links of graph is 1180 links. From this graph we clustering the node to two class, positive adjective and negative adjective, which is shown in in Fig. 10. The blue line is for positive adjective class and the red line is for negative adjective class. The percentage of positive adjective is $(\frac{1112}{68})100\% = 94.23\%$, it is shown that the positive adjective class is 94.23% from the total adjective word. The comparison link betwen positive adjective and negative adjective is shown in Table 1. with total link show in Table 1 The first scenario aims to show the performance of clustering. Accuracy was measured to see the performance of prediction with Accuracy. The result was shown in the table below. There were use basic preprocessing at this first scenario because it just wanted to show the real result of baseline method to build a prediction model and measure the accuracy value. The experiments were run 5 and 10 times. The lower data set provides the lower accuracy in cross fold validation with $k = 5$.

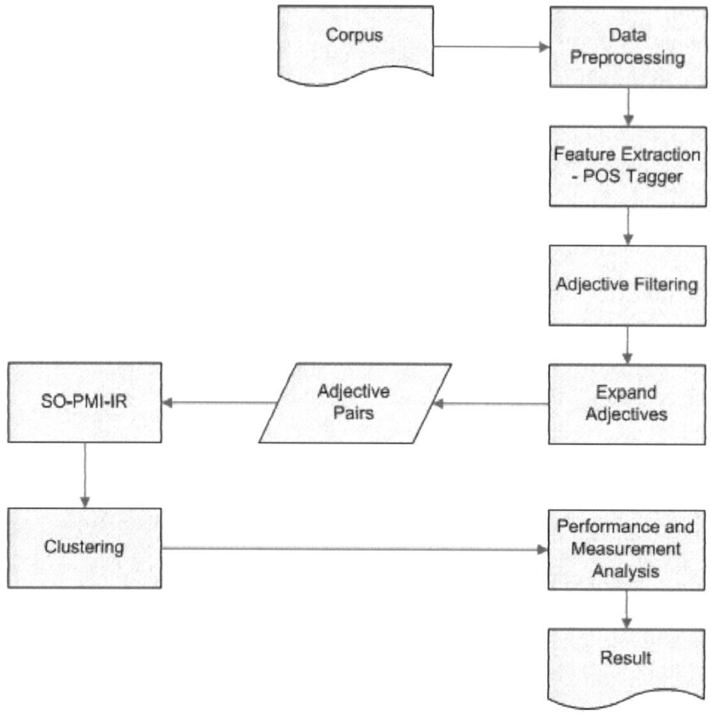

Figure 8. Scenario 2

Table 1. Total link 1st scenario

No	word	Total links
1	Positive Adjective	1112
2	Negative Adjective	68

Table 2. Accuracy of Scenario 1 k Fold=5

Data Set	K Cross Fold (%)					Accuracy (%)
	1	2	3	4	5	
100%	65.97	63.61	63.57	65.92	65.29	64.87
75%	64.15	65.25	62.88	62.03	65.25	63.92
50%	65.64	64.74	68.59	64.36	67.95	66.26
25%	67.95	62.51	66.67	59.56	58.97	62.33

The average accuracy of testing was 60% to 68%. The result of this scenario still had a small value of accuracy. The small value of accuracy was caused by some adjective word that can not be detected because they had affixed. So in the scenario 3, we added some preprocessing technique.

4.2. Scenario 2

In this scenario, the total links of graph is 1644 links. From this graph we clustering the node to two class, positive adjective and negative adjective, which is shown in in Fig. 11. The blue line is for positive adjective class and the red line is for negative adjective class. Compared with the graph of first scenario, in this scenario the link of graph is more than link in first scenario

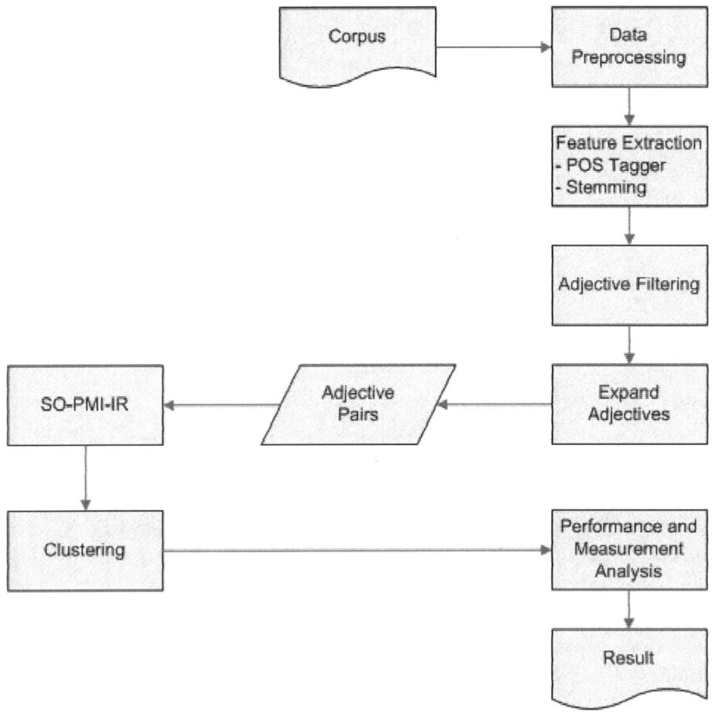

Figure 9. Scenario 3

because the effect of PMI. The percentage of positive adjective is $(\frac{1518}{126})100\% = 92.33\%$, it is shown that the positive adjective class is 92.33% from the total adjective word. The comparison link betwen positive adjective and negative adjective is shown in Table 3. with total link show in Table 3 The second scenario aims to show the performance SO-PMI-IR with basic preprocessing

Table 3. Total link 2nd scenario

No	word	Total links
1	Positive Adjective	1518
2	Negative Adjective	126

and clustering . The result was shown in the table below.

Table 4. Accuracy of Scenario 2 k Fold=5

Data Set	K Cross Fold (%)					Accuracy (%)
	1	2	3	4	5	
100%	72.29	72.93	70.38	65.92	64.65	69.24
75%	72.03	64.83	75.00	69.07	67.37	69.66
50%	69.23	64.10	66.67	76.92	66.03	68.59
25%	69.23	69.23	64.10	47.44	57.69	61.54

The second scenario showed the performance of the clustering using data with preprocessing and combined witg SO-PMI-IR. The experiments were run 5 and 10 times. Compared to without SO-PMI-IR implementation in scenario 1, overall accuracy slighly increased 2%, compared to without SO-PMI-IR. The accuracy without SO-PMI-IR is 68% and with SO-PMI-IR is 70%

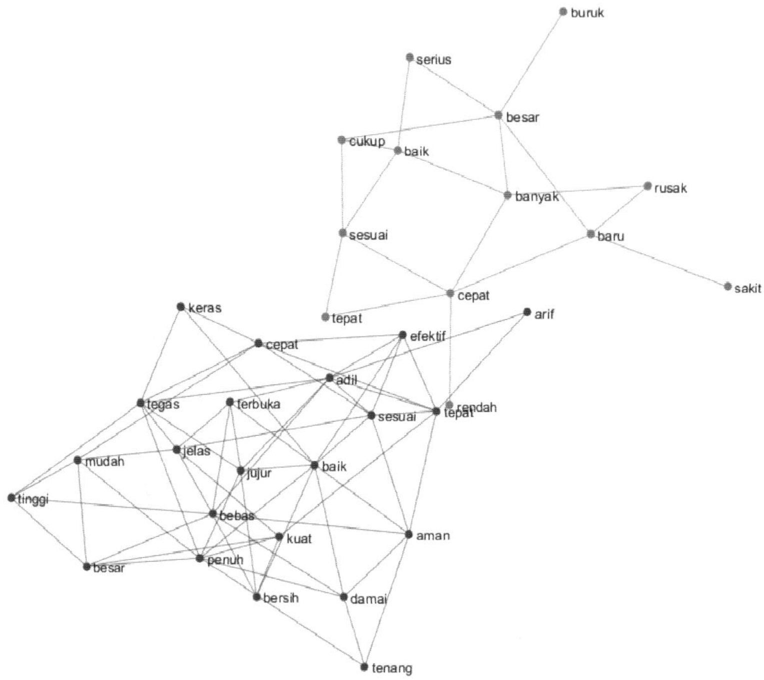

Figure 10. graph and cluster of 1st scenario.

4.3. Scenario 3

The graph of 3rd scenario is shown in Fig. 12. In this scenario, the total links of graph is 3132 links. From this graph we clustering the node to two class, positive adjective and negative adjective. The blue line is for positive adjective class and the red line is for negative adjective class. Compared with the graph of first scenario and second scenario, in this scenario the link of graph is the highest because the effect of PMI and preprocessing. The comparison link betwen positive adjective and negative adjective is shown in Table 5. with total link show in Table 5 The third scenario aims to show the performance of proposed preprocessing combine with

Table 5. Total link 3rd scenario

No	word	Total links
1	Positive Adjective	2315
2	Negative Adjective	817

SO-PMI-IR and clustering. The result was shown in the table below.

Compared to scenario 1, in overall proposed preprocessing method combined with SO-PMI-IR improves accuracy by 8%. Compared to scenario 2, proposed preprocessing combined with SO-PMI-IR improves accuracy by 6%. From the table above shown that the lower data set provide the lower accuracy in cross fold validation with $k = 5$. It is indicates that the proposed method has better performances with higher data set.

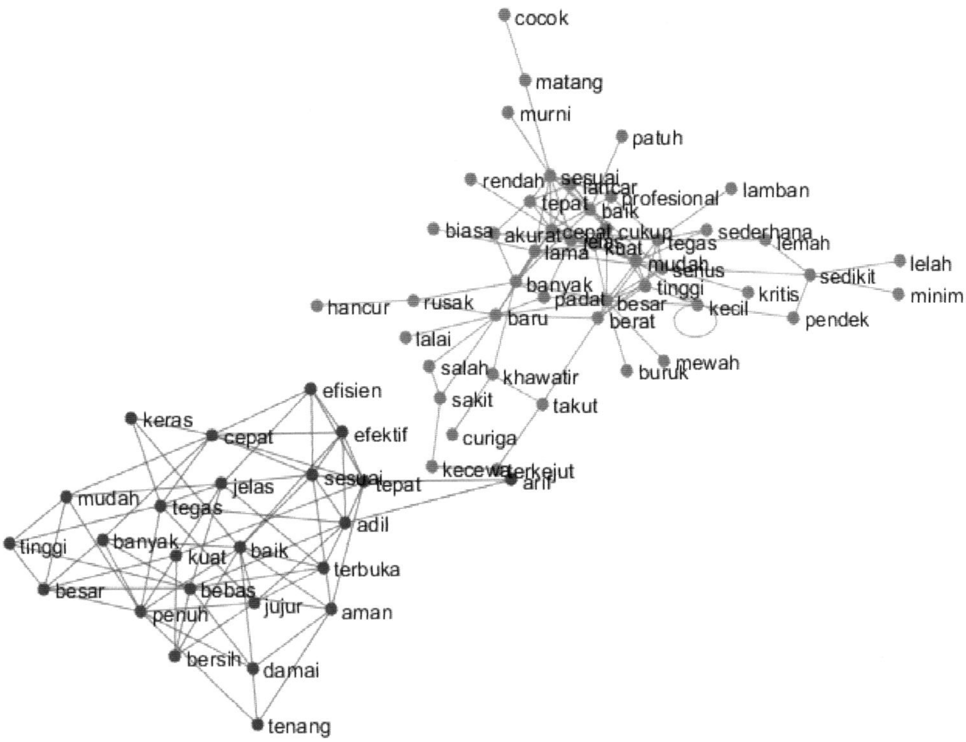

Figure 11. graph of 2nd scenario.

Table 6. Accuracy of Scenario 3 k Fold=5

Data Set	K Cross Fold (%)					Accuracy (%)
	1	2	3	4	5	
100%	75.90	72.30	70.14	64.57	69.96	70.58
75%	79.33	69.95	71.88	68.03	64.18	70.67
50%	77.70	67.99	67.99	71.94	69.06	70.94
25%	74.64	70.29	72.46	60.87	68.84	69.42

4.4. Summary of findings

From section 4, we resume the summary of accuracy in all scenarios, these is shown in Figs. 13 and 14.

From Figs. 13 and 14, its shown that lower dataset has lower accuracy. Compared by each scenario, the best scenario is third scenario.

5. Conclusions

This research proposed the combination of graph clustering and PMI the classificate the adjective word from adjective pairs. The performance of classification models was compared between developed by not using preprocessing, without preprocessing and using combined graph clustering and PMI, and using preprocessing and using combined graph clustering and PMI. From the experiment result, some conclusions of this research are as follows:

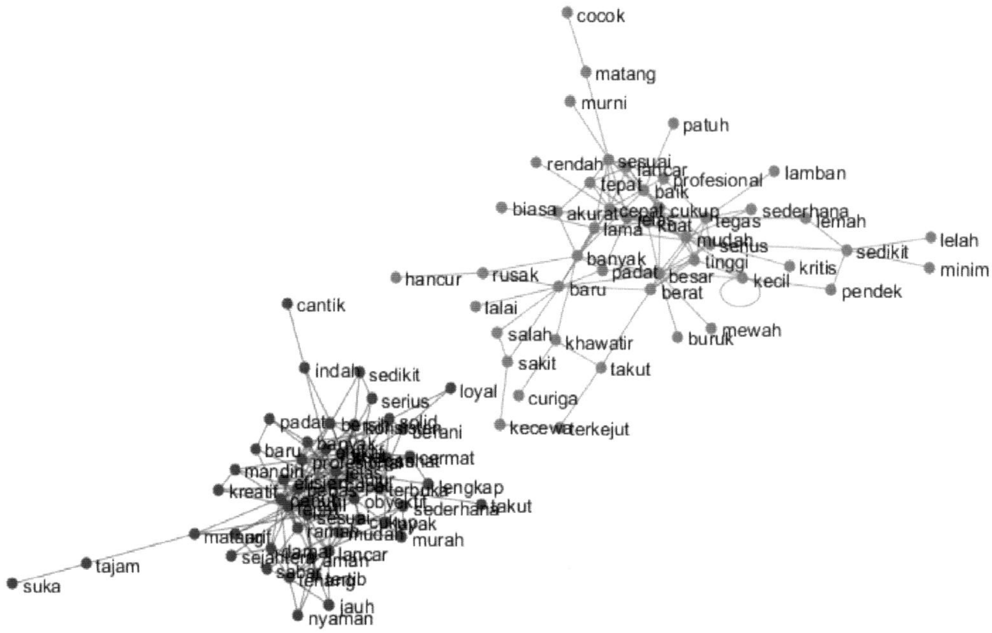

Figure 12. graph and cluster of 3rd scenario.

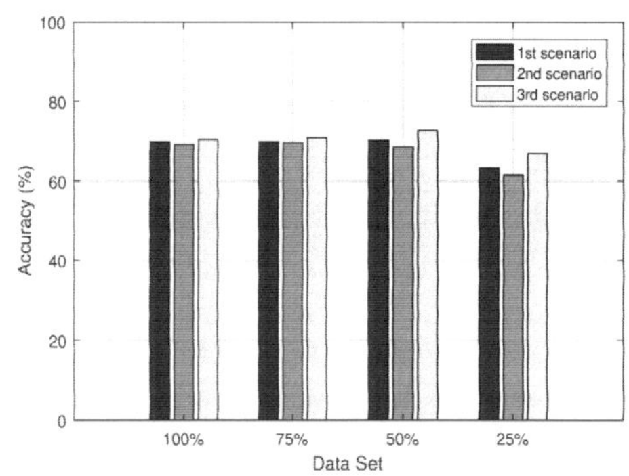

Figure 13. accuracy of cross fold k=5.

1. There are difference of system results and human labeling. This can be happen because every human being has a different behavior and understanding in giving the sentiments of a lexicon.
2. The lower data set provides the lower accuracy in cross fold validation with $k = 5$ and $k = 10$. It is indicates that the proposed method has better performances with higher data set.
3. The implementation of combined graph clustering and PMI is not improved the performance of accuracy. Compared to the third scenario, which is the best performance of all scenario, the effect of preprocessing makes better performance then without using preprocessing.

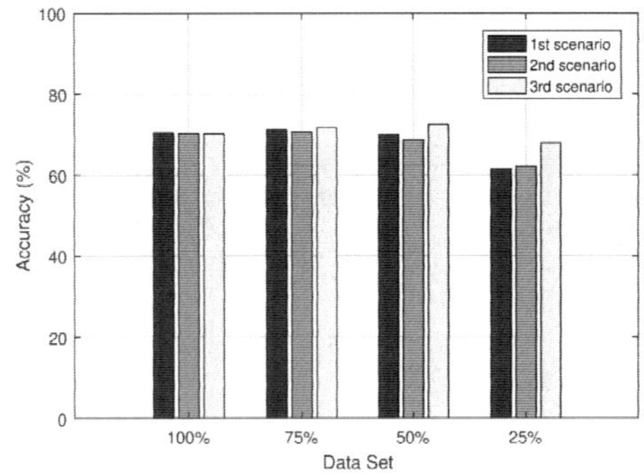

Figure 14. accuracy of cross fold k=10.

4. The overall accuracy result is 72% . These results indicate that not too many successful results have been generated by the system with Indonesian Corpus. This result occurs because the terms in the default dictionary are not complete or not fully in accordance with the needs of Indonesian vocabulary. This can also happen because there has been no special treatment for some features.

References

[1] B. Pang, L. Lee, and S. Vaithyanathan. Thumbs up?: Sentiment Classification Using Machine Learning Techniques. In *EMNLP02:Proc. Of The ACL-02 conf on Empirical methods in natural language processing*, Pages 79-86, ACL, 2002.

[2] T. OKeefe and O. Katrinska. Feature Selection and Weighting Methods in Sentiment Analysis. In *Proceedings of the Australasian document computing symposium*, 2009.

[3] V. Hatzivassiloglou and K.R. Predicting the semantic orientation of adjective. In *Proceeding of the 35th annual meeting on Association for Computational Linguistic*, pages 174-181, Doi:10.3115/976909.979640, 1997

[4] S. Vazquez, M. Padro, N. Bel and J. Gonzalo. Automatic Extraction of Polar Adjectives For The Creation of Polarity Lexicons. In *Proceeding of COLING 2012: Posters:24th International Conference on Computational Linguistics COLING 2012*, Pages 1271-1280, ACL, 2012

[5] P. Turney and M.L. Littman. Unsupervised learning of semantic orientation from a hundred-billion-word corpus. nparc.cisti-icist.nrc-cnrc.gc.ca. 2002

[6] F.Rashel, A. Luthfi, A. Dinakaramani and R. Manurung. Building an Indonesian Rule-Based Part-of-Speech Tagger. In *International Conference on Asian Language Processing (IALP)*, 2014

[7] A. Dinakaramani, F. Rashel, A. Luthfi and R. Manurung. Designing an Indonesian Part of Speech Tagset and Manually Tagged Indonesian Corpus. In *International Conference on Asian Language Processing (IALP)*, 2014

[8)] F. Tala, J. Kamps, K. Muller, and M. de Rijke. The Impact of stemming on Information Retrieval in Bahasa Indonesia(Abstract). In *14th Meeting of Computational Linguistics in the Netherlands (CLIN-2003)*, 2003

[9] K.W. Church and P. Hanks. Word Association Norms, Mutual Information and Lexicography. In *Proceedings of the 27th Annual Conference of the Association of Computational Linguistics (pp. 76-83)*, 1989

[10] P.D. Turney. Mining the Web for synonyms: PMI-IR versus LSA on TOEFL. In *Proceedings of the Twelfth European Conference on Machine Learning Berlin: Springer-Verlag*,2001

[11] B. Liu. Sentiment Analysis and Opinion Mining. In *Chicago: Department of Computer Science*, 2012.

[12] M. Taboada, et al. Lexicon-Based Methods for Sentiment Analysis. In *Association for Computational Linguistics Volume 37, Number 2*, 2010.

[13] L.V. Fausett. Fundamental of Neural Network: Architectures, Algorithm, And Application. In *Prentice Hall, 1st edition. ISBN-13: 978-0133341867.*

[14] M.S.H. Wicaksono. Unsupervised Approach for Sentiment Analysis on Indonesian Movie Review In *6th Conference of Indonesian Students Association in Korea (CISAK)*, 2013

[15] R.F. Manurung Machine Learning-based Sentiment Analysis of Automatic Indonesian Translations of English

Movie Reviews In *Proceedings of the International Conference on Advanced Computational Intelligence and Its Applications (ICACIA)*, 2008.

[16] Y. Muhamad. Analisis Sentimen pada Dokumen Berbahasa Indonesia dengan Pendekatan Support Vector Machine. In *P*hD thesis, Bina Nusantara University.

[17] I. Sunni and D. Widyantoro. Analisis Sentimen dan Ekstraksi Topik Penentu Sentimen pada Opini Terhadap Tokoh Publik. In *Jurnal Sarjana Institut Teknologi Bandung Bidang Teknologi Elektro dan Informatika, vol. 1, no. 1, pp. 200-206*, 2002.

[18] E. Terra, and C.L.E. Clarke. Frequency Estimates for Statistical Word Similarity Measures. In *Proceedings of the 2003 ACL Main Paper pp 165-172*, 2003 .

[19)] Y. Matsuo, K. Uchiyama, T. Sakaki, and M. Ishizuka Graph Based Word Clustering Using a Web-search Engine. In *Proceedings of the 2006 Conference on Empirical Methods in Natural Language Processing (EMNLP)*, 2006 .

Measuring e-Commerce service quality from online customer review using sentiment analysis

Puspita Kencana Sari[1], Andry Alamsyah[2], Sulistyo Wibowo[3]

School of Economics and Business Telkom University, Bandung, Indonesia
puspitakencana@telkomuniversity.ac.id[1], andrya@telkomuniversity.ac.id[2], sulistyo.wib@gmail.com[3]

Abstract. The biggest e-Commerce challenge to understand their market is to chart their level of service quality according to customer perception. The opportunities to collect user perception through online user review is considered faster methodology than conducting direct sampling methodology. To understand the service quality level, sentiment analysis methodology is used to classify the reviews into positive and negative sentiment for five dimensions of electronic service quality (e-Servqual). As case study in this research, we use Tokopedia, one of the biggest e-Commerce service in Indonesia. We obtain the online review comments about Tokopedia service quality during several month observations. The *Naïve Bayes* classification methodology is applied for the reason of its high-level accuracy and support large data processing. The result revealed that personalization and reliability dimension required more attention because have high negative sentiment. Meanwhile, trust and web design dimension have high positive sentiments that means it has very good services. The responsiveness dimension have balance sentiment positive and negative.

1. Introduction

Recently, Internet technology make people can easily generate online contents and able to share information quickly, such as in the review site, e.g: TrustedCompany, Tripadvisor; and in social media e.g: Facebook, Twitter. User generated content sites that review about a product or service from a company can be made as one of the valuable data sources for the company. The activity of data mining from those sites can be an important insight for the company.

Data analytics is an interdisciplinary field with the objective to find pattern on the data with contributions from many fields, such as statistics, machine learning, information search, pattern recognition, and bioinformatics [1]. This activity is widely used in many fields, such as in retail, finance, telecommunications, and media. The several techniques or model for data analytics including classification, prediction, clustering, outlier detection, association rules, sequence analysis, time series analysis, and text mining, as well as new approach such as social network analysis and sentiment analysis [2]. Research in data analytics or specifically data mining, has successfully led to methods, tools and algorithm to solve user's problems using many data type to apply on the following applications, such as business intelligent, predictive model, and decision support system.

To understand their service quality level, most companies use conventional methods to obtain data using sampling and/or questionnaire. This approach has limitation in terms of taking longer time and higher operation cost. We propose a different approach to analyze service quality level using user generated content data from online review site. Because of unstructured nature of the data, we use sentiment analysis method [2] to classify service quality dimensions according to users' perception. By using this method, we can determine the perceived e-commerce service quality.

In this paper, we use Tokopedia, one of the largest e-commerce in Indonesia as our case study to be evaluated its service quality. Between 2016-2017, Tokopedia has 50 million website visitors' and sales number achieve 16.5 million every month [3]. The general purpose of the study is to determine the services quality of e-commerce site using Tokopedia users' review on the TrustedCompany, an online review site. The objective of this research is to know how consumer perception based on data from TrustedCompany using service quality measurement, which is e-service quality. The results of this study can be used as an evaluation for Tokopedia about their service quality.

2. Theoretical Background

Service quality (servqual) is a model to assess services based on its performance's perception. There are five general service quality dimension: *tangibles, reliability, responsiveness, assurance,* and *empathy* [4]. There are some issues in servqual model usage in some industry, one of the issue is in handling service of electronic business or using electronic media [4]. The conceptual model to understand and to improve the service quality in electronics form is called e-Servqual [5]. The new model suggests that e-Servqual incorporate new dimension according to the scale required to measure the quality of e-commerce services:

Table 1. e-Servqual Dimensions [5]

Web Design	Web site design describes the appeal that user interface design presents to customers.
Reliability	Reliability represents the ability of the web site to fulfill orders correctly deliver promptly and keep personal information secure.
Responsiveness	Customers expect online stores to respond to their inquiries promptly
Trust	Trust is defined as customer willingness to accept vulnerability in an online transaction and based on their positive expectations regarding online store behaviors
Personalization	Personalization involves individualized attention personal thank you notes from online stores and the availability of a message area for customer questions or comments

Sentiment analysis is a field of study that analyzes people's opinions, sentiments, evaluations, appraisals, attitudes, and emotions to entities such as products, services, organizations, individuals, problems, events, topics, and their attributes [2]. Generally, there are three levels of sentiment analysis, they are document level, sentence level, entity and aspect level [6]. Sentiment analysis performed automatically to classify opinions into classes by contained words. Multiclass classification is a classification method which has more than two target classes.

Pre-processing is preparation step to process data before entering the classification process [7]. Data resulted from pre-processing step is required to guarantee model fitness and result accuracy. The target class is all e-Servqual dimensions. The details of pre-processing step are as follows.

1) Tokenization process : to divide sentence into words, phrase, or symbol.
2) Filtering process : to remove stopwords or unnecessary text attributes such as *'dan', 'atau', 'dari', 'untuk'*.
3) *Stemming* process : to change words into their word stem, or basic such as *'berjualan'* to *'jual', 'membeli'* to *'beli'*. We distinguish the stopwords dictionary in *Bahasa* between binary classification, positive-negative sentiment classification case and Multiclass Classification, and five dimension e-Servqual classification case.

Table 2. Pre-processing Example

Raw Data	barang kurang update terbaru dan saat mau pembayaran membingungkan jd gk beli dech
Tokenization	barang, kurang, update, terbaru, dan, saat, mau pembayaran, membingungkan, jd, gk beli, dech,
Filtering Stopword	Barang kurang update terbaru pembayaran membingungkan gk beli
Stemming	Barang kurang update baru bayar bingun tidak beli

After finishing the pre-processing procedure, each word given weight with *TF-IDF* methods. The *TF-IDF* is a method to weighting words which is combination step between the *Term Frequency*, and the *Inverse Document Frequency*. *TF-IDF* give higher weight to words that have low frequency of occurrence in some documents, and simultaneously, has high frequency of occurrence in a document [8]. This ensures that words with high *TF-IDF* values can be used as representative examples of the documents in which they originate, and words that often appear in documents, will be given a low weight.

Naïve Bayes Classifier (NBC) is a statistical classification method that can be used to predict class membership probabilities, such as the probability that a tuple is assigned belonging to a specific class [9]. The NBC's advantages are easy to implement, requires little training data for parameters estimation, and higher accuracy in many studies [10].

3. Methodology

In this research, we use dataset collected comes from an online review website namely Trustedcompany which reviews e-Commerce sites in *Bahasa*. The contents are all reviews regarding *Tokopedia*. The data collection period is from January 2017 until April 2017. As the result, we get total 609 data reviews. The data that has been taken is cleaned through the pre-processing process. Unnecessary attributes are discarded, we only keep data that compatible with research needs. After we get the data, we construct machine learning model to automate the classification process with help of *Rapidminer* software.

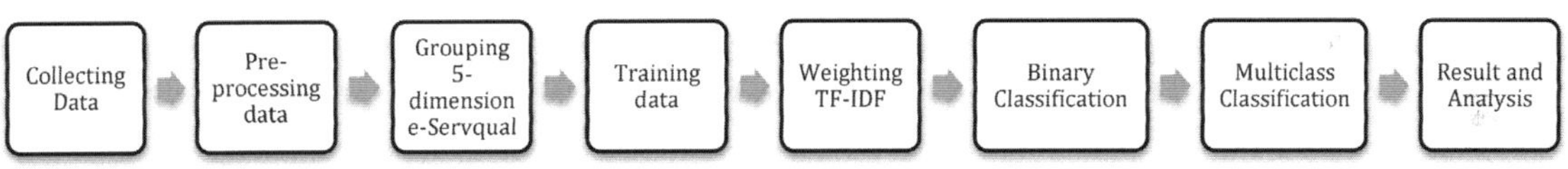

Figure 1. Research Workflow

We classify user review based on five e-Servqual dimensions and sentiment analysis. The five dimensions e-Servqual classification example can be seen in Table 3. Meanwhile, Table 4 is sentiment analysis classification example. To measure model accuracy, we use several measurement, such as: accuracy, recall, precision. We also calculate the confusion matrix [11] shown in Table 5. *TP (True Positive)* is the positive tuples that were correctly labelled by the classifier, *FP (False Positive)* is the negative tuples that were incorrectly labelled as positive, and *TN (True Negative)* is the negative tuples that were correctly labelled by the classifier, and *FN (False Negative)* is the positive tuples that were mislabelled as negative.

Table 3. Example of e-Servqual Dimension

Reviews	*eServqual Dimension*
Sulit berhubungan dengan produsen. utk komunikasipun kadang berbelit2 dan lama.	*Personalization*
Barang sesuai pesanan. Kiriman tidk pernah terlambat	*Reliability*
Pedagang disana dikirimin pesan pada gak responsive.	*Responsiveness*
Pembelian barang terjamin, aman transaksi belanjanya, dan mudah.	*Trust*
webnya sudah sangat baik dan nyaman.	*Web Design*

Table 4. Sentiment Analysis Classification Example

Reviews	*Sentiment*

Hari ini sangat mengecewakan tokopedia. Status sdh di kirim 2 jam yg lalu tp msh blm di terima. Padahal di kirim by gosend.	*Negative*
Bagus banget buat segala kebutuhan one stop solutions	*Positive*

Table 5. Confusion Matrix

	***True* Positive**	***True Negative***
***False* Positive**	42 (TP)	2 (FP)
***False* Negative**	8 (FN)	48 (TN)

The equation of accuracy, recall, precision formula is as follows

$$Accuracy = \frac{TP+TN}{TP+FN+TN+FP} \quad (1)$$

$$Precision = \frac{TP}{TP+FP} \quad (2)$$

$$Recall = \frac{TP}{TP+FN} \quad (3)$$

From formula (1) we get Accuracy result is 90%, from formula (2) we get Precision is 90,5%, and from formula (3) we get Recall is 90%. Indicating that machine learning model using NBC method successfully classified the data.

4. Result and Analysis

From 609 review data, we classify 321 positive sentiment and 288 negative sentiment, shown in Table 6. For the case of multiclass classification of e-Servqual dimension, the result is shown in Table 7. From table 6 data is classified based on e-Servqual dimensions per sentiment category, the result proportion sentiment on each dimension is shown in Table 8.

There are 138 data in reliability dimension. There are not too significant difference between negative (46%) and positive (54%) sentiment. From five dimensions of e-Servqual, two dimension, (Reliability and Responsiveness) have sentiment value almost equal. Meanwhile, other two-dimensions (Trust and Personalization) have contrary sentiment value. Therefore, the whole sentiment value almost balance. This case shown that consumers perception about service quality from website have balance value although Tokopodia has some weaknesses, but also has some strengthness.

Negative sentiment about the Reliability dimension have topics such as; goods are not met buyer expectations and not sent by the seller. The Personalization dimension has the most negative sentiments (79%) among the other dimensions. Topic that often reviewed by *Tokopedia* users is regarding customer service performance to handle complaints. The Trust dimension has the most positive sentiment (92%) among other dimensions. Topics that are often reviewed regarding about transactions security, as *Tokopedia* implements "*escrow*" service as guarantee. The Web Design dimension also has dominant positive sentiment (66%). It means that *Tokopedia* user interface is good. Most users praise the usability, the look, and the easiness to navigate the website. The Responsiveness dimension has a balanced value, 50% for positive sentiment and 50% for negative sentiment.

Table 6. Binary Classification Result

	Number	*Percentage (%)*
Sentiment Positive	321	53
Sentiment Negative	288	47
Total	609	100

Table 7. Multiclass Classification Result

Dimension	*Number*	*Percentage (%)*

Reliability	138	23
Personalization	139	23
Responsiveness	121	20
Trust	112	18
Web Design	99	16
TOTAL	609	100

Table 8. Sentiment Proportion of each e-Servqual Dimension

Dimension	*Number*	*Sentiment Positive (%)*	*Sentiments Negative(%)*
Reliability	138	46	54
Personalization	139	21	79
Responsiveness	121	50	50
Trust	112	92	8
Web Design	99	66	34
Total	609	321	288

5. Conclusion

Sentiment analysis using Naïve Bayes Classifier method perform well to classify sentiment and issue regarding e-commerce e-Servqual from online review site. Data collection through online source is more appropriated for the research object, since the object also conduct their business through online modus. We considered this approach are more effective and efficient. Moreover, the methods to process unstructured data such as text is available and robust. However, this procedure need to be investigated thoroughly as we collect the data during specific duration. In order to get more representative result we need to conduct another source data such as social media (Twitter and Facebook) to verify the service quality.

The implication from this research is company can evaluate their electronic service quality from consumer perception faster and cheaper because it exploit data which has existed from website review such as Trustedcompany or review in their website. Difference with conventional way, company have to collect data using questioners to evaluate service quality.

Based on the results, the overall sentiment about *Tokopedia* service quality is positive. We conclude *Tokopedia* service quality is good enough. Although the margin is not too high between positive and negative sentiment. On the e-Servqual dimensions, the personalization dimension should get serious attention because they are dominated by negative sentiments.

References

[1] Alamsyah, A., Rahma, W., & Irawan, H. (2015). *Sentiment Analysis Based On Appraisal Theory For Marketing Intelligence In Indonesia's Mobile Phone Market. Journal of Theoretical and Applied Information Technology (2015)*

[2] Zhao, Y. (2013). R Data Minning: Examples and Case Studies. *R and Data Mining*, 2012

[3] Tempo.co. (2016). *Tokopedia Jadi Situs E-Commerce No 1 di Indonesia.* [online]. https://bisnis.tempo.co/read/news/2016/08/17/092796716/tokopedia-jadi-situs-e-commerce-nomor-1-di-indonesia

[4] Zeithaml, V.A. (2002), "Service excellent in electronic channels", Managing Service Quality, Vol. 12 No. 3, pp. 135-8.

[5] Lee, Gwo-Guang., & Lin, Hsiu-Fen. (2005). *Customer Perceptions of e-service quality in online shopping, 33(2). International Journal of Retail& Distribution Management.*

[6] Liu, B. (2012). *Sentiment Analysis and Opinion Mining. Synthesis Lectures on Human Language Technologies*, *5*(1), 1–167.

[7] Kiz, Aysun Kapucugil., dan Özdağoğlu, Güzin. (2015). Text Mining As A

Supporting Process for VoC Clarification. *Alphanumeric Journal*, 3(1), 025–040. Retrieved from The Journal of Operations Research, Statistics, Econometrics and Management Information Systems

[8] Alamsyah, A., & Saviera, F. (2017). *A Comparison of Indonesia's E-Commerce Sentiment Analysis for Marketing Intelligence Effort (case study of Bukalapak, Tokopedia and Elevenia). The* 8th International Conference on Sustainable Collaboration in Business, Technology, Information and Innovation (2017)

[9] Moens, M. F., Li, J., dan Chua, T.-S. (2014). Mining User Generated Content. Boca Raton: CRC Press.

[10] Han, J., Micheline K. & Jian P. (2012). Data Mining: Concepts and Techniques (3rd Ed.).Waltham: Morgan Kauffman Publishers

[11] Manning, Christopher D., Raghavan, Prabhakar., & Schütze, Hinrich. (2009). *An Introduction to Information Retrieval.* Cambridge University Press

Automatic sentence extraction for the detection of scientific paper relations

Y Sibaroni[1], S S Prasetiyowati[2], M Miftachudin[3]

[1,2,3]Computational Science, School of Computing, Universitas Telkom

[1]yuliant@telkomuniversity.ac.id, [2]srisuryani@telkomuniversity.ac.id, [3]Muhammad.Miftachudin55@gmail.com

Abstract. The relations between scientific papers are very useful for researchers to see the interconnection between scientific papers quickly. By observing the inter-article relationships, researchers can identify, among others, the weaknesses of existing research, performance improvements achieved to date, and tools or data typically used in research in specific fields. So far, methods that have been developed to detect paper relations include machine learning and rule-based methods. However, a problem still arises in the process of sentence extraction from scientific paper documents, which is still done manually. This manual process causes the detection of scientific paper relations longer and inefficient. To overcome this problem, this study performs an automatic sentences extraction while the paper relations are identified based on the citation sentence. The performance of the built system is then compared with that of the manual extraction system. The analysis results suggested that the automatic sentence extraction indicates a very high level of performance in the detection of paper relations, which is close to that of manual sentence extraction.

1. Introduction

Relations between scientific papers are forms of an interrelationship between one paper and another. Such relations are very useful and important in the fields of research and education. By observing the relationships between the existing scientific articles, one can obtain a description of recent research in particular fields as well as those which can be developed further. In addition, scientific paper relations can be used to investigate the development trend in future research. Manually, the forms of such paper relations can be obtained through an in-depth analysis of a collection of scientific papers.

Large number of worldwide produced scientific papers which reach millions of papers annually will certainly complicate the process of manually identified relations [1]. Therefore, the need for the process of identifying relations between papers automatically becomes more urgent.

In the case of citation-based relations, the actual applications actually have been created. Some websites which serve as a paper recommender system have been available, such as scholar.google.com, www.acm.org, www.sciencedirect.com, and citeseerx.ist.psu.edu. Websites may provide recommendations of desired scientific papers on the basis of similarity relations or citation relations. Similarity relations in question mean the level of content similarity between scientific papers.

Developing an automatic paper recommendation system takes several stages. In a supervised learning-based stage, the process includes file reading and sentence splitting based on a collection of pdf scientific papers, creation of a scientific paper dataset and class annotation, feature identification and extraction, model learning, and, finally, model evaluation [2]. All of these steps (except class annotation) should be done automatically to be effective for large data applications.

One of the problems that arise in those studies is the process of automatic file reading and sentence splitting in which some sentences cannot be separated completely. An improper sentence separation can certainly affect the identification of the whole paper relations. So far, no studies have been done on research paper relations which measure the effect of automatic extraction error on the performance of scientific paper relation identification.

This study was aimed to develop an automatic sentence extraction process for detecting scientific paper relations and, at the same time, to measure the effectiveness of the automatic sentence extraction compared to the manual sentence extraction. Therefore, this study seeks to obtain a system which is able to automatically separate sentences in scientific papers whose results are expected to approach the results of manual sentence extraction. The result of this study is very important to know when we want to develop business applications of papers' relation detection based on input pdf-type papers that come from the internet.

2. Related work

The relations between scientific papers are formed when the papers have similar contents or when a paper cites other papers. The first is known as a *content-based approach* while the latter is known as a *citation analysis-based approach* [3]. This citation-based approach is further developed into a citation context approach. A citation context is defined as a collection of words or phrases in a sentence which quotes or discusses other papers [2]. This citation context-based approach provides a more detailed relation form of the citation analysis approach which only contains similarity relations. Some examples of paper relations which can be developed based on this approach include *extend, criticize, compare,* and *improve* [4].

The research focus in each approach is different. In the content-based approach, the research focus of the majority of researchers is on developing better similarity measurements. Several similarity measurements developed include *identity measures* [5], *Jaccard coefficient* [6] and *cosine similarities* [7]. *Identity measures* are used to determine the co-derivative (plagiarism) of an article based on a collection of article database. *Jaccard coefficient* is used to measure the similarity of text in an article, while cosine similarities are used to measure the power level of co-citation between two sentences in an article.

In the citation analysis approach, the research focus is actually also on developing similarity measurements like the previous approach. However, in this approach, the similarity measurements are calculated according to the occurring citations. Some basic similarity measurements include *bibliographic coupling, co-citation analysis, reference list, and cited by* [8], [9]. In this approach, the similarity of a paper is not only defined by its content, but also by its citation pattern. The citation pattern in question includes citation sequence, citation location in a paragraph or chapter, and so on. Some of the more complex citation-based similarity measurements developed by researchers include *in-text citation frequency analysis* (ICFA) and *in-text citation distance analysis* (ICDA) by Gipp et al.[8], *Citation Proximity Analysis* (CPA) and *Citation Order Analysis* (COA) by Gipp and Bell [9] and *Global Relation Strength* (GRS) by Liang et.al [10].

In the citation context approach, the research focus is on detecting paper relations based on a citation context contained in citation sentences. Teufel et al. classify citation context into 12 types [11] in which these 12 categories are sentences which contain inter-paper relationships. Teufel et al. use a supervised machine learning as the classification method. The development of an explicit paper relation scheme based on this citation context is then performed by Wang. Based on the results of citation context analysis, Wang divides paper relations into extend, criticize, compare, and improve [4]. The method used by Wang to classify paper relations is the rule-based method. In both studies, the main feature used for classifying or determining paper relations is the cue phrase feature based on the citation sentence identified previously.

The existence of paper relations, therefore, depends on the results of citation sentence identification. In relation to the process of citation sentence identification, some studies are found to be very relevant. Two of them are those conducted by Sugiyama et al. [12] and Widyantoro et.al.[13]. Both studies use the supervised machine learning method for the classification process. The accuracy of this method is largely defined by the reliability of the used features. Sugiyama et al. [12] use 6 main features for the classification process, including *unigram, bigram, position, proper Noun, previous* and *next-sentence,* and *orthographic*, while Widyantoro et al. [13] use citation styles-related features. Widyantoro et al. use four forms of citation style to identify the citation sentence.

Unfortunately, in the implementation, Sugiyama et al. only use the "accuracy" measure in determining system performance based on the proposed feature even though their research focus is on investigating the extent to which the proposed feature is able to identify the citation sentence appropriately. The use of "accuracy" measure becomes less precise because the obtained value is greatly influenced by the system performance in identifying non-citation sentence as well. The use of citation styles-related features is even more promising to be used for identifying *citation sentence.*

3. Identification of citation-based paper relations

The existence of a paper relation can be first identified using citation and subsequently be detected using a particular cue phrase [4]. The accuracy of paper relation identification significantly depends on the accuracy of citation detection in every sentence in scientific papers. There are four types of scientific paper relations that can be identified according to the citation context existing in citation sentence, including extend, criticize, compare, and improve. These four types of paper relation also intersect with the citation function grouping of a citation sentence contained in the 12 categories as performed by [14].

4. The proposed system designs

The system design of learning classifier model for detecting paper relations conducted in this study can be seen in Figure 1. The process begins with Reading Data from a collection of pdf papers that will be processed. The data used in this study is 30 pdf English language articles that are obtained from the internet for free. The data reading process utilizes pdfBox library so that the pdf files can be read as text files. The process is also combined with the Sentence Splitting process, which is an automatic extraction of sentences from a parent article using the breakIterator function. The whole sentences obtained from the reading and separation process are later stored in a single database. The stored database (citation dataset) contains information on id_file, id_sentence, and sentence content. The number of sentences generated from this automatic extraction process is as many as 5878.

For the learning process, an annotation must be done for each sentence as generated from the separation process. This annotation means class labelling process for each sentence as a "citation" or "non-citation" class. The annotation process generates 874 citation sentences and 5004 non-citation sentences. The number of the automatically generated sentences is slightly different from the number of manually generated sentences considering that the accuracy of breakIterator function has not reached 100%. After the annotation process is finished, the next is feature extraction, i.e. a process to generate citation style-based features. The dataset and the resulting features are then used for the learning process. The classifier model used in this study includes Support Vector Machine (SVM) and Naïve Bayes (NB).

The evaluation method used in this study is 5-fold Cross Validation based on id-paper. The determination of this evaluation method aims to make the formation of a model more objective than cross validation based on random sentences. This is because in a paper, there will likely be several sentences that have similar writing patterns. The performance of relation identification through automatic sentence extraction process is then compared with the performance of relation identification through manual sentence extraction process. In addition to comparing the number of citation sentences

IOP Conf. Series: Journal of Physics: Conf. Series **971** (2018) 012054 doi:10.1088/1742-6596/971/1/012054

identified, the performance measurement is also done by considering the F-Measure value obtained from each process. The comparison of these two processes will identify the accuracy of automatic extraction process in the detection of paper relations.

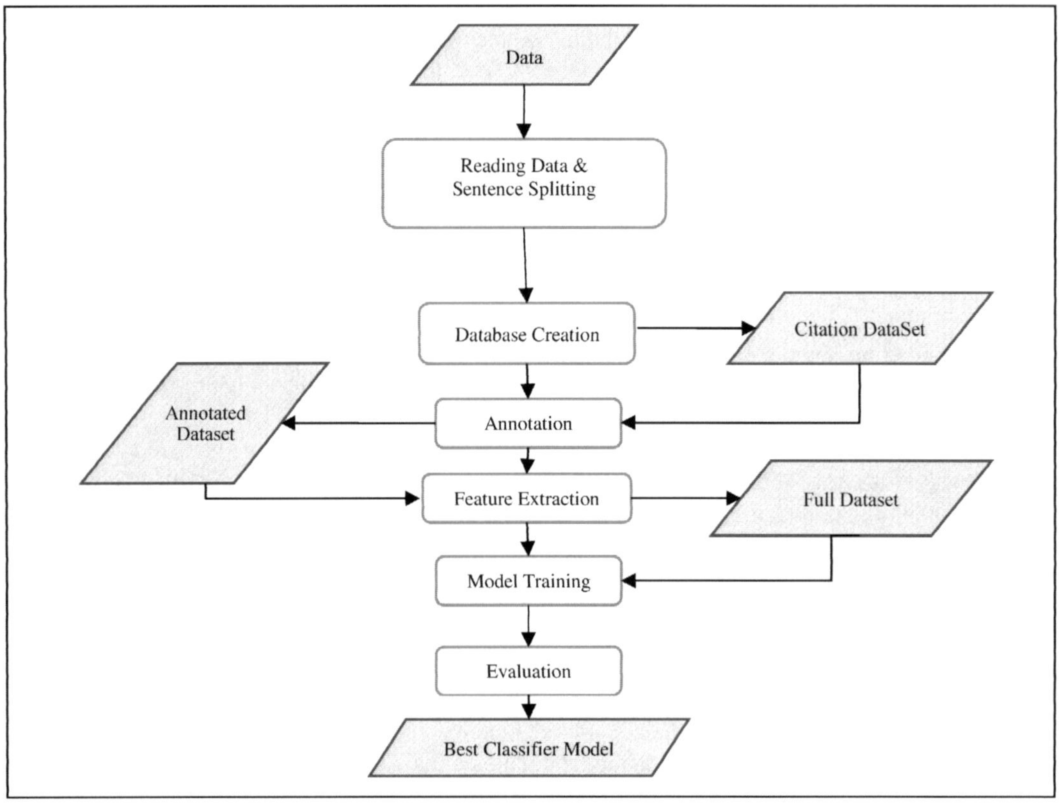

Figure 1. Model learning of relation identification

5. Features used

The accuracy of citation identification is defined by the use of appropriate features. In this study, the citation feature used to detect citation includes four citation style features as proposed by [13]. We use 4 features in [13] because these four features provide good results. The four citation style features along with the regular expression and its feature explanation can be seen in Table 1.

Table 1. Features for citation detection [13]

Regular Expressions	Name of Feature	Description
m/(\(\d{4}\))/g	Textual Syntactic	Declare patterns in the form of: "digits with 4 numbers" flanked by brackets Example: .... Sibaroni (2015)
m/(\([A-Z][^\(]+\d{4}\))/g	Textual Parenthetical	Patterns of letters, punctuation, and numbers consisting of 4 digits appear in sequence. Example: Ahuja, Magnanti, and Orlin, 1995
(m/(\[\d{1,3}\])/g)	Numbered 1	Pattern: There are 1 to 3 digit numbers in a square bracket Example: .... and PolyBench [5,7]
(m/(\[\d{1,3}.*\])/g	Numbered 2	Pattern: There are 1 to 3 digit numbers (and can be repeated more 1 time) in an square bracket

The performance of citation sentence identification through automatic sentence extraction is then compared with that of citation sentence identification through manual sentence extraction. The number of citation sentences in the data set, which is generated through manual sentence splitting, serves as the reference to the performance measurement of automatic sentence splitting. The evaluation method used is a 5-cross validation method based on id-paper, while the classifiers used are Naïve Bayes and Support Vector Machine (SVM).

6. Experimental results and evaluation results

The results of citation sentence identification through manual sentence extraction are shown in Table 2, while the ones of citation sentence identification through automatic sentence extraction are presented in Table 3. The system performance is evaluated by taking into account the ratio of the number of successfully detected citations between automatic and manual sentence extraction. Furthermore, the F-Measure values of the two processes are also compared.

Table 2. Citation identification performance for manual sentence extraction

Actual/prediction	Naïve Bayes / SVM	
	citation	non-citation
citation	632	264
non-citation	34	5738

Based on the experimental results as shown in Table 2, the SVM and NB classifiers generate the same results in detecting citations contained in all sentences obtained using manual extraction. As many as 632 citation sentences are identified from a total of 896 citation sentences. Meanwhile, the F-Measure calculation results in 81%, which means the performance is quite good.

Table 3. Citation identification performance for automatic sentence extraction

Actual/prediction	Naïve Bayes / SVM	
	citation	non-citation
citation	616	258
non-citation	74	4930

According to the experimental results as illustrated in Table 3, the SVM and NB classifiers give the same result in detecting citations contained in all sentences obtained using automatic extraction. As many as 616 citation sentences are identified from a total of 874 citation sentences.

The number of citation sentences generated from automatic extraction is slightly different from that generated from manual extraction. The calculation of F-Measure results in 79%, which means the performance is quite good, even approaching the automatic extraction performance (81%). The comparison between the number of citation sentences obtained through automatic extraction and that obtained through manual extraction generates a value of 97.5%.

Based on the analysis results, it can be concluded that the process of automatic sentence extraction using the breakIterator function gives very satisfactory results from which the results of citation sentence detection are close to those of citation sentence detection using manual sentence extraction.

7. Conclusion and future work

In general, the obtained results are very satisfactory. The proposed of automatic sentence extraction system by utilizing the library function available in java programming language for the process of relation detection provides excellent performance. This means the development of business applications for the detection of paper relations based on this extraction system is very feasible to do.

Further research that can be done is to conduct experiments on a larger scale on scientific article data with larger size and more varied fields. The text quality of scientific papers to be processed should also be considered because it will affect the obtained file reading results. The identification process of paper relations, including *extend, criticize, compare,* and *improve* by directly inputting pdf paper data can also be done for the detection of a more specific paper relation. The encouraged to develop appropriate features to identify each type of paper relation is the next research challenge that can be tried to do. Furthermore, an experiment also needs to be conducted in other classifiers in addition to Naïve Bayes and SVM to obtain the best system performance.

References

[1] M. Ware and M. Mabe, "The stm report," *STM Rep.*, no. September, p. 68, 2009.

[2] Y. Sibaroni, D. H. Widyantoro, and M. L. Khodra, "Extend Relation Identification in Scientific Papers Based On Supervised Machine Learning," in *The 2016 International Conference on Advanced Computer Science and Information Systems (ICACSIS 2016)*, 2016.

[3] Y. Sibaroni, D. H. Widyantoro, and M. L. Khodra, "Survey on Research Paper 's Relations," *Int. Conf. Inf. Technol. Syst. Innov.*, 2015.

[4] W. Wang, P. Villavicencio, and T. Watanabe, "Analysis of reference relationships among research papers, based on citation context," *Int. J. Artif. Intell. Tool*, vol. 21, no. 2, pp. 1–24, 2012.

[5] T. C. Hoad and J. Zobel, "Methods for Identifying Versioned and Plagiarised Documents," *JASIST*, vol. 54, pp. 203–215, 2003.

[6] H. Xu, E. Martin, and A. Mahidadia, "Contents and time sensitive document ranking of scientific literature," *J. Informetr.*, vol. 8, no. 3, pp. 546–561, 2014.

[7] Y. Kyung Jeong, M. Song, and Y. Ding, "Content-based author co-citation analysis," *J. Informetr.*, vol. 8, no. 1, pp. 197–211, 2014.

[8] B. Gipp, J. Beel, and C. Hentschel, "Scienstein : A Research Paper Recommender System," *Proc. Int. Conf. Emerg. Trends Comput. (ICETiC'09), pages 309–315, Virudhunagar (India), January 2009. Kamaraj Coll. Eng. Technol. India, IEEE*, no. January, pp. 309–315, 2009.

[9] B. Gipp and J. Beel, "Identifying related Documents for Research Paper Recommender by CPA and COA," *Proc. World Congr. Eng. Comput. Sci. 2009 Vol I WCECS 2009, Oct. 20-22, 2009, San Fr. USA*, vol. I, pp. 20–23, 2009.

[10] Y. Liang, Q. Li, and T. Qian, "Finding Relevant Papers Based on Citation Relations," *Lect. Notes Comput. Sci.*, vol. 6897, no. Department of Computer Science, City University of Hong Kong, pp. 403–414, 2011.

[11] S. Teufel, a Siddharthan, and D. Tidhar, "An annotation scheme for citation function," *Proc. 7th SIGdial Work. Discourse Dialogue*, pp. 80–87, 2006.

[12] K. Sugiyama, T. Kumar, M.-Y. Kan, and Ramesh C. Tripathi, "Identifying Citing Sentences in Research Papers Using Supervised Learning," *978-1-4244-5651-2/10/$26.00 ©2010 IEEE*, pp. 67–72, 2010.

[13] D. H. Widyantoro and I. Amin, "Citation Sentence Identification and Classification for Related Work Summarization," *Int. Conf. Adv. Comput. Sci. Inf. Syst. (ICACSIS), Jakarta, Indones.*, 2014.

[14] S. Teufel and D. Tidhar, "Automatic classification of citation function," *Proc. EMNLP-06, Sydney, Aust.*, 2006.

POS-Tagging for informal language (study in Indonesian tweets)

Endang Suryawati, Devi Munandar, Dianadewi Riswantini, Achmad Fatchuttamam Abka, Andria Arisal

Research Center for Informatics, Indonesian Institute of Sciences

endang@informatika.lipi.go.id

Abstract. This paper evaluates Part-of-Speech Tagging for the formal Indonesian language can be used for the tagging process of Indonesian tweets. In this study, we add five additional tags which reflect to social media attributes to the existing original tagset. Automatic POS tagging process is done by stratified training process with 1000, 1600, and 1800 of annotated tweets. It shows that the process can achieve up to 66.36% accuracy. The experiment with original tagset gives slightly better accuracy (67.39%) than the experiment with five additional tags, but will lose important informations which given by the five additional tagset.POS-Tagging for Informal Language (Study in Indonesian Tweets).

1. Introduction

Twitter is one of the most popular information sharing media. This microblog is continuously producing data. This data are in the form of tweets, messages sent using Twitter, which has a limit of the number of characters for each tweet. It is an interesting task for enthusiasts in the field of computational linguistic to obtain useful information from this huge data. The information are usually about public opinion on social, politics, health, education, disaster, etc. In an effort to extract information derived from tweets, it is necessary to understand the pattern of tweets.

Tweets are known to have informal language styles. This requires special attention in the process of recognizing every word compared to formal language. The main problems encountered in informal language, such as in tweets, are the use of non-standard words and abbreviation as the effect of character limitation for each message. Sometimes, tweet's writer also use ellipsis and emoticon to show expression. Tweets also can be a mixing between different local languages and dialects with the formal/official language.

One of the fundamental process when extracting information from twitter data is part-of-speech (POS) tagging. POS tagging determine tags of words in tweets based on the syntactic analysis. The studies about POS tagging for twitter data have been performed using both rule-based and statistical-based approaches. The addition of Twitter data that tends to be informal to a formal language corpus can decrease the performance of a model. It becomes a challenge to be able to handle the problem of extracting Twitter data pattern, in addition to building the tweets corpus.

Gimpel et al. [1] developed POS tagging for English using conditional random field (CRF) method by training 1827 manually annotated tweets. The results shows better performance than Stanford POS tagger. The research by Gimpel et al. inspired another group of researcher to develop POS tagging for tweets in Dutch [2]. Avontuur et al. [2] use SoNaR's standard twitter collection added with five special tags to form a golden corpus. The tagger used is existing Frog tagger combined with post-processing module. In addition to English, POS tagging for Twitter data has been developed for several other languages such as Myanmar [3], Japanese [4], Irish [5], Indian [6] and Arabic [7]. In these studies, various treatments for informal language have been conducted to provide better results.

Indonesian tweets that has variety of language styles also requires special handling. Indonesian tweets is strongly influenced by foreign languages and local languages that have diverse dialects. Although there is no specific research about Indonesian tweets corpus for POS tagging, however some research have taken advantage of POS tagging for Indonesian tweets to support natural language processing (NLP) applications, namely event extraction [8][9]. The research shows an attempt to correct deficiencies when handling out of vocabulary (OOV) problem.

In order to achieve good performance of NLP applications, the handling of problem in tweets should be done in pre-processing stage. In an NLP application that utilizes POS tagger and named-entity recognition (NER) for extracting information, improper OOV handling will worsen the performance. The poor performance of POS tagger will decrease NER accuracy and further reduce overall system performance [10]. Therefore, special handling in developing POS tagger for informal language, such as in twitter data, is needed.

In general, there are two ways to develop POS tagging for Twitter data. The first way is to build a tweets corpus by manually annotating each word, then building a tagger model based on the corpus [8]. The second way is to automatically annotate each word in twitter data collection using existing formal language tagger and make corrections if needed, then the tagger is retrained to get new model for tweets data [1][2]. The retraining process will recur in accordance with the addition of training data that has been corrected, so that the performance continue to increase.

In this paper, POS tagger for Indonesian tweets is developed using the second way, that is utilizing existing POS tagger for formal language to automatically perform annotations on twitter data collection. The goal is to prove that POS tagger for the formal language can be used to annotate twitter data collection, with the addition of five new tags related to twitter attribute and the use of informal words. Word embeddings generated by using word2vec [11] and glove [12] are used to represent word in addition to random method. The random word embeddings is used as the baseline.

The next sections of this paper is managed as follows. Chapter 2 describes issues on twitter data for Indonesian and foreign languages (non-Indonesian) and how to handle them. Chapter 3 describes the experiments (corpus and method used). The results and discussion are outlined in Chapter 4 and Chapter 5 is the conclusion of this paper.

2. Twitter Sentence Pattern

2.1. The Problem of Twitter Data

The informal pattern often used in tweets are abbreviation, misspelled word, use of emoticons, use of interjection words, dialect variation and noisy data. Tweets also have orthographic forms with many variants. Indonesian tweets has more irregular pattern. In addition to have such an informal pattern, Indonesian tweets is influenced by foreign languages, local languages, and social languages. Social languages are often referred as the *alay* language. The *alay* language contributes to the addition of OOV lists that degrade model performance.

To deal with the informal pattern of Twitter data, A Le et al. [13] grouped informal language pattern into six groups of patterns, namely abbreviation, interjection, foreign word, blending, emoji, and emoticons. In the foreign word group, sometimes also found the form of abbreviation, such as the word "ht" which means "hot topic". In the blending group, there is an example of mixing between Indonesian tweets and English such as "*gagal* move on", which means failed to move on. Although this approach is intended for sentiment analysis applications, it can be adopted for POS tagging for the normalization process at the pre-processing stage.

In twitter data also found words derived from local languages, such as *nak*, *ora*, *piye*, and *kabare*. In addition, there is an unusual form of the loop like *tetembak* which means firing a shot. The influence of *alay* language is also strong enough to affect the pattern of Indonesian tweets, word such as *cucok* which means fit or suitable and a number of other *alay* words.

2.2. *Handling The Problem of Twitter Data*

In general the problem in Twitter data can be addressed either through the selection of appropriate tagging methods, the selection of appropriate word representation (feature), or the process of normalizing the informal pattern into a formal pattern. According to Avontuur et al. [2], there are two approaches to handle problem in twitter data pattern. The first approach is to revise or normalize the text of tweets from its informal form to formal form. The second approach attempt to adopt NLP tools that address twitter data issues. In addition to the choice of methods, features, and normalization, it is also necessary to have a large number of data.

Owoputi et al. [14] tried to improve POS tagging performance developed by Gimple et al. [1] by addressing the issue of informal pattern in Twitter data. This is done by the addition of cluster-based features and the level-token name lists feature. Word clustering groups informal words based on their variants, for example, the words "lololol", "loool", and "lmbo" will be in different groups with words "gonna", "gunna", "gonne" and "goona".

Other researchers propose POS tagging with bootstrapping techniques to reduce sparsity data [15]. It contribute in a module that is part of GATE TwitE toolkit for processing social media text. Kaji et al. [4] proposed a combination of model with lexical normalization to overcome orthographic diversity. Gosh et al. [6] used various pre-processing modules (word lists and some word features) and post-processing with logical reasoning to annotate words that fail to be annotated by the model.

Another approach is by choosing the right method of tagging. Some of these methods have been used in POS tagging for Twitter data research such as Conditional Random Field (CRF) [1] [16] [17] and Neural Network [17]. Some tools that have been used in POS tagging for Twitter data research are Stanford [5] [6] [17], GATE [17], AMIRA [7] and MADA [7], NLTK, OpenNLP, CoreNLP, Pattern, TweetNLP, TwitterNLP and TwitIE [18].

3. Experimental Setup

First we performed an automatic annotation using existing POS tagger for formal Indonesia [19], with addition of five new tags for Twitter data. The data annotation is done semi-automatically, where the tagger annotates new data automatically and the result of the annotation is corrected manually. The resulting data is added to the training data to rebuild the model. The model is trained multiple times, with data amounts of 1000, 1600 and 1800. This technique is referred to as active learning. The normalization process does not use normalized word lists but utilizes the Kateglo dictionary and Kamus Besar Bahasa Indonesia, to find the equivalent word of the informal language manually.

3.1. *Word Embeddings*

The construction of word embeddings started with preparation of unlabeled corpus to build a vocabulary. Unlabeled corpus is built from 101,800 Indonesian tweets. Pre-processing of the unlabeled text corpus produces vocabulary of 178,736 words. Each method of word embeddings (random, CBOW, Skip-Gram, and Glove) are trained using the vocabulary. The training process will produce a word vector for every word that is in the vocabulary. This word vector is used as input in the tagging process. Some important parameters used in this experiment are vocabulary size, context window size and vector dimension. We use vector dimension of 500 and the context window size is 2.

3.2. *Tagset*

The tagset used are an available 23 tags [19] with addition of five new tags related to the twitter attributes (bold texts in Table 1). These five tags (twitter attributes) are HASH, DISC, AT, URL and EMO. HASH is a tag used for word starts with "#". DISC is a tag used for "RT" and ":" in a re-tweets sentence. AT is a tag used for word starts with the "@". Tag of URL is used for word that indicate an URL. EMO is used for word that contain emotion expressions. The addition of these five new tags refers to the five twitter tags used by Avontuur et al. [2]. These five new tags are used to accommodate five twitter attributes. These new tags (classes) can be used as the keyword for the information searching process from twitter data. Meanwhile, informal words will be kept classified as the original twenty-three class, even though these informal words will be automatically classified as the NNP class before they are manually corrected for the retraining process.

Table 1. Indonesian tagset for social media

Tag	Description
CC	Coordinating Conjunction
CD	Cardinal Number
OD	Ordinal Number
DT	Determiner / article
FW	Foreign Word
IN	Preposition
JJ	Adjective
MD	Modal and auxiliary verb
NEG	Negation
NN	Noun
NNP	Proper Noun
NND	Classifier, partitive, and measurement noun
PR	Demontrative Pronoun
PRP	Personal pronoun
RB	Adverb
RP	Particle
SC	Subordinating Conjunction
SYM	Symbol
UH	Interjection
VB	Verb
WH	Question
X	Unknown
Z	Punctuation
AT	**Mention ('@')**
DISC	**Discourse Maker ('RT')**
HASH	**Hastag ('#')**
URL	**Uniform Resources Locator ('http')**
EMO	**Emoticon ('☺', '☹')**

3.3. POS Tagging with Neural Network
The model used to perform the tagging process is Feedfroward Neural Network with one hidden layer. This model has been built on the previous research for the formal Indonesian language trained on online media corpus [19]. In our research, the model is trained using semi-automatic labeled corpus, that created from 1000, 1600 and 1800 Indonesian tweets. The labeled corpus used in this training process is divided into 64% for training, 16% for validation and 20% for testing. The use of original tagset (without five additional tags) is intended to build a tagger model which used as the baseline.

4. Result and Discusion
From previous research, it has been concluded that almost all methods of word embedding (CBOW, Skip-Gram, and Glove) give good performance (roughly 93 %) for tagging process of Indonesian language sentences from online media, except random method that gives a fairly low performance [19]. Table 2 shows the accuracy for each embeddings method which produced from 1000 tweets as training data. Tables 3 and Table 4 show the same measurement which produced from 1600 tweets and 1800 tweets as training data.

Table 2. Tagging accuracy (%) with 1000 tweets

Random		CBOW		Skip-gram		Glove	
Testing	Validation	Testing	Validation	Testing	Validation	Testing	Validation
64.87	65.37	43.19	42.08	43.05	42.85	46.05	46.22

Table 3. Tagging accuracy (%) with 1600 tweets

Random		CBOW		Skip-gram		Glove	
Testing	Validation	Testing	Validation	Testing	Validation	Testing	Validation
64.03	65.11	64.94	64.83	**65.69**	65.86	63.91	63.91

Table 4. Tagging accuracy (%) with 1800 tweets

Random		CBOW		Skip-gram		Glove	
Testing	Validation	Testing	Validation	Testing	Validation	Testing	Validation
63.41	64.01	65.61	64.63	**66.36**	65.81	64.14	63.54

Table 5. Tagging Accuracy (%) with 1800 Tweets without the five additional tags

Random		CBOW		Skip-gram		Glove	
Testing	Validation	Testing	Validation	Testing	Validation	Testing	Validation
64.46	65.10	67.19	66.78	**67.39**	68.05	66.87	67.06

Based on the accuracy shown in table 2, random method gives the best performance among the other embeddings methods with accuracy of 64.87%. CBOW, Skip-Gram, and Glove give lower accuracy (approximately 43% - 46%). Table 3 shows an increase in performance almost on all methods caused by an increase in the amount of training data from 1000 tweets to 1600 tweets. The same result occurs when the amount of training data is increased to 1800 (66.36%), where almost all methods give performance improvement, except for Random which has a slight decrease in performance (see Table 4). This proves that an increase in the amount of data accompanied by increasing data pattern can improve performance. Based on this experiment Skip-Gram outperformed other methods in 1600 and 1800 training data. This indicate that the use of word embeddings can support the improvement of tagger performance.

The resulting accuracy of POS tagger for twitter data is smaller than the accuracy of POS tagger for formal Indonesian Language [19]. The main reason is the amount of data pattern is still small to support the performance of tagger. The source of twitter data is also obtained randomly (common account), not derived from specific accounts. The obtained data is mainly a private conversation that contains more noisy data.

As can be seen in Table 5, our experiment in POS tagging without the five additional tags (the original tagset) gives slightly better accuracy than the experiment with the five additional tagset

(Table. 4). It is can be inferred that classification with smaller number of category gives better accuracy. It is because the five additional tags to classify Twitter attributes ('#', '@', 'http') which dominate in tweets are correctly classified as NNP class. Other Twitter attributes ('RT', 'emoticons') are correctly classified as SYM class. Even though they are correctly classified, in this case, removing those five additional tags leads to missing important informations that relate with social media case. This experiment further emphasizes that adding the five additional tags does not significantly reduce POS tagging accuracy, but also keep important features of social media messages.

5. Conclusion

Part-of-speech tagging for the formal Indonesian language can be used for tagging of Indonesian tweets with the additional of five new tags. Approach to handle informal language patterns by matching a word with Kateglo dictionary (a simple normalization) helps model to recognize the word. Several word embedding techniques have been evaluated for part-of-speech tagging of social media messages (tweets) in Indonesian language. The experiment shows that the using of word embedding is an alternative technique for POS tagging with a promising results (66.36 %). This is because the word embeddings can capture syntactic and semantic information of word based on nearby words. The characteristic of word embeddings is suitable to recognize and predict tweets patterns.

For the future, in order to improve the performance of this tagger, it is necessary to add a large number of training data which brings many new patterns. In the normalization process, it needs to add informal word list. Data sources should derive from the specific twitter accounts as needed to avoid irregular informal patterns and noisy data. Unknown words also need to be treated separately and included in model learning iteration (active learning). It is recommended to attempt another word embeddings methods, such as TweetNLP, TwitterNLP, and TwitIE to find the best word embeddings method for Indonesian tweets.

This work is a study that aimed to enrich the Indonesian language corpus by creating a high quality informal language corpus. In addition to increasing amount of training data and selecting appropriate word embedding methods, it is possible to apply other types of neural networks such as CNN, RNN or other types of neural networks in order to improve the tagger performance.

References

[1] K. Gimpel, N. Schneider, B. O'Connor, D. Das, D. Mills, J. Einstein, M. Heilman, D. Yogatama, J. Flanigan, and N. A. Smith 2011 Part-of-Speech for Twitter: Annotation, Features, and Experiments *Proceedings of the 49th Annual Meeting of the Association for Computational Linguistics* Portland, Oregon pp. 42-47

[2] T. Avontuur, I. Balemans, L. Elshof, N. V. Noord, M. V. Zanen 2012 Developing a part-of-speech tagger for Dutch tweets *Computational Linguistics in the Netherlands Journal* Vol 2, pp. 34-51

[3] B. O'Connor, C. Dyer, K. Gimpel and N. A. Smith 2011 Normalization of Myanmar Grammatical Categories for Pat-of-Speech Tagging *International Journal of Computer Applications* Vol 36 No. 1 pp. 10-17

[4] N. Kaji and M. Kitsuregawa 2014 Accurate Word Segmentation and POS Tagging for Japanese Microblogs: Corpus Annotation and Joint Modeling with Lexical Normalization *Proceedings of the 2014 Conference on Empirical Methods in Natural Language Processing (EMNLP)* pp. 99-109

[5] T. Lynn, K. Scannel, and E. Maguire 2015 Minority Language Twitter: Part-of-Speech tagging and Analysis of Irish Tweets Proceedings of the ACL pp. 1-8

[6] S. Lynn, S. Ghosh, and D. Das 2016 Part-of-Speech Tagging of Code-Mixed Social Media Text Proceedings of the second Workshop on Computational Approaches to Code Switching, pp. 90-97

[7] F. Albogami, and A. Ramsay 2016 Fast and Robust POS Tagger for Arabic Tweets Using Agreement-based Botstrapping Proceedings of the Tenth International Conference on

IOP Conf. Series: Journal of Physics: Conf. Series **971** (2018) 012055 doi:10.1088/1742-6596/971/1/012055

Language Resources Workshop on Computational Approaches to Code Switching and Evaluation pp. 90-97
[8] F. Muhammad, and A. Purwanti 2016 Handling Out of Vocabulary in Supervised Event Extraction on Indonesian Tweets *International Conference on Data and Software Engineering*
[9] F. Muhammad, and M. L. Kodra 2015 Event Information Extraction from Indonesian Tweets using Conditional Random Field *Advanced Informatics: Concepts, Theory and Applications (ICAICTA)*
[10] A. Purwarianti, A. Andhika, A. F. Wicaksono, I. Afif, F. Ferdiany 2016 InaNLP: Indonesia Natural Language Processing Toolkit *Proceedings of International Conference on Advanced Informatics: Concept, teori and Application*
[11] T. Mikolov, K. Chen, G. Corrado, and J,. Dean 2013 Efficient estimation of word representations in vector space *arXiv* prepint arXiv:1301.3781
[12] J. Pennington, R. Socher, and C. D. Manning 2014 Glove: Global vectors for word representation *EMNLP* vol. 14, pp. 1532
[13] T. A. Le, D. Moeljadi, Y. Miura, and T. Ohkuma 2016 Sentiment Analysis for Low Resource Languages: A Study on Informal Indonesian Tweets *Proceedings of the 12th Workshop on Asian Language Resources*
[14] O. Owoputi, B. O'Connor, C. Dyer, K. Gimpel, N. Schneider, and N. A. Smith 2013 Improved Part-of-Speech Tagging for Online Conversational Text with Word Cluster *Proceedings of NAACL*
[15] L. Derzynski, A. Ritter, S. Clark, and K. Bontcheva 2013 Twitter Part-of-Speech Tagging for All: Overcoming Sparse and Noisy Data
[16] C. Li, and Y. Liu 2015 Joint POS Tagging and Text Normalization for Informal Text *Proceedings of the 24th International Joint Conference on Artificial Intelligence*
[17] T. Gui, Q. Zhang, HM. Peng and X. Huang 2017 Part-of-Speech for Twitter with Adversarial Neural Network *Proceedings of the 2017 Conference on Empirical Methods in Natural Language Processing*
[18] A. Pinto, H. G. Oliveira, and A. O. Alves 2016 Comparing the Performance of Different NLP Toolkits in Formal and Social Media Text *SLATE*
[19] A. F. Abka 2016 Evaluating The Use of Word Embeddings for Part-of-Speech Tagging in Bahasa Indonesia *Proceedings of International Conference on Computer, Control and Informatics: Concept, Teori and Application*

An implementation of support vector machine on sentiment classification of movie reviews

I M Yulietha, S A Faraby, Adiwijaya, and W C Widyaningtyas

School of Computing Telkom University

irenemys@student.telkomuniversity.ac.id, saidalfarby@telkomuniversity.ac.id, adiwijaya@telkomuniversity.ac.id, windawinda@student.telkomuniversity.ac.id

Abstract. With technological advances, all information about movie is available on the internet. If the information is processed properly, it will get the quality of the information. This research proposes to the classify sentiments on movie review documents. This research uses Support Vector Machine (SVM) method because it can classify high dimensional data in accordance with the data used in this research in the form of text. Support Vector Machine is a popular machine learning technique for text classification because it can classify by learning from a collection of documents that have been classified previously and can provide good result. Based on number of datasets, the 90-10 composition has the best result that is 85.6%. Based on SVM kernel, kernel linear with constant 1 has the best result that is 84.9%

1. Introduction

Information needs are driven by the abundance of technology and research. There is so much information, so it necessary to present information but not lose the value of the information [1]. With the development of web, more people that can write their opinion about a product or service. Some of great website allows the user to contribute, modify, and increase the content [2]. Every internet user can express their opinion about something freely. One of them is movie review.

Nowadays everyone can give their opinion in social network [3] [4]. As well as movie review [4]. By the technology, information about movie review are available on internet. It hard to find the correct information that user wants. If the information is processed we can get the value and the quality of the information. It will allow users to get movie quality easily by movie review classification.

From this movie review, there is a challenge to handle negation word in review. Negation word is statement that has denial or reverses the value. Usually negation word uses 'not'. Negation word may affect movie review because it can change the value of sentiment. Positive words can turn negative by negation word.

Sentiment analysis used to see the opinion or tendency of opinion about a movie is positive or negative. This sentiment analysis research was built using a machine learning approach known as Support Vector Machine and devoted to movie review in English language. Machine learning is used because it can be used in every area of computerization such as text analyze, image processing, finance, search and recommendation engine, speech understanding, etc. Machine learning in this case is used to classify texts based on positive or negative user opinion.

2. Literature Review

There are several research that have been done in sentiment analysis, some of them are Hidayatullah and Azhari in 2014 by using Naïve Bayes Classifier and Support Vector Machine to analyze data tweet of presidential vote in 2014. Support Vector Machine have better result than Naïve Bayes. Songbo Tan et al (2008) compared Naïve Bayes, centroid classifier, K-Nearest Neighbor (KNN), winnow classifier and Support Vector Machine on dataset in China language. The result is Support Vector Machine have the best result. By those research Support Vector Machine have better accuracy, so the author uses Support Vector Machine to classify movie reviews.

Another research that have been conducted in sentiment analysis classification is Maraes et al that has analyzed about movie review and product from Amazon used Support Vector Machine and artificial Neural Network. Fatimah Wulandini and Anto Satriyo Nugroho have analyzed Text Classification using Support Vector Machine for Web Mining Based Spation Temporal Analysis of the Spread of Tropical Disease by comparing four classify methods. Zhang et al in 2011 has analyzed sentiment classification in restaurant reviews on internet in Canton language using Naïve Bayes and Support Vector Machine. Setyo Budi in 2017 has analyzed about movie review from Cornell Computer Science as dataset using K-Means method. There are several classification method commonly used for text classification or speech classification, that is Multinomial Naïve Bayes [5], Bayesian Network [6], and Hidden Markov Model [7] [8].

Sentiment analysis is one of computer science to identify emotion, assessment, attitude, opinion, evaluation of a product, services, organization, individual, public figure, topic, event, or other activity. The goal is to categorize someone opinion as a sentiment positive or sentiment negative [12]. In this paper, we propose a method for classifying sentiments on the document movie reviews. The dataset source is https://www.cs.cornell.edu/people/pablo/movie-review-data/. From that source, there are 1000 reviews with positive sentiment and 1000 reviews with negative sentiment. Data has been labeled positive and negative [9].

Support Vector Machine (SVM) that developed by Boser and Guyon, Vapnik in 1992 is a classification technique that aims to find the hyperplane with the largest margin [20]. The good hyperplane is obtained with the maximal of margin value. Margin is a distance between hyperplane with support vector (closest point with hyperplane).

3. System Design

In this research, we will build system that can classify sentiment on movie review. The system described in figure 1.

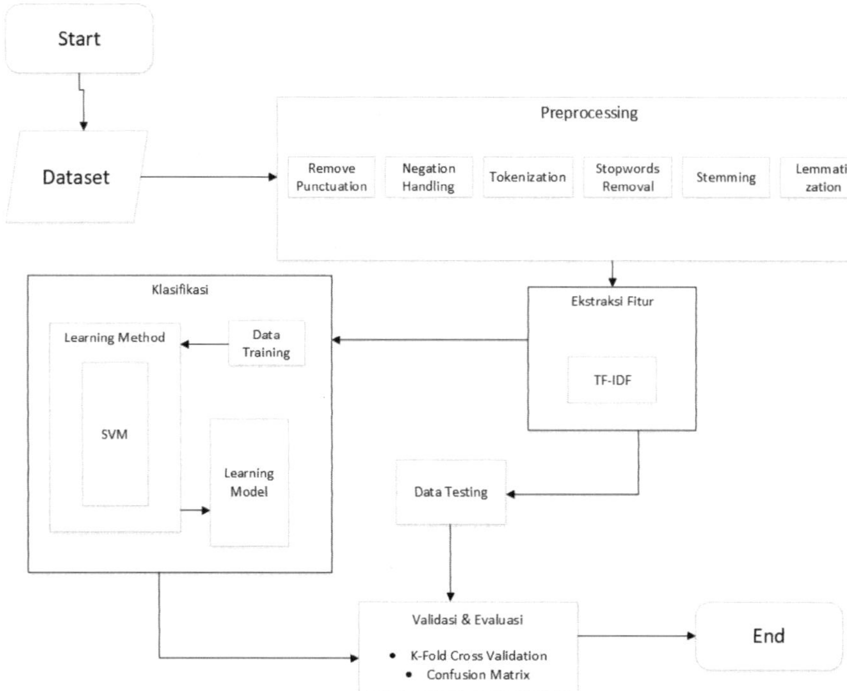

Figure 1 System Design

The purpose of this research is to analyze :

a. Comparison of the amount of training data and testing data to classification model
b. Influence of pre-processing and Negation Handling to Support Vector Machine method in movie review

In validation stage is used k-fold cross validation method. In k-fold cross validation data divided into k parts, that is D_1, D_2,...D_k and each D has same quantity of data. It will be iterated k times, which in each i iteration will be specified as testing data. Another data will be used as training data to get the classification model. Evaluation is to measure the performance of the system. The result will evaluated by predictive classification table, the Confusion Matrix. Confusion Matrix [22] is table that can be used to determine performance of classification model [23]. The example of confusion matrix can be seen in Table 1.

Table 1 Predictive Classification.

		Actual Class	
		+	-
Predicted Class	+	True Positive (TP)	False Positive (FP)
	-	False Negative (FN)	True Negative (TN)

From table 1, there is document categorize in a searching process [24]:

1. True Positive (TP) is system prediction result is positive and the target result is positive.
2. True Negative (TN) is system prediction result is negative and the target result is negative.

3. False Positive (FP) is system prediction result is positive but the target result is negative.
4. False Negative (FN) is system prediction result is negative but the target result is positive.

4. Data and Analysis

a. Influence of Number of Dataset

Based on the result, more number of data given to the training data, the accuracy higher. That is because the classification model can handle more diversity of testing data so the model can classify better. Testing result with composition data between training and testing 90:10 have higher accuracy, ie 85.6% with 1800 training data and 200 testing data. In composition 50% training data and 50% testing data has smallest value of F1-Score that is 82.7%, the number of training data is less than before so the classification model can't handle more diversity on training data but because of good training data characteristic (data have difference to classified) so the model that have been built is good.

b. Influence of SVM Kernel Function

Comparison of SVM kernel uses three different kernels. That is linear, Radial Basic Function (RBF) and polynomial. The results are given in the Table 2.

Table 2. Linear Kernel Result.

No.	C Value	*F-1Score Average* (%)
1.	0.01	32,7%
2.	0.1	38,3%
3.	1	**84,9%**
4.	10	83,6%
5.	100	83,3%
6.	1000	83,7%

Based on the experiment. Table 2 shows that the comparison of accuracy values on SVM Linear with C as constant, F1-Score 84.9% with C=1.0 is the best result. In this scene, C affects the accuracy of testing data classification. Greater the value of C then accuracy result is smaller. This is because trade-off between margin and error is bigger. The value of results is obtained from average of all fold.

Table 3. RBF Kernel Result.

C	γ					
	0.01	0.1	1	10	100	1000
0.01	32,4%	32,2%	32,1%	32,8%	32,4%	32,6%
0.1	32,6%	32,5%	32,4%	32,6%	32,8%	32,2%
1	32,5%	79,7%	83,6%	33,5%	33,4%	32,6%
10	79,3%	83,8%	83,9%	32,7%	32,8%	32,7%
100	83,3%	83,8%	83,8%	32,7%	32,7%	32,7%
1000	83,6%	83,5%	**84,3%**	33,4%	32,9%	32,2%

Based on the experiment. Table 3 shows that F1-Score from RBF Kernel with C and γ, the best result of kernel RBF classification is 84.3% with C = 1000 and $\gamma = 1.0$. The value of C and γ can affect accuracy of testing data classification. The value of C is the most commonly used in all SVM kernels. The function of γ is to determine proximity level between two points so it easier to find the consistent hyperplane. Based on the result, the smaller value of gamma then the F1-Score value is smaller so it needs larger C parameter.

Table 4. Polynomial Kernel Result.

d	C					
	0.01	0.1	1	10	100	1000
1	32,5%	**32,8%**	32,7%	32,7%	32,1%	31,8%
2	32,2%	32,3%	32,4%	32,7%	32,7%	32,6%
3	32,2%	32,2%	32,7%	32,1%	32%	31,4%

Based on the experiment. Table 4 shows that the F1-Score value of polynomial kernel classification with C and d (degree). The best result of polynomial kernel classification was 32.8% with d = 1 and C = 0.1. For the optimal result d may change. Based on the character of the data, in this testing data d =1 optimal result. The function of d is to help mapping data from input space to higher dimension in feature space, so in the new dimension can be found a consistent hyperplane.

c. Influence of Pre-processing and Negation Handling

In this research, pre-processing step is to compare between stemming and lemmatization. Stemming and lemmatization have different word-cutting processes. Stemming works by eliminate any prefixes, inserts, suffixes, or combination of prefix and suffix. But lemmatization works by change each word to the base of word.

Table 5 Lemmatization and Stemming Result.

Fold	F1-Score	
	Lemmatization	Stemming
1	85.8%	84.1%
2	84.8%	83.9%
3	84.1%	84.8%
4	83.6%	83.8%
5	86.2%	83.1%
Average	84.9%	83.9%

Based on the experiment. Table 5 shows that lemmatization value is higher than stemming. Lemmatization has better value because each word is converted into base word and make the true feature in each document increase, that cause the extracted feature increase too. This research suggests that lemmatization tends to produce better classification performance. Another reason that causes stemming process has less feature than lemmatization because there are difference of word cutting process. Evaluation result of stemming is reduced because there is missing information from the word that has been stemmed.

In this research is comparing the influence of using negation handling and without negation handling. This process finds negation word, the example are no, not, etc. If in a sentence contain negation word (no or not), then words after negation word are marked as negation word.

Based on the experiment. Table 6 shows that the best value of F1-Score is by using negation handling, that is 84.4%. Without using negation handling the F1-Score is 82.4%. The difference between negation handling and without negation handling is 0.024. Negation handling can increase performance of sentiment analysis in sentences. The performance increase because of a word in different sentences context can detect negation synthetic that may affect the class prediction result on movie review that used.

Table 6. Influence of Negation Handling

Fold	Without Negation Handling			Using Negation Handling		
	Precision	Recall	F-1 Score	Precision	Recall	F-1 Score
1.	79,4%	81,1%	80,3%	86,5%	78,1%	82,1%
2.	85,2%	79%	82,4%	84,8%	89,6%	87,2%
3.	85,4%	83,2%	84,3%	87,5%	85,5%	86,5%
4.	85,1%	87,3%	83,4%	85,4%	86,6%	86%
5.	83,6%	79,8%	81,7%	80,6%	84%	82,2%
Average			82,4%			84,8%

5. Conclusion

Base on the research, the conclusion are more data that used as training data, the F1-Score result is better for classify. In this case the best result is 85.6% with 90% data as training data and 10% data as testing data. In linear separable and non-linear separable testing, the better F1-Score is 84.9% using linear separable. In this case used linear kernel, RBF kernel, and polynomial kernel in SVM. The result is linear kernel has the best result of F1-Score is 84.9%. Using negation handling give effect in classification. Classification with negation handling has better result. In stemming and lemmatization testing, lemmatization F1-Score is better than stemming because the difference in process cut-off word. For the next research, the different selection feature is expected to improve the accuracy result for the example on polynomial accuracy result.

References

[1] B. Liu, "*Sentiment Analysis: A Multi-Faceted Problem*," IEEE Intelligent Systems, 2010.

[2] K. Yessenov and S. Misailovic, "*Sentiment Analysis of Movie Review Comments*," 2009.

[3] C. L. Liu, W. Hsaio, C. H. Lee and E. Jou, "*Movie Rating and Review Summarization in Mobile Environment*," IEEE Transactions on Systems Man, and Cybernetics, Part C (Applications and Reviews), 2012.

[4] V. Chandani, R. S. Wahono and Purwanto, "*Komparasi Algoritma Klasifikasi Machine Learning Dan Feature Selection pada Analisis Sentimen Review Film*," Journal of Intelligent Systems, vol. 1, 2015.

[5] R. A. Aziz, M. S. Mubarok and Adiwijaya, "*Klasifikasi Topik Pada Lirik Lagu Dengan Metode Multinomial Naive Bayes*," Indonesia Symposium on Computing (IndoSC), 2016.

[6] A. H. R. Z.A, M. S. Mubarok and Adiwijaya, "*Learning Struktur Bayesian Networks Menggunakan Novel Modified Binary Differential Evolution Pada Klasifikasi Data*," Indonesia Symposium on Computing (IndoSC*)*, 2016.

[7] I. N. Yulita, H. L. The and Adiwijaya, "*Fuzzy Hidden Markov Models For Indonesian Speech Classification*," JACIII, vol. 16(3), pp. 381-387, 2012.

[8] U. N.Wisesty, M. Mubarok and Adiwijaya, "*A Classification Of Marked Hijaiyah Letters' Pronunciation Using Hidden Markov Model*," AIP Conference Proceedings 1867, vol. 020036, 2017.

[9] E. Simoudis, "*Reality Check for Data Mining*," IEEE Expert: Intelligent Systems and Their Application, vol. 11, 1996.

[10] M. Hu and B. Liu, "*Mining and Summarizing Customer Review*," 2004.

[11] Irawati, "*Optimisasi Parameter Support Vector Machine (SVM) Menggunakan Algoritme Genetika*," 2010.

[12] B. Liu, "*Sentiment Analysis And Opinion Mining*," 2012.

[13] R. Feldman and J. Sanger, "*The Text Mining HandBook: Advanced Approaches in Analyzing Unstructured Data*," 2007.

[14] M. S. Mubarok, Adiwijaya and M. D. Aldhi, "*Aspect-based Sentiment Analysis to Review Products Using Naive Bayes*," AIP Conference Proceedings 1867, vol. 020060, 2017.

[15] C. C. Aggarwal and C. Zhai, "*A Survey of Text Clustering*," IBM T.J. Watson Research Center, pp. 78-128, 2012.

[16] Adiwijaya, *Aplikasi Matriks dan Ruang Vektor*, Yogyakarta: Graha Ilmu, 2014.

[17] Adiwijaya, *Matematika Diskrit dan Aplikasinya*, Bandung: Alfabeta, 2016.

[18] J. Han and M. Kamber, "*Data Mining: Concepts and Techniques*," 2006.

[19] O. Maimon and L. Rokach, " *Data Mining and Knowledge Discovery Handbook*," 2010.

[20] P.-N. Tan, M. Steinbach and V. Kumar, *Introduction to Data Mining*, Boston: Pearson Addison Wesley, 2006.

[21] A. S. Nugroho, A. B. Witarto and D. Handoko, *Support Vector Machine - Teori dan Aplikasinya dalam Bioinformatika*, Penerbit IlmuKomputer.Com, 2003.

[22] D. Xhemali, C. J. Hinde and R. G. Stone, "*Naive Bayes vs. Decision Trees vs. Neural Networks in the Classification of Training Web Pages*," International Journal of Computer Science, no. 4(1), pp. 16-23, 2009.

[23] L. Hamel, *The Encyclopedia of Data Warehousing and Mining*, 2nd Edition, Idea Group Publishers, 2008.

[24] S. Bird, E. Klein and E. Loper, *Natural Language Processing with Python*, 2009.

[25] N. M. S. Hadna, P. I. Santosa and W. W. Winarno, "*Studi Literatur Tentang Perbandingan Metode Untuk Proses Analisis Sentimen Di Twitter*," Seminar Nasional Teknologi Informasi dan Komunikasi 2016 (SENTIKA 2016), 2016.

[26] P. Chaovalit and L. Zhou, "*Movie Review Mining: a Comparison between Supervised and Unsupervised Classification Approaches,*" Proceedings of the 38th Hawaii International Conference on System Sciences, 2005.

The Language Grid: supporting intercultural collaboration

T Ishida

Department of Social Informatics, Kyoto University, Yoshida-Honmachi, Sakyo-ku, Kyoto 606-8501, Japan

ishida@i.kyoto-u.ac.jp

A variety of language resources already exist online. Unfortunately, since many language resources have usage restrictions, it is virtually impossible for each user to negotiate with every language resource provider when combining several resources to achieve the intended purpose. To increase the accessibility and usability of language resources (dictionaries, parallel texts, part-of-speech taggers, machine translators, etc.), we proposed the *Language Grid* [1]; it wraps existing language resources as atomic services and enables users to create new services by combining the atomic services, and reduces the negotiation costs related to intellectual property rights [4]. Our slogan is "language services from language resources." We believe that *modularization with recombination* is the key to creating a full range of customized language environments for various user communities.

The first step is to wrap each language resource as an atomic language service. Such services are not restricted to a simple function accessing the language resource. For example, the parallel text service can return a high quality translation of a sentence that is similar to the input sentence. Wrapping machine translators is straightforward, but even human interpreters can be wrapped as translation services. Users do not have to distinguish machine from human translation services other than by their quality: machine translators can provide faster services while human interpreters return high quality translations.

The second step is to combine atomic language services to create new services [2]. For example, to translate Japanese sentences into Vietnamese, we need to cascade Japanese-English and English-Vietnamese translation services, because currently there is no direct translator handling Japanese to Vietnamese. Furthermore, if we need to replace words output by machine translators with the technical words in dictionaries specific to the service domain, we can use part-of-speech taggers to divide the input sentences into parts and then identify the technical words. If we have different types of translators, then we need a service to determine which one is the best. In spite of all these efforts, if the quality of translation is still insufficient, we may then call upon human translation services. The above scenario introduces what we call the composite service workflow.

The Language Grid allows users to register services to share them. Major stakeholders in forming the service grid fall into three categories: *service providers*, *service consumers*, and *grid operators*. Service providers register language services to the service grid, while service consumers invoke registered language services as needed for their intercultural activities. Grid operators manage and control language resources and services. For the institutional design of the Language Grid, we have to consider the following contexts of the stakeholders:

- How to protect the intellectual property rights of service providers and to motivate them to register services with the service grid. To this end, service providers should be allowed to specify for what purposes their services can be used.

- How to encourage the widest variety of activities of service consumers to increase their satisfaction. To this end, service consumers should be allowed to run application systems that employ the services permitted for such use.
- How to reduce the load of grid operators while allowing them to globally extend their service grids. To this end, *federated operation* is to be facilitated, where several operators collaboratively operate their service grids by connecting them in a peer-to-peer fashion [5].

We have organized the Language Grid project based on collaboration among service providers in various universities, research institutes and for-profit companies, and service consumers in non-profit and non-governmental organizations [3]. Action research methodologies are being employed during the project. Software development, applications in real communities, and institutional design for federated operation are all related, and thus are being performed in parallel. As a result, the Language Grid has become one of the most advanced infrastructures for services computing. As of September 2017, 177 groups in 23 countries are participating by sharing 230 language services through the federated operation of Language Grid servers located in Kyoto, Jakarta, Bangkok and Xinjiang.

The Language Grid runs on the *service grid server software*, which is registered with GitHub. Note that language services are often used together application-domain-specific services: healthcare, agriculture, marketing etc. Our service grid server software is general enough to permit sharing of not only language services but also services in various domains. For example, to support patients who are using local languages, we can combine the Language Grid with a healthcare service grid. Federated operation can be applied to connect different types of service grids. We are now entering the era of IoT. Future application domains will scale the server network to adapt to the growing IoT environment. A pioneering work that analyzes such future service networks appears in [6].

Acknowledgement
This research was partially supported by a Grant-in-Aid for Scientific Research (A) (17H00759, 2017-2020) from Japan Society for the Promotion of Science (JSPS).

References

[1] Ishida T 2006 Language Grid: an infrastructure for intercultural collaboration *IEEE/IPSJ Symp. on Applications and the Internet* pp 96-100 (invited paper).
[2] Lin D, Shi C and Ishida T 2012 Dynamic service selection based on context-aware QoS *IEEE Int. Conf. on Services Computing* pp 641-648.
[3] Ishida T ed. 2011 *The Language Grid: Service-Oriented Collective Intelligence for Language Resource Interoperability* (Springer Science & Business Media).
[4] Ishida T 2016 Intercultural collaboration: a brief history *Int. Conf. on Principle and Practices in Multi-Agent Systems* pp 3-19 (invited paper).
[5] Murakami Y, Tanaka M, Lin D and Ishida T 2012 Service grid federation architecture for heterogeneous domains *IEEE Int. Conf. on Services Computing* pp 539-546.
[6] Lhaksmana K M, Murakami Y and Ishida T 2016 Analysis of large-scale service network tolerance to cascading failure *IEEE IoT J.* **3**:6 pp 1159-1170.